BIOLOGICAL INVADERS IN INLAND WATERS: PROFILES, DISTRIBUTION, AND THREATS

INVADING NATURE -
SPRINGER SERIES IN INVASION ECOLOGY
Volume 2

Series Editor: JAMES A. DRAKE
University of Tennessee,
Knoxville, TN, U.S.A.

Biological invaders in inland waters: Profiles, distribution, and threats

Edited by

FRANCESCA GHERARDI

Springer

A C.I.P. Catalogue record for this book is available from the Library of Congress.

ISBN 978-1-4020-6028-1 (HB)
ISBN 978-1-4020-6029-8 (e-book)

Published by Springer,
P.O. Box 17, 3300 AA Dordrecht, The Netherlands.

www.springer.com

Printed on acid-free paper

Contents

Contributors

Maria C. Abete

Experimental Zooprofilactic Institute of Piedmont, Liguria and Valle d'Aosta, Via Bologna 148, 10154 Torino, Italy
mariacesarina.abete@izsto.it

Michael J. Adams

USGS Forest and Rangeland Ecosystem Science Center, 3200 SW Jefferson Way, Corvallis, OR 97331, USA
michael_adams@usgs.gov

Paloma Alcorlo

Department of Ecology, Autonomous University of Madrid, c/ Darwin 2, 28049 Madrid, Spain
paloma.alcorlo@uam.es

Kęstutis Arbačiauskas

Institute of Ecology, Vilnius University, Akademijos 2, LT-08412 Vilnius-21, Lithuania
arbas@ekoi.lt

Mária Balážová-L'avrinčíková

Comenius University, Faculty of Natural Sciences, Department of Ecology, Mlynská dolina B-2, 84215 Bratislava, Slovakia
lavrincikova@fns.uniba.sk

Angel Baltanás

Department of Ecology, Autonomous University of Madrid, c/ Darwin 2, 28049 Madrid, Spain
angel.baltanas@uam.es

Silvia Barbaresi

Department of Animal Biology and Genetics, University of Firenze, Via Romana 17, 50125 Firenze, Italy
silvia.barbaresi@unifi.it

Jean-Nicolas Beisel

LIEBE, CNRS, Campus Bridoux, Paul Verlain University, Metz, Avenue du Général Delestraint, 57070 Metz, France
beisel@univ-metz.fr

Nadezhda A. Berezina

Laboratory of Freshwater and Experimental Hydrobiology, Zoological Institute, Russian Academy of Sciences, Universitetskaya emb. 1, 199034 St. Petersburg, Russia
nber@zin.ru

João M. Bernardo

Department of Ecology, University of Évora, Colégio Luis Verney, Rua Romão Ramalho, 7000-671 Évora, Portugal
jmb@uevora.pt

Sandro Bertolino

DI.VA.P.R.A., Laboratory of Entomology and Zoology, Via L. da Vinci 44, 10095 Grugliasco, Italy
sandro.bertolino@unito.it

Marco Bodon

Department of Environmental Sciences, University of Siena, Via Mattioli 4, 53100 Siena, Italy
mabodon@tin.it

Dani Boix

Institute of Aquatic Ecology, University of Girona, Campus Montilivi, 17071 Girona, Spain
dani.boix@udg.es

Miguel Angel Bravo-Utrera

Biological Reserve of Doñana, 21760, Matalascañas, Huelva, Spain
mabravo@ebd.csic.es

J. Anne Brown

School of Biosciences, Hatherly Laboratories, Prince of Wales Road, University of Exeter, Exeter, EX4 4PS, UK
j.a.brown@exeter.ac.uk

Antonella Carosi

Department of Cellular and Environmental Biology, University of Perugia, Via Elce di Sotto, 06123 Perugia, Italy
faunaittica@provincia.terni.it

Sandra Casellato

Department of Biology, University of Padova, Via U. Bassi 58/BIG, 35121 Padova, Italy
sandra.casellato@unipd.it

Florence W. Chege

Eastern Africa Regional Programme, IUCN – The World Conservation Union, P.O. Box 68200, Nairobi, 00200, Kenya
florence.chege@iucn.org

Loke Ming Chou

Department of Biological Sciences, National University of Singapore, 14 Science Drive 4, 117543, Singapore
dbsclm@nus.edu.sg

Simone Cianfanelli

Museum of Natural History, Zoological Section "La Specola", University of Firenze, Via Romana 17, 50125 Firenze, Italy
simone.cianfanelli@unifi.it

Miguel Clavero

Institute of Aquatic Ecology, University of Girona, Campus Montilivi, 17071 Girona, Spain
miguel.clavero@udg.es

Gordon H. Copp

Centre for Environment, Fisheries & Aquaculture Science (CEFAS), Salmon & Freshwater Team, Pakefield Road, Lowestoft, NR33 0HT, UK
gordon.copp@cefas.co.uk

Massimiliano Corboli

Department of Cellular and Environmental Biology, University of Perugia, Via Elce di Sotto, 06123 Perugia, Italy
emmeci@angelfire.com

María Crehuet

Department of Ecology, Autonomous University of Madrid, c/ Darwin 2, 28049 Madrid, Spain
maria.crehuet@uam.es

Simon Devin

LIEBE, CNRS, Campus Bridoux, Paul Verlain University, Metz, Avenue du Général Delestraint, 57070 Metz, France
devin@univ-metz.fr

Yury Yu. Dgebuadze

Severtsov Institute of Ecology and Evolution of the Russian Academy of Science, Leninsky Prospect 33, 117071 Moscow, Russia
dgebuadze@sevin.ru

Ambrosius J. M. Dörr

Department of Cellular and Environmental Biology, University of Perugia, Via Elce di Sotto, 06123 Perugia, Italy

Antonia C. Elia
Department of Cellular and Environmental Biology, University of Perugia, Via Elce di Sotto, 06123 Perugia, Italy
elia@unipg.it

Sílvia Fernandes
Department of Ecology, University of Évora, Colégio Luis Verney, Rua Romão Ramalho, 7000-671 Évora, Portugal
smariaf@hotmail.com

Andrew A. Filippov
Faculty of Biology and Soil Science, St. Petersburg State University, Universitetskaya nab. 7-9, 199034 St. Petersburg, Russia
filippov@ecopro.spb.ru

John Foster
Environment Agency, Saxon House, Little High Street, Worthing, West Sussex, BN11 1DH, UK
john.foster@environment-agency.gov.uk

Michael G. Fox
Environmental and Resource Studies Program and Department of Biology, Trent University, Peterborough, Ontario, K9J 7B8, Canada
mfox@trentu.ca

Leopold Füreder
River Ecology and Freshwater Fauna Laboratory, Institute of Ecology, Leopold-Franzens University, Technikerstr. 25, 6020 Innsbruck, Austria
leopold.fuereder@uibk.ac.at

Elda Gaino
Department of Cellular and Environmental Biology, University of Perugia, Via Elce di Sotto, 06123 Perugia, Italy
gaino@unipg.it

Emili García-Berthou
Institute of Aquatic Ecology, University of Girona, Campus Montilivi, 17071 Girona, Spain
emili.garcia@udg.es

Piero Genovesi

National Institute of Wild Fauna (INFS), Via Ca' Fornacetta 9, 40064 Ozzano Emilia, Italy
piero.genovesi@infs.it

Francesca Gherardi

Department of Animal Biology and Genetics, University of Firenze, Via Romana 17, 50125 Firenze, Italy
francesca.gherardi@unifi.it

Lucia Ghetti

Umbria Region, Servizio Programmazione Forestale, Faunistico-Venatoria ed Economia Montana, Via M. Angeloni 61, 06124 Perugia, Italy
lghetti@yahoo.it

Stephan Gollasch

GoConsult, Grosse Brunnenstr. 61, 22763 Hamburg, Germany
http://www.gollaschconsulting.de

Danièle Guinot

National Museum of Natural History, Department of Habitats and Aquatic Populations, P.O. Box 53, 57 rue Cuvier, F-75231 Paris cedex 05, France
guinot@mnhn.fr

Simona Gumuliauskaitė

Centre for Ecology and Environmental Sciences, Vilnius University, Čiurlionio 21/27, LT-07119 Vilnius-10, Lithuania
simona.gumuliauskaite@gf.vu.lt

David Harper

Department of Biology, University of Leicester, University Road, Leicester, LE1 7RH, UK
dmh@leicester.ac.uk

Leif-Matthias Herborg

Great Lakes Institute for Environmental Research, University of Windsor, Windsor, Ontario, N9B 3P4, Canada
lherborg@uwindsor.ca

David M. Holdich

East Midlands Environmental Consultants, The Old Ragged School, Brook Street, Nottingham, NG1 1EA, UK
david.holdich@ntlworld.com

Geoffrey W. Howard

Eastern Africa Regional Programme, IUCN – The World Conservation Union, P.O. Box 68200, Nairobi, 00200, Kenya
geoffrey.howard@iucn.org

Maria Ilhéu

Departament of Ecology, University of Évora, Colégio Luis Verney, Rua Romão Ramalho, 7000-671 Évora, Portugal
milheu@uevora.pt

Stanislav Katina

Comenius University, Faculty of Mathematics, Physics and Informatics, Department of Probability and Mathematical Statistics, Mlynská dolina, 84248 Bratislava, Slovakia
katina@fmph.uniba.sk

Brenton Knott

School of Animal Biology (M092), The University of Western Australia, 35 Stirling Highway, Crawley, W.A. 6009, Australia
bknott@cyllene.uwa.edu.au

Vladimír Kováč

Comenius University, Faculty of Natural Sciences, Department of Ecology, Mlynská dolina B-2, 84215 Bratislava, Slovakia
kovac@fns.uniba.sk

Tisza Lancioni
Department of Cellular and Environmental Biology, University of Perugia, Via Elce di Sotto, 06123 Perugia, Italy
t.lancioni@arpa.umbria.it

Giovanni La Piana
Department of Biology, University of Padova, Via U. Bassi 58/BIG, 35121 Padova, Italy
giovannilapiana@virgilio.it

Daniel Larson
Department of Environmental Assessment, Swedish University of Agricultural Sciences, P.O. Box 7050, 75007 Uppsala, Sweden
daniel.larson@ma.slu.se

Anderson O. Latini
UNILESTEMG, University Centre of Eastern Minas Gerais, CCB Campus Bom Retiro, Rua Bárbara Heliodora, 725 Bairro Bom Retiro, Ipatinga, 35160-215 MG, Brazil
aolatini@bol.com.br

Massimo Lorenzoni
Department of Cellular and Environmental Biology, University of Perugia, Via Elce di Sotto, 06123 Perugia, Italy
lorenzon@unipg.it

Elisabetta Lori
Museum of Natural History, Zoological Section "La Specola", University of Firenze, Via Romana 17, 50125 Firenze, Italy
elisabetta.lori@unifi.it

Jessica Lynas
School of Animal Biology (M092), The University of Western Australia, 35 Stirling Highway, Crawley, W.A. 6009, Australia
jlynas@cyllene.uwa.edu.au

Hugh J. MacIsaac
Great Lakes Institute for Environmental Research, University of Windsor, Windsor, Ontario, N9B 3P4, Canada
hughm@uwindsor.ca

Alessio Mengoni
Department of Animal Biology and Genetics, University of Firenze, Via Romana 17, 50125 Firenze, Italy
alessio.mengoni@unifi.it

Dan Minchin
Marine Organism Investigations, 3 Marina Village, Ballina, Killaloe, County Clare, Ireland
minchin@indigo.ie

Carlos Montes
Department of Ecology, Autonomous University of Madrid, c/ Darwin 2, 28049 Madrid, Spain
carlos.montes@uam.es

Jim R. Muirhead
Great Lakes Institute for Environmental Research, University of Windsor, Windsor, Ontario, N9B 3P4, Canada
muirhe1@uwindsor.ca

Pek Kaye Abigayle Ng
Department of Biological Sciences, National University of Singapore, 14 Science Drive 4, 117543, Singapore
g0306158@nus.edu.sg

Pierre Y. Noël
National Museum of Natural History, UMR CNRS-UPMC-MNHN 5178, Department of Habitats and Aquatic Populations, P.O. Box 53, 57 rue Cuvier, F-75231 Paris cedex 05, France
pnoel@mnhn.fr

Ruth M. O'Riordan

Department of Biological Sciences, National University of Singapore, 14 Science Drive 4, 117543, Singapore & Department of Zoology, Ecology and Plant Science, University College Cork, Distillery Fields, North Mall, Cork, Ireland
r.oriordan@ucc.ie

Vadim E. Panov

Faculty of Biology and Soil Science, St. Petersburg State University, Universitetskaya nab. 7-9, 199034 St. Petersburg, Russia
rbic@zin.ru; vpanov@aquaticinvasions.ru

Christopher A. Pearl

USGS Forest and Rangeland Ecosystem Science Center, 3200 SW Jefferson Way, Corvallis, OR 97331, USA
christopher_pearl@usgs.gov

Giovanni Pedicillo

Department of Cellular and Environmental Biology, University of Perugia, Via Elce di Sotto, 06123 Perugia, Italy
pedicillog@yahoo.it

Miguel Petrere Jr

São Paulo State University, Júlio de Mesquita Filho, UNESP, Biosciences Institute, Av. 24 A n.1515, Bela Vista, P.O. Box 199, Rio Claro, SP, 13506–900, Brazil
mpetrere@rc.unesp.br

Manfred Pöckl

Vienna Ecology Centre, Department of Freshwater Ecology, University of Vienna; State Government of Lower Austria Landhausplatz 1, 3109 St. Pölten, Austria
manfred.poeckl@noel.gv.at

Marino Prearo

Experimental Zooprofilactic Institute of Piedmont, Liguria and Valle d'Aosta, Via Bologna 148, 10154 Torino, Italy
marino.prearo@izsto.it

Neil F. Ramsay

Department of Biological Sciences, National University of Singapore, 14 Science Drive 4, Singapore 117543 & Department of Zoology, Ecology and Plant Science, University College Cork, Distillery Fields, North Mall, Cork, Ireland
r.ramsay@ucc.ie

Riccardo Scalera

Via Torcegno 49 V1 A2, 00124 Roma, Italy
scaleric@tin.it, riccardo.scalera@alice.it

Dawn M. Scott

School of Biosciences, Hatherly Laboratories, Prince of Wales Road, University of Exeter, Exeter, EX4 4PS, UK
d.m.scott@exeter.ac.uk

Tamara A. Shiganova

Shirshov Institute of Oceanology of the Russian Academy of Science, 36 Nakhimov Avenue, 117997 Moscow, Russia
shiganov@ocean.ru

Catherine Souty-Grosset

UMRCNRS 6556 Génétique et Biologie des Populations de Crustacés, University of Poitiers, 40 Avenue du recteur Pineau, 86022 Poitiers cedex, France
catherine.souty@univ-poitiers.fr

Andrew Storey

School of Animal Biology (M092), The University of Western Australia, 35 Stirling Highway, Crawley, W.A. 6009, Australia
awstorey@cyllene.uwa.edu.au

Gabrielle Thiébaut

LIEBE, CNRS, Campus Bridoux, Paul Verlain University, Metz, Avenue du Général Delestraint, 57070 Metz, France
thiebaut@univ-metz.fr

Jozef Tomeček

Comenius University, Faculty of Natural Sciences, Department of Ecology, Mlynská dolina B-2, 84215 Bratislava, Slovakia
tomecek@sopsr.sk

Alessandra Visentin

Department of Biology, University of Padova, Via U. Bassi 58/BIG, 35121 Padova, Italy
alessandravisentin@libero.it

Eva Willén

Department of Environmental Assessment, Swedish University of Agricultural Sciences, P.O. Box 7050, SE-750 07 Uppsala, Sweden
eva.willen@ma.slu.se

Rod W. Wilson

School of Biosciences, Hatherly Laboratories, Prince of Wales Road, University of Exeter, Exeter, EX4 4PS, UK
r.w.wilson@exeter.ac.uk

David A. Wright

University of Maryland, Center for Environmental Science, Chesapeake Biological Laboratory, Solomons, MD 20688, USA
wright@cbl.umces.edu

Foreword

Q: Why are some environments more vulnerable to invasion than others?
A: Environments? In the first volume of this series you asked which species traits coincide with good invaders ... now environments!
Q: Sure, if traits are not terribly powerful predictors alone, I thought that perhaps species traits and kinds of systems might somehow covary. What do you think?
A: Well that depends.
Q: Depends on what?
A: That depends too.

Freshwater environments are islands of sorts. Lakes, rivers, streams and wetlands are uniquely bounded and discrete at one readily apparent scale of observation, the basin or channel boundary. It comes as no surprise, then, that ecological thought has been so strongly influenced by research conducted within the confines of systems so easily circumscribed. Yet, aquatic and terrestrial habitats are inexorably coupled such that this boundary, while substantial at some scales or levels of organization, vanishes at others. This fact is well illustrated by the cascading or indirect effects induced by non-native species that readily traverse the interface between land and water.

Biological invasions represent the latest threat to the integrity of freshwater ecosystems worldwide, systems that are already impacted in massive fashion by human activity. Based on the number of documented and potential extinctions, the freshwater fauna of North America are experiencing an extinction rate that is five times that of the terrestrial environment.[i] While unrelenting habitat modification and resource exploitation directly accounts for many of these species losses, exotic species have surely played a role in what is an unprecedented

[i] Ricciardi, A. and J.B. Rasmussen. 1998. Extinction rates of North American Freshwater Fauna. Conserv. Biol. 13: 1220–1222.

episode of extinction in real time. Humans, however, are not safe in their role as the dominant purveyors of extinction. Ecosystem modification is often accompanied by increased susceptibility to invasion, and once established, exotic species are fully capable of changing all the rules driving system organization.

This is a broad-based volume, crafted with the widest possible brush strokes. Francesca Gherardi set out to create a volume that not only addressed the phenomenon of biological invasions in freshwater systems, but that also reflects the very breadth of contemporary approaches employed to understand the threats posed by the global movement of species. Here, the reader will find specificity and generality, application as well as theory, along with the socio-political implications and response to a global crisis. In a very real sense, this volume represents *everything invasions*.

James Drake
Series Editor

Preface

Nowadays we live in a very explosive world, and while we may not know where or when the next outburst will be, we might hope to find ways of stopping it or at any rate damping down its force.

Charles Elton 1958

In the past few decades, it has become clear to scientists and policy-makers that the human-mediated introduction of species – meaning the deliberate or accidental introduction into the wild of microbes, fungi, plants, and animals, including genetically modified organisms (GMOs) outside their natural range of distribution – is the main driver of biodiversity change (Sala *et al.* 2000). Acting often in concert with other anthropogenic alterations to the environment, such as changes in land use, climate, nitrogen deposition, and atmospheric CO_2, the effects on global biodiversity are expected to increase quickly with time in both extent and intensity.

Changes in the natural distribution of species should not, in general, be viewed as abnormal events (Lodge 1993). They are commonplace in nature, often occurring over the course of geological times in association with climate change (Graumlich and Davis 1993). But only rare events, usually associated with unusual climatic conditions such as storms, may induce the dispersal of species to habitats previously beyond their natural dispersal capabilities (MacIsaac *et al.* 2001). Human actions are more frequent and powerful. Such actions have greatly increased the temporal rate at which species disperse and the distances they traverse, accomplishing in a few decades something that could have never been accomplished by the means of natural events alone (Lodge 1993).

Since their earliest migrations, humans have contributed to the spread of organisms, always carrying them and their propagules over long distances. But the frequency of human-induced introductions of species and the consequent risks associated with them have augmented exponentially in the recent past in concert with the fast growth of the human population and with the rapid escalation of our potentials to alter the environment. Large numbers of people are today traveling faster and farther, and more and more goods and materials are being traded among nations and continents (Pimentel *et al.* 2002), creating a "New Pangaea" (Mooney 1998 cited by Rosenzweig 2001). All these factors combined have produced burgeoning rates of non-indigenous species (NIS) in every ecosystem that has been monitored. Over 480,000 NIS have been introduced into the varied ecosystems on earth (Pimentel *et al.* 2002) and have come to dominate about 3% of the ice-free surface over the last 500 years (Mack 1985). Their prevalence is exacerbated by climatic changes that in their turn favor the natural spread and proliferation of NIS (Dukes and Mooney 1999, Carlton 2000, Cowx 2002). The combined effect of the spread of cosmopolitan species and the extinction or range contraction of regional and endemic indigenous organisms often results in the "mingling" of the taxonomic composition of once disparate biota (Olden *et al.* 2004). This phenomenon is inevitably leading to the "homogenization" (McKinney and Lockwood 1999) or "McDonaldization" of the biosphere (Lövei 1997) that will characterize, it has been said, the forthcoming "Homogocene" era (Orians 1994 cited by Rosenzweig 2001).

Indeed, several introduced species have been beneficial to humans; species such as corn, wheat, rice, plantation forests, domestic chicken, cattle, and others provide now more than 98% of the world's food supply (Ewel *et al.* 1999, Pimentel *et al.* 2002). Many cause minimal environmental impact, as predicted by the oft-cited "tens rule" (Williamson and Fitter 1996). So, the fraction of the introduced species that cause problems is small, but their impact could be very serious. These species have the potential of becoming numerically and ecologically prominent; they spread from the point of introduction and are often able to dominate indigenous populations and communities (Kolar and Lodge 2001); they may profoundly and adversely affect indigenous species, ecosystem processes, economic interests, and public health (e.g. Ricciardi *et al.* 1998). In sum, they may turn out to be invasive. Their effects that justify alarm include biodiversity loss at the level of species, large reduction in the lower (genetic) and higher (generic) levels of biodiversity, changes in ecosystem functions, alteration of the ecosystem services provided to humans, aesthetic modifications of landscapes, direct costs to industries, damage to crops and forests, and the spread of human diseases, such as HIV and West Nile virus (Mack *et al.* 2000). Also, deliberate introductions made to solve local or regional problems may be responsible for serious ecological and economic consequences, the so-called Frankenstein Effect (Moyle *et al.* 1986). The costs they inflict form a hidden but onerous "tax" on many goods and services and the damages they cause are often irrevocable: biological invaders act as biological pollutants that, unlike chemicals, reproduce and spread autonomously, over great distances, and can adapt to changing conditions. Their impacts may be continuously increasing over time, even when their introduction ceases.

Since the 1980s, studies of NIS have expanded greatly, resulting in a flood of scientific publications and in the foundation of two invasion-focused journals, *Diversity and Distribution* (Blackwell Publishing) in 1998 and *Biological Invasions* (Kluwer-Springer) in 1999, this growth reflecting the rise in popularity that the discipline of invasion biology has gained as an appealing area of research among ecologists. The overall number of published articles appears, however, to be significantly biased towards terrestrial invaders; invasive events occurring in freshwater systems have been most often neglected or analyzed in a few regional contexts or for a small number of paradigmatic species.

In this book, the identity, distribution, and impact of freshwater NIS will be examined, as well as the dynamics of their invasion. Rather than providing a broad and comprehensive review of the issue, *Biological Invaders in Inland Waters* focuses on old and new invaders and also raises questions and opens perspectives that will be of stimulus for further research. Inland waters will be taken here as meaning rivers, lakes, and reservoirs. Coastal lagoons, saline lakes, estuaries, and low salinity seas such as the Baltic, will be mentioned when appropriate.

The ultimate, ambitious purpose of the book is to help define a more general framework for our knowledge of invasions in fresh waters. Such a framework will be indispensable to the planning of a science-based management program. Inspiration for this effort came from the International Workshop, "Biological Invasions in Inland Waters" (INWAT), held in Florence (Italy) between 5 and 7 May 2005 and made possible by support from Ente Cassa di Risparmio di Firenze, the Italian Ministry of University and Scientific Research (MIUR), the University of Florence, the Provinces of Arezzo, Firenze, Grosseto, Pisa, and Pistoia, and the International Association of Astacology. The INWAT Workshop was a necessary addendum of the final meeting (Florence, 2–5 May 2005) of the European network CRAYNET ("European crayfish as keystone species – linking science, management and economics with sustainable environmental quality", coordinator Catherine Souty-Grosset) (Fifth EU Framework).

This volume benefited greatly from the collaboration – and patience – of the authors and the numerous reviewers. Heartfelt thanks are directed to Jim Drake, who was the first supporter of the project, to Geoff Sanders for his careful linguistic revision, to Claudia Angiolini for her editorial help, and to Suzzanne Mekking and Martine van Bezooijen at Springer for their ability to transform a dream in a printed volume.

F.G.
Florence, Italy
November 2006

REFERENCES

Carlton, J. T. 2000. Global change and biological invasions in the oceans. Pages 31–53 in H. A. Mooney and R. J. Hobbs, editors. Invasive species in a changing world. Island Press, Covelo, CA.

Cowx, I. G. 2002. Analysis of threats to freshwater fish conservation: past and present challenges. Pages 201–220 in M. J. Collares-Pereira, I. G. Cowx, and M. M. Coelho, editors. Conservation of freshwater fishes: options for the future. Blackwell Science, Oxford, UK.

Dukes, J. S. and H. A. Mooney. 1999. Does global change increase the success of biological invaders? Trends in Ecology and Evolution **14**, 135–139.

Elton, C. S. 1958. The ecology of invasions by animals and plants. University of Chicago Press, Chicago, IL, Edition 2000.

Ewel, J. J., D. J. O'Dowd, J. Bergelson, C. C. Daehler, C. M. D'Antonio, D. Gomez, D. R. Gordon, R. J. Hobbs, A. Holt, K. R. Hopper, C. E. Hughes, M. Lahart, R. R. B. Leakey, W. G. Lee, L. L. Loope, D. H. Lorence, S. M. Louda, A. E. Lugo, P. B. Mcevoy, D. M. Richardson, and P. M. Vitousek. 1999. Deliberate introductions of species: research needs. BioScience **49**, 619–630.

Graumlich, L. J. and M. B. Davis. 1993. Holocene variation in spatial scales of vegetation pattern in the upper Great Lakes. Ecology **74**, 826–839.

Kolar, C. S. and D. M. Lodge. 2001. Progress in invasion biology: predicting invaders. Trends in Ecology and Evolution **16**, 199–204.

Lodge, D. M. 1993. Biological invasions: lessons for ecology. Trends in Ecology and Evolution **8**, 133–137.

Lövei, G. L. 1997. Global change through invasion. Nature (London) **388**, 627–628.

MacIsaac, H. J., I. A. Grigorovich, and A. Ricciardi. 2001. Reassessment of species invasion concepts: the Great Lakes basin as a model. Biological Invasions **3**, 405–416.

Mack, R. N. 1985. Invading plants: their potential contribution to population biology. Pages 127–142 in J. White, editor. Studies on plant demography. Academic, London, UK.

Mack, R. N., D. Simberloff, W. M. Lonsdale, H. Evans, M. Clout, and F. A. Bazzaz. 2000. Biotic invasions: causes, epidemiology, global consequences, and control. Ecological Applications **10**, 689–710.

McKinney, M. L. and J. L. Lockwood. 1999. Biotic homogenization: a few winners replacing many losers in the next mass extinction. Trends in Ecology and Evolution **14**, 450–453.

Moyle, P. B., H. W. Li, and B. A. Barton. 1986. The Frankenstein Effect: impact of introduced fishes on native fishes in North America. Pages 415–426 in R. H. Strond, editor. Fish culture in fisheries management. American Fisheries Societies, Bethesda, MD.

Olden, J. D., N. L. Poff, M. R. Douglas, M. E. Douglas, and K. D. Fausch. 2004. Ecological and evolutionary consequences of biotic homogenization. Trends in Ecology and Evolution **19**, 18–24.

Pimentel, D., S. McNair, J. Janecka, J. Wightman, C. Simmonds, C. O'Connell, E. Wong, L. Russel, J. Zern, T. Aquino, and T. Tsomondo. 2002. Economic and environmental threats of alien plant, animal, and microbe invasions. Pages 307–329 in D. Pimentel, editor. Biological invasions: economic and environmental costs of alien plant, animal, and microbe species. CRC Press, Boca Raton, FL.

Ricciardi, A., R. J. Neves, and J. B. Rasmussen. 1998. Impending extinctions of North American freshwater mussels (Unionida) following the zebra mussel (*Dreissena polymorpha*) invasion. Journal of Animal Ecology **67**, 613–619.

Rosenzweig, M. 2001. The four questions: what does the introduction of exotic species do to diversity? Evolutionary Ecology Research **3**, 361–367.

Sala, O. E., F. S. Chapin III, J. J. Armesto, R. Berlow, J. Bloomfield, R. Dirzo, E. Huber-Sanwald, L. F. Huenneke, R. B. Jackson, A. Kinzig, R. Leemans, D. Lodge, H. A. Mooney, M. Oesterheld, N. L. Poff, M. T. Sykes, B. H. Walker, M. Walker, and D. H. Wall. 2000. Global biodiversity scenarios for the year 2100. Science **287**, 1770–1774.

Williamson, M. and A. Fitter. 1996. The varying success of invaders. Ecology **77**, 1666–1670.

F.G. wishes to thank the friends and colleagues who acted as reviewers:

Michael Adams, Angelo Antonio Agostinho, Carles Alcaraz, Fernando Alonso Gutiérrez, Giorgio Bavestrello, João Bernardo, Roberto Berti, Giuseppe Boitani, Loïc Bollache, Marco Bologna, Hugo Coops, Gordon Copp, Stephen Corn, Claudia Corti, Alan Covich, Ian Cowx, Bill Daniels, Jaimie Dick, Natham Dorn, Cédric d'Udekem d'Acoz, Frauke Ecke, Jim Fetzner, Jordi Figuerola, John Foster, Michael Fox, Emili Garcia-Berthou, Cristina Giacoma, Folco Giusti, Stephan Gollasch, Joanna Grabowska, Michal Grabowski, Manuela Gualtieri, Joseph Heller, David Holdich, Pierre Horwitz, Jeoffry Howard, Mark Hutchinson, John Jackson, Krzysztof Jazdzewski, Ladd Johnson, Frank Kligenstein, Vladimír Kováč, Daniel Larson, Thomas Lauer, Erni Leppäkoski, Sandro Lovari, Luca Luiselli, Hugh MacIsaac, Alexandra Marçal Correia, Tracy McCollin, Marit Mjerde, Noel Morrissy, Ilona Musko, Erin Muths, Stefan Nehring, Annamaria Nocita, Francesco Nonnis-Marzano, Sergej Olenin, Vadim Panov, Gad Perry, George Phiri, Manfred Pöckl, Mark Ponniah, Giovanni Principato, Wolfang Rabitsch, Philip Rainbow, Sanjeevi Rajagopal, Julian Reynolds, Nicoletta Riccardi, Karl-Otto Rothhaupt, Jordi Sala, Giuseppe Simone, Patrik Stenroth, Gabrielle Thiébaut, Pietro Giorgio Tiscar, Marco Vannini, Marcel Veldhuis, Karen Veverica, Anna Vila-Gispert, Hartwell Welsh, Dudley Williams, Sergio Zerunian.

Introduction

We must make no mistake: we are seeing one of the great historical convulsions in the world's fauna and flora.

Charles Elton (1958)

Chapter one

Biological invasions in inland waters: an overview

Francesca Gherardi

INTRODUCTION

The value of inland waters to humankind is obviously infinite and the induced changes in the goods and services they provide have a strong impact on human welfare. Lakes/rivers and wetlands currently contribute 20% to the estimated annual global value of the entire biosphere amounting to US$33 trillion per year (Costanza *et al.* 1997). These elevated numbers may justify the present general concern about the increasing degradation of freshwater systems, associated with the rapid extinction rate of their biodiversity – in some cases even matching that of tropical forests (Ricciardi and Rasmussen 1999).

Together with other anthropogenic sources of disturbance, such as the impoundment of rivers (e.g. dams and weirs, water removal), water quality deterioration (e.g. pollution, eutrophication, acidification), habitat degradation and fragmentation (e.g. channelization and land use change), and over-exploitation, the introduction of non-indigenous species (NIS) into fresh waters is today regarded as the main driver of biodiversity change (Millennium Ecosystem Assessment 2005). The effects of such a driver has been estimated to be greater in freshwater than in terrestrial ecosystems (Sala *et al.* 2000). This is particularly apparent in lakes where biological invaders have been recognized as one of the greatest causes of species extinctions (Lodge 2001).

Francesca Gherardi, Biological invaders in inland waters: Profiles, distribution, and threats, 3–25.

THE VULNERABILITY OF INLAND WATERS TO INVASIONS

Inland waters have been the theatres of spectacular biological invasions. Well-known cases are the introduction of the Nile perch *Lates niloticus* (Linnaeus) into Lake Victoria followed by the elimination of about 200 species of haplochromine cichlids (Craig 1992), the alteration of the Laurentian Great Lakes communities and ecosystems by sea lamprey *Petromyzon marinus* Linnaeus, zebra mussel *Dreissena polymorpha* (Pallas), and other invaders (MacIsaac *et al.* 2001), and the complete domination of lowland rivers in the western USA by non-indigenous fish and invertebrates (Moyle and Light 1996a). In several freshwater systems, other less celebrated dramas are however ongoing with the intervention of several, previously unsuspected actors, such as *Lepomis gibbosus* (Linnaeus) (Chapter 15) and *Carassius auratus* (Linnaeus) (Chapter 13) among fish, *Dikerogammarus villosus* (Sowinsky) (Chapters 12 and 27), *Gmelinoides fasciatus* (Stebbing) (Chapter 26), and *Pontogammarus robustoides* (Sars) (Chapter 25) among crustaceans, and *Rana catesbeiana* Shaw among amphibians (Chapters 7 and 38). This confirms that invasions by NIS are pervasive and highly diffused phenomena in fresh waters but also that our predictive ability may be weak. Meanwhile, other apparently harmless NIS are spreading (see potamid crabs in southern France, Chapter 3).

The reasons that freshwater systems are vulnerable to NIS are several, including the higher intrinsic dispersal ability of freshwater species compared with terrestrial organisms (Beisel 2001). Lakes and some streams are comparable to islands in that their geographic isolation has led to local adaptation with the evolution of many endemisms and sometimes to a low biodiversity (Lodge 1993). The extensive introduction of organisms in inland waters, either inadvertent (e.g. via ship ballast, artificial/natural canals, or estuarine saline-bridges, Chapters 17, 21, and 22; as parasites of other introduced species, such as the oomycete *Aphanomyces astaci*, Chapter 6) or deliberate (e.g. stocking of fish and crayfish, Chapters 20 and 31; intentional releases of pets or farm organisms, Chapters 8 and 9), is a direct consequence of the intensity with which humans utilize these systems for recreation, food sources, and commerce (Rahel 2000, Ricciardi 2001). Human-mediated dispersal of crustacean zooplankton, for instance, might exceed the natural rate by up to 50,000-fold (Hebert and Cristescu 2002). And the frequency of species invasions in freshwater systems is likely to continue to grow commensurate with enhanced global commerce and human exploitation of these communities.

Finally, freshwater systems are subject, especially at higher latitudes, to altered seasonal temperature regimes due to global climatic warming and, especially in developed countries, to strong human disturbance. In fact, many NIS are migrating to new areas where the climate has warmed, such as some introduced warm-water fish [e.g. *Micropterus salmoides* (Lacepède), *Lepomis macrochirus* Rafinesque, and *Lepomis cyanellus* (Rafinesque)] that are spreading in North America into higher latitudes and altitudes (Eaton and Scheller 1996,

Chapter 35). Disturbed ecosystems and communities attract biological invasions more than pristine systems; disturbance results in the resharing of space and energy resources that are available to indigenous and non-indigenous species and may open new vacant niches for the most adaptable and tolerant invaders (Ross *et al.* 2001).

The vulnerability of inland waters to biological invasions is a cause of the complete domination of vast waterscapes in certain regions by NIS, such as water hyacinth [*Eichornia crassipes* (Martius) Solms] in many tropical lakes and rivers (Chapter 10) and the red swamp crayfish [*Procambarus clarkii* (Girard)] in several waterbodies of southern Europe (Gherardi 2006, Chapter 2). Xenodiversity may be extraordinarily high in, for instance, large rivers of developed countries that usually host dozens to hundreds of NIS (Mills *et al.* 1996, Chapter 12). The Hudson River, for instance, contains more than 100 species of non-indigenous fish, vascular plants, and large invertebrates, a considerable fraction of which are ecologically important – such as Eurasian watermilfoil (*Myriophyllum spicatum* Linnaeus), purple loosestrife (*Lythrum salicaria* Linnaeus), zebra mussel (*D. polymorpha*), Atlantic rangia (*Rangia cuneata* Gray), and common carp (*Cyprinus carpio* Linnaeus) (Strayer *et al.* 2005). Some taxa are particularly affected by species introductions: the New Zealand fish fauna contains 30 NIS (53% of the total) (Vitousek *et al.* 1997); isolated islands often have more non-indigenous than indigenous fish species (Hawaii: 19 *vs.* 6), but also continental areas have relatively large numbers of non-indigenous fish species (California: 42 *vs.* 76, Brazil: 76 *vs.* 517; references in Vitousek *et al.* 1996); at least 76 fish species belonging to 21 families have been introduced into European fresh waters (Lehtonen 2002), of which 51 have become established.

Species originating from diverse biogeographical areas now coexist in several basins; in the Rhine, indigenous crustaceans [*Gammarus pulex* (Linnaeus)] occur with North American species [*Gammarus tigrinus* Sexton and *Orconectes limosus* (Rafinesque)], Mediterranean species (the freshwater shrimp *Atyaephyra desmaresti* Millet), and Ponto-Caspian species [*Gammarus roeseli* Gervais and *Dikerogammarus villosus* (Sowinsky)] (Beisel 2001). Biotic homogenization is constantly increasing; freshwater fish similarity among the States of the USA amounts today to 7% (Rahel 2000) and some of them, such as Arizona and Montana, which previously had no fish species in common, now share more than 30 species. National borders are obviously irrelevant and they provide no barrier to the natural dispersal of NIS: the Nile perch released in Tanzania spread to other countries bordering Lake Victoria (Welcomme 1988) and *P. clarkii* introduced into Spain invaded Portugal via the common hydrographic basins (Gherardi 2006).

Some freshwater systems function as "hotspots" where NIS accumulate. The Great Lakes system contains over 145 non-indigenous invertebrates, pathogens, algae, fish, and plants, with approximately 75% originated from Eurasia of which 57% are native to the Ponto-Caspian region (Mills *et al.* 1993,

Ricciardi and Rasmussen 1998, MacIsaac *et al.* 2001). These species include a wide array of taxa, such as mussels [*D. polymorpha, Dreissena bugensis* (Andrusov)], amphipods (*Echinogammarus ischnus* Stebbing), cladocerans [*Cercopagis pengoi* (Ostroumov)], harpacticoid copepods [*Nitocra incerta* (Richard) and *Schizopera borutzkyi* (Monchenko)], and fish [*Neogobius malanostomus* (Pallas)], *Proterorhinus marmoratus* (Pallas), *Gymnocephalus cernuus* (Linnaeus)]. It has been ascertained that Ponto-Caspian species reach the Great Lakes in ballast along five shipping "corridors" (MacIsaac *et al.* 2001). Four of these corridors require the first transfer of species via rivers and canals to ports in the North and Baltic seas that, in their turn, function as "hubs", acting as the donor for other ships that transport these species in secondary invasions to ports in North America and, potentially, in East Asia, San Francisco Bay, and Australia (e.g. Cohen and Carlton 1998, Ruiz *et al.* 2000).

Finally, many freshwater invaders are moved among biogeographic regions within continents and are transported among continents in association with economic activity and trade globalization that benefit millions worldwide (Lodge and Shrader-Frechette 2003). The inevitable tension between two often competing goals – increasing economic activity and protecting the environment from invasive species – make it difficult to justify the need for decision-makers to contain the spread of these species and to mitigate the environmental risks they pose. For instance, a number of issues has been raised in favor of the outcomes of introducing crayfish (Gherardi 2006, Chapter 28). First, in the absence of indigenous species, invasive crayfish were claimed to occupy vacant niches, constituting the unique large macro-consumer within polluted or eutrophicated waters, where the native fauna has already been severely decimated (Gherardi *et al.* 2000). The second claim is that they constitute abundant prey for rare or threatened birds and mammals, like several Ardaeidae and the otter (e.g. Barbaresi and Gherardi 2000, Rodríguez *et al.* 2005). Third, from a socioeconomic perspective, introduced crayfish have contributed to: (1) the restoration of traditional habits, e.g. by crayfishing in Sweden and Finland (Kirjavainen and Sipponen 2004); (2) economic benefits for local crayfishermen, e.g. the Spanish netsmen; (3) diversification of agriculture to include astaciculture, e.g. by crayfish farmers in Britain and in Spain; and (4) increased trade between countries inside Europe as well as between European and extra-European countries (Ackefors 1999).

THE CURRENT STATE OF RESEARCH

Studies on the identity, distribution, and impact of freshwater NIS and on the dynamics of their invasion have increased exponentially since the 1990s, resulting in a flood of publications particularly abundant in the last decade (Fig. 1). This pattern of growth in the literature is a reflection of the rise in popularity that invasion biology has gained as an appealing area of research

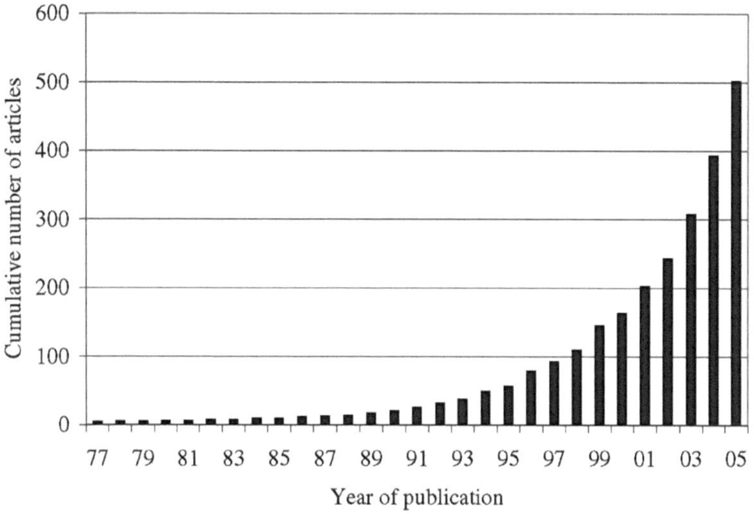

Fig. 1 Cumulative number of publications dealing with non-indigenous species (total number = 502). Papers were identified via keywords from Biosis analyzed between 1967 and December 2005.

among ecologists (Kolar and Lodge 2001). The overall number of published articles is, however, significantly biased towards terrestrial invaders (Fig. 2). This is not surprising: terrestrial systems are the most visible and accessible habitats for humans and, as such, have received the preponderance of ecological attention.

The majority of studies of freshwater invaders has been conducted in North America, and mostly in the Great Lakes (Fig. 3A), and centered on animals (74%) more than on plants (20%), whereas a small fraction (6%) analyzed both kingdoms simultaneously. Of all the animal taxa, fish, particularly salmonids, have received the greatest scientific attention (Fig. 3B), as the result of their

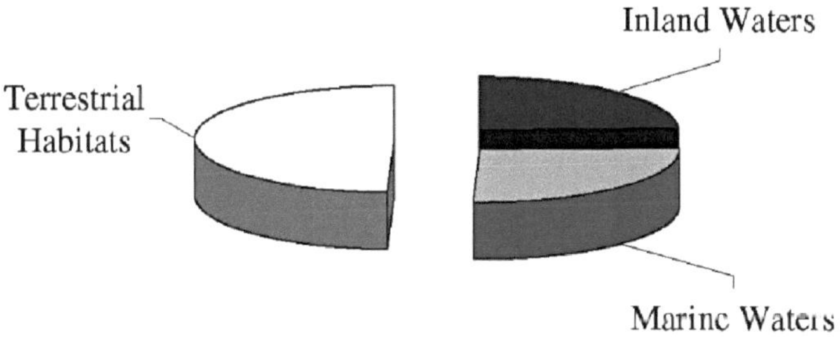

Fig. 2 Frequency distribution of research articles published in the journal *Biological Invasions* (Springer) since 1999 distinguished among habitats (total number = 354).

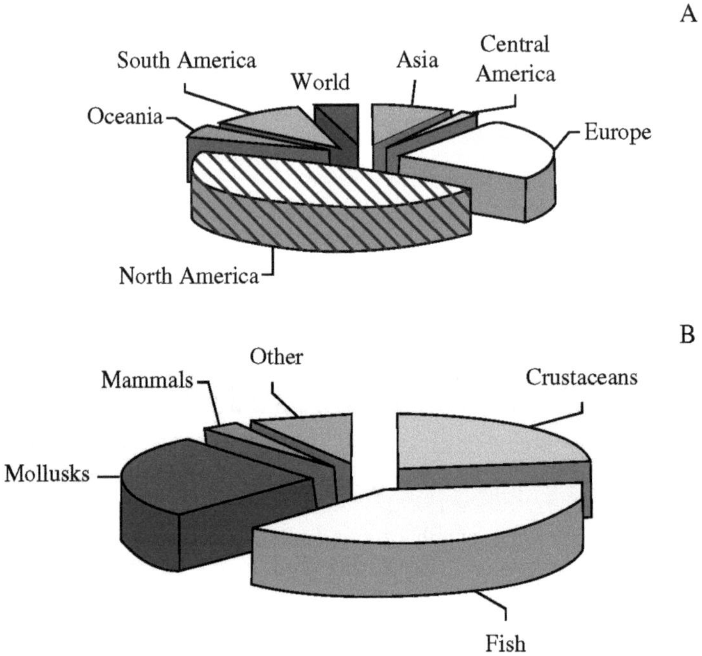

Fig. 3 Frequency distributions of research articles published in the journal *Biological Invasions* (Springer) since 1999 dealing with inland waters distinguished among continents (A) and animal taxa (B) (total number = 69).

perceived ecological role in aquatic food webs and their economic importance to humans. In the other taxa, dreissenids among mollusks and crayfish among crustaceans had been most often studied. Only recently has been the taxonomic coverage of freshwater invaders broadened. New privileged study animals have been Eleotridae and Poeciliidae among fish (e.g. Bedarf *et al.* 2001, Laha and Mattingly 2006, Pusey *et al.* 2006); the Asian clam [*Corbicula fluminea* (Muller)], the golden mussel [*Limnoperna fortunei* (Dunker)], and the golden apple snail [*Pomacea canaliculata* (Lamarck)] among mollusks (e.g. Darrigran 2002, Carlsson and Lacoursière 2005, Boltovskoy *et al.* 2006, Oliveira *et al.* 2006, Yusa *et al.* 2006); and the spiny waterflea (*Bythotrephes longimanus* Leydig) and cladocerans among crustaceans (e.g. Çelik *et al.* 2002, Shurin and Havel 2002, Branstrator *et al.* 2006).

ARE GENERALIZATIONS POSSIBLE?

A consequence of the concentrated interest on terrestrial biomes and of the limited geographic and taxonomic breadth in fresh waters is that traditional invasion paradigms have been mostly derived from terrestrial studies and have been rarely tested in aquatic organisms (Beisel 2001). However, a number of generalizations about freshwater invasion is emerging today. They derive from two main approaches (Moyle and Light 1996b): the first analyzes case studies of invaders and their distribution (e.g. *P. clarkii*, Chapter 4; mollusks, Chapter 5; amphibians and reptiles, Chapter 7; plants, Chapter 11), whereas the second approach aims to extend recent developments in ecological theory to freshwater invaders (e.g. Chapters 19 and 23), in which the focus has been mainly directed to the interaction between the invader and the target community and to the biological characteristics of both the invading species and the ecosystem being invaded. Based on the examples provided by the recent literature, a list of 15 general statements characterizing some of the known events of biological invasion in inland waters can be drawn, as follows.

(1) The establishment and spread success of freshwater NIS often exceeds the 10% value predicted by the "tens rule" (Williamson 1996).

This is seen in the results obtained by Jeschke and Strayer (2005), who analyzed the introductions of vertebrates between Europe and North America (USA and Canada). Using corrected data for unrecorded introductions, the authors showed that, of the 220 and 713 fish species native to Europe and North America respectively, 11% and 6% have been introduced into Europe and North America, respectively, and 36% and 49% have become established after slightly longer than a decade, while 56% and 63% of the established fish had spread and become invasive. On a more global scale, Ruesink (2005) used a database of 1,424 intentional international transfers of freshwater fish and found that up to 64% of the introduced fish became established and 22% of the established cases had exerted a documented impact (i.e. changes in food availability, habitat structure, nutrient dynamics, or top-down trophodynamics).

(2) Propagule pressure is often a major predictor of the establishment of freshwater organisms.

Recent findings showed that the large number of propagules present in an inoculating population, such as the thousands of zebra and quagga mussels carried in the ballast of cargo ships, and the frequencies of sequential inoculations, such as multiple introductions of "desired" species, are positively correlated with invasion success (Lonsdale 1999, Kolar and Lodge 2001, Mack *et al.* 2000, Ricciardi 2001). For instance, as showed by Ruesink (2005), introduced fish species were more likely to establish when humans intended their establishment (76%) rather than when fish were cultivated or used with no explicit desire for naturalization (57%).

(3) The often elevated propagule pressures may explain the several instances in which introduced populations appear to be immune from bottlenecks – usually depleting genetic variation (the "genetic paradox").

Aquatic organisms introduced in large numbers via ballast or subject to multiple introductions can carry a large fraction of the genetic variability of their source populations or bring genetic races from different parts of their native range (Stepien *et al.* 2002). Hence, many colonizers arrive with a high phenotypic and genetic diversity. This adds to other features that may favor their adaptability to the recipient areas, such as the fast acquisition of genetic variability after their arrival that results from such sources as hybridization with closely related organisms, epistasis (i.e. an interaction in which one gene influences the expression of another), or the potential for chromosomal restructuring by inversion, translocation, or duplication (Cox 2004).

(4) Failures of NIS to establish derive most often from their inability to meet the "environmental resistance" on the part of the recipient community – the different regimes of temperature, current, water chemistry, or abiotic resources.

Several examples from different taxa support this statement. Moyle and Light (1996a), for instance, showed that freshwater fish invading North American basins are likely to become established when abiotic conditions are appropriate, regardless of the biota already present. The narrow thermal tolerance of *C. fluminea* may explain its absence from most of the Great Lakes system (Ricciardi 2001), whereas salinity of that system was too low to allow for the successful reproduction of the Chinese mitten crab *Eriocheir sinensis* (Milne Edwards), notwithstanding the frequent introductions of this latter species over the past decades in ship ballast (MacIsaac 1999). The general harshness of the environment may reduce the ability of non-indigenous fish to invade. An example is Eagle Lake, California, which is a highly alkaline (pH: 8–9) terminal lake containing only four indigenous fish species (Moyle and Light 1996b). Any attempt to introduce fish failed in the long term, including the introduction in the early 20th century of the largemouth bass, *M. salmoides*.

(5) As a consequence of (4), success in the establishment of freshwater invaders may depend on a close match between their physiological requirements and the environmental characteristics of the system being invaded.

Species from nearby areas are more likely to be successful invaders than those from more distant locations, as found for fish species in North America that are most likely to be successful if they are adapted to the local, highly seasonal, hydrological regime of the recipient environment (Moyle and Light 1996b). For instance, two species of Cyprinidae introduced into the Pecos River, Texas, USA from nearby areas became established because the artificial flow regime of the recipient river closely resembled that of their native streams (Bestgen *et al.* 1989). Convergent salinity conditions in donor and recipient ecosystems played a key role in the success of invaders in the Great Lakes (MacIsaac *et al.* 2001). An additional prerequisite for successful invasion that allows a species to survive transportation (e.g. in ballast; Bailey *et al.* 2004) and to become established in a recipient area is its euryoeciousness, i.e. its ability to tolerate wide environmental conditions (Ricciardi and Rasmussen 1998). For instance, the range of salinity in which a species can live and reproduce provides a reliable basis for discrimination between invasive and non-invasive North American and East European gammarid amphipods (Devin and Beisel 2007). In general, areas with a wide salinity gradient, like the Baltic Sea, may offer a range of hospitable conditions for invaders, functioning as hot spots of xenodiversity (Leppäkoski *et al.* 2002). Finally, the increased ionic content of large European rivers as the result of pollution has allowed salt tolerant species to spread in new river basins in recent decades (Ketelaars *et al.* 1999).

(6) Demographic factors (sexual precocity, fecundity, and number of generations per year; Lodge 1993, Chapter 12), biological plasticity (Chapters 14 and 16), and/or the ability to overcome the biotic resistance posed by the recipient community (the complex of native predators, parasites, pathogens, and competitors, and previously introduced species) may be neither essential nor sufficient for freshwater species to become invasive.

Usually, *r*-selected crayfish (e.g. *P. clarkii*) rather than K-selected species [(e.g. *Austropotamobius pallipes* (Lereboullet)] have a high probability of spreading. But fecundity and number of generations per year often are not sufficient to explain why the amphipod *G. pulex* has invaded the streams in Northern Ireland where it outcompetes *G. duebeni* Liljeborg (Devin and Beisel 2007). Often, the magnitude of an invader's impact may be predicted by its "taxonomic distinctiveness" within the recipient community (Ricciardi and Atkinson 2004). Indeed, the lack of evolutionary experience with the invader – meaning the absence of competitors, predators, or parasites and the inability to respond to them with an appropriate behavior – may predispose communities to be altered by invasions (e.g. Diamond and Case 1986). For instance, eliminations of indigenous species by *D. polymorpha* have rarely been reported from the invaded European lakes whose native fauna was previously exposed to *Dreissena* during the Pleistocene era (Ricciardi *et al.* 1998). Similarly, the introduced tilapiine species, *Oreochromis niloticus* (Linnaeus), is one of the few fish species to persist in Lake Victoria in large numbers in face of the Nile perch invasion as the result of its past evolutionary experience with similar predators (Moyle and Light 1996b). Sometimes, the interaction between two species that do not share an evolutionary history leads to the positive outcome for one of the two in a form of "evolutionary release" (Schlaepfer *et al.* 2005). For instance, indigenous prey (e.g. anuran tadpoles and metamorphs) may be unable to recognize introduced predators (e.g. the introduced *R. catesbeiana* in the western USA; Rosen and Schwalbe 2002) and their style of preying. As a result, the naïve predators are released from the difficulties of finding a prey (Kiesecker and Blaustein 1997); they may dispose of a high availability of food during the establishment phase of their invasion, which is a precondition of their fast spread. The phenomenon of evolutionary release might explain the paradox of why invasive species sometimes enjoy a competitive advantage over locally adapted species, although there would be a priority effect for residents (e.g. Shea and Chesson 2002, Schlaepfer *et al.* 2005). Also a reduced attack from natural enemies (predators and parasites) encountered outside their natural range gives some species the ability to spread and to become invasive, as predicted by the "enemy release hypothesis" (ERH) (e.g. Torchin *et al.* 2003). However, generalizations about the role of the naïveté of introduced species and of their potential lack of enemies in the recipient community may not be possible in freshwater systems. Several examples, in fact, provide contrasting evidence. The sea lamprey, *P. marinus*, eliminated large fish from Lake Michigan even if some of these species [e.g. the lake trout *Salvelinus namaycush* (Walbaum)] coexisted with the lamprey in other lakes where the species have been together for thousands of years (Moyle 1986). In addition, the ERH has been verified in relatively few organisms and subject to limited criticism (Colautti *et al.* 2004), being only one of the several hypotheses that can explain the abundance and/or the impact of a given invader (Enemy Inversion Hypothesis, climatic variables, selection for "invasive" genotypes, human disturbance, etc.).

(7) In fresh waters, species-rich communities may be as vulnerable to invasion as less speciose, less biologically "sophisticated" communities.

This statement contrasts with one of the most well-established generalizations in the invasion literature since Elton (1958), i.e. that communities with high diversity and complexity are the least susceptible to invasion because of the strength of the community interactions (e.g. Lodge 1993, Levine and D'Antonio 1999, Kennedy *et al.* 2002, Shea and Chesson 2002). Numerous examples at both the global and local levels demonstrate that often the opposite occurs in freshwater systems (Moyle 1986, Ross *et al.* 2001). Jeschke and Strayer (2005), for instance, showed that there is no clear difference in the probability of fish species becoming established and spreading once introduced from Europe to North America and *vice versa*, suggesting that, at the global level, the biota of North America do not offer more resistance to invaders, notwithstanding that it is less disturbed by humans and species-richer than Europe. At a local level, Lake Victoria, which once contained the richest endemic fish communities on the planet, was devastated by a single invader, the Nile perch *L. niloticus*, because it encountered neither predation nor competition from indigenous fish (Goldschmidt *et al.* 1993). Mississipi River, once the most speciose of all temperate rivers, has been invaded by several non-indigenous fish, including common carp *C. carpio*, goldfish *C. auratus*, grass carp *Ctenopharyngodon idella* (Valenciennes), striped bass *Morone saxatilis* (Walbaum), rainbow smelt *Osmerus mordax* (Mitchill), rainbow trout *Oncorhyncus mykiss* Walbaum, and white catfish *Ictalurus catus* (Linnaeus) (Burr and Page 1986). Similarly, the zebra mussel has become established at high densities throughout the Mississippi River basin, which contains the world's richest endemic assemblage of freshwater mussels (Ricciardi *et al.* 1998).

The above examples, however, contrast with the results obtained by employing disturbance treatments in pond zooplankton communities (Smith and Shurin 2006). Shurin (2000) found that reducing the abundance of indigenous species allowed four times as many invaders to establish and to obtain 16 times greater total abundance, therefore showing that in some instances local interactions may be strong enough to exclude a large fraction of potential invaders.

(8) Often, freshwater NIS, instead of interfering with one another, facilitate each other's establishment and/or continued existence, and therefore increase the likelihood and the magnitude of their ecological impact, as predicted by the phenomenon of "invasional meltdown" (Simberloff and Von Holle 1999, Simberloff 2006).

Invasive species may facilitate further invasions by direct effects – providing benefits to another invader – and indirect effects – reducing an invader's enemies or enhancing its prey (Ricciardi 2001). Strong experimental evidence was provided by Adams *et al.* (2003), who showed that introduced fish facilitated invasion by the bullfrog (*R. catesbeiana*) in western North America by preying on native macroinvertebrates (such as dragonfly nymphs) that otherwise precluded establishment or severely limited the numbers of frogs. In the Great Lakes, mutualistic, commensal, and asymmetric exploitative interactions facilitated the survival and population growth of many invaders (Ricciardi 2001). *Dreissena polymorpha* facilitated colonization by several invertebrate NIS (Ricciardi *et al.* 1998) because it increased the surface area and spatial heterogeneity, creating settling sites, providing refuge, and trapping sediment and biodeposits. Additionally, *Dreissena* generates filtration currents that are exploited by other invertebrates (Stewart and Haynes 1994) and it is a food source for several introduced fish, i.e. white bass [*Morone chrysops* (Rafinesque)] and round goby [*Neogobius melanostomus* (Pallas)] (French 1993). Its invasion also increased the spread of Eurasian watermilfoil (MacIsaac 1996). The reduction of piscivores by the parasite sea lamprey was an indirect effect that paved the way for invasion by the planktivore alewife

[*Alosa pseudoharengus* (Wilson)] followed by *Oncorhynchus* spp. (Moyle 1986). Finally, sequential invasions by Ponto-Caspian species completed the life cycle of parasitic organisms, such as the trematode *Bucephalus polymorphus* (Baer) of western Europe origin. The introduction of the first intermediate host (the zebra mussel) of the trematode and its definitive host [the pikeperch *Stizostedion lucioperca* (Linnaeus)] allowed it to spread, causing high mortality in its secondary intermediate hosts, the indigenous cyprinids (Combes and Le Brun 1990).

(9) There is still a poor state of knowledge of whether invasive species are the "drivers" of the extinction of indigenous populations or species, or merely the "passengers" along for the environmental ride (MacDougall and Turkington 2005).

Local and global extinctions frequently overlap invasions in space and time. For instance, the loss of genetically distinct populations of unionids in North America has been accelerated by a factor of 10 after the invasion of zebra mussels (Ricciardi *et al.* 1998). Of the 40 fish species known to have become extinct since 1890 in North America, 27 were negatively affected by the introduction of NIS (Wilcove and Bean 1994). However, the dominance of NIS might be an indirect consequence of habitat modifications that by themselves lead to both indigenous species loss and NIS invasion (Gurevitch and Padilla 2004, Didham *et al.* 2005). For instance, the unionid declines began before the introduction in the mid-1980s of zebra mussels and were caused by several stressors, such as habitat destruction and deterioration resulting from water diversion, erosion, an increase in eutrophication (which causes periods of anoxia), pesticides, loss of host fish for parasitic unionid larvae, historic harvesting for the button industry and harvesting for the pearl industry (reviewed in Gurevitch and Padilla 2004). In some instances, successful freshwater invaders have been integrated without eliminations of species from the communities being invaded. In Lake Malawi, East Africa, 12 species of bottom-feeding haplochromine cichlids from one part of the lake were introduced into another part of the lake in which they were absent without any apparent changes in the abundance of the local species (Trendall 1988). After the completion of the Panama Canal in 1914 and the consequent creation of a freshwater corridor between the Rio Chagres on the Caribbean slope and the Rio Grande on the Pacific slope of the Isthmus of Panama, the freshwater fish assemblages of previously isolated drainage basins were enabled to interchange. However, no cases of local extinctions were recorded but species richness increased by 10% in the Rio Chagres and 22% in the Rio Grande (Smith *et al.* 2004). And there are documented cases (see statement 15) of indigenous species that, given enough time, learn or evolve the ability to escape the "evolutionary trap" caused by an invasive species (Schlaepfer *et al.* 2005).

(10) Among the diverse ways that introduced species threaten the existence of indigenous species (e.g. predation, parasitism, vectoring of pathogens, and competition; Mack *et al.* 2000), the most underestimated is hybridization with indigenous species (Olden *et al.* 2004).

Hybridization is thought to alter the integrity of the endemic gene pools of unionids, crayfish, and fish; it produces hybrid swarms that eliminate indigenous taxa often in a very short time frame (Perry *et al.* 2002). Examples are several, especially in fish. Within a 4-year period following its introduction, the non-indigenous pupfish, *Cyprinodon variegates* Lacépède, was involved in a large-scale introgressive hybridization event with the endemic *Cyprinodon pecosensis* Echelle and Echelle in New Mexico, USA (Echelle and Connor 1989). Similarly, anadromous populations of wild brown trout (*Salmo trutta* Linnaeus) were highly introgressed by stocking with hatchery fish and eventually reduced their fitness (Hansen 2002).

(11) Introduced species have effects at multiple ecological levels in freshwater systems (Simon and Townsend 2003, Chapter 24), but a few studies, mostly focused on salmonids (Simon and Townsend 2003), have analyzed this multifaceted impact (Parker *et al.* 1999).

At the level of individual organisms, invaders may alter the behavior of native species, influencing habitat use and foraging. At the population and community levels, they may induce changes in the abundance or distribution of other species and affect both direct and indirect interactions among populations, respectively. Finally, at the ecosystem level, invaders may change the pathways and magnitude of movements of energy and nutrients.

(12) A first strong danger posed by freshwater invaders to native biota arises if they are either macro-enemies (predators or grazers) or micro-enemies (pathogens or parasites) (Williamson 1996).

Specifically, the NIS posed on the top of the food web or those that are generalized predators (Williamson 1996) are likely to produce marked effects on ecosystem processes (see the case of *P. clarkii*; Chapters 29 and 30) because their impact can "cascade" through the entire food web, altering both ecosystem processes and the behavior of the indigenous species (Townsend 1996). This is especially true in aquatic systems, in which trophic cascades appear to be more common than in terrestrial biomes (Strong 1992). For instance, in California the most successful fish invaders have been piscivores and omnivores (Moyle and Light 1996a), while detritivorous fish seemed to have little effect on indigenous fish assemblages (Power 1990).

(13) A second strong danger posed by freshwater invaders to native biota is their role of "ecosystem engineers", i.e. species that "directly or indirectly control the availability of resources to other organisms by causing physical state changes in biotic or abiotic materials" (Jones *et al.* 1994).

Engineering organisms may cause physical modifications to the environment and influence the maintenance or creation of habitats. Their ecological effects on other species occur because of physical state changes caused, either directly or indirectly, by the engineer, and because, as engineers, they affect the control and use of resources by other species. Zebra mussels have all the properties of ecosystem engineers. They change the characteristics of biotic and abiotic environments by their presence and activities, especially their feeding and filtering. These changes are system-wide, affecting species composition, species interactions, community structure, and ecosystem properties (Karatayev *et al.* 2002, Chapters 32 and 33).

(14) In freshwater systems, as in other biomes, invaders may be subject to evolutionary changes that influence several life history characteristics (Cox 2004).

Once established, NIS are freed from the constraints of the gene flow from their parent population and from the biotic pressures of former enemies, they are subject to altered selection pressures, and they impose strong new evolutionary pressures on the indigenous species. Substantial evolution may take place over relatively short timescales (Carroll and Dingle 1996). For instance, following its introduction to Pacific rivers in North America, the American shad, *Alosa sapidissima* (Wilson), evolved geographic changes in its life history patterns in less than a century (summarized in Dingle 1980). In its native rivers in eastern North America, fecundity and the yearly number of spawns vary as a function of latitude, with reduced clutch size and increased repeat spawning more prevalent in northern versus southern rivers. Fish from Pacific rivers are 32–77% (*vs.* 20–40% from Atlantic rivers) repeat spawners; their age at maturity varies from 3.3 to 3.8 years for males (*vs.* 4 years) and from 4 to 4.5 years for females (*vs.* 4.6 years), and their mean lifetime fecundities range from 321,000 to 500,000 eggs (*vs.* 300,000–350,000). These variations are a function of latitude and water temperature, reflecting rapid post-invasion evolution under selection by local environmental conditions. The western mosquitofish [*Gambusia affinis* (Baird and Girard)] native to North America has shown rapid genetic changes in several locations into which it has been introduced (references in Cox 2004); a period of about 70 generations was sufficient to induce adaptive changes in the life history of the populations introduced into Hawaii, whereas, in the populations introduced into thermal springs in Nevada, changes in body fat content and size at maturity required about 110–165 generations.

(15) In freshwater systems there is growing evidence for adaptive evolutionary responses by indigenous species to NIS and for the influence of such responses on the community dynamics (Lambrinos 2004).

Declines in native populations may be ephemeral if indigenous species are genetically variable in their susceptibility to NIS and can evolve in response to invasion. Alternatively, lack of the ability to evolve in the face of strong selection from invaders can cause extinction (see statement 9). For instance, in 70 years *Rana aurora* Baird and Girard has acquired the ability to recognize the chemical cues emitted by its new predator, the introduced bullfrog *R. catesbeiana*, and behaves accordingly, by reducing their foraging activity and increasing their refuge use, whereas frogs from uninvaded ponds do not change their behavior when presented with bullfrogs (Kiesecker and Blaustein 1997). Evolutionary adaptation may also involve habitat and resource use, leading to the phenomenon of character displacement (i.e. increased difference in quantitative characters of two or more species in areas of syntopy compared to areas of allopatry) (Strauss *et al.* 2006). In the Great Lakes, the bloater [*Coregonus hoyi* (Milner)] is one of the indigenous fish that survived competition with the introduced alewife for zooplankton. Following the explosion of the alewife populations in the 1960s, the bloater shifted its diet from small zooplankton to larger benthic prey (Crowder and Binkowski 1983, Crowder and Crawford 1984). In less than 20 years, this shift was accompanied by an adaptive change of its feeding apparatus that showed a decrease by about 15% in the number of gill rakers (Crowder 1984). Similarly, in North American lakes where bluegill sunfish (*L. macrochirus*) have been absent, the pumpkinseed (*L. gibbosus*) exhibits pelagic and littoral ecotypes as adaptations to the diet of zooplankton (for pelagic forms) and of benthic arthropods and mollusks (for littoral forms). In the lakes where bluegills have been introduced, the pumpkinseed populations exhibited exclusively littoral ecotypes, thus restricting its feeding activity to littoral areas in response to the competition with the dominant newcomer (Robinson *et al.* 2000).

CONCLUSIONS

In recent years, a rising awareness of the economic and ecologic costs caused by invasions in fresh waters has encouraged more proactive research and this has increased our understanding of invasive processes in aquatic systems. Notwithstanding some obvious limitations derived from the relatively small taxonomic coverage of invasion studies and the prevailing focus on certain systems, some general issues regarding freshwater invaders can be raised. First and foremost, predicting the likelihood of the success of a freshwater invader or predicting the invasibility of an aquatic system depends on a detailed understanding of the characteristics of the invader and of the system that is being invaded (Moyle and Light 1996b). But both are likely to be idiosyncratic and complex at the local level, which makes it difficult to apply some generalized theories of invasion biology. As observed by Simberloff (2006), this is part of the larger problem that "ecology is fundamentally an idiographic science" (p. 917): we will need a large catalogue of case studies in order to generate the level of understanding required to deal with many of the environmental problems (Simberloff 2004).

The "tens rule" does not hold for invasion processes in fresh waters. Introduction is a critical step, so the most effective means of minimizing the adverse impact of freshwater invaders is to prevent species transport in the first place. Once introduced, several species have a high potential to establish, and, once established, eradication is often impossible and mitigation and control are difficult and expensive, if possible at all (Chapters 34, 36, 37, and 38). The successful establishment of a species is positively related to propagule pressure (Chapter 18). A consequence is that the probability of establishment might be lessened by reducing both the number of individuals accidentally released via commerce-related activities and the frequency of such releases. The importance of propagule pressure also alerts us about the need to construct effective legislative barriers against the introduction of "desirable" species that might turn out to be "Frankensteins" (Moyle *et al.* 1986).

The most likely "monsters" in fresh waters are those species whose physiological requirements closely match with the environmental characteristics of the recipient system or those species able to tolerate a wide range of environmental conditions. With some exceptions, the biotic resistance exerted by the recipient community, including its richness in species and complexity, seems to be less effective in countering the establishment of freshwater invaders, while an *r*-selected strategy is only in some cases a prerequisite for a species to become invasive. Similarly, the naïveté of introduced species and the assumed lack of enemies in the recipient community cannot explain *per se* freshwater invasions. Most freshwater communities are not saturated with species (Cornell and Lawton 1992), but instead are capable of supporting greater numbers of them if the pool of potential colonists and the rate of colonization from the pool is increased (Gido and Brown 1999). Introduced species, in fact, often facilitate each other's establishment and/or their continued existence, therefore increasing the likelihood and the magnitude of the global ecological impact inflicted by biological invasions.

NIS exert multiform effects on the recipient community, most often acting simultaneously at multiple ecological levels. They may pose threats to indigenous species, populations, and genes, and may induce changes to individuals, populations, communities, and ecosystems. The most dangerous species are parasites but also predators or omnivores that may produce trophic cascades in the recipient community, and ecosystem engineers that may cause physical modifications of the environment and may influence the maintenance or creation of habitats. All these recognized impacts of invaders represent, however, only the "tip" of an ecological and evolutionary iceberg (Palumbi 2001). In inland waters, as in the other biomes, the introduction of species may interact with habitat destruction and degradation, overexploitation of plants and animals, and global climate change to create an "evolutionary revolution" (Cox 2004). And empirical data and theories are urgently needed to enable prediction, understanding, and management of the acute and chronic effects of species invasions (Strayer *et al.* 2006).

REFERENCES

Ackefors, H. 1999. The positive effects of established crayfish. Pages 49–61 in F. Gherardi and D. M. Holdich, editors. Crayfish in Europe as alien species: How to make the best of a bad situation? A. A. Balkema, Rotterdam, The Netherlands.

Adams, M. J., C. A. Pearl, and R. B. Bury. 2003. Indirect facilitation of an anuran invasion by non-native fishes. Ecology Letters **6**, 343–351.

Barbaresi, S. and F. Gherardi. 2000. The invasion of the alien crayfish *Procambarus clarkii* in Europe, with particular reference to Italy. Biological Invasions **2**, 259–264.

Bailey, S. A., I. C. Duggan, C. D. A. Van Overdijk, T. H. Johengen, D. F. Reid, and H. J. MacIsaac. 2004. Salinity tolerance of diapausing eggs of freshwater zooplankton. Freshwater Biology **49**, 286–295.

Bedarf, A. T., K. R. McKaye, E. P. Van Den Berghe, L. J. L. Perez, and D. H. Secor. 2001. Initial six-year expansion of an introduced piscivorous fish in a tropical Central American lake. Biological Invasions **3**, 391–404.

Beisel, J.-N. 2001. The elusive model of a biological invasion process: time to take differences among aquatic and terrestrial ecosystems into account? Ethology, Ecology & Evolution **13**, 193–195.

Bestgen, K. R., S. P. Platania, J. E. Brooks, and D. L. Probst. 1989. Dispersal and life history traits of *Notropis girardi* (Cypriniformes, Cyprinidae) introduced into the Pecos River, New Mexico. American Midland Naturalist **122**, 228–235.

Boltovskoy, D., N. Correa, D. Cataldo, and F. Sylvester. 2006. Dispersion and ecological impact of the invasive freshwater bivalve *Limnoperna fortunei* in the Río de la Plata watershed and beyond. Biological Invasions **8**, 947–963.

Branstrator, D. K., M. E. Brown, L. J. Shannon, M. Thabes, and K. Heimgartner. 2006. Range expansion of *Bythotrephes longimanus* in North America: evaluating habitat characteristics in the spread of an exotic zooplankter. Biological Invasions **8**, 1367–1379.

Burr, B. M. and L. M. Page. 1986. Zoogeography of fishes in the lower Ohio-Upper Mississippi Basin. Pages 287–324 in C. H. Hocutt and E. O. Wiley, editors. The zoogeography of North American freshwater fishes. Wiley, New York, NY.

Carlsson, N. O. L. and J. O. Lacoursière. 2005. Herbivory on aquatic vascular plants by the introduced golden apple snail (*Pomacea canaliculata*) in Lao PDR. Biological Invasions **7**, 233–241.

Carroll, S. P. and H. Dingle. 1996. The biology of post-invasion events. Biological Conservation **78**, 207–214.

Çelik, K., J. E. Schindler, W. J. Foris, and J. C. Knight. 2002. Predator-mediated coexistence of exotic and native crustaceans in a freshwater lake? Biological Invasions **4**, 451–454.

Cohen, A. N. and J. T. Carlton. 1998. Accelerating invasion rate in a highly invaded estuary. Science **279**, 555–558.

Colautti, R. I., A. Ricciardi, I. A. Grigorovich, and H. J. MacIsaac. 2004. Is invasion success explained by the enemy release hypothesis? Ecology Letters **7**, 721–733.

Combes, C. and N. Le Brun. 1990. Invasions by parasites in continental Europe. Pages 285–296 in F. Di Castri, A. J. Hansen and M. Debussche, editors. Biological Invasions in Europe and the Mediterranean Basin. Kluwer Academic Publishers, Dordrecht, The Netherlands.

Cornell, H. V. and J. H. Lawton. 1992. Species interactions, local and regional processes, and limits to the richness of ecological communities: a theoretical perspective. Journal of Animal Ecology **61**, 1–12.

Costanza, R., R. d'Arge, R. de Groot, S. Farber, M. Grasso, B. Hannon, K. Limburg, S. Naeem, R. V. O'Neill, J. Paruelo, R. G. Raskin, P. Sutton, and M. van den Belt. 1997. The value of the world's ecosystem services and natural capital. Nature **387**, 253–260.

Cox, G. W. 2004. Alien species and evolution. Island Press, Washington, DC.

Craig, J. F. 1992. Human induced changes in the composition of fish communities in the African Great Lakes. Reviews in Fish Biology and Fisheries **2**, 93–124.

Crowder, L. B. 1984. Character displacement and habitat shift in a native cisco in southeastern Lake Michigan: evidence for competition? Copeia **1984**, 878–883.

Crowder, L. B. and F. P. Binkowski. 1983. Foraging behaviors and the interaction of alewife, *Alosa pseudoharengus*, and bloater, *Coregonus hoyi*. Environmental Biology of Fishes **8**, 105–113.

Crowder, L. B. and H. L. Crawford. 1984. Ecological shifts in resource use by bloaters in Lake Michigan. Transactions of the American Fisheries Society **113**, 694–700.

Darrigran, G. 2002. Potential impact of filter-feeding invaders on temperate inland freshwater environments. Biological Invasions **4**, 145–156.

Devin, S. and J.-N. Beisel. 2007. Biological and ecological characteristics of invasive species: a gammarid study. Biological Invasions **9**, 13–24.

Diamond, J. and T. J. Case. 1986. Overview: introductions, extinctions, exterminations, and invasions. Pages 65–79 in J. Diamond and T. J. Case, editors. Community Ecology. Harper & Row, New York, NY.

Didham, R. K., J. M. Tylianakis, M. A. Hutchison, R. M. Ewers, and N. J. Gemmell. 2005. Are invasive species the drivers of ecological change? Trends in Ecology and Evolution **20**, 470–474.

Dingle, H. 1980. The ecology and evolution of migration. Pages 1–101 in S. A. Gauthreaux Jr., editor. Animal migration, orientation, and navigation. Academic Press, New York, NY.

Eaton, J. G. and R. M. Scheller. 1996. Effects of climatic warming on fish thermal habitat in streams of the United States. Limnology and Oceanography **41**, 1110–1115.

Echelle, A. A. and P. J. Connor. 1989. Rapid, geographically extensive genetic introgression after secondary contact between two pupfish species (Cyprinodon, Cyprinodontidae). Evolution **43**, 717–727.

Elton, C. S. 1958. The Ecology of invasions by animals and plants. The University of Chicago Press, Chicago, IL, Edition 2000.

French, J. R. P. I. 1993. How well can fishes prey on zebra mussels in eastern North America? Fisheries **18**, 13–19.

Gherardi, F. 2006. Crayfish invading Europe: the case study of *Procambarus clarkii*. Marine and Freshwater Behaviour and Physiology **39**, 175–191.

Gherardi, F., A. Raddi, S. Barbaresi, and G. Salvi. 2000. Life history patterns of the red swamp crayfish, *Procambarus clarkii*, in an irrigation ditch in Tuscany, Italy. Crustacean Issues **12**, 99–108.

Gido, K. B. and J. H. Brown. 1999. Invasion of North American drainage by alien fish species. Freshwater Biology **42**, 387–399.

Goldschmidt, T., F. Witte, and J. Wanink. 1993. Cascading effects of the introduced Nile perch on the detritivorous/phytoplanktvorous species in the sublittoral areas of Lake Victoria. Conservation Biology **7**, 686–700.

Gurevitch, J. and D. K. Padilla. 2004. Are invasive species a major cause of extinctions? Trends in Ecology and Evolution **19**, 470–474.

Hansen, M. H. 2002. Estimating the long-term effects of stocking domesticated trout into wild brown trout (*Salmo trutta*) populations: an approach using microsatellite DNA analysis of historical and contemporary samples. Molecular Ecology **11**, 1003–1015.

Hebert, P. D. N. and M. Cristescu. 2002. Genetic perspective on invasions: the case of the Cladocera. Canadian Journal of Fisheries and Aquatic Sciences **59**, 1229–1234.

Jeschke, J. M. and D. L. Strayer. 2005. Invasion success of vertebrates in Europe and North America. Proceedings of the National Academy of Sciences USA **102**, 7198–7202.

Jones, C. G., J. H. Lawton, and M. Shachak, 1994. Organisms as ecosystem engineers. Oikos **69**, 373–386.

Karatayev, A. Y., L. E. Burlakova, and D. K. Padilla. 2002. Impacts of zebra mussels on aquatic communities and their role as ecosystem engineers. Pages 433–446 in E. Leppäkoski, S. Gollasch and S. Olenin, editors. Invasive aquatic species of Europe: distribution, impacts and management. Kluwer Academic Publishers, Dordrecht, The Netherlands.

Kennedy, T. A., S. Naeem, K. M. Howe, J. M. H. Knops, D. Tilman, and P. Relch. 2002. Biodiversity as a barrier to ecological invasion. Nature **417**, 636–638.

Ketelaars, H. A. M., F. E. Lambregts-van de Clundert, C. J. Carpentier, A. J. Wagenvoort, and W. Hoogenboezem. 1999. Ecological effects of the mass occurrence of the Ponto-Caspian invader, *Hemimysis anomala* G.O. Sars, 1907 (Crustacea: Mysidacea), in a freshwater storage reservoir in the Netherlands, with notes on autoecology and new records. Hydrobiologia **394**, 233–248.

Kiesecker, J. M. and A. R. Blaustein. 1997. Population differences in responses of red-legged frogs (*Rana aurora*) to introduced bullfrogs. Ecology **78**, 1752–1760.

Kirjavainen, J. and M. Sipponen. 2004. Environmental benefit of different crayfish management strategies in Finland. Fisheries Management and Ecology **11**, 213–218.

Kolar, C. S. and D. M. Lodge. 2001. Progress in invasion biology: predicting invaders. Trends in Ecology and Evolution **16**, 199–204.

Laha, M. and H. T. Mattingly. 2006. Identifying environmental conditions to promote species coexistence: an example with the native Barrens topminnow and invasive western mosquitofish. Biological Invasions **8**, 719–725.

Lambrinos, J. G. 2004. How interactions between ecology and evolution influence contemporary invasion dynamics? Ecology **85**, 2061–2070.

Lehtonen, H. 2002. Alien freshwater fishes of Europe. Pages 153–161 in E. Leppäkoski, S. Gollasch and S. Olenin, editors. Invasive aquatic species of Europe: distribution, impact and management. Kluwer Academic Publishers, Dordrecht, The Netherlands.

Leppäkoski, E., S. Gollasch, and S. Olenin. 2002. Alien species in European waters. Pages 1–6 in E. Leppäkoski, S. Gollasch and S. Olenin, editors. Invasive Aquatic Species of Europe: Distribution, impact and management. Kluwer Academic Publishers, Dordrecht, The Netherlands.

Levine, J. M. and C. M. D'Antonio. 1999. Elton revisited: a review of evidence linking diversity and invisibility. Oikos **87**, 15–26.

Lodge, D. M. 1993. Biological invasions: lessons for ecology. Trends in Ecology and Evolution **8**, 133–137.

Lodge, D. M. 2001. Lakes. Pages 277–313 in F. S. Chapin III, O. E. Sala and E. Huber-Sannwald, editors. Global biodiversity in a changing environment: scenarios for the 21st century. Springer, New York, NY.

Lodge, D. M. and K. Shrader-Frechette. 2003. Nonindigenous species: ecological explanation, environmental ethics, and public policy. Conservation Biology **17**, 31–37.

Lonsdale, W. M. 1999. Global patterns of plant invasions and the concept of invasibility. Ecology **80**, 1522–1536.

MacDougall, A. S. and R. Turkington. 2005. Are invasive species the drivers or passengers of change in degraded ecosystems? Ecology **86**, 42–55.

MacIsaac, H. J. 1996. Potential abiotic and biotic impacts of zebra mussels on the inland waters of North America. American Zoologist **36**, 287–299.

MacIsaac, H. J. 1999. Biological invasions in Lake Erie: past, present, and future. In State of Lake Erie (SOLE) – past, present, and future. Pages 305–322 in M. Munawar and T. Edsall, editors. Ecovision World Monograph Series, Backhuys, Leiden, The Netherlands.

MacIsaac, H. J., I. A. Grigorovich, and A. Ricciardi. 2001. Reassessment of species invasion concepts: the Great Lakes basin as a model. Biological Invasions **3**, 405–416.

Mack, R. N., D. Simberloff, W. M. Lonsdale, H. Evans, M. Clout, and F. A. Bazzaz. 2000. Biotic invasions: causes, epidemiology, global consequences, and control. Ecological Applications **10**, 689–710.

Millennium Ecosystem Assessment. 2005. Ecosystem and human well-being: biodiversity synthesis. World Resources Institute, Washington, DC.

Mills, E. L., R. M. Dermott, E. F. Roseman, D. Dustin, E. Mellina, D. B. Conn, and A. P. Spidle. 1993. Colonization, ecology, and population structure of the "quagga" mussel (Bivalvia: Dreissenidae) in the lower Great Lakes. Canadian Journal of Fisheries and Aquatic Sciences **50**, 2305–2314.

Mills, E. L., D. L. Strayer, M. D. Scheuerell, and J. T. Carlton. 1996. Exotic species in the Hudson River basin –a history of invasions and introductions. Estuaries **19**, 814–823.

Moyle, P. B. 1986. Fish introductions into North America: patterns and ecological impact. Pages 27–43 in H. A. Mooney and J. A. Drake, editors. Ecology of biological invasions of North America and Hawaii. Springer, New York, NY.

Moyle, P. B., H. W. Li, and B. A. Barton. 1986. The Frankenstein Effect: impact of introduced fishes on native fishes in North America. Pages 415–426 in R. H. Strond, editor. Fish culture in fisheries management. American Fisheries Societies, Bethesda, MD.

Moyle, P. B. and T. Light. 1996a. Fish invasions in California: do abiotic factors determine success? Ecology **76**, 1666–1670.

Moyle, P. B. and T. Light. 1996b. Biological invasions of fresh water: empirical rules and assembly theory. Biological Conservation **78**, 149–161.

Olden, J. D., N. L. Poff, M. R. Douglas, M. E. Douglas, and K. D. Fausch. 2004. Ecological and evolutionary consequences of biotic homogenization. Trends in Ecology and Evolution **19**, 18–24.

Oliveira, M. D. de, A. M. Takeda, L. F. de Barros, D. S. Barbosa, and E. K. de Resende. 2006. Invasion by *Limnoperna fortunei* (Dunker, 1857) (Bivalvia, Mytilidae) of the Pantanal wetland, Brazil. Biological Invasions **8**, 97–104.

Palumbi, S. 2001. Humans as the world's greatest evolutionary force. Science **293**, 1786–1790.

Parker, I. M., D. Simberloff, W. M. Lonsdale, K. Goodell, M. Wonham, P. M. Kareiva, M. H. Williamson, B. Von Holle, P. B. Moyle, J. E. Byers, and L. Goldwasser 1999. Impact: toward a framework for understanding the ecological effects of invaders. Biological Invasions **1**, 3–19.

Perry, W. L., D. M. Lodge, and J. L. Feder. 2002. Importance of hybridization between indigenous and nonindigenous freshwater species: an overlooked threat to North American biodiversity. Systematic Biology **51**, 255–275.

Power, M. E. 1990. Effects of fish in river food webs. Science **250**, 811–814.

Pusey, B., D. Burrows, A. Arthington, and M. Kennard. 2006. Translocation and spread of piscivorous fishes in the Burdekin River, north-eastern Australia. Biological Invasions **8**, 965–977.

Rahel, F. J. 2000. Homogenization of fish faunas across the United States. Science **288**, 854–856.

Ricciardi, A. 2001. Facilitative interactions among aquatic invaders: is an "invasional meltdown" occurring in the Great Lakes? Canadian Journal of Fisheries and Aquatic Sciences **58**, 2513–2525.

Ricciardi, A. and S. K. Atkinson. 2004. Distinctiveness magnifies the impact of biological invaders in aquatic ecosystems. Ecology Letters **7**, 781–784.

Ricciardi, A., R. J. Neves, and J. B. Rasmussen. 1998. Impending extinctions of North American freshwater mussels (Unionida) following the zebra mussel (*Dreissena polymorpha*) invasion. Journal of Animal Ecology **67**, 613–619.

Ricciardi, A. and J. B. Rasmussen. 1998. Predicting the identity and impact of future biological invaders: a priority for aquatic resource managers. Canadian Journal of Fisheries and Aquatic Sciences **55**, 1759–1765.

Ricciardi, A. and J. B. Rasmussen. 1999. Extinction rates of North American freshwater fauna. Conservation Biology **13**, 1220–1222.

Robinson, B. W., D. S. Wilson, and A. S. Margosian. 2000. A pluralistic analysis of character release in pumpkinseed sunfish (*Lepomis gibbosus*). Ecology **81**, 2799–2812.

Rodríguez, C. F., E. Bécares, M. Fernández-Aláez, and C. Fernández-Aláez. 2005. Loss of diversity and degradation of wetlands as a result of introducing exotic crayfish. Biological Invasions **7**, 75–85.

Rosen, P. C. and C. R. Schwalbe. 2002. Widespread effects of introduced species on reptiles and amphibians in the Sonoran Desert region. Pages 220–240 in B. Tellman, editor. Invasive exotic species in the Sonoran region. University of Arizona Press and the Arizona-Sonora Desert Museum, Tucson, AR.

Ross, R. M., W. A. Lellis, R. M. Bennett, and C. S. Johnson. 2001. Landscape determinants of nonindigenous fish invasions. Biological Invasions **3**, 347–361.

Ruesink, J. L. 2005. Global analysis of factors affecting the outcome of freshwater fish introductions. Conservation Biology **19**, 1883–1893.

Ruiz, G. M., P. W. Fofonoff, J. T. Carlton, M. J. Wonham, and A. H. Hines. 2000. Invasion of coastal marine communities in North America: apparent patterns, processes, and biases. Annual Review of Ecology and Systematics **31**, 481–531.

Sala, O. E., F. S. Chapin III, J. J. Armesto, R. Berlow, J. Bloomfield, R. Dirzo, E. Huber-Sanwald, L. F. Huenneke, R. B. Jackson, A. Kinzig, R. Leemans, D. Lodge, H. A. Mooney, M. Oesterheld, N. L. Poff, M. T. Sykes, B. H. Walker, M. Walker, and D. H. Wall. 2000. Global biodiversity scenarios for the year 2100. Science **287**, 1770–1774.

Schlaepfer, M. A., P. W. Sherman, B. Blossey, and M. C. Runge. 2005. Introduced species as evolutionary traps. Ecology Letters **8**, 241–246.

Shea, K. and P. Chesson. 2002. Community ecology theory as a framework for biological invasions. Trends in Ecology and Evolution **17**, 170–176.

Shurin, J. B. 2000. Dispersal limitation, invasion resistance, and the structure of pond zooplankton communities. Ecology **81**, 3074–3086.

Shurin, J. B. and J. E. Havel. 2002. Hydrologic connections and overland dispersal in an exotic freshwater crustacean. Biological Invasions **4**, 431–439.

Simberloff, D. 2004. Community ecology: is it time to move on? American Naturalist **163**, 787–799.

Simberloff, D. 2006. Invasional meltdown 6 years later: important phenomenon, unfortunate metaphor, or both? Ecology Letters **9**, 912–919.

Simberloff, D. and B. Von Holle. 1999. Positive interactions of nonindigenous species: invasional meltdown? Biological Invasions **1**, 21–32.

Simon, K. S. and C. R. Townsend. 2003. Impacts of freshwater invaders at different levels of ecological organization, with emphasis on salmonids and ecosystem consequences. Freshwater Biology **48**, 982–994.

Smith, S. A., G. Bell, and E. Bermingham. 2004. Cross-Cordillera exchange mediated by the Panama Canal increased the species richness of local freshwater fish assemblages. Proceedings of the Royal Society of London B **271**, 1889–1896.

Smith, S. A. and J. B. Shurin. 2006. Room for one more? Evidence for invasibility and saturation in ecological communities. Pages 423–447 in M. W. Cadotte, S. M. McMahon and T. Fukami, editors. Conceptual ecology invasion biology: reciprocal approaches to nature. Springer, Dordrecht, The Netherlands.

Stepien, C. A., C. D. Taylor, and K. A. Dabrowska. 2002. Genetic variability and phylogeographical patterns of a nonindigenous species invasion: a comparison of exotic vs. native zebra and quagga mussel populations. Journal of Evolutionary Biology **15**, 314328.

Stewart, T. W. and J. M. Haynes. 1994. Benthic macroinvertebrate communities of southwestern Lake Ontario following invasion of *Dreissena*. Journal of Great Lakes Research **20**, 479–493.

Strauss, S. Y., J. A. Lau, and S. P. Carroll. 2006. Evolutionary responses of natives to introduced species: what do introductions tell us about natural communities? Ecology Letters **9**, 357–374.

Strayer, D. L., E. A. Blair, N. F. Caraco, J. J. Cole, S. Findlay, W. C. Nieder, and M. L. Pace. 2005. Interactions between alien species and restoration of large-river ecosystems. Archiv für Hydrobiologie, Supplement **155**/1–4, 133–145.

Strayer, D. L., V. T. Eviner, J. M. Jeschke, and M. L. Pace. 2006. Understanding the long-term effects of species invasions. Trends in Ecology and Evolution **21**, 645–651.

Strong, D. R. 1992. Are trophic cascades all wet? Differentiation and donor-control in speciose ecosystems. Ecology **73**, 747–754.
Torchin, M. E., K. D. Lafferty, A. P. Dobson, V. J. McKenzie, and A. M. Kuris. 2003. Introduced species and their missing parasites. Nature **421**, 628–630.
Townsend, C. R. 1996. Invasion biology and ecological impacts of brown trout *Salmo trutta* in New Zealand. Biological Conservation **78**, 13–22.
Trendall, J. 1988. The distribution and dispersal of introduced fish at Thumbi West Island in Lake Malawi, Africa. Journal of Fish Biology **33**, 357–369.
Vitousek, P. M., C. M. D'Antonio, L. D. Loope, M. Rejmánek, and R. Westbrooks. 1997. Introduced species: A significant component of human-caused global change. New Zealand Journal of Ecology **21**, 1–16.
Vitousek, P. M., C. M. D'Antonio, L. D. Loope, and R. Westbrooks. 1996. Biological invasions as global environmental change. American Scientist **84**, 468–478.
Welcomme, R. I. 1988. International introductions of inland aquatic species. Fisheries technical paper 294. Food and Agriculture Organization, Rome, Italy.
Wilcove, D. S. and M. J. Bean. 1994. The big kill: declining biodiversity in America's lakes and rivers. Environmental Defense Fund, Washington, DC.
Williamson, M. 1996. Biological invasions. Chapman & Hall, London, UK.
Yusa, Y., N. Sugiura, and T. Wada. 2006. Predatory potential of freshwater animals on an invasive agricultural pest, the apple snail *Pomacea canaliculata* (Gastropoda: Ampullariidae), in southern Japan. Biological Invasions **8**, 137–147.

Distribution of invaders

Since the Age of Exploration began, there has been a drastic breaching of biogeographic barriers that previously had isolated the continental biota for millions of years. We are now developing a whole new cosmopolitan assemblage of organisms across the surface of the Earth with large consequences not only for the functioning of ecosystems but also for the future evolutionary trajectory of life.

Harold Mooney and Elsa Cleland (2001)

Chapter two

Invasive crustaceans in European inland waters

David M. Holdich and Manfred Pöckl

INTRODUCTION

At least 52,000 species of crustaceans have been described, although many more probably exist (Martin and Davis 2001). They are amongst the most prolific macroinvertebrates in the aquatic environment, both in terms of numbers and species diversity, but they do not usually cause public concern unless they are large and become invasive, e.g. the red king crab, *Paralithodes camtschaticus* Samouelle, in Norway, the Chinese mitten crab, *Eriocheir sinensis*, in Germany and the UK, and the red swamp crayfish, *Procambarus clarkii*, in African lakes (Chapter 4).

The Global Invasive Species Database (http://www.issg.org/database) lists three crustaceans in its world's worst 100 invasive non-indigenous species (NIS), i.e. the green crab, *Carcinus maenas* Linnaeus; the fishhook waterflea, *Cercopagis pengoi*; and the Chinese mitten crab, *E. sinensis*. However, in the 'Global Strategy on Invasive Alien Species' (McNeely *et al.* 2001) crustaceans are not dealt with, save for a brief mention of non-indigenous crayfish escaping from a London fish market.

Many aquatic crustaceans produce planktonic larvae or resistant propagules (Panov *et al.* 2004) and consequently can be moved great distances, either naturally or by human-mediated means, e.g. they are the commonest faunal component in ballast water of ships (Panov *et al.* 2004). Some attach themselves to solid surfaces or construct tubes on such surfaces, which may then become mobile, e.g. ships' hulls and oil platforms, whilst others burrow into softer materials such as wood – these habits can result in the crustacean being transported outside its home range, and even transcontinentally. Others have

Francesca Gherardi, Biological invaders in inland waters: Profiles, distribution, and threats, 29–75.

been translocated for economic reasons such as aquaculture and to enhance fish production, and for the pet and restaurant trades, and have subsequently become established in the wild. The majority of crustacean introductions have been recorded for the marine and estuarine environments, and these have been well documented (e.g. Carlton 1996, Ruiz *et al.* 1997, Rodríguez and Suárez 2001); in general less attention has been paid to introductions into inland waters (Welcomme 1988, Gherardi and Holdich 1999, Leppäkoski *et al.* 2002b), with the exception of fish (Welcomme 1991, Lehtonen 2002).

Despite the large number of crustacean species present in the aquatic environment, relatively few have become established outside their natural range due to accidental or deliberate introductions into European waters. Most of those that have become established occur in the marine and estuarine environments, but a growing number of species are becoming established in inland waters, mainly amphipods from the Ponto–Caspian basin and North America, and crayfish from North America. In this review, details are given of invasive crustaceans that have become established in European inland waters in recent times. Although all groups are dealt with, particular attention is given to the amphipods and decapods as they are currently having the most impact. In total, three species of Branchiopoda, four species of Copepoda, one species of Branchiura, and 46 species of Malacostraca (5 Mysida, 21 Amphipoda, 4 Isopoda, and 15 Decapoda [two Caridea (prawns), six Brachyura (crabs), and nine Astacida (crayfish)]) are listed in Tables 1–3. In the majority of cases it is difficult to assess whether or not an invasive species is having a high impact, but when this is known then it is highlighted in the tables.

In this review, inland waters will be taken as meaning rivers, lakes, and reservoirs. Coastal lagoons, saline lakes, estuaries, and low salinity seas, such as the Baltic, will be mentioned as appropriate. Although the Baltic is, to quote Leppäkoski *et al.* (2002a), "a sea of invaders", particularly for invasive crustaceans, it has been well covered elsewhere, e.g. Jaźdźewski and Konopacka (2002), Leppäkoski *et al.* (2002a, b, c) and Telesh and Ojavear (2002). However, the following facts are of interest. The Ponto–Caspian branchiopod, *Evadne anonyx* Sars, is widespread in the Baltic but cannot tolerate freshwater (V. E. Panov 2006, personal communication) so is unlikely to invade inland waters. The North American copepod, *Acartia tonsa* Dana, is widespread in Europe, particularly in the Baltic, but does not appear to have entered inland waters. The New Zealand barnacle, *Elminius modestus* Darwin, is also widespread in coastal waters, but does not occur in inland waters.

The classification of Crustacea used in this review is mainly based on that of Martin and Davis (2001). However, the higher taxonomic categories other than family have not been given a name, e.g. class, infraorder, order, etc., as in many cases there still seems to be disagreement over the correct terminology. For example, some workers refer to the Cladocera as a suborder (Martin and Davis 2001), whilst others call them a superorder (V. E. Panov 2006, personal communication). Many workers still use the term Mysidacea, whilst Martin and

Table 1 Non-indigenous Amphipoda in European inland waters (those with a + are considered to be highly invasive).

Scientific name, authority, family	Natural range	Introduced range
Malacostraca, Amphipoda		
+*Chelicorophium curvispinum* (G. O. Sars) (Corophiidae)	Ponto–Caspian basin	Widespread in western Europe, including British Isles
Chelicorophium robustum (Bousfield and Hoover) (Corophiidae)	Ponto–Caspian basin	Poland, Germany
Corophium sowinskyi Martynov (Corophiidae)	Ponto–Caspian basin	Poland, Czech Republic
+*Crangonyx pseudogracilis* Bousfield (Crangonyctidae)	USA	British Isles, The Netherlands
Dikerogammarus bispinosus Martynov (Gammaridae)	Ponto–Caspian basin	Hungary, Austria, (poorly known?)
Dikerogammarus haemobaphes (Eichwald) (Gammaridae)	Ponto–Caspian basin	Hungary, Austria, Germany, Poland
+*Dikerogammarus villosus* (Sowinsky) (Gammaridae)	Ponto–Caspian basin	Hungary, Austria, Germany, The Netherlands, France, Italy
Echinogammarus berilloni (Catta) (Gammaridae)	Iberian Peninsula	France, Belgium, Luxembourg, Germany, The Netherlands
+*Echinogammarus ischnus* (Stebbing) (Gammaridae)	Ponto–Caspian basin	Poland, Germany, The Netherlands, Austria, Hungary
Echinogammarus trichiatus (Martynov) (Gammaridae)	Ponto–Caspian basin	Bavarian Danube, Rhine, Austr. Danube
Echinogammarus warpachowskyi (G. O. Sars) (Gammaridae)	Ponto–Caspian basin	Ukraine, Lithuania, Poland
+*Gammarus pulex* (Linnaeus) (Gammaridae)	Most European countries	New water bodies within Europe, including Ireland
Gammarus roeseli Gervais (Gammaridae)	Balkan Penins, Asia Minor	Central Europe, France, Germany
+*Gammarus tigrinus* Sexton (Gammaridae)	USA	British Isles, Germany, The Netherlands, southern Baltic
+*Gmelinoides fasciatus* (Stebbing) (Gammaridae)	Russia (Lake Baikal)	Lakes Ladoga, Onega, Peipsi, R. Neva and Neva estuary
Hyalella azteca (Saussure) (Hyalellidae)	USA	Not yet self-established, high risk
Obesogammarus crassus (G. O. Sars) (Gammaridae)	Ponto–Caspian basin	Lithuania, Poland
+*Obesogammarus obesus* (G. O. Sars) (Gammaridae)	Ponto–Caspian basin	Austrian and Bavarian Danube
Orchestia cavimana Heller (Orchestidae)	Ponto–Caspian basin	Widespread, including England
+*Pontogammarus robustoides* (G. O. Sars) (Gammaridae)	Ponto–Caspian basin	Poland, Germany

Table 2 Non-indigenous Decapoda (Pleocyemata) in European inland waters (those with a + are considered to be highly invasive).

Scientific name, authority, family	Natural range	Introduced range
Malacostraca, Caridea		
Atyaephyra desmaresti (Millet) (Atyidae)	Southern Europe	France (1843), Belgium (1888), The Netherlands (1916), Germany (1932), R. Main (1983), Main-Danube Canal (1990), Bavarian Danube (1997), Austrian Danube (1999)
Palaemon macrodactylus Rathbun (Palaemonidae)	North-east Asia	England (2004)
Malacostraca, Brachyura		
Callinectes sapidus M.J. Rathbun (Portunidae)	USA	Widespread (France, 1901)
+*Eriocheir sinensis* H. Milne Edwards (Varunidae)	SE Asia	Widespread (Germany, 1912)
Potamon sp. (Potomidae)	Unknown	France (1985)
Potamon fluviatile (Herbst) (Potomidae)	Italy, Balkans, Greece	France (early 19th century)
Potamon ibericum tauricum (Czerniavsky) (Potomidae)	Turkey	France (1960s)
Rhithropanopeus harrisii Maitland (Xanthidae)	USA	Widespread (The Netherlands, 1874)
Malacostraca, Astacida		
+*Astacus leptodactylus* Eschscholtz (Astacidae)	Ponto–Caspian basin	Widespread
Cherax destructor Clark (Parastacidae)	Australia	Spain (Spain, 1983), possibly Switzerland
Orconectes immunis (Hagen) (Cambaridae)	USA	Germany (Germany, 1997)
+*Orconectes limosus* (Rafinesque) (Cambaridae)	USA	Widespread (Germany, 1890)
+*Orconectes rusticus* (Girard) (Cambaridae)	USA	France (France, 2005*)
+*Orconectes virilis* (Hagen) (Cambaridae)	North America	The Netherlands (The Netherlands, 2005*)
+*Pacifastacus leniusculus* (Dana) (Astacidae)	North America	Widespread (Sweden, 1960s)
+*Procambarus clarkii* (Girard) (Cambaridae)	USA	Widespread (Spain, 1970s)
+*Procambarus* sp. (Cambaridae)	USA	Germany, The Netherlands (Germany, 2003)

* reported in that year but probably present earlier

Table 3 Other non-indigenous crustaceans in European inland waters (those with a + are considered to be highly invasive).

Scientific name, authority, (family in brackets)	Natural range	Introduced range
Branchiopoda, Cladocera		
+*Bythotrephes longimanus* Leydig (Cercopagididae)	Widespread in Europe	Widespread – further invasions in Europe
+*Cercopagis pengoi* (Ostroumov) (Cercopagididae)	Ponto–Caspian basin	Widespread, including Baltic Sea
Cornigerius maeoticus (Pengo) (Podonidae)	Ponto–Azov	Widespread, including Baltic Sea
Copepoda, Calanoida		
+*Acanthocyclops americanus* (Marsh) (Cyclopoida)	North America	Britain (19th century), now widespread in Europe
Apocyclops panamensis (Marsh)	Central America	Italy (1990s)
Boeckella triarticulata (G. M.Thompson) (Calanoida)	Australia	Italy (1992)
Eudiaptomus gracilis (G. O. Sars) (Calanoida)	Europe, but not Italy	Italy (late 1980s)
Branchiura, Arguloida		
Argulus japonicus Thiele (Argulidae)	South-east Asia	Spain (1921), Germany (1935), France, Italy, Poland, UK (1990)
Malacostraca, Isopoda		
Asellus communis Say (Asellota)	USA	England
+*Jaera istri* Veuille (Asellota)	Ponto–Caspian basin	Austrian Danube (1934), Bavarian Danube (1967), Main–Danube Canal (1993), R. Main (1994), R. Rhine delta (1997), R. Rhine (2003), R. Elbe (1999)
Proasellus coxalis (Asellota)	Southern Europe	Widespread
Proasellus meridionalis Racovitza (Asellota)	Southern Europe	Widespread

Table 3 Continued.

Scientific name, authority, (family in brackets)	Natural range	Introduced range
Malacostraca, Mysida		
+*Hemimysis anomala* G. O. Sars (Mysidae)	Ponto–Caspian basin	Baltic Sea countries (1990s), The Netherlands (1998), Austrian Danube (1999), Belgium (2000), England (2004), Germany (2004), France (2006)
+*Katamysis warpachowskyi* G. O. Sars (Mysidae)	Ponto–Caspian basin	Hungarian, Slovakian and Austrian Danube (2001)
+*Limnomysis benedeni* Czerniavsky (Mysidae)	Ponto–Caspian basin	Budapest (1947), Vienna (1973), Bavarian Danube (1993), Main–Danube Canal (1998), R. Rhine (1998), R. Rhine delta (1998)
Paramysis intermedia (Czerniavsky) (Mysidae)	Ponto–Caspian basin	Baltic countries
Paramysis lacustris (Czerniavsky) (Mysidae)	Ponto–Caspian basin	Baltic countries

Davis (2001) call them the Mysida. Martin and Davis (2001) discussed the conflicting views about the terminology used for crayfish and admit that the one they have used is misleading, i.e. Superfamilies Astacoidea and Parastacoidea in the Infraorder Astacidea, as the crayfish are now considered to be monophyletic (Crandall *et al.* 2000, Scholtz 2002) and yet the two superfamilies are given equal rank with the three other superfamilies in the infraorder. K. Crandall (2006, personal communication) is of the opinion that the crayfish should not be elevated to their own infraorder, e.g. the Astacida, as suggested by Scholtz and Richter (1995) (see also Scholtz 2002 and Taylor 2002), until more studies are carried out, and that the original classification of H. H. Hobbs Jr (see, e.g. Hobbs Jr 1988) should be retained for the time being. However, in this review the classification used by Taylor in Holdich (2002a) is used, i.e. the crayfish are in the Infraorder Astacida with two superfamilies as noted above. Ahyong (2006) in a recent analysis of homarid phylogeny also places the crayfish in the Astacida.

INVASIVE CRUSTACEANS – ORIGINS, SPREAD, AND IMPACT

Background

The invasion of European inland waters by crustaceans has been mainly on three fronts: introductions (a) from North America, Australia, and Asia; (b) from one European region to another; and (c) from the Ponto–Caspian Basin by three routes. These are: (1) northern invasion corridor – Volga–Baltic inland waterway; (2) central invasion corridor – Dnieper–Vistula–Oder–Elbe–Rhine; and (3) southern invasion corridor – the Danube River connection with the Rhine basin (Fig. 1). Many of the species using these invasion corridors have become established in the low salinity Baltic Sea and its associated gulfs, but have moved by natural diffusion or aided by ships through these freshwater corridors to get there.

Invasive crustaceans have either been introduced intentionally or unintentionally, or in some cases have made their own way from one region to another via canals and rivers, and during floods. Van der Velde *et al.* (2000) and Bernaurer and Jansen (2006) note that the River Rhine has many invasive crustaceans (e.g. mysids, amphipods, isopods, and decapods) that have migrated there via the Main–Danube Canal from the River Danube, which itself contains a number of Ponto–Caspian species. Anthropomorphic effects in the R. Rhine has raised salt and temperature levels, thus making conditions favourable for species that originally lived in estuarine or brackish water. However, Kelleher et al. (2000b) point out that water quality in the lower R. Rhine has in fact improved since the restoration plan initiated after the Sandoz chemical spill in 1986, and whilst this is making conditions favourable for the return of some indigenous species, it is also attracting increasing numbers of NIS. Similarly, the

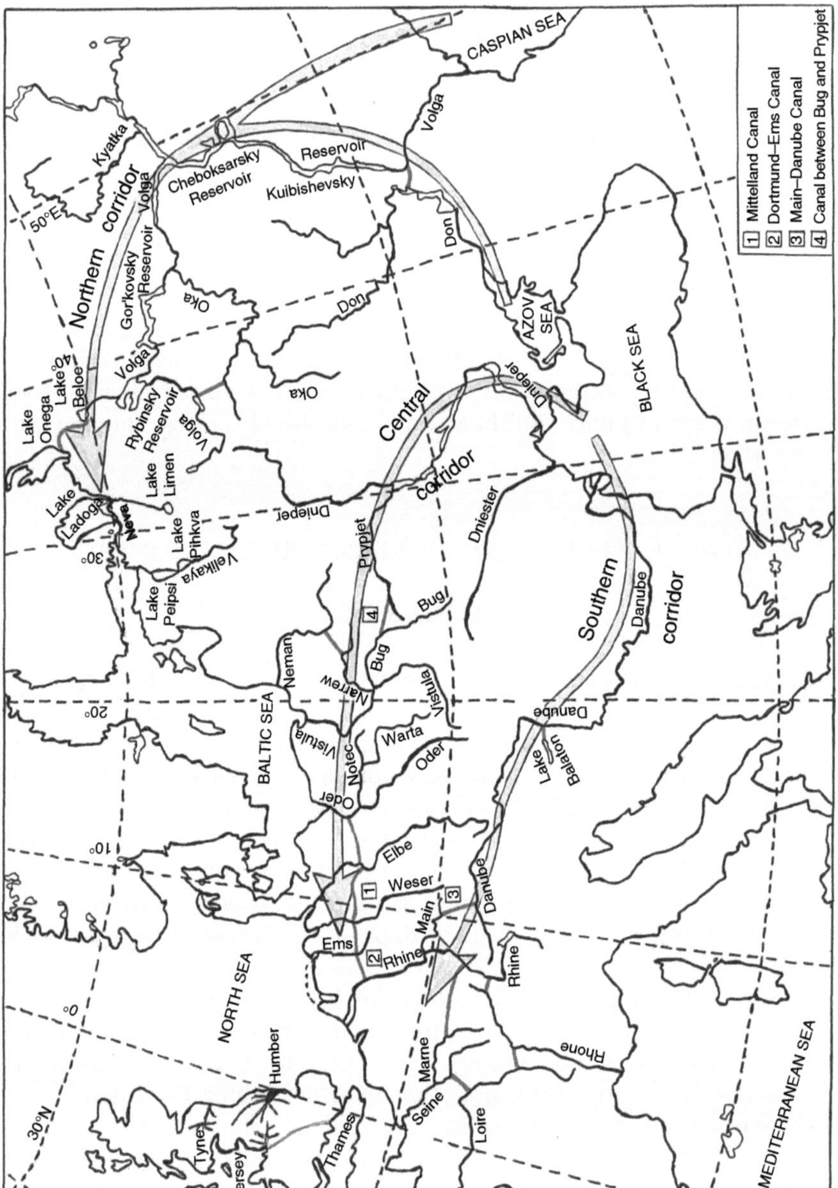

Fig. 1 Map of Europe showing the main inland waterways (rivers and canals), and the three main migration corridors (arrows), i.e. northern, central, and southern, used by Ponto–Caspian species. (Redrawn from Jazdzewski 1980, Bij de Vaate *et al.* 2002)

ability of many freshwater crustacean species, including crayfish (Firkins and Holdich 1993, Holdich *et al.* 1997), to tolerate elevated temperature and salt levels increases their chances of becoming established in new areas. Jaździewski and Konopacka (2002) suggest that the recent massive invasion of Ponto–Caspian species into central and western Europe may be due to the increasing ionic content of large European rivers, caused by agricultural and industrial inputs.

Intentional introductions include those for aquaculture (e.g. crayfish), human food (e.g. crabs and crayfish), fish food (e.g. mysids, amphipods, crayfish), pet trade (e.g. crayfish), management (e.g. crayfish for weed clearance), and stock enhancement (e.g. crayfish). Unintentional introductions have occurred via ballast water (e.g. branchiopods, copepods, isopods, amphipods, mysids, decapods), stocking of fish (e.g. branchiurans, decapods), attachment to mobile surfaces such as ships' hulls (e.g. tube-dwelling amphipods), entanglement in nets (e.g. decapods), floating weed and fouled mobile surfaces (a possible route for many species), fish bait (e.g. decapods), dumping of pets or excess stock (e.g. decapods), and perhaps even via predators such as birds, including waterfowl (Niethammer 1950, Segerstrale 1954). Anglers often use invasive crayfish species as bait and this can result in what is known as 'bait-bucket' introductions, which is a particular problem in North America, where the invasive rusty crayfish, *Orconectes rusticus*, has been spread northwards into Canada by this means, displacing indigenous crayfish species along the way (Lodge *et al.* 2000a, b). In Europe recreational anglers sometimes introduce crayfish such as the North American spiny-cheek crayfish, *Orconectes limosus*, in the belief that it will increase fish production (Holdich and Black 2007). This may be the case, but after a time the presence of large numbers of crayfish can have a detrimental impact on the fishing activity itself as well as on the freshwater environment (see below). Examples of those making their own way can be found in most of the invasive crustacean groups, but because they are relatively large, perhaps most noticeable are the decapods (e.g. the Chinese mitten crab, *E. sinensis*, and the narrow-clawed crayfish, *Astacus leptodactylus*).

Many other accidental introductions must also have occurred, but they have either not become established, or not had any noticeable impact. In some cases the introduction becomes established, but remains very localized, even though it may have been present for decades, as in the case of the North American isopod, *Asellus communis*, which only occurs in one isolated lake in England (Gledhill *et al.* 1993, Harding and Collis 2006). The situation is very fluid, with new records for non-indigenous invasive species being discovered on a frequent basis. For example, the Ponto–Caspian mysid, *Hemimysis anomala*, which has been introduced into a number of European countries as fish food, has suddenly appeared in central England (Holdich *et al.* 2006). Also, populations of the North American crayfish, *Orconectes virilis*, have been found recently in the Netherlands, and populations of *O. rusticus* have appeared in one region of France (Souty-Grosset *et al.* 2006).

Some crustaceans introduced via the various routes mentioned above have done equally well or better in their new environments, once they have become established. In some cases there have been positive effects through aquaculture, stock enhancement, and recreational activities (e.g. crayfish, see Ackefors 1999, Westman 2002), and this has encouraged secondary introductions (e.g. the North American red swamp crayfish, *P. clarkii*, and signal crayfish, *Pacifastacus leniusculus*). In others there have been negative effects through competition with indigenous species (e.g. branchiopods, copepods, mysids, amphipods, crayfish), transmission of disease (e.g. crayfish plague), and physical damage to the freshwater environment and its biota (e.g. crabs and crayfish, see Holdich 1999).

The majority of crustacean groups have invasive representatives in European inland waters, although amphipods provide the greatest number. In terms of publications, the majority are on amphipods and branchiopods (J. T. A. Dick 2006, personal communication), although invasive brachyuran crabs and crayfish have attracted a lot of attention in recent years (Gherardi and Holdich 1999, Gollasch 1999, Herborg *et al.* 2003). In their review of the anthropogenic dispersal of decapod crustaceans in the aquatic environment, Rodríguez and Suárez (2001) list 58 marine species that have been dispersed from their natural distribution areas, with 51 of these occurring in European waters. They list an additional eight freshwater and estuarine non-crayfish decapod species, only two of which (*E. sinensis* and *Rhithropanopeus harrisii*) have become established in European waters. They also list 20 crayfish species, including six that have become established in Europe, although this number has now increased (Souty-Grosset *et al.* 2006).

Taxonomic survey

Amphipoda (Table 1)

During the last few decades, numerous previously unrecorded amphipod species have been observed in European inland waters, but there is not enough space in this book to give all the immigration details for each of these species. Therefore only those that have had major effects in their new territories, by displacing indigenous species and/or changing the aquatic community including food web interactions, are dealt with below.

Amphipods have been introduced deliberately to boost secondary productivity and hence yields for the fishing industry. Leppäkoski *et al.* (2002a, b) state that more than 30 amphipod species from the Caspian complex have been introduced for this purpose. Especially in the former Soviet Union, new reservoirs, lakes, and any kind of waterbody were inoculated with species that promised high reproductive capacity. Canals that connected previously separated catchments offered an opportunity to invade new territories by passive and active anthropogenic vectors like navigation and transport in ballast water tanks. Some amphipods can leave the water and migrate at least a short distance over land (e.g. *Gammarus duebeni* Lilljeborg, Gledhill *et al.* 1993), like invasive North American crayfish in Europe. Nevertheless, the occurrence of amphipod species in many isolated waterbodies had been a mystery until Niethammer (1950) and Segerstrale (1954) proved the role of waterfowl in the transport of gammarids and other freshwater invertebrates. The latter showed experimentally that *Gammarus lacustris* Sars could become attached to the plumage and feathers of a mallard and remained in this position even after the wing has been taken out of the water. The attachment is mainly effected by pereopods 3–7, the last, claw-like segment of which is hooked into the plumage. The curved position, typical of the amphipod when out of water, prevents rapid desiccation of the gills. Thus, it may be possible for amphipods to be carried huge distances over land by this means and reach isolated bodies of water.

Most of the non-indigenous amphipod species in Continental Europe originate from the Ponto–Caspian basin. However, two well-established species in western Europe, *Gammarus tigrinus* and *Crangonyx pseudogracilis*, originated from the USA. A third species complex, the Mexican freshwater shrimp, *Hyalella azteca*, from the USA, Mexico, Central America, and the Caribbean, is sold intensively in the aquarium trade: it is kept by many aquarists and in garden ponds (Proßeckert 2001), and it is just a matter of time before it establishes self-sustaining populations in the wild. In Lake Ladoga, Lake Onega, Lake Peipsi, and the Neva estuary (Russia), *Gmelinoides fasciatus* from the Siberian Lake Baikal established dense populations as a result of introduction trials that had been very common in Soviet Fisheries management programmes. *Gammarus roeseli* that originates from the Balkan Peninsula and Asia Minor invaded larger rivers of the lower parts of central Europe, its western border being the eastern parts of France. *Echinogammarus berilloni* originated from the Iberian Peninsula and has invaded France, Belgium, the Netherlands, Luxembourg, and parts of Germany.

Although *Gammarus pulex* is indigenous to Europe, it has a very wide geographical range, stretching from eastern Siberia and China westwards to the British Isles, although it is absent from Norway (Pinkster 1972) and parts of Scotland (Gledhill *et al.* 1993). It has been introduced into some waters in Northern Ireland (where it is not indigenous) and more widely in Britain, supposedly to stock angling waters to enhance fish production (Strange and Glass 1979). It has also recently been introduced to the Irish Republic (McLoughlin *et al.* 2000). At several sites on the western seaboard of Britain, the indigenous *Gammarus duebeni celticus* Stock and Pinkster was supposed to have been displaced by competition with incoming *G. pulex* (Hynes 1954), but Sutcliffe (1967) found no evidence to support this. In Germany, fishery managers favoured the spreading of *G. pulex* (Haempel 1908) and *Gammarus fossarum* Koch (which had been regarded as a subspecies of *G. pulex* by many workers at these times) in any suitable body of water. Lough Neagh in Northern Ireland has been invaded by three non-indigenous amphipods: *G. tigrinus* and *C. pseudogracilis* from North America, and *G. pulex* from Europe, which have come into contact with the sole indigenous species, *G. duebeni celticus*, that is still present. Various studies have shown that *G. pulex* in Ireland is capable of replacing *G. duebeni* by competitive exclusion (Dick *et al.* 1990a, b, 1993, Dick 1996, MacNeil *et al.* 1999, McLoughlin *et al.* 2000, J. D. Reynolds 2006, personal communication). *Gammarus duebeni* is also common in parts of NW France, but is now extinct in Normandy due to interactions with the expanding *G. pulex* (Piscart *et al.* 2006). In Brittany, a recent study has revealed a decline of the endangered *G. d. celticus* since 1970 due to changes in environment and interference from indigenous *G. pulex*, which is expanding its range (Piscart *et al.* 2007).

Since its discovery in the London area in the 1930s, *C. pseudogracilis* has become widespread in most of England and Wales, and has spread northwards into Scotland (Gledhill *et al.* 1993). Similarly, since *C. pseudogracilis* was recorded from a pond in Dublin (Holmes 1975), it has become widespread in both Northern Ireland and the Irish Republic (Dick *et al.* 1999). It was discovered in the Netherlands in 1979 (Platvoet *et al.* 1989). It is likely that it will spread further in continental Europe – it was discovered in the R. Rhine in 1992 (Bernaurer and Jansen 2006). Notes on the ecology of this species are given in Gledhill *et al.* (1993). The species inhabits any kind of waterbody, from fresh to brackish and clean to organically enriched.

Gammarus tigrinus, which originates from the Atlantic seaboard of North America, was introduced by unknown means into Britain, perhaps early in the 20th century (Sexton 1939), where it thrived in areas where the water was salty due to mining pollution. In 1957, specimens from Wyken Slough near Coventry were deliberately introduced into the Rivers Weser and Werra (also the Elbe, Ems, and Schlei) in Germany where indigenous gammarids have disappeared due to pollution (Bulnheim 1985). *Gammarus tigrinus* thrives in polluted, slightly saline waters and is a source of food for fish. By 1964 it had become common in the IJselmeer and northern parts of the Netherlands (Pinkster *et al.* 1977). It is now widespread throughout the lowlands of western Europe and has become one of the dominant macroinvertebrates in many catchments, where it has outcompeted indigenous species (Fries and Tesch 1965). It is also known from coastal lagoons in the southern Baltic (Leppäkoski *et al.* 2002a). Its distribution in Britain is summarized by Gledhill *et al.* (1993), and its spread through the Netherlands is documented by Pinkster *et al.* (1977, 1980, 1992), Pinkster and Platvoet (1983), and Platvoet *et al.* (1989). During rapid colonization in the 1960s and 1970s, *G. tigrinus* displaced the indigenous *G. pulex* from many freshwater habitats, and the indigenous *G. duebeni* and *Gammarus zaddachi* Sexton from brackish water habitats. When the salt-enriched River Erewash was breached near a series of water-filled gravel pits in the English Midlands in the 1980s to allow further gravel extraction, *G. tigrinus* quickly colonized the gravel pits and became the dominant macro-invertebrate amongst the marginal vegetation for a number of years. However, due to a decline in the coal mining industry in the area, the river became less saline and so did the gravel pits, resulting in a dramatic (although not complete) decline in the NIS (D. M. Holdich 1995, personal observation).

Fig. 2 The tube-dwelling Ponto–Caspian amphipod, *Chelicorophium curvispinum*, from the Morava River, the border stream of Austria and Slovakia, between Zwerndorf and Baumgarten, Lower Austria. (Photo: W. Graf and A. Schmidt-Kloiber)

The tube-dwelling amphipod *Chelicorophium curvispinum* (Fig. 2) originates from large rivers discharging into the Black and Caspian seas, e.g. Volga, Dnieper, Dniester, Danube, etc., and clearly dispersed through the central corridor into the Baltic and North Sea drainage systems. The earliest report (1912) of the corophiid outside its natural range was from the Spree–Havel system near Berlin where it was described as *Corophium devium* (Wundsch 1912). It was also found in the Mittelland Canal and Dortmund–Ems Canal in 1956 and 1977, respectively (Van den Brink *et al.* 1989). *Chelicorophium curvispinum* was first recorded in Britain in the early 1930s by Crawford (1935) from the River Avon at Tewkesbury, and from then onwards it was reported widely as occurring in the interconnected canals and rivers of the English Midlands, as well as in other river systems, e.g. the R. Stour in SE England (Buckley *et al.* 2004). It was most likely introduced to Britain by ships sailing from ports of the Elbe estuary (Harris 1991). It is now present in Ireland in the R. Shannon and R. Erne systems (Lucy *et al.* 2004). In the Austrian Danube it has been known at least from the 1960s as far as the German border at Passau (Vornatscher 1965). The Main–Danube Canal where it was found in 1993 has been colonized by *C. curvispinum* from two directions, i.e. from the rivers Rhine (1987) and Main (1988) and from the Upper Danube (1959). The adults range in length from 2.5 to 7.0 mm. They filter suspended particles from the water column for the construction of tubes on solid substrates in which they live, giving them some shelter against predation. Soon after being recorded in the middle and lower Rhine in 1987 (Van den

Brink *et al.* 1989, Schöll 1990a), its numbers have increased explosively, and densities of well above 100,000 m^{-2} (maximum approximately 750,000 m^{-2}) have been recorded, which is much higher than numbers recorded in other rivers. The overgrowth of stones by the tubes of these animals can bind mud with a dry weight of up to 1.044 g m^{-2} and thus completely change the habitat (Van der Velde *et al.* 1998), causing direct environmental impacts over a distance of 200–500 km in the Rhine (Van den Brink and Van der Velde 1991) due to: (a) competition for space; (b) competition for food; and (c) changes in food web interactions. Larva of the zebra mussel *Dreissena polymorpha* Pallas, also a successful invader from the Ponto–Caspian basin, need bare hard substrates on which to settle, which may not be available because of the tube-building activities on such surfaces by the corophiid invader. Other filter feeders, such as the invading *D. polymorpha* and the indigenous species such as *Hydropsyche contubernalis* McLachlan (a caseless caddisfly larvae), chironomid larvae, and zooplankton species may be outcompeted. Eel and perch were found to shift their diet because of the invasion by *C. curvispinum*, which provided a new source of food (Kelleher *et al.* 1998). Dutch workers have found that *C. curvispinum* breeds from April to September, producing three generations a year – one more than related corophiid species (Rajagopal *et al.* 1998).

Specimens of *Chelicorophium robustum* were sampled in the R. Main in 2003 in the States of Bavaria and Hessen, being the first records of this species in Germany (Bernerth and Stein 2003, Berthold and Kaiser 2004). It was also recorded in the R. Rhine in 2004 (Bernaurer and Jansen 2006). Compared with *C. curvispinum*, the newly recorded species is easily detected by the large body size of adult specimens, i.e. 9 mm. A further spread in European inland waters is expected.

Migration patterns of *Corophium sowinskyi* are unclear because it is difficult to distinguish it from *C. curvispinum*. The species originates from the Danube, Dnieper, Volga, Don, and Dniester rivers (Mordukhai-Boltovskoi 1979). Records of this species in the Czech Republic indicate that the southern corridor could become the most obvious route for its range extension.

Dikerogammarus haemobaphes, originating from the Ponto–Caspian basin, was reported in the early 1960s from the Austrian Danube near Vienna by Vornatscher (1965) and in 1992 for the Bavarian stretch of the Danube (Tittizer 1996). During the 1980s it was the most abundant species in stony sediments (Pöckl 1988, 2002). It was probably the first amphipod species to invade the R. Rhine system via the southern corridor (Schleuter *et al.* 1994). For the first time in the Baltic Sea basin, the species was recorded in Poland in 1997, and its range expansion was reported by Jaźdźewski and Konopacka (2000). Its life history is presently being studied in the Vistula River where it is multivoltine, with three generations per year and high fecundity (Bacela and Konopacka 2005b).

Specimens of *Dikerogammarus villosus* (Fig. 3), which can reach a male maximum length of almost 30 mm, were not found in the Austrian Danube

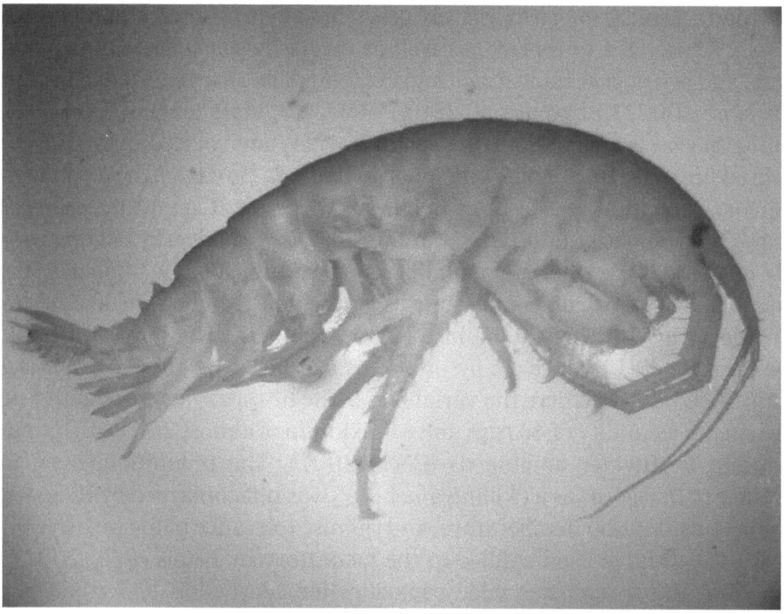

Fig. 3 The Ponto–Caspian amphipod, *Dikerogammarus villosus*, from the Austrian Danube at Linz, Upper Austria. Although this species dominates the community by number and biomass, other amphipod species do occur. (Photo: W. Graf)

before 1989 (Nesemann *et al.* 1995), and examples for different colour variants in live animals are given in that paper. It was demonstrated, however, that the different colour types cannot be differentiated at the allozyme level (Müller *et al.* 2002). The species was not found in the Bavarian Danube before 1992. It used the southern corridor and was sampled from the lower Rhine in the Netherlands (Bij de Vaate and Klink 1995). *Dikerogammarus villosus* is reported to be a successful invader by competition and predation: *D. haemobaphes* is rarely found in the Rhine system since the arrival of *D. villosus*, which has successfully invaded via the Rhône system (Müller and Schramm 2001) and the large rivers in northern Germany (Grabow *et al.* 1998), as well as the Moselle and other French hydrosystems (Devin *et al.* 2001). *Dikerogammarus haemobaphes* on the other hand is actively expanding in Poland (Jaźdźewski and Konopacka 2000). In the Netherlands, Dick and Platvoet (2000) have found that *D. villosus* is having a marked impact on the indigenous *G. duebeni*, as well as the non-indigenous *G. tigrinus*, and they predict that it will further reduce amphipod diversity in a range of freshwater habitats in Europe. *Dikerogammarus villosus* also occurs in several lakes, e.g. Traunsee and R. Traun, Austria (O. Moog 2003, personal communication), Lake Constance, Germany (K. O. Rothhaupt 2003, personal communication), Lake Garda, Italy (Casellato *et al.* 2005), where it is partially replacing the indigenous *Echinogammarus stammeri*

(Karaman). *Dikerogammarus villosus* preys heavily on other amphipod species, which it is thought to replace, as well as on *Asellus aquaticus* Linnaeus, insect larvae, and fish eggs (Chapter 27), and even small fish are reported to be attacked (Dick *et al.* 2002, La Piana *et al.* 2005). However, Platvoet (2005) also showed that the species is able to nourish itself by a wide range of feeding methods, such as shredding, grazing, collecting micro- and macro-algae, coprophagy, and carnivory, and that the feeding habits are dependent on water temperature and the micro-distribution of food organism. *Dikerogammarus villosus* is apparently less predatory when a population is well established in comparison to the phase when it is rapidly increasing its individual numbers in a new habitat (Van Riel *et al.* 2005). The life history and population dynamics of *D. villosus* have been studied intensively by one of us (M. Pöckl) in the Austrian Danube during 2002–2004, where the variability in life history and reproductive output with a mean fecundity of 43 eggs and a maximum of almost 200 were found to be unique in freshwater amphipods (Pöckl 2007). The reputation given in the literature of *D. villosus* as a "killing machine" was not confirmed by these studies. Oxygen consumption, temperature, and salinity tolerance of the invasive amphipod *D. villosus* have been studied in the laboratory by Bruijs *et al.* (2001), who found that the species has wide capacities for adaptation and could possibly survive ballast water exchange and thus develop large populations in temperate areas on a global scale.

Müller and Schramm (2001) reported that a third riverine *Dikerogammarus* sp., *D. bispinosus*, has colonized the middle and upper R. Danube (Austrian stretch at Linz from 1998). Their genetic analyses demonstrate the clear species status of this taxon, which formerly had been described as a subspecies of *D. villosus* by Martynov (1925) from the lower Dnieper. The lack of hybrid genotypes indicates a reproductive isolation among *D. haemobaphes*, *D. villosus*, and *D. bispinosus* in a syntopic population from the Hungarian Danube near Szob (Müller *et al.* 2002). The dispersal behaviour of *D. bispinosus* may be species-specific as with *D. haemobaphes* and *D. villosus*.

Echinogammarus ischnus belongs to the group of Ponto–Caspian amphipods that have advanced farthest north-westwards, reaching the systems of the North and Baltic seas. In 1928, it was recorded for the first time from the Vistula below Warsaw (Jarocki and Demianowicz 1931), and has probably passed through the Rivers Dnieper, Pripet, the Pripet-Bug Canal, and the R. Bug. Using a similar pathway, the Neman–Pripet canal, *E. ischnus* had reached the lower R. Neman by about 1960 (Gasjunas 1965, 1968 in Jażdżewski 1980). Herhaus (1978) discovered the species in the Dortmund–Ems canal. The well-developed canal systems joining the Vistula, Oder, Elbe, and Weser rivers seem to have been its most probable route (Jażdżewski 1980). Between 1979 and 1981 specimens of *E. ischnus* were sampled in the Mittellandkanal (Herbst 1982). In the late 1980s the species was observed from the Rhine–Herne Canal and the Weser–Dattel Canal (Schöll 1990b). At about the same time, *E. ischnus* was also found in the Mecklenburgian and Pommeranian lakes (Jażdżewski and Konopacka 1990, Köhn and Waterstraat

1990), indicating that in western Europe it is colonizing habitats that are comparable to those of its natural Ponto–Caspian distribution area, where it occurs in several inshore Black Sea lakes (Jaźdźewski 1980). From the lower Rhine in Germany, a density of 100 m^{-2} of hard substrate has been reported (Schöll 1990b), and in 1991 and 1992 specimens were recorded from the lower Rhine delta in the Netherlands (Van den Brink *et al.* 1993). The development of a dense population of *E. ischnus* in the Rhine may have been hindered by *G. tigrinus*, as well as the mass abundance of *C. curvispinum* and *D. villosus*. In the Austrian stretch of the Danube, *E. ischnus* had not been reported to occur in the 1960s (Vornatscher 1965), but during the 1980s dense populations were encountered on stony substrate (Pöckl 1988). Via the Bavarian Danube (1989) the species migrated to the Main–Danube Canal (1995) (Van der Velde *et al.* 1998). Köhn and Waterstraat (1990) suggested that *E. ischnus* is closely associated with clumps of *D. polymorpha* in Lake Kummerow, Germany.

Echinogammarus berilloni originates from Mediterranean rivers, and adult males can reach a body length of 22 mm. When true estuarine species are absent (e.g. *G. zaddachi*, *Gammarus chevreuxi* Sexton), as in north-western Spain, it is able to penetrate into estuarine regions. According to Pinkster (1993), it is a typical species of middle courses of streams and rivers, and has never been found in the upper reaches. It is an active migrant, which has made use of canals to reach a large area of western Europe. Presently, it is found in the Rhine from Düsseldorf to Basel where it occurs in low densities, and in the Mosel (Moselle), a tributary of the Rhine. Meyer *et al.* (2004) reported that in temporary waters of a karstic system in western Germany, *Gammarus* species dominated, but in permanent downstream sections *E. berilloni* almost completely replaces *G. pulex* and *G. fossarum*. The occurrence of *E. berilloni* in the estuarine parts of some river systems in north-western Spain indicates that the species can stand high salinities as well as considerable changes in salinity. It also can withstand a high amount of organic pollution and high temperatures, and has been sampled in some parts of Spain at temperatures of up to 31°C (Pinkster 1993).

Outside its original natural distribution area in the Ponto–Caspian basin, *Echinogammarus trichiatus* was recorded for the first time in 1996 in the Bavarian Danube (Weinzierl *et al.* 1997), and three years later at a distance of 120 km from the first record in the Upper Danube. In 2000 and 2001, the species was sampled in the Upper and Lower Rhine, respectively (Podraza *et al.* 2001), which means that it must have spread through the Main–Danube Canal, using the southern invasive corridor. In 1998, *E. trichiatus* was also found in the Austrian Danube (H. Nesemann 1999, personal communication), which is later than the German record. Distribution with the stream flow is obviously easy, but the distribution of this species is still largely unknown and may be scattered.

Echinogammarus wurpachowskyi originates from the brackish parts of the Caspian Sea and the deltas and estuaries of many Ponto–Caspian rivers. Some 40 years ago the species became one of the main objects of Soviet acclimatization enterprises. It was introduced into reservoirs and lakes in the Ukraine, and in Lithuania in the Neman River drainage area (Kaunnasskoe Reservoir). It penetrated into artificial reservoirs in the Dnieper River and into the Kuronian Lagoon (Jaźdźewski 1980). This small species (adult males reaching a maximum of 6.5 mm) is very tolerant to both varying salinity and temperature conditions and has spread over large parts of eastern Europe.

Since 1994, the stout and small *Obesogammarus obesus* from the Ponto–Caspian basin, which swims in an upright position (it is not laterally compressed like most gammarids), is known to occur in the Austria Danube, and has in some parts developed high densities (M. Pöckl 1994, personal observation), occupying a position after *C. curvispinum*, *D. villosus*, and *E. ischnus* in abundance. In 1995 it was recorded from the Bavarian Danube (Weinzierl *et al.* 1996), and an estimated density of 3,300 m^{-2} was reported. The invasion of the R. Rhine is expected to occur via the Main–Danube Canal in the near future. In October 2004, the species was recorded in the R. Rhine near Koblenz, Germany. The sampling site was located approximately 0.5 km away from the main river, quite close to a sports boat marina. Additional records from the same location in 2005 and 2006 indicate that this species may have become established in the central section of the R. Rhine (Nehring 2006). It can be speculated that *O. obesus* will extend its range within the European river and canal system in the near future.

Obesogammarus crassus was intentionally introduced in the 1960s into the Kaunas Reservoir (Lithuania), in the Neman River, and in several aquatic habitats along the Baltic coast of the former Soviet Union (Jaźdźewski 1980). From the Neman River, the species colonized the Kuronian Lagoon. Recently, *O. crassus* was observed from the Vistula Lagoon (Jaźdźewski *et al.* 2002). Westward dispersal has been the result of offshore transportation (in ballast water) via the Baltic Sea, indicating the northern dispersal route. However, part of the central corridor is considered to be a potential second corridor.

The first record of *Pontogammarus robustoides* in Germany dates from 1994 when it was found in the Peene (Rudolph 1997). From there it may have used the Hohenstaaten–Friedrichthaler Wasserstraße, the R. Oder, the Oder–Havel Canal, the Havel Canal, and the R. Elbe to reach the Mittellandkanal where it was sampled at Wolfsburg in 1998 (Martens *et al.* 1999, Tittizer *et al.* 2000). Like the other Ponto–Caspian species, *C. curvispinum*, *E. ischnus*, and *P. robustoides* probably also used the central corridor to penetrate westwards, and clearly not the southern one via the R. Danube. The average body length of mature specimens was 11.15 mm, ranging from 4.5 to 21.0 mm, and the smallest ovigerous females were 8.5 mm long. The mean brood size for all gravid females was 64.5 and varied from 11 to 185, and the egg number was exponentially correlated to female body length. These traits in life history determine the success of *P. robustoides* as a potential invader (Bacela and Konopacka 2005a).

Before the 1960s, the distribution area of the Baikalian amphipod, *G. fasciatus*, was limited to basins of Siberian Rivers (Angara, Lena, Yenisay, Irtysch, Pyasina, Tunguska, Selenga, Barguzin). In the former Soviet Union it was considered to be a suitable species for intentional introduction to enhance fish production in lakes and reservoirs, because of its high environmental plasticity and general high abundances within its native range. During the 1960s and 1970s, hundreds of millions of *G. fasciatus* specimens were introduced into 22 lakes and reservoirs outside its native range in Siberia and European Russia (Panov and Berezina 2002). In European Russia, *G. fasciatus* was introduced intentionally into Gorkovskoe Reservoir in the R. Volga basin, several Karelian Isthmus lakes located close to the western shore of Lake Ladoga, and Lake Ilmenin in the Lake Ladoga basin. *Gammarus fasciatus* invaded the western and northern shores of Lake Ladoga, some 18,400 km^2 in area, in the late 1980s (Panov 1996) and by the 1990s it had successfully colonized the whole littoral zone of this largest European lake. In the 1990s, from Lake Ladoga via the R. Neva, *G. fasciatus* penetrated into the Neva estuary, the largest estuary in the Baltic Sea (3,600 km^2). In 1996, the species was found in the Neva Bay and by 2001 it had established successfully in the coastal zone of the estuary (Berezina and Panov 2003). In 2001, *G. fasciatus* established self-sustaining populations along the western shore of Lake Onega. Berezina (Chapter 26) discusses the changes in the littoral communities of large lakes caused by introduction of *G. fasciatus*. In Lake Peipsi, *G. fasciatus* was introduced accidentally at the beginning of the 1970s during several attempts to enrich the native population of *G. lacustris* G.O. Sars by addition of specimens of this species from Siberian populations. These introductions were "contaminated" because the material released (several million specimens) contained a mixture of *G. fasciatus* (1–2% in density) and *G. lacustris*. The accidentally introduced *G. fasciatus* survived and were first observed in Lake Peipsi in 1972. By 1990 it had become established in the whole littoral zone of this lake (Berezina 2004). Two decades ago, the indigenous amphipod species *G. lacustris* was common in Lakes Ladoga and Onega as well as the freshwater parts of the Neva estuary. After invasion by the Baikalian amphipod, the indigenous *G. lacustris* has disappeared from many habitats. Moreover, the density of the freshwater isopod, *Asellus aquaticus*, was found to be dependent on the density of the Baikalian amphipod, decreasing significantly at localities with more than 500 *G. fasciatus* m^{-2}.

Astacida (crayfish) (Table 2)

Approximately 600 species of freshwater crayfish belonging to three families (Astacidae, Cambaridae, and Parastacidae) have been described and new species are being described on a regular basis, particularly from the Americas and Australasia (Taylor 2002, Fetzner 2005). However, there are only five indigenous crayfish species in Europe, all belonging to the Astacidae (Holdich 2002b, 2003, Souty-Grosset *et al.* 2006). After the last glaciation some 10,000 years ago, these crayfish species gradually colonized Europe by natural diffusion, either from glacial refugia or from the Ponto–Caspian basin. Subsequently, at least four of the five species have been translocated by man, or have migrated via man-made structures such as canals, to an extent that often makes it difficult to determine their origins. However, molecular genetic studies are being used to gradually unravel their origins, and what some countries consider to be their indigenous species appear to have been probably introduced. For example, the white-clawed crayfish, *Austropotamobius pallipes* (Lereboullet) *sensu lato*, in England and Ireland has been introduced on several separate occasions from France (Grandjean *et al.* 1997, Gouin *et al.* 2003), and the same species was introduced into Spain from Italy (Machino and Holdich 2006). When these events occurred, however, is not known, though probably in the last 1,000 years. Similarly, the noble crayfish, *Astacus astacus* (Linnaeus), which is now a treasured gastronomic icon, was introduced into Sweden and Norway in the Middle Ages (Machino and Holdich 2006).

When these species were introduced into the fresh waters of new areas and became established they must have been invasive and had an impact on the pre-existing biota. This is often a fact that is overlooked and is particularly relevant to such a keystone species as a crayfish, which can have a considerable impact when introduced into a waterbody that has not experienced it before. A case in point is *A. leptodactylus sensu lato*, which is indigenous to the Ponto–Caspian basin, but which has spread naturally via rivers and canals into northern and eastern Europe, and has been introduced into western Europe for commercial purposes. This crayfish is highly fecund and can grow to a very large size (up to 500 g wet weight) and reach very high densities. As a consequence of this, it can become the dominant animal in a waterbody, displacing other crayfish species if they are present (Souty-Grosset *et al.* 2006). As with other European crayfish it does not carry crayfish plague, but is susceptible to it (see below).

From the middle of the 19th century, a disease now commonly known as crayfish plague entered the waters of the Po Valley in Italy and gradually spread throughout Europe, killing off many populations of indigenous crayfish (Holdich 1999, 2003). The ranges of indigenous crayfish such as *A. astacus* and *A. pallipes* in western Europe were particularly affected and are still being compromised today (Souty-Grosset *et al.* 2006). Crayfish plague is indigenous to North America, and all those North American crayfish that have been studied are carriers of the oomycete causing it.

As crayfish were a valuable commodity in Europe in the 19th century, to boost European stocks steps were taken to introduce a North American crayfish species that was immune to the disease, i.e. *O. limosus* (see below). Two further species, i.e. *P. leniusculus* and *P. clarkii*, were introduced in the 1960s and 1970s respectively, to improve stocks further (see below). Their spread throughout Europe (see below) has only made the situation worse for the indigenous species, particularly as they are superior competitors (Holdich 1999), although there have also been some commercial, management, and recreational benefits (Ackefors 1999). These three species are further dealt with below as they have the widest distribution of invasive crayfish occurring in European inland waters. The Australian crayfish, *Cherax destructor*, was also introduced for commercial purposes in the 1980s, but it is restricted to Spanish waters, although it is imported live for restaurants in other countries and is cultivated in Italy (Souty-Grosset *et al.* 2006).

A second wave of crayfish introductions occurred in the late 20th and early 21st centuries (see Table 2), but unlike the four species mentioned above it seems likely that most were imported for the pet trade, and were subsequently released or escaped into natural waters. At the present time, although they can be considered invasive, their range is very limited (see Souty-Grosset *et al.* 2006 for further details). Of particular concern is the marbled crayfish, *Procambarus* sp., of unknown origin and species, which has been made widely available through the aquarium trade in recent years, and which now occurs in the wild in Germany and the Netherlands. This crayfish is parthenogenetic and can produce large numbers of offspring in a short space of time (Vogt *et al.* 2004, Seitz *et al.* 2005, Souty-Grosset *et al.* 2006). Considering the number of crayfish species available through the aquarium trade this source of invasive crayfish is likely to be a continuing problem as owners want to get rid of their pets as they grow too large or breed too rapidly.

As mentioned above, the first non-indigenous crayfish to be introduced into Europe from another continent was *O. limosus* (Fig. 4). After its introduction to Germany in 1890, secondary introductions were made into other parts of Germany and into Poland and France, in an attempt to make up for losses of

Fig. 4 The North American spiny-cheek crayfish, *Orconectes limosus*, and its burrows. This has become well-established in continental Europe since its introduction into Germany in 1890, but has only recently invaded England. Adults from Clifton Pond, Nottingham, England. (Photo: J. Black)

A. astacus through crayfish plague (Souty-Grosset *et al.* 2006). It also spread naturally through rivers and canals and is now probably the commonest crayfish in continental Europe, occupying at least 20 countries. Out of 300 lakes recently examined in NE Germany, 214 were found to have *O. limosus*. In Poland, populations of *O. limosus* increased from 57 in 1959 to at least 1,383 by 2004. It is gradually spreading eastwards in Europe and recently has been found in Croatia (Maguire 2003) and Serbia (Karaman and Machino 2004); it is likely to spread into Bulgaria, Romania, and the Ukraine via the R. Danube before long (Machino and Holdich 2006). It has been implicated in the demise of indigenous crayfish populations through competition and crayfish plague. Its large numbers and burrowing activity are likely to have a marked effect on the freshwater environment. However, *O. limosus* has never fulfilled its role as a replacement for *A. astacus* from the gastronomic point of view, as other species are much preferred. It is commonly used as fish bait and this has led to its introduction into new sites and countries such as England, where it has built up large populations in a short space of time (Holdich and Black 2007).

As indicated by the number of contributions relating to its biology in this volume, *P. clarkii* attracts a lot of attention because of its invasive capabilities. It was introduced to southern Spain in 1973 for aquacultural purposes but soon became widely established in the wild and is now present in 13 European countries, including islands in the Azores and Canaries (Holdich 2002b, Souty-Grosset *et al.* 2006). Although it has brought undoubted benefits to the Spanish economy through its harvesting and export, mainly to Scandinavian countries (Ackefors 1999), its environmental impact caused by burrowing and high consumption of both plant (including rice seedlings) and animal matter can be striking. For example, prior to 1996, Chozas Lake in NW Spain used to harbour a rich community in its clear, shallow waters. *Procambarus clarkii* was then introduced and its activities caused the waters to become turbid (Rodriguez *et al.* 2005). This was followed by a 99% reduction in plant cover, 71% loss of macroinvertebrates, 83% reduction in amphibian species, 52% reduction in waterfowl, and plant-eating birds such as ducks were also reduced by 75%. However, carnivorous birds increased their presence after the introduction of the crayfish. In addition to such effects, *P. clarkii* is also a carrier of crayfish plague and is thought to be responsible for the decline in indigenous crayfish species in a number of countries, e.g. Italy and Spain.

Pacifastacus leniusculus, is the only member of the Astacidae to be introduced into Europe (Lewis 2002). As the indigenous crayfish fauna in Europe all belong to this same family, many aspects of their biology are similar. *Pacifastacus* is the only genus of the Astacidae in North America, but like members of the other family, the Cambaridae, it too carries crayfish plague. *Pacifastacus leniusculus* was first released into Swedish waters in 1960, to replenish stocks of crayfish with an ecological and gastronomic homologue replacing *A. astacus*, which had been badly affected by crayfish plague. It soon became a popular species for stocking and culture, and as a result of secondary introductions (both from Sweden and North America) had become established in 24 European countries, from the UK across to eastern Europe by 2005, making it the most widespread non-indigenous crayfish (Souty-Grosset *et al.* 2006). In most of the countries into which it has been introduced, it has become established in the wild either as a result of escapes or deliberate seeding of waters. *Pacifastacus leniusculus* is most established in Sweden where it occurs in approximately 3,000 localities. Although it was welcomed in many countries by aquaculturists, particularly in Sweden and Finland, conservationists were concerned about the dangers of introducing a large, aggressive, highly fecund, fast-growing species into the freshwater environment, especially where indigenous crayfish, which are susceptible to the effects of crayfish plague, were still present. The fears of conservationists have proved true, whilst *P. leniusculus* has not provided the huge improvement in stocks that was predicted. Ironically, it is in the UK, which does not have a modern tradition for eating crayfish and which had good healthy stocks of its so-called (see above) indigenous species, *A. pallipes*, where problems are most acute since the introduction of *P. leniusculus* for aquacultural purposes in the 1970s. Despite a raft of legislation being drafted to protect the indigenous species and measures taken to try and stop the spread of the NIS, *A. pallipes* may well become extinct in a few decades (Sibley 2003, Holdich *et al.* 2004, Holdich and Pöckl 2005). This is due to a combination of the effects of crayfish plague and the superior competitive abilities of *P. leniusculus*. In addition, the burrowing activity (Fig. 5) of *P. leniusculus*, coupled with its insatiable appetite, is

Fig. 5 River bank collapse caused by the burrowing activities of the North American signal crayfish, *Pacifastacus leniusculus*, in the Gaddesby Brook, Leicestershire, England. (Photo: P. J. Sibley)

having both a physical and biological impact on both lentic and lotic environments (Guan and Wiles 1997, Sibley 2000). Attempts at eradicating *P. leniusculus* have so far proved futile despite the removal of many thousands of adults at a number of sites, e.g. in Scottish rivers (Collins 2006). As Holdich *et al.* (1999) predicted in their review of eradication methods, the only sure way is to use biocides, and this method has been successfully trialled in the UK, but it is only of use in enclosed waterbodies (Peay *et al.* 2006).

Caridea (Table 2)

Van der Velde *et al.* (2000) note that the freshwater river prawn, *Ataephyra desmaresti* (Atyidae), which originates from southern Europe, has spread widely throughout Europe via canals that connect European river basins. It was first observed in 1843 near Paris, 1888 in Belgium, 1916 in the Netherlands, 1925 in Metz, and 1929 in Strasbourg (Thienemann 1950). The first German record dates from 1932, from a backwater of the Lower R. Rhine near Rees. Thereafter it occurred at several sites of the dense canal network in north-western Germany, and after passing the Mittellandkanal it was found near Hannover in 1936. Several records were noted from the Rhine–Rhône Canal, the R. Saar at Saarbrücken, and the R. Mosel at Merl (summarized by Kinzelbach 1972). It was observed in the Lower R. Main in 1983 (Nesemann 1984), from where it migrated upstream in this river, entered the Main Canal (Heuss *et al.* 1990), passed the highest point of the Main–Danube Canal (Wittmann 1995), went downstream, was recorded in 1997 in the Bavarian Danube (Weinzierl *et al.* 1997), and in 1998 in the Austrian Danube (Moog *et al.* 1999). It therefore can be expected to occur in Slovakia, Hungary, and further downstream in the R. Danube in the near future. It feeds on micro-organisms, algae, plants, detritus, live and dead animals. *Ataephyra desmaresti* has a wide tolerance to temperature and salinity ranges, and lacks planktonic larvae, which are all useful attributes to possess when colonizing new habitats (Steffen 1939, Fidalgo 1989b). Adult males can reach a body length of 16–27 mm, females a maximum of 35 mm. The number of eggs is reported to vary between 100 and 1,400 (Fidalgo 1989a, b). Normally, the lifespan is 12–14 months, but under unfavourable conditions juveniles grow at a slower rate, reach sexual maturity in the second year and can live for three years. It is not known how many successive broods can be produced by an individual female.

The first record for the oriental prawn, *Palaemon macrodactylus* (Palaemonidae), for Europe was reported by Ashelby *et al.* (2004). Although this introduced species is widespread in the western USA, it has only colonized one location in Europe so far. It is thought to have been introduced into the R. Orwell estuary on the eastern coast of England some time between mid-2000 and late-2001, where it is now common and breeding. Transport in ballast water seems the most likely route of its introduction. It has since been found in the adjacent R. Stour estuary (Ashelby *et al.* 2004). *Palaemon macrodactylus* is extremely hardy and is known to live in salinities as low as 1.0 ppt in California. It is probable that aided by shipping this species will spread in European coastal waters and could enter inland waters.

Brachyura (Table 2)

The indigenous freshwater crab, *Potamon fluviatile*, occurs in Italy, the Balkans and Greece. Its behaviour has been studied by Italian workers (Barbaresi and Gherardi 1997, Gherardi *et al.* 1999). In competitive situations with the white-clawed crayfish, *A. pallipes*, it is usually the crab that wins as it has higher levels of aggression and strength. At present the two species have mainly different distributional ranges but where these overlap they never share the same water-body. They may have had a common distribution in the past, but competitive exclusion by the crab has meant that the crayfish has been pushed into less favourable habitats (Barbaresi and Gherardi 1997). The same may happen with some populations of indigenous crayfish species in France, where three non-indigenous *Potamon* species (see Table 2) have been introduced and become established there, one since the early 19th century (see Chapter 3).

The catadromous North American blue crab, *Callinectes sapidus* Rathbun, is euryhaline and eurythermic, and in its natural range migrates down rivers to reproduce in the sea. It was introduced into the Netherlands in 1932, probably in ballast water (Adema 1991). It has been recorded in a number of other European countries, e.g. France (Goulletquer *et al.* 2002), and is breeding in the eastern Mediterranean (Froglia 2005). It has also been recorded in the Marmara (Ozturk 2002), Black and Azov seas (Gomoiu *et al.* 2002). It does not appear to have extended its range into inland waters very far. Often, only single specimens are found, e.g. one was recorded from a river on the eastern coast of England in 1982 (Gledhill *et al.* 1993).

One of the commonest non-indigenous crab species is the North American mud crab, *R. harrisii*, which is found in a number of estuaries and coastal lagoons throughout Europe, including in the Mediterranean and Adriatic (Froglia 2005), as well as the Black and Azov seas since 1932 (Gomoiu *et al.* 2002), and also the Caspian Sea (Aladin *et al.* 2002). It is also present in Wales (UK) (Minchin and Eno 2002). It is thought to have been introduced via ballast water into the Netherlands in the 19th century (Adema 1991). It has been recorded in the lower R. Rhine, but only in low numbers (Van der Velde *et al.* 2000). In Poland, Jaźdźewski and Konopacka (2000) noted that this species attained very high densities in brackish waters in the 1950s and 1960s and became a major component of the zoobenthos, although its numbers decreased after that time. It does not appear to penetrate far into inland waters.

The most invasive of the non-indigenous crabs is the catadromous Chinese mitten crab, *E. sinensis* (Fig. 6), from SE Asia, where it has been recorded as far as 1,400 km upstream in China (Gollasch 1999). It is considered a delicacy in the Far East where it supports a billion dollar industry (Herborg *et al.* 2005), but

Fig. 6 An ovigerous female Chinese mitten crab, *Eriocheir sinensis*, from the River Thames (November 2005), London, England. *Eriocheir sinensis* was first recorded in Germany in 1912 and is now widely spread in European inland waters and estuaries. (Photo: P. Hurst)

in Europe it only tends to be eaten by Chinese immigrants. It is thought to have reached Europe via ballast water and was first recorded in the River Aller (a tributary of the R. Weser in Germany) in 1912 (Peters and Panning 1933, Gollasch 1999). Like the blue crab it migrates down rivers to breed after which it dies, and then as juveniles migrates upstream again in large numbers, taking 3–5 years to become sexually mature in Europe (Schubert 1935 in Herborg *et al.* 2005). In its migrations it can move across land to get around weirs (Rettig 2000 in Puky *et al.* 2005, Herborg *et al.* 2003). It is euryhaline and eurythermic and is capable of moulting in freshwater.

Despite being introduced in 1912, the range expansion of *E. sinensis* was not reported until 1927 when it migrated from Germany via the Kiel Canal into the Baltic Sea (Peters 1938 in Herborg *et al.* 2003), from where it reached Russia and Finland by 1933. It reached the Netherlands in 1929 and spread throughout the country (Van der Velde *et al.* 2000), France in 1930, Belgium in 1933, and England in 1935. It had migrated 700 km up the R. Elbe to Prague and 512 km along the R. Rhine by 1932, and by 1934 it occurred 464 km up the R. Oder as far as Breslau (Robbins *et al.* 2000, Herborg *et al.* 2003). Although it was known from the French coast as far as Le Havre in 1943, there appears to have been a secondary introduction, probably via ballast water or associated with oyster cultivation, to the R. Gironde region (1954–1960), from where *E. sinensis* reached the Mediterranean coast via canals in 1959 (Herborg *et al.* 2003). It is known from the Austrian (Rabitsch and Schiemer 2003) and Serbian (Karaman and Machino 2004) stretches of the Danube and has recently (2003) been found in the Hungarian part (Puky *et al.* 2005). It has also been reported from the White Sea (Berger and Naumov 2002) and the first record for Europe's largest lake, Lake Ladoga in Russia, was found in 2005 (Panov 2006). It was discovered in the Black Sea (Gomoiu *et al.* 2002) and Azov Sea in 1997 (Murina and Antonovsky 2001 in Herborg *et al.* 2003), and from the River Tazeh Bekandeh that drains into the Caspian Sea, Iran in 2002 (Robbins *et al.* 2006). Recently, in the west, it has been found in Waterford Harbour on the south-eastern coast of the Irish Republic (J. D. Reynolds 2006, personal communication).

According to Jaźdźewski and Konopacka (2000), *E. sinensis* is less of a problem in most of Europe than it used to be in the 1920–1940s. In the 1930s and 1940s lack of competition and an abundant food supply led to them becoming so abundant in Germany that millions of juveniles were caught during their upstream migration in 1936, but subsequently pollution led to a reduction in the crab's food supply and the crab itself (Gollasch 1999). Due to recent improvements in the water quality and a consequent increase in food supplies of some European rivers, *E. sinensis* is becoming abundant again and, in 1998, 75,000 crabs were taken by hand in only two hours in the River Elbe, where it is once again causing problems due to its migratory habit (Gollasch 1999, S. Gollasch 2006, personal communication). The main problems associated with *E. sinensis* are its burrowing habit that may endanger flood defences, and the fact that it reaches high densities, thus competing with indigenous species for food, including crayfish (Robbins *et al.* 2000), as well as endangering navigation. It also interferes with recreational and commercial fishing by taking bait and interfering with nets (Herborg *et al.* 2003, 2005).

Recent studies have mainly dealt with its spread in the UK, where it was first observed in the River Thames in 1935 but remained at low numbers until the 1990s when numbers escalated (Robbins *et al.* 2000), possibly as a result of improving water quality (Herborg *et al.* 2005). In England, *E. sinensis* spread along the coasts at an average rate of 78 km per year in the period 1976–1999, but this increased dramatically to 448 km per year in the period 1997–1999, which is similar to the rate of spread along the Baltic coast in 1928–1935 (Herborg *et al.* 2005). The upstream spread was 16 km per year in 1973–1998 with a marked increase to 49 km per year in 1995–1998. There are concerns about the impact that it will have on the structure of river banks and the indigenous fauna, including crayfish.

Other taxa (Table 3)

Space does not permit a full review of the other taxa of Crustacea that have invaded European inland waters, but brief details are given below and a tentative list is given in Table 3.

Branchiopoda. The spiny water flea, *Bythotrephes longimanus*, is indigenous to lakes throughout the Palearctic, ranging from the British Isles to the Bering Sea. Sikes (2002) has summarized what is known of this species as an invader. *Bythotrephes longimanus* has been found in areas where it was not previously known, including the North American Great Lakes in 1982, where it is thought to have been introduced with ballast water from a transoceanic ship originating from St. Petersburg harbour (Russia), where it is common in the nearby Lake Lagoda. *Bythotrephes longimanus* reproduces rapidly by parthenogenesis and its ability to produce sexual eggs allows it to survive adverse environmental conditions. It is a dominant predator of zooplankton in the summer months. Van der Velde *et al.* (2000) have found that *B. longimanus* is common in water storage reservoirs, lakes, and rivers in the Netherlands, and that it has a marked effect on other zooplankton. They also mention that it is found in the catchment area of the R. Rhine in the Alps, north-eastern Germany, Poland, Belarus, the Baltic States, Scandinavia, and the British Isles. They suggest that it has reached the Netherlands and other countries by long distance dispersal through the transport of resting eggs by rivers and/or birds. Panov *et al.* (2006) stated that the predatory Ponto–Caspian cladocerans, *C. pengoi*, *E. anonyx*, and *Cornigerius maeoticus* have moved through to the Baltic via the Volga–Baltic waterway (northern invasion corridor), probably in ships' ballast water, and that, although they are warm-water species, they have the potential to become established in a wide range of inland and coastal water ecosystems in temperate zones. Rodionova and Panov (2006) noted that *E. anonyx* increased ten fold in the eastern Gulf of Finland between 2000 and 2004. Litvinchuk and Maximova (2005) have studied the biology of *E. anonyx* and *Cornigerius maeoticus maeoticus* Pengo in the Baltic Sea and found them living with the indigenous branchiopod fauna. *Cercopagis pengoi* and *C. maeoticus* are known to occur in freshwater reservoirs in the Ponto–Caspian basin as well as those associated with the R. Volga, and it is probably only a matter of time before they are recorded from similar habitats further west. According to Panov *et al.* (2006), most Ponto–Caspian onychopod cladocerans are euryhaline and can survive in relatively high salinities as well as in freshwater (with the exception of *E. anonyx*). The invasion of the Baltic by these species has been helped by climate changes and the intensive shipping activity along the corridor. They suggest that Ponto–Caspian cladocerans should be considered as "high risk" invasive species because of their potential for range expansion and the impact they have on the recipient ecosystem. Indeed, Telesh and Ojaveer (2002) have found that *C. pengoi* in the Baltic Sea has a marked impact on the zooplankton community as well as pelagic food webs involving planktivorous fish species. They suggest that the dietary overlap with young planktivorous fish may lead to a decline in food sources for fish such as herring and sprat, although this is compensated to some extent by the fact that the fish can feed on the branchiopod.

Copepoda. Until recently few invasive Copepoda had been reported, but N. Riccardi and G. Rossetti (2006, personal communication) have found the calanoid, *Eudiaptomus gracilis*, in lowland waters of northern Italy. Although this species is indigenous and widespread in Europe, it was not known in Italy until the 1980s and it now appears to be having an adverse effect on the indigenous *Eudiaptomus padanus* Burckhardt in northern Italy at least. Ferrari *et al.* (1992) have recorded the Australian calanoid, *Boeckella triarticulata*, from fishponds in northern Italy, and Baldaccini *et al.* (1997) have reported the occurrence of the Central American cyclopoid, *Apocyclops panamensis*, from Lake Massaciuccoli (a brackish shallow lake) in Tuscany (Italy). Alekseev *et al.* (2002) have reported that a cyclopoid, *Acanthocyclops americanus*, now occurs in Belgium at densities of 40,000 m^{-2}. They mention that this species rapidly expanded across Europe and Asia in the 20th century after its introduction from North America into Britain in the 19th century. It is now found as far east as reservoirs on the Rivers Volga and Dnieper, where it is the dominant pelagic animal in the summer months (Alekseev and Kosova 1977 in Alekseev *et al.* 2002).

Branchiura. The fish-louse, *Argulus japonicus*, has a worldwide distribution having being moved with farmed fish stocks, e.g. koi carp, *Cyprinus carpio* Linnaeus from the Orient (Rushton-Mellor 1992, Lester and Roubal 1999). It is common wherever goldfish are found. Its distribution overlaps that of *Argulus foliaceus* Linnaeus but generally occurs in warmer water. *Argulus japonicus* was first discovered in Europe in 1921 (Spain) and has since been found in Germany, France, Italy, Poland, and Slovakia (G. Boxshall 2006, personal communication) on many fish, including species *Carassius*, *Cyprinus*, *Esox*, *Perca*, *Tinca*, and *Scardinius*. The first record for the UK was in 1990 on koi and mirror carp in English ponds (Rushton-Mellor 1992), and it has since spread to indigenous fish populations in southern England (G. Boxshall 2006, personal communication).

Isopoda. Only isopods belonging to the Asellota have invaded European inland waters. *Asellus communis* from North America was mentioned earlier. *Proasellus coxalis*, originating from the western Mediterranean, southern Italy, Sicily, and the Aegean Sea, reached the R. Rhine via southern France, through the Rhône, Saône, Doubs, and the Rhine-Rhône Canal. It is chiefly distributed in streams and rivers of northern Germany: Ems, Saale, Ruhr, Weser, Aller, and Elbe. It is seldom found in the upper Rhine in southern Germany. *Proasellus meridionalis* was previously distributed in western Europe, and was recorded in the 1930s and 1940s in France and England. It made use of the Rhône-Saône-Seine Canal and the Rhine-Rhône Canal for its further spread. The density of this isopod in the Rivers Saar and Rhine, however, is not high. Van der Velde *et al.* (2000) list *P. coxalis* and *P. meridionalis* as having invaded the R. Rhine in the Netherlands.

The isopod genus *Jaera* Leach (Family Janiridae) has been revised by Veuille (1979) who described *Jaera istri* as a new species using morphological characters. The type locality was near Kladovo on the R. Danube near the Iron Gate. At that time, *J. istri* was endemic to the R. Danube from Romania to the Austrian–German border. The distribution of *Jaera sarsi* (Valkanov) is limited to brackish waters in Bulgaria, where it colonizes the supra-littoral zone of the

Black Sea and adjacent areas. It has been described from Lake Gebedze and Lake Schabla. The literature before the revision of Veuille (1979) has obviously listed only *J. sarsi* but it is certain that the purely freshwater populations from the middle and upper stretches of the R. Danube can be referred to as *J. istri*. The oldest Austrian record of the species is from the R. Danube in Vienna, dating back to the year 1934 (Strouhal 1939). In the Bavarian section of the R. Danube, *J. istri* was observed in 1967 (Kothé 1968). After the opening of the R. Danube-Main-R. Rhine Canal, thus joining two previously separated catchments to create a new navigation route between the North Sea and the Black Sea in 1992, *J. istri* has taken the southern corridor to invade many waters in the west of the European Continent. Muskó *et al.* (2005) stated that the littoral zone of Lake Balaton in Hungary has been invaded by *J. istri*. It was found in 1993 in the Main-Danube Canal (Tittizer 1997), in 1994 in the R. Main (Schleuter and Schleuter 1995), and in 1996 in the middle section of the R. Rhine (Schöll and Banning 1996). The Rhine delta was colonized in 1997 (Kelleher *et al.* 2000a), where this lithophilous isopod species inhabits solid substrates. In 1999, *J. istri* was found in the R. Elbe, having used the central corridor for further range extension into the north-eastern part of Europe (Schöll and Hardt 2000). The species, which is salt tolerant, may be spread by means of vessels. The food of this small (1.98 mm) isopod, which can reach mean densities of 2,814 m^{-2}, and highest maximum densities of 5,110 m^{-2} (Kelleher *et al.* 2000b), consists of algae, plant remains, and detritus. Its abundance on all sides of a stone, either sheltered or unsheltered, is fairly similar, in contrast to amphipods such as *E. ischnus* and *D. villosus*, which are found mainly on more sheltered areas and crevices of stones. The life history and reproductive behaviour have not been studied.

Mysida. Species belonging to the mysid genera, *Hemimysis*, *Limnomysis* and *Paramysis* are amongst those crustaceans that have been deliberately introduced from the Ponto–Caspian basin to eastern European countries as fish food (Borodich and Havlena 1973, Ketelaars *et al.* 1999, Arbaciauskas 2002). *Limnomysis benedeni* is a euryhaline mysid species and tolerates a salinity of 6.5‰. Lentic environments with aquatic vegetation and tree roots are preferred. Originally, the species was endemic to the coastal waters of the Black and Caspian seas and can be found several hundred kilometres upstream in rivers discharging into both seas. The species have been intentionally introduced into several habitats along the Baltic coast of the former Soviet Union, and in Lake Balaton, Hungary, for the enhancement of fish production (Bij de Vaate *et al.* 2002). However, Muskó *et al.* (2005) noted that it is only found sporadically in that lake. In 1947, *L. benedeni* was found in the R. Danube in the vicinity of Budapest (Dudich 1947), in 1973 in an ox bow lake of the Austrian Danube near Schönau (Weish and Türkay 1975), in 1993 in the Bavarian Danube (Wittmann 1995), and in 1998 in the Main–Danube Canal (Reinhold and Tittizer 1998). However, by 1998 the species had already reached the middle R. Rhine and the Rhine delta (Kelleher *et al.* 1999, Ketelaars *et al.* 1999). This

Fig. 7 The Ponto–Caspian mysid, *Hemimysis anomala*. Widely spread in continental Europe as a result of stockings to enhance fish production since the middle of the 20th century, but a recent invader in central England. Specimen figured taken from a large swarm in the National Water Sports Centre, Holme Pierrepont, Nottingham, England. (Photo: M. Winter and L. Rippon)

well documented immigration clearly indicates the southern corridor of the migration route for the westward range extension of *L. benedeni*. As Wittmann (1995) has most frequently caught the species in harbours, he suggests that the major vector of migration is shipping. *Hemimysis anomala* (Fig. 7) is a euryhaline mysid from the coastal regions and lagoons associated with the Black, Azov, and Caspian seas, as well as extending upstream into rivers. In the middle 20th century it was introduced into a number of reservoirs in the former USSR, and from these to Lithuania from where it spread to the Baltic Sea (Ketelaars *et al.* 1999). Subsequently, it was recorded from the R. Rhine catchment in 1997, then in the R. Main in 1998. In 1999, it was recorded from the R. Danube (Wittmann et al. 1999). It reached the Netherlands via the Main–Danube Canal or invaded from the Baltic via ballast water (Faasse 1998, Van der Velde *et al.* 2000). It has also been recorded in Belgium (Verslycke *et al.* 2000) and most recently in France (Dumont 2006) and Germany (Bernauer and Jansen 2006). It is a voracious predator and also an omnivorous feeder, and its adverse impact on zooplankton and algae in a freshwater storage reservoir in the Netherlands has been documented by Ketelaars *et al.* (1999). This species has made a sudden appearance in central England, although it is not known how it entered the country. It is presently most common in a large recreational lake that hosts international rowing events (Holdich *et al.* 2006). In 2001, *Katamysis warpachowskyi*, a further mysid shrimp from the Ponto–Caspian basin, was recorded for the Hungarian, Slovakian, and Austrian stretches of the R. Danube (Wittmann 2002). Although the relict mysid, *Mysis relicta* Lovén, is indigenous to Europe it was introduced into lakes in Norway and Sweden as fish food. Subsequently, in Swedish lakes many cladocerans disappeared and this led to reduced growth of Arctic char (*Salvelinus alpinus* Linnaeus) (Langeland *et al.* 1991 in Ketelaars *et al.* 1999). The dramatic impact that the stocking of *M. relicta* in Flathead Lake (North America) has had at all trophic levels, including top carnivores, is provided by Spencer *et al.* (1991).

CONCLUSIONS

As noted by Aladin *et al.* (2002) in relation to the Caspian Sea, all resident species can be described as invaders, the only difference being the time of introduction, with the most ancient invaders now being regarded as indigenous. The same can be said for the inland waters of Europe, which were invaded by a wide variety of species after the last glaciation some 10,000 years ago or were colonized by glacial relics. In this review of invasive crustaceans, only recent invaders are considered. Leppäkoski *et al.* (2002a) noted that the "North American" barnacle, *Balanus improvisus* Darwin, was first recorded in Europe in 1844, although P. Rainbow (2006, personal communication) doubts that it is a North American species, but occurs naturally on both sides of the Atlantic. The Chinese mitten crab, *E. sinensis*, is considered to be the first recorded case (1912) of a species being transported between continents in ballast water (Carlton 1985). The narrow-clawed crayfish, *A. leptodactylus*, might well be the first recorded case of a Ponto–Caspian crustacean invading northern Europe. Huxley (1881) stated that, "the invading *Astacus leptodactylus* is everywhere overcoming and driving out *Astacus nobilis* in the struggle for existence, apparently in virtue of its more rapid multiplication." He was referring to the displacement of *A. astacus* in the White Sea region of Russia, and the fact that *A. leptodactylus* had probably reached this region via canals connecting its rivers to the R. Volga. The spiny-cheek crayfish, *O. limosus*, introduced into Germany in 1890, is certainly the first example of a crustacean being introduced from North America for stocking purposes (Holdich 2002b, Machino and Holdich 2006, Holdich and Black 2007).

Introductions usually increase biodiversity, but this can be at a cost to the indigenous fauna. For example, the UK had a single indigenous species of crayfish before the 1970s, but by 2004 there were five other established crayfish species, imported deliberately for aquaculture, restaurant, bait, and pet trades, and all with the potential to harm the indigenous species through competition, and transmission of disease in the case of North American species (Holdich *et al.* 2004, Holdich and Pöckl 2005). Many countries associated with the R. Danube and R. Rhine and their various connecting canals, have seen a large increase in the number of mostly Ponto–Caspian species, particularly amphipods, cladocerans and mysids, inhabiting their inland waters over the last few decades, e.g. the Netherlands and Germany (Van der Velde *et al.* 2000, Bernauer and Jansen 2006), which have been found to have an adverse impact on the indigenous biota. However, there can also be benefits, e.g. Kelleher *et al.* (2000b), pointed out that many non-indigenous amphipods in the R. Rhine now form a large part of the diet of macrozoobenthivorous fish. These invaders have either diffused naturally or been aided by human activities such as shipping and inoculation of waters to enhance fish production.

For crustaceans, the trends outlined in this review are set to become worse. A case in point is the R. Rhine, which amongst European rivers is probably the best documented (see above). Most recently, Bernauer and Jansen (2006) reported that NIS made up 74% of the total number of organisms collected from ship-based samples and 85% from cooling water intake (of which 64% were *D. villosus*). They recorded 17 species of non-indigenous crustaceans in the upper R. Rhine, which was just over 50% of the non-indigenous macroinvertebrate species found there. It is thought that most of these species have made their way into the R. Rhine via the Main–Danube canal, which first opened in 1992. Bernauer and Jansen (2006) have shown that the macroinverebrate community of the upper R. Rhine has been severely altered by the invasion of several highly successful NIS (mostly crustaceans) that has resulted in the elimination or population decline of some of the indigenous species.

The largest crustacean invaders, the Chinese mitten crab and the North American crayfish, still have parts of Europe to conquer. The former is gradually moving round Britain, and has recently entered Irish waters, as well as spreading into eastern Europe and the Near East, whilst the latter is moving into eastern Europe, often aided by humans with aquacultural interests. Ireland is of particular interest when considering invasive species (J. D. Reynold 2006, personal communication). It was cut off from continental Europe before 9000 BP, and as a consequence freshwater species had special difficulties in bridging the more saline seas. The amphipod *G. duebeni* perhaps invaded from the sea in peri-glacial times of low salinity to become widespread in freshwaters, with *G. lacustris* invading large lakes. Most of the invasive species in continental European inland waters have yet to arrive in Ireland, and for larger species such as crayfish this is in part due to strict legislation on imports (Reynolds 1997).

It is virtually impossible to eliminate an established non-indigenous aquatic macroinvertebrate in anything but a small, enclosed waterbody (Holdich *et al.* 1999, Peay *et al.* 2006). The aim of eradication is to completely remove the invasive species, whereas control aims for its reduction over time. Eradication is best attempted in the early stages of invasion. However, many invasive species are not noticed until they become established, e.g. the case of the crayfish, *O. virilis*, in the Netherlands (Souty-Grosset *et al.* 2006) and the mysid, *H. anomala*, in England (Holdich *et al.* 2006). Unless an invasive species can be seen by the relevant authorities to be causing economic or physical harm to the aquatic environment, then they are unlikely to be prepared to spend large sums of money on eradication programmes. Control is also very difficult if only trappable crustaceans are removed as the lack of large individuals may result in smaller cohorts growing more rapidly. Hundreds of thousands of non-indigenous crayfish have been removed from some rivers in Britain at great expense but this has had little impact (Collins 2006). No amount of legislation (Holdich and Pöckl 2005) will prevent a child tipping a pet crayfish into a lake after it has outgrown its tank. If such a crayfish is parthenogenetic, as has been found in the case of *Procambarus* sp. (Scholtz *et al.* 2003, Vogt *et al.* 2004, Seitz *et al.* 2005), then it only needs one individual to start a population. It is also very difficult to get the message over to recreational anglers that introducing live non-indigenous crustaceans as food to increase fish production is not a good idea. One huge problem concerns the pet trade, which some European countries appear unwilling to do anything about (Holdich and Pöckl 2005). Extensive lists of North American crayfish are available to European aquarists. Potentially invasive crustaceans are even traded on eBay!

As pointed out by Gollasch and Leppäkoski (1999), all invasive species should be treated as "guilty until proved innocent", as there is no way of exactly predicting how a NIS will behave in a new habitat (Leppäkoski *et al.* 2002b). However, it seems to be the case that the majority of introduced crustaceans have some of the characteristics of *r*-selected species (short life cycles, high fecundity, fast growth), whilst the established indigenous species are more K-selected (Lindqvist and Huner 1999, Van der Velde *et al.* 2000). Hopefully, continued education, vigilance, and prosecutions may eventually lead people to realize the dangers of intentional introductions of invasive species, but unintentional introductions will be impossible to stop. All that can be hoped for is that education will lead to increased vigilance, so that when NIS are reported to the relevant authorities they may try and do something about them rather than procrastinating for years until it is too late, as is usually the case (Holdich and Pöckl 2005). Certainly, in the UK, press and TV coverage have raised public awareness of the dangers of introduced crustacean species. In recent years, the UK attention has been on crayfish, but currently it has become focused on the Chinese mitten crabs with headlines such as "Crab that eats riverbanks brings flood threat" (Daily Mail, 17 November, 2005).

ACKNOWLEDGEMENTS

The authors wish to thank Christopher Ashelby, Roger Bamber, Geoff Boxshall, Keith Crandall, Jamie Dick, Leonor Fidalgo, Stephan Gollasch, Elisabeth Licek,

Vadim Panov, Philip Rainbow, Julian Reynolds, Nicoletta Riccardi, Roni Robbins, David Sutcliffe, and Karl Wittman for providing useful information for this review; the following for permission to use the photographs: W. G. Graf and A. Schmidt-Kloiber (Fig. 2), W. Graf (Fig. 3), John Black (Fig. 4), Peter Sibley (Fig. 5), Natural History Museum Photo Unit, London (Fig. 6), and Lesley Rippon (Fig. 7); and Julian Reynolds and David Sutcliffe for critically reading the manuscript.

REFERENCES

Ackefors, H. 1999. The positive effects of established crayfish introductions in Europe. Pages 49–61 in F. Gherardi and D. M. Holdich, editors. Crayfish in Europe as alien species. How to make the best of a bad situation? A. A. Balkema, Rotterdam, The Netherlands.

Adema, J. P. H. M. 1991. De krabben van Nederland en België (Crustacea, Decapoda, Brachyura). Nationaal Natuurhistorish Museum, Leiden, **XI**, 1–244 (34 maps and 79 figures).

Ahyong, S. T. 2006. Phylogeny of the clawed lobsters (Crustacea: Decapoda: Homarida). Zootaxa **1109**, 1–14.

Aladin, N. V., I. S. Plotnikov, and A. A. Filippov. 2002. Invaders in the Caspian Sea. Pages 351–359 in E. Leppäkoski, S. Gollasch, and S. Olenin, editors. Invasive aquatic species of Europe. Kluwer Academic Publishers, Dordrecht, The Netherlands.

Alekseev, V., E. Fefilova, and H. J. Dumont. 2002. Some noteworthy free-living copepods from surface freshwater in Belgium. Belgian Journal of Zoology **132**, 133–139.

Arbaciauskas, K. 2002. Ponto–Caspian amphipods and mysids in the inland waters of Lithuania: history of introduction, current distribution and relations with native malacostracans. Pages 104–115 in E. Leppäkoski, S. and S. Olenin, editors. Invasive aquatic species of Europe. Kluwer Academic Publishers, Dordrecht, The Netherlands.

Ashelby, C. W., T. M. Worsfold, and C. H. J. M. Fransen. 2004. First records of the oriental prawn *Palaemon macrodactylus* (Decapoda: Caridea), an alien species in European waters, with a revised key to British Palaemonidae. Journal of the Marine Biological Association, UK **84**, 1041–1050.

Bacela, K. and A. Konopacka. 2005a. The life history of *Pontogammarus robustoides*, an alien amphipod species in Polish waters. Journal of Crustacean Biology **25**, 190–195.

Bacela, K. and A. Konopacka. 2005b. Invasive by life history? – The case of *Dikerogammarus haemobaphes* in the Vistula river. Page 21 in abstracts for Biological Invasions in Inland Waters (INWAT) workshop, Firenze, Italy, May 5–7, 2005.

Baldaccini, G. N., P. Ercolini, and M. Mattioli. 1997. Eutrofizzazione del Lago di Massaciuccoli. Composizione ed evoluzione temporale delle communità zooplanctonica e macrobenthonica. Pages 289–346 in Lago di Massaciuccoli. **13** Ricerche finalizzare al risanamento, Febbraio, 1997.

Barbaresi, S. and F. Gherardi. 1997. Italian freshwater decapods: exclusion between the crayfish *Austropotamobius pallipes* (Faxon) and the crab *Potamon fluviatile* (Herbst) Bulletin Français de la Pêche et de la Pisciculture **347**, 731–747.

Berezina, N. A. 2004. Causes, characteristics and consequences of non-indigenous amphipod species dispersal in aquatic ecosystems of Europe. Pages 254–268 in Biological invasions in aquatic and terrestrial ecosystems. KMK Scientific Press, Moscow, Russia. (In Russian with English summary)

Berezina, N. A. and V. E. Panov. 2003. Establishment of new gammarid species in the eastern Gulf of Finland (Baltic Sea) and their effects on littoral communities. Proceedings of the Estonian Academy of Sciences in Biology and Ecology **52**, 284–304.

Berger, V. J. A. and A. D. Naumov. 2002. Biological invasions of the White Sea. Pages 235–239 in E. Leppäkoski, S. Gollasch, and S. Olenin, editors. Invasive aquatic species of Europe. Kluwer Academic Publishers, Dordrecht, The Netherlands.

Bernauer, D. and W. Jansen. 2006. Recent invasions of alien macroinvertebrates and loss of native species in the Upper Rhine, German. Aquatic Invasions **1**, 55–71.

Bernerth, H. and S. Stein. 2003. *Crangonyx pseudogracilis* and *Corophium robustum* (Amphipoda), zwei neue Einwanderer im hessischen Main sowie Erstnachweise für Deutschland von *C. robustum*. Lauterbornia **48**, 57–60.

Berthold, E. and I. Kaiser. 2004. Weitere Funde von *Crangonyx pseudogracilis* und *Chelicorophium robustum* (Amphipoda) im Main. Lauterbornia **50**, 15–16.

Bij de Vaate, A. and A. G. Klink. 1995. *Dikerogammarus villosus* Sowinsky (Crustacea: Gammaridae) a new immigrant in the Dutch part of the Lower Rhine. Lauterbornia **20**, 51–54.

Bij de Vaate, A., K. Jażdżewski, H. A. M. Ketelaars, S. Gollasch, and G. Van der Velde. 2002. Geographical patterns in range extension of Ponto–Caspian macroinvertebrate species in Europe. Canadian Journal of Fisheries and Aquatic Sciences **59**, 1159–1174.

Borodich, N. D. and F. K. Havlena. 1973. The biology of mysids acclimatized in the reservoirs of the Volga River. Hydrobiologia **42**, 527–539.

Bruijs, M. C. M., B. Kelleher, G. Van der Velde, and A. Bij de Vaate. 2001. Oxygen consumption, temperature and salinity tolerance of the invasive amphipod *Dikerogammarus villosus*: indicators of further dispersal via ballast water transport. Archiv für Hydrobiologie **152**, 633–646.

Buckley, P., Dussart, G., and J. A. Trigwell. 2004. Invasion and expansion of Corophiidae (Amphipoda) in the Stour estuary (Kent, UK). Crustaceana **77**, 425–433.

Bulnheim, H-P. 1985. Genetic differentiation between natural populations of *Gammarus tigrinus* (Crustacea, Amphipoda) with reference to its range extension in European continental waters. Archiv für Hydrobiologie **102**, 273–290.

Carlton, J. T. 1985. Transoceanic and interoceanic dispersal of coastal marine organisms: the biology of ballast water. Oceanography and Marine Biology **23**, 313–371.

Carlton, J. T. 1996. Pattern, process, and prediction in marine invasion technology. Biological Conservation **78**, 97–106.

Casellato, S., G. La Piana, L. Latella, and S. Ruffo. 2005. *Dikerogammarus villosus* (Sowinsky 1894), a new invasive species in Garda Lake (Northern Italy). Pages 26–27 in abstracts for Biological Invasions in Inland Waters (INWAT) workshop, Firenze, Italy, May 5–7, 2005.

Collins, V. 2006. £100,000 battle against alien crayfish ends in defeat. Crayfish NEWS – International Association of Astacology Newsletter **28**, 10–11.

Crandall, K. A., J. Harris, and J. W. Fetzner Jr. 2000. The monophyletic origin of freshwater crayfish estimated from nuclear and mitochondrial DNA sequences. Proceedings of the Royal Society of London B **267**, 1679–1686.

Crawford, G. I. 1935 *Corophium curvispinum* GO Sars, var. *devium* Wundsch, in England. Nature (London) **136**, 685.

Devin, S., J-N. Beisel, V. Bachmann, and J-C. Moreteau. 2001. *Dikerogammarus villosus* (Amphipoda: Gammaridae): another invasive species newly established in the Moselle river and French hydrosystems. Annales de Limnologie **37**, 21–28.

Dick, J. T. A. 1996. Post-invasion amphipod communities of Lough Neagh, Northern Ireland: influences of habitat selection and mutual predation. Journal of Animal Ecology **65**, 756–767.

Dick, J. T. A., R. W. Elwood, and D. E. Irvine. 1990a. Displacement of the native Irish freshwater amphipod *Gammarus duebeni* by the introduced *Gammarus pulex*. Irish Naturalists' Journal **23**, 313–316.

Dick, J. T. A., D. E. Irvine, and R. W. Elwood. 1990b. Differential predation by males on moulted females may explain competitive displacement of *Gammarus duebeni* by *G. pulex* (Amphipoda). Behavioural Ecology and Sociobiology **26**, 41–45.

Dick, J. T. A., C. McNeil, and R. Anderson. 1999. The distribution of *Crangonyx pseudogracilis* Bousefield 1958 (Crustacea: Amphipoda) in Northern Ireland, with notes on its ecology and behaviour. Irish Naturalists' Journal **26**, 236–240.

Dick, J. T. A., I. Montgomery, and R. W. Elwood. 1993. Replacement of the indigenous amphipod *G. duebeni* by the introduced *G. pulex*: differential cannibalism and mutual predation. Journal of Ecology **62**, 79–88.

Dick, J. T. A. and D. Platvoet. 2000. Invading predatory crustacean *Dikergammarus villosus* eliminates both native and exotic species. Proceedings of the Royal Society, London B **267**, 977–983.

Dick, J. T. A., D. Platvoet, and D. W. Kelly. 2002. Predatory impact of the freshwater invader *Dikerogammarus villosus* (Crustacea: Amphipoda). Canadian Journal of Fisheries and Aquatic Sciences **59**, 1078–1084.

Dudich, E. 1947. Die höheren Krebse (Malacostraca) der Mitteldonau. Fragmenta Faunistica Hungarica **10**, 125–232.

Dumont, S. 2006. A new invasive species in the north-east of France, *Hemimysis anomala* G. O. Sars, 1907 (Mysidacea). Crustaceana **79**, 1269–1274.

Faasse, M. A. 1998. The Pontocaspian mysid *Hemimysis anomala* Sars, 1907, new to the fauna of the Netherlands. Bulletin Zoölogisch Museum **16**, 73–76.

Ferrari, I., A. Farabegoli, A. Pugnetti, and E. Stella. 1992. The occurrence of a calanoid Australian species, *Boeckella triarticulata* (Thompson), in fish ponds in Northern Italy. Verhandlungen der Internationalen Vereingung für Limnologie **24**, 2822–2827.

Fetzner, J. W. 2005. A productive year for describing new crayfish species. Crayfish NEWS, International Association of Astacology Newsletter **27**, 1, 3.

Fidalgo, M. L. 1989a. Some additional observations on the population dynamics of the freshwater *Atyaephyra desmaresti* Millet (Decapoda: Natantia) in Crestuma/Lever reservoir (River Douro, Portugal). Publicazioni Instituto Zoologico Dr Augusto Nobre **214**, 1–14.

Fidalgo, M. L. 1989b. Biology of the freshwater shrimp *Atyaephyra desmaresti* Millet (Decapoda: Natantia) in the River Douro, Portugal. I. Life cycle and individual growth. Archiv für Hydrobiologie **116**, 97–106.

Firkins, I. and D. M. Holdich. 1993. Thermal studies with three species of freshwater crayfish. Freshwater Crayfish **9**, 241–248.

Fries, G. and F. W. Tesch. 1965. Der Einfluss des Massenvorkommens von *Gammarus tigrinus* Sexton auf Fische und niedere Tierwelt in der Weser. Archiv für Fischereiwissenschaften **16**, 133–150.

Froglia, C. 2005. The spreading of alien decapod crustaceans in marine and freshwater habitats in Italy. Page 192 in abstracts, Sixth International Crustacean Congress, Glasgow, UK, July 18–22, 2005.

Gherardi, F., G. Baldaccini, P. Ercolini, S. Barbaresi, G. De Luise, D. Mazzoni, and M. Mori. 1999. The situation in Italy. Pages 107–128 in F. Gherardi and D. M. Holdich, editors. Crayfish in Europe as alien species: How to make the best of a bad situation? A. A. Balkema, Rotterdam, The Netherlands.

Gherardi, F. and D. M. Holdich, editors. 1999. Crayfish in Europe as alien species: How to make the best of a bad situation? A. A. Balkema, Rotterdam, The Netherlands.

Gledhill, T., D. W. Sutcliffe, and W. D. Williams. 1993. British Freshwater Crustacea Malacostraca: a Key with Ecological Notes. Freshwater Biological Association Scientific Publication No. 52, Ambleside.

Gollasch, S. 1999. *Eriocheir sinensis* (Milne-Edwards, 1854), the Chinese mitten crab. Pages 55–61 in S. Gollasch, D. Minchin, H. Rosenthal, and M. Voigt, editors. Exotics across the ocean. Case histories on introduced species: their general biology, distribution, range expansion and impact. Logos Verlag, Berlin, Germany.

Gollasch, S. and E. Leppäkoski. 1999. Initial risk assessment of alien species in Nordic coastal waters. Nord 1998, **8**. Nordic Council of Ministers, Copenhagen, Denmark.

Gomoiu, M-T., B. Alexandrov, N. Shadrin, and T. Y. Zaitsev. 2002. The Black Sea – a recipient, donor and transit area for alien species. Pages 341–350 in E. Leppäkoski, S. Gollasch, S. and S. Olenin, editors. Invasive aquatic species of Europe. Kluwer Academic Publishers, Dordrecht, The Netherlands.

Gouin, N., F. Grandjean, S. Pain, C. Souty-Grosset, and J. Reynolds. 2003. Origin and colonization history of the white-clawed crayfish, *Austropotamobius pallipes*, in Ireland. Heredity **9**, 70–77.

Goulletquer, P., G. Bachelet, P. G. Sauriau, and P. Noël. 2002. Open Atlantic coast of Europe – a century of introduced species into French waters. Pages 276–290 in E. Leppäkoski, S. Gollasch, and S. Olenin, editors. Invasive aquatic species of Europe. Kluwer Academic Publishers, Dordrecht, The Netherlands.

Grabow, K., T-O. Eggers, and A. Martens 1998 *Dikerogammarus villosus* Sowinsky (Crustacea: Amphipoda) in navigable canals and rivers in northern Germany. Lauterbornia **33**, 103–107.

Grandjean, F., C. Souty-Grosset, and D. M. Holdich. 1997. Mitochondrial DNA variation in four British populations of the white-clawed crayfish *Austropotamobius pallipes pallipes*: implications for management. Aquatic Living Resources **10**, 121–126.

Guan, R-Z. and P. R. Wiles. 1997. Ecological impact of introduced crayfish on benthic fishes in a British lowland river. Conservation Biology **11**, 641–647.

Haempel, O. 1908. Über die Fortpflanzung und künstliche Zucht des gemeinen Flohkrebses (*Gammarus pulex* L. und *fluviatilis* R.). Allgemeine Fischerei Zeitung **33**, 86–141.

Harding, P. T. and G. M. Collis. 2006. The occurrence of *Asellus communis* Say, 1818 (Crustacea, Isopoda) at Bolam lake, Northumberland. Bulletin of the British Myriapod and Isopod Group **21**, 8–11.

Harris, R. R. 1991. Amphipod also invades Britain. Nature (London) **354**, 194.

Herborg, L.-M., S. P. Rushton, A. S. Clare, and M. G. Bentley. 2003. Spread of the Chinese mitten crab (*Eriocheir sinensis* H. Milne Edwards) in continental Europe: analysis of a historical data set. Hydrobiologia **503**, 21–28.

Herborg, L.-M., S. P. Rushton, A. S. Clare, and M. G. Bentley. 2005. The invasion of the Chinese mitten crab (*Eriocheir sinensis*) in the United Kingdom and its comparison to continental Europe. Biological Invasions **7**, 959–968.

Herbst, V. 1982. Amphipoden in salzbelasteten niedersächsischen Oberflächengewässern. Gewässer und Abwässer **68/69**, 35–40.

Herhaus, K. F. 1978. Die ersten Nachweise von *Gammarus tigrinus* Sexton, 1939, und *Chaetogammarus ischnus* (Stebbing 1906) (Crustacea, Amphipoda, Gammaridae) im Einzugsgebiet der Ems und ihre verbreitungsgeschichtliche Einordnung. Natur und Heimat **38**, 71–77.

Heuss, K., W-D. Schmidt, and H. Schödel. 1990. Die Verbreitung von *Atyaephyra desmaresti* (Millet) (Crustacea, Decapoda) in Bayern. Lauterbornia **6**, 85–88.

Hobbs Jr, H. H. 1988. Crayfish distribution, adaptive radiation and evolution. Pages 52–82 in D. M. Holdich and R. S. Lowery, editors. Freshwater crayfish: biology, management and exploitation. Croom Helm, London, UK.

Holdich, D. M. 1999. The negative effects of established crayfish introductions. Pages 31–47 in F. Gherardi and D. M. Holdich, editors. Crayfish in Europe as alien species: How to make the best of a bad situation? A. A. Balkema, Rotterdam, The Netherlands.

Holdich, D. M., editor. 2002a. The biology of freshwater crayfish. Blackwell Science Publishers, Oxford, UK.

Holdich, D. M. 2002b. Present distribution of crayfish in Europe and some adjoining countries. Bulletin Français de la Pêche et de la Pisciculture **367**, 611–650.

Holdich, D. M. 2003. Crayfish in Europe – an overview of taxonomy, legislation, distribution, and crayfish plague outbreaks. Pages 15–34 in D. M. Holdich and P. J. Sibley, editors. Management & conservation of crayfish. Proceedings of a conference held in Nottingham on 7th November, 2002. Environment Agency, Bristol, UK.

Holdich, D. M. and J. Black. 2007. The spiny-cheek crayfish, *Orconectes limosus* (Rafinesque, 1817) [Crustacea: Decapoda: Cambaridae], digs into the UK. Aquatic Invasions **2**, 1–16.

Holdich, D. M., S. Gallagher, L. Rippon, P. Harding, and R. Stubbington. 2006. The invasive Ponto–Caspian mysid, *Hemimysis anomala*, reaches the UK. Aquatic Invasions **1**, 4–6.

Holdich, D. M., R. Gydemo, and W. D. Rogers. 1999. A review of possible methods for controlling nuisance populations of alien crayfish. Pages 245–270 in F. Gherardi and D. M. Holdich, editors. Crayfish in Europe as alien species: How to make the best of a bad situation? A. A. Balkema, Rotterdam, The Netherlands.

Holdich, D. M., M. M. Harlioğlu, and I. Firkins. 1997. Salinity adaptations of crayfish in British waters with particular reference to *Austropotamobius pallipes*, *Astacus leptodactylus* and *Pacifastacus leniusculus*. Estuarine, Coastal & Shelf Science **44**, 147–154.

Holdich, D. M. and M. Pöckl. 2005. Does legislation work in protecting vulnerable species? Bulletin de Français de la Pêche et de la Pisciculture **376–377**, 809–827.

Holdich, D. M., P. J. Sibley, and S. Peay. 2004. The white-clawed crayfish – a decade on. British Wildlife **15**, 153–164.

Holmes, J. M. C. 1975. *Crangonyx pseudogracilis* Bousfield, a freshwater amphipod new to Ireland. Irish Naturalist's Journal **18**, 225–226.

Huxley, T. H. 1881. The crayfish. An introduction to the study of zoology. 3rd edn. Kegan Paul, London, UK. (The MIT Press, Cambridge, Massachusetts republished this book in 1974).

Hynes, H. B. N. 1954. The ecology of *Gammarus duebeni* Lilljeborg and its occurrence in fresh water in western Britain. Journal of Animal Ecology **23**, 38–84.

Jarocki, J. and A. Demianowicz. 1931. Über das Vorkommen des pontokaspischen Amphipoden *Chaetogammarus tenellus* (G.O. Sars) in der Wistula (Weichsel). Bulletin of the Academy of Polish Science, Cracovie (**B II**), 513–530.

Jażdżewski, K. 1980. Range extensions of some Gammaridean species in European Inland waters caused by human activity. Crustaceana Supplement **6**, 84–107.

Jażdżewski, K. and A. Konopacka. 1990. Nowe, interesujace stanowisko Ponto-Kaspijskiego kielza *Echinogammarus ischnus* (Stebbing, 1898) (Crustacea, Amphipoda) w Polsce. Prezegląd Zoologiczny **34**, 101–111.

Jażdżewski, K. and A. Konopacka. 2000. Immigration history and present distribution of alien crustaceans in Polish waters. Pages 55–64 in J. C. von Vaupel Klein and F. R. Schram, editors. The biodiversity crisis and Crustacea. A. A. Balkema, Rotterdam, The Netherlands.

Jażdżewski, K. and A. Konopacka. 2002. Invasive Ponto–Caspian species in waters of the Vistula and Oder basins and the southern Baltic Sea. Pages 384–398 in E. Leppäkoski, S. Gollasch, and S. Olenin, editors. Invasive aquatic species of Europe. Kluwer Academic Publishers, Dordrecht, The Netherlands.

Jażdżewski, K., A. Konopacka, and M. Grabowski. 2002. Four Ponto–Caspian and one American gammarid species (Crustacea, Amphipoda) recently invading Polish waters. Contributions to Zoology **71**, 115–122.

Karaman, I. and Y. Machino. 2004. Occurrence of the spiny-cheek crayfish (*Orconectes limosus*) and the Chinese mitten crab (*Eriocheir sinensis*) in Serbia. Crayfish NEWS, International Association of Astacology Newsletter **26**, 11, 19–20.

Kelleher, B., P. J. M. Bergers, F. W. B. Van den Brink, P. S. Giller, G. Van der Velde, and A. Bij de Vaate. 1998. Effects of exotic amphipod invasions on fish diet in the Lower Rhine. Archiv für Hydrobiologie **143**, 363–382.

Kelleher, B., A. Bij de Vaate, M. Swarte, A. G. Klink, and G. Van der Velde. 2000a. Identification, invasion and population development of the Ponto–Caspian isopod *Jaera istri* Veuille (Janiridae) in the lower Rhine, The Netherlands. Beaufortia **50**, 89–94.

Kelleher, B., G. Van der Velde, P. S. Giller, and A. Bu de Vaate. 2000b. Dominant role of exotic invertebrates, mainly Crustacea, in diets of fish in the lower River Rhine. Pages 35–46 in J. C. von Vaupel Klein and F. R. Schram, editors. The biodiversity crisis and Crustacea. A. A. Balkema, Rotterdam, The Netherlands.

Kelleher, B., G. Van der Velde, K. J. Wittmann, M. A. Faasse, and A. Bij de Vaate. 1999. Current status of the freshwater Mysidae in the Netherlands with records of *Limnomysis benedini* Czerniavsky 1882, a Ponto–Caspic species in the Dutch Lower Rhine branches. Bulletin Zoölogisch Museum Universiteit Amsterdam **16**, 89–94.

Ketelaars, H. A. M., F. E. Lambregts-van de Clundert, C. J. Carpentier, A. J. Wagenvoort, and W. Hoogenboezem. 1999. Ecological effects of the mass occurrence of the Ponto-Caspian invader, *Hemimysis anomala* G.O. Sars, 1907 (Crustacea: Mysidacea), in a freshwater storage reservoir in the Netherlands, with notes on its autecology and new records. Hydrobiologia **394**, 233–248.

Kinzelbach, R. 1972. Einschleppung und Einwanderung von Wirbellosen in Ober- und Mittelrhein. Mainzer Naturwissenschaftliches Archiv **11**, 109–150.

Köhn, J. and A. Waterstraat. 1990. The amphipod fauna of Lake Kummerow (Mecklenburg, German Democratic Republic) with reference to *Echinogammarus ischnus* Stebbing, 1899. Crustaceana **58**, 74–82.

Kothé, P. 1968. *Hypania invalida* (Polychaeta sedentaria) and *Jaera sarsi* (Isopoda) ertmals in der deutschen Donau. Archiv für Hydrobiologie Supplement **34**, 88–114.

La Piana, G., A. Visentin, and S. Casellato. 2005. *Dikerogammarus villosus*: a danger for fish too! Page 47 in abstracts for Biological Invasions in Inland Waters (INWAT) workshop, Firenze, Italy, May 5–7, 2005.

Lehtonen, H. 2002. Alien freshwater fish of Europe. Pages 153–161 in E. Leppäkoski, S. Gollasch, and S. Olenin, editors. Invasive aquatic species of Europe. Kluwer Academic Publishers, Dordrecht, The Netherlands.

Leppäkoski, E., S. Gollasch, P. Gruszka, H. Ojaveer, S. Olenin, and V. Panov. 2002a. The Baltic – a sea of invaders. Canadian Journal of Fisheries and Aquatic Science **59**, 1175–1188.

Leppäkoski, E., S. Gollasch, and S. Olenin, editors. 2002b. Invasive aquatic species of Europe. Kluwer Academic Publishers, Dordrecht, The Netherlands.

Leppäkoski, E., S. Gollasch, and S. Olenin. 2002c. Alien species in European waters. Pages 1–6 in E. Leppäkoski, S. Gollasch, and S. Olenin, editors. Invasive aquatic species of Europe. Kluwer Academic Publishers, Dordrecht, The Netherlands.

Lester, R. J. G. and F. R. Roubal. 1999. Phylum Arthropoda Branchiura. Pages 542–549 in P. T. K. Woo, editor. Fish diseases and disorders, Vol. 1: Protozoan and metazoan infections. CABI Publishing, New York, NY.

Lewis, S. D. 2002. *Pacifastacus*. Pages 511–540 in D. M. Holdich, editor. Biology of freshwater crayfish. Blackwell Science, Oxford, UK.

Lindqvist, O. V. and J. V. Huner. 1999. Life history characteristics of crayfish: what makes some of them good colonizers? Pages 23–30 in F. Gherardi and D. M. Holdich, editors. Crayfish in Europe as alien species: How to make the best of a bad situation? A. A. Balkema, Rotterdam, The Netherlands.

Litvinchuk, L. and O. B. Maximova. 2005. New Ponto–Caspian invaders *Evadne anonyx* and *Conigerius maeoticus maeoticus* (Polyphemoidea, Cladocera) in the Baltic Sea. Page 49 in abstracts for Biological Invasions in Inland Waters (INWAT) workshop, Firenze, Italy, May 5–7, 2005.

Lodge, D. M., C. A. Taylor, D. M. Holdich, and J. Skurdal. 2000a. Nonindigenous crayfishes threaten North American freshwater biodiversity. Fisheries **25**, 7–20.

Lodge, D. M., C. A. Taylor, D. M. Holdich, and J. Skurdal. 2000b. Reducing impacts of exotic crayfish introductions. Fisheries **25**, 21–23.

Lucy, F., D. Minchin, J. M. C. Holmes, and M. Sullivan. 2004 First records of the Ponto–Caspian amphipod *Chelicorophium curvispinum* (Sars, 1895) in Ireland. Irish Naturalists' Journal **27**, 461–464.

Machino, Y. and D. M. Holdich. 2006. Distribution of crayfish in Europe and adjacent countries: updates and comments. Freshwater Crayfish **15**, 292–323.

MacNeil, C., J. T. A. Dick, and R. W. Elwood. 1999. The dynamics of predation on *Gammarus* spp. (Crustacea: Amphipoda). Biological Reviews of the Cambridge Philosophical Society **74**, 375–395.

Maguire, I. 2003. Appearance of *Orconectes limosus* in Croatia. Crayfish NEWS – International Association of Astacology Newsletter **25**, 7.

Martens, A., T. O. Eggers, and K. Grabow. 1999. Erste Funde von *Pontogammarus robustoides* (Sars) im Mittellandkanal (Crustacea: Amphipoda). Lauterbornia **35**, 39–42.

Martin, J. W. and G. E. Davis. 2001. An updated classification of the recent Crustacea, No. **39** Science Series, Natural History Museum of Los Angeles County, CA.

Martynov, A. W. 1925. Gammaridae des unteren Laufes des Dnjepr. Arbeiten der Ukainischen wissenschaftlich-praktischen Staats-Station des Schwarzen und Asow Meeres **B 1**, 133–153.

McLoughlin, N., J. Reynolds, J. Lucey, and M. McGarrigle. 2000. The biogeography and current status of *Gammarus duebeni* Liljeborg and *Gammarus pulex* (L.) (Crustacea, Amphipoda) in freshwater in the republic of Ireland. Bulletin of the Irish Biogeographical Society **24**, 142–152.

McNeely, J. A., H. A. Mooney, L. E. Neville, P. J. Schei, and J. K. Waage. 2001. A Global Strategy on Invasive Alien Species. IUCN Gland, Switzerland, and Cambridge, UK.

Meyer, A., Kashek, N., and E. I. Meyer. 2004. The effect of low flow and stream drying on the distribution and relative abundance of the alien amphipod, *Echinogammarus berilloni* (Catta, 1878) in a karstic stream system (Westphalia, Germany). Crustaceana **77**, 909–922.

Minchin, D. and C. Eno. 2002. Exotics of coastal and inland waters of Ireland and Britain. Pages 267–275 in E. Leppäkoski, S. Gollasch, and S. Olenin, editors. Invasive aquatic species of Europe. Kluwer Academic Publishers, Dordrecht, The Netherlands.

Moog, O., H. Nesemann, H. Zitek, and A. Melcher. 1999. Erstnachweis der Süßwassergarnele *Atyaephyra desmaresti* (Millet, 1831) (Decapoda) in Österreich. Lauterbornia **35**, 67–70.

Mordukhai-Boltovskoi, F. D. 1979. Composition and distribution of Caspian fauna in the light of modern data. Internationale Revue der Gesamten Hydrobiologie **64**, 1–38.

Müller, J. and S. Schramm. 2001. A third *Dikerogammarus* invader is located in front of Vienna. Lauterbornia **41**, 49–52.

Müller, J. C., S. Schramm, and A. Seitz. 2002. Genetic and morphological differentiation of *Dikerogammarus* invaders and their invasion history in Central Europe. Freshwater Biology **47**, 2039–2048.

Muskó, I. B., C. Balogh, A. P. Tóth, E. Varga, and G. Lakatos. 2005. Seasonal changes of Malacostraca in the stony littoral zone of Lake Balaton (Hungary) in a drought and the next generation period. Page 197 in Abstracts, Sixth International Crustacean Congress, Glasgow, UK, July 18–22, 2005.

Nehring, S. 2006. The Ponto–Caspian amphipod *Obesogammarus obesus* (Sars, 1894) arrived in the Rhine River via the Main–Danube Canal. Aquatic Invasions **1**, 148–153.

Nesemann, H. 1984. Die Zehnfußkrebse (Crustacea, Decapoda) der Untermainaue im Jahre 1983. Hessische Faunistische Briefe **4**, 63–69.

Nesemann, H., M. Pöckl, and K. J. Wittmann. 1995. Distribution of epigean Malacostraca in the middle and upper Danube (Hungary, Austria, Germany). Miscellanea Zoologica Hungarica **10**, 49–68.

Niethammer, G. 1950. Zum Transport von Süßwassertieren durch Vögel. Zoologischer Anzeiger **151**, 41–42.

Ozturk, B. 2002. The Marmara Sea, a link between the Mediterranean and the Black sea. Pages 337–340 in E. Leppäkoski, S. Gollasch, and S. Olenin, editors. Invasive aquatic species of Europe. Kluwer Academic Publishers, Dordrecht, The Netherlands.

Panov, V. E. 1996. Establishment of the Baikalian endemic amphipod *Gmelinoides fasciatus* in Lake Ladoga. Hydrobiologia **322**, 187–192.

Panov, V. E. 2006. First record of the Chinese mitten crab, *Eriocheir sinensis* H Milne Edwards 1853 (Crustacea: Brachyura: Varunidae) from Lake Ladoga, Russia. Aquatic Invasions **1**, 28–31.

Panov, V. E. and N. A. Berezina. 2002. Invasion history, biology and impacts of the Baikalian amphipod *Gmelinoides fasciatus*. Pages 96–103 in E. Leppäkoski, S. Gollasch, and S. Olenin, editors. Invasive aquatic species of Europe. Kluwer Academic Publishers, Dordrecht, The Netherlands.

Panov, V. E., P. I. Krylov, and N. Riccardi. 2004. Role of diapause in dispersal and invasion by aquatic invertebrates. Journal of Limnology **63** (Supplement 1), 56–69.

Panov, V. E., N. V. Rodionova, P. V. Bolshagin, and E. A. Bychek. 2006. Invasion biology of Ponto–Caspian onychopod cladocerans (Crustacea: Cladocera: Onychopoda). Hydrobiologia, in press.

Peay, S., P. D. Hiley, P. Collen, and I. Martin. 2006. Biocide treatment of ponds in Scotland to eradicate signal crayfish. Bulletin de Français de la Pêche et de la Pisciculture **380–381**, 1363–1379.

Peters, N. and A. Panning. 1933. Die chinesische Wollhandkrabbe (*Eriocheir sinensis* H. Milne-Edwards) in Deutschland. Zoologischer Anzeiger **101**, 267–271.

Pinkster, S. 1972. On members of the *Gammarus pulex*-group (Crustacea-Amphipoda) from western Europe. Bijdragen Tot de Dierkunde **42**, 164–191.

Pinkster, S. 1993. A revision of the genus *Echinogammarus* Stebbing, 1899, with some notes on related genera (Crustacea, Amphipoda). Memorie del Museo Civico di Storia Naturale (II Series) Sezione Scienze della Vita (A. Biologia) **10**, 1–185.

Pinkster, S., J. Dielman, and D. Platvoet. 1980. The present position of *Gammarus tigrinus* Sexton, 1939, in the Netherlands, with description of a newly discovered amphipod species, *Crangonyx pseudogracilis* Bousfield, 1958 (Crustacea, Amphipoda). Bulletin Zoölogisch Museum, Universiteit Amstderdam **7**, 33–45.

Pinkster, S. and D. Platvoet. 1983. Further observations on the distribution and biology of two alien amphipods, *Gammarus tigrinus* Sexton, 1939, and *Crangonyx pseudogracilis* Bousfield, 1958, in the Netherlands (Crustacea, Amphipoda). Bulletin of the Zoöligisch Museum Universiteit van Amsterdam **9**, 154–164.

Pinkster, S., M. Scheepmaker, D. Platvoet, and N. Broodbakker. 1992. Drastic changes in the amphipod fauna (Crustacea) of Dutch inland waters during the last 25 years. Bijdragen tot de Dierkunde **61**, 193–204.

Pinkster, S., H. Smit, and N. Brandse-de Jong. 1977. The introduction of the alien amphipod *Gammarus tigrinus* Sexton, 1939, in the Netherlands and its competition with indigenous species. Crustaceana Supplement **4**, 91–105.

Piscart, C., A. Manach, G. Copp, and P. Marmonier. 2007. Distribution and microhabitats of native and non-native gammarids (Amphipoda, Crustacea) in Brittany, with particular reference to the endangered endemic sub-species *Gammarus duebeni celticus*. Journal of Biogeography **34**, 524–533.

Platvoet, D. 2005. Temperature dependent feeding in *Dikerogammarus villosus* Sowinsky. Page 60 in abstracts for Biological Invasions in Inland Waters (INWAT) workshop, Firenze, Italy, May 5–7, 2005.

Platvoet, D., M. Scheepmaker, and S. Pinkster. 1989. The position of two introduced amphipod crustaceans, *Gammarus tigrinus* and *Crangonyx pseudogracilis* in the Netherlands during the period 1987–1988. Bulletin Zoöligish Museum, Universiteit van Amsterdam **11**, 197–202.

Pöckl, M. 1988. Bestimmungsschlüssel für Peracarida der österreichischen Donau (Crustacea, Malacostraca). Wasser und Abwasser **32**, 89–110.

Pöckl, M. 2002. Krebstiere: Flohkrebse (Crustacea: Amphipoda: Corophiidae und Gammaridae). Pages 273–284 in F. Essl and W. Rabitsch, editors. Neobiota in Österreich. Federal Environment Agency.

Pöckl, M. 2007. Strategies of a successful new invader in European fresh waters: fecundity and reproductive potential of the Ponto–Caspian amphipod *Dikerogammarus villosus* in the Austrian Danube, compared with the indigenous *Gammarus fossarum* and *G. roeseli*. Freshwater Biology **52**, 50–63.

Podraza, P., T. Ehlert, and P. Roos. 2001. Erstnachweis von *Echinogammarus trichiatus* (Crustacea: Amphipoda) im Rhein. Lauterbornia **41**, 129–131.

Proßeckert, B. 2001. *Hyalella azteca* – ein Krebschen erobert die Welt. Aquaristik aktuell **7–8**, 36–37.

Puky, M., J. D. Reynolds, and P. Schád. 2005. Native and alien Decapoda species in Hungary: distribution, status, conservation importance. Bulletin de Français de la Pêche et de la Pisciculture **376–377**, 553–568.

Rabitsch, W. and F. Schiemer. 2003. Chinesische Wollhandkrabbe (*Eriocheir sinensis*) in der österreichischen Donau. Österreichs Fischerei **56**, 61–65.

Rajagopal, S., G. Van der Velde, B. G. P. Pfaffen, and A. Bij de Vaate. 1998. Growth and production of *Corophium curvispinum* G.O. Sars, 1895 (Amphipoda), an invader in the Lower Rhine. Pages 3–33 in J. C. von Vaupel Klein and F. R. Schram, editors. The biodiversity crisis and Crustacea. A. A. Balkema, Rotterdam, The Netherlands.

Reinhold M. and T. Tittizer. 1998. *Limnomysis benedeni* Czerniavsky (Crustacea: Mysidacea), ein pontokaspisches Neozoon im Main-Donau-Kanal. Lauterbornia **33**: 37–40.

Reynolds, J. D. 1997. The present status of freshwater crayfish in Ireland. Bulletin Français de la Pêche et de la Pisciculture **347**, 693–700.

Robbins, R. S., M. Sakari, S. N. Baluchi, and P. F. Clarke. 2006. The occurrence of *Eriocheir sinensis* H Milne Edwards 1853 (Crustacea: Brachyura: Varunidae) from the Caspian Sea region, Iran. Aquatic Invasions **1**, 32–34.

Robbins, R. S., B. D. Smith, P. S. Rainbow, and P. F. Clarke. 2000. Seasonal changes (1995–1997) in the population structure of Chinese mitten crabs, *Eriocheir sinensis* (Decapoda, Brachyura, Grapsidae) in the Thames at Chelsea, London. Pages 343–350 in J. C. von Vaupel Klein and F. R. Schram, editors. The biodiversity crisis and Crustacea. A. A. Balkema, Rotterdam, The Netherlands.

Rodionova, N. V. and V. E. Panov. 2006. Establishment of the Ponto–Caspian predatory cladoceran *Evadne anonyx* in the eastern Gulf of Finland, Baltic Sea. Aquatic Invasions **1**, 7–12.

Rodriguez, C. F., E. Bécares, M. Fernández-Aláez, and C. Fernández-Aláez. 2005. Loss of diversity and degradation of wetlands as a result of introducing exotic crayfish. Biological Invasions **7**, 75–85.

Rodríguez, G. and H. Suárez. 2001. Anthopogenic dispersal of decapod crustaceans in aquatic environments. Interciencia **26**, 282–288.

Rudolph, K. 1997. Zum Vorkommen des Flohkrebses *Pontogammarus robustoides* im Peene-Mündungsgebiet. Natur und Museum **127**, 306–312.

Ruiz, G. M., J. T. Carlton, E. D. Grosholz, and A. H. Hines. 1997. Global invasions of marine and estuarine habitats by non-indigenous species: Mechanisms, extent, and consequences. American Zoologist **37**, 621–632.

Rushton-Mellor, S. 1992. Discovery of the fish-louse, *Argulus japonicus* Thiele (Crustacea Branchiura) in Britain. Aquaculture & Fisheries Management **23**, 269–271.

Schleuter, M. and A. Schleuter. 1995. *Jaera istri* (Veuille) (Janiridae, Isopoda) aus der Donau erreicht über den Main-Donau-Kanal den Main. Lauterbornia **21**, 177–178.

Schleuter, M., A. Schleuter, S. Potel, and M. Banning. 1994. *Dikerogammarus haemobaphes* (Eichwald, 1841) (Gammaridae) aus der Donau erreicht über den Main-Donau-Kanal den Main. Lauterbornia **19**, 155–159.

Schöll, F. 1990a. Zur Bestandssituation von *Corophium curvispinum* Sars im Rheingebiet. Lauterbornia **5**, 67–70.

Schöll, F. 1990b. Erstnachweis von *Chaetogammarus ischnus* Stebbing im Rhein. Lauterbornia **5**, 71–74.

Schöll, F. and M. Banning. 1996. Erstnachweis von *Jaera istri* (Veuille) (Janiridae, Isopoda) im Rhein. Lauterbornia **25**, 61–62.

Schöll, F. and D. Hardt. 2000. *Jaera istri* (Veuille) (Janiridae, Isopoda) erreicht die Elbe. Lauterbornia **38**, 99–100.

Scholtz, G. 2002. Phylogeny and evolution. Pages 30–52 in D. M. Holdich, editor. Biology of freshwater crayfish. Blackwell Science, Oxford, UK.

Scholtz, G., A. Braband, L. Tolley, A. Reimann, B. Mittmann, C. Lukhaup, F. Steuerwald, and G. Vogt. 2003. Parthenogenesis in an outsider crayfish. Nature (London) **421**, 806.

Scholtz, G. and S. Richter. 1995. Phylogenetic systematics of the reptantian Decapoda (Crustacea, Malacostraca. Zoological Journal of the Linnean Society **113**, 289–328.

Segerstrale, S. G. 1954. The freshwater amphipods, *Gammarus pulex* (L.) and *Gammarus lacustris* G.O. Sars, in Denmark and Fennoscandia – a contribution to the late- and post-glacial immigration history of the fauna in Northern Europe. Societas Scientiarum Fennica, Commentationes Biologica XV, **1**, 1–91.

Seitz, R., K. Vilpoux, U. Hopp, S. Harzsch, and G. Maier. 2005. Ontogeny of the Marmorkrebs (marbled crayfish): a parthenogenetic crayfish with unknown origin and phylogenetic position. Journal of Experimental Zoology **303A**, 393–405.

Sexton, E. W. 1939. On a new species of *Gammarus* (*G. tigrinus*) from Droitwich District. Journal of the Marine Biological Association UK **23**, 543–552.

Sibley, P. 2000. Signal crayfish management in the River Wreake catchment. Pages 95–108 in D. Rogers and J. Brickland, editors. Crayfish conference Leeds. Environment Agency, Leeds, UK.

Sibley, P. J. 2003. The distribution of crayfish in Britain. Pages 64–72 in D. M. Holdich and P. J. Sibley, editors. Management & conservation of crayfish. Environment Agency, Bristol, UK.

Sikes, B. A. 2002. Spiny water flea (*Bythotrephes longimanus* (Leydig 1860). Invader of the Month. Institute for Biological Invasions, June, 1–18.

Steffen, G. F. 1939. Untersuchungen über Morphologie, Lebensweise und Verbreitung von *Atyaephyra desmaresti* Millet (Decapoda, Natantia, Atyidae). Ph.D. thesis, University of Berlin, Germany.

Strange, C. D. and G. B. Glass. 1979. The distribution of freshwater gammarids in Northern Ireland. Proceedings of the Royal Irish Academy **79B**, 145–153.

Strouhal, H. 1939. Einige bemerkenswerte Vorkommen von Wirbellosen, besonders Isopoden, in der Ostmark. Festschrift für Prof. Dr. E. Strand **5**, 68–80.

Souty-Grosset, C., D. M. Holdich, P. Y. Noël, J. D. Reynolds, and P. Haffner, editors. 2006. Atlas of Crayfish in Europe. Muséum national d'Histoire naturelle, Paris, France (Patrimoines naturels, 64).

Spencer, C. N., B. R. McClelland, and J. A. Stanford. 1991. Shrimp stocking, salmon collapse, and eagle displacement. BioScience **41**, 14–21.

Sutcliffe, D. W. 1967. A re-examination of observations on the distribution of *G. duebeni* Lilljeborg in relation to the salt content in fresh water. Journal of Animal Ecology **36**, 579–597.

Taylor, C. 2002. Taxonomy and conservation of native crayfish stocks. Pages 236–257 in D. M. Holdich, editor. Biology of freshwater crayfish. Blackwell Science, Oxford, UK.

Telesh, I. V. and H. Ojaveer. 2002. The predatory water flea *Cercopagis pengoi* in the Baltic Sea: invasion history, distribution and implications to ecosystem dynamics. Pages 62–65 in E. Leppäkoski, S. Gollasch and S. Olenin, editors. Invasive aquatic species of Europe. Kluwer Academic Publishers, Dordrecht, The Netherlands.

Thienemann, A. 1950. Verbreitungsgeschichte der Süßwassertierwelt Europas. Die Binnengewässer **18**, 1–809.

Tittizer, T. 1996. Vorkommen und Ausbreitung aquatischer Neozoen in den europäischen Bundeswasserstraßen. Pages 49–86 in Gebietsfremde Tierarten – Auswirkungen auf einheimische Arten, Lebensgemeinschaften und Biotope – Situationsanalyse. S. Gebhardt, R. Kinzelbach and H. Schmidt-Fischer, editors. Ecomed Verlag, Landsberg, Germany.

Tittizer, T. 1997. Ausbreitung aquatischer Neozoen (Makrozoobenthos) in den europäischen Wasserstraßen, erläutert am Beispiel des Main-Donau-Kanals. Pages 113–134 in Güteentwicklung der Donau, Rückblick und Perspektiven. Schriftenreihe des Bundesamtes für Wasserwirtschaft (Wien) **4**.

Tittizer, T., F. Schöll, M. Banning, A. Haybach, and M. Schleuter. 2000. Aquatische Neozoen im Makrobenthos der Binnenwasserstraßen Deutschlands. Lauterbornia **39**, 1–72.

Van den Brink, F. W. B., G. Van der Velde and A. Bij de Vaate. 1989. A note on the immigration of *Corophium curvispinum* Sars, 1895 (Crustacea: Amphipoda) into the Netherlands via the River Rhine. Bulletin Zoölogisch Museum, Universiteit Amsterdam **11**, 211–213.

Van den Brink, F. W. B. and G. Van der Velde. 1991. Amphipod invasion on the Rhine. Nature **352**, 576.

Van den Brink, F. W. B., B. G. P. Pfaffen, F. M. J. Oosterbroeck, and G. Van der Velde. 1993. Immigration of *Echinogammarus* (Stebbing, 1899) (Crustacea: Amphipoda) into the Netherlands via the lower Rhine. Bulletin Zoölogisch Museum, Universiteit Amsterdam **13**, 167–170.

Van der Velde, G., S. Rajagopol, B. Kelleher, I. B. Muské, and A. B. de Vaate. 2000. Ecological impact of crustacean invaders: General considerations and examples from the Rhine River. Pages 3–33 in J. C. Von Vaupel Klein and F. R. Schram, editors. The biodiversity crisis and Crustacea. A. A. Balkema, Rotterdam, The Netherlands.

Van der Velde, G., S. Rajagopal, F. W. B. Van den Brink, B. Kelleher, B. P. G. Paffen, A. J. Kempers, and A. Bij de Vaate. 1998. Ecological impact of an exotic amphipod invasion in the River Rhine. Pages 159–169 in P. H. Nienhuis, R. S. E. W. Leuven, and A. M. J. Ragas, editors. New concepts for sustainable management of river basins. Backhuys Publishers, Leiden, The Netherlands.

Van Riel, M. C., G. Van der Velde, and A. Bij de Vaate. 2005. Trophic relationships in the Lower Rhine food web during invasion and after establishment of *Dikerogammarus villosus*. Page 67 in abstracts for Biological Invasions in Inland waters (INWAT) workshop, Firenze, Italy, May 5–7, 2005.

Verslyche, T., C. Janssen, K. Lcok and J. Mees. 2000. First occurrence of the Pontocaspian invader *Hemimysis anomala* (Sars, 1907) in Belgium (Crustacea: Mysidacea). Belgian Journal of Zoology **130**, 157–158.

Veuille, M. 1979. L' evolution du genre *Jaera* Leach (Isopodes; Asellotes) et ses rapports avec l'histoire de la Méditerranée. Bijdragen tot de Dierkunde **49**, 195–217.

Vogt, G., L. Tolley, and G. Scholtz. 2004. Life stages and reproductive components of the Marmorkrebs (marbled crayfish), the first parthenogenetic decapod crustacean. Journal of Morphology **261**, 286–311.

Vornatscher, J. 1965. Amphipoda, Teil VIII f. Catalogus Faunae Austriae. Österreichische Akademie der Wissenschaften. Österreichische Staatsdruckerei, 1–3.

Weinzierl, A., S. Potel, and M. Banning. 1996. *Obesogammarus obesus* (Sars 1894) in der oberen Donau (Amphipoda, Gammaridae). Lauterbornia **26**, 87–89.

Weinzierl A., G. Seitz, and R. Thannemann. 1997. *Echinogammarus trichiatus* (Amphipoda) und *Atyaephyra desmaresti* (Decapoda) in der bayerischen Donau. Lauterbornia **31**, 31–32.

Weish, P. and M. Türkay. 1975. *Limnomysis benedeni* in Österreich mit Betrachtungen zur Besiedlungsgeschichte (Crustacea: Mysidacea). Archiv für Hydrobiologie Supplement **44**, 480–491.

Welcomme, R. L. 1988. International introductions of inland aquatic species. FAO Fisheries Technical paper 294. FAO, Rome, Italy.

Welcomme, R. L. 1991. International introductions of freshwater fish species into Europe. Finnish Fisheries Research **12**, 11–18.

Westman, K. 2002. Alien crayfish in Europe: negative and positive impacts and interactions with native crayfish. Pages 76–95 in E. Leppäkoski, S. Gollasch and S. Olenin, editors. Invasive aquatic species of Europe. Kluwer Academic Publishers, Dordrecht, The Netherlands.

Wittmann, K. J. 1995. Zur Einwanderung potamophiler Malacostraca in die obere Donau: *Limnomysis benedeni* (Mysidacea), *Corophium curvispinum* (Amphipoda) und *Atyaephyra desmaresti* (Decapoda). Lauterbornia **20**, 77–85.

Wittmann, K. J. 2002. Weiteres Vordringen pontokaspischer Mysidacea (Crustacea) in die mittlere und obere Donau: Erstnachweise von *Katamysis warpachowskyi* für Ungarn, die Slowakei und Österreich mit Notizen zur Biologie und zum ökologischen Gefährdungspotential. Lauterbornia **44**, 49–63.

Wittmann, K. J., J. Theiss, and M. Banning. 1999. Die Drift von Mysdiacea und Decapoda und ihre Bedeutung für die Ausbreitung von Neozoen im Main-Donau-System. Lauterbornia **35**: 55–66.

Wundsch, H. H. 1912. Eine neue Spezies des Genus *Corophium* Latreille aus dem Müggelsee bei Berlin. Zoologischer Anzeiger **39**, 729–738.

Chapter three

Non-indigenous freshwater crabs in France: a new occurrence of a potamid near Nice

Pierre Y. Noël and Danièle Guinot

INTRODUCTION

The presence of a crab in the Cagne River, in south-eastern France near Nice (Alpes-Maritimes), has been known since about 1985 (M. Pascal 2000, personal communication). A brachyuran crab of the genus *Potamon* Savigny, family Potamidae Ortmann, referred to by locals as the "blue crab", was first observed in 2000 (J.-L. Teyssié 2002, personal communication) and then collected in this river in 2000 and the following years. The occurrence of the crab was confirmed in 2005 by new field observations by one of us (P. Noël) and also documented by canyoning enthusiasts and fishermen (Barla 2005, Teyssié 2005). This crab is totally dependent on freshwater, even if it is located not very far from the sea (about 10 km); it shows a direct development without free-swimming larvae and has no recognizable close relatives in the sea.

The species has been presumably introduced with Turkish imported crayfish, *Astacus leptodactylus* (Eschscholtz) in the 1960s and 1970s, but its precise origin in Turkey is unknown. There is no doubt that the crab introduced into France belongs to the genus *Potamon* (type species: *Potamon fluviatile* Savigny, junior subjective synonym of *Cancer potamios* Olivier; nec *Cancer fluviatilis* Herbst), the only genus which occurs around the Mediterranean Basin, in North Africa and

Francesca Gherardi, Biological invaders in inland waters: Profiles, distribution, and threats, 77–90.

in the Middle East. It is widely accepted that two species of freshwater crabs of the genus *Potamon* occur in the Mediterranean region. The first species, *Potamon (Eutelphusa) fluviatile* (Herbst), has been known since antiquity and occurs in the western Mediterranean: Italy, Greece, Macedonia to the northern Peloponnese, western Ionian and Aegean Islands, Dalmatia, Montenegro, Albania, and Malta (d'Udekem d'Acoz 1999). The Italian specimen figured by Rondelet (1555: 208) was selected as lectotype by Holthuis (1962: 238). The second species, *Potamon (Potamon) potamios* (Olivier), is present in eastern Europe and the Near East: SE Balkans, Greek mainland and south-eastern Aegean Islands, S Russia, Turkey to Iran and Sinai Peninsula, Egypt (Holthuis 1962, Bott 1967, 1970). A specimen collected in 1980 from the Golan Heights was subsequently designated by Pretzmann (1983b: 380) as the neotype to be in concordance with a possible topotypic region along the route followed by Olivier during his travels (Olivier 1804).

The specific identification of the Cagne crab was difficult because several *Potamon* species, subspecies, and even infrasubspecific categories are present in Turkey (Pretzmann 1962, 1984), in Greece (Pretzmann 1980, 1988), and in the eastern region. Brandis *et al.* (2000) demonstrated the high variability of the species. The Cagne crab is provisionally assigned with reservation to *Potamon (Pontipotamon) ibericum* (Bierberstein). *Potamon* aff. *ibericum* is very close to another French introduced crab found in the Hérault River in the 1990s (Charmantier 1992, 1993a, b, Defontaines and Bayle 1993).

The material examined and used for identification is deposited in institutions abbreviated as follows: MNHN, Muséum national d'Histoire naturelle, Paris; NHMW, Naturhistorisches Museum, Vienna. The abbreviations used are as follows: G1, first male pleopod or gonopod; G2, second male pleopod or gonopod; Mxp3, third maxilliped; P1, cheliped; P1–P5, first to fifth pereopods (P2–P5, walking legs). Measurements given are carapace length × carapace width (CW), in millimetres (mm), and indicate the maximum dimensions.

Material examined from the Cagne River

- *Potamon (Pontipotamon)* aff. *ibericum*, Alpes-Maritimes, Cagne River, near Lubiane tributary, coll. August 2000, C. Jourdain and E. Vigneux leg.17.11.2000: 1 male 30.2 × 38.0 mm (MNHN-B 29917).
- *Potamon (P.)* aff. *ibericum*, Alpes-Maritimes, Cagne River, "Le Poutaouchoun", Michel Pascal coll. November 2000 and leg. 21.11.2000: 1 male 30.3 × 38.0 mm (MNHN-B 29915).
- *Potamon (P.)* aff. *ibericum*, Alpes-Maritimes, Cagne River, "Le Poutaouchoun", Michel Pascal coll. 2003 and leg. 08.09.2003: 1 male 36.2 ×45.5 mm, 1 female 29.7 × 37.4 mm (MNHN-B 29916).
- *Potamon (P.)* aff. *ibericum*, M. Pascal coll. 2003 and leg. October 2004: 4 males 35.2 × 43.3 mm, 32.0 × 40.0 mm, 28.1 × 35.7 mm (plus one not measured), 2 females 31.9 × 39.1 mm, 31.7 × 40.3 mm (MNHN-B 29618).

- *Potamon* (*P.*) aff. *ibericum*, P. Noël coll. and leg. 21.07.2005: 3 males 15.7 × 19.6 mm, 15.6 × 19.4 mm, 15 × 18.8 mm, 2 females 24.4 × 31.3 mm, 1 juv. soft, damaged (MNHN-B 29918).

Material examined from the Hérault River

- *Potamon* (*Pontipotamon*) *ibericum tauricum* (Czerniavsky), Hérault River, St-Guilhem-le-Désert, G. Charmantier coll. 07.10.1991/01.02.1992, det. and leg.: 2 males 26.1 × 31.3 mm, 25.1 × 30.4 mm, 1 female 25.3 × 31.5 mm (MNHN-B 22365).
- *Potamon* (*P.*) *ibericum*, Hérault River, St-Guilhem-le-Désert, G. Charmantier coll., det. and leg. 26.03.1994: 1 male 26.9 × 33.2 mm, 1 female 29 × 31 mm (MNHN-B 27888).
- *Potamon* (*P.*) *ibericum*, Hérault River, St-Guilhem-le-Désert, P. Noël coll. 28.04.1992 and det. and leg.: 1 female (MNHN-B 29914).

Comparative material

- *Potamon* (*Pontipotamon*) *ibericum tauricum*, Turkey, Smyrne [Izmir] region, Mélès River, on banks and under stones, Mission zoologique H. Gadeau de Kerville, May 1912, G. Pretzmann det. 1964: about 50 specimens (MNHN-B 5131 and B 5189); Turkey, Smyrne [Izmir] region, Mélès River, Mission zoologique H. Gadeau de Kerville, May 1912, G. Pretzmann det. 1964: about 10 specimens (MNHN-B 5184).
- *Potamon* (*P.*) *ibericum*, Bulgaria, J.P. Gasc coll. May 1970, D. Brandis det. 1996: 3 specimens (MNHN-B 26939).
- *Potamon* (*P.*) *ibericum tauricum* Natio *bithynensis*, Türkei, W. Gerede (38 km NW Kizilcalium), Pretzmann, Radda & Konetschnig leg. 1967, Pretzmann det. 1976 (NHMW 3971 *pro parte*) [Brandis *et al.* (2000). det. *P. (P.) ibericum*].
- *Potamon* (*P.*) *ibericum tauricum* Natio *cappadociensis*, Turkey, Hynat-Deres, Bach estl. Ordu, Pretzmann leg. 1967 and det. 1976 (NHMW 3990 *pro parte*) [Brandis *et al.* (2000). det. *P.* (*P.*) *ibericum*].
- *Potamon* (*Pontipotamon*) *ibericum meandris* Pretzmann, "Cotypus", Meandre, Pretzmann det. 1963 (NHMW 3244) [Brandis *et al.* (2000). det. *P.* (*P.*) *ibericum*].
- *Potamon* (*Potamon*) *potamios hippocratis* Ghighi, Turkey, zwischen Kale und Mugla, Quellen vor Paß NW Mugla, Pretzmann leg. 1969 and det. 1976 (NHMW 4093) [Brandis *et al.* (2000) det. *Potamon* (*Potamon*) *rhodium* (Parisi)].
- *Potamon* (*Potamon*) *potamios hippocratis* Natio *antalyensis*, Turkey, 30 km NW Antalya, Pretzmann leg. 1969 and det. 1976 (NHMW 4103) [Brandis *et al.* (2000) det. *P.* (*P.*) *potamios*].

- *Potamon* (*P.*) *potamios hippocratis* Natio *egerdiri*, Turkey, SW Burdur, Pretzmann leg. 1965 and det. 1976 (NHMW 4111 *pro parte*) [Brandis *et al.* (2000) det. *P.* (*P.*) *potamios*].
- *Potamon (Potamon) potamios karamani* Pretzmann Natio *karamani*, Turkey, Tschagat-Tichai, Taurus, coll. Fahringer 1922 (NHMW 3138) [Brandis *et al.* (2000) det. *P.* (*P.*) *potamios*].
- *Potamon* (*Centropotamon*) *hueceste hueceste* Pretzmann, Turkey, Sereflikochisar, O.Zufluß zum Tuz Göl [W.Tuzgöl?], Pretzmann leg. 1972 and det. 1976 (NHMW 4133).
- *Potamon* (*Centropotamon*) *magnum vangoelium* Pretzmann, Turkey, Cucurca (Vil.Hakkari), 1600 m, Eiselt & Bilek leg. 1968, Pretzmann det. 1976 (NHMW 3885) [Brandis et al. (2000) det. *Potamon* (*Orientopotamon*) *persicum* Pretzmann].

In addition to the collected material, videos by P. Noël and colour photographs were examined for some of the descriptions given below.

The Cagne River and crab localization

The Cagne River is a coastal, 24.6 km long river that flows on limestone from the slope of the Cheiron Mountain near Coursegoules at 950 m altitude. It follows some narrow canyons and ends in the Mediterranean, not very far from Nice. The upper and lower part of the river can dry up, especially during summer. The water quality is good in the higher part of the river (trout are present in the brook part of the river). The calcium and carbonate content is probably high, since gravels, rocks, and even crabs tend to be covered by. Water quality worsens in the lower part, owing to the presence of a wastewater treatment plan in Vence; bathing is forbidden in the lower river.

The Cagne crab population seems to be well established at present. A population was first found near a place named "imprimerie Trulli" (J.-L. Teyssié 2000, personal communication) in winter 1999–2000 and in April 2000. It was previously recorded in the nineties from "Le Poutaouchoun" (P. Ponte 2000, personal communication), upstream of an old bridge between Vence and La Gaude (Fig. 1). Initial surveys have been limited to the portion from the bridge to an ancient barrage, situated around 600 m upstream. A few surveys in the upper waters first gave negative results. New investigations carried out on the lower part of the canyon revealed the presence of the crab there. It appears that the crab population is only located along some 4–5 km of the central part of the river course between "Le Riou" (north) and "Les Salles" (south), where the river rarely dries up.

Morphological description (Fig. 1)

Relatively large size. Carapace flat, smooth, lateral borders ornamented with a few blunt, flat tubercules in some specimens. Frontal border slightly sinuous.

Fig. 1 (Upper-right) The Cagne River and the ancient bridge at the place called "Poutaouchoun". (Photograph P. Noël) (Lower-right) *Potamon* aff. *ibericum* female *in situ*: the original specimen shows blue colour. (Photograph P. Noël) (Upper-left) *Potamon* aff. *ibericum*: the original specimen shows dark brown carapace, bluish coloration on Mxp3 and greenish tints due to green algae. (Courtesy of M. Pascal) (Lower-left) *Potamon (Eutelphusa) fluviatile* shown on the "Fontana del Porcellino" in Firenze. (Photo: P. Noël) [Colour photographs available on the website: http://www.mnhn.fr/]

Orbital margins smooth or slightly toothed. Anterolateral borders relatively short, regularly converging, only weakly and regularly serrated. Postfrontal cristae moderately strong. Suborbital field varying from weakly to moderately tuberculated. Chelipeds markedly heterochelous; larger chela (right) may be stout; fingers elongated, variably gaping; cutting edges with several pointed teeth separated by smaller interteeth. Pereopods 2–4 rather short and stout. Male abdomen slenderly triangular; segment 6 short and wide; telson rather short. G1 and G2 of the typical *Potamon* shape. Terminal joint of G1 spindle-like, slender, its tip reaching button of abdominal maintaining apparatus (press-button); flexible zone between subterminal and terminal joints broadened, projecting toward sternum. G2 as long as G1.

The colour of specimens found in the wild matched perfectly with the surrounding environment, the common name "blue crab" referring mostly to the colour of adult P1, and the blue colour being noticed especially among large crabs when wet or immersed in water. The blue colour is hardly seen when crabs wander on river banks, outside water. Adults, i.e. specimens of CW from 15–20 mm, with a dark brown dorsal carapace; depressed zones lighter; ventral surface pale cream, with brownish marks on abdomen (mostly proximal part). Telson pale violet. Mxp3 dark in its flattened anterior half. Both chelipeds markedly bluish-violet dorsally, fingers tips whitish; ventral side may be lighter in colour. Walking legs dark brown, most distal part darker, with bluish tints in some instances; a distinct pattern often present in small individuals (may vanish with growth), with slightly marbled (alternate dark and light patches) walking legs. No significant sex colour differences for individuals of similar size. Very young individuals (CW less than 10–15 mm) paler, with translucent parts of carapace. Tiny individuals almost colourless; recently moulted individuals with bright colours. Colours fading with time due to the presence of epibiotes (microscopic green algae, bacteria, etc.) on cuticle.

Some of the largest individuals show an unusual phenomenon giving them a peculiar greyish coloration. Since these individuals live in water with a high amount of carbonates and since they moult less often than younger ones, patches of calcium carbonate tend to develop on their cuticules. When these patches peel, the epicuticule is removed, giving the crab a diseased appearance (leprosis).

Remarks

The Cagne crab belongs to *Potamon* (*Pontipotamon*) as recognized by Pretzmann (1962: 228) and redefined by Brandis *et al.* (2000: 22, 55). It is closely related to *P. ibericum* as shown by the gonopod morphology, i.e. G1 with a spindle-shaped, elongated, terminal joint (Brandis *et al.* 2000: Fig. 6). A characteristic, perhaps distinctive, feature is the markedly broadened extremity of the subterminal joint on its mesial side (D. Brandis 2005, personal communication). We have not found definite characters distinguishing the Cagne and the Hérault crabs, which suggests a common origin for both *Potamon* introduced into France. Examination of more specimens seems necessary to exclude eventual morphological variations and to avoid the uncertainty surrounding the status of the two introduced crabs. A Turkish origin is the most probable hypothesis, in view of the morphological characters and available information. It is worth noting that the 16S sequence of the Cagne crab was found to be "exactly the same as that of the Hérault crab, while there are few differences from the *P. ibericum* group from Central Anatolia" (D. Brandis 2006, personal communication). The identity of both introduced "French" freshwater crabs deserves further investigation.

According to Brandis *et al.* (2000: 25), the indigenous *P. ibericum*, which include *P. ibericum tauricum*, have constant gonopod characters from Greece to the coastal regions of the Caspian Sea and show morphological structures without intrapopulational or regional variations.

Colour photographs of the Cagne crab were published recently in newspapers dealing with "la Cagne blue crab" (Barla 2005, Teyssié 2005). According to Charmantier (1993a) *P. (Pontipotamon) ibericum tauricum* are dark brown on carapace, brownish to pale yellow on ventral surface, violet on chelipeds and partly on Mxp3. The colour (when including the species pattern range) is viewed as a reliable criterion in the systematics of brachyuran crabs, but it is not or rarely mentioned for freshwater crabs. Coloration of *Potamon* species has rarely been reported, especially the blue colours, mainly because blue pigments (carotenoproteins) are destroyed by ethanol. The colour of the cheliped tips, used as a diagnostic character for two South African species of *Potamonautes* (see Daniels *et al.* 1998), was later considered doubtful (Gouws *et al.* 2002). We have nevertheless considered it useful to describe the colour of the "Cagne crab", which is consistent in the available material and in both sexes.

Process of introduction

The origin of the newly discovered *Potamon* in France seems to be the fish breeding installation (which is no longer working). The crab was presumably imported together with the Turkish crayfish *A. leptodactylus* when small crabs were observed in the crayfish lots that were imported (from Kerevitas, Istanbul) between 1975 and 1983. The crab was not recorded in the fish inventories of July 2004, made in the vicinity downstream from an ancient fish breeding plan (at about 500 m). It is now known at the entrance of the gorges (about 6 km downstream from the fish breeding plan). A population of the indigenous white-clawed crayfish, *Austropotamobius pallipes* Lereboullet spp. complex, is present upstream.

Behaviour and ecology

At night, individuals of *P.* aff. *ibericum* wander on river banks for feeding; some, especially large ones, can be caught by hand in the morning, facing the current on the boulders. Usually at daylight, crabs shelter under stones, in crevices, among submerged roots, or within decaying leaves. There is no evidence of the digging of holes or burrows on the river banks. Mating seems to occur at the end of August as observed in 2003 and 2004. Associated fauna includes fish like the brown trout (*Salmo trutta fario* Linnaeus), Mediterranean barbel (*Barbus meridionalis* Risso, now *Barbus graellsii* Steindachner), European eel (*Anguilla anguilla* Linnaeus), amphibians (unidentified tadpoles), and many insects.

Taxonomy

It is commonly assumed that freshwater crabs show strong similarities in carapace and pereopod shapes. Identification based on these external characters is unreliable and unstable because of their high variability, even within populations. Alternatively, freshwater taxonomists (Pretzmann 1962, 1980, 1982, 1983a, b, c, 1984, 1987, 1988, 1990, 1993, Bott 1967, 1970) have focused their attention on the morphology of gonopods, because of the presumably weak external selection pressure on these organs. Slight differences in G1 morphology have been used. For example, Pretzmann (1962, 1980, 1983a, b, 1993) described many subspecies with infrasubspecific forms ("natio") in *P. potamios*. These taxonomic categories have been considered to have a highly questionable justification and therefore have been synonymized under *P. potamios*, while waiting for morphometric and genetic results (Brandis *et al.* 2000: 3, 50, Table 1). Similarly, subspecies of *Potamon (Pontipotamon) ibericum* (*P. tauricum; P. meandris* Pretzmann; *P. albanicum* Starobogatov and Vassilenko) as well as its infrasubspecific categories (see Pretzmann 1983b, c, 1988) have been synonymized with the nominal species (Brandis *et al.* 2000: 25, Table 1).

The topology of the gonopods could be markedly variable within high level monophyletic groups (Rodríguez 1982, 1992, Ng 1988, Cumberlidge 1999). The characters that are significant for the identification of freshwater crabs and informative for phylogenetic reconstruction thus remain problematical. Assigning an organism to a particular species is further complicated by the occurrence of sibling species, expressing only limited discriminating morphological or genetic characteristics. Today, molecular sequences provide data for the recognition of the genetic differentiation among distinct species and also among populations, in particular for those that are geographically intermediate (Daniels *et al.* 2003). It is increasingly clear that presumably "good" species of freshwater crabs, at the moment insufficiently isolated, may continue to exchange genetic material through hybridization (introgression) (Gouws *et al.* 2002, Daniels *et al.* 2002).

This is particularly true for the different lineages of freshwater crabs that occupied the tropical and subtropical regions in the Upper Cretaceous and the Paleogene. *Potamon* arrived from the east in central and southern Europe in the Upper Miocene and established in the circum-Mediterranean region (Pretzmann 1972, 1982, 1983a, b, c, 1987, 1990). The present subgenera and species developed as the result of isolation caused by the separation of islands. In the Pleistocene, freshwater crabs survived in small refuges in southern Europe, and the modern subspecies developed. During more humid periods, gene-introgressions occurred, and interrupting gene-change caused different subsequent differentiations (Pretzmann 1987: 21, 22, Fig. 3).

The recent revision of Brandis *et al.* (2000) demonstrated the difficulties in the identification of potamid freshwater crabs. Only the examination of considerable material from different areas allows for the characterization of morphotypes, features of carapace and chelipeds being variable. In the absence of any morphological transition between the distinct morphotypes, each of these morphotypes proved to be confined to a specific river system or geographic region, with copulatory structures appearing to be the only stable characters (Brandis *et al.* 2000: 6, 50, Fig. 18; see also Brandis *et al.* 1999, von Sternberg *et al.* 1999).

In the new classification of the European and Middle East species proposed by Brandis *et al.* (2000), *Potamon* was divided into four subgenera, each belonging to a particular river drainage system: *Potamon*, *Eutelphusa* Pretzmann, *Pontipotamon* Pretzmann, and *Orientopotamon* Pretzmann.

Freshwater taxonomists have to face many problems, including the unresolved question of the monophyly of the freshwater Brachyura and the higher taxonomy (Martin and Davis 2002), as well as the delimitation of genera and species with new, significant, and consistent morphological characters, such as the thoracic sternum (Yeo 2000, Yeo and Ng 2004), the endophragmal system (von Sternberg and Cumberlidge 1999), and the sternal button that holds the abdomen (D. Guinot 2006, unpublished data).

Introductions of freshwater crabs into mainland France

Freshwater brachyuran crabs are not known as indigenous to France. Nevertheless, the present record in the Cagne River is the third occurrence of Potamidae in France.

The first case of a non-indigenous freshwater crab in France, which dates back to the end of the 18th century, was not accidental and it concerned the same region as the present species. The crab which has been deliberately acclimatized in south-east France at this time was probably *P.* (*E.*) *fluviatile*. This European freshwater indigenous species is found in Italy (less than 300 km from Nice), where it is very common, and illustrated on a fountain in Florence (Fig. 1). This crab is also found on the Greek mainland and in the Balkans. Risso (1819: 504) first indicated that freshwater brachyuran crabs were acclimatized in southern France and were used as food before the French Revolution (1789). Nice (Nizza) was Italian until 1793. Risso's record was mentioned by Desmarest (1825: 127, 128, as *Thelphusa fluviatilis* Pretzmann), who wrote "Au rapport de M. Risso, on avait transporté et acclimaté son espèce [the Italian *Potamon*] aux environs de Nice, il y a trente ans environ", that corresponds to the end of the 18th century. Risso (1827: 14, as *P. fluviatilis*) also explained

> Je ne cite ici cette espèce que pour engager les propriétaires du midi de la France qui ont dans leurs jardins des ruisseaux ou des réservoirs d'eaux vives, d'acclimater ces crabes, comme l'avait fait, il y a plusieurs années, le M. le comte Audiberti. Il les avoit tellement multipliés en peu d'années, qu'on en rencontrait dans tous les endroits de son jardin, et que ces potamophiles étaient devenus un assez bon comestible.

(see Holthuis 1977: 68). The origin of the crabs possessed by le Comte Audiberti is not known, but it is obvious that these animals were introduced. The Italian *P. fluviatilis* was known as source of human food for a long time, this edible crab being sold in the markets in Rome (Aldrovandi 1606: 206, as *Cancer fluviatilis* Matthioli; see Grmek and Guinot 1965: 55). It is not known if the acclimatized freshwater crabs in south-east France were really used as food before the French Revolution. Both mentions in the above text of "combustible" (combustible) and "comestible" (edible) are somewhat confusing. Perhaps the crabs were used both for food and as combustible in a time of scarcity, unless these two different terms may be the result of a typographical error. Bouvier (1940: 273, as *Potamon edulis* Linnaeus) stated that freshwater crabs had not been recorded in France for a long time.

The second record in France concerns the Hérault River near Montpellier. The crab was identified as *Potamon (Pontipotamon) ibericum* (see Charmantier 1992) or *P.* (*P.*) *ibericum tauricum* (see Charmantier 1993a, b). Charmantier (1992: 624) presented the possibility of a recent introduction with imported Turkish crayfish. There is no recent evidence of the presence of this species so it seems that it has recently become less common.

The third record is the present one in the Cagne River, at about 300 km from the Hérault River. The hypothesis that some potamid species occur in other French rivers but still undetected cannot be excluded. As far as we know, no other similar introduction has been reported in Europe. In northern European countries, where living crayfish could be imported for human consumption, freshwater crabs could not become acclimatized to cold environments. It is difficult to say if, in the long term, *P. ibericum* or related freshwater crabs could spread in other rivers of France. The two instances in France (Hérault and the Cagne River) are far from being invasive.

An invasive crab in France was *Eriocheir sinensis* H. Milne Edwards which lives predominantly in freshwater but migrates seawards for the breeding season. This species, however, belongs to the Varunidae H. Milne Edwards and not to the Potamidae. After a period of invasion, its populations decreased drastically in France where it seems to be rare at present.

Crab and crayfish interactions

Potamon fluviatile and the indigenous crayfish *A. pallipes* are found in different distribution ranges but live sympatrically in Europe; a coexistence with competitive interactions may have occurred during the Pleistocene (Pretzmann 1987, Laurent 1988). In Italian freshwater systems, *P. fluviatilis* and *A. pallipes* never share the same stream or pond, suggesting a sharp segregation of the two species (Barbaresi and Gherardi 1997). More recently, agonism and interference competition was studied in freshwater decapods with a focus on an invasive crayfish *Procambarus clarkii* Girard, and two indigenous species *Austropotamobius italicus* Faxon, and *P. fluviatile* (Gherardi and Cioni 2004).

In southern France, *A. pallipes* and *P.* aff. *ibericum* share the same stream. The introduced *P.* aff. *ibericum* is probably restricting the crayfish upstream where it was and is still present, in areas isolated by waterfalls. This suggests competitive exclusion (although the crab may not have been established in the Cagne River for a long time). The potential problems to biodiversity deriving from a non-indigenous freshwater crab in the two French rivers are not known. This would need further investigation, as far as French populations are concerned.

ACKNOWLEDGEMENTS

We are greatly indebted to Michel Pascal, head of Brigade départementale du Conseil Supérieur de la Pêche, Nice, who collected most part of the material studied here and provided many precious information. Thanks to Jean-Louis Teyssié (Agence internationale de l'énergie atomique, Monaco) for earliest information, and to Jean-Luc Cerutti, President of the Saumonnée Vençoise, Vence. Special thanks are due to Peter C. Dworschak (Naturhistorisches Museum, Vienna, Austria) for the loan of a part of the European freshwater material deposited in his institution. We gratefully acknowledge the cooperation of Dirk Brandis, specialist of the Potamidae, for his advice about the Cagne River crab, in particular in providing preliminary results of DNA sequences. Thanks also to Rémi Chappaz (Université Aix Marseille I, EA Evolution Génome Environnement), and to Peter Castro (California State Polytechnic University, Pomona, California) who corrected our English text and contributed pertinent suggestions. The authors would like to express their thanks to Cédric d'Udekem d'Acoz and Francesca Gherardi for their valuable comments, and to Alain Foubert for his help in preparing Fig. 1.

REFERENCES

Aldrovandi, U. 1606. De reliquis animalibus exanguibus libri quatuor post mortem ejus editi; nempe de mollibus, crustaceis, testaceis et zoophytis. J. P. Bellagambam, Bologna, Italy.

Barbaresi, S. and F. Gherardi. 1997. Italian freshwater decapods: exclusion between the crayfish *Austropotamobius pallipes* (Faxon) and the crab *Potamon fluviatile* (Herbst). Bulletin Français de la Pêche et de la Pisciculture **347**, 731–747.

Barla, C. 2005. Les crabes de la Cagne: la naissance d'une espèce. Le Pêcheur de France **267**, 90–91.

Bott, R. 1967. Potamidae (Crustacea Decapoda) aus Afghanistan, Westasien und dem Mittelmeerraum. (Eine Revision der Untergattung *Potamon* s. str.). Videnskabelige Meddelelser fra Dansk Naturhistorisk Forening (København) **130**, 7–43.

Bott, R. 1970. Die Süsswasserkrabben von Europa, Asien, Australien und ihre Stammesgeschichte. Abhandlungen der Senckenbergischen Naturforschenden Gesellschaft (Frankfurt) **526**, 1–338.

Bouvier, E.-L. 1940. Décapodes marcheurs. Faune de France, Lechevalier et Fils, Paris, France.

Brandis, D., V. Storch, and M. Türkay. 1999. Morphology and function of the copulatory system in freshwater crabs of the genus *Potamon*. Journal of Morphology **239**, 157–166.

Brandis D., V. Storch, and M. Türkay. 2000. Taxonomy and zoogeography of the freshwater crabs of Europe, North Africa, and the Middle East (Crustacea: Decapoda: Potamidae). Senkenbergiana biologica **80**, 5–56.

Charmantier, G. 1992. Occurrence of freshwater crabs, genus *Potamon*, in southern France. Journal of Crustacean Biology **12**, 620–626.

Charmantier, G. 1993a. *Potamon (Pontipotamon) ibericum tauricum* (Czerniavsky, 1884). Pages 41–42 in E. Vigneux, P. Keith, and P. Noël, editors. Atlas préliminaire des crustacés décapodes d'eau douce de France. Coll. Patrimoines Naturels Vol. **14**, SFF BIMM-MNHN CSP Min. Env., Paris, France.

Charmantier, G. 1993b. Potamid crabs in southern France. Page 28 in Actes de la première conférence européenne sur les crustacés. Paris, 31 août–5 septembre 1992. P. Noël, editor. Editions du Muséum national d'Histoire naturelle, Paris, France.

Cumberlidge, N. 1999. The freshwater crabs of west Africa. Family Potamonautidae. Collection Faune et Flore Tropicales **36**, 1–382.

Daniels, S. R., G. Gouws, B. A. Stewart, and M. Coke. 2003. Molecular and morphometric data demonstrate the presence of cryptic lineages among freshwater crabs (Decapoda: Potamonautidae: *Potamonautes*) from the Drakensberg Mountains, South Africa. Biological Journal of the Linnean Society **78**, 129–147.

Daniels, S. R., B. A. Stewart, and M. J. Gibbons. 1998. *Potamonautes granularis* sp. nov. (Brachyura, Potamonautidae), a new cryptic species of river crab from the Olifants river system, South Africa. Crustaceana **71**, 885–903.

Daniels, S. R., B. A. Stewart, G. Gouws, M. Cunningham, and C. A. Matthee. 2002. Phylogenetic relationships of the southern African freshwater crab fauna (Decapoda: Potamonautidae: *Potamonautes*) derived from multiple data sets reveal biogeographic patterning. Molecular Phylogenetics and Evolution **25**, 511–523.

Defontaines, P. and P. Bayle. 1993. Contribution du grand-duc d'Europe *Bubo bubo* à la découverte dans l'Hérault d'un crabe d'eau douce du genre *Potamon*. Alauda **61**, 58–60.

Desmarest, A.-G. 1825. Considérations générales sur la classe des crustacés, et description des espèces de ces animaux, qui vivent dans la mer, sur les côtes, ou dans les eaux douces de la France. F. G. Levrault, Paris, France.

Gherardi, F. and A. Cioni. 2004. Agonism and interference competition in freshwater decapods. Behaviour **141**, 1297–1324.

Gouws, G., S. R. Daniels, B. A. Stewart, and P. Cook. 2002. Allozyme electrophoresis demonstrate the presence of a species boundary in freshwater crabs (Decapoda: Potamonautidae). Journal of Natural History **36**, 1199–1222.

Grmek, M. D. and D. Guinot. 1965. Les Crabes chez Ulysse Aldrovandi : un aperçu critique sur la Carcinologie du XVIe siècle. Vie et Milieu Supplement **19**, 45–64.

Holthuis, L. B. 1962. Forty-seven genera of Decapoda (Crustacea); proposed addition to the official list Z N (S) 1499. Bulletin of Zoological Nomenclature **19**, 232–253.

Holthuis, L. B. 1977. The Mediterranean decapod and Stomatopod crustacea. A Risso's published works and manuscripts. Annales du Muséum d'Histoire naturelle de Nice **5**, 37–88.

Laurent, P.-J. 1988. *Austropotamobius pallipes* and *A. torrentium*, with observations on their interactions with other species in Europe. Pages 341–364 in D. M. Holdich and R. S. Lowery, editors. Freshwater crayfish: biology, management and exploitation. Croom Helm, London, UK.

Martin, J. W. and G. E. Davis. 2002. An updated classification of the recent Crustacea. Natural History Museum of Los Angeles County, Science series **39**, 1–124.

Ng, P. K. L. 1988. Freshwater crabs and prawns of Peninsular Malaysia and Singapore. Shinglee, National University of Singapore, Singapore.

Olivier, A. G. 1804. Voyage dans l'Empire Othoman, l'Egypte et la Perse, fait par ordre du gouvernement pendant les six premières années de la République, vol. **2**, Paris, France.

Pretzmann, G. 1962. Die mediterranen und vorderasiatischen Süsswasserkrabben. Annalen des Naturhistorischen Museums in Wien **65**, 205–240.

Pretzmann, G. 1972. Bericht über die dritte nach Anatolien durchgeführt zoologische Sammelreise. Annalen des Naturhistorischen Museums in Wien **76**, 747–751.

Pretzmann, G. 1980. Potamiden aus Griechenland (leg Malicky, leg Pretzmann). Annalen des Naturhistorischen Museums in Wien **83**, 667–672.

Pretzmann, G. 1982. Die westmediterranen Süsswasserkrabben. Quaderni del Laboratorio di Tecnologia della Pesca **3**, 363–366.

Pretzmann, G. 1983a. Ergebnisse einiger Sammelreisen nach Vorderasien 7. Die Süswasserkrabben der Turkei. Annalen des Naturhistorischen Museums in Wien **84** (B), 281–300.

Pretzmann, G. 1983b. Die Süsswasserkrabben der Mittelmeerinseln und der westmediterranen Länder. Annalen des Naturhistorischen Museums in Wien **84** (B), 369–387.

Pretzmann, G. 1983c. Ergebnisse einiger Sammelreisenim Mittelmeergebiet. Annalen des Naturhistorischen Museums in Wien **85** (B), 113–116.

Pretzmann, G. 1984. Zur Verbreitung der Potamiden in der Türkei. Annalen des Naturhistorischen Museums in Wien **86** (B), 275–277.

Pretzmann, G. 1987. A contribution to a historic analysis of Mediterranean freshwater decapods chorology. Investigación Pesquera **51** Supplement 1, 17–25.

Pretzmann, G. 1988. *Potamon (Pontipotamon)* auf Chios. Annalen des Naturhistorischen Museums in Wien **90** (B), 179–182.

Pretzmann, G. 1990. Speziation der mediterranen Süsswasserkrabben. Sitzungsberichte-Österreichische Akademie der Wissenschaften Mathematisch-naturwissensschaftliche Klasse Abt 1 **198** (1 bis 4), 13–22.

Pretzmann, G. 1993. Speziation trends in potamids. Bios **1**, 9–17.

Risso, A. 1819. Dictionnaire des Sciences Naturelles, t **XIII.** F. G. Levrault, Paris, 504 pp.

Risso, A. 1827. Histoire naturelle des principales productions de l'Europe méridionale et particulièrement de celles des environs de Nice et des Alpes Maritimes, vol. 5. Histoire naturelle des crustacés des Alpes Maritimes; Description de quelques myriapodes, scorpionides, arachnides et acarides habitant les Alpes Maritimes; Liste des principaux insectes des Alpes Maritimes; Catalogue de vers intestinaux . . . et observations relatives à plusieurs d'entre eux; Description des principaux radiaires vivants ou fossiles existants dans les Alpes Maritimes; Tableau des zoophytes les plus ordinaires. F. G. Levrault, Paris, France.

Rodríguez, G. 1982. Les Crabes d'eau douce d'Amérique. Famille des Pseudothelphusidae. Faune Tropicale, ORSTOM, Paris, France.

Rodríguez, G. 1992. The freshwater crabs of America. Family Trichodactylidae and supplement to the family Pseudothelphusidae. Faune Tropicale, ORSTOM, Paris, France.

Rondelet, G. 1555. Universae aquatilium historiae pars altera, cum veris ipsorum imaginibus. Lugduni [Gallorum], Lyon, France.

Sternberg von, R. and N. Cumberlidge. 1999. Autapomorphies of the endophragmal system in trichodactylid freshwater crabs (Crustacea: Decapoda: Eubrachyura). Journal of Morphology **256**, 23–28.

Sternberg R. von, N. Cumberlidge and G. Rodríguez. 1999. On the marine sister groups of the freshwater crabs (Crustacea: Decapoda: Brachyura). Journal of Zoological Systematics and Evolution Research **37**, 19–38.

Teyssié, J.-L. 2005. Le monde des truites. Pêche Mouche **47**, 74–76.

Udekem d'Acoz, C. d' 1999. Inventaire et distribution des crustacés décapodes de l'Atlantique nord-oriental, de la Méditerranée et des eaux continentales adjacentes au nord de 25 °N. Collection Patrimoines Naturels **40**, Muséum national d'Histoire Naturelle, Paris, France, 384 pp.

Yeo, D. C. J. 2000. Revision of the freshwater crabs (Crustacea: Decapoda: Brachyura: Potamoidea and Gecarcinucoidea) of Indochina. Ph.D. thesis, National University of Singapore, Singapore.

Yeo, D. C. J. and P. K. L. Ng. 2004. Recognition of two subfamilies in the Potamidae Ortmann, 1896 (Brachyura, Potamidae) with a note on the genus *Potamon* Savigny, 1816. Crustaceana **76**, 1289–1235.

Chapter four

Status and ecosystem interactions of the invasive Louisianan red swamp crayfish Procambarus clarkii *in East Africa*

John Foster and David Harper

INTRODUCTION

There are no indigenous crayfish in continental Africa although there are indigenous crayfish on the island of Madagascar (Hobbs 1988). However, various non-indigenous North American and Australian crayfish have been introduced to continental Africa since the 1970s, notably the Louisianan red swamp crayfish *Procambarus clarkii* (Girard). This is a relatively large, prolific, aggressive, burrowing crayfish (Hobbs *et al.* 1989 quoted in Holdich 1999) and it is well documented as an invasive species worldwide (Holdich 1999). Concern about the impact of exotic crayfish on aquatic ecosystems in South Africa (Mikkola 1996) is mirrored for fresh waters in East Africa where *P. clarkii* has established itself and is spreading (Howard and Matindi 2003).

The status of *P. clarkii* in Kenya and the Lake Victoria catchment was determined by reference to the literature, Nairobi Museum Records, and personal communications from scientists and riparian owners. The actual or probable impacts of *P. clarkii* in Kenya and the Lake Victoria catchment were

Francesca Gherardi, Biological invaders in inland waters: Profiles, distribution, and threats, 91–101.

ascertained from the literature, including field studies by the authors (Foster and Harper 2006a, b).

UGANDA

Water bodies in Uganda are illustrated in Fig. 1. *Procambarus clarkii* was introduced to Uganda in East Africa in the 1960s. In 1966, *P. clarkii* was cultured at Fisheries Resources Research Institute/National Agricultural Research Organisation's

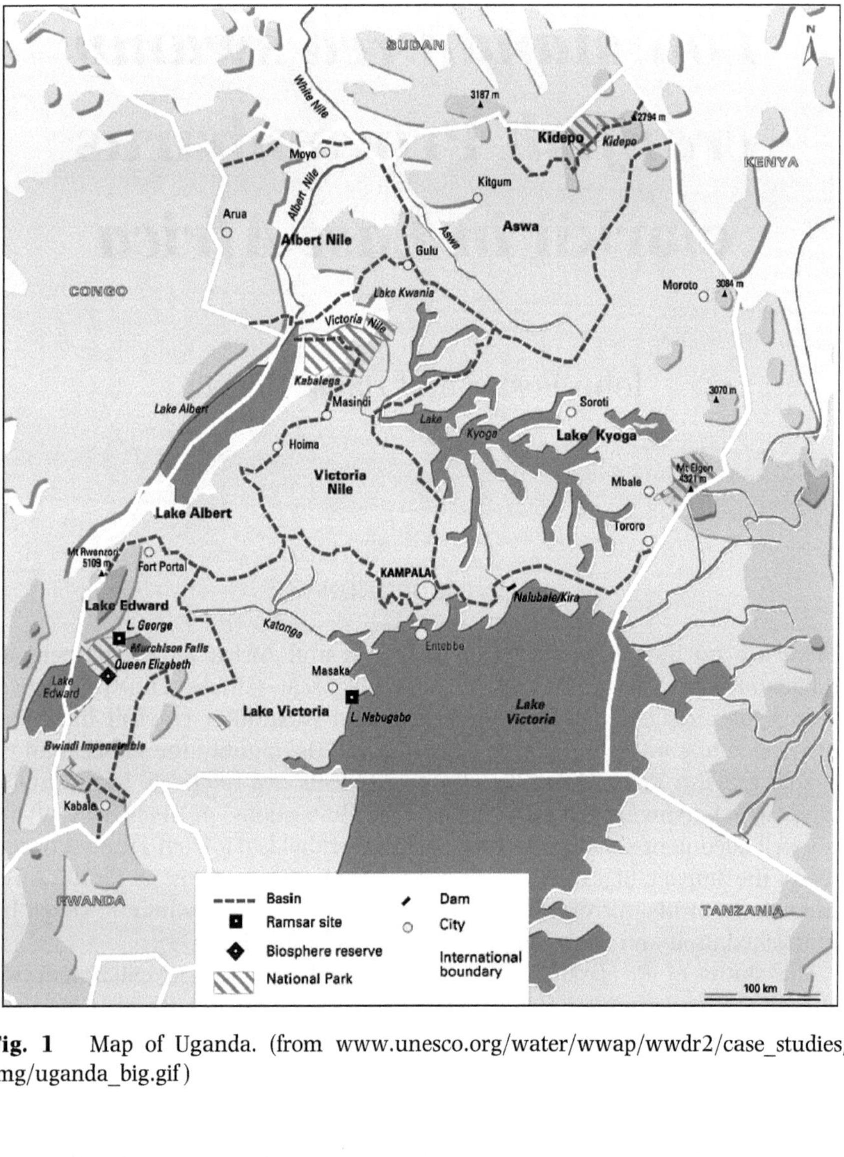

Fig. 1 Map of Uganda. (from www.unesco.org/water/wwap/wwdr2/case_studies/img/uganda_big.gif)

ponds at Kajjansi near Entebbe and Lake Victoria (Lowery and Mendes 1977) where it is still present (W. Daniels 2006, personal communication).

The species is established in Lake Bunyonyi in SW Uganda where it is exploited for the local restaurant trade. Lake Bunyonyi has no indigenous fish species but was stocked over the years with numerous local and foreign species. Lake Bunyonyi is a deep volcanic barrier lake which flows into the Ruhumba river which, in turn, flows into the Ruhumba swamps (but not to Lake Victoria) (Foster 2005). Water from Lake Bunyonyi flows through rock formations, not as an overflow. *Procambarus clarkii* may be quite widespread but under recorded in Uganda and may even have colonized the periphery of Lake Victoria in Uganda (W. Daniels 2006, personal communication).

Furthermore, anecdotal records suggest that *P. clarkii* may be established in the River Kagera which enters Lake Victoria on the Uganda–Tanzania border. The river originates in up country Rwanda and Burundi close to the Ruzizi River which flows into Lake Tanganyika (G. Howard 2005, personal communication). This presents a possible colonization route for *P. clarkii* into Lake Tanganyika in the long term.

KENYA

Water bodies in Kenya are illustrated in Fig. 2. *Procambarus clarkii* was introduced from Uganda to various parts of Kenya including the eastern basin of Lake Naivasha between 1966 and 1970 to enhance the commercial fisheries in the lakes and dams (Parker 1974, Lowery and Mendes 1977, Mikkola 1996). A commercial fishery was opened for *P. clarkii* in Lake Naivasha in 1975 (Mikkola 1996) and by 1977 the species was prevalent throughout the lake (Oluoch 1990). The status of *P. clarkii* in Lake Naivasha has been reviewed by Oluoch (1990), Harper *et al.* (2002), and Foster and Harper (2006a). The status of *P. clarkii* in the Rivers Gilgil, Malewa, and Karati flowing into Lake Naivasha from 1999 to 2003 is discussed by Foster and Harper (2006b).

Lake Naivasha (0°45′S, 36°20′E) is located in the Eastern Rift Valley at 1,890 m above sea level, approximately 100 km north-west of Nairobi. The lake is freshwater with two main rivers, the Gilgil and the Malewa draining into the northern perimeter of the lake in addition to the minor ephemeral River Karati; the lake has a subterranean inflow and outflow (Ase 1987). The recent biological history of the lake is reviewed by Harper *et al.* (1990).

The Naivasha Basin is bounded by the Aberdare Mountains to the east and the Mau Escarpment to the west. About 90% of the discharge into Lake Naivasha derives from the Malewa River (1,730 km^2 catchment), which receives its water from the Kinangop Plateau and the Aberdares. Much of the remaining inflow is from the River Gilgil (420 km^2 catchment) which drains the Bahati Highlands to the north of the Elmenteita–Nakuru basin, although a significant proportion of the Gilgil's water is abstracted for irrigation (Barnard

Fig. 2 Map of Kenya. (from www.unesco.org/water/wwap/wwdr2/case_studies/img/kenya_big.gif)

and Biggs 1988). A map of Lake Naivasha is illustrated in Foster and Harper (2006a), while a map of the rivers flowing into the lake is illustrated in Barnard and Biggs (1988).

Although *P. clarkii* had colonized the entire area of Lake Naivasha by 1977 (Oluoch 1990), it was only recorded from the lower reaches of the Rivers Gilgil and Malewa flowing into Lake Naivasha since 1999 (Foster and Harper 2006b). There is also anecdotal evidence that the crayfish has been introduced to ponds in the Malewa catchment to control leeches which feed on livestock and horses (S. Higgins 2003, personal communication). In October 2005, *P. clarkii* and guppies were recorded at the Njunu Springs at an altitude of about 2,300 m at Lake Ol Bolossat near Nyahururu in the Aberdare Mountains (www.nature-kenya.org 2006). Lake Ol Bolossat and its adjacent swamps (in the headwaters of the Ewaso Narok which joins the Ewaso Ng'iro North River) have no indigenous fish species.

Since 1974, *P. clarkii* has been recorded from the Athi/Galana river system which flows into the Indian Ocean in Kenya (Lowery and Mendes 1977, Nairobi Museum records) and was common in the Karen Pools in the suburbs of Nairobi in 1973 (Nairobi Museum Records). It is established in the Nairobi River (K. M. Mavuti, Nairobi University, 2003, personal communication) and in the Ewaso Ng'iro river system flowing off Mount Kenya to the swamps of the arid zones of northern Kenya (Lowery and Mendes 1977). *Procambarus clarkii* has been introduced into various farm dams and into various ditches, streams, and rivers draining these dams across Kenya (Lowery and Mendes 1977). The species is also present in some small high altitude tropical man-made reservoirs in the Kenyan Eastern Rift Valley (Mwaura *et al.* 2002), including Gathanje reservoir which has a fairly reliable fishery for it (Mwaura 2006).

Prior to 1977, *P. clarkii* was introduced into the catchment area of the Nzoia River draining to Lake Victoria from north-west Kenya (Lowery and Mendes 1977). In 1991, the species was recorded in abundance at Eldoret on the Eldoret river system by Mr M. D. MacDonald (Nairobi Museum records) who noted:

> This exotic was abundant in the rivers. While looking for amphibians there I saw one hundred or so. Earlier whilst looking for *Charmaeles ellioti* some boys passed by and asked me if I'd like to see some scorpions from the river. I told them scorpions didn't live in the river. It was only later that I realised that they had been referring to the crayfish.

The River Eldoret also flows into Lake Victoria from north-west Kenya.

BIOLOGICAL CONTROL OF SCHISTOSOMIASIS AND ANOPHELINE MOSQUITOES BY CRAYFISH

Procambarus clarkii has been deliberately introduced to certain aquatic locations in Kenya to combat the debilitating parasitic disease schistosomiasis by eating the parasite's snail vector (Mkoji *et al.* 1992); it may, in any event, inadvertently control schistosomiasis if it colonizes a schistosomiasis infected water body. Under certain environmental conditions, *P. clarkii* exerts a significant impact on the transmission of human schistosomiasis at locations in Kenya (Mkoji *et al.* 1999a).

Laboratory studies have indicated that *P. clarkii* has the ability to consume the aquatic life stages of the malaria-carrying anopheline mosquitoes and may therefore cause decreases in pathogen-transmitting mosquito populations in areas of Kenya where it has become established (Mkoji *et al.* 1999b). Thus there may be an incentive to introduce *P. clarkii* to those areas of Africa where schistosomiasis is endemic (such as the Sudan) in order to attempt to combat the debilitating parasitic disease in the human population. However, Lodge *et al.* (2005) noted that *P. clarkii* reduced populations of slow moving benthic invertebrates including snails, chironomid larvae, and oligochaetes in laboratory mesocosms and that water lilies disappeared from a pond that *P. clarkii* were introduced to. They state that, given the large impacts of freshwater crayfish on indigenous aquatic invertebrate and macrophyte communities, promotion of *P. clarkii* as a biological control agent should not proceed without additional assessment of the risks posed to indigenous African biota including fish. They conclude that, if freshwater crayfish colonized the large natural lakes of East Africa, globally important freshwater biodiversity resources might be at risk.

ECOLOGICAL IMPACTS OF *PROCAMBARUS CLARKII*

The ecological, economic, and social effects of the impacts of invasive species on waters and wetlands can be dramatic. Two classic examples of this are the well documented colonization of Lake Victoria by the exotic floating weed water hyacinth *Eichhornia crassipes* (Mart.) Solms (Howard and Matindi 2003) and the deliberately introduced Nile perch *Lates niloticus* Linnaeus.

Some non-indigenous crayfish can profoundly impact aquatic ecosystems (Hobbs *et al.* 1989, Holdich 1999, Lodge and Hill 1994, Chapter 28) and can affect the dynamics and biodiversity of the invaded community (Gherardi and Holdich 1999); such impacts may be negative and ecologically disastrous in the long term. These effects may include displacement of indigenous crab (Foster and Harper 2006b) and crayfish species. There may also be transfer of disease; consumption of fish eggs, large quantities of aquatic macroinvertebrates and macrophytes; damage to production in rice fields by eating rice shoots and burrowing into the banks; displacement of amphibians and possible physical damage to irrigation structures and banks of rivers and lakes by burrowing crayfish species (Holdich 1999). The Kenyan rice growing agricultural sector may be impacted by the spread of *P. clarkii* and its burrowing behaviour may also cause adverse impacts within the country.

There is an inverse relationship between *P. clarkii* abundance and the distribution and abundance of floating leaved and submerged aquatic plants in Lake Naivasha (Harper *et al.* 2002) as well as a ‘special’ relationship between water hyacinth *E. crassipes* and *P. clarkii* (Foster and Harper 2006a). *Procambarus clarkii* has been observed at densities in excess of 500 m^{-2} (juveniles and adults) in floating water hyacinth mats on Lake Naivasha (Harper *et al.* 2002). It is likely that these water hyacinth mats, which normally fringe the papyrus reeds at the edge of the lake, play an important role in *P. clarkii* recruitment and population dynamics. Water hyacinth mats were abundant in the littoral zone in 1999 as were *P. clarkii*. In 2001 and 2002, the water hyacinth mats were extensively damaged by an non-indigenous coleopteran, *Cyrtobagus eichhorniae* (Warner), which had been introduced as a biological control agent for *E. crassipes* in 1995 (Harper and Mavuti 2004). Throughout 2001 and 2002, adult *C. eichhorniae* were found in *E. crassipes*. The destruction and lack of recovery of the water hyacinth mats, coupled with their importance to *P. clarkii* recruitment may explain a *P. clarkii* population crash which occurred concurrently in Lake Naivasha (Foster and Harper 2006a).

The related dynamics of *P. clarkii*, floating leaved, and submerged aquatic plants (notably the invasive *E. crassipes* which is abundant about the periphery of Lake Victoria) may follow similar patterns to those observed in Lake Naivasha if *P. clarkii* establishes itself in Lake Victoria or other East African lakes.

Periodic fluctuations in *P. clarkii* populations in Lake Naivasha may also be related to periodic droughts which decimate key littoral habitat along the lake shore (Foster and Harper 2006a) and floods which induce breeding in *P. clarkii*. The droughts and floods are related to the 'El Niño' event and afflict the East African highlands on a regular basis (Vincent *et al.* 1979, Hay *et al.* 2002).

The indigenous freshwater crab *Potamonautes loveni* (Colosi) was recorded in the Rivers Gilgil and Malewa flowing into Lake Naivasha in the 1980s (Barnard and Biggs 1988). In field studies from 1999 to 2003, *P. loveni* was only recorded in these rivers in the absence of the invasive *P. clarkii* (Foster and Harper 2006b). It may be that *P. clarkii* is eliminating *P. loveni* through some mechanism when contact occurs. The possible impact of the non-indigenous crayfish *P. clarkii* on indigenous freshwater crabs has worrying implications for freshwaters in the rest of Kenya and in the Lake Victoria catchment where *P. clarkii* has been introduced and where there were or are populations of indigenous freshwater crabs, some of the species of which are still "new to science".

The African clawless otter, *Aonyx capensis* (Davis), occurs in Kenya and in the Lake Victoria catchment and will feed on both non-indigenous crayfish and indigenous freshwater crabs. *Procambarus clarkii* has been observed in abundance in the spraints of the African clawless otter at Lake Naivasha (J. Foster 1999, personal observation). Interactions between the African clawless otter, crayfish, and freshwater crabs in the Ewaso Ng'iro river system, Kenya have been studied by Ogada (2006). Crayfish have supplanted the indigenous freshwater crabs in the Ewaso Ng'iro river system. The crayfish are the primary food source of the African clawless otter, but this resource varies seasonally due to the increased exposure of crayfish to other predators such as baboons, genets, herons, and monitor lizards. A stable prey for the African clawless otters (freshwater crabs) has been replaced by an unstable prey (crayfish). This is leading to seasonal variation in otter behaviour and a predicted local extinction of otters. A similar situation is likely to exist in respect of the marsh mongoose, *Atilax paludinosus* (Cuvier). Low ambient water temperature may be a limiting factor on the spread of *P. clarkii* in the high altitude mountainous areas of East Africa.

Procambarus clarkii may have other impacts on river and lake ecology in East Africa, including likely undocumented impacts on aquatic invertebrates. It is exposed to a range of predator types such as aquatic invertebrates, predatory fish, amphibians, reptiles, birds, and mammals, and will interact with them (Foster and Slater 1995). *Procambarus clarkii* is an important component of the diet of American largemouth bass, *Micropterus salmoides* (Lacepède), in Lake Naivasha and it is also fed upon by cormorants, fish eagles, wading birds such as ibises, and mammals such as the marsh mongoose (Harper *et al.* 2002). Crayfish can have effects on food webs by direct and indirect trophic effects (Nyström 2002).

IMPACT ON COMMERCIAL FISHERIES

Procambarus clarkii detrimentally affected established commercial finfish fisheries in Lake Naivasha by attacking fish caught in commercial nets and damaging the nets by entangling their bodies in the mesh of the nets (Lowery and Mendes 1977). This scenario may be repeated in respect of the substantial commercial finfish fisheries of Lake Victoria, if or when *P. clarkii* becomes established there. It has been demonstrated that *P. clarkii* may out-compete tilapia for food, at least under experimental conditions (Brummett and Alon 1994) and significant potentially adverse impacts may occur in the two lakes.

Procambarus clarkii can form the basis of substantial new commercial fisheries in its own right, as has been the case at Lake Naivasha since 1975, and may become the case at Lake Victoria. Catches of several hundred tonnes per annum of *P. clarkii* were exported live, mainly to Europe (predominantly Sweden and Germany) until 1981 when catches peaked at 500 tonnes or about 19 million adult crayfish. Then, the European Union imposed a temporary ban on the import of live crayfish from Kenya due to fears concerning an outbreak of 'cholera' in East Africa (Foster *et al.* 2001).

This led to a collapse of the Lake Naivasha commercial crayfish fishery and bankruptcy for some of the businesses involved. The internal Kenyan market for crayfish is small and mainly limited to tourist outlets in the Naivasha and Nairobi areas and some local consumption in the Naivasha area. Since 1981 catches have been well below observed potential maximum sustainable yields, averaging about 20–60 tonnes per annum or about 0.75–2.25 million adult crayfish per annum (Foster *et al.* 2001).

Despite this, efforts to establish commercial crayfish fisheries in Kenya have been hampered by the fact that many Kenyans do not eat crayfish, nor do they have economic or technical means to catch or export them properly (Mikkola 1996).

CONCLUSIONS

Procambarus clarkii is now abundant in freshwater bodies across Kenya and also occurs in Uganda. It has impacted the aquatic ecology of those waters and it may eliminate indigenous African freshwater crabs when it comes into competition with them. It supports a variable commercial fishery on Lake Naivasha but does not seem to be exploited elsewhere in Kenya. It has been used, with some success, as a biological control agent for the parasitic disease, schistosomiasis, in Kenya.

The crayfish has been recorded from the Nzoia and Eldoret rivers draining to Lake Victoria from north-west Kenya (Lowery and Mendes 1977, Nairobi Museum records) and it may already have escaped into Lake Victoria from ponds at Kajjansi, Uganda (W. Daniels 2006, personal communication) The colonization routes for *P. clarkii* into Lake Victoria clearly exist. Lake Victoria may be subject to ecological perturbations and changes to its fisheries associated with colonization by *P. clarkii* in the foreseeable future. The pattern and effects of colonization of Lake Victoria by *P. clarkii* may be similar to that observed in Lake Naivasha since 1970.

ACKNOWLEDGEMENTS

This study is part of a wider research project at Naivasha undertaken by the Universities of Leicester, UK, and of Nairobi, Kenya. We are grateful to the Office of the President of the Government of Kenya for permission to undertake research. We wish to thank the Earthwatch Volunteers for their invaluable assistance in survey work and the Nairobi Museum for permission to examine their crustacean collection. The views expressed are those of the authors and are not necessarily those of their parent organisations or employers.

REFERENCES

Ase, L. E. 1987. A note on the water budget of Lake Naivasha – especially the role of *Salvina molesta* Mitch. and *Cyperus papyrus* L. Geografiska Annaler **69** A, 415–429.

Barnard, P. C. and J. Biggs. 1988. Macroinvertebrates in the catchment streams of Lake Naivasha, Kenya. Revue d'Hydrobiologie Tropicale **21**, 127–134.

Brummet, R. E. and N. C. Alon. 1994. Polyculture of Nile tilapia (*Oreochromis niloticus*) and Australian red claw crayfish (*Cherax quadricarinatus*) in earthen ponds. Aquaculture **122**, 47–54.

Foster, J. 2005. Freshwater crayfish in Lake Bunyonyi, Uganda. Crayfish News, International Association of Astacology Newsletter **27**, 8.

Foster, J. and D. M. Harper. 2006a. The Alien Louisianan Red Swamp Crayfish *Procambarus clarkii* Girard in Lake Naivasha, Kenya 1999–2003. Freshwater Crayfish **15**, 195–202.

Foster, J. and D. M. Harper. 2006b. Status of the alien Louisianan red swamp crayfish *Procambarus clarkii* Girard and the native African freshwater crab *Potamonautes loveni* in rivers of the Lake Naivasha catchment, Kenya. Freshwater Crayfish **15**, 189–194.

Foster, J. and F. M. Slater. 1995. A global review of crayfish predation with observations on the possible loss of *Austropotamobius pallipes* in the Welsh Wye due to crayfish plague. Freshwater Crayfish **8**, 589–613.

Foster, J., D. K. Mbogo, and D. M. Harper. 2001. The Louisianan red swamp crayfish *Procambarus clarkii* in Lake Naivasha, Kenya. Crayfish News, International Association of Astacology Newsletter **23**, 9–11.

Gherardi, F. and D. M. Holidich, editors. 1999. Crayfish in Europe as alien species. How to make the best of a bad situation? A. A. Balkema, Rotterdam, The Netherlands.

Harper, D. M. and K. M. Mavuti. 2004. Lake Naivasha, Kenya: Ecohydrology to guide the management of a tropical protected area. Ecohydrology and Hydrobiology **4**, 287–305.

Harper, D. M., K. M. Mavuti, and S. M. Muchiri. 1990. Ecology and management of Lake Naivasha, Kenya, in relation to climatic change, alien species' introductions and agricultural development. Environmental Conservation **17**, 328–336.

Harper, D., A. C. Smart, S. Coley, S. Schmitz, R. North, C. Adams, P. Obade, and K. Mbogo. 2002. Distribution and abundance of the Louisiana red swamp crayfish *Procambarus clarkii* Girard at Lake Naivasha, Kenya between 1987 and 1999. Hydrobiologia **488**, 57–71.

Hay, S. L., J. Cox, D. J.D. Rogers, S. E. Randolph, D. I. Stern, G. D. Shanks, M. F. Myers, and R. W. Snow. 2002. Climate change and the resurgence of malaria in the East African highlands. Nature (London) **415**, 905–909.

Hobbs, H. H. Jr. 1988. Crayfish distribution, adaptive radiation and evolution. Pages 52–82 in D. M. Holdich and R. S. Lowery, editors. Freshwater crayfish: biology, management and exploitation. Croom Helm, London, UK.

Hobbs, H. H., J. P. Jass, and J. V. Huner. 1989. A review of global crayfish introductions with particular emphasis on two North American species (Decapoda, Cambaridae). Crustaceana **56**, 299–316.

Holdich, D. M. 1999. The negative effects of established crayfish introductions. In G. W. Howard and S. W. Matindi. 2003. Alien invasive species in African wetlands: some threats and solutions. IUCN, IUCN Eastern Africa Regional Office, Nairobi, Kenya, 15 pp.

Howard, G. W. and S. W. Matindi. 2003. Alien invasive species in African wetlands: some threats and solutions. IUCN, IUCN Eastern Africa Regional Office, Nairobi, Kenya, 15 pp.

Lodge, D. M. and A. M. Hill. 1994. Factors governing species composition, population size and productivity of cool-water crayfishes. Nordic Journal of Freshwater Research **69**, 111–136.

Lodge, D. M., S. K. Rosenthal, K. M. Mavuti, W. Muohi, P. Ochieng, S. S. Stevens, B. N. Mungai, and G. M. Mkoji. 2005. Louisiana crayfish (*Procambarus clarkii*) (Crustacea: Cambaridae) in Kenyan ponds: non-target effects of a potential biological control agent for schistosomiasis. African Journal of Aquatic Science **30**, 119–124.

Lowery, R. S. and A. J. Mendes. 1977. *Procambarus clarkii* in Lake Naivasha, Kenya, and its effects on established and potential fisheries. Aquaculture **11**, 111–121.

Mikkola, H. 1996. Alien freshwater crustacean and indigenous mollusc species with aquaculture potential in eastern and southern Africa. South African Journal of Aquatic Science **22**, 90–99.

Mkoji, G. M., B. V. Hofkin, B. N. Mungai, J. H. Kihara, F. K. Mungai, J. H. Ouma, D. K. Koech, and E. S. Loker. 1992. Control of natural populations of schistosome transmitting snails by the crayfish, *Procambarus clarkii*, in temporary man-made ponds in Kenya. Pages 264–271 in Aquaculture and Schistosomiasis. National Academy Press, Washington, DC.

Mkoji, G. M., B. V. Hofkin, A. M. Kuris, A. Stewart-Oaten, B. N. Mungai, J. H. Kihara, F. Mungai, J. Yundu, J. Mbui, J. R. Rashid, C. H. Kariuki, J. H. Ouma, D. K. Koech, and E. S. Loker. 1999a. Impact of the crayfish *Procambarus clarkii* on *Schistosoma haematobium* transmission in Kenya. American Journal of Tropical Medicine and Hygiene **61**, 751–759.

Mkoji, G. M., T. G. Boyce, B. N. Mungai, R. S. Copeland, B. V. Hofkin, and E. S. Loker. 1999b. Predation of aquatic stages of *Anopheles gambiae* by the Louisiana red swamp crawfish (*Procambarus clarkii*). Journal of the American Mosquito Control Association **15**, 69–71.

Mwaura, F. 2006. Some aspects of water quality characteristics in small shallow tropical man-made reservoirs in Kenya. African Journal of Science and Technology **7**, 82–96.

Mwaura, F., K. M. Mavuti, and W. N. Wamicha. 2002. Biodiversity characteristics of small high-altitude tropical man-made reservoirs in the Eastern Rift Valley, Kenya. Lakes and Reservoirs: Research and Management **7**, 1–12.

Nyström, P. 2002. Ecology. Pages 192–224 in D. M. Holdich, editors. Biology of Freshwater Crayfish. Blackwell Science, Oxford, UK.

Ogada, M. O. 2006. Effects of the Louisiana crayfish invasion on the African clawless otter in the Ewaso Ng'iro ecosystem, Kenya. Abstract for the 14th International Conference on Aquatic Invasive Species, May 2006, Key Biscayne, FL.

Oluoch, A. O. 1990. Breeding biology of the Louisiana red swamp crayfish *Procambarus clarkii* Girard in Lake Naivasha, Kenya. Hydrobiologia **208**, 85–92.

Parker, I. S. C. 1974. The status of the Louisiana red swamp crayfish *Procambarus clarkii* Girard in Lake Naivasha. Wildlife Services Report to the Fisheries Department of the Kenyan Government commissioned by Messrs Ndoto, Nairobi, Kenya, 47 pp.

Vincent, C. E., T. D. Davies, and A. K. C. Beresford. 1979. Recent changes in the level of Lake Naivasha, Kenya, as an indicator of equatorial westerlies over East Africa. Climatic Change **2**, 175–189.

Chapter five

Non-indigenous freshwater molluscs and their distribution in Italy

Simone Cianfanelli, Elisabetta Lori, and Marco Bodon

INTRODUCTION

Since the second half of the 19th century, there have been reports of the presence of non-indigenous freshwater molluscs in Italy, though they were not yet recognized as "alien". Most species of molluscs were introduced into Italy more recently, in the second half of the 20th century, as the result of the development of commercial routes and the intensification of intercontinental traffic (Fig. 1). A complete list of non-indigenous freshwater mollusc species was not published until recently (Cianfanelli *et al.* 2007), though certain species were already considered "alien" in the "Checklist delle specie della fauna italiana" (Bodon *et al.* 1995, 2005a, 2005b; Castagnolo 1995; Manganelli *et al.* 1995, 1998) and in other articles (Lori *et al.* 2005). Eleven species of non-indigenous freshwater molluscs, differing in invasiveness (Fig. 2), are currently known (Table 1). Using data from the literature and unpublished records from field research, we created a data bank that enabled mapping of the distribution of non-indigenous molluscs. To show collection sites, UTM maps (10 km grid) were used (Fig. 3a–h, Fig. 4a–d).

Invasiveness varies widely between mollusc species and depends on their biology, vectors, availability of ecological niches, compatibility with new habitats, and habitat integrity. It is almost impossible to eradicate invasive species once they have successfully colonized a new environment. The best defence is therefore prevention (Genovesi and Shine 2004). Our aim here is to contribute

Francesca Gherardi, Biological invaders in inland waters: Profiles, distribution, and threats, 103–121.

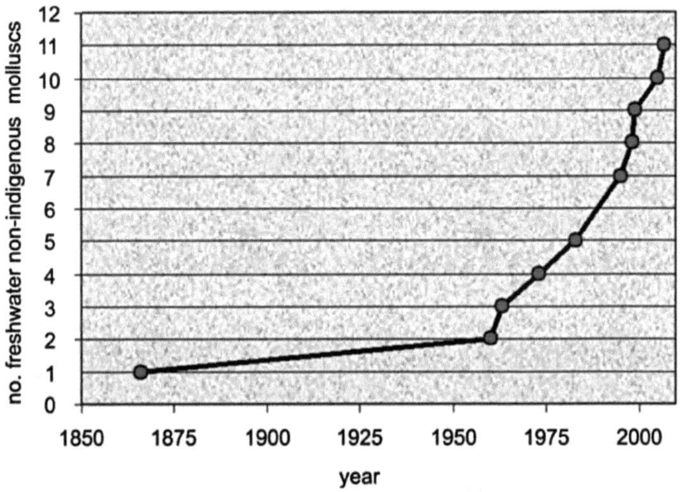

Fig. 1 Temporal trend in the number of non-indigenous freshwater mollusc species introduced into Italy. Dates for each species denote the year of the first published report of their occurrence.

to the knowledge of the Italian status and to suggest measures to prevent the spread of the non-indigenous malacofauna.

NON-INDIGENOUS MOLLUSCS: STATE-OF-THE-ART

Some non-indigenous species (NIS), such as *Melanoides tuberculata* (O. F. Müller) and *Helisoma duryi* (Wetherby), propagate slowly but their impact on the ecosystem is only apparently negligible. Though these two species have been reported in Italy for at least 10 years, their distribution still seems to be limited. Their low invasiveness is probably due to environmental factors that prevent their spread (Fig. 2).

Melanoides tuberculata is a large prosobranch (shell length up to about 5 cm) from tropical and subtropical regions, introduced by the popularity of aquaria. Found for the first time in Italy in 1984 in the marshes of Comacchio, its distribution is currently limited to four sites in northern and central Italy (Bodon *et al.* 1995, 2005b). In southern Tuscany (Fig. 3a), its high population density (S. Cianfanelli, E. Lori, and M. Bodon 2005), that also derives from its parthenogenetic mode of reproduction, is a threat for the Italian endemic *Melanopsis etrusca* Brot, a species whose distribution is limited to a few sites (Cianfanelli *et al.* 1991, Bodon *et al.* 2005b) particularly in hot springs (Manganelli *et al.* 2000). The already critical status of *M. etrusca* is expected to deteriorate irreversibly due to the competition with the NIS. *Melanopsis*

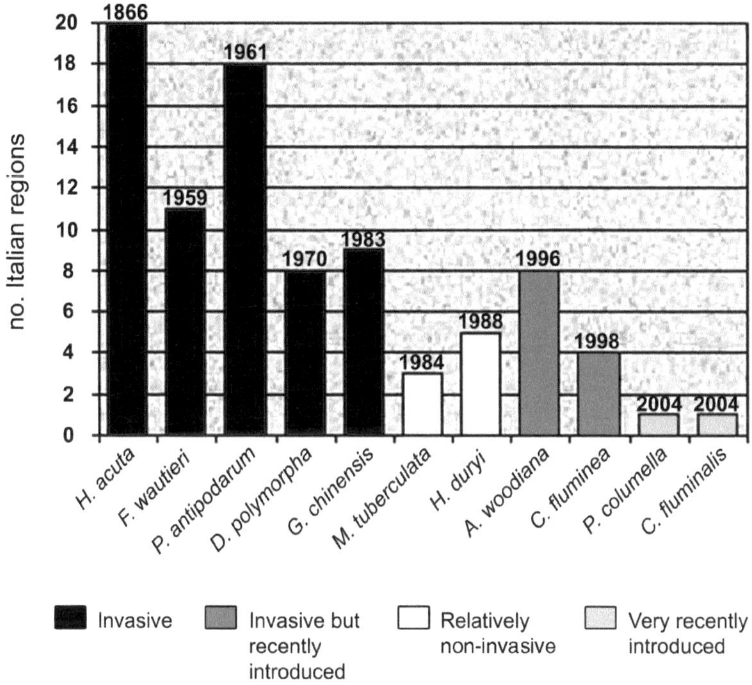

Fig. 2 Number of Italian regions in which non-indigenous freshwater molluscs are present. The species, reported in chronological order of finding (top of columns), are divided into four groups showing their current state of invasiveness in Italy (invasive, i.e. NIS spreading from the point of introduction and becoming abundant; invasive but introduced recently, i.e. from 1990 to 2000; relatively non-invasive; introduced very recently, i.e. after 2000).

tuberculata has been reported in nine other European countries (Austria, France, Germany, Hungary, Malta, Netherlands, Poland, Slovak Republic, and Spain including Canary Islands; Falkner *et al.* 2001, Girardi 2003, Piechocki *et al.* 2003, Bank 2005).

Helisoma duryi is a medium-sized pulmonate with planospiral shell (maximum diameter: about 2.5 cm) from North America, first found in Italy in 1988 in the Lake of Albano (Giusti *et al.* 1995, Manganelli *et al.* 1995, Alexandrowicz 2003, Mienis 2004b). Today it is known in five sites between Liguria, Apulia, and Sicily (Fig. 3g). Its presence is caused by release of aquarium specimens or introduction of fish.

The North American *Helisoma anceps* (Menke), another planorbid (maximum diameter of the shell: about 2 cm) similar to *H. duryi*, has been reported in Italy (Fig. 3f). Considered congeneric with *H. duryi* (both are sometimes attributed to the genus *Planorbella* Haldeman), it was identified in 1963 in a single site in Tuscany, the River Frigido (Henrard 1968) (the snail reported by Zettler and

Table 1 List of freshwater molluscs introduced into Italy, with the date of the first collection in Italy, references to the published report of their occurrence, and Italian regions where it occurs today.

Freshwater non-indigenous molluscs in italy	first collection	first published report	Current distribution
Melanoides tuberculata (O.F. Müller)	1984	Bodon *et al.* 1995	Emilia-Romagna, Tuscany, Latium
Potamopyrgus antipodarum (J.E. Gray)	1961	Berner 1963	Piedmont, Aosta Valley, Lombardy, Trentino-Alto Adige, Veneto, Friuli-Venezia Giulia, Emilia-Romagna, Liguria, Tuscany, Umbria, Marches, Latium, Abruzzo, Molise, Campania, Apulia, Calabria, Sicily
Haitia acuta (Draparnaud)	1866 or before	Issel 1866 (as *Physa pisana*)	Piedmont, Aosta Valley, Lombardy, Trentino-Alto Adige, Veneto, Friuli-Venezia Giulia, Emilia-Romagna, Liguria, Tuscany, Umbria, Marches, Latium, Abruzzo, Molise, Campania, Apulia, Basilicata, Calabria, Sicily, Sardinia
Pseudosuccinea columella (Say)	2004	This paper	Liguria
Gyraulus (Gyraulus) chinensis (Dunker)	1983 or before	Meier-Brook 1983	Piedmont, Lombardy, Veneto, Emilia-Romagna, Liguria, Tuscany, Umbria, Molise, Calabria
Helisoma duryi (Wetherby)	1988	Giusti *et al.* 1995; Manganelli *et al.* 1995	Liguria, Tuscany, Latium, Apulia, Sicily
Ferrissia wautieri (Mirolli)	1959	Mirolli 1960	Piedmont, Lombardy, Veneto, Emilia-Romagna, Liguria, Tuscany, Umbria, Latium, Campania, Apulia, Sardinia
Anodonta woodiana (Lea)	1996	Manganelli *et al.* 1998	Piedmont, Lombardy, Veneto, Emilia-Romagna, Tuscany, Umbria, Marches, Latium
Dreissena polymorpha (Pallas)	1970	Giusti and Oppi 1973	Piedmont, Lombardy, Trentino-Alto Adige, Veneto, Emilia-Romagna, Tuscany, Umbria, Molise
Corbicula fluminalis (O.F. Müller)	2004	Lori *et al.* 2005	Trentino-Alto Adige
Corbicula fluminea (O.F. Müller)	1998	Fabbri and Landi 1999	Piedmont, Lombardy, Veneto, Emilia-Romagna

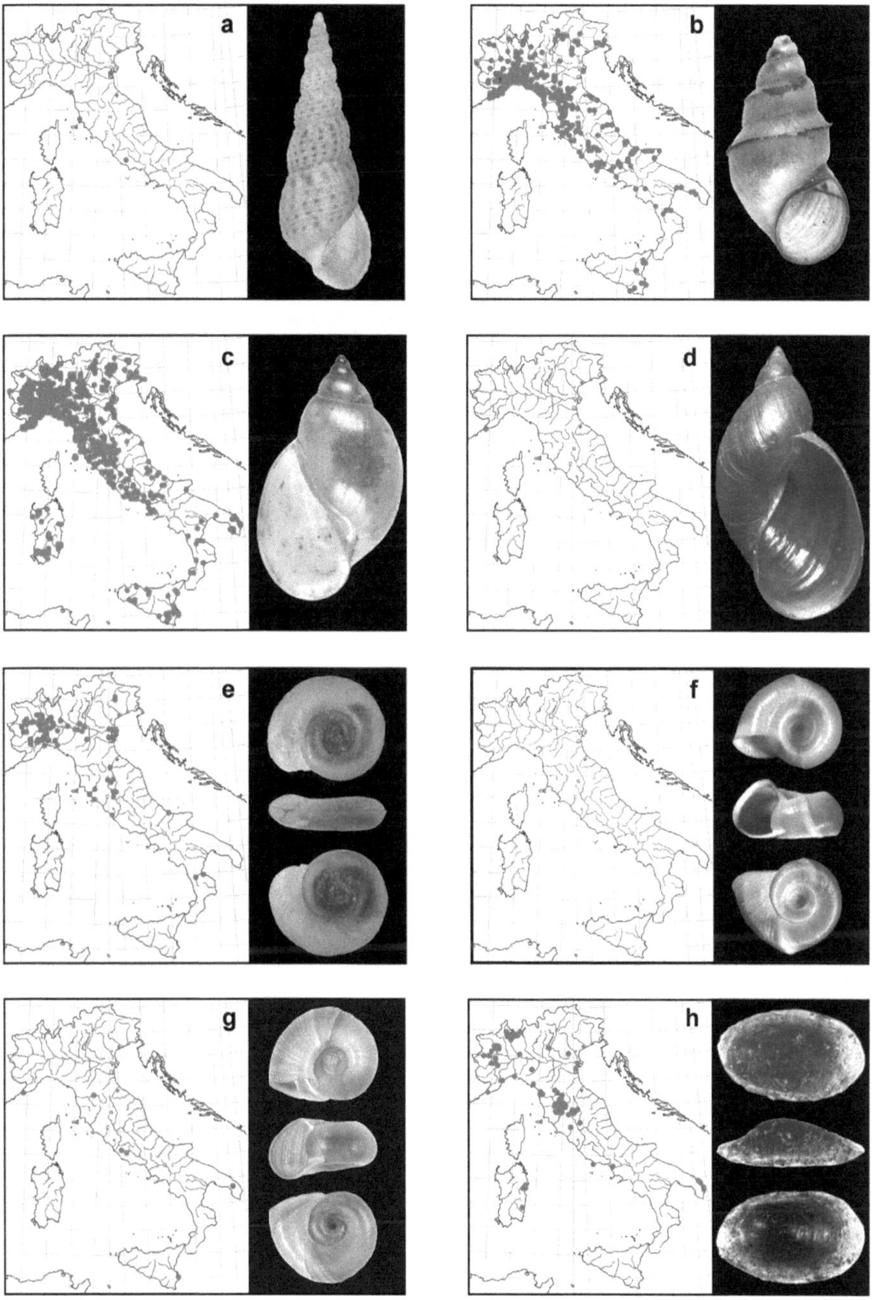

Fig. 3 Distribution and shell of: a – *Melanoides tuberculata*, b – *Potamopyrgus antipodarum*, c – *Haitia acuta*, d – *Pseudosuccinea columella*, e – *Gyraulus chinensis*, f – *Helisoma anceps*, g – *Helisoma duryi*, h – *Ferrissia wautieri*.

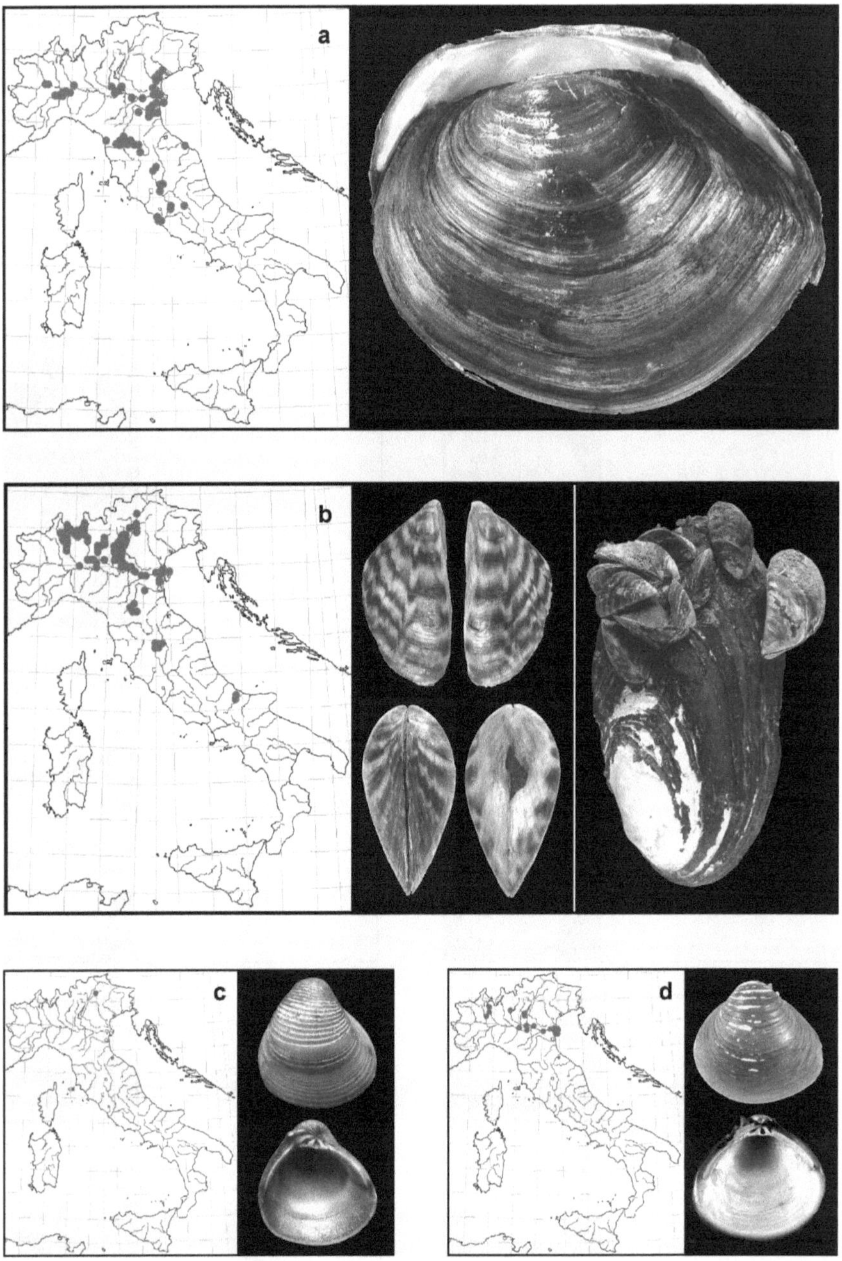

Fig. 4 Distribution and shell of: a – *Anodonta woodiana*, b – *Dreissena polymorpha*; the picture with specimens on *Microcondylaea compressa* suggests the damage that *D. polymorpha* can cause when settled on unionids, c – *Corbicula fluminalis*, d – *Corbicula fluminea*.

Richard 2003 from Siracusa was actually *H. duryi*). The fact that its determination has not been verified and its current absence from the site originally reported made it impossible to include this species in the present list. Unlike the congeneric *H. duryi*, which is found in 13 European countries (Austria, Denmark, France including Corsica, Germany, Hungary, Iceland, Italy, Latvia, Malta, Poland, Portugal [only Madeira], Spain including Balearic Islands and Canary Islands, and United Kingdom; Giusti *et al.* 1995, Vimpère 2004, Bank 2005, Greke 2005), *H. anceps* has been reported from natural environments only in Italy and very recently from Lake Prespa (Albania, Greece, Macedonia; Eröss *et al.* 2005).

The pulmonate *Pseudosuccinea columella* (Say) and the bivalve *Corbicula fluminalis* (O. F. Müller) have only recently been reported in Italy, whereas two other bivalves, *Corbicula fluminea* (O.F. Müller) and *Anodonta woodiana* (Lea) (assigned by some authors to the genus *Sinanodonta* Modell), have been known for about 10 years (Fig. 2). To judge by their success in other countries, the latest two seem to be invasive (Malavasi *et al.* 1999, Hubenov 2001, Mienis 2004a). This is confirmed by their rapid spreading in Italy into many degraded environments in the north: if not contained, they are expected to spread further.

Pseudosuccinea columella has a fusiform shell of medium size (height up to about 2 cm) and comes from North America (Zilch 1959). It has been introduced into many European countries: Austria, Greece, Hungary, Spain including Balearic and Canary Islands, and Switzerland (Turner *et al.* 1998, Falkner *et al.* 2001, Anderson 2004, Reischütz and Reischütz 2004, Bank 2005). In Italy, it was recorded for the first time in 2004 in a single site in Liguria (Hanbury Botanical Gardens, Ventimiglia; Fig. 3d). Since the site was a botanic garden, introduction was presumably associated to the importation of ornamental aquatic plants. The species has a rather limited distribution in Europe so far, probably due to unfavourable environmental conditions. In fact, in warmer climates it is known to spread rapidly and is now present in many other countries of the new world, as well as in Australia (Smith and Stanisic 2006), Cuba (Gutiérrez *et al.* 2001), and Hawaii (Cowie 1998).

Corbicula fluminalis is a medium-sized bivalve (valves up to about 2.5 cm long) from South Asia, reported in Italy for the first and only time in 2004 (Lori *et al.* 2005) in the lake at the Passo di Lavazzè (Cavalese, Trentino) (Fig. 4c). The limited European spread suggests that it is less invasive than the congeneric *C. fluminea*. However, it has also been reported from eight other European countries (Belgium, France, Germany, Hungary, Luxembourg, Netherlands, Portugal, and Switzerland; Swinnen *et al.* 1998, Turner *et al.* 1998, Csányi 1999, Falkner *et al.* 2001, Araujo 2005), though some of these reports are dubious because of the somewhat uncertain identity of the European populations.

Corbicula fluminea is similar to the latter in form and size (valves up to about 2 cm long); it is native to the south-eastern Asia and now widespread in many European countries (Austria, Belgium, Bulgaria, Czech Republic, France, Germany, Hungary, Luxembourg, Netherlands, Portugal, Romania, Spain,

Switzerland, and United Kingdom) and in other continents (North, Central and South America, and Hawaii) (Mouthon 1981, Mienis 1991, Swinnen *et al.* 1998, Turner *et al.* 1998, Csányi 1999, Beran 2000, Bij de Vaate and Hulea 2000, Falkner *et al.* 2001, Hubenov 2001, Vanden Bossche 2002, Chevallier 2003, Cadée and Soes 2004, Teodósio *et al.* 2004, Van Peursen 2004, Araujo 2005). First found in Italy in 1998 (Fabbri and Landi 1999), today it is present in various sites in the Po basin, northern Italy (Malavasi *et al.* 1999, Nardi and Braccia 2004, Bodon *et al.* 2005a; Fig. 4d). It is expected to spread further along rivers, since in a few years it has appeared with large established populations in many places where it was previously unknown.

Among the species recently introduced into Italy, *A. woodiana* is the largest freshwater bivalve (valve length of up to about 30 cm) and the fastest spreading. Indeed, since the first report in 1996, it has colonized eight Italian regions, mainly the hydrographic basins of the Po, Adige, Piave, Reno, Arno, and Tiber rivers (Manganelli *et al.* 1998, Bodon *et al.* 2005a, Solustri and Nardi 2006; Fig. 4a). This rapid spread suggests that there will be population explosions in many parts of northern and central Italy. Originally from East Asia, *A. woodiana* is now found in 14 European countries: Austria, Belgium, Bulgaria, Czech Republic, France, Germany, Greece, Hungary, Poland, Romania, Serbia, Slovak Republic, Slovenia, and Ukraine (Petró 1984, Sárkány-Kiss 1986, Girardi and Ledoux 1989, Guelmino 1992, Protasov *et al.* 1994, Košel 1995, Beran 1997, Reischütz and Reischütz 2000, Tappenbeck 2000, Yurishinets and Korniushin 2001, Sablon 2002, Albrecht *et al.* 2006, Hubenov 2006). Indiscriminate introductions of fish seem to be the main causes of its dispersal. The larval forms (glochidia) of this species, like all the unionids, are parasites of fish gills and the mollusc is therefore introduced together with fish restocking. *Anodonta woodiana* could seriously threaten the survival of populations of some indigenous unionids, already threatened by pollution, excessive water intake, and cementification of river banks. Competition with other indigenous species, especially other *Anodonta*, some populations of which are already showing disquieting signs of rarefaction (Fabbri and Landi 1999, Niero 2003), has been observed. Recent reports of other NIS of *Anodonta* in Italy (Cisotto 2003) have to be attributed to *A. woodiana*.

Other NIS with a large distribution in Italy and found in many collecting sites include: *Potamopyrgus antipodarum* (Gray), *Haitia acuta* (Draparnaud), *Gyraulus* (*Gyraulus*) *chinensis* (Dunker), *Ferrissia wautieri* (Mirolli), and *Dreissena polymorpha* (Pallas) (Fig. 2).

Potamopyrgus antipodarum is a prosobranch gastropod with small conical shell (height of up to about 7 mm), introduced into Europe from New Zealand at the end of the 19th century. It was first reported in Italy in 1961 (Berner 1963) and in about 40 years it has colonized all regions except Sardinia (Favilli *et al.* 1998, Bodon *et al.* 2005b; Fig. 3b). Various factors, such as euryoeciousness, parthenogenetic reproduction, dispersal with fish restocking, and transport during monitoring of water courses, have facilitated its fast spread. Population densities of up to 800,000 m^{-2} have been reported (Adam 1942, Lucas 1959,

Réal 1973, Falniowski 1987). It may cause the disappearance of other freshwater molluscs (Doby *et al.* 1966, Berner 1971, Albaret *et al.* 1981, Hershler *et al.* 1994). In Europe, it is the most widespread non-indigenous prosobranch species; only Iceland and some eastern countries (Albania, Bulgaria, and Former Yugoslavia; Grossu 1986, Fischer 1994, Falkner *et al.* 2001, Bank 2005) have escaped invasion. However, in the latter countries, presumed absence could be due to a lack of recent field observations.

Haitia acuta, a basommatophoran pulmonate with medium-sized, sinistral, ovate shell (height of up to about 17 mm), is common and abundant in lotic and lentic environments. It was introduced into Europe from North America (Taylor 2003); its first report in Italy dates back to Issel (1866), who described it as *Physa pisana*. Perusal of historical malacological collections demonstrates that its introduction was one of the causes of the gradual rarefaction of the indigenous basommatophore *Physa fontinalis* (Linnaeus) (Manganelli *et al.* 2000). For example, the malacological collection of the Museum of Natural History of Florence includes many shells of *P. fontinalis* collected since 1857 from areas where the species now no longer exists; after 1868, the first shells of *H. acuta* appeared, becoming increasingly numerous and from many parts of Italy. *Haitia acuta* is currently present in all 20 Italian regions, including highly polluted water bodies, often forming large populations (Feliksiak 1939, Saraceni 1971, Moretti *et al.* 1979, Melone 1981; Fig. 3c). In Europe, it is found almost everywhere, except in Bosnia-Herzegovina, Finland, Iceland, Norway, Yugoslavia, and Baltic countries except Lithuania (Feliksiak 1939, Falkner *et al.* 2001, Velkovrh 2001, Anderson 2003, 2005, Lobato Paraense and Pointier 2003, Bank 2005, Kantor and Sysoev 2005, Zettler *et al.* 2005). Absence from certain countries may be due to lack of recent data and lack of careful field observations.

Gyraulus chinensis is a basommatophoran pulmonate with small planospiral shell (max. diameter: about 5 mm) native to Asia. First reported in Italy by Meier-Brook (1983), it has found a congenial environment in rice fields. It has colonized north-western Italy where rice is cultivated intensively (Fig. 3e). Its spread could be however underestimated because of its small size and its similarity with some congeneres. In Europe, it has also been reported in Austria, France, Germany, Netherlands, Portugal, and Spain (Falkner *et al.* 2001, Albuquerque de Matos 2004, Bank 2005).

Ferrissia wautieri is a small basommatophoran pulmonate with limpet-like shell (length of up to about 4 mm), now widespread in much of Italy, with populations that apparently do not cause impact to the environment. Its distribution could be underestimated because its small size and mimesis make it elusive. Hubendick (1972) suspects it to be a NIS accidentally introduced into Italy, whereas Falkner *et al.* (2002) consider it cryptogenic. *Ferrissia wautieri* is nevertheless an entity that has not yet been classified definitively. Some authors proposed *Ferrissia clessiniana* (Jickeli) as a senior synonym of *F. wautieri* (Hubendick 1970, Falkner *et al.* 2002), while others used the synonym *F. fragilis* (Tryon) for some East European populations (Walther *et al.* 2006). We prefer conserving the name given by Mirolli, due to persisting doubts about the identity of southern European populations. Identified for the first time in Italy in 1959 from the lake of Mergozzo and in an aquarium supplied with water from Lake Maggiore (Mirolli 1960), its distribution includes 10 continental regions of Italy and Sardinia (Girod *et al.* 1974, Castagnolo *et al.* 1982, Talenti and Cianfanelli 1989, Baldaccini and Papasogli 1990, Ferreri 1995, Manganelli *et al.* 1995; Fig. 3h). It is frequent in lentic waters, often in contaminated, dystrophic conditions. Its distribution includes other 20 European countries (Albania, Austria, Belgium, Bulgaria, Czech Republic, France including Corsica, Germany, Greece, Hungary, Latvia, Luxembourg, Netherlands, Poland, Romania, Slovak Republic, Slovenia, Spain including Balearic Islands, Switzerland, Ukraine, and United Kingdom; Grossu 1987, Dhora and Welter-Schultes 1996, Falkner and Proschwitz 1998, Hubenov 1998, Falkner *et al.* 2001, Anderson 2004, Bank 2005).

Dreissena polymorpha is a medium-sized bivalve with mytiloid shell (valve length of up to about 4 cm) and with free larvae (veliger), of Ponto-Caspian origin. It was first reported in Italy in 1970 in Lake Garda (Franchini 1976), where it was almost certainly transported attached to the hulls of boats from Germany (Giusti and Oppi 1973). It has so far been reported from 8 Italian regions in 14 natural lakes, 6 artificial lakes, and 2 coastal wetlands (marshes of Comacchio and Sacca del Canarin), where its colonization is only marginal (Bodon *et al.* 2005a, Cianfanelli *et al.* 2007; Fig. 4b). It has already been monitored in 21 natural and artificial riverine water courses, almost always downstream of lentic environments. Most of the collection sites are in the north (42 water bodies) and those in central Italy (5 water bodies) are limited to Tuscany (Florence and Pistoia; Lori and Cianfanelli 2006), Umbria (Perugia), and Molise (Campobasso). It is present in four northern hydrographic basins (Po, Adige, Brenta, and Reno) and three in central Italy (Arno, Tiber, and Biferno), but occupies vast areas only in the Po basin. The altitude of collecting sites is between sea level and 842 m; sites above 500 m are almost all artificial or artificially regulated lakes. Because of its invasiveness, the high economic costs inflicted to several European countries and to the USA, and its property of bioindicator (it accumulates and transfers micro-contaminants such as DDT, heavy metals, PCBs, and other xenobiotics: Camusso *et al.* 2001, Binelli *et al.* 2004, Ricciardi *et al.* 2004), *D. polymorpha* is the most widely studied and monitored non-indigenous mollusc. The chronology of its colonization of Italy has been reconstructed from a number of reports (Cianfanelli *et al.* 2007): it took 7 years for this species to conquer the hydrological network from Lake Garda to the mouth of the Po and 30 years to populate nearly all of the lower Po plain. In a few years, *D. polymorpha* will certainly spread to the Venetian part of the same plain, especially the basins of the Adige and Brenta rivers. As in the case of other particularly invasive species, future measures can only hope to control and contain this expansion. The situation for central and southern Italy is different, because there seems to be still time to act with success. The spread of

Dreissena in the rest of Italy could be prevented if correct plans of intervention are adopted, and measures to avoid new introductions and possibly to eradicate small populations are implemented. The species is found everywhere in Europe except Cyprus, Iceland, Norway, Portugal, and Former Yugoslavia except Macedonia (Zhadin 1952, Ghenciu *et al.* 1980, Maassen 1980, Milkov 1983, Lyakhnovich *et al.* 1982, Brezeanu *et al.* 1986, Dhora and Welter-Schultes 1996, Hubenov 1998, Falkner *et al.* 2001, Korniushin *et al.* 2002, Araujo 2005, Kantor and Sysoev 2005).

CONCLUSIONS

In most cases, introductions of NIS are caused by man. They are nearly always accidental, though there are some significant examples of introductions related to commercial activity. This is the case of *A. woodiana,* introduced into Tuscany not only unintentionally but also specifically for the production of artificial pearls (Berni *et al.* 2004).

Measures to prevent deliberate introduction would be easy to develop case by case, though local entrepreneurs and maximization of profits may be at odds with correct management of the fauna. Measures to prevent accidental introductions seem, on the contrary, more difficult to be taken. Molluscs are largely introduced through practices related to fish management and aquaria. First, molluscs are introduced as temporary parasites of fish or as occasional guests of containers for fish transport. It is therefore necessary to avoid introducing fish from infested environments and prudentially also those from different hydrographic basins. Second, molluscs are introduced into the natural environment when aquaria are emptied. In aquaria, they are often raised as living filters (e.g. bivalves such as *Anodonta* spp.) or as cleaners of algae growing on the aquarium walls that gastropods, such as *M. tuberculata* and *Helisoma* spp., scrape with their radula. Their spread into nearby environments may be rapid and may be aided by natural factors. For example, it seems that specimens of *P. antipodarum* and *D. polymorpha* ingested by birds or fish may go through the digestive tract unharmed and are excreted elsewhere, or they may be transported in mud on the feet or feathers of migratory birds (Haynes *et al.* 1985). Prevention is also difficult in the case of introductions related to plant nurseries (*P. columella*) and farming (*G. chinensis*), and dispersal may be rapid.

Another means of mollusc species dispersal is the lack of precise criteria for the analysis of water. In order to limit further damage to the aquatic ecosystems it is important to take all precautions to limit the spread of any NIS to other hydrographic basins. Water body management should involve all possible measures to avoid accidental introductions into uncolonized environments. It is therefore necessary to avoid simultaneous monitoring of networks that include infested and uninfested waters. Alternatively, measures should be taken to prevent contamination, such as disinfection of equipment and personal articles that come into contact with the waters or with the substrate (nets, sampling instruments, boots), before entering other waters for monitoring purposes. With regard to species considered good bioindicators or useful indicators for monitoring water quality, such as *D. polymorpha*, all types of introduction into adjacent environments should be avoided, including those for study purposes. The only exception would be for environments already infested with populations coming from the same water body. Transport of specimens of NIS from a colonized basin may also be due to vectors such as recreational boats. The practice of checking equipment and hulls is a rule that may lead to positive effects (Minchin *et al.* 2002).

Protection of the biodiversity of indigenous species thus depends primarily on careful precautions to prevent introduction of NIS and secondarily on efforts to block their spread. Specific laws, kept up to date, are therefore necessary to ensure and regulate conservation and detailed control. Most of all, an information campaign should be directed to those who manage and enjoy water, at all levels.

ACKNOWLEDGEMENTS

We thank Folco Giusti and Giuseppe Manganelli for bibliographical research and many helpful comments to improve this paper. We also thank Saulo Bambi for technical assistance and Helen Ampt for the translation. This research was partly financed by the University of Siena and by the Museum of Natural History, Zoological Section "La Specola", University of Florence (Italy).

REFERENCES

Adam, W. 1942. Note sur les Gastéropodes. 11. Sur la répartition et la biologie de *Hydrobia jenkinsi* Smith en Belgique. Bulletin du Musée Royal d'Histoire Naturelle de Belgique **18**, 1–18.

Albaret, J. L., P. Orecchia, P. Lanfranchi, H. Picot, and Ch. Bayssade-Dufour. 1981. *Potamopyrgus* et *Bulins* en Corse (Octobre 1980). Annales de Parasitologie Humaine et Comparée **56**, 559–562.

Albrecht, C., D. Lohfink, and R. Schultheiß. 2006. Dramatic decline and loss of mollusc diversity in long-lived lakes in Greece. Tentacle **14**, 11–13.

Albuquerque de Matos, R. M. 2004. Non-marine testaceous Gastropoda of continental Portugal and Berlengas Islands. I. Catalogue and bibliography. Arquivos do Museu Bocage, Nova Série **4**, 1–158.

Alexandrowicz, S. W. 2003. *Planorbella duryi* (Wetherby, 1879) from the crater-lake Albano (Central Italy). Folia Malacologica **11**, 89–93.

Anderson, R. 2003. *Physella (Costatella) acuta* Draparnaud in Britain and Ireland – Its taxonomy, origins and relationships to other introduced Physidae. Journal of Conchology **38**, 7–21.

Anderson, R. 2004. *Pseudosuccinea columella* (Say) and other additions to the fauna of Menorca. Journal of Conchology **38**, 323.

Anderson, R. 2005. An annotated list of the non-marine mollusca of Britain and Ireland. Journal of Conchology **38**, 607–637.

Araujo, R., editor. 2005. Mollusca: Bivalvia. Fauna Europaea version 1.2 (7 March 2005). 07/10/2006. http://www.faunaeur.org/

Baldaccini, G. N. and G. L. Papasogli. 1990. Note sulla malacofauna del territorio versiliese (Toscana). – *Ferrissia wautieri* "complex" (Mirolli) (Mollusca, Pulmonata) nella Macchia Lucchese. Bollettino Malacologico **25**, 355–359.

Bank, R. A., editor. 2005. Mollusca: Gastropoda. Fauna Europaea version 1.2 (7 March 2005). 07/10/2006. http://www.faunaeur.org/

Beran, L. 1997. First record of *Sinanodonta woodiana* (Mollusca: Bivalvia) in the Czech Republic. Acta Societatis Zoologicae Bohemicae **61**, 1–2.

Beran, L. 2000. First record of *Corbicula fluminea* (Mollusca: Bivalvia) in the Czech Republic. Acta Societatis Zoologicae Bohemicae **64**, 1–2.

Berner, L. 1963. Sur l'invasion de la France par *Potamopyrgus jenkinsi* (Smith). Archiv für Molluskenkunde **92**, 19–29.

Berner, L. 1971. La régulation des naissances chez *Potamopyrgus jenkinsi* (Smith) en function des ressources alimentaires. 93° Congrès national des Sociétés savantes, Tours, 1968, Sciences **2**, 391–397.

Berni, P., S. Bitossi, M. Salvato, M. Orlandi, J. Salviati, M. Silvestri, P. Megale, P. Orlandi, and R. Billiard. 2004. Valorizzazione del territorio attraverso produzioni alternative di perle di acqua dolce di elevata qualità. Con tecniche di policoltura eco-sostenibile. International Workshop "Tinca e acquacoltura nelle acque interne", Ceresole d'Alba, Italy, 179–185.

Binelli, A., F. Ricciardi, and A. Provini 2004. Present status of POP contamination in Lake Maggiore (Italy). Chemosphere **57**, 27–34.

Bij de Vaate, A. and O. Hulea. 2000. Range extension of the Asiatic clam *Corbicula fluminea* (Müller 1774) in the River Danube: first record from Romania. Lauterbornia **38**, 23–26.

Bodon, M., S. Cianfanelli, G. Manganelli, L. Castagnolo, E. Pezzoli, and F. Giusti. 2005a. Mollusca Bivalvia. Pages 83–84 in S. Ruffo and F. Stoch, editors. Checklist e distribuzione della fauna italiana. Memorie del Museo Civico di Storia Naturale di Verona, 2. serie, Sezione Scienze della Vita 16, Verona, Italy.

Bodon, M., S. Cianfanelli, G. Manganelli, E. Pezzoli, and F. Giusti. 2005b. Gastropoda Prosobranchia ed Heterobranchia Heterostropha. Pages 79–81 in S. Ruffo and F. Stoch, editors. Checklist e distribuzione della fauna italiana. Memorie del Museo Civico di Storia Naturale di Verona, 2. serie, Sezione Scienze della Vita 16, Verona, Italy.

Bodon, M., G. Manganelli, L. Favilli, and F. Giusti. 1995. 14. Prosobranchia Archaeogastropoda Neritimorpha (generi 013–014); Prosobranchia Caenogastropoda Architaenioglossa (generi 060–065); Prosobranchia Caenogastropoda Neotaenioglossa p.p. (generi 070–071, 077, 095–126); Heterobranchia Heterostropha p.p. (genere 294). Pages 1–60 in A. Minelli, S. Ruffo and S. La Posta, editors. Checklist delle specie della fauna italiana. Calderini, Bologna, Italy.

Brezeanu, G. H., L. Gruia, and A. Petgu. 1986. Die Strukturdynamik einer population von *Drcissena polymorpha* Pallas unter den Bedingungen gesteuerten Wachstums im Stausee "Eisernes Tor". Revue Roumaine de Biologie, Série de Biologie Animale **31**, 145–148.

Cadée, G. C. and D. M. Soes. 2004. *Corbicula fluminea* (Müller, 1774) (Bivalvia, Corbiculidae) eaten by oystercatchers. Basteria **68**, 87–90.

Camusso, M., R. Balestrino, and A. Binelli. 2001. Use of zebra mussel (*Dreissena polymorpha*) to assess trace metal contamination in the largest Italian subalpine lakes. Chemosphere **44**, 263–270.

Castagnolo, L. 1995. 17. Bivalvia (specie d'acqua dolce generi 063–065, 128, 132–134). Pages 1–21 in A. Minelli, S. Ruffo and S. La Posta, editors. Checklist delle specie della fauna italiana. Calderini, Bologna, Italy.

Castagnolo, L., G. Manganelli, and F. Giusti. 1982. *Ferrissia wautieri* (Mirolli) (Mollusca. Pulmonata) nella Val di Farma (Siena – Grosseto: Italia centrale). Bollettino Malacologico **18**, 267–284.

Chevallier, H. 2003. Nouvelles données sur la propagation en France des *Corbicula* Mergerl von Mühlfeld, 1811 (Bivalvia: Corbiculidae). Documents Malacologiques **4**, 11–14.

Cianfanelli, S., E. Lori, and M. Bodon. 2007. *Dreissena polymorpha*: current status of knowledge about the distribution in Italy (Bivalvia: Dreissenidae). In G. Van der Velde, S. Rajagopal and A. Bij de Vaate, editors. The Zebra Mussels in Europe. Backhuys Publishers, Leiden, The Netherlands, in press.

Cianfanelli, S., E. Talenti, and M. Calcagno. 1991. Le stazioni di *Melanopsis dufouri* Férussac, 1823 (Gastropoda, Prosobranchia) in Italia. Quaderni del Museo di Storia Naturale di Livorno **10**, 59–76.

Cisotto, P. 2003. Primo rinvenimento in Italia di *Anodonta suborbiculata* Say, 1831 e di *Anodonta grandis* Say, 1829. Lavori della Società Veneta di Scienze Naturali **28**, 11–15.

Cowie, R. H. 1998. Patterns of introduction of non-indigenous non-marine snails and slugs in the Hawaiian Islands. Biodiversity and Conservation **7**, 349–368.

Csányi, B. 1999. Spreading invaders along the Danubian highway: first record *Corbicula fluminea* (O. F. Müller, 1774) and *C. fluminalis* (O. F. Müller, 1774) in Hungary (Mollusca: Bivalvia). Folia Historico Naturalia Musei Matraensis **23**, 343–345.

Dhora, D. and F. W. Welter-Schultes. 1996. List of species and atlas of the non-marine molluscs of Albania. Schriften zur Malakozoologie **9**, 90–197.

Doby, J. M., A. Chabaud, G. Mandahl-Barth, B. Rault, and H. Chevallier. 1966. Extension en Corse du mollusque gastropode *Potamopyrgus jenkinsi* (Smith, 1889) (Hydrobiidae). Bulletin du Muséum National d'Histoire Naturelle, 2 Série **37**, 833–843.

Domagała, J., A. M. Łabęcka, M. Pilecka-Rapacz, and B. Migdalska. 2004. *Corbicula fluminea* (O. F. Müller, 1774) (Bivalvia: Corbiculidae) – a species new to the Polish malacofauna. Folia Malacologica **12**, 145–148.

Eröss, Z. P., Z. Fehér, and A. Hunyadi. 2005. Invasion of a North American alien. *Planorbella anceps* (Menke, 1830) (Mollusca: Gastropoda: Planorbidae), in the ancient Lake Prespa. Tentacle **13**, 6–7.

Fabbri, R. and L. Landi. 1999. Nuove segnalazioni di molluschi, crostacei e pesci esotici in Emilia-Romagna e prima segnalazione di *Corbicula fluminea* (O. F. Müller, 1774) in Italia (Mollusca Bivalvia, Crustacea Decapoda, Osteichthyes Cypriniformes). Quaderno di Studi e Notizie di Storia Naturale della Romagna **12**, 9–20.

Falkner, G., R. A. Bank, and T. Proschwitz. 2001. Clecom-project Check-list of the nonmarine Molluscan Species-group taxa of the States of Northern, Atlantic and Central Europe (CLECOM I). Heldia **4**, 1–76.

Falkner, G. and T. Proschwitz. 1998. A record of *Ferrissia (Pettancylus) clessiniana* (Jickeli) in Sweden, with remarks on the identity and distribution of the European *Ferrissia* species. Journal of Conchology **36**, 39–40.

Falkner, G., T. E. J. Ripken, and M. Falkner. 2002. Mollusques continentaux de France. Liste de Référence annotée et Bibliographie. Publications Scientifiques du M.N.H.N., Paris, France.

Falniowski, A. 1987. Hydrobioidea of Poland (Prosobranchia: Gastropoda). Folia Malacologica **1**, 1–122.

Favilli, L., G. Manganelli, and M. Bodon. 1998. La distribuzione di *Potamopyrgus antipodarum* (Gray, 1843) in Italia e in Corsica (Prosobranchia: Hydrobiidae). Atti della Società italiana di Scienze naturali e del Museo Civico di Storia Naturale di Milano **139**, 23–55.

Feliksiak, S. 1939. *Physa acuta* Draparnaud in den Fabrikteichen von Łódź und ihre allgemeine Verbreitung. Fragmenta Faunistica Musei Zoologici Polonici **4**, 243–259.

Ferreri, D. 1995. Molluschi, Irudinei e Turbellari Tricladi delle acque dolci della provincia di Lecce. Thalassia Salentina **21**, 29–49.

Fischer, W. 1994. Süsswassermollusken aus Zypern. Nachrichtenblatt der Estern Voralberger Malakologischen Gesellschaft **2**, 47–48.

Franchini, D. A. 1976. Sostituzione degli insediamenti malacologici autoctoni ad opera della *Dreissena polymorpha* (Pallas) in una stazione del Benaco (Torri del Benaco, VR). Lavori del Simposio sui Molluschi terrestri e dulcicoli dell'Italia Settentrionale, Mantova, 10–11/5/1975, Lavori della Società Malacologica Italiana **13**, 89–95.

Genovesi, P. and S. Shine. 2004. European strategy on invasive alien species. Nature and environment, No. 137, Council of Europe Publishing, Strasbourg, France.

Ghenciu, V. A., A. Munteanu-Ghenciu, and I. Balanescu. 1980. Contributions to the knowledge of geographical expansion in Moldavia of the species *Dreissena polymorpha* Pallas. Studii si Comunicari Muzeul de Stiintele Naturii **10–12**, 25–26.

Girardi, H. 2003. Note sur la presence de *Melanoides tuberculatus* (O.F. Müller, 1774) dans une exurgence karstique à Bagnères-de-Bigorre (Hautes-Pyrénées, France). Documents Malacologiques **4**, 15–17.

Girardi, H. and J. C. Ledoux. 1989. Présence d'*Anodonta woodiana* (Lea) en France (Mollusques, Lamellibranches, Unionidae). Bulletin mensuel de la Société Linnéenne de Lyon **58**, 286–291.

Girod, A., E. Bona, and A. Freddi. 1974. Nuovi dati sulla distribuzione di *Ferrissia wautieri* (Mirolli) in Lombardia e nel Canton Ticino. Atti della Società italiana di Scienze naturali e del Museo Civico di Storia naturale di Milano **115**, 243–250.

Giusti, F. and E. Oppi. 1973. *Dreissena polymorpha* (Pallas) nuovamente in Italia. (Bivalvia, Dreissenidae). Memorie del Museo Civico di Storia Naturale di Verona **20**, 45–49.

Giusti, F., G. Manganelli, and P. J. Schembri. 1995. The non-marine molluscs of the Maltese Islands. Monografie XV Museo Regionale di Scienze Naturali, Torino, Italy.

Greke, K. 2005. [Mollusca]. In: NOBANIS North European and Baltic Network on Invasive Alien Species (1 August 2005). 07/10/2006 http://www.nobanis.org/speciesInfo.asp?taxaID=1606

Grossu, A. V. 1986. Gastropoda Romaniae 1. Editura Litera, Bucharest, Romania.

Grossu, A. V. 1987. Gastropoda Romaniae 2. Editura Litera, Bucharest, Romania.

Guelmino, J. 1992. *Anodonta woodiana woodiana* Lea, 1834 (Mollusca: Bivalvia) prvi nalaz u donjem toku Tise. Matica Srpska, Novi Sad, Serbia.

Gutiérrez, A., G. Perera, M. Yong, and L. Wong. 2001. The effect of isolation on the life-history traits of *Pseudosuccinea columella* (Pulmonata: Lymnaeidae). Memorias do Instituto Oswaldo Cruz **96**, 577–581.

Haynes, A., B. J. R. Taylor, and M. E. Varley. 1985. The influence of the mobility of *Potamopyrgus jenkinsi* (Smith E. A.) (Prosobranchia: Hydrobiidae) on its spread. Archiv für Hydrobiologie **103**, 497–508.

Henrard, J. B. 1968. On the occurrence of *Helisoma anceps* (Menke) in Italy. Basteria **32**, 2–3.

Hershler, R., T. J. Frest, E. J. Johannes, P. A. Bowler, and F. G. Thompson. 1994. Two new genera of hydrobiid snails (Prosobranchia: Rissoidea) from the Northwestern United States. Veliger **37**, 221–243.

Hubendick, B. 1970. Studies on Ancylidae. The Palearctic and Oriental Species and Formgroups. Acta regiae Societatis Scientiarum et Litterarum Gothoburgensis Zoologica **5**, 1–51.

Hubendick, B. 1972. The European fresh-water limpets (Ancylidae and Acroloxidae). Informations de la Société Belge de Malacologie, Série 1 **8–9**, 109–128.

Hubenov, Z. 1998. Caspian Relict Species of Mollusca. Appendix 9 in C. Meine, editor. Bulgaria's Biological Diversity: Conservation Status and Needs Assessment Biodiversity Support Program. World Wildlife Found. 07/10/2006 http://www.worldwildlife.org/bsp/publications/europe/bulgaria/bulgaria.html

Hubenov, Z. 2001. Corbiculidae – a new family to the Bulgarian recent malacofauna (Mollusca: Bivalvia). Acta Zoologica Bulgarica **53**, 61–66.

Hubenov, Z. 2006. *Anodonta (Sinanodonta) woodiana* (Lea, 1834) (Mollusca: Bivalvia: Unionidae) – a new invasive species for the Bulgarian malacofauna. Acta Zoologica Bulgarica **58**, 37–42.

Issel, A. 1866. Catalogo dei molluschi raccolti nella Provincia di Pisa. Memorie della Società Italiana di Scienze Naturali **II**, 1–38.

Kantor, Y. I. and A. V. Sysoev. 2005. Catalogue of molluscs of Russia and adjacent countries. KMK Scientific Press, Moscow, Russia.

Korniushin, A. V., N. J. Janovich, and R. K. Melnichenko. 2002. Artenliste der Süßwassermuscheln der Ukraine, mit Bemerkungen über taxonomischen Status, Verbreitung und Gefährdungskategorien einiger Arten und Formen. Pages 463–478 in M. Falkner, K. Groh and M. C. D. Speight, editors. Collectanea Malacologica. ConchBooks, Hackenheim, Germany.

Košel, V. 1995. The first record of *Anodonta woodiana* (Mollusca, Bivalvia) in Slovakia. Acta Facultatis Rerum Naturalium Universitatis Comeniae **39**, 3–7.

Łabęcka, A. M., J. Domagała, and M. Pilecka-Rapacz. 2005. First record of *Corbicula fluminalis* (O. F. Müller, 1774) (Bivalvia: Corbiculidae) in Poland. Folia Malacologica **13**, 25–27.

Lobato Paraense, W. and J. P. Pointier. 2003. *Physa acuta* Draparnaud, 1805 (Gastropoda: Physidae): a study of topotypic specimens. Memorias do Instituto Oswaldo Cruz **98**, 513–517.

Lori, E., M. Bodon, and S. Cianfanelli. 2005. Molluschi continentali alieni in Italia: presenza e distribuzione. Notiziario S.I.M. **23**, 71. http://www.sim-online.it/

Lori, E. and S. Cianfanelli. 2006. New records of *Dreissena polymorpha* (Pallas, 1771) (Mollusca: Bivalvia: Dreissenidae) from Central Italy. Aquatic Invasions **1**, 281–283. http://www.aquaticinvasions.ru/2006/AI_2006_1_4_Lori_Cianfanelli.pdf

Lucas, A. 1959. Les *Hydrobia* (Bythinellidae) de l'Ouest de la France. Journal de Conchyliologie **99**, 3–14.

Lyakhnovich, V. P., S. I. Gavrilov, A. Y. Karataev, I. V. Karataeva, and T. I. Nyakhaeva. 1982. Long-term changesin the macrobenthos of Lukoml'skoe Lake. Vyestsi Akademii Navuk Byelarusi SSR Syeryya Biyalahichnykh Navuk **1**, 91–93.

Maassen, W. J. M. 1980. De zoetwatermollusken van het Ohrid – en Prespameer en hun omgeving. De Kreukel **16**, 81–98.

Malavasi, D., R. Fabbri, and A. Bernardoni. 1999. Prima segnalazione nel tratto medio del fiume Po di *Corbicula fluminea* Mueller (Mollusca Bivalvia Corbiculidae). Pianura **11**, 183–185.

Manganelli, G., M. Bodon, S. Cianfanelli, L. Favilli, and F. Giusti. 2000. Conoscenza e conservazione dei molluschi non marini italiani: lo stato delle ricerche. Bollettino Malacologico **36**, 5–42.

Manganelli, G., M. Bodon, L. Favilli, L. Castagnolo, and F. Giusti. 1998. Checklist delle specie della fauna d'Italia, molluschi terrestri e d'acqua dolce. Errata ed addenda, 1. Bollettino Malacologico **33**, 151–156.

Manganelli, G., M. Bodon, L. Favilli, and F. Giusti. 1995. 16. Gastropoda Pulmonata. Pages 1–60 in A. Minelli, S. Ruffo and S. La Posta, editors. Checklist delle specie della fauna italiana. Calderini, Bologna, Italy.

Meier-Brook, C. 1983. Taxonomic studies on *Gyraulus* (Gastropoda: Planorbidae). Malacologia **24**, 1–113.

Melone, G. 1981. La malacofauna del fiume Po a Caorso Isola Serafini. Rivista di Idrobiologia **20**, 255–268.

Mienis, H. K. 1991. Some remarks concerning Asiatic clams invading Europe with a note on sample of *Corbicula fluminea* (Müller, 1774) from Trapani, Sicily. Notiziario SIM **9**, 137–139.

Mienis, H. K. 2004a. Additional information concerning the conquest of Europe by the invasive Chinese Pond Mussel *Sinanodonta woodiana* 9. News from Belgium, Italy, Romania and Serbia. Ellipsaria **6**, 8–9.

Mienis, H. K. 2004b. A graveyard of *Planorbella duryi* forma *seminole* on the shores of Lake Albano, Italy. Ellipsaria **6**, 12–13.

Milkov, D. L. 1983. Charniere rudimentaire chez les jeunes *Dreissena polymorpha* Pallas et son importance taxonomique. Acta Zoologica Bulgarica **22**, 82–85.

Minchin, D., F. Lucy, and M. Sullivan. 2002. Zebra mussel: impacts and spread. Pages 135–146 in E. Leppäkoski, S. Gollasch and S. Olenin, editors. Invasive Aquatic Species of Europe. Distribution, impacts and management, Kluwer Academic Publishers, Dordrecht, The Netherlands.

Mirolli, M. 1960. Morfologia, biologia e posizione sistematica di *Watsonula wautieri*, n.g., n.s. (Basommatophora, Ancylidae). Memorie dell'Istituto Italiano di Idrobiologia **12**, 121–163.

Moretti, G. P., F. Cianficconi, and F. Tucciarelli. 1979. Bilancio ecologico e geonemico delle biocenosi alle confluenze del F. Tevere in Umbria. Lavori della Società Italiana di Biogeografia Nuova Serie **6** (1976), 525–566.

Mouthon, J. 1981. Sur la présence en France et au Portugal de *Corbicula* (Bivalvia, Corbiculidae) originaire d'Asie. Basteria **45**, 109–116.

Nardi, G. and A. Braccia. 2004. Prima segnalazione di *Corbicula fluminea* (O.F. Müller, 1774) per il Lago di Garda (provincia di Brescia) (Mollusca, Bivalvia, Corbiculidae). Bollettino Malacologico **39**, 181–184.

Niero, I. 2003. Sulla presenza in Veneto e centro Italia di *Anodonta woodiana woodiana* (Lea, 1834) (Mollusca, Bivalvia). Bollettino del Museo Civico di Storia Naturale di Venezia **54**, 29–33.

Petró, E. 1984. The occurrence and presence of *Anodonta woodiana woodiana* (Lea 1834) in Hungary. Állattani Közlemények **71**, 189–191.

Piechocki, A., B. Wawrzyniak-Wydrowska, and B. Zdanowski. 2003. *Melanoides tuberculatus* (O. F. Müller, 1774) (Orthogastropoda: Thiaridae), a gastropod species new for the fauna of Poland. Folia Malacologica **11**, 39–41.

Protasov, A. A., S. A. Afanasjev, and B. Zdanowski. 1993. Natural self-purification systems of Konin lakes. Komunikaty Rybackie **6**, 6–9.

Réal, G. 1973. Répartition en France de *Potamopyrgus jenkinsi* (E. A. Smith, 1889). Haliotis **3**, 199–204.

Reischütz, A. and P. L. Reischütz. 2000. Beiträge zur Kenntnis der Molluskenfauna Niederösterreichs 17. Die Chinesische Teichmuschel *Sinanodonta woodiana* (Lea, 1834) in Österreich. Nachrichtenblatt der Ersten Vorarlberger Malakologischen Gesellschaft **8**, 67.

Reischütz, A. and P. L. Reischütz. 2004. Hellenikà pantoía, 10: *Pseudosuccinea columella* (Say 1817) im Nomos Florina (NW-Makedonien, Griechenland). Nachrichtenblatt der Ersten Vorarlberger Malakologischen Gesellschaft **12**, 9.

Ricciardi, F., A. Binelli, and A. Provini. 2004. Utilizzo di biomarker in *Dreissena polymorpha* come sistema di monitoraggio rapido: il caso dei grandi laghi subalpini. Pages 9 in R. Casagrandi and P. Melià, editors. Atti del XIII Congresso S.It.E. Como, Italy. Ecologia quantitativa: metodi sperimentali, modelli teorici, applicazioni. Atti 27.

Sablon, R. 2002. Exotic mussel species invasions in Belgian freshwater systems (Mollusca, Bivalvia). Bulletin de l'Institut Royal des Sciences Naturelles de Belgique, Biologie, **72**.

Saraceni, C. 1971. Biologia ed ecologia delle comunità macrobentoniche del fiume Bardello. Memorie dell'Istituto Italiano di Idrobiologia **27**, 61–111.

Sárkány-Kiss, A. 1986. *Anodonta woodiana* (Lea, 1834) a new species in Romania (Bivalvia, Unionacea). Travaux du Museum National d'Histoire Naturelle "Grigore Antipa" **28**, 15–17.

Smith, B. J. and J. Stanisic. 2006. Mollusca. In CSIRO – Commonwealth Scientific and Industrial Research Organisation – AICN – Australian Insect Common Names – version 1.53 (16 June 2005). 07/10/2006. http://www.ento.csiro.au/aicn/name_s/b_3491.htm

Solustri, C. and G. Nardi. 2006. Una nuova stazione di *Anodonta woodiana woodiana* (Lea, 1834) nell'Italia centrale (Bivalvia, Unionidae). Quaderno di Studi e Notizie di Storia Naturale della Romagna **23**, 1–8.

Swinnen, F., M. Leynen, R. Sablon, L. Duvivier, and R. Vanmaele. 1998. The Asiatic clam *Corbicula* (Bivalvia: Corbiculidae) in Belgium. Bulletin de l'Institut Royal de Sciences Naturelles de Belgique, Biologie **68**, 47–53.

Talenti, E. and S. Cianfanelli. 1989. Due nuove stazioni di *Ferrissia wautieri* (Mirolli, 1960) in Toscana. Bollettino Malacologico **25**, 251–254.

Tappenbeck, L. 2000. *Sinanodonta woodiana*, die 'Chinesische Teichmuschel' – eine neue Großmuschel in Deutschland. Halophila – Mitteilungsblatt Fachgruppe Faunistik und Ökologie Staßfurt **41**, 11.

Taylor, D. W. 2003. Introduction to Physidae (Gastropoda: Hygrophila); biogeography, classification, morphology. Revista de Biología Tropical **51** Supplement 1, 1–287.

Teodósio, J., M. A. Chícaro, and L. Chícaro. 2004. The Asiatic clam *Corbicula fluminea* (Müller) in the Guadiana basin. World Congress of Malacology, Perth, Western Australia 11–16 July 2004, Western Australian Museum, abstracts, 146.

Turner, H., J. G. J. Kuiper, N. Thew, R. Bernasconi, J. Rüetschi, M. Wüthrich, and M. Gosteli. 1998. Atlas der Mollusken der Schweiz und Liechtensteins. Fauna Helvetica 2.CSCF und SEG, Neuchâtel.

Vanden Bossche, J. P. 2002. First records and fast spread of five new (1995–2000) alien species in the River Meuse in Belgium: *Hypania invalida, Corbicula fluminea, Hemimysis anomala, Dikerogammarus villosus* and *Crangonyx pseudogracilis*. Bulletin de l'Institut Royal de Sciences Naturelles de Belgique **72**, Supplément 19, 73–78.

Van Peursen, A. D. P. 2004. Oproep om vindplaatsen van *Corbicula*'s in Nederland te melden. Spirula **340**, 96.

Velkovrh, F. 2001. Analiza stanja biotske raznovrstnosti za področje kopenskih in Sladkovodnih Mehkužcev. Pages 74–79 in ARSO. Ekspertne študije za Pregled stanja biotske raznovrstnosti in krajinske pestrosti v Sloveniji. Ljubljana, Republika Slovenija.

Vimpère, J. 2004. Introduction en France continentale d'un gastéropode d'eau douce originaire de Floride: *Planorbella duryi* (Wetherby, 1879), (Mollusca: Gastropoda: Pulmonata). Le Naturaliste Vendéen **4**, 127–130.

Walther A. C., T. Lee, J. B. Burch, and D. Ó Foighil. 2006. Confirmation that the North American ancylid *Ferrissia fragilis* (Tryon, 1863) is a cryptic invader of European and East Asian freshwater ecosystems. Journal of Molluscan Studies **72**, 318–321.

Yurishinets, V. I. and A. V. Korniushin. 2001. The new species in the fauna of Ukraine *Sinanodonta woodiana* (Bivalvia, Unionidae), its diagnostics and possible ways of introduction. Vestnik Zoologii **35**, 79–84.

Zettler, M. L. and D. Richard. 2003. Kurze Bemerkungen über Süßwassermollusken Siziliens unter besonderer Berücksichtigung von *Theodoxus meridionalis* (Philippi, 1836). Malakologische Abhandlungen **21**, 29–38.

Zettler, M. L., A. Zettler, and D. Daunys. 2005. Bemerkenswerte Süßwassermollusken aus Litauen. Aufsammlungen vom September 2004. Malakologische Abhandlungen **23**, 27–40.

Zhadin, V. I. 1952. Mollusks of Fresh and Brackish Waters of the USSR. Keys to the Fauna of the USSR published by the Zoological Institute of the Academy of Sciences of the USSR, 46. Translated from Russian 1965, Israel Program for Scientific Translations, Jerusalem, Israel.

Zilch, A. 1959. Euthyneura 2. Pages 1–834 in O. H. Schindewolf, editor. W. Wenz. Gastropoda. Handbuch der Paläozoologie 6. Gebrüder Borntraeger, Berlin, Germany.

Chapter six

Non-indigenous animal species naturalized in Iberian inland waters

Emili García-Berthou, Dani Boix, and Miguel Clavero

INTRODUCTION

Invasions by human-introduced non-indigenous species (NIS) are one of the main threats to biodiversity and a driving force of global change (Vitousek *et al.* 1997, Mack *et al.* 2000, Clavero and García-Berthou 2005). The Iberian Peninsula (IP) is a hotspot of biodiversity (Médail and Quézel 1999) and a knowledge of the invasive species inhabiting it is essential for conservation and environmental management. Naturalized vertebrates and plants in the IP have received considerable attention (see e.g. Vilà *et al.* 2001, Pleguezuelos 2002, Sobrino *et al.* 2002, Lloret *et al.* 2004, Alcaraz *et al.* 2005), but its invasive invertebrates are very poorly known. Although there are many records of some invertebrate invasive species, particularly crustaceans, there are very few available reviews of selected taxa of invertebrate invaders in the IP (e.g. Espadaler and Collingwood 2001). The aim of this chapter is to review the animal species naturalized in Iberian inland waters, including vertebrates and free-living and parasitic invertebrates. As usual, the taxonomy and biogeography of vertebrate species are much better known than for invertebrates, so our data for invertebrates should be regarded as a preliminary check-list. Similarly, the parasites of non-commercial aquatic species are poorly studied and the data in the IP mostly come from studies of the eel, *Anguilla anguilla* (Linnaeus), thus certainly underestimating the range of introduced parasites (Blanc 1997, 2001). We feel, however, that it is important to provide such a

Francesca Gherardi, Biological invaders in inland waters: Profiles, distribution, and threats, 123–140.

first check-list because many of the invertebrates involved are nowadays common in the IP and for many of them it is largely unknown even by biologists that they are not indigenous to the IP. Increasing the awareness on the introduced status and current distribution of these species is essential to reduce their spread and impact.

We compiled animal species cited (by March 2006) as currently naturalized in Iberian inland waters from the scientific literature and unpublished Spanish Ph.D. theses (http://teseo.mec.es/teseo/). We included species from estuaries and saline coastal lagoons but excluded purely marine taxa and terrestrial animal species not strictly linked to aquatic ecosystems. We list invertebrate and vertebrate species introduced by humans and currently naturalized, i.e. species that reproduce and sustain populations in the wild without human intervention (see e.g. Richardson *et al.* 2000, Pyšek *et al.* 2004). A few uncertain cases are listed in a separate table. The introduced origin of parasite invertebrates is particularly uncertain but we followed Blanc (1997, 2001), who has recently provided a comprehensive list of aquatic parasites introduced to Europe, together with their native distribution.

NATURALIZED ANIMALS IN IBERIAN INLAND WATERS

The invertebrate and vertebrate species naturalized in Iberian inland waters are listed in Tables 1 and 2, respectively. A few cases, for which it is not clear whether the species is indigenous to the IP or whether they have established, are listed in Table 3. We found 45 invertebrate and 28 vertebrate species certainly naturalized at present in Iberian inland waters.

Among the 45 invertebrates, 12 were parasites (mostly Platyhelminthes flatworms), mainly of freshwater fish and introduced to Europe from Asia with common carp (*Cyprinus carpio*), goldfish (*Carassius auratus*), or Japanese eel (*Anguilla japonica* Temminck and Schlegel) (see Blanc 1997, 2001); several of the parasites have now been recorded on fish species indigenous or endemic to the IP (see references in Table 1). The remaining 33 invertebrates were free-living species, mostly crustaceans (18 species) or molluscs (6 species). Most of the 28 vertebrates were fish (23 species), and there was no aquatic bird naturalized and only one amphibian and one reptilian species.

The continent of origin was significantly different between vertebrates and invertebrates (independence test; $\chi^2 = 37.1$, df $= 7$, $P < 0.0005$) because most naturalized vertebrates were native to the rest of Europe (43% of the 28 species) or North America (29%), origins that in turn were rare among invertebrates (0 and 12%, respectively), which predominantly came from Asia (38%). There was no significant variation in origin between free-living and parasitic invertebrates ($\chi^2 = 6.5$, df $= 5$, $P = 0.26$) or between crustaceans and molluscs ($\chi^2 = 3.8$, df $= 5$, $P = 0.59$).

Table 1 Non-indigenous invertebrate species naturalized in inland waters of the Iberian Peninsula (IP). When a species is probably an old introduction, but the introduction date is largely unknown, a question mark is given as the first record date. The habitats for non-parasite species are coded as: 1, streams and rivers (excluding estuaries); 2, lakes and reservoirs; 3, ponds and pools; 4, rice fields; and 5, estuarine or saline waters.

GROUP Species	Indigenous distribution	First record in the IP	Habitat/ Parasite	References for the IP	Other references
CNIDARIA					
Cordylophora caspia (Pallas)	Ponto–Caspian	2001	1,2,5	Escot *et al.* 2003, Solà 2004	Schuchert 2004
Craspedacusta sowerbyi Lankester	Asia	1968	1,2	Margalef 1974, Ferreira 1985	
Haliplanella lineata (Verrill)	Pacific coast	1996	5	Cuesta *et al.* 1996	Gollasch and Riemann-Zürneck 1996
PLATYHELMINTHES TURBELLARIA					
Dugesia tigrina (Girard)	North America	?	1,2	Baguñà *et al.* 1980	Young and Reynoldson 1999
PLATYHELMINTHES MONOGENEA					
Dactylogyrus anchoratus (Dujardin)	Asia	?	parasite	Sánchez Suárez 2002	Blanc 1997, 2001
Gyrodactylus cyprini Diarova	Asia	?	parasite	Lacasa Millán 1992	Blanc 1997, 2001
Gyrodactylus katharineri Malmberg	Asia	?	parasite	Gutiérrez Galindo and Lacasa Millán 1999	Blanc 1997, 2001
Gyrodactylus salaris Malmberg	Baltic	?	parasite	Bakke *et al.* 2002	Blanc 1997, 2001
Pseudodactylogyrus anguillae (Yin and Sproston)	Asia and Australia	?	parasite	San Martín Outeiral 2004, Gómez-Juaristi and Salvador 2006	Blanc 1997, 2001
Pseudodactylogyrus bini (Kikuchi)	Asia and Australia	?	parasite	Orts Muñoz 1993, Gómez-Juaristi and Salvador 2006	Blanc 1997, 2001

Table 1 Continued.

GROUP Species	Indigenous distribution	First record in the IP	Habitat/ Parasite	References for the IP	Other references
PLATYHELMINTHES TREMATODA					
Phyllodistomum folium (Olfers, 1816)	Ponto–Caspian?	2004	parasite	Peribáñez *et al.* 2006	
PLATYHELMINTHES CESTODA					
Bothriocephalus acheilognathi Yamaguti [= *Bothriocephalus opsarichthydis* (Yeh)]	Asia	?	parasite	Lacasa Millán 1992	Blanc 1997, 2001
NEMATODA					
Anguillicola crassus Kuwahara, Niimi and Itagaki	Asia	?	parasite	Gallastegi *et al.* 2002, Maillo *et al.* 2005	Blanc 1997, 2001
MOLLUSCA GASTROPODA					
Gyraulus chinensis (Dunker)	Asia	1979	2,4	Brown *et al.* 1998	Anderson 2005
Physella acuta (Draparnaud)	North America	1845	1,2,4	Vidal-Abarca and Suárez 1985	Anderson 2003
Potamopyrgus antipodarum (Gray) [= *P. jenkinsi* (Smith)]	New Zealand	1951	1,5	Vidal-Abarca and Suárez 1985	Paavola *et al.* 2005
MOLLUSCA BIVALVIA					
Corbicula fluminea (Müller)	Asia, Africa, and Australia	1981	1,3,5	Vidal-Abarca and Suárez 1985, Escot *et al.* 2003	McMahon 2000
Dreissena polymorpha (Pallas)	Ponto–Caspian	1880	1,2	Vidal-Abarca and Suárez 1985, Altaba *et al.* 2001	
Mytilopsis leucophaeta (Conrad)	Gulf of Mexico	1993	1,2,5	Escot *et al.* 2003	Laine *et al.* 2006
ANNELIDA					
Branchiura sowerbyi Beddard	Asia	1970s	1,2,5	Prat 1980	Brinkhurst and Jamieson 1971
Xironogiton victoriensis Gelder and Hall	North America	1996	parasite	Gelder 1999	Ohtaka *et al.* 2005

Ficopomatus enigmaticus (Fauvel) (= *Mercierella enigmatica* Fauvel)	Indian Ocean	1924	5	Rioja 1924, Fischer-Piette 1951	
CRUSTACEA BRANCHIOPODA					
Artemia franciscana (Kellog)	North, Central, and South America	1980s	5	Amat *et al.* 2005	
Wlassicsia pannonica Daday	Eurasia	1990s	4	Martinoy *et al.* 2006	Leoni *et al.* 1999
CRUSTACEA OSTRACODA					
Dolerocypris sinensis Sars	Asia	1986	3,4	Forés *et al.* 1986, Baltanás *et al.* 1996	Rossi *et al.* 2003
Ilyodromus viridulus (Brady)	Australia and New Zealand	1996	4	Baltanás *et al.* 1996	Rossi *et al.* 2003
Cypris sp. (= *Cypris subglobosa* Sowerby)	America, Africa, and Asia?	1986	2,4	Forés *et al.* 1986, Baltanás *et al.* 1996	Whatley *et al.* 2003
Isocypris beauchampi (Paris)	Africa	1976	2,4	Armengol 1976, Baltanás *et al.* 1996	Rossi *et al.* 2003
Stenocypris major (Baird)	Asia	1986	3,4	Forés *et al.* 1986, Baltanás *et al.* 1996	Rossi *et al.* 2003
Strandesia vavrai (Müller)	Africa	1983	4	Paulo and Moutinho 1983, Baltanás *et al.* 1996	
Strandesia vinciguerrae (Masi)	Africa	1986	4	Forés *et al.* 1986, Baltanás *et al.* 1996	
Tanycypris sp.	Asia	1988	3,4	Forés 1988, Baltanás *et al.* 1996	Rossi *et al.* 2003
CRUSTACEA COPEPODA					
Acartia tonsa Dana	North and South America	1990s	5	Sobral 1985, Frisch *et al.* 2005	
Lernaea cyprinacea Linnaeus	Asia		parasite	Moreno *et al.* 1986, Gutiérrez-Galindo and Lacasa-Millán 2005	Blanc 1997, 2001

Table 1 Continued.

GROUP Species	Indigenous distribution	First record in the IP	Habitat/ Parasite	References for the IP	Other references
CRUSTACEA BRANCHIURA					
Argulus japonicus Thiele	Asia	1921	parasite	http://www.fauna-iberica.mncn.csic.es/	Blanc 1997, 2001
CRUSTACEA MYSIDACEA					
Synidotea laticauda Benedict	North Pacific and South Atlantic coasts	1996	5	Cuesta *et al.* 1996	
CRUSTACEA DECAPODA					
Cherax destructor Clark	Australia	1983	5	Gutiérrez-Yurrita *et al.* 1999	
Eriocheir sinensis (Milne Edwards)	Asia	?	5	Cuesta *et al.* 2006	
Pacifastacus leniusculus (Dana)	North America	1974	1	Habsburgo-Lorena 1978	
Palaemon macrodactylus Rathbun	Asia	1999	5	Cuesta *et al.* 2004	
Procambarus clarkii (Girard)	North America	1974	1,2,4,5	Habsburgo-Lorena 1978	
Rhithropanopeus harrisii (Gould)	North-west Atlantic	1991	5	Cuesta *et al.* 1991	
INSECTA					
Aedes albopictus (Skuse)	Asia	2004	3	Aranda *et al.* 2006	Eritja *et al.* 2005
Stenopelmus rufinasus Gyllenhal	Asia	2003	1,2,4,5	Dana and Viva 2006	
Trichocorixa verticalis (Fieber)	Atlantic coast of America	1997	3,5	Günther 2004, Sala and Boix 2005	Hutchinson 1931

Table 2 Non-indigenous vertebrate species naturalized in inland waters of the Iberian Peninsula (IP). The habitats are coded as: 1, rivers (excluding estuaries); 2, lakes and reservoirs; 3, rice fields; and 4, estuarine or saline waters.

GROUP Species	Indigenous distribution	First record in the IP	Habitat	References for the IP
PISCES				
Abramis bjoerkna (Linnaeus)	Europe	1995	1,2	Doadrio 2002
Abramis brama (Linnaeus)	Europe	2004	2	Benejam *et al.* 2005
Alburnus alburnus (Linnaeus)	Europe	1992	1,2	Doadrio 2002
Ameiurus melas (Rafinesque)	North America	1910	1,2	Doadrio 2002
Carassius auratus Linnaeus	Asia	17th century	1,2,3	Doadrio 2002
Cobitis bilineata Canestrini	Europe	2002	1	Doadrio 2002
Cyprinus carpio Linnaeus	Eurasia	17th century	1,2	Doadrio 2002
Esox lucius Linnaeus	Europe	1949	1,2	Doadrio 2002
Fundulus heteroclitus (Linnaeus)	North America	1970	4	Doadrio 2002
Gambusia holbrooki (Girard)	North America	1920	1,2,3,4	Doadrio 2002
Herichthys facetum (Jenyns)	South America	1985	1,2	Doadrio 2002
Hucho hucho (Linnaeus)	Europe	1970	1	Doadrio 2002
Lepomis gibbosus (Linnaeus)	North America	1910	1,2	Doadrio 2002
Micropterus salmoides (Lacepède)	North America	1955	1,2	Doadrio 2002
Oncorhynchus mykiss (Walbaum)	North America	19th century	1,2	Doadrio 2002
Perca fluviatilis Linnaeus	Europe	1975	1,2	Doadrio 2002
Poecilia reticulata Peters	South America	2000	1,4	Doadrio 2002
Pseudorasbora parva (Temminck and Schlegel)	Asia	2001	1	Caiola and Sostoa 2002
Rutilus rutilus (Linnaeus)	Europe	1910	1,2	Doadrio 2002
Salvelinus fontinalis (Mitchill)	North America	19th century	1,2	Doadrio 2002
Sander lucioperca (Linnaeus)	Europe	1975	1,2	Doadrio 2002
Scardinius erythrophthalmus (Linnaeus)	Europe	1910	1,2	Doadrio 2002
Silurus glanis L.	Europe	1974	1,2	Doadrio 2002
AMPHIBIA ANURA				
Discoglossus pictus Otth	Africa	1900	1	Pleguezuelos 2002
REPTILIA CHELONIA				
Trachemys scripta (Schoepf)	America	1985	1,2,3,4	Pleguezuelos 2002
MAMMALIA				
Mustela vison Schreber	North America	1978	1,2,3,4	Ruiz-Olmo *et al.* 1997, Palomo and Gisbert 2002
Myocastor coypus Molina	South America	1970	1	Palomo and Gisbert 2002
Ondatra zibethicus (Linnaeus)	North America	2002	1	Elosegi 2004

Table 3 Animal species possibly introduced to inland waters of the Iberian Peninsula (IP), but with uncertain status. Some species are cryptogenic (Carlton 1996), i.e. it is very difficult to know whether they are indigenous or introduced; the other species have been reported in the wild but it is uncertain whether they have established permanent populations (naturalized).

GROUP Species	Possibly indigenous to the IP	Uncertain establishment	References
MOLLUSCA GASTROPODA			
Ferrissia wautieri (Mirolli) [= *F. clessiniana* (Jickeli)]	yes	no	Anderson 2005
CRUSTACEA DECAPODA			
Austropotamobius italicus (Faxon) / *Austropotamobius pallipes* (Lereboullet)	yes	no	Grandjean *et al.* 2001
PISCES			
Acipenser baeri Brandt	no	yes	Elvira and Almodóvar 2001
Aphanius fasciatus (Valenciennes)	no	yes	Doadrio 2002
Ctenopharyngodon idella (Valenciennes)	no	yes	J. M. Queral 2005, personal communication
Ictalurus punctatus (Rafinesque)	no	yes	Doadrio 2002
Oncorhynchus kisutch (Walbaum)	no	yes	Doadrio 2002
Tinca tinca (Linnaeus)	yes	no	Doadrio 2002
AMPHIBIA			
Bufo mauritanicus Schlegel	no	yes	Pleguezuelos 2002
Rana catesbeiana Shaw	no	yes	Pleguezuelos 2002
Rana ridibunda, Rana kl. *esculenta, Rana lessonae*	no	yes	Arano *et al.* 1995, García-París *et al.* 2004
REPTILIA CHELONIA			
Pelodiscus sinensis (Wiegmann)	no	yes	Pleguezuelos 2002
AVES			
Aix galericulata (Linnaeus)	no	yes	GAE 2006
Anser erythropus (Linnaeus)	no	yes	GAE 2006
Branta canadensis (Linnaeus)	no	yes	GAE 2006
Oxyura jamaicensis (Gmelin)	no	yes	GAE 2006
MAMMALIA			
Castor fiber Linnaeus	yes	yes	Ceña *et al.* 2004

The main habitat also differed between vertebrates and invertebrates (independence test; $\chi^2 = 22.8$, df $= 4$, $P < 0.0005$), because the former were mostly present in streams and rivers (26 of the 28 species were present in streams and rivers) or lakes and reservoirs, whereas several invertebrates were only present in estuaries/saline waters (e.g. several decapod crustaceans introduced into the Guadalquivir River through ballast water) or in rice fields (namely ostracods).

The mechanism of introduction is obviously also different for invertebrates and vertebrates, because most of the former are accidental introductions (e.g. Asian ostracods in rice fields, ballast water, etc.), whereas most fish species have been introduced intentionally (nowadays illegally). Therefore, naturalized vertebrates and invertebrates showed opposite patterns, with the former (mostly fish) intentionally introduced from the rest of Europe or North America to Iberian streams and reservoirs and most invertebrates originating from Asia and accidentally introduced to estuaries or rice fields.

UNCERTAIN CASES

We found four species for which it is uncertain whether the species is indigenous to the IP and 13 species that they may not have established (Table 3). An interesting case illustrating both the lack of knowledge on invasive species and the power of modern genetic techniques is the crayfish of the *Austropotamobius pallipes* species complex. Until the 1980s the populations in the IP were generally regarded as an endemic species or subspecies in strong decline due to the introduction of the oomycete *Aphanomyces astaci* Schikora with North American crayfish (Martínez *et al.* 2003). Grandjean *et al.* (2000) showed that two species (*A. pallipes* and *Austropotamobius italicus*) could be distinguished within the species complex and that Spanish populations were very close to some Italian populations, so they might be of anthropogenic origin, as already proposed by Albrecht (1983), and should be regarded as *A. italicus*. Grandjean *et al.* (2001) demonstrated a drastic bottleneck in Spanish populations but discussed several potential mechanisms alternative to the hypothesis of introduction by humans. With further genetic analyses, Trontelj *et al.* (2005) supported the anthropogenic origin for the Spanish populations but did not find unequivocal separation between *A. pallipes* and *A. italicus* (but see also Schulz and Grandjean 2005).

These genetic techniques might also prove useful for tench [*Tinca tinca* (Linnaeus)] in the IP. Tench is indigenous to many parts of Europe but considered introduced into Italy (Bianco 1998) and Portugal (Almaça 1995). This latter country shares its largest river basins (Duero, Tajo, and Guadiana rivers) with Spain. There are doubts about its indigenous status in Spain (Doadrio 2002). In fact, Gómez Caruana and Díaz Luna (1991) considered it introduced into the IP around the 17th century. There are records of tench stocking by monks in Spanish and Portuguese ponds several centuries ago (Almaça 1995, García-Berthou and Moreno-Amich 2000). As far as we know, no phylogeographic study on tench has been performed, in contrast to many other European cyprinids, although they could be most helpful in clarifying its native distribution.

A similar, more solved example of "cryptogenic" species (see Carlton 1996) is the case of the freshwater snail *Physella acuta* (Draparnaud). This species was first described from Europe (Drapanaud, 1805), namely from the River

Garonne, near Bordeaux (France). This species is widely distributed in the IP and the rest of western Europe (Vidal-Abarca and Suárez 1985) and inhabits all types of fresh waters. It has been generally regarded as indigenous to continental Europe (Haas 1929, Germain 1930, Macan and Cooper 1977, Girod *et al.* 1980, Vidal-Abarca and Suárez 1985) and its presence in North America was not reported until the 1990s (Wu *et al.* 1997). Nowadays, three types of evidence indicate, however, that *P. acuta* is indigenous to North America and not to Europe: (i) the lack of records of *Physella* shells from European sediments older than the 18th century (Lozek 1964); (ii) recent studies using internal morphology comparisons (Anderson 2003) and reproductive isolation experiments (Dillon *et al.* 2002) showing that at least one *Physella* species from North America [*Physella heterostropha* (Say)] is actually *P. acuta*; and (iii) some historical data of the cotton trade between France and the United States in the 18th century that could explain the arrival of this species to the River Garonne, where it was first observed (Anderson 2003).

The case of *P. acuta* illustrates the importance of historical data and the fossil record as tools for the identification of old introductions by man. Fossil records have been very helpful to establish the introduced nature of ostracods and suggest that dispersal by man of many other invertebrates is very old and has been generally neglected (McKenzie and Moroni 1986, Rossi *et al.* 2003).

The other group of species in Table 3 are species that have been reported in the wild but it is uncertain whether they have established. There are several other NIS that have been recorded in the wild (see e.g. Elvira and Almodóvar 2001, Pleguezuelos 2002) but have certainly not established permanent populations.

ECOLOGICAL IMPACT

The ecological impact of most of these NIS is largely unknown with a few exceptions. The red swamp crayfish, *Procambarus clarkii*, has altered the functioning and structure of many aquatic ecosystems in the IP reducing macrophytes and associated species, among other impacts (Geiger *et al.* 2005, Rodríguez *et al.* 2005, Chapter 28). The eastern mosquitofish (*Gambusia holbrooki*) has been experimentally demonstrated to affect endemic cyprinodontiform fishes [*Aphanius iberus* (Valenciennes) and *Valencia hispanica* (Valenciennes)] by resource and interference competition (Rincón *et al.* 2002). The zebra mussel (*Dreissena polymorpha*) is one of the best known invasive species and, although it is a very old introduction into Portugal, only recently has it been introduced to Spain through the Ebro River, where it is widespread nowadays and might affect the endangered giant pearl mussel, *Margaritifera auricularia* Spengler (Altaba *et al.* 2001). The zebra mussel is still not widespread in the IP, but it will probably be fostered by the illegal, poorly controlled introduction and translocations of fish that are still very frequent.

The polychaete *Ficopomatus enigmaticus* (Fauvel) is very abundant in some Spanish coastal lagoons and probably profoundly affects its ecosystem functioning because it builds large reef-like aggregates (Schwindt and Iribarne 1998). Many piscivorous fish have been introduced into the IP and some unique ecosystems such as Lake Banyoles have been profoundly altered and are nowadays completely dominated by NIS (García-Berthou and Moreno-Amich 2000).

The distribution, abundance, and impact of introduced parasites in the IP is largely unknown but some species such as *Lernaea cyprinacea* are widespread (Moreno *et al.* 1986, Gutiérrez-Galindo and Lacasa-Millán 2005) and several of them have now been recorded on endemic fish species (see references in Table 1). The swimbladder nematode *Anguillicola crassus*, which was transferred from its indigenous host (the Japanese eel, *A. japonica*) to the European eel (*A. anguilla*), can severely impair swimbladder function (and thus possibly spawning migration) and has caused mortalities in both farmed and wild populations in the presence of other stressors (Kirk 2003). Similarly to the case of crayfish plague, Gozlan *et al.* (2005) have recently shown that the topmouth gudgeon, *Pseudorasbora parva*, an Asiatic cyprinid highly invasive in Europe and recently introduced to the IP, carries a pathogen that strongly affects indigenous cyprinids.

Given the enormous impact of the few well-investigated invasive species, the considerable number of introduced species, and the presence in the IP of many endemic species of plants (Médail and Quézel 1999), freshwater fish (Doadrio 2002), and amphibians (Pleguezuelos *et al.* 2002), the overall potential impact of these naturalized species is enormous and should be urgently investigated. The room for management and educational improvement by public administrations to prevent further introductions and translocations and to reduce the spread of invasive species is even larger. We hope this paper will contribute to the improved understanding and control of invasive species in European waters.

ACKNOWLEDGEMENTS

We thank J.F. Gutiérrez-Galindo, F. Padrós, and anonymous reviewers for helpful comments. This study was financially supported by the Spanish Ministry of Science and Technology (REN2003–00477) and the Ministry of Universities, Research and Information Society (DURSI), Government of Catalonia (Distinction Award for university research 2004 to EGB).

REFERENCES

Albrecht, H. 1983. Besiedlungsgeschichte und ursprünglich holozäne Verbreitung der europäischen Flußkrebse (Decapoda: Astacidae). Spixiana **6**, 61–77.

Alcaraz, C., A. Vila-Gispert, and E. García-Berthou. 2005. Profiling invasive fish species: the importance of phylogeny and human use. Diversity and Distributions **11**, 289–298.

Almaça, C. 1995. Fish species and varieties introduced into Portuguese inland waters. Publicações avulsas do Museu Bocage, Lisboa, Portugal.

Altaba, C. R., P. J. Jiménez, and M. A. López. 2001. El temido mejillón cebra empieza a invadir los ríos españoles desde el curso bajo del río Ebro. Quercus **118**, 50–51.

Amat, F., F. Hontoria, O. Ruiz, A. J. Green, M. I. Sánchez, J. Figuerola, and F. Hortas. 2005. The American brine shrimp as an exotic invasive species in the western Mediterranean. Biological Invasions **7**, 37–47.

Anderson, R. 2003. *Physella* (*Costatella*) *acuta* Draparnaud in Britain and Ireland – its taxonomy, origins and relationships to other introduced Physidae. Journal of Conchology **38**, 7–21.

Anderson, R. 2005. An annotated list of the non-marine mollusca of Britain and Ireland. Journal of Conchology **38**, 607–638.

Aranda, C., R. Eritja, and D. Roiz. 2006. First record and establishment of the mosquito *Aedes albopictus* in Spain. Medical and Veterinary Entomology **20**, 150–152.

Arano, B., G. A. Llorente, M. García-París, and P. Herrero. 1995. Species translocation menaces Iberian waterfrogs. Conservation Biology **9**, 196–198.

Armengol, J. 1976. Crustáceos acuáticos del Coto de Doñana. Oecologia aquatica **2**, 93–97.

Baguñà, J., E. Saló, and R. Romero. 1980. Les planàries d'aigües dolces a Catalunya i les Illes Balears. I. Clau sistemàtica i distribució geogràfica.Butlletí de la Institució Catalana d'Història Natural **45**, 15–30.

Bakke, T. A., P. D. Harris, and J. Cable. 2002. Host specificity dynamics: observations on gyrodactylid monogeneans. International Journal for Parasitology **32**, 281–308.

Baltanás, A., B. Beroiz, and A. López. 1996. Lista faunística y bibliográfica de los ostrácodos no-marinos (Crustacea, Ostracoda) de la Península Ibérica, Islas Baleares e Islas Canarias. Asociación Española de Limnología, Madrid, Spain, 71 pp.

Benejam, L., J. Carol, C. Alcaraz, and E. García-Berthou. 2005. First record of the common bream (*Abramis brama*) introduced to the Iberian Peninsula. Limnetica **24**, 273–274.

Bianco, P. G. 1998. Freshwater fish transfers in Italy: history, local modification of fish composition, and a prediction on the future of native populations. Pages 165–197 in I. J. Cowx, editor. Stocking and introductions of fishes. Fishing News Books, Blackwell, Oxford, UK.

Blanc, G. 1997. L'introduction des agents pathogènes dans les écosystèmes aquatiques: aspects théoriques et réalités. Bulletin Français de la Pêche et de la Pisciculture **344/345**, 489–513.

Blanc, G. 2001. Introduction of pathogens in European aquatic ecosystems: attempt of evaluation and realities. Pages 37–56 in A. Uriarte and B. Basurco, editors. Environmental impact assessment of Mediterranean aquaculture farms. CIHEAM-IAMZ, Zaragoza, Spain.

Brinkhurst, R. O. and B. G. M. Jamieson. 1971. Aquatic Oligochaeta of the world. Oliver and Boyd, Edinburgh, UK, 860 pp.

Brown, D. S., M. A. A. Gracio, and C. Meier-Brook C. 1998. The Asian freshwater snail *Gyraulus chinensis* (Dunker, 1848) (Planorbidae) in West Africa and Europe. Journal of African Zoology **112**, 203–213.

Caiola, N. and A. de Sostoa. 2002. First record of the Asiatic cyprinid *Pseudorasbora parva* in the Iberian Peninsula. Journal of Fish Biology **61**, 1058–1060.

Carlton, J. T. 1996. Biological invasions and cryptogenic species. Ecology **77**, 1653–1655.

Ceña, J. C., I. Alfaro, A. Ceña, U. Itoitz, G. Berasategui, and I. Bidegain. 2004. Castor europeo en Navarra y La Rioja. Galemys **16**, 91–98.

Clavero, M. and E. García-Berthou. 2005. Invasive species are a leading cause of animal extinctions. Trends in Ecology and Evolution **20**, 110.

Cuesta, J. A., J. E. García-Raso, and J. I. González-Gordillo. 1991. Primera cita de *Rhithropanopeus harrisii* (Gould, 1814) (Crustacea, Decapoda, Brachyura, Xanthidae) en la Península Ibérica. Boletín del Instituto Español de Oceanografía **7**, 149–153.

Cuesta, J. A., E. González-Ortegón, P. Drake, and A. Rodríguez. 2004. First record of *Palaemon macrodactylus* Rathbun, 1902 (Decapoda, Caridea, Palaemonidae) from European waters. Crustaceana **77**, 377–380.

Cuesta, J. A., E. González-Ortegón, A. Rodríguez, F. Baldó, C. Vilas, and P. Drake. 2006. The decapod crustacean community of the Guadalquivir Estuary (SW Spain): seasonal and inter-year changes in community structure. Hydrobiologia **557**, 85–95.

Cuesta, J. A., L. Serrano, M. R. Bravo, and J. Toja. 1996. Four new crustaceans in the Guadalquivir river estuary (SW Spain), including an introduced species. Limnetica **12**, 41–45.

Dana, E. D. and S. Viva. 2006. *Stenopelmus rufinasus* Gyllenhal 1836 (Coleoptera: Erirhinidae) naturalized in Spain. The Coleopterists Bulletin **60**, 41–42.

Dillon, R. T., A. R. Wethington, J. M. Rhett, and T. P. Smith. 2002. Populations of the European freshwater pulmonate *Physa acuta* are not reproductively isolated from American *Physa heterostropha* or *Physa integra*. Invertebrate Biology **121**, 226–234.

Doadrio, I., editor. 2002. Atlas y libro rojo de los peces continentales de España. Ministerio de Medio Ambiente, Madrid, Spain, 347 pp.

Draparnaud, J. P. R. 1805. Histoire naturelle des mollusques terrestres et fluviatiles de la France. Paris, 164 pp.

Elosegi, M. M. 2004. Observación de una rata almizclera, *Ondatra zibethicus* (Linnaeus, 1766) en Ezkurra (Navarra). Galemys **16**, 63–64.

Elvira, B. and A. Almodóvar. 2001. Freshwater fish introductions in Spain: facts and figures at the beginning of the 21st century. Journal of Fish Biology **59**, 323–331.

Eritja, R., R. Escosa, J. Lucientes, E. Marquès, R. Molina, D. Roiz, and S. Ruiz. 2005. Worldwide invasion of vector mosquitoes: present European distribution and challenges for Spain. Biological Invasions **7**, 87–97.

Escot, C., A. Basanta, F. Cobo, and M. A. Gonzalez. 2003. Sobre la presencia de *Mytilopsis leucophaeata* (Conrad, 1831) (Bivalvia, Dreissenacea, Dreissenidae) en el río Guadalquivir (sur de la Peninsula Ibérica). Graellsia **59**, 91–94.

Espadaler, X. and C. A. Collingwood. 2001. Transferred ants in the Iberian Peninsula (Hymenoptera, Formicidae). Nouvelle Revue d'Entomologie **17**, 257–263.

Ferreira, M. T. 1985. Occurrence of the freshwater medusa *Craspedacusta sowerbyi* Lank. 1880 (Hidrozoa: Olindiidae) in Portuguese reservoirs. Boletim da Sociedade Portuguesa de Ciências Naturais **22**, 41–46.

Fischer-Piette, E. 1951. Stations de l'annélide tubicole *Mercierella enigmatica* Fauvel sur la côte Nord d'Espagne. Bulletin du Laboratorie Maritime de Dinard **34**, 7–9.

Forés, E. 1988. Els ostràcodes dels arrossars del Delta de l'Ebre: sistemàtica, ecologia i distribució geogràfica. Butlletí de la Institució Catalana d'Història Natural **45**, 15–30.

Forés, E., M. Menéndez, and F. A. Comín. 1986. Contribución al conocimiento de Crustáceos y Rotíferos del Delta del Ebro. Miscellània Zoològica **10**, 105–111.

Frisch, D., H. Rodríguez-Pérez, and A. J. Green. 2005. Invasion of artificial ponds in Doñana Natural Park, southwest Spain, by an exotic estuarine copepod. Aquatic Conservation-Marine and Freshwater Ecosystems **15**, 483–492.

GAE (Grupo de Aves Exóticas de SEO/BirdLife). 2006. Aves invasoras en España: lista de especies en las categorías C y E. http://www.seo.org/programa_ficha.cfm?idPrograma=17.

Gallastegi, I., A. Rallo, and M. F. Mulcahy. 2002. A report of *Anguillicola crassus* from Spain. Bulletin of the European Association of Fish Pathologists **22**, 283–284.

García-Berthou, E. and R. Moreno-Amich. 2000. Introduction of exotic fish into a Mediterranean lake over a 90-year period. Archiv für Hydrobiologie **149**, 271–284.

García-París, M., A. Montori, and P. Herrero. 2004. Amphibia, Lissamphibia. Pages 640 in M. A. Ramos *et al.*, editors. Fauna Ibérica, 24. Museo Nacional de Ciencias Naturales, Madrid, Spain.

Gelder, S. R. 1999. Zoogeography of branchiobdellidans (Annelida) and temnocephalidans (Platyhelminthes) ectosymbiotic on freshwater crustaceans, and their reactions to one another in vitro. Hydrobiologia **406**, 21–31.

Geiger, W., P. Alcorlo, A. Baltanás, and C. Montes. 2005. Impact of an introduced Crustacean on the trophic webs of Mediterranean wetlands. Biological Invasions **7**, 49–73.

Germain, L. 1930. Mollusques terrestres et fluviatiles. In Faune de France **21/22**. Lechevalier, Paris, France, 897 pp.

Girod, A., I. Bianchi, and M. Mariani. 1980. Gasteropodi, 1 (Gastropoda: Pulmonata, Prosobranchia: Neritidae, Viviparidae, Bithyniidae, Valvatidae). In S. Ruffo, editor. Guide per il riconoscimento delle specie animali delle acque interne italiane, 7. Consiglio Nazionale delle Richerche, Verona, Italy, 86 pp.

Gollasch, S. and K. Riemann-Zürneck. 1996. Transoceanic dispersal of benthic macrofauna: *Haliplanella lineata* (Verrill, 1898) (Anthozoa, Actinaria) found on a ship's hull in a ship yard dock in Hamburg Harbour, Germany. Helgoländer Meeresuntersuchungen **50**, 253–258.

Gómez Caruana, F. and J. L. Díaz Luna. 1991. Guía de peces continentales de la Península Ibérica . Penthalon, Madrid, Spain.

Gómez-Juaristi, M. and A. Salvador. 2006. Anguila – *Anguilla anguilla*. In L. M. Carrascal and A. Salvador, editors. Enciclopedia virtual de los vertebrados españoles. Museo Nacional de Ciencias Naturales, Madrid, Spain. http://www.vertebradosibericos.org/

Gozlan, R. E., S. St Hilaire, S. W. Feist, P. Martin, and M. L. Kent. 2005. Disease threat to European fish. Nature (London) **435**, 1046.

Grandjean, F., D. J. Harris, C. Souty-Grosset, and K. A. Crandall. 2000. Systematics of the European endangered crayfish species, *Austropotamobius pallipes* (Decapoda: Astacidae). Journal of Crustacean Biology **20**, 522–529.

Grandjean, F., N. Gouin, C. Souty-Grosset, and J. Diéguez-Uribeondo. 2001. Drastic bottlenecks in the endangered crayfish species *Austropotamobius pallipes* in Spain and implications for its colonization history. Heredity **86**, 431–438.

Günther, H. 2004. *Trichocorixa verticalis verticalis* (Fieber), eine nearktische Ruderwanze in Europa. Mitteilungen des Internationalen Entomologischen Verereins **29**, 45–49.

Gutiérrez-Galindo, J. F. and M. I. Lacasa-Millán. 1999. Monogenea parásitos de Cyprinidae en el río Llobregat (NE de España) (*Barbus haasi* Petersen, 1925). Revista AquaTIC **7**. http://www.revistaaquatic.com/aquatic/art.asp?t=h&c=69

Gutiérrez-Galindo, J. F. and M. I. Lacasa-Millán. 2005. Population dynamics of *Lernaea cyprinacea* (Crustacea: Copepoda) on four cyprinid species. Disease of Aquatic Organisms **67**, 111–114.

Gutiérrez-Yurrita, P. J., J. M. Martinez, M. Ilheu, M. A. Bravo-Utrera, J. M. Bernardo, and C. Montes. 1999. The status of crayfish populations in Spain and Portugal. Pages 161–192 in F. Gherardi and D. Holdich, editors. Crayfish in Europe as alien species: How to make the best of a bad situation? A. A. Balkema, Rotterdam, The Netherlands.

Haas, F. 1929. Fauna malacológica terrestre y de agua dulce de Cataluña. Treballs del Museu de Ciències Naturals **13**, 1–491.

Habsburgo-Lorena, A. S. 1978. Present situation of exotic species of crayfish introduced into Spanish continental waters. Freshwater Crayfish **4**, 175–184.

Hutchinson, G. E. 1931. On the occurrence of *Trichocorixa* Kirkaldy (Corixidae, Hemiptera-Heteroptera) in salt water and its zoo-geographical significance. The American Naturalist **65**, 573–574.

Kirk, R. S. 2003. The impact of *Anguillicola crassus* on European eels. Fisheries Management and Ecology **10**, 385–394.

Lacasa Millán, M. I. 1992. Introducción a la parasitofauna de peces ciprínidos de la cuenca media del río Llobregat. Ph.D. dissertation, Universidad Autònoma de Bacelona, Barcelona, Spain.

Laine, A. O., J. Mattila, and A. Lehikoinen. 2006. First record of the brackish water dreissenid bivalve *Mytilopsis leucophaeata* in the northern Baltic Sea. Aquatic Invasions **1**, 38–41.

Leoni, B., M. Cotta-Ramusino, and F. G. Margaritora. 1999. Seasonal succession of Cladocerans in a rice field in Italy. Hydrobiologia **391**, 241–247.

Lloret, F., F. Medail, G. Brundu, and P. E. Hulme. 2004. Local and regional abundance of exotic plant species on Mediterranean islands: are species traits important? Global Ecology and Biogeography **13**, 37–45.

Lozek, V. 1964. Quartärmollusken der Tschechoslowakie. Rozpravy ústredního ústavu geologického **31**, 1–374.

Macan, T. T. and R. D. Cooper. 1977. A key to the British fresh- and brackish-water gastropods. Freshwater Biological Association, Ambleside, UK, 46 pp.

Mack, R. N., D. Simberloff, W. M. Lonsdale, H. Evans, M. Clout, and F. A. Bazzaz. 2000. Biotic invasions: causes, epidemiology, global consequences, and control. Ecological Applications **10**, 689–710.

Maillo, P. A., M. A. Vich, H. Salvado, A. Marques, and M. P. Gracia. 2005. Parasites of *Anguilla anguilla* (L.) from three coastal lagoons of the River Ebro delta (Western Mediterranean). Acta Parasitologica **50**, 156–160.

Margalef, R. 1974. Ecología. Omega, Barcelona, Spain, 951 pp.

Martinez, R., E. Rico, and E. Alonso. 2003. Characterisation of *Austropotamobius italicus* (Faxon, 1914) in a central Spain area. Bulletin Français de la Pêche et de la Pisciculture **370–371**, 43–56.

Martinoy, M., D. Boix, J. Sala, S. Gascón, J. Gifre, A. Argerich, R. Barrera, S. Brucet, A. Badosa, R. López-Flores, M. Méndez, J. M. Utgé, and X. D. Quintana. 2006. Crustacean and aquatic insect assemblages in the Mediterranean coastal ecosystems of Empordà Wetlands (NE Iberian Peninsula). Limnetica **25**, 53–70.

McKenzie, K. G. and A. Moroni. 1986. Man as an agent of crustacean passive dispersal via useful plants – exemplified by ostracoda *ospiti esteri* of the Italian ricefields ecosystem – and implications arising therefrom. Journal of Crustacean Biology **6**, 181–198.

McMahon, R. F. 2000. Invasive characteristics of the freshwater bivalve *Corbicula fluminea*. Pages 315–343 in R. Claudi and J. H. Leach, editors. Nonindidenous Freshwater Organisms: Vectors, Biology and Impacts. Lewis Publishers. Boca Raton, FL.

Médail, F. and P. Quézel. 1999. Biodiversity hotspots in the Mediterranean basin: setting global conservation priorities. Conservation Biology **13**, 1510–1513.

Moreno, O., C. Granado, and F. Garcia-Novo. 1986. Variabilidad morfológica de *Lernaea cyprinacea* (Crustacea: Copepoda) en el embalse de Arrocampo (Cuenca de Tajo: Caceres). Limnetica **2**, 275–270.

Ohtaka, A., S. R. Gelder, T. Kawai, K. Saito, K. Nakata, and M. Nishino. 2005. New records and distributions of two North American branchiobdellidan species (Annelida: Clitellata) from introduced signal crayfish, *Pacifastacus leniusculus*, in Japan. Biological Invasions **7**, 149–156.

Orts Muñoz, M. E. 1993. Parásitos y parasitismo de la anguila *Anguilla anguilla* L. Ph.D. dissertation, University of Valencia, Valencia, Spain.

Paavola, M., S. Olenin, and E. Leppäkoski. 2005. Are invasive species most successful in habitats of low native species richness across European brackish water seas? Estuarine Coastal and Shelf Science **64**, 738–750.

Palomo, L. J. and J. Gisbert. 2002. Atlas de los mamíferos terrestres de España . Ministerio de Medio Ambiente, Madrid, Spain.

Paulo, L. F. and M. Moutinho. 1983. Systématique et distribution de Ostracodes au Portugal. Publicaçoes do Instituto de Zoologia Dr Augusto Nobre **173**, 1–32.

Peribáñez, M. A., M. L. Elrío, M. J. Gracia, D. Fernández de Luco, J. A. Castillo, J. Lucientes, and I. Cia. 2006. *Phyllodistomum folium* (Trematoda: Gorgoderidae) infecting zebra mussels (*Dreissena polymorpha*) in the Ebro River, Spain. Parasitology International **55**, 143–145.

Pleguezuelos, J. M. 2002. Las especies introducidas de anfibios y reptiles. Pages 501–532 in J. M. Pleguezuelos, R. Márquez, and M. Lizana, editors. Atlas y Libro Rojo de los Anfibios y Reptiles de España. Dirección General de Conservación de la Naturaleza/ Asociación Herpetológica Española, Madrid, Spain.

Pleguezuelos, J. M., R. Márquez, and M. Lizana, editors. 2002. Atlas y Libro Rojo de los Anfibios y Reptiles de España. Dirección General de Conservación de la Naturaleza/ Asociación Herpetológica Española, Madrid, Spain, 587 pp.

Prat, N. 1980. Bentos de los embalses españoles. Oecologia Aquatica **4**, 3–43.

Pyšek, P., D. M. Richardson, M. Rejmbnek, G. Webster, M. Williamson, and J. Kirschner. 2004. Alien plants in checklists and floras: towards better communication between taxonomists and ecologists. Taxon **53**, 131–143.

Richardson, D. M., P. Pyšek, M. Rejmánek, M. G. Barbour, F. D. Panetta, and C. J. West. 2000. Naturalization and invasion of alien plants: concepts and definitions. Diversity and Distributions **6**, 93–107.

Rincón, P. A., A. M. Correas, F. Morcillo, P. Risueño, and J. Lobón-Cerviá. 2002. Interaction between the introduced eastern mosquitofish and two autochthonous Spanish toothcarps. Journal of Fish Biology **61**, 1560–1585.

Rioja, E. 1924. La Mercierella enigmatica Fauvel, serpúlido de agua salobre, en España. Boletin de la Real Sociedad Espanola de Historia Natural **24**, 160–169.

Rodríguez, C. F., E. Bécares, M. Fernández-Aláez, and C. Fernández-Aláez. 2005. Loss of biodiversity and degradation of wetlands as a result of introducing exotic crayfish. Biological Invasions **7**, 75–85.

Rossi, V., G. Benassi, M. Veneri, C. Bellavere, P. Menozzi, A. Moroni, and K. G. McKenzie. 2003. Ostracoda of the Italian ricefields thirty years on: new synthesis and hypothesis. Journal of Limnology **62**, 1–8.

Ruiz-Olmo, J., S. Palazón, F. Bueno, C. Bravo, I. Munilla, and R. Romero. 1997. Distribution, status and colonization of the American mink *Mustela vison* in Spain. Journal of Wildlife Research **2**, 30–36.

Sala, J. and D. Boix. 2005. Presence of the nearctic water boatman *Trichocorixa verticalis verticalis* (Fieber, 1851) (Heteroptera, Corixidae) in the Algarve region (S Portugal). Graellsia **61**, 31–36.

San Martín Outeiral, S. 2004. Helmintos parásitos de *Anguilla anguilla* L. en las rías de Arousa y Ferrol. Ph.D. dissertation, University of Santiago de Compostela, Santiago de Compostela, Spain.

Sánchez Suárez, I. 2002. Contribución al conocimiento de la parasitofauna de la anguila europea (*Anguilla anguilla* L., 1758). Ph.D. dissertation, Universidad Complutense de Madrid, Madrid, Spain.

Schuchert, P. 2004. Revision of the European athecate hydroids and their medusae (Hydrozoa, Cnidaria): families Oceanidae and Pachycordylidae. Revue Suisse de Zoologie **111**, 315–369.

Schulz, H. K. and F. Grandjean. 2005. Roundtable session 3, Phylogeny of European crayfish; improving the taxonomy of European crayfish for a better conservation. Bulletin Français de la Pêche et de la Pisciculture **376–377**, 829–836.

Schwindt, E. and O. O. Iribarne. 1998. Reef of *Ficopomatus enigmaticus* (Polychaeta, Serpulidae) in the Mar Chiquita Coastal Lagoon, Argentina. Bolletí de la d'Història Natural de les Balears **41**, 35–40.

Sobral, P. 1985. Distribuição de *Acartia tonsa* Dana no estuário do Tejo e sua relação com *Acartia clausi* Giesbrecht. Boletim do Instituto Nacional de Investigaçao das Pescas, Lisbon **13**, 61–75.

Sobrino, E., M. Sanz-Elorza, E. D. Dana, and A. González-Moreno. 2002. Invasibility of a coastal strip in NE Spain by alien plants. Journal of Vegetation Science **13**, 585–594.

Solà, C. 2004. Impacte de l'accident miner d'Aznalcóllar sobre el riu Guadiamar. Recuperació de la comunitat de macroinvertebrats i bioacumulació de metalls pesants. Ph.D. dissertation, University of Barcelona, Barcelona, Spain, 248 pp.

Trontelj, P., Y. Machino, and B. Sket. 2005. Phylogenetic and phylogeographic relationships in the crayfish genus *Austropotamobius* inferred from mitochondrial COI gene sequences. Molecular Phylogenetics and Evolution **34**, 212–226.

Vidal-Abarca, C. and M. L. Suárez. 1985. Lista faunística y bibliográfica de los moluscos (Gasterópodos & Bivalvia) de las aguas continentales de la Península Ibérica e Islas Baleares. Asociación Española de Limnología, Madrid, Spain, 190 pp.

Vilà, M., E. García-Berthou, D. Sol, and J. Pino. 2001. Survey of the naturalised plants and vertebrates in peninsular Spain. Ecologia Mediterranea **27**, 55–67.

Vitousek, P. M., H. A. Mooney, J. Lubchenco, and J. M. Melillo. 1997. Human domination of Earth's ecosystems. Science **277**, 494–499.

Young, J. O. and T. B. Reynoldson. 1999. Continuing dispersal of freshwater triclads (Platyhelminthes; Turbellaria) in Britain with particular reference to lakes. Freshwater Biology **42**, 247–262.

Whatley, R. C., S. Bajpai, and J. E. Whittaker. 2003. The identity of the non-marine ostracod *Cypris subglobosa* Sowerby from the intertrappean deposits of Peninsular India. Palaeontology **46**, 1281–1296.

Wu, S. K., R. D. Oesch, and M. E. Gordon. 1997. Missouri aquatic snails. Missouri Department of Conservation, Jefferson, MO, 97 pp.

Chapter seven

An overview of the natural history of non-indigenous amphibians and reptiles

Riccardo Scalera

INTRODUCTION

In his *Naturalis Historia* (77 AD), Pliny the Elder wrote: "*Mirium rerum naturam non solum alia aliis dedisse terris animalia, sed in eodem quoque situ quaedam aliquis locis negasse*" [It is a remarkable fact that nature not only assigned different countries to different animals, but that, even in the same country, it denied certain species to particular localities] (book VIII 83). Pliny the Elder, an erudite natural philosopher and encyclopaedist, could not imagine that, as a side-effect of what we currently call globalization, an ever increasing number of animals and plants would have been moved from one place to another outside their natural range. It is somehow an odd connection that ancient Romans were among the main early actors in fostering the movement of species within the European and Mediterranean regions. Besides several species of mammals and birds introduced for food or hunting, Romans probably also contributed to the movement of reptiles. For example, at that time, pond turtles (i.e. *Emys orbicularis* Linnaeus) were already kept as pets, as were various land-dwelling tortoises, *Testudo* Linnaeus spp. Indeed, ancient Romans were not the very first people contributing to the spread of non-indigenous species, because many introductions are known at least since the Neolithic (Kraus 2003), especially in the Mediterranean region (Pleguezuelos 2002). Thus, introductions probably started centuries before Pliny's time, but certainly since then, a growing number of species has been involved in this global reshuffling. The result is that today about 270 species of amphibians and reptiles

Francesca Gherardi, Biological invaders in inland waters: Profiles, distribution, and threats, 141–160.

are known to exist in countries outside their natural range (Lever 2003) and an unknown number has been subject to other small scale translocations.

In general, despite the increasing interest in non-indigenous amphibians and reptiles (NIAR) (Lever 2003), these taxa have been subordinate in the literature to other species, possibly because most of them are not perceived to be as urgent a threat as others (Kraus and Campbell 2002). As held by Kiesecker (2003), the main constraint in understanding the role of NIAR in the worldwide decline of amphibians has been the lack of their recognition as an important global problem. As a consequence, the long-term effects of the spread of NIAR are largely neglected. Even the Global Invasive Species Database, developed by the IUCN/SSC Invasive Species Specialist Group (ISSG) and available at www.issg.org/database, is not yet as exhaustive for amphibians and reptiles as one would expect: it includes only four amphibians – the cane toad, *Bufo marinus* (Linnaeus), *Eleutherodactylus coqui* Thomas, the American bullfrog, *Rana catesbeiana* Shaw, and the African clawed frog, *Xenopus laevis* (Daudin) – and three reptiles – the brown tree snake, *Boiga irregularis* Merrem, the brown anole, *Norops sagrei* Duméril and Bibron, and the red-eared slider, *Trachemys scripta elegans* (Wied) – out of a total of 284 non-indigenous species introduced worldwide.

As shown on the following pages, besides the best known case studies of some notable taxa like *B. marinus*, *R. catesbeiana*, and *T. s. elegans*, whose spread represents a recognized threat at the global level, there is a number of local situations that are greatly overlooked. This chapter aims at stimulating attention to the large number of worldwide introductions of herpetofauna. A special emphasis is given to those amphibians and reptiles occurring in freshwater ecosystems or somehow linked to inland waters, such as frogs, salamanders, some snakes and lizards (e.g. monitors), and freshwater turtles.

ARE NIAR DISREGARDED?

At the global level, the main comprehensive review of NIAR is the recent book by Lever (2003), which describes 83 species of amphibians and 185 species of reptiles, providing information on relevant threats, taken from about 1,450 sources. It clearly represents the first reference book on this topic though, given its ambitious scope, it has been criticized for not being fully exhaustive (Lovich 2005) and for not providing an analysis of the data reported (Hailey 2005).

Indeed, an exhaustive overview of the worldwide distribution of NIAR is not easy: the real extent of their occurrence, considering both recent and ancient introductions, still needs to be clearly evaluated, and the literature on the topic is far from adequate. At a regional scale, other than North America, little attention has been paid to them. Even in Australia (M. Hutchinson 2006, personal communication) and in most European countries, where non-indigenous species are clearly perceived as a major threat for both nature conservation and human welfare, the current knowledge on these taxa is not exhaustive.

The areas primarily affected by the colonization of NIAR are quite scattered throughout the planet, with an undoubted world "supremacy" held by North America. In Florida alone some 40 naturalized species of amphibians and reptiles are present (Meshaka *et al.* 2004, Smith 2006). In Europe, the number of NIAR is likely to fall between the number of species established in North America (more than 50; McCoid and Kleberg 1995, Lever 2003) and the five – surprisingly few – occurring in Australia (Hutchinson 2001, Bomford 2003). However, their exact number is not yet known because comprehensive studies have never been carried out, notwithstanding the long history of introductions in this region and the rich knowledge accumulated on its faunistic and ecological features. The European situation is a good example with which to highlight the general gap in information on NIAR. It is not clear why the scientific community has the tendency to disregard the subject. For instance, the main reference in Europe, the Atlas of Amphibians and Reptiles (Gasc *et al.* 1997), treats all taxa at the same level, so that the American bullfrog, *R. catesbeiana*, is described among the other species of the European fauna without adequate emphasis on its "alien" status. Two other species, the Mauritanian toad, *Bufo mauritanicus* Schlegel, and the red-eared slider, *T. s. elegans*, are only mentioned in additional notes. Of course, the main aim of the atlas was to show the state of the art concerning the distribution of European species. But does this justify the lack of emphasis on NIAR?

In general, from a mere faunistic point of view, it seems that in Europe comprehensive studies dealing with introduced species, or at least with those considered invasive, have not been felt as a priority at either the continental or the national level. Herpetofauna atlases for countries and islands of the European and Mediterranean regions only rarely include specific contributions dedicated to NIAR. An exception is the Iberian Peninsula, which has produced publications which integrate the typical distributional and taxonomical treatise of an atlas with specific sections dealing with NIAR, both at the national (Mateo 1997, Barbadillo *et al.* 1999, Pleguezuelos 2002) and the local scales (i.e. Catalonia; Llorente *et al.* 1995).

The situation is similar for island ecosystems. Although indigenous species are known to be particularly affected by the presence of introduced species, special attention to NIAR has been paid only in the two Spanish archipelagos: the Balearic Islands, which with 13 species introduced out of a total of 16, show the highest degree of colonization of non-indigenous taxa, and the Canary Islands, which host a rich herpetofauna composed of 14 indigenous and six introduced species (Pleguezuelos 2002).

ON "THE ORIGIN OF SPECIES"

The way species are actively or passively introduced beyond their natural range through human agency is considered a main issue for the understanding of biological invasions. A detailed analysis of the invasion patterns and main pathways for NIAR, particularly those introduced into the USA, has been carried out by Kraus (2003).

In general, species may be introduced either deliberately or unintentionally. The importance of each pathway in herpetofauna introductions varies geographically and over the years (Kraus 2003). Also the origin of NIAR has changed with time. In particular, there are regions where species which have been subject to translocations in ancient times are often part of the indigenous herpetofauna of neighbouring areas – as reported for Spain (Pleguezuelos 2002) – or nearby islands – as suggested for Florida (Butterfield *et al.* 1997). In recent times, such local translocations have been supplemented by those regarding species whose origin may be from other biogeographical regions. Such introductions are mostly linked to the international food and pet trade.

A poor understanding of the distribution of the species and the unavailability of a definitive assessment concerning their phylogeographic status might explain why not much attention has been paid to NIAR so far. In particular, the origin of species translocated in ancient times can be difficult to assess, especially in regions with a long history of the movement of people and goods, such as the Mediterranean.

Today, the origin of species translocated in ancient times can be effectively investigated through modern genetic analysis, using the same techniques that are commonly used for the purpose of systematic and biogeographical studies. The adoption of such techniques in NIAR research could help elucidate their phylogeography. For instance, genetic analyses have been carried out for Mediterranean freshwater turtles of the genus *Emys* Duméril A. (Fritz *et al.* 2005), which have been subject to translocations since ancient times, and to confirm the status of the introduced viperine snake, *Natrix maura* Linnaeus in Menorca, Spain (Guicking *et al.* 2006). In the latter case, molecular data strongly support the evidence of its recent introduction from France, justifying the implementation of strict management measures to reduce its impact on the endemic midwife toad, *Alytes muletensis* (Sanchíz and Adrover).

The results of these studies are thus quite encouraging and suggest the need for increasing the use of genetic tests in the future, so as to rely on more objective data for identification and assignment to likely sources of origin of the introduced species and populations.

TRAVELLERS BY CHANCE

In 1998, the first amphibian species in the evolutionary history of the Galapagos, the Fowler's snouted tree frog, *Scinax quinquefasciatus* (Fowler), indigenous to the Pacific lowlands of Colombia and Ecuador, started spreading on Isabela (Snell and Rea 1999). Apparently, this small tree frog, now considered a leading conservation problem for the survival of indigenous arthropods, reached the archipelago accidentally, stowing away on cargo ships connecting the Galapagos to the Ecuadorian coast. But is it only by chance that a brand new class of vertebrate invaded Melville's *Enchanted Isles*?

Since ancient times, non-indigenous species have been introduced unintentionally around the globe as a consequence of the necessary movement of people and goods. Indeed many species can disperse passively as "hitch-hikers", taking advantage of the various means of transport put at their disposal by human activities linked to international trade (McNeely 1997). Examples include shipping containers, ships, planes, trains, trucks, and cars, where such species can easily hide in commercial commodities, packing material, nursery stock, food, wood, and other goods.

In general, such means of transport are not very suitable for sensitive animals which may lack the basic needs to survive for long distances, as all those species strictly linked to freshwater ecosystems, amphibians in particular. However, there are a few notable exceptions, like *S. quinquefasciatus*, which managed to reach the Galapagos after sailing for about 1,000 km in a ship's cargo from the Ecuadorian coast. Of course, travelling as stowaway is easier for species whose adaptation to hostile environments makes travelling without food and in extreme conditions more feasible (i.e. snakes, lizards, and geckos). For instance, many non-indigenous snakes arrive continuously in Hawaii not only through the smuggling of pet animals, but also introduced as cargo stowaways (Kraus and Cravalho 2001).

Accidental introductions can also be linked to military activities. For example, Bruno and Maugeri (1990) report that in ancient times snakes were used to frighten enemies, i.e. during Roman assaults. Some introduced populations of snakes in the Mediterranean islands could be linked to such events, as suggested by Pleguezuelos (2002) for the origin of *N. maura* in the Balearic, Spain (Guicking *et al.* 2006).

INTENTIONAL INTRODUCTIONS

Intentional introductions of herpetofauna have been linked to human activities such as biological control, farming, aquaculture, and various "scientific" or ornamental purposes. However, the purpose of a release could be sometimes multifaceted, and the difference between intentional and unintentional transport could be a mere shade of meaning. As a consequence, a certain degree of uncertainty is possible on the reasons behind a species introduction. This is especially true for ancient introductions where clues of active or passive transport by human agency, such as fossil remains or historical documents, are often unavailable.

Ranid frogs are typical edible species that have experienced several introductions throughout the world for human consumption, often linked to the restaurant trade. The most common case is certainly the American bullfrog,

R. catesbeiana. A "gourmet" indigenous to North America, the bullfrog is generally considered the most suitable for aquaculture and is frequently farmed for commercial production in countries outside its native range. The species is now spreading in several countries and islands worldwide (for a review see IUCN *et al.* 2006), mainly as a consequence of escapes from breeding facilities (but also from garden ponds), or following intentional releases aimed at establishing wild populations to be regularly harvested. However, several other species have been introduced for the same purpose, though on a smaller spatial scale. For example, the marsh frog, *Rana ridibunda* Pallas, an indigenous species of eastern Europe and Asia, has been introduced into several European countries outside its natural range, including Italy (Sindaco *et al.* 2006), Switzerland (Wittenberg 2005), the UK (Zeisset and Beebee 2003), France (Pagano *et al.* 2003), Belgium (Percsy and Percsy 2002), and a number of places in Asia (Kuzmin 2006). Also Mediterranean islands have been affected by introductions of frogs for human consumption. For example, the pool frog, *Rana lessonae* Camerano, and the edible frog, *Rana* klepton *esculenta* Linnaeus, were translocated from the Italian peninsula to Sardinia (Scalera 2003, Sindaco *et al.* 2006).

Biological control has been the main cause of introductions for the cane toad, *B. marinus*, an indigenous species to South America wrongly considered effective in controlling insect pest on crops, and now established in several countries throughout the world, notably Australia (Lever 2001, Kiesecker 2003). Unfortunately, there is no evidence that *B. marinus* has successfully contributed to pest control in Australia, and is now to be considered a pest itself in its introduced range. Introductions carried out for pest control include the coqui frog, *E. coqui* (Kraus and Campbell 2002), and the poison arrow frog, *Dendrobates auratus* (Girard), in Hawaii (Kraus *et al.* 1999), and the Iberian green frog, *Rana perezi* Seoane, in the Balearics, Spain (Pleguezuelos 2002). Reptiles have also been released as biological control agents. In the Marshall islands (northwest equatorial Pacific), the mangrove monitor, *Varanus indicus* Daudin, which is indigenous to nearby areas, was released to control populations of rats *Rattus* Fischer sp. on military bases (Spennemann 1997). Also in Spain, the Mediterranean pond turtle, *Mauremys leprosa* Schweigger, was introduced for controlling slugs in gardens (see Pleguezuelos 2002).

Deliberate introductions occur particularly as a side-effect of the pet industry (Kraus 2003). In this context, aquarium and vivarium hobbyists keeping amphibians and reptiles in captivity play a major role in fostering the lucrative trade of a growing number of species. Such movements carry the inherent risk of escape or abandonment and the potential establishment of wild self-sustaining populations of NIAR. The alarming successful establishment of naturalized populations of red-eared sliders in several countries throughout the world is clearly linked to: (a) the massive numbers produced in commercial farming activities, both within and outside its native range; and (b) the huge trade of live specimens aimed at supplying either the pet trade or the demand for human consumption and traditional Chinese medicine (Hoover 1998, van Dijk *et al.*

2000; Chapter 8). Likewise, the pond turtle, *E. orbicularis*, has often been the subject of scattered introductions in the European and Mediterranean regions, as a consequence of being a popular pet since ancient times, besides being consumed as food (Pleguezuelos 2002). As a consequence, its natural distribution pattern is clearly affected by introductions, as documented by Fritz *et al.* (2005) in various parts of Italy.

The introductions carried out for research purposes are certainly the most unusual. For instance, non-indigenous populations of the Ambrosi's cave salamander, *Speleomantes ambrosii* (Lanza), and the Italian cave salamander, *Speleomantes italicus* (Dunn), were released in a cave outside their natural range in northern Italy, to verify the possibility of interbreeding (Scalera 2001, Sindaco *et al.* 2006). Similarly, cave salamanders, *Speleomantes* Dubois spp., are known to have been introduced as an experiment in the French Pyrenees (Pascal *et al.* 2006).

Another peculiar case of introduction is related to the use of amphibians in medicine: the African clawed frog, *X. laevis*, was used until recently as a test for human pregnancy and is still common in biology research laboratories and in the pet trade. Because of these uses, it has been introduced from its sub–Saharan African range to the USA (McCoid and Kleberg 1995), Chile (Lobos and Measey 2002), and some European countries, such as the UK (Measey and Tinsley 1998), France (Pascal *et al.* 2006), and Italy (Sindaco *et al.* 2006).

ECOLOGICAL IMPACT

Although in most cases the adverse ecological effects of NIAR are not well known or may be going unnoticed, mainly because of a lack of specific studies, there is evidence of potential threats to indigenous species as a consequence of competition for food and cover and as a result of predatory dynamics. Thus, the occurrence of NIAR may lead to the loss of indigenous species, and changes in community structures and function. Invasive NIAR are also considered as one of the main factors explaining the global decline in many amphibian populations (Kiesecker 2003).

Of course, the effects of species introduced recently can be more apparent than for those of taxa introduced in ancient times, provided that specific researches are envisaged and carried out (but, in general, the impact has been mostly assessed on an empirical basis so far). Species naturalized in the distant past are likely to be now in balance with the extant biological communities, even though damage may well have occurred in the past, so that their impact is not easy to assess.

In the recent years, particular attention has been paid to the process through which regionally distinct, indigenous communities are gradually replaced by locally expanding, cosmopolitan, non-indigenous communities. This process is called biotic homogenization (*sensu* McKinney and Lockwood 1999) and is the result of three interacting processes, the introduction of species, the extinction of indigenous species, and the alteration of pristine habitats (Rahel 2002). On a continental scale, in North America, the average homogenization level in amphibians and reptiles is expected to be higher than in mammals and birds (particularly in southern US) as reported by Olden *et al.* (2006) after investigating the quantitative relation between the biotic homogenization, human population size, and urbanization. This evidence has been confirmed by Smith (2006), who published the result of a specific study on NIAR in Florida.

The homogenization process is likely to be a very widespread phenomenon. In Europe, for instance, there is a growing concern for the spread of the red-eared slider, *T. s. elegans*, a freshwater turtle indigenous to North America currently introduced in several countries, which is silently replacing the few remnant populations of the European pond turtle, *E. orbicularis*. Some authors suggest that such turtles may compete for food, nesting sites, and basking places (Cadi and Joly 2003), although further studies are needed to confirm the impact of such interactions (see Luiselli *et al.* 1997). However, *T. s. elegans* is also known to endanger other species of turtles at the local level, like the Caspian pond turtle, *Mauremys caspica* Gmelin, in Cyprus (Hadjichristophorou 1999).

Several naturalized species are opportunistic feeders with a wide trophic niche, a feature that increases their fitness in a wide range of ecological situations. This is clearly the case for the cane toad, *B. marinus* (Kiesecker 2003) and the American bullfrog, *R. catesbeiana* (Albertini and Lanza 1987), whose diet seems to reflect habitat rather than food preference. The same applies to *R. perezi*, an endemic of the Iberian peninsula, which has been introduced both in the Balearic Islands, where it represents a threat for *A. muletensis*, and in the Canary islands, where predation upon the endemic lizard *Gallotia galloti* Oudart has been recorded (Pleguezuelos 2002). Moreover, studies have been carried out on *B. marinus* (Smith 2005) and *R. catesbeiana* (Kupferberg 1997), which show that even their larvae may adversely impact indigenous tadpoles as a result of interspecific competition.

Non-indigenous snakes can be a major cause of extinction. For instance, in the Balearic Islands, Spain, introduced *N. maura* is known to represent a serious threat to the endangered endemic *A. muletensis* in Mallorca, and was probably involved (with the support of introduced populations of *E. orbicularis*) in the extinction of the endemic *Alytes talaioticus* (Sanchiz and Alcover) – now considered a synonym of *A. muletensis* (see Martínez-Solano *et al.* 2004) – in Menorca (Pleguezuelos 2002).

Also non-indigenous monitors are considered dangerous predators for indigenous wildlife, e.g. in Florida, where the Nile monitor, *Varanus niloticus* Linnaeus, has been recently introduced (Enge *et al.* 2004).

Although some effects are obvious, others might be subtler, yet still of great concern. In Australia, the cane toad, *B. marinus*, is known to poison indigenous predators, besides competing for habitat and food resources with other reptiles and amphibians. Phillips *et al.* (2003), analysing the potential impact of *B. marinus*, determined that about 30% of terrestrial frog-eating snakes are potentially at risk from ingestion of toad-toxins. The cane toad is also thought to contribute to the spread of pathogens that could infect indigenous amphibians (Kiesecker 2003). In North America, other interesting cases of ecologically relevant interactions between indigenous species and NIAR have been reported. Pearl *et al.* (2005) observed the occurrence of interspecific amplexus between each of the two indigenous frogs, the red-legged frog, *Rana aurora* Baird and Girard, and the Oregon spotted frog, *Rana pretiosa* Baird and Girard, and the introduced *R. catesbeiana*, which could have negative demographic consequences for the indigenous ranids (i.e. reducing numbers of males available to couple with conspecifics during their breeding periods). On the other hand, *R. catesbeiana* seems to benefit from the presence of some other NIAR occurring in the same ecosystem. For instance, in western North America the non-indigenous bluegill *Lepomis macrochirus* Rafinesque is facilitating the invasion of *R. catesbeiana* by lowering the abundance of indigenous dragonfly nymphs, which are one of the few predators of the unpalatable bullfrog tadpoles (Adams *et al.* 2003). As emphasized by Simberloff and Von Holle (1999), such positive interactions between non-indigenous species should receive greater attention, because they can be at least as common as detrimental ones.

Reporting some positive effects of introduced species could be misleading, but it is worth mentioning that there are indigenous species, like some ground-nesting birds, that may benefit from the *B. marinus* induced reductions in the numbers of predators, both indigenous and non-indigenous (van Dam *et al.* 2002). Misunderstandings could arise if laypersons read this information without taking into proper consideration the overall negative effects documented for this and other NIAR.

GENETIC EFFECTS

The process of homogenization can extend across all levels of biological organization (Rahel 2002), including the genetic level. For example, non-indigenous species may hybridize with closely related indigenous taxa. Hybridization may cause loss of diversity in genetically different and locally adapted populations and species through genetic introgression, and may interfere with the natural evolutionary processes.

European waterfrogs are characterized by a complex hybridogenetic gametogenesis of the hybrids (Schultz 1969). They are currently receiving increasing attention, especially in relation to the serious ecological and genetic consequences which could negatively interfere with the clonal reproduction typical of this group (Plenet *et al.* 2005). In general, in natural populations where *R.* klepton *esculenta* is a natural hybrid between *R. ridibunda* and *R. lessonae*, the proportion of parental species and hybrids seems to be dependent on the environment (Uzzell and Berger 1975). Vorburger and Reyer (2003) have documented a genetic mechanism of species replacement, according to which the introduction of *R. ridibunda* could unbalance the proportion of the hybrids and even replace the indigenous waterfrogs, *R. lessonae* and *R.* klepton *esculenta*, in several areas of central Europe. This threat may concern several populations of waterfrogs throughout Europe: indeed, so far, non-indigenous populations of *R. ridibunda* are known in Italy (Sindaco *et al.* 2006), Spain (Pleguezuelos 2002), Switzerland (Wittenberg 2005), and France (Plenet *et al.* 2005).

The potential effects of such introductions were clearly overlooked, perhaps because there is uncertainty in whether genetic pollution represents an actual risk for the survival of the affected species (Pagano *et al.* 2003). This might explain why it is only recently that some populations introduced into Italy and formerly considered to be *R. ridibunda* were recognized as being *Rana kurtmuelleri* Gayda, a species indigenous to the Balkans, south-eastern Europe (Scalera 2003, Sindaco *et al.* 2006). A similar situation is emerging also in France where, as reported by Pagano *et al.* (2003), several other non-indigenous waterfrogs have been introduced, but have not yet been identified.

Waterfrogs are not the only taxon affected by this problem. For example, hybridization also occurs between the indigenous great crested newt, *Triturus cristatus* (Laurenti), and the Italian crested newt, *Triturus carnifex* (Laurenti), which has been introduced into Switzerland (Wittenberg 2005) and the UK (Inskipp 2003). In Switzerland, the non-indigenous *T. carnifex* is also replacing the other species, though the exact mechanism is not yet known (Wittenberg 2005). The possibility of genetic contamination has also been reported in some populations of *E. orbicularis* in Italy, following the translocation of specimens belonging to different subspecies (Fritz *et al.* 2005).

Evolutionary changes in introduced species are also being investigated and the results are interesting from a conservation perspective. In France, Schmeller *et al.* (2005) found a higher genetic variability in introduced populations of *R. ridibunda* than in indigenous ones, due to the mixed origin of the populations. Similarly, Zeisset and Beebee (2003), analysing the population dynamics and genetics of the introduced populations of *R. ridibunda* in Britain, found that, despite starting with few founders, significant bottleneck effects were undetectable, presumably because of rapid population expansions immediately after translocations.

Evolutionary processes have also been invoked to explain genetic changes in morphological features. The possibility of genetic drift in an introduced species has been documented for *B. marinus* in Australia, where some westerly populations have longer legs than those distributed further east. This may be due to natural selection for features that suit them for long-distance dispersal (Phillips *et al.* 2006).

NEGATIVE EFFECTS ON ANIMAL AND HUMAN HEALTH

Other than inflicting ecological and genetic harm, NIAR may represent a potential vector of new pathogens, some of which might even threaten human health. The spread of new pathogens can affect indigenous species in a way similar to that of non-indigenous predators. The introduction of NIAR via farming and pet trade is considered the main vector of pathogens and diseases among indigenous species, and is likely to be involved in the global amphibian decline (Laurance *et al.* 1996, Kiesecker 2003).

Chytridiomycosis is one of the most alarming diseases. It is caused by the zoosporic fungus *Batrachochytrium dendrobatidis* Longcore, Pessier and Nichols, which has been associated with a number of amphibian declines and extinctions in geographically disparate parts of the world (see Ron 2005). Strong evidence linking this cutaneous fungal infection to species extinctions was found for the Australian sharp-snouted day frog *Taudactylus acutirostris* (Andersson) (Daszak *et al.* 2003, Schloegel *et al.* 2006). This disease is now recorded in several regions throughout the world, except Asia, apparently as a consequence of the spread of an increasing number of amphibians in trade, which may act as a vector (see Weldon *et al.* 2004). The American bullfrog, *R. catesbeiana*, is clearly among those species which could play a key role in the global dissemination of this pathogen (Mazzoni *et al.* 2003, Hanselmann *et al.* 2004). But infections of this chytrid fungus have also been found in the invasive *E. coqui* in Hawaii (Beard and O'Neill 2005). This indigenous species of Puerto Rico could therefore contribute to the spread of the chytrid fungus to geographic areas where it does not yet exist. Outbreaks of *B. dendrobatidis* are already implicated in an estimated 67% of about 110 species of *Atelopus* (Duméril and Bibron) that have disappeared in the American tropics, and global warming is considered a key contributing factor (Pounds *et al.* 2006).

But for humans one of the greatest health risks associated to amphibians and reptiles is related to the fact that many species are known to be a vector of *Salmonella* Lignieres, a genus of bacillus responsible for severe gastroenteritis, typhoid, and septicaemia, often with serious complications including even meningitis (Mermin *et al.* 2004). As documented by a rich medical literature accumulated in the last 30 years on this topic, many species commonly kept as pets could therefore place their owners, particularly children, at risk of dangerous illness following direct contact with infected animals. Mermin *et al.* (2004) have assessed that reptile and amphibian exposure is associated with about 6% of the approximately 1.24 million sporadic human *Salmonella* infections that occur annually in the USA. As a preventive measure, since 1975, the USA decided to ban the domestic trade of turtles with a carapace length of less than four inches (see Code of Federal Regulations, Title 21 – Sec. 1 240.62 Turtles intrastate and interstate requirements. 21CFR1240.62). It is worth mentioning that the ban did not affect the exports, and therefore USA bred turtles – particularly *T. s. elegans* – have continued to be spread throughout the world, as well with their questionable "shipment" of parasites. Of course, pets are not the only source of contamination: free-ranging introduced species can also be a significant vector of salmonellosis, especially when commonly associated with human habitations, and are progressively extending their range, as shown by specific surveys on *B. marinus* in Australia (O'Shea *et al.* 1990, Speare 1990).

Another aspect of the health risk posed by NIAR to human safety is linked to the threat due to poisonous species. For example, the poisonous skin secretions of *B. marinus* are known to be deadly to humans (Lever 2001) and household pets (McCoid and Kleberg 1995). Another kind of health hazard associated with cane toad regards the use of skin extract and toxin in traditional medicine and as a drug (van Dam *et al.* 2002). Concern may also arise in relation to the spread of the poison arrow frog *D. auratus* in Hawaii, an indigenous species from tropical America whose extremely toxic skin secretions are considered lethal also to humans (they are used to tip the hunting arrows of the indigenous people). However, dendrobatids toxicity seems dependant on compounds sequestered from arthropod prey in their native habitat, which are presumably absent from Hawaii, and therefore the toxicity of the introduced populations is likely to be greatly reduced compared to the toxicity of the founding individuals (Wright 2001).

ECONOMIC AND SOCIAL IMPACT

The "ecological homogenization" of the world, resulting in global McEco-systems (Enserink 1999), is a side-effect of globalization. The introduction of non-indigenous species is considered an externalized cost of international trade, but nonetheless current economics rarely account for the long-term global change it may cause (McNeely 1997). Only a minor percentage of all species introduced beyond their natural range are likely to become naturalized, and in turn only a few of them are likely to become invasive (Williamson 1996). Nevertheless, once a species has become invasive, the relative economic impact is likely to be considerable. The importance of assessing the actual or potential economic impact of non-indigenous species relies on the fact that threats to biodiversity overlapping with threats to human activities are more likely to be taken seriously into consideration by the public opinion, thus fostering political support for the implementation of proper management programmes.

But quantitative data concerning the cost of impacts and management of NIAR are rare, and in general refer to local situations. The main available figures only relate to the most renowned species, like *B. marinus* and *R. catesbeiana*. In Australia, expensive initiatives which required precise cost/benefit assessments have been planned or are being undertaken against *B. marinus*. This species, besides the ecological impact, causes economic losses to some human activity, like apiculture (for the species is a predator of the European honey bee, *Apis mellifera* Linnaeus; Phillips *et al.* 2003). Although the economic impact of cane toads has not been calculated, there are figures which might help understand the problem. According to the Standing Committee on Agriculture (2005) mapping the impacts of the cane toad on biodiversity in Kimberley

(Western Australia) would cost the government some AUS$ 600,000. Other AUS$ 3 million were also allocated for finding a biological control solution to toads, in addition to funding for other specific research programmes. On the other hand, McLeod (2004) estimated that research cost is AUS$ 0.5 million per annum, while figures on ordinary management cost (quarantine checks and public awareness and response) are unavailable. It is also calculated that the construction of a 6 km exclusion fence across the Cobourg Peninsula neck (Northern Territory, Australia) would cost AUS$ 3.6–5.7 million, with an additional expense for annual maintenance in the range of AUS$ 0.4–0.9 million (Brook *et al.* 2004).

In Europe, figures are only available for local attempts of species eradication. For instance, Reinhardt *et al.* (2003) tried to determine the cost to control *R. catesbeiana* in Germany. In this country the presence of the bullfrog was limited to a few populations. However, the foreseen annual cost to implement control measures on only five ponds (mainly by means of electrofishing) is €270,000. Reinhardt *et al.* (2003) also underlines that the total cost would rise to €4.4 billion (and obviously the ecological harm would likewise increase commensurately) in the event that this species spreads throughout Germany. In the UK, south-east England, early efforts to eradicate the first breeding bullfrog population cost some US$ 29,000 (Inskipp 2003).

CONCLUSIONS

In 1839, Charles Darwin, reflecting on the creatures met in the Galapagos islands being remarkably unafraid of humans, wrote in his *Journal of Researches into the Geology and Natural History of the various Countries Visited by H. M. S. Beagle:* "We may infer from these facts, what havoc the introduction of any new beast of prey must cause in a country, before the instincts of the indigenous inhabitants have become adapted to the stranger's craft or power". Darwin's concern likely referred mostly to mammals, birds, and invertebrates, the main taxa which at that time started the reshuffling process which the peculiar Galapagos faunistic communities have been experiencing so far. Of course, the man who so greatly contributed to the fame of the "enchanted islands" could not imagine that, not long after his visit, a small tree frog would have spread in the Galapagos islands, ready to jeopardise indigenous arthropods.

Something is happening to the evolutionary history of the world biota that we cannot control and we are only beginning to understand. The introduction of NIAR is altering the composition and the ecology of original biological communities in space and time. The alarming spread, particularly of cosmopolitan species, and the naturalization of new taxa are expected to continue in the future with an increasing trend towards globalization, possibly exacerbated by major environmental perturbations. For instance, climate changes (Mooney 1996,

Bright 1998) were probably involved in the successful colonization of *S. quinquefasciatus* in the Galapagos (Snell and Rea 1999) and may be related to outbreaks of chytridiomycosis in Central America (Pounds *et al.* 2006). Of course, also minor environmental modifications (Kiesecker 2003) could provide new opportunities to the spread of NIAR, for instance contributing to the connectivity of distant places and creating a more homogeneous habitat. Indeed, in Australia roads facilitate the dispersal of *B. marinus* (Seabrook and Dettmann 1996). On the other hand, it is interesting to notice that, at least in Florida, none of the colonizations that occurred so far originated in natural habitats and most NIAR have remained only in disturbed sites (Butterfield *et al.* 1997).

The spread of NIAR has clearly been overlooked so far, but, following the increased understanding of their negative effects, it is likely that the perception will change and management strategies to deal with their presence and to prevent further introductions, particularly those accidentally caused by keeping so many species as pets, will be implemented. In this context, developing adequate laws and regulations would be essential (see also Chapter 37).

From a conservation point of view, a proper knowledge of the origin of a NIAR (particularly when confusion arises to whether it is actually introduced or just overlooked autochthonous) can be very important, for instance when planning a management strategy. The story of the pool frog, *R. lessonae*, in Britain shows the implications of attributing a wrong status to a species from a conservation perspective. This species was traditionally considered as occurring in Britain only as a result of ancient introductions, but recent findings suggest that indigenous populations also occurred, before disappearing in the 1990s (Beebee *et al.* 2005). As a result, the conservation status of this species is now totally overturned, *R. lessonae* being Britain's most endangered amphibian, currently targeted even by reintroduction programmes. This is not an isolated case. In France there seems to be an analogous situation (Pagano *et al.* 2003). Also in Switzerland, the reintroduction of *E. orbicularis* carried out with specimens of unknown origin shows the high risk of genetic introgression which might occur in areas occupied also by indigenous populations, when little attention is paid on the choice of the taxa to be used (Wittenberg 2005).

It is clear that the scientific community should dedicate more efforts on finding solutions to face the spread of NIAR. The stimulation of further studies would certainly help find a solution for this global challenge. Focusing on scientific research to mitigate the impact on indigenous species can give interesting results. For instance, in the future it could be possible to reduce the impact of chytridiomycosis through bacterial species recently isolated on the skins of some Australian amphibians which inhibit the growth of fungi, including *B. dendrobatidis* (Harris *et al.* 2006).

Dedicated communication campaigns and other measures aimed at raising awareness in public opinion and among policy makers would certainly prevent further introductions. The production of reader friendly publications on this topic would involve academics and the general public in the process of clarifying the geographical distribution of problem species, as well as in the work of early detection of newly established species.

Meanwhile, waiting for an adequate reply from the scientific communities and the competent authorities, nature is doing its best to overcome this threat. For instance, in Australia both body size and toxicity of *B. marinus* – and therefore its impact on indigenous predators – are considered to be decreasing with time (Phillips and Shine 2005). On the other hand, there are indigenous snakes in Australia, such as the keelback *Tropidonophis mairii* Gray, which seem capable of adaptively responding to the toad invasion by increasing toxin resistance (Phillips *et al.* 2004). Thus, although most introductions prove to be irreversible, there are clues suggesting that some indigenous species are learning to live with the threats posed by invasive species. From the perspective of threatened indigenous species, this is clearly the best they can do, without adequate human support.

REFERENCES

Adams, M. J., C. A. Pearl, and R. B. Bury. 2003. Indirect facilitation of an anuran invasion by non-native fishes. Ecology Letters **6**, 343–351.

Albertini, G. and B. Lanza. 1987. *Rana catesbeiana* Shaw, 1802 in Italy. Alytes **6**, 117–129.

Barbadillo, L. J., I. Lacomba, V. Perez Mellado, V. Rancho, and L. F. Lopez Jurado. 1999. Anfibios y Reptiles de la peninsula Iberica, Baleares y Canarias. Editorial Geo Planeta, Spain, 419 pp.

Beard, K. H. and E. M. O'Neill. 2005. Infection of an invasive frog *Eleutherodactylus coqui* by the chytrid fungus *Batrachochytrium dendrobatidis* in Hawaii. Biological Conservation **126**, 591–595.

Beebee, T., J. Buckley, I. Evans, J. Foster, A. Gent, C. Gleed-Owen, G. Kelly, G. Rowe, C. Snell, J. Wycherley, and I. Zeisset. 2005. Neglected native or undesirable alien? Resolution of a conservation dilemma concerning the pool frog *Rana lessonae*. Biodiversity and Conservation **14**, 1607–1626.

Bomford, M. 2003. Risk assessment for the import and keeping of exotic vertebrates in Australia. Bureau of Rural Sciences, Canberra, Australia, 135 pp.

Bright, C. 1998. Life out of bounds: bioinvasion in a borderless world. The Worldwatch Environmental Alert Series. Earthscan Publications, London, UK, 287 pp.

Brook, B. W., P. J. Whitehead, and J. K. Dingle. 2004. Potential cane toad short to medium term control techniques – the biological feasibility and cost of exclusion as a mitigating control strategy. Report to the Department of Heritage and Environment, Darwin, Australia.

Bruno, S. and S. Maugeri. 1990. Serpenti d'Italia e d'Europa. Mondadori, Italy. 224 pp.

Butterfield, B. P., W. E. Meshaka Jr, and C. Guyer. 1997. Nonindigenous amphibians and reptiles. Pages 123 138 in D. Simberloff, D. C. Schmitz, and T. C. Brown, editors. Strangers in paradise. Impact and management of nonindigenous species in Florida. Island Press, Washington, DC.

Cadi, A. and P. Joly. 2003. Competition for basking places between the endangered European pond turtle (*Emys orbicularis galloitalica*) and the introduced red-eared slider (*Trachemys scripta elegans*). Canadian Journal of Zoology **81**, 1392–1398.

Daszak, P., A. A. Cunningham, and A. D. Hyatt. 2003. Infectious disease and amphibian population declines. Diversity and Distributions **9**, 141–150.

Enge, K. M., K. L. Krysko, K. R. Hankins, T. S. Campbell, and F. W. King. 2004. Status of the Nile monitor (*Varanus niloticus*) in south-western Florida. South-eastern Naturalist **3**, 571–582.

Enserink, M. 1999. Biological invaders sweep in. Science **285**, 1834–1836.

Fritz, U., T. Fattizzo, D. Guicking, S. Tripepi, M. G. Pennisi, P. Lenk, U. Joger, and M. Wink. 2005. A new cryptic species of pond turtle from southern Italy, the hottest spot in the range of the genus *Emys* (Reptilia, Testudines, Emydidae). The Norwegian Academy of Science and Letters – Zoologica Scripta **2005**, 1–21.

Gasc, J. P., A. Cabela, J. Crnobrnja-Isailovic, D. Dolmen, K. Grossenbacher, P. Haffner, J. Lescure, H. Martens, J. P. Martinez Rica, H. Maurin, M. E. Oliveira, T. S. Sofianidou, M. Veith, and A. Zuiderwijk. 1997. Atlas of Amphibians and Reptiles in Europe. Societas Europaea Herpetologica, Muséum National d'Histoire Naturelle, Paris, France, 494 pp.

Guicking, D., R. A. Griffiths, R. D. Moore, U. Joger, and M. Wink. 2006. Introduced alien or persecuted native? Resolving the origin of the viperine snake (*Natrix maura*) on Mallorca. Biodiversity and Conservation **15**, 3045–3054.

Hadjichristophorou, M. 1999. Control and eradication of non-native terrestrial vertebrates in Cyprus. Pages 43–47 in Proceedings of the Workshop on the control and eradication of non-native terrestrial vertebrates. CE, Environmental. Encounters, no. 41, Strasbourg, France.

Hailey, A. 2005. Naturalized reptiles and amphibians of the world. Christopher Lever. Book review. Applied Herpetology **2**, 86–88.

Hanselmann, R., A. Rodrýguez, M. Lampo, L. Fajardo-Ramos, A. Alonso Aguirre, A. Marm Kilpatrick, J. P. Rodrýguez, and P. Peter Daszak. 2004. Presence of an emerging pathogen of amphibians in introduced bullfrogs *Rana catesbeiana* in Venezuela. Biological Conservation **120**, 115–119.

Harris, R. N., T. Y. James, A. Lauer, M. A. Simon, and A. Patel. 2006. Amphibian pathogen *Batrachochytrium dendrobatidis* is inhibited by the cutaneous bacteria of amphibian species. EcoHealth **3**, 53–56.

Hoover, C. 1998. The US role in the international live reptile trade: amazon tree boas to Zululand dwarf chamaleons. TRAFFIC North America. Washington, DC.

Hutchinson, M. 2001. Tealeaves and tumbling dice – Predicting animal invasions. Proceedings of Exotic Reptile and Amphibian Conference October 2001, Barossa Valley, Australia. On line at http://www.venomsupplies.com/exoticconference/conferencewelcome.html

Inskipp, C. 2003. Making a lasting impression. The impact of the UK's wildlife trade on the world's biodiversity and people. TRAFFIC International, UK, 74 pp.

IUCN, Conservation International Nature Serve. 2006. *Rana catesbeiana* – American Bullfrog, Bullfrog. Global Amphibian Assessment. <www.globalamphibians.org>. Accessed on 29 April 2006.

Kiesecker, J. M. 2003. Invasive species as a global problem. Toward understanding the worldwide decline of amphibians. Pages 113–126 in R. D. Semlitsch, editors. Amphibian conservation. Smithsonian, Washington, DC.

Kraus, F. 2003. Invasion pathways for terrestrial vertebrates. Pages 68–92 in J. Carlton, G. Ruiz, and R. Mack, editors. Invasive species: Vectors and management strategies. Island Press, Washington, DC.

Kraus, F. and E. W. Campbell. 2002. Human-mediated escalation of a formerly eradicable problem: the invasion of Caribbean frogs in the Hawaiian Islands. Biological Invasions **4**, 327–332.

Kraus F., Campbell E. W., Allison A., and Pratt T. 1999. *Eleutherodactylus* frog introductions to Hawaii. Herpetological Review **30**, 21–25.

Kraus, F. and D. Cravalho. 2001. The risk to Hawaii from snakes. Pacific Science **55**, 409–417.

Kupferberg, S. J. 1997. Bullfrog (*Rana catesbeiana*) invasion of a California river: the role of larval competition. Ecology **78**, 1736–1751.

Kuzmin, S. L. 2006. *Rana ridibunda*. AmphibiaWeb. Information on amphibian biology and conservation. [web application]. 2006. Berkeley, CA. Available: http://amphibiaweb.org/. (Accessed: Apr 29, 2006).

Laurance, W. F., K. R. McDonald, and R. Speare. 1996. Epidemic disease and the catastrophic decline of Australian rain forest frogs. Conservation Biology **10**, 406–413.

Lever, C. 2001. The cane toad: the history and ecology of a successful colonist. Westbury Publishing, West Yorkshire, UK, 230 pp.

Lever, C. 2003. Naturalized reptiles and amphibians of the world. Oxford University Press, Oxford, UK, xx + 318 pp.

Llorente, G. A., A. Montori, X. Santos, and M. A. Carretero. 1995. Atlas dels Amfibis i Reptils de Catalunya i Andorra. Edicions El Brau, Barcelona, Spain, 192 pp.

Lobos, G. and G. J. Measey. 2002. Invasive populations of *Xenopus leavis* (Daudin) in Chile. Herpetological Journal **12**, 163–168.

Lovich, J. 2005. Introduced reptiles and amphibians of the world: unwanted exotic species. Diversity and Distributions **11**, 591.

Luiselli, L., M. Capula, D. Capizzi, E. Filippi, M. Trujillo Jesus, and C. Anibaldi. 1997. Problems for conservation of Pond Turtles (*Emys orbicularis*) in Central Italy: is the introduced Red-eared Turtle (*Trachemys scripta*) a serious threat? Chelonian Conservation and Biology **2**, 417–419.

Martínez-Solano, I., H. A. Gonçalves, J. W. Arntzen, and M. García-París. 2004. Phylogenetic relationships and biogeography of midwife toads (Discoglossidae: *Alytes*). Journal of Biogeography **31**, 603–618.

Mateo, J. A. 1997. Las especies entroducidas en la Peninsula Iberica, Baleares, Canarias, Madeira y Azores. In J. M. Pleguezuelos, editor. Distribucion y biogeographia de los amphibios y reptiles en España y Portugal. Universidad de Granada – Asociacion Herpetologica Española. Monografias de Herpetologia **3**, 465–469.

Mazzoni, R., A. A. Cunningham, P. Daszak, A. Apolo, E. Perdomo, and G. Speranza. 2003. Emerging pathogen of wild amphibians in frog (*Rana catesbeian*a) farmed for international trade. Emerging Infectious Diseases **9**, 995–998.

McCoid, M. J. and C. Kleberg. 1995. Non-native reptiles and amphibians. Pages 433–437 in E. T. LaRoe, G. S. Farris, C. E. Puckett, P. D. Doran, and M. J. Mac, editors. Our living resources: A report to the nation on the distribution, abundance, and health of US plants, animals and ecosystems. Department of the Interior, National Biological Service, Washington, DC.

McKinney, M. L. and J. L. Lockwood. 1999. Biotic homogenization: a few winners replacing many losers in the next mass extinction. Trends in Ecology and Evolution **14**, 450–453.

McLeod, R. 2004. Counting the cost: Impact of invasive animals in australia, 2004. Cooperative Research Centre for Pest Animal Control, Canberra, Australia, 82 pp.

McNeely, J. A. 1997. IUCN and alien invasive species: a framework for action. Pages 3–9 in D. A. R. Clayton and O. L. Gerry, editors. Conserving vitality and diversity. Proceedings of the World Conservation Congress Workshop on alien invasive species. IUCN, North American Wetlands Conservation Council, Canadian Wildlife Service. Ottawa, Canada.

Measey, G. J. and R. C. Tinsley. 1998. Feral *Xenopus leavis* in South Wales. Herpetological Journal **8**, 23–27.

Mermin, J., L. Hutwagner, D. Vugia, S. Shallow, P. Daily, J. Bender, J. Koehler, R. Marcus, F. J. Angulo, and Emerging Infections Program FoodNet Working Group. 2004. Reptiles, amphibians, and human *Salmonella* infection: a population-based, case-control study. Clinical Infectious Diseases. **38**, 253–261.

Meshaka Jr, W. E., B. P. Butterfield, and J. B. Hauge. 2004. The exotic amphibians and reptiles of Florida. Krieger Publishing, Melbourne, FL, 166 pp.

Mooney, H. A. 1996. Biological Invasion and Global Change. Pages 123–126 in O. T. Sandlund, P. J. Schei, and A. V. Viken, editors. Proceedings of the Norway/UN Conference on Alien Species. Trondheim, Norway.

O'Shea, P., R. Speare, and A. D. Thomas. 1990. Salmonellas from the Cane Toad, *Bufo marinus*. Australian Veterinary Journal **67**, 310.

Olden, G. D., N. LeRoy Poff, and M. L. McKinney. 2006. Forecasting faunal and floral homogenization associated with human population geography in North America. Biological Conservation **127**, 261–271.

Pagano, A., A. Dubois, D. Lesbarrères, and T. Lodé. 2003. Frog alien species: a way for genetic invasion? Comptes Rendus Biologies **326**, 85–92.

Pascal, M., O. Lorvelec, and J.-D. Vigne. 2006. Invasions Biologiques et Extinctions: 11000 ans d'histoire des Vertébrés en France. Coédition Belin – Quæ, Paris, France, 350 pp.

Pearl, C. A., M. P. Hayes, R. Haycock, J. D. Engler, and J. Bowerman. 2005. Observations of interspecific amplexus between western North American ranid frogs and the introduced American bullfrog (*Rana catesbeiana*) and an hypothesis concerning breeding interference: American Midland Naturalist **154**, 126–134.

Percsy, C. and N. Percsy. 2002. Evolution des populations indigènes et introduites de grenouilles "vertes" en Brabant wallon. Bulletin de l'Institut royal des Sciences naturelles de Belgique, Biologie **72** (Supplement), 213–217.

Phillips, B. L., G. P. Brown, and R. Shine. 2003. Assessing the potential impact of cane toads on Australian snakes. Conservation Biology **17**, 1738–1747.

Phillips, B. L., G. P. Brown, and R. Shine. 2004. Assessing the potential for an evolutionary response to rapid environmental change: invasive toads and an Australian snake. Evolutionary Ecology Research **6**, 799–811.

Phillips, B. L., G. P. Brown, K. J. Webb, and R. Shine. 2006. Invasion and the evolution of speed in toads. Nature (London) **439**, 803.

Phillips, B. L. and R. Shine. 2005. The morphology, and hence impact, of an invasive species (the cane toad, *Bufo marinus*): changes with time since colonisation. Animal Conservation **8**, 407–413.

Pleguezuelos, J. M. 2002. Las especies introducidas de Anfibios y Reptiles. In J. M. Pleguezuelos, R. Marquez and M. Lizana, editors. Atlas y Libro Rojo de los Amphibios y Reptiles de España. Dirección General de Conservación de la Naturaleza – Asociación Herpetológica Española, Madrid, Spain, 584 pp.

Plenet, S., D. S. Schmeller, and P. Joly. 2005. Consequences of species introduction in a hybridogenetic complex. Pages 60–61 in abstract Volume for the International workshop INWAT: Biological Invasions in Inland Waters. Firenze, Italy, May 5–7, 2005.

Pounds, A., M. R. Bustamante, L. A. Coloma, J. A. Consuegra, M. P. L. Fogden, P. N. Foster, E. La Marca, K. L. Masters, A. Merino-Viteri, R. Puschendorf, S. R. Ron, G. A. Sánchez-Azofeifa, C. J. Still, and B. E. Young. 2006. Widespread amphibian extinctions from epidemic disease driven by global warming. Nature (London) **439**, 161–167.

Rahel, F. J. 2002. Homogenization of freshwater faunas. Annual Review of Ecology and Systematics **33**, 291–315.

Reinhardt, F., M. Herle, F. Bastiansen, and B. Streit. 2003. Economic impact of the spread of alien species in Germany. Federal Environmental Agency (Umweltbundesamt), Berlin, Germany, 229 pp.

Ron, S. R. 2005. Predicting the distribution of the amphibian pathogen *Batrachochytrium dendrobatidis* in the New World. Biotropica **37**, 209–221.

Scalera, R. 2001. Invasioni biologiche. Le introduzioni di vertebrati in Italia: un problema tra conservazione e globalizzazione. Collana Verde, 103. Corpo Forestale dello Stato. Ministero delle Politiche Agricole e Forestali, Roma, Italy, 368 pp.

Scalera, R. 2003. Anfibi e rettili italiani. Elementi di tutela e conservazione. Collana Verde, 104. Corpo Forestale dello Stato. Ministero delle Politiche Agricole e Forestali, Rome, Italy, 232 pp.

Schloegel, L. M., J. M. Hero, L. Berger, R. Spear, K. McDonald, and P. Daszak, 2006. The decline of sharp-snouted day frog (*Taudactylus acutirostris*): The first documented case of extinction by infection in a free-ranging wildlife species? EcoHealth **3**, 35–40.

Schmeller, D. S., P. Joly, and S. Plenet. 2005. Genetic diversity of an introduced frog. Pages 62–63 in abstract volume for the International workshop INWAT: Biological Invasions in Inland Waters. Firenze, Italy, May 5–7, 2005.

Schultz, R. J. 1969. Hybridization, unisexuality and polyploidy in the teleost Poeciliopsis (Poeciliidae) and other vertebrates. American Naturalist **103**, 605–619.

Seabrook, W. A. and E. B. Dettmann. 1996. Roads as activity corridors for cane toads in Australia. Journal of Wildlife Management **60**, 363–368.

Simberloff, D. and B. Von Holle. 1999. Positive interactions of non-indigenous species: invasional meltdown? Biological Invasions **1**, 21–32.

Sindaco, R., G. Doria, E. Mazzetti, and F. Bernini, editors. 2006. Atlante degli Anfibi e dei Rettili d'Italia / Atlas of Italian Amphibians and Reptiles. Societas Herpetologica Italica, Edizioni Polistampa, Firenze, Italy, 791 pp.

Smith, K. G. 2005. Effects of non-indigenous tadpoles on native tadpoles in Florida: evidence of competition. Biological Conservation **123**, 433–441.

Smith, K. G. 2006. Patterns of non-indigenous herpetofaunal richness and biotic homogenization among Florida counties. Biological Conservation **127**, 327–335.

Snell, H. L. and S. Rea. 1999. El Niño 1997–1998 en Galápagos: Se puede estimar 120 años de variaciones climáticas con estadisticas de 34? Pages 65–71 in P. Ospina and E. Munoz, editors. Informe Galapagos 1998–1999. Fundación Natura, Quito, Ecuador.

Speare, R. 1990. Review of the diseases of the Cane Toad, *Bufo marinus*, with comments on biological control. Australian Wildlife Research **17**, 87–410.

Spennemann, D. H. R. 1997. Distribution of rat species (*Rattus* spp.) on the atolls of the Marshall Islands: Past and present dispersal. Atoll Research Bulletin No. **446**, 18 pp.

Standing Committee on Agriculture. 2005. Taking control: a national approach to pest animals: inquiry into the impact on agriculture of pest animals. House of Representatives, Standing Committee on Agriculture, Fisheries and Forestry, The Parliament of the Commonwealth of Australia, Canberra, Australia, 248 pp.

Uzzell, T. and L. Berger. 1975. Electrophoretic phenotypes of *Rana ridibunda, Rana lessonae* and their hybridogenetic associate *Rana esculenta*. Proceedings of the Academy of Natural Sciences of Philadelphia **127**, 13–24.

van Dam, R. A., D. Walden, and G. Begg. 2002. A preliminary risk assessment of cane toads in Kakadu National Park. Scientist Report 164. Supervising Scientist, Darwin N.T., Australia, 92 pp.

van Dijk, P. P., B. L. Stuart, and A. G. J. Rhodin, editors. 2000. Asian Turtle Trade: Proceedings of a Workshop on Conservation and Trade of Freshwater Turtles and Tortoises in Asia. Chelonian Research monographs, No. 2, 164 pp.

Vorburger, C. and H.-U. Reyer. 2003. A genetic mechanism of species replacement in European waterfrogs? Conservation Genetics **4**, 141–155.

Weldon, C., L. H. du Preez, A. D. Hyatt, R. Mullera, and R. Speare. 2004. Origin of the amphibian chytrid fungus. Emerging Infectious Diseases **10**, 2100–2105.

Williamson, M. 1996. Biological invasions. Chapman & Hall, London, UK.

Wittenberg, R., editor. 2005. An inventory of alien species and their threat to biodiversity and economy in Switzerland. CABI Bioscience Switzerland Centre, Report to the Swiss Agency for Environment, Forests and Landscape, Delémont, Switzerland.

Wright, A. 2001. Introduced species summary project: green and black poison dart frog or poison arrow frog (*Dendrobates auratus*). http://www.columbia.edu/itc/cerc/danoff-burg/invasion_bio/inv_spp_summ/Dendrobates_auratus.htm. (9 December 2002)

Zeisset, I. and T. J. C. Beebee. 2003. Population genetics of a successful invader: the marsh frog *Rana ridibunda* in Britain. Molecular Ecology **12**, 639–646.

Chapter eight

The red-eared slider (Trachemys scripta elegans) in Asia: a review

Neil F. Ramsay, Pek Kaye Abigayle Ng, Ruth M. O'Riordan, and Loke Ming Chou

INTRODUCTION

This chapter reviews what is currently known about the status of red-eared sliders in Asia, as well as the potential impact of this non-indigenous species (NIS) on the native Asian biota. It includes published literature, information from websites, as well as mentioning ongoing research where known.

The red-eared terrapin or slider, *Trachemys scripta elegans* (Wied) is native to the south-eastern United States. Estimates of the annual trade in hatchlings vary from 3–4 million (Platt and Fontenot 1992), 4–7 million (Warwick 1991), 8 million (Williams 1999), to 43.6 million from the USA between 1998 and 2002 and 52 million between 1989 and 1997 (Telecky 2001). The popularity of this species has been influenced by crazes amongst children coinciding with cartoons featuring Teenage Mutant Ninja (Hero in the UK) Turtles. Following on from its popularity in the international pet trade, individuals have been released (as discarded pets or for religious reasons) in many places outside their natural range around the world, including a number of Asian countries. *Trachemys scripta elegans* is in the list of the top 100 of the world's worst invasive NIS drawn up by the World Conservation Union IUCN (Global Invasive Species Database, http://www.issg.org/database) and is considered a major threat to

Francesca Gherardi, Biological invaders in inland waters: Profiles, distribution, and threats, 161–174.

indigenous aquatic flora and fauna. In 1975, the US Food and Drug Administration banned the domestic sale of terrapins less than four inches (~12 cm) in length (which some children put in their mouths) because they were causing an estimated 300,000 cases of salmonellosis annually (Williams 1999). According to Williams (1999), at that time most of the estimated 8 million hatchlings annually exported to 60 nations were infected with salmonella (ranched terrapins are fed slaughterhouse offal rich in salmonella). Slider ranching is also an important activity in countries other than the USA. A number of salmonella infections in humans have been traced back to pet terrapins, most of which were *T. s. elegans*. Several authorities believe that they should be considered to be potential vectors of salmonellosis: care should be taken to prevent water in which a pet terrapin has been kept from coming into contact with kitchen utensils or food (Newbery 1984). In people, salmonella causes diarrhoea, fever, and nausea, and can lead to more serious complications such as blood poisoning, meningitis, or death. The most serious cases are found in infants and people with weak immune systems (Salzberg 2000). The importation of red-eared terrapins into New Zealand was banned by the Department of Agriculture due to the potential human health risk (Robb 1980). Conversely, concerns have been raised regarding the depletion of *T. s. elegans* in their natural habitats in southern Louisiana, due to over collection of adults as breeding stock for farms (Warwick *et al.* 1990).

THE GLOBAL SITUATION

Introductions of red-eared sliders due to releases and/or escapes from the pet trade have been reported in Guam (Mariana Islands), Taiwan, Korea, Japan, Malaysia, Singapore, Thailand, Indonesia, Sri Lanka, New Zealand, Israel, Arabia, Bahrain, South Africa, Brazil, Panama, Bermuda, Italy, Spain, Britain, France, Guadeloupe, Guyana, Martinique, Polynesia, and Reunion, as well as in North America outside its natural range (Newbery 1984, Bouskila 1986, Uchida 1989, Ernst 1990, McCoid 1992, Platt and Fontenot 1992, Daniels 1994, da Silva and Blasco 1995, Moll 1995, Ota 1995, Luiselli *et al.* 1997, Servan and Arvy 1997, Chen and Lue 1998, Thomas and Hartnell 2000).

Although the red-eared slider is now found on every continent except Antarctica (Salzberg 2000), the ecological effects of introductions of *T. s. elegans* have been poorly documented (Platt and Fontenot 1992). Most research on its ecology and biology has been in its native temperate regions (e.g. Cagle 1944a, 1944b, 1946, 1950). This species is generally diurnal, feeding mainly in the morning and frequently basking on shores, logs, or while floating, during the rest of the day (Morreale and Gibbons 1986). At night, it sleeps lying on the bottom or resting on the surface near brush piles and hummocks (Ernst and Barbour 1972); however males may move overland at night. Aggressive interactions during basking among four species of emydid terrapins have been observed (Lindeman 1999). Cagle (1946) correlated basking, feeding, and courtship with temperature. It is thought that the terrapins do not feed beyond the extremes of the temperature range of 10–37 °C and consequently do not grow. Juvenile red-eared sliders are mainly carnivorous, eating tadpoles, insects, snails, and spiders, but adults are opportunistic omnivores, consuming almost any food item available, including small fish, amphibians, water plants, and various molluscs (Newbery 1984, Parmenter and Avery 1990). They feed at any time of the day but usually in the early morning and late afternoon (Newbery 1984).

With its broad ecological tolerances, omnivorous diet, and dispersal ability, there is the potential for establishing breeding populations in many areas of the world but little research has been carried out yet. In some countries, where it has been introduced, red-eared sliders have been said to compete with indigenous species for food and basking spots (Salzberg 2000). There is some preliminary evidence that introduced *T. scripta*, now common in Bermuda, are eating mosquito fish (*Gambusia*) as well as a variety of local snails (Davenport *et al.* 2003). In almost all countries where they have been introduced, there are already indigenous freshwater chelonians.

Before focusing on Asia, we will first briefly examine some of the research on this species in areas where it has been introduced, which are pertinent to the Asian situation.

RESEARCH OUTSIDE ASIA

Although there has been some research carried out on the possible impacts of sliders in Europe, there is however to date no hard evidence of threats to the indigenous species *Emys orbicularis* (Linnaeus), *Mauremys leprosa* (Schweigger), and *Mauremys caspica* (Gmelin), or to freshwater ecosystems. However, in an experimental set-up in France, Cadi and Joli (2003) found sliders outcompeting *E. orbicularis* for basking sites. Again in France, Servan and Arvy (1997) reported that *T. scripta* was widely distributed and reproducing in three regions where the European pond turtle *E. orbicularis* occurred and Cadi *et al.* (2004) confirmed production of both sexes from nests incubated in the wild. A comparison of biological parameters with *E. orbicularis* showed that the red-eared slider was bigger, had a more precocious reproduction, the eggs were larger, the young heavier, and the populations more numerous than those of the European pond turtle (Servan and Arvy 1997). The minimum length of males of the red-eared slider at maturity is less than that of the *E. o. orbicularis*, which explains the precocious maturity of the red-eared slider: two to five years for *T. s. elegans* (Cagle 1950) versus six to 16 years for *E. o. orbicularis* (Servan and Arvy 1997). Morreale and Gibbons (1986) and da Silva and Blasco (1995) suggested that breeding populations of *T. scripta* would become established in south-western Spain, an area of habitats and climate similar to parts of its native range, and so there is the potential for competition between *T. scripta* and the indigenous species (*M. leprosa* and *E. orbicularis*). In Valencia there is evidence of reproduction; nest sites and hatchlings (Sancho *et al.* 2005, N. F. Ramsay and R. M. O'Riordan 2005, personal observation). In the 1990s, da Silva and Blasco (1995) warned that, if the range of *T. scripta* expands, a displacement of the indigenous species can be expected and that this event would have especially deleterious consequences for *E. orbicularis*, as it is far more endangered and scarce in Estremadura than is *M. leprosa* (da Silva 1993). In 1997, the then 16-member European Union banned the import of red-eared sliders on the grounds that they were having a deleterious effect on the indigenous European pond terrapin (*E. orbicularis*).

In Israel, *T. scripta* is believed to compete with *M. caspica* (Bouskila 1986), while in South Africa it is suspected that *T. scripta* has displaced the native range of *Pelomedusa subrufa* (Lacépède) through competition. Sliders have adapted completely to the seasonal changes. Instead of breeding between March and September (the normal spring and summer months in North America), they breed during the South African equivalent (late August to February). The reproductive success of animals kept in large open pits on the Transvaal high veldt has been very good, suggesting that reproductive success will also be good for those individuals released into the natural environment. In Queensland, Australia, the red-eared slider became a declared Class 1 pest species in 2003 [Queensland Land Protection (Pest and Stock Route Management) Act 2002]. Class 1 pests are those species that have the potential to cause adverse economic, environmental, or social impacts.

IN ASIA

Asia is the world's most speciose region for tortoises and terrapins as well as having the greatest percentage of threatened species, with more than 75% Critically Endangered, Endangered, or Vulnerable, and 91% on the IUCN Red list (Turtle Conservation Fund 2002). For obvious reasons, research effort and funding has focused on the indigenous species which are both often poorly known and highly endangered due to habitat loss and overcollection, and for aquatic species increasing industrial, agricultural, and domestic pollution of waterbodies. The introduction of NIS, perhaps carrying novel diseases and parasites, as well as being potential competitors, may pose another threat to their already precarious survival. Little research has been carried out in Asia on non-indigenous terrapins, including impacts of red-eared sliders on indigenous species, although there is ongoing research in Singapore (see below).

Commercial farming of species, particularly *Pelodiscus sinensis* (Wiegmann), can lead to other problems. When there is a slump in the market then farms go out of business as in Thailand and Malaysia in 2000 (CITES 2003). It is perhaps not unreasonable to assume that when a farm goes bankrupt unsold animals are released. Unless there is effective bio-security, farms also act as reservoirs of disease from wastewater, escapes, vermin, etc.

Unlike in the EU, where the import of the red-eared slider was banned in 1997, it is still imported into many Asian countries. Whereas the trade in live freshwater and terrestrial chelonians in Europe and North America is almost entirely driven by the demands of the pet trade, in Asia there are multiple markets. There are the local traditional ones, for food especially the soft-shelled Trionychidae species; medicine, e.g. Three-striped Box Terrapin *Cuora trifasciata* (Bell) can fetch US$2,000 on the black market, with material from the plastron of this species believed to be a cure for cancer (Guynup 2004); religious reasons (release for karma); and a growing internal pet trade. Demands from all of these potential markets within Asia are rising with a burgeoning middle class and consumer-driven populace.

Singapore

The red-eared slider is the only reptile species legally sold in Singapore in the pet trade, with the numbers imported peaking in 2005 at over 587,852 animals in that year (Lye Fong Keng, Agri-Food and Veterinary Authority, Singapore, 2006, personal communication). Table 1 shows the total numbers of imports and those from the United States of this species in the last few years, with a clear increase after a drop in 2003. A number of other chelonian species have been found illegally for sale in pet shops (ACRES 2005, Goh and O'Riordan 2007). In the wild, in Singapore, at least 11 species of terrapins have been recorded, both indigenous and NIS (Lim and Chou 1990, Lim and Lim 1992, Chou 1995, Teo and Rajathurai 1997). One of the indigenous species, the Mangrove or River Terrapin *Batagur baska* (Gray), is ranked as Critically Endangered by the IUCN and was first on the list of the World's Top 25 Most Endangered Turtles released by the Turtle Conservation Fund in 2002, while two other species found in Singapore, the Spiny or Spiny Hill Terrapin, *Heosemys spinosa* (Gray), and the Giant Soft-shell Turtle, *Pelochelys cantorii* Gray, are categorized as Endangered. Furthermore, about 5% of Singapore's reptiles have become extinct over the last 183 years (Brooks *et al.* 2003). By far, the most frequently observed chelonian in Singapore waterbodies is the red-eared slider. However, other NIS have also been found in them, including the Chinese Striped-neck, *Ocadia sinensis* (Gray), the Cooter, *Pseudemys* sp., Chinese Softshell, *P. sinensis*, and the Pig-nosed or Fly river turtle, *Carettochelys insculpta* Ramsay (P.K.A.Ng 2005, personal observation). The Chinese Softshell, *P. sinensis*, is the only species of terrapin permitted for import into Singapore for food. So, as in many parts of Asia, in Singapore there are three separate but sometimes overlapping trades in terrapins; the pet trade; for human consumption; and release for religious reasons. In Singapore, releases of introduced terrapins have gone on for decades (Lim and Lim 1992) for religious reasons, and when red-eared sliders become too large or aggressive to be kept as pets.

Table 1 Numbers and origins of red-eared sliders imported into Singapore (2001–2005).

Year	Numbers from US	Total Numbers
2001	284,000	301,245
2002	266,604	269,904
2003	147,363	149,863
2004	388,236	389,036
2005	522,502	587,852

(Source: Agri-Food and Veterinary Authority, Singapore)

Teo and Rajathurai (1997) mentioned 87 records of sliders in their survey of the nature reserves of Singapore and noted that this species was well established in the reservoirs and lakes of Singapore. They commented that fortunately at that time it had not established itself in forest streams, but it is known that red-eared sliders may move overland (Cagle 1944a) especially if a habitat becomes unsuitable, if there is overcrowding, for mate-seeking, or for securing food, and they may move up to 1.6 km from the nearest water to nest (Cagle 1950). Sulaiman (2002) noted that a major concern for conservationists in Singapore was that sliders may outcompete local species, such as the Spiny Terrapin and the Malayan box terrapin, *Cuora amboinensis* (Daudin). Sliders have been recorded laying eggs in the Botanic Gardens (Teo and Rajathurai 1997) and nesting behaviour has been observed there (T. W. H. Tan 2003, personal communication) and at the Night Safari of Singapore Zoological Gardens (T. M. Leong 2003, personal communication), but it is unknown whether the eggs were viable. In Singapore, there has as yet been no published research to show if *T. s. elegans* and other introduced terrapins are successfully reproducing in the wild and whether there are effects on indigenous species, e.g. if the reproductive period of *T. s. elegans* overlaps with indigenous species, there may be potential competition for nest sites. Sulaiman (2002) noted that a female slider can produce up to a dozen eggs twice a year and may out-breed the Malayan box terrapin that lays two eggs each time.

Since 2002, scientists at the National University of Singapore have been undertaking a comprehensive research programme on the biology of red-eared sliders in Singapore. The aspects under study are: (1) diet and whether there is overlap with indigenous terrapins or other species; (2) surveys of waterbodies for the presence and population size of both indigenous and non-indigenous terrapins; (3) estimates of the population size and structure, in particular for evidence of recruitment (habitat preferences of NIS are being compared with those of indigenous species); (4) the activity patterns of red-eared sliders are being examined and an ethogram determined (the occurrence of courtship and nesting behaviour are being recorded and the timing of reproduction and egg-laying, the number of eggs produced, and the site of nesting are being examined; interactions with other species, terrapins and others, including aggression, competition for food and basking sites and predation are recorded); (5) the parasite and disease status are also being examined; (6) comparative studies of how red-eared sliders and two "local" species respond to food items, accelerative ability and food capture/handling methods by video-recording and kinematic analysis (c.f. Davenport *et al.* 1992) have been completed (Davenport 2005, personal communication).

The aim at the end of the research programme is to establish if there is an impact by sliders on the indigenous fauna and to provide a programme for long-term monitoring. Data from this research will be of use for resource management by the National Parks Board (the Singapore body responsible for parks and catchment areas) and to educate the public with respect to the potential problems that can result from releasing NIS into the environment.

Thailand

Adult sliders are abundant in all ponds in parks and temples in Bangkok (Jenkins 1995, Cox *et al.* 1998), and have been released into reservoirs and canals and captured in the wild, north of Bangkok (cited in Jenkins 1995). They have also been described as 'common' in Lumphini Park in southern Thailand (Ransdale 2001). There is also some commercial production and export of sliders but this is not thought to be significant (CITES 2003).

China

In China, a range of North American species are being farmed largely for local demand with 500,000 sliders being produced over three years. Recent exports of sliders to China from the USA were 4.65 million in 1998, 4.71 million in 1999, and 7.5 million in 2000. China has now stopped the import of sliders less than 10 cm long (CITES 2003). Surveys of some animal markets found sliders for sale in Chengdu and Kunming. At Qingshiqiao, 740 individuals of 11 species were recorded with sliders making up 95% (91% of these were hatchlings), at Huaniao 529 'turtles' were on sale with 98% being sliders (97% hatchlings) (Shi 2000). A small number of sliders were on sale on Hainan Island in 2002 (Shi 2004).

Hong Kong

The presence of sliders in the wild has been recorded by the Hong Kong Reptile and Amphibian Society (www.hkas.com). Surveys of Kau Sai Chau, Sai Kung by Dahmer *et al.* (2001) found a new record for a slider in 2000 compared with a 1993 survey (Lau and Dudgeon 1999).

Malaysia

Sharma (1994) reported that in Peninsular Malaysia *T. scripta* hatchlings are commonly sold in pet shops in Penang, Perak (Ipoh and Taiping), Melaka, Terengganu, and Kuala Lumpur. According to Lim and Das (1999), *T. s. elegans* is widespread in the wild in both rural and suburban areas in Peninsular Malaysia; however, these authors do not mention it occurring in Borneo (see under Indonesia). At the Batu Caves near Kuala Lumpur, adult semi-captive sliders have been seen in ponds (Jenkins 1995). The authors know that some research is currently being undertaken on the sliders in Sabah.

Vietnam

Slider hatchlings have been seen on sale in Hanoi's Dong Xuan Market for the last few years, but have only recently been seen in the waterways. One was first discovered in Hoan Kiem Lake in Hanoi in 1997 by Professor Ha Dinh Duc of Hanoi University and, in 2003, more than a dozen juveniles and two adults were observed (Turtle Conservation Indochina 2003). The red-eared sliders, reportedly shipped in as hatchlings from Thailand, appear to have found their way into the lake as releases for religious reasons. Perhaps a hundred or more turtles are released into the lake by Hanoians each year as part of their tradition. Prof. Ha Dinh Duc noted that the Buddhist tradition of releasing wildlife has resulted in 12 species of turtles thus far being recorded in the lake, including *Indotestudo elongata* (Blyth), *Manouria impressa* (Günther), *Pyxidea mouhotii* (Gray), *O. sinensis*, and *P. sinensis* (Turtle Conservation Indochina 2003). In 2004, Prof. Ha Dinh Duc said that there had been no formal research into the impact of the red-eared slider on Hoan Kiem Lake's indigenous wildlife, but felt it was clear that there would be negative consequences as water levels fall, and called for detailed research into the consequences that NIS would have on genetic diversity and the lake's ecosystem. According to Turtle Conservation Indochina (2003), there is no clear evidence that the red-eared sliders are eaten in Vietnam or shipped to China, although larger individuals are occasionally observed in Ho Chi Minh and Hanoi markets. Red-eared sliders are likely to establish a foothold in Hoan Kiem Lake and possibly other places, as the lake offers suitable nesting, and it is likely that releases will continue to augment existing numbers in the future.

Republic of Korea

There are records of red-eared sliders from a number of areas of Korea. Sliders were originally imported into Korea in the 1970s for Buddhist release ceremonies and later as pets. There has been an estimate of 6.5 million animals imported up until when their import was banned in late 2001(Soh Ji-young 2003).

Japan

Uchida (1989) wrote of the then current status of non-indigenous terrapins in Japan. More recently, a survey of 802 sites in 46 prefectures, conducted by the Nature Conservation Society of Japan in 2003, found 5,966 'turtles' and 90% of these were NIS, being mostly species common in Taiwan and North and South America. Sliders made up 62% (3,708) of all turtle records (Turtle and Tortoise Newsletter 2004, Templado 2005). There are also records from Okinawa (Ota 1995), while according to Brazil (2005) sliders can be found in every prefecture. The Invasive Alien Species Act was enacted in Japan at the start of June 2005. It prohibits the importation, sale, raising, and release into the wild of 37 NIS. Included in the list is the North American snapping turtle (*Chelydra serpentina* Linnaeus), and pet owners had until 1 December 2005 to register their turtle with the Environmental Ministry. However, the red-eared slider, of which up to a million are imported into Japan each year, is not listed. According to Templado's (2005) article, fear of a mass slider release is one of the reasons why this species was not included in the list.

Indonesia

The Asian Turtle Conservation Network has listed red-eared sliders from Sumatra, Java, Kalimantan (Borneo), Sulawesi, and Irian Jaya (Hendrie and Vazquez 2004). The Irian News reports that the WWF have an additional record for Manokwari in Irian Jaya in 2004 (Irian News 2004).

Taiwan

Lue and Chen (1996) found *T. scripta* to be the second most abundant turtle of all the rivers surveyed in Taiwan. Subsequently, these authors suggested that the wide ecological tolerance and dietary habits of sliders may cause impacts on indigenous chelonians in Taiwan (Chen and Lue 1998). Although the introduction of sliders may unfavourably affect indigenous fauna, only limited data are available regarding the status of its populations, as well as its relationship with indigenous organisms in Taiwan. In the Taipei Botanical Garden, released individuals of the sliders have almost eradicated the vegetation (water lilies) in a pond. The release of non-indigenous freshwater chelonians is banned in Taiwan, but the law is very difficult to enforce and some sliders are released through Buddhist Mercy Ceremonies. Severinghaus and Chi (1999) commented that in Taiwan prayer released birds are usually wild caught, while the turtles and fishes tend to be captive bred NIS, such as *T. scripta* and carp. The import into Taiwan of reptiles as pets is now banned by the government (Chen and Lue 1998).

We have not found any published information on red-eared sliders in India, Cambodia, Lao PDR, or Myanmar. Except for the record of its occurrence mentioned in Servan and Arvy (1997) we have no other data for Sri Lanka.

CONCLUSIONS

There is a growing awareness across much of Asia of the potential problems of NIS as well as of the trade in wild-caught chelonians. There is however a lack of research in most countries on any potential effects, negative or otherwise, of sliders and other non-indigenous chelonians on indigenous species. Indeed, the CITES (2003) report on the trade in chelonians has argued that since *T. scripta* evolved in a region with a diverse range of other terrapin species sharing the habitat, that although opportunistic, it generally does not exclude other species. If there is competition with other terrapins, it is more likely to be in temperate regions where basking becomes more important. The report suggested that it is unlikely that sliders will establish dominant populations in hill or forest streams nor in large rivers and reservoirs, but is more likely to establish in lowland, vegetated, slow-moving, or static waterbodies, e.g. in canals, ponds, and lakes. Luiselli *et al.* (1997) have emphasized that the introduction of NIS should always be strongly discouraged and that severe preventative measures should be adopted by each responsible government to reduce such activities. Releases into the wild can have far-reaching and harmful consequences for natural ecosystems (Newbery 1984) and it is necessary to educate the public about the potential deleterious effects. In Spain, steps are being taken to eliminate sliders (da Silva and Blasco 1995). Several Asian countries have now banned the importation of red-eared sliders and/or have attempted to restrict the trade in wild species. Unlike other tropical regions, there is a sizeable, local (i.e. Asian) demand for chelonians. The cultural significance has to be treated sensitively to avoid complaints of interference with religious beliefs and traditional medicine. This demonstrates the need for local, culturally sensitive educators preferably from the same communities in combination with scientific research. According to Jenkins (1995) and to the Asian Turtle Trade Working Group (2000), keeping chelonians as pets in South-east Asia was much less prevalent than in Europe and North America, but is not uncommon and is increasing. In the European Union, the banning of the sale of *T. s. elegans* has resulted in the importation of another subspecies, *T. s. scripta*, the yellow-bellied slider. It is therefore likely that a similar scenario of releases and escapes will result in another slider in the wild, while there is still not enough research on the potential impact of *T. s. elegans*. The CITES (2003) report commented that only continued monitoring of non-indigenous "freshwater turtle distribution combined with ecological studies of turtle communities in Asia and beyond can provide answers and suggest methods for active management" of non-indigenous freshwater turtle populations.

ACKNOWLEDGEMENTS

We would like to thank the Agri-Food and Veterinary Authority of Singapore for providing the data on imports of red-eared sliders into Singapore, the

National Parks Board and Public Utilities Board of Singapore for access to the various water bodies, as well as the National University of Singapore for funding the ongoing research in Singapore.

REFERENCES

ACRES. 2005. http://www.acres.org.sg/Press%20releases/22Jul05.html. Accessed 18/01/06.

Asian Turtle Trade Working Group. 2000. Conclusions and recommendations. Pages 148–155 in P. P. van Dijk, B. L. Stuart, and A. G. J. Rhodin, editors. Asian turtle trade: proceedings of a workshop on conservation and Trade of freshwater turtles and tortoises in Asia, Phnom Penh, Cambodia, 1–4 December 1999. Chelonian Research Monographs 2.

Bouskila, A. 1986. On the danger of the red eared terrapin, *Chrysemys scripta*, in natural habitats in Israel. Hardun **3**, 63.

Brazil, M. 2005. Natural Numbers Game. Wild Watch, Isles of Biodiversity. The Japan Times. http://search.japantimes.co.jp/print/features/enviro2005/fe20050217mb.htm. Accessed 18/02/06.

Brooks, B.W., N. S. Sodhi, and P. K. Ng. 2003. Catastrophic extinctions follow deforestation in Singapore. Nature **424**, 420–423.

Cadi, A., V. Delmas, A.-C. Prévot-Julliard, P. Joly, C. Pieau, and M. Girandot. 2004. Successful reproduction of the introduced slider turtle (*Trachemys scripta elegans*) in the South of France. Aquatic Conservation: Marine and Freshwater Ecosystems **14**, 237–246.

Cadi, A. and P. Joly. 2003. Competition for basking places between the endangered European pond terrapin (*Emys orbicularis galloitalica*) and the introduced red-eared slider (*Trachemys scripta elegans*). Canadian Journal of Zoology **81/8**, 1392–1398.

Cagle, F. R. 1944a. Home range, homing behaviour, and migration in turtles. Museum of Zoology, University of Michigan Miscellaneous Publications **16**, 1–34.

Cagle, F. R. 1944b. A technique for obtaining turtle eggs for study. Copeia **1**, 60.

Cagle, F. R. 1946. The growth of the slider turtle, *Pseudemys scripta elegans*. American Midland Naturalists **36**, 685–729.

Cagle, F. R. 1950. The life history of *Pseudemys scripta troostii* (Holbrook). Ecological Monographs **20**, 31–54.

Chen, T.-H. and K.-Y. Lue. 1998. Ecological notes on feral populations of *Trachemys scripta elegans* in Northern Taiwan. Chelonian Conservation and Biology **3**, 87–90.

Chou, L. M. 1995. Amphibians and reptiles. Pages 145–150 in S. C. Chin, R. T. Corlett, Y. C. Wee, and S. Y. Geh, editors. Rain Forest in the City: Bukit Timah Nature Reserve, Singapore. Garden's Bulletin Singapore Supplement **3**, 145–150.

CITES. 2003. Trade in tortoises and freshwater turtles. Development of mid- and long-term conservation measures for tortoises and freshwater turtles. AC19 Doc. 15.2 (Rev. 1).

Cox, M. J., P. P. van Dijk, J. Nabhitabhata, and K. Thirakhupt. 1998. A photographic guide to the snakes and other reptiles of Peninsular Malaysia, Singapore, and Thailand. New Holland, London, UK, 144 pp.

da Silva, E. 1993. Distribución de los Emídios *Mauremys leprosa* Schw. (1812) y *Emys orbicularis* L. (1758) de la provincia de Badajoz: factores que puderian influir en sus áreas de ocupacion. Doñana Acta Vertebrata **20**, 260–266.

da Silva, E. and M. Blasco. 1995. *Trachemys scripta elegans* in Southwestern Spain. Herpetological Review **26**,133–134.

Dahmer, D. T., H. K. Kwokand, and M. L. Felley. 2001. New reptile and amphibian records for Kau Sai Chau, Sai Kung. Porcupine **23**, 1.

Daniels, T. P. 1994. Red-eared Terrapin at Loch Ardinning, Milngavie (V.C.86). The Glasgow Naturalist **22**, 430–431.

Davenport, J., K. Black, G. Burnell, T. Cross, S. Culloty, S. Ekaratne, B. Furness, M. Mulcahy, and H. Thetmeyer. 2003. Aquaculture: the ecological issues. The British Ecological Society, Ecological Issues Series, Blackwell, Oxford, UK, 89 pp.

Davenport, J., M. Spikes, S. M. Thornton, and B. O. Riley. 1992. Crab-eating in the diamond-back terrapin *Malaclemys terrapin* (Latreille): dealing with dangerous prey. Journal of the Marine Biological Association of the United Kingdom **72**, 835–848.

Ernst, C. H. 1990. Systematics, taxonomy, variation and geographic distribution of the slider turtle. Pages 57–67 in J. W. Gibbons, editor. Life history and ecology of the slider turtle. Smithsonian Institution Press, Washington, DC.

Ernst, C. H. and R. W. Barbour. 1972. Turtles of the United States. University Press of Kentucky, Lexington, KY, 347 pp.

Global Invasive Species Database http://www.issg.org/database/species/search.asp?st=sss&sn=&rn=Korea%2C%20 Republic%20of&ri=18376&hci=-1&ei=-1&fr=1&sts=sss. Accessed 20/03/06.

Goh, T. Y. and R. M. O'Riordan. 2007. Are tortoises and freshwater turtles still traded illegally as pets in Singapore? Oryx **41**, 97–100.

Guynup, S. 2004. Turtle Triage. Wildlife Conservation, online edition. http:/www.wildlifeconservation.org/wcm-home/wcm-article/7191105. Accessed 20/02/06.

Hendrie, D. and R. I. Vazquez. 2004. Tortoises and Freshwater Turtles of Indonesia. Asian Turtle Conservation Network, PO Box 222, No. 2/C5, Tap the DH Ngoai Thuong, Hanoi, Vietnam. November 27, 2004.

Irian News. 2004. WWF discovers Red Ear Slider in Manokwari, Indonesian's Irian Jaya Province. Tuesday May 25, 2004.

Jenkins, M. D. 1995. Tortoises and Freshwater Turtles: The trade in Southeast Asia. Traffic International United Kingdom, 50 pp.

Lau, M. W. N. and D. Dudgeon. 1999. Composition and distribution of Hong Kong amphibian fauna. Memoirs of the Hong Kong Natural History Society **22**, 1–80.

Lim, K. K. P. and L. M. Chou. 1990. The Herpetofauna of Singapore. Pages 50–59 in L. M. Chou and P. K. L. Ng, editors. Essays in Zoology. Department of Zoology, National University of Singapore, Singapore.

Lim, B. L. and I. Das. 1999. Turtles of Borneo and Peninsular Malaysia. Natural History Publications (Borneo) Sdn. Bhd, Kota Kinabalu, Sabah, Malaysia, 151 pp.

Lim, K. K. P. and F. L. K. Lim. 1992. A guide to the Amphibians and Reptiles of Singapore. Singapore Science Centre, Singapore, 160 pp.

Lindeman, P. V. 1999. Aggressive interactions during basking among four species of emydid turtles. Journal of Herpetology **33**, 214–219.

Lue, K. Y. and T. H. Chen. 1996. The distribution and conservation of fresh water turtles in Taiwan. Council of Agriculture. (In Chinese)

Luiselli, L., M. Capula, D. Capizzi, E. Philippi, V. TruilloJesus, and C. Anibaldi. 1997. Problems for conservation of pond turtles (*Emys orbicularis*) in central Italy: is the introduced red-eared turtle (*Trachemys scripta*) a serious threat? Chelonian Conservation and Biology **2**, 417–419.

McCoid, M. J. 1992. Geographic distribution: *Trachemys scripta elegans*. Herpetological Review **23**, 26.

Moll, E. O. 1995. The turtle *Trachemys scripta* and the pet trade. Aliens **2**, 3.

Morreale, S. J. and J. W. Gibbons. 1986. Habitat suitability index models: Slider turtle. US Fish and Wildlife Service Biological Report 82 (10.125), 14 pp.

Newbery, R. 1984. The American red-eared terrapin in South Africa. African Wildlife **38**, 186–189.

Ota, H. 1995. A review of introduced reptiles and amphibians of the Ryukyu Archipelago Japan. Island Studies in Okinawa **13**, 63–78.

Parmenter, R. R. and H. W. Avery. 1990. The feeding ecology of the slider turtle. Pages 257–266 in J. W. Gibbons, editor. Life history and ecology of the slider turtle. Smithsonian Institute Press, Washington, DC.

Platt, S. G. and L. W. Fontenot. 1992. The red-eared slider, *Trachemys scripta* (Wied) in South Korea. Bulletin of the Chicago Herpetological Society **27**, 113–114.

Queensland Land Protection (Pest and Stock Route Management) Act 2002. http://www.nrm.qld.gov.au/about/documents/annual_report/performance/turtle.html Accessed 22/03/06.

Ransdale, N. 2001. www.surfbirds.com/mb/Trip Reports/thailand.html. Accessed 16/03/06.

Robb, J. 1980. New Zealand amphibians and reptiles in colour. Collins, Auckland, New Zealand, 119 pp.

Salzberg, A. 2000. The cage papers. Salmonella and reptiles. The Norway rat of the turtle world. Reptile and Amphibian Hobbyist **5**, 84.

Sancho, V., I. Lacomba, F. Ramia, J. V. Bataller, and A. Forteza. 2005. Reproduction of *Trachemys scripta elegans* in Valencia Region (East Spain). Abstract from the 4th International Symposium on *Emys orbicularis* Valencia, Spain, 6–10 June 2005, p. 10.

Servan, J. and C. Arvy. 1997. Introduction de la tortue de Floride *Trachemys scripta* en France. Un nouveau compétiteur pour les espèces de tortues d'eau douce européennes. Bullétin Français de la Pêche et de la Pisciculture **344–345**, 173–177.

Severinghaus, L. L. and L. Chi. 1999. Prayer animal release in Taiwan. Biological Conservation **89**, 301–304.

Sharma, D. S. K. 1994. Tortoise and freshwater turtle trade and utilisation in Peninsular Malaysia. TRAFFIC Southeast Asia Field Report No 15 unpublished, 34 pp.

Shi, H. T. 2000. Results of turtle market surveys in Chengdu and Kunming. Turtle and Tortoise Newsletter **6**, 15b–16.

Shi, H. T. 2004. Observations of wildlife trade in Wenchang City, Hainan Island, China. Turtle and Tortoise Newsletter **8**, 12–13.

Soh Ji-young. 2003. Ministry declares war on destroyers of environment. http://times.hankooki.com/1page/nation/200306/kt2003062718025011970.htm. Accessed 20/01/2006.

Sulaiman, S. 2002. American ex-pets pushing out locals. The Sunday Times (Singapore). September 29, 30 pp.

Telecky, T. M. 2001. United States import and export of live turtles and tortoises. Turtle and Tortoise Newsletter **4**, 8–13.

Templado, L. 2005. Alien species push around locals. http://www.aquaticaliens.org/news_detail.php?News_Id=3. Accessed 10/10/2005.

Teo, R. C. H. and S. Rajathurai. 1997. Mammals, reptiles and amphibians in the nature reserves of Singapore- Diversity, Abundance and Distribution. Proceedings of the Nature Reserves Survey Seminar. Garden's Bulletin Singapore **49**, 353–425.

Thomas, M. and P. Hartnell. 2000. An occurrence of a red-eared turtle (*Trachemys scripta elegans*) in the Waikato River at Hamilton, New Zealand. Herpetofauna (Sydney) **30**, 15–17.

Turtle & Tortoise Newsletter. 2004. Over 60% of turtles in Japan are foreign species. Issue 8: 32 Dec. 2004/Japan For Sustainability Newsletter and International Wildlife Coalition, 16 December 2003.

Turtle Conservation Fund. 2002. A Global Action Plan for Conservation of Tortoises and Freshwater Turtles. Strategy and Funding Prospectus 2002–2007. Washington, DC: Conservation International and Chelonian Research Foundation, 30 pp.

Turtle Conservation Indochina. 2003 Newsletter of the Turtle Conservation and ecology Program (TCEP) 2.

Uchida, I. 1989 The current status of feral turtles of Japan. Anima **205**, 80–85.

Warwick, C, 1991. Conservation of red-eared terrains *Trachemys scripta elegans*: threats from international pet and culinary markets. Testudo **3**, 34–44.

Warwick, C., C. Steedman, and T. Holford. 1990. Ecological implications of the Red-eared turtle trade. The Texas Journal of Science **42**, 419–422.

Williams, T. 1999. The Terrible Turtle Trade. Audubon **101/ 44**, 46–48.

Chapter nine

Semiaquatic mammals introduced into Italy: case studies in biological invasion

Sandro Bertolino and Piero Genovesi

INTRODUCTION

Several semiaquatic mammal species introduced into European freshwater ecosystems are very successful colonizers and rapid invaders. There are four species established in the regional wetlands: the coypu (*Myocastor coypus* Molina), the muskrat (*Ondatra zibethicus* Linnaeus), the American mink (*Mustela vison* Schreber), and the Canadian beaver (*Castor canadensis* Kuhl). These species were imported into Europe to be farmed or were intentionally released into the wild with the aim of harvesting their furs. The Canadian beaver was erroneously released in Finland in an attempt to recover the extinct population of European beavers exterminated by overhunting in the 19th century (Nummi 1996).

Three of them (coypu, muskrat, and mink) are present in Italy with a different status and pose different and complex challenges for preventing further introductions, eradicating key populations, and mitigating the impact of the populations already established in the wild. The principles of a national strategy on the non-indigenous semiaquatic species in Italy could thus provide examples for addressing other species and countries.

Francesca Gherardi, Biological invaders in inland waters: Profiles, distribution, and threats, 175–191.

Despite the wide differences in the ecological niches of the species considered in this review (the mink is a strictly carnivore predator specialized for predation on vertebrates; the muskrat and the coypus are grazers, feeding on several aquatic plant species), all the species share several similar biological traits: (1) all are strictly linked to aquatic habitats, are good swimmers and fast colonizers, able to rapidly occupy vacant suitable habitats; (2) have a large original range (American mink and muskrat inhabit most of North America; the coypu is widespread in South America); (3) feeding niches are very broad, being all able to adapt their diets to the local availability of prey/plants; (4) are larger than most similar indigenous European species – American mink is larger than the European mink *Mustela lutreola* (Linnaeus), while no medium-large aquatic rodents are present in Italy and in southern Europe, since the beaver is extinct in all this area.

On the basis of the biological traits and distribution of the species, and taking into account the potential impacts they may cause, we discuss the main elements for a national Italian policy on non-indigenous semiaquatic mammal species.

THE SPECIES

The coypu is a rodent native to South America that has been imported for fur farming to Europe, Asia, Africa, and North America (Lever 1985, Carter and Leonard 2002). Individuals escaped from the farms or released into the wild established populations along riverbanks and in wetlands of many countries. In Europe, the coypu is widespread from Spain to Romania and from Italy to Germany (Mitchell-Jones *et al.* 1999, Carter and Leonard 2002). The species is considered a pest because of the damage produced by feeding on natural vegetation and crops and for its burrowing activity that undermines riverbanks and dikes (Table 1).

The muskrat is a rodent native to North America, extending its range from Canada to the Gulf of Mexico, the Rio Grande, and Colorado. The present distribution of muskrat outside its native range is the widest for any introduced vertebrate, apart from commensal rats and mice; in Eurasia, the species now ranges from Atlantic to Pacific coast (Long 2003). The introduction outside its original range was a consequence of the release or escape of specimens from breeding farms followed by their natural expansion.

The muskrat was first introduced into Czech Republic near Prague in 1905; here five animals originated two million of individuals in 10 years (Nummi 2002). Its present distribution in Europe ranges from France to Russia and from Italy and Romania to Scandinavia and Baltic countries. In some European countries, the species is responsible for damage to riverbanks and to cereal crops, and changes the composition of aquatic plants (Table 1) with significant impacts on the invertebrate fauna (Nummi *et al.* 2006). There is also some

Table 1 Impacts of three introduced semiaquatic mammals on ecosystem and human activities.

Introduced species	Threats/damage	Source
Damage to vegetation		
Myocastor coypus	Impact on natural aquatic vegetation by feeding	Boorman and Fuller 1981, Foote and Johnson 1993, Bertolino *et al.* 2005
Ondatra zibethicus	Effects on the abundance of some plants and on species dominance	Danell 1996, 1977
Threats to the indigenous fauna		
Myocastor coypus	Destruction of bird nests; predations on eggs	Scaravelli 2002, Tinarelli 2002
Ondatra zibethicus	Impact on invertebrate fauna by changing vegetation structure; competition with water vole; suggested impact on freshwater mussels and crayfish	Danell 1996, Nummi *et al.* 2006
Mustela vison	Competition with European mink and polecat; negative impact on water vole, other rodents and ground-nesting birds	Sidorovich *et al.* 1999, Sidorovich and Macdonald 2001, Macdonald *et al.* 2002, Nordström *et al.* 2002, 2003, Banks *et al.* 2004
Other impacts		
Myocastor coypus	Burrowing activity weakening riverbanks and dikes	Carter *et al.* 1999, Panzacchi *et al.* 2007
Ondatra zibethicus	Burrowing activity weakening riverbanks and dikes (low impact)	Danell 1996
Mustela vison	Predation on poultry, reared game birds, and fisheries	Harrison and Symes 1989, Moore *et al.* 2000, Sheail 2004

evidence of competition with the water vole *Arvicola terrestris* (Linnaeus) and negative impacts on other aquatic species.

The American mink is a predator native to North America, where it is broadly distributed except in Mexico, southern areas of the USA, and north of the Arctic Circle. The first American minks were imported to Europe for fur trade in the 1920s and at present breeding farms are located in various countries. In eastern Europe this predator has also been intentionally released into the wild in the 1930s and 1940s to give rise to populations that could then be harvested for the fur trade. Established populations of the species are present in most part of Europe, especially in the north-eastern countries, in the British Islands, and in the Mediterranean countries (Bonesi and Palazon 2007). The mink poses severe impacts on a number of indigenous species, in particular the water vole, ground-nesting birds, and waterfowl (Table 1). In some cases, its predation caused the complete breeding failure of invaded colonies of terns and gulls (Craik 1997). It is considered a major threat to the endemic European mink (Sidorovich *et al.* 1999) and competes for space and food with the polecat *Mustela putorius* Linnaeus. It also can cause significant damage to poultry runs, reared game birds, and fisheries.

THE ITALIAN SITUATION

The distribution of the coypu in Italy (Fig. 1a) had a sharp increase in recent years, passing from scattered to the present widespread range, with two more or less separated populations: one in northern Italy – from the Po Valley

Fig. 1a Distribution of the coypu in Italy (Cocchi and Riga 2000).

and along the Adriatic coast as far as Abruzzo – and the second along the Tyrrhenian coast of Tuscany and Latium. In southern Italy and in the major islands, the presence of the species is still localized (Cocchi and Riga 2001), although in southern Sardinia it is already quite widespread. The species range is still in expansion where ecological and environmental conditions are favorable; on the basis of a suitability habitat model, the present range could further increase in the future by 2.5–3.3 times (Ottaviani cited in: Panzacchi *et al.* 2007).

The coypu can compromise the integrity of the sloping embankments of irrigation canals through its burrowing activity, and is suspected to have contributed to the causes of flooding, with major economic losses (Panzacchi *et al.* 2007). The species can also cause locally important economic damage by feeding on crops, such as sugar beets and maize. The impact of the coypu on wetlands through feeding on aquatic vegetation can be also severe (Reggiani *et al.* 1993, Cocchi and Riga 2001, Bertolino *et al.* 2005). It destroys nests and preys on eggs of several aquatic birds, including some endangered species (Scaravelli 2002, Tinarelli 2002). It has been hypothesized that the species has a role in the epidemiology of leptospirosis (Michel *et al.* 2001), although it is less important for the spread of the bacteria in the environment compared to other species as rats (V. Guberti 1999, personal communication).

A recent national survey on the economic losses due to the coypu in Italy showed that in 6 years (1995–2000) damage to the riverbanks exceeded €10 millions and impact on agriculture reached €935,138 (Panzacchi *et al.* 2007). At present, the eradication from Italy is considered impractical, because the population is well established and widespread, and permanent control is the most common management policy. In the period covered by the survey (1995–2000), control activities involved the removal of 220,688 coypus and a cost of €2,614,408; however, this effort did not seem to have successfully contained either the expansion of the rodent or its damage at a national level (Panzacchi *et al.* 2007). However, local experiences indicated that well-planned control programs can slow down coypu population increase or manage to eradicate isolated populations (Velatta and Ragni 1991, Bertolino *et al.* 2005).

The muskrat is confined to very few wetlands of north-eastern Italy. Cases of natural expansion of the Slovenian population have been observed, causing the establishment of the species in some localities of Friuli-Venezia Giulia (Lapini and Scaravelli 1993). The present distribution in the country is still very limited and no impacts are recorded so far (Fig. 1b).

The first breeding farms of American minks in the country were established in the 1950s; nowadays there are less than 30 farms, mostly in central and north-eastern Italy. Feral populations originated by individuals that escaped from captivity or were released are recorded in four areas: Friuli (Bon *et al.* 1995), Ronco River (Emilia-Romagna region; D. Scaravelli 2002,

Fig. 1b Distribution of the muskrat in Italy (Spagnesi and De Marinis 2002).

personal communication), Aniene valley (Latium region; Angelici *et al.* 1998), and Flumendosa River (Sardinia Island; A. Deiana 2001, personal communication) (Fig. 1c). It is not yet clear which populations are self-sustaining (reproduction has only been observed in Latium), but data of its constant presence reported for Sardinia and Ronco River seem to confirm that there are several naturalized populations in the country.

Fur farms are often the objects of raids by animal liberation activists who enter the captive-centers during the night, opening cages and releasing the animals. In Italy, in the last 6 years more than 30,000 individuals were released in this way in five areas (Table 2). Most of the animals died from predation, starvation, or were killed by cars, killed illegally, or recaptured. It is known that animals adapted to captivity usually show reduced fitness under natural conditions (Tufto 2001); however, it cannot be excluded that some of the released individuals gave rise to small populations that went undetected.

Fig. 1c Distribution of the American mink in Italy (Andreotti *et al.* 2001).

Table 2 American mink released in Italy from fur farms by the animal liberation activists in 2000–2005.

Province	Region	Year	Animals released
Parma	Emilia-Romagna (NE Italy)	?	5,000
Forlì-Cesena	Emilia-Romagna (NE Italy)	2001	3,000
Treviso	Veneto (NE Italy)	2002	5,000
Ferrara	Emilia-Romagna (NE Italy)	2003	20,000
Padova	Veneto (NE Italy)	2005	200

THE REASON OF A SUCCESS

The three species here considered are among the most successful invaders in Europe; the mink and the coypu are included in the IUCN list of the "100 worst alien species" (Lowe *et al.* 2000). According to Ehrlich (1989), successful invaders are species able to cross major barriers – in this case with the help of humans – rapidly establish viable populations and expand both in number and range in the new habitats relatively quickly. There are several biogeographic, ecological, biological, physiological, and genetic attributes which can be used to identify a potential invader. These were reviewed by Ehrlich (1989) for vertebrates and applied to muskrat by Danell (1996). According to Ehrlich (1989), successful invaders have broad ecological amplitude; this assumption has also been supported by Vázquez (2005) who proposed a "niche breadth-invasion success" hypothesis, suggesting that generalists are more successful invaders than specialists. Non-indigenous species more adapted to occupy human-modified habitats are more likely to establish into the wild (Sol *et al.* 2002). Also *r*-selected species often make successful invaders (Saether 1988).

Coypu, muskrat, and mink have most of the attributes of successful invaders, as they: (1) have large native ranges; (2) often present consistent populations; (3) are rapid dispersers; (4) have broad diet and good behavioral flexibility; (5) show short generation times; (6) are partially gregarious; (7) pregnant females can colonize new areas; (8) have larger size than local congeneric; and (9) are all able to colonize human-modified habitats (Table 3).

Table 3 Attributes of successful vertebrate invaders (adapted from Ehrlich 1989, Danell 1996) and how these apply to the semiaquatic mammals introduced into Italy.

Attributes of successful invaders	Muskrat	Coypu	American mink
Large native range	+	+	+
Abundant in original range	+	+	+
Mobile	+	+	+
Broad diet	+	+	+
Short generation lines	+	+	+
Much genetic variability	?*	?*	?*
Gregarious	+	+	–
Female able to colonize alone	+	+	
Larger than most relatives	+	+	+
Associated with man	–	–	–
Able to function in a wide range of environmental conditions	+	+	+
Able to colonize human-modified habitats	+	+	

* Founder animals originated from breeding farms; repeated releases may have avoided decrease in genetic variability

CONTROL *VS*. ERADICATION

Eradication of non-indigenous species is globally acknowledged as a key management option for mitigating the impacts caused by biological invasions (Wittenberg and Cox 2001, Genovesi 2005). Many invasive introduced species have been eradicated worldwide, managing in this way to prevent the impacts they cause to biological diversity and economy (Simberloff 2002, Genovesi 2005). However, eradication and control in freshwater ecosystems are often much more difficult than in terrestrial environments, and in fact no eradication of amphibians, reptiles, plants, or invertebrates have been completed so far in Europe (Chapter 34).

The main – not to say the only – exception to the difficulty of controlling non-indigenous species in aquatic environments is the removal of mammal species, considering that several eradication projects of introduced semiaquatic mammals have been successfully completed in Europe (Table 4), including the eradication of the muskrat and the coypu from the UK and of the American mink at the local scale, such as its removal from the island of Hiimaa, Estonia (Genovesi 2005), and in the Outer Hebryds (Hebridean Mink Project 2006). The eradication campaign against the coypus in England during the period 1981–1992 is considered one of the most successful eradication projects carried out in Europe and the largest completed on mainland in this region (Gosling 1989, Genovesi 2005).

Eradication is most applicable when populations are still small, and thus is more likely to succeed when realized within a short period after the introduction (i.e. Bomford and O'Brien 1995). Once a non-indigenous species has widely established in the wild and population size becomes large, eradication usually becomes expensive and technically complex (Genovesi 2000). In this case, a permanent control campaign is an alternative option, and in fact the three mammal species here considered are controlled in many countries to reduce

Table 4 Eradication programs carried out in Europe on American mink, coypu, and muskrat. (Sources: Gosling and Baker 1989, Genovesi 2005, Bonesi and Palazon 2007)

Species	Country	Region	Year	Outcome
Mustela vison	Estonia	Hiiumaa Island	1998–1999	Successful
	Finland	Islands in the Baltic Sea	1993–2001	Successful
	Great Britain		1964–1970	Not successful
Myocastor coypus	Great Britain	West Anglia	1981–1989	Successful
Ondatra zibethicus	Great Britain	Scotland, Shropshire, Sussex, Surrey	1932–1937	Successful

Table 5 European countries where the American mink and the coypu are controlled or hunted. (Sources: Carter and Leonard 2002, Bonesi and Palazon 2007; Website: http://www.nwrc.usgs.gov/special/nutria/index.htm)

Species	Country	Method	Year of starting
Mustela vison	Belarus	Control	1992
	Iceland	Hunting	1940s
	Lithuania	Hunting	1980s
	UK – Itchen	Control	1990s
	UK – Thames	Control	2002
	UK – Western Isles	Control/Eradication	2001
Myocastor coypus	Austria	Control	1935
	France	Control	1974
	Germany	Control/Hunting	?
	Italy	Control	?
	The Netherlands	Control	?

their impacts. In Table 5, a list of countries in Europe where mink and coypu are controlled or hunted is reported from two recent reviews published on these species (Carter and Leonard 2002, Bonesi and Palazon 2007).

However, permanent control can be very expensive, and the cost/benefit ratio of this management option should be carefully evaluated before starting a campaign. For example, in Italy the number of coypus removed during control activities in the year 2000 alone ($n = 64{,}338$) almost doubled the number of animals removed in the entire eradication campaign in England ($n = 34{,}822$). The costs paid in Italy for the management (damage and control operations) of the coypu in the year 2000 only, accounted for about 75% of the overall costs of the eradication completed in England and the cost/benefits of the coypu policy in Italy is thus debatable (Panzacchi *et al.* 2007). As a consequence, whenever it is technically feasible, eradication is the best option in comparison to permanent control, because it is definitive and does not require permanent removal efforts and the standing costs of management.

TOWARD A NATIONAL STRATEGY TO PREVENT AND MITIGATE THE IMPACTS OF SEMIAQUATIC MAMMALS IN ITALY

The best strategy for preventing the negative consequences caused by biologic invasions is based on a hierarchical approach that comprises: (1) prevention of non-indigenous species introduction; (2) in the case that prevention fails, prompt eradication of the introduced species; and (3) when eradication is not feasible, spatial containment and/or population control campaign (Wittenberg and Cock 2001, Genovesi and Shine 2004). In the case of the semiaquatic mammals introduced into Italy, prevention should be focused on (1) strictly regulating the existing fur farm facilities, (2) banning farming in not yet invaded areas, (3) rapidly respond to illegal releases, and (4) contain arrival through spread of populations established in neighboring countries. Eradication campaigns should become the key management option when prevention fails, or in the case of isolated populations already established in the country. Large-scale containment or a control campaign should be planned on solid science, only after an in-depth analysis of the cost/benefit ratio.

Legal

All the considered species are automatically protected under the Italian legislation (Law 157/1992), which does not distinguish between indigenous and non-indigenous species and does not include a clear reference to eradication. The legal framework should be revised in order to ensure that legal status of introduced species is compatible with rapid response and mitigation measures. DPR 357 (modified and integrated by DPR 120) has introduced a general ban on the release into the wild of non-indigenous species; however, the practical interpretation of such ban remains unclear, and enforcement is very complex. Clear guidelines for the implementation of this ban and for a clarification of the responsibilities following the introduction of this legal tool are thus urgently needed.

Prevention

The pathway of introduction for the three species is fur farming (both through accidental escapes and intentional releases by animal liberation activists); prevention should thus focus on stricter rules on farm facilities, in order to prevent further escapes, and restrictions to farming in critical areas (most vulnerable habitats, critical areas for expansion, isolated areas not yet invaded). Fencing and security of existing farms should be verified and improved, and stricter criteria of fencing should be imposed when issuing licenses to farms. It is critical that authorization to farms is conditioned to the capacity of the competent authorities (Provincial Administrations) to verify and periodically control the adequacy of the facilities. Muskrats and coypus do not have any commercial interest at present, and stricter rules for farming these rodents are thus not a priority. The involvement of farmers is critical, and a voluntary code of conduct should be developed. A memorandum of understanding with the AIAV (Italian Association of Mink Breeders) should be established.

Rapid response

Rapidity and efficacy of response to new escapes can be increased by defining clear protocols for the competent authorities in case of escapes or releases. Also in this case, the involvement of farmers is an important element of a response strategy; a system of rapid alert after escapes or attacks should be developed. Lists of competent authorities and telephone numbers should be provided to farmers. Decision process for response should be streamlined in order to ensure that capture of animals is started within 2–3 h after escape. Competent authorities in areas with mink farms should develop contingency plans and collect basic equipments (traps, baits, transport cages, nets) for recapturing animals and transporting these to the farms. Staff should be trained at trapping and handling mink. Agreements with farmers should also include support for careful evaluation of damage to fences and support for immediate repairs.

In order to set up a system of rapid alert in case of new introductions or of expansions of the species into new areas, it would be important to involve key groups (i.e. angling association, public fish departments, etc.), producing and circulating informative documentations to enhance rapid identification of the species and providing contacts for collecting and evaluating observation data.

Eradication

Priority should be given to eradicate populations more likely to expand into new areas and to isolated populations. Particularly important should be the eradication of the small populations of the muskrat in north-eastern Italy, because this could prevent very high costs in the future. Such eradication should then be followed by a constant rapid detection and response system aimed at containing new arrivals of muskrats from Slovenia.

Eradication of the mink population recently discovered in Sardinia is particularly urgent, as the removal of this isolated population could prevent severe impact to the biodiversity of the island that hosts many important bird nesting areas and important endemic amphibian species. Mink populations recorded in northern and central Italy have not yet started to expand and the feasibility of local eradications should thus be evaluated by the competent local authorities and then rapidly enforced when appropriate.

The widespread distribution of the coypu in Italy makes an overall eradication an impracticable option. However, similarly to the case of the American mink, eradication of the isolated populations of Sardinia and Sicily is of critical importance for preventing high costs and severe impacts to the wetlands of these islands.

Control

When eradication is not feasible – as in the case of the coypu – the need and efficacy of applying permanent control actions should be evaluated. Control should not be started only as a response – because damage is recorded – prior to assessing the cost/benefit ratio of the activities. Control policies should be planned at an adequate, biologically sound spatial scale, taking into account the potential counteracting effects of immigration. Efforts should be concentrated in the most vulnerable areas in terms of biodiversity and human activities. Limiting factors for the species in southern Europe should be identified and management should be focused accordingly. Efficacy of management should be constantly evaluated, with particular reference to cost/benefit, and activities should be reviewed accordingly.

International cooperation

The case of the muskrat, colonizing Italy by its natural spread from the neighboring Slovenia, shows the importance of international coordination and cooperation. In parallel with the eradication of the Italian population, it would be important that Slovenia controls the larger population living in its territory and keep Italy informed on the expansion patterns occurring toward the border. On the other hand, the technical experiences gathered in Italy – for example, on the efficacy and cost/benefit of the coypu control – should be rapidly circulated to the other European countries sharing similar problems, as for example Spain (Panzacchi *et al.* 2007).

CONCLUSIONS

Freshwater ecosystems of Europe – hosting an important portion of the regional biodiversity – are particularly vulnerable to biological invasions. The implementation of more stringent policies for preventing and mitigating the impacts caused by non-indigenous species in these ecosystems is urgent and critical for preserving European biological diversity. Among freshwater non-indigenous species, semiaquatic mammals are particularly harmful. In this chapter, we have identified the main elements of a policy on the three species present in Italy – the coypu, the muskrat, and the American mink – all efficient invaders that threaten indigenous species and ecosystems, and affect human activities.

Priority should be given to prevention, by revising and implementing the legal framework, strictly regulating fur farms, banning farms in critical areas, and imposing more efficient fencing devices on farmers. Prompt eradication in case of escapees and removal of key isolated populations should be the basis of management. Eradication of the muskrat before it starts to spread is a priority for the near future. Involvement of the fur farmers is an important element for prevention and rapid response for new invasions. A better coordination with neighboring countries is another key element, as shown by the recommended eradication of the muskrat in north-eastern Italy that could be undermined by lack of action in Slovenia.

At the European level, it seems important to work toward a restriction of import and trade of all the three species here considered, taking into account that such a measure is legally justified on the basis of a science-based risk assessment.

The case of the semiaquatic non-indigenous mammals present in Italy provides a practical example of the elements to be developed in a national strategy on invasive species. The general approach described for the three non-indigenous mammals follows the guiding principles developed under the Convention on Biological Diversity, and the approach defined by the European Strategy on Invasive Alien Species: prevention of new unwanted introductions should be the first line of defense; prompt identification and early eradication are the best options when prevention fails; control or containment should only be undertaken when eradication is impracticable or fails.

ACKNOWLEDGEMENTS

We thank the editor for the invitation to contribute to this volume. Sandro Lovari and two anonymous referees provided helpful comments on previous drafts of the manuscript.

REFERENCES

Andreotti, A., N. Baccetti, A. Perfetti, M. Besa, P. Genovesi, and V. Guberti. 2001. Alien mammals and birds in Italy: survey, impacts on biodiversity and management guidelines. Quaderni di Conservazione della Natura, 2. Ministero Ambiente – Istituto Nazionale per la Fauna Selvatica, Italy, 189 pp. (In Italian with English summary)

Angelici, F. M., L. Luiselli, and L. Ruggiero. 1998. Preliminary data on the diet of a naturalised population of American mink (*Mustela vison domestica*) in Latium. Page 90 in Proceedings of the 2nd Italian Theriological Congress, Varese, Italy, 28–30 ottobre 1998. (In Italian)

Banks, P. B., M. Nordstrom, M. Ahola, and E. Korpimaki. 2005. Variable impacts of alien mink predation on birds, mammals and amphibians of the Finnish archipelago: a long-term experimental study. Proceedings of the IX International Mammalogical Congress, Sapporo, Japan.

Bertolino, S., A. Perrone, and L. Gola. 2005. Effectiveness of coypu control in small Italian wetland areas. Wildlife Society Bulletin **33**, 714–720.

Bomford, M. and P. O'Brien. 1995. Eradication or control for vertebrate pests? Wildlife Society Bulletin **23**, 249–255.

Bon, M., P. Paolucci, F. Mezzavilla, R. De Battisti, and E. Vernier. 1995. The Mammal atlas of Venetia. Lavori della Società Veneziana di Scienze Naturali, Supplemento **21**, 1–136. (In Italian)

Bonesi, L. and S. Palazon. 2007. The American mink in Europe: status, impacts, and control. Biological Conservation **134**, 470–483.

Boorman, L. A. and R. M. Fuller. 1981. The changing status of reedswamp in the Norfolk broads. Journal of Applied Ecology **18**, 241–269.

Carter, J., A. L. Foote, and L. A. Johnson-Randall. 1999. Modelling the effects of nutria (*Myocastor coypus*) on wetland loss. Wetlands **19**, 209–219.

Carter, J. and B. P. Leonard. 2002. A review of the literature on the worldwide distribution, spread of, and efforts to eradicate the coypu (*Myocastor coypus*). Wildlife Society Bulletin **30**, 162–175.

Cocchi, R. and F. Riga. 2001. Guidelines for the control of Coypu (*Myocastor coypus*). Quaderni di Conservazione della Natura, 5, Ministero Ambiente and Istituto Nazionale per la Fauna Selvatica, Italy, 189 pp. (In Italian with English summary)

Craik, C. 1997. Long-term effects of North American Mink *Mustela vison* on seabirds in western Scotland. Bird Study **44**, 303–309.

Danell, K. 1977. Short-term plant succession following the colonization of a northern Swedish lake by the muskrat, *Ondatra zibethica*. Journal of Applied Ecology **14**, 933–947.

Danell, K. 1996. Introductions of aquatic rodents: lessons of the muskrat *Ondatra zibethicus* invasion. Wildlife Biology **2**, 213–220.

Ehrlich, P. 1989. Attributes of invaders and the invading processes: vertebrates. Pages 315–328 in J. A. Drake, editor. Biological Invasions: A Global Perspective. Wiley, Chichester, UK.

Foote, A. L. and L. A. Johnson. 1993. Plant strand development in Louisiana coastal wetlands: coypu grazing effects on plant biomass. Pages 265–271 in Proceedings of the 13th Annual Conference of the Society of Wetland Scientists, New Orleans. Society of Wetland Scientists, Utica, NY.

Genovesi, P. 2000. Guidelines for eradication of terrestrial vertebrates: a European contribution to the Invasive Alien Species Issue. Council of Europe, tpvs65e-2000.

Genovesi, P. 2005. Eradications of invasive alien species in Europe: a review. Biological Invasions **7**, 127–133.

Genovesi, P. and C. Shine. 2004. European strategy on invasive alien species. Nature and environment, n. 137. Council of Europe Publishing, Strasbourg, France.

Gosling, L. M. 1989. Extinction to order. New Scientist 4 March 1989, 44–51.

Gosling, L. M. and S. J. Baker. 1989. The eradication of muskrats and coypus from Britain. Biological Journal of the Linnean Society **38**, 39–51.

Harrison, M. D. K. and R. G. Symes. 1989. Economic damage by feral American mink (*Mustela vison*) in England and Wales. Pages 242–250 in R. J. Putman, editor. Mammals as pests. Chapman & Hall, London, UK.

Hebridean Mink Project. 2006. Web site: http://www.snh.org.uk/scottish/wisles/intro.asp.

Lapini, L. and D. Scaravelli. 1993. Preliminary data on the muskrat *Ondatra zibeticus* (Linneo, 1766) in North-eastern Italy (Mammalia, Rodentia, Arvicolidae). Pages 249–252 in M. Spagnesi and E. Randi, editors. Proceedings of the VII Congress of the Association A. Ghigi for the Biology and Conservation of Vertebrates. Supplemento Ricerche della Biologia della Selvaggina 21, Bologna, Italy. (In Italian)

Lever, C. 1985. Naturalized mammals of the world. Longman, Harlow, Essex, UK.

Long, J. L. 2003. Introduced mammals of the world: their history, distribution and influence. CAB International, Wallingford, UK.

Lowe, S., M. Browne, S. Boudjelas, and M. De Poorter. 2000. 100 of the world's worst invasive alien species. A selection from the Global Invasive Species Database. Published by ISSG as special lift-out in Aliens 12, December 2000, 12 pp.

Michel, V., N. Ruvoen-Clouet, A. Menard, C. Sonrier, C. Fillonneau, F. Rakotovao, J. P. Ganiere, and G. Andre-Fontaine. 2001. Role of the coypu (*Myocastor coypus*) in the epidemiology of leptospirosis in domestic animals and humans in France. European Journal of Epidemiology **2**, 111–121.

Mitchell-Jones, A. J., G. Amori, W. Bogdanowicz, B. Kryštufek, P. J. H. Reijnders, F. Spitzenberger, M. Stubbe, J. B. M. Thissen, V. Vohralík, and J. Zima. 1999. The atlas of European mammals. Academic Press, London, UK.

Moore, N., P. A. Robertson, and J. N. Aegerter. 2000. Feasibility study into the options for management of mink in the Western Isles. CSL, MAFF, London, UK.

Macdonald, D. W., V. E. Sidorovich, E. I. Anisomova, N. V. Sidorovich, and P. J. Johnson. 2002. The impact of American mink *Mustela vison* and European mink *Mustela lutreola* on water voles *Arvicola terrestris* in Belarus. Ecography **25**, 295–302.

Nordström, M., J. Hogmander, J. Nummelin, J. Laine, N. Laanetu, and E. Korpimaki. 2002. Variable responses of waterfowl breeding populations to long-term removal of introduced American mink. Ecography **25**, 385–394.

Nordström, M., J. Hogmander, J. Nummelin, J. Laine, N. Laanetu, and E. Korpimaki. 2003. Effects of feral mink removal on seabirds, waders and passerines on small islands of the Baltic Sea. Biological Conservation **109**, 359–368.

Nummi, P. 1996. Wildlife introductions to mammal deficient areas: the Nordic countries. Wildlife Biology **2**, 221–226.

Nummi, P. 2002. Introduced semiaquatic mammals and birds in Europe. Pages 162–172 in E. Leppäkoski, S. Olenin, and S. Gollash, editors. European invasive aquatic species: distributions, impacts and management. Kluwer Academic Publishers, Dordrecht, The Netherlands.

Nummi, P., V. M. Väänänen, and J. Malinen. 2006. Alien grazing: indirect effects of muskrats on invertebrates. Biological Invasions **8**, 993–999.

Panzacchi, M., S. Bertolino, R. Cocchi, and P. Genovesi. 2007. Cost/benefit analysis of two opposite approaches to pest species management: permanent control of *Myocastor coypus* in Italy versus eradication in East Anglia (UK). Wildlife Biology **13**, in press.

Reggiani, G., L. Boitani, S. D'Antoni, and R. De Stefano. 1993. Biology and control of the coypu in the Mediterranean area. Supplemento Ricerche dei Biologi della Selvaggina, Bologna **21**, 67–100.

Saether, B. E. 1988. Pattern of covariation between life-history traits of European birds. Nature (London) **331**, 616–617.

Scaravelli, D. 2002. *Myocastor* problem: some considerations from the experience in the province of Ravenna. Pages 25–28 in Proceedings of the National Congress "The management of alien species in Italy; the case of the coypu and the red swamp crayfish". Firenze, Italy, 24–25 October 2002. (In Italian)

Sheail, J. 2004. The mink menace: The politics of vertebrate pest control. Rural History **15**, 207–222.

Sidorovich, V., H. Kruuk, and D. W. Macdonald. 1999. Body size, and interactions between European and American mink (*Mustela lutreola* and *M. vison*) in Eastern Europe. Journal of Zoology **248**, 521–527.

Sidorovich, V. and D. W. Macdonald. 2001. Density dynamics and changes in habitat use by the European mink and other native mustelids in connection with the American mink expansion in Belarus. Netherlands Journal of Zoology **51**, 107–126.

Simberloff, D. 2002. Today Tiritiri Matangi, tomorrow the world! Are we aiming too low in invasives control? Pages 4–12 in D. Veitch and M. Clout, editors. Turning the Tide: the Eradication of Invasive Species. IUCN SSC Invasive Species Specialist Group. IUCN, Gland, Switzerland/Cambridge, UK.

Sol, D., S. Timmermans, and L. Lefebvre. 2002. Behavioural flexibility and invasion success in birds. Animal Behaviour **63**, 495–502.

Spagnesi, M. and A. M. De Marinis, editors. 2002. Italian mammals. Quaderni della Conservazione della Natura, 14, Ministero dell'Ambiente – Istituto Nazionale Fauna Selvatica, Italy. (In Italian)

Tinarelli, R. 2002. The impact of the coypu in the wetlands of Emilia Romagna and some considerations on the measures for its control. Pages 39–40 in Proceedings of the National Congress "The management of alien species in Italy; the case of the coypu and the red swamp crayfish". Firenze, Italy, 24–25 October 2002. (In Italian)

Tufto, J. 2001. Effects of releasing maladapted individuals: a demographic-evolutionary model. American Naturalist **158**, 331–340.

Vázquez, D. 2005. Exploring the relationship between niche breadth and invasion success. Pages 307–322 in M. W. Cadotte, S. M. McMahon, and T. Fukami, editors. Conceptual ecology and invasions biology: reciprocal approaches to nature. Springer, Dordrecht, The Netherlands.

Velatta, F. and B. Ragni. 1991. The coypu population in Lake Trasimeno. Population parameters and numerical control. Pages 311–326 in M. Spagnesi and S. Toso, editors. Atti del II Congresso Nazionale dei Biologi della Selvaggina, Supplemento Ricerche di Biologia della Selvaggina **19**. (In Italian)

Wittenberg, R. and M. Cock. 2001. Invasive Alien Species: a Toolkit of Best Prevention and Management Practices. GISP/CAB International, Wallingford, UK.

Chapter ten

Invasions by plants in the inland waters and wetlands of Africa

Geoffrey W. Howard and Florence W. Chege

INTRODUCTION

Africa is a large continent (near 30 million km^2) with 53 countries including several island States. There are many hundreds of small rivers and lake basins within those countries, while most of the large drainage basins include at least two countries – up to as many as 13 (UN/Water Africa 2006). Freshwater wetlands are similarly widespread and variable in size with many of the larger swamps and floodplains extended into at least two countries. A classic example is the Nile River Basin, which encompasses parts of ten countries in eastern Africa, the Horn of Africa, and northern Africa. It includes lakes and impoundments at all altitudes, and has a great variety of wetland types, many of which stretch across those national boundaries (Howard 2007).

Most of Africa is tropical, being between the two tropics at 32° latitude North and South. North African countries on the Mediterranean coast are more temperate as are parts of South Africa and its near neighbours. Despite a great variety of rainfall patterns and ecosystem types, there are very few natural barriers to the movement of aquatic plant propagules between the States of Africa (apart from the island States), which is significant when describing and discussing the distribution and spread of invasive species. Some species of tropical non-indigenous invasive plants have become widespread in the waters and wetlands of Africa as a result of both natural and man-enhanced movement. While introductions of non-indigenous species (NIS) to Africa have most

Francesca Gherardi, Biological invaders in inland waters: Profiles, distribution, and threats, 193–208.

commonly been brought about by intentional imports, trade, travel, transport, and tourism, their spread across Africa has often been by the natural movement of aquatic plants within the major drainage basins – after interbasin transfers by a variety of means. The political boundaries of nation States are hardly barriers to this spread so that the likelihood of countries being able to prevent spread is limited, once a potentially invasive freshwater species becomes established within tropical Africa (Howard 2004). Temperate species are able to spread at the northern and southern ends of the continent, but the mainly tropical areas in between have kept these two extremes largely separate with regard to aquatic plant invasions by mainly temperate species. However, climate change is beginning to affect African aquatic systems such that temperate and tropical areas are less separated – which increases the chances of colonization by invasive species across previously separated systems.

This introduction describes the background to invasions by non-indigenous plants in the fresh waters and wetlands of Africa in comparison to those in Europe. Similar movement of species is enabled by drainage systems between countries in Europe, but where water systems are more managed and where species are seen as non-indigenous if they come from outside national borders. The concept of "alien invasive species" in Africa is further confounded by increasing examples of invasions by plants that are truly indigenous to the waters and wetlands of some part or parts of Africa. These will be treated somewhat separately below, but there is need for use of the concept of "nativeness" as referring to an ecosystem or habitat rather than to a country within the waters and wetlands of Africa. Increasingly of late, there are more examples of waters and wetlands being invaded by plants that are indigenous to Africa – but not necessarily indigenous to a particular country or ecosystem within Africa. Within a country or even an ecosystem, it is clear that, when a system is modified by human intervention or natural disaster, species that are indigenous to that system can become invasive because of changes in the nature of that system. Thus, when African water and wetland systems are significantly affected by changes in water regime, changes in water quality, changes in vegetation and herbivorous fauna, or changes in topography and landscape, plant species that were previously indigenous and part of the native vegetation may react to the disturbance and so become invasive. We will describe some examples of this later.

TYPES OF PLANTS THAT BECOME INVASIVE

The term "aquatic plants" is deliberately not used in this context as we wish to place emphasis on the ecosystems invaded rather than on the taxonomy or physiology of the plants concerned. Higher plants (Spermatophyta) that inhabit fresh waters in tropical Africa range from facultative aquatics (hydrophytes) that are totally dependent upon the presence of freshwater and are unable to survive without it, to those that rely upon (or tolerate) inundation for some part of the annual cycle. Thus, we have followed the system of Mitchell (1985a) and divided the plants that are the subject of this discussion into three broad types: floating plants, emergent plants, and submerged plants with the middle group including some species that are occasionally dependent upon (or tolerant of) inundation for their growth and spread. Few species are entirely limited to one category as they often have habitat adaptations to allow survival with changing water availability. For example, the notorious water hyacinth, while being primarily a floating plant, can survive as a rooted plant in very shallow water or with only subsurface inundation. These categories are more useful for the consideration of detection and for the management of invasions than as tight biological boundaries.

The range of species considered describes the range of macrophyte types, situations, threats, and management challenges for species that invade the waters and wetlands of Africa, and is not meant to be all-inclusive. Tables 1 and 2 list the species mentioned with some records of their range or potential as invaders. Information about the plant species that invade African waters is limited and skewed in distribution across species, countries, and ecosystems. The most widespread invaders have an extensive literature that describes their distribution and impacts and sometimes their control, while others are barely known and so are infrequently mentioned. Some countries, notably South Africa, have many data and publications about invasive plants with excellent references for recognition and management. Other countries are beginning to assemble national listings and distributions of invasive species but have published little. Some information is available from international publications and websites, but there are few that focus on the continent of Africa.

Information for Tables 1 and 2 and for the discussion has been derived mainly from personal observations, records, and communications over the last 30 years. Additional information has been derived from Howard (1985), Mitchell (1985a), Pieterse and Murphy (1990), Cronk and Fuller (2001), Henderson and Cilliers (2002), Howard and Matindi (2003), Weber (2003), and the following websites: Aquatic, Wetland and Invasive Plant Database, Global Invasive Species Database, Global Invasive Species Information Network, Smithsonian Environment Research Center – Aquatic Invasions Research Directory. These tables include species that are regarded by the authors (and others) as invasive and do not include all those that are known in Africa to be "aquatic weeds" in general or in specific situations relating to agriculture, water management and use, human health, and development.

FLOATING PLANTS

Mitchell (1985b) used the term "obligate acropleustopyhte" to typify the floating water plants that cannot grow well in other situations and listed ten or so species indigenous to Africa as well as four genera of the "duckweeds"

Table 1 Major non-indigenous invasive species in the waters and wetlands of Africa. Data from listed websites, personal communications, and observations, 1975–2006.

Invasive species and type	Higher plant family	Origin	Records in Africa as invasive	Recorded in Africa as potentially invasive
Floating plants				
Eichhornia crassipes	Pontederiaceae	South America	Africa-wide	
Pistia stratiotes	Araceae	South America	Africa-wide	
Salivinia molesta	Pteridophyta Salviniaceae	South America	Africa-wide	
Azolla filiculoides	Pteridophyta Azollaceae	South America	Africa-wide	
Emergent plants				
Mimosa pigra	Leguminosae Mimosoidea	Central and South America	Flood plains and riverine wetlands in tropical Africa	Riverine habitats from where it can move to flood plains
Myriophyllum aquaticum (Vell.) Verdc.	Haloragaceae	South America	South Africa, (eastern Africa)	Many wetlands and water gardens across Africa
Myriophyllum spicatum Linnaeus	Haloragaceae	Asia, Europe, North Africa	South Africa	Central Africa
Pontederia cordata Linnaeus	Pontederiaceae	North, central South America	South Africa	Southern and eastern Africa

Sesbania punicea	Leguminosae Papilionoidea	South America	Southern Africa	Eastern Africa
Arundo donax Linnaeus	Poaceae	Mediterranean	South Africa	Eastern Africa
Canna indica	Cannacea	Pan-tropical (Asia, Americas now in Africa)	South Africa	Eastern and southern Africa
Ipomoea carnea fistulosa (Mart. Ex Choisy) D.F.Austin	Convolvulaceae	Central and South America	South Africa	
Rorippa nasturtium-aquaticum (L.) Hayek = *(Nasturtium officinale)*	Brassicaceae	Europe	South Africa	Central, eastern and southern Africa
Submerged and floating-leaved				
Hydrilla verticillata (L.f.) Royle	Hydrocharitaceae	Asia, Americas, now Africa		North and southern Africa
Hydrocleys nymphoides	Butomaceae	South America		Eastern Africa
Ceratophyllum demersum Linnaeus	Ceratophyllaceae	Worldwide (Tropical and temperate)		West, central, southern, and eastern Africa

Table 2 Examples of indigenous African plants that are, or may become, invasive in the waters and wetlands of Africa. Data from Henderson and Cilliers (2002), personal communications, observations and records 1975–2006.

Species or genera and type	Higher plant family	Origin	Records in Africa as invasive	Recorded in Africa as potentially invasive
Floating plants				
Lemna Linnaeus., *Spirodela* Schleid, *Wolffia* Horkel ex Schleid *Pseudowolffia* Hartog and Plas, *Wolffiella* (Hegelm) Hegelm	Lemnaceae "Duckweeds"	Africa	Africa	Africa
Emergent plants				
Hydrocotyle ranunculoides L. f.	Apiaceae (Umbelliferae)	Africa (naturalized in USA and Europe)	Isolated invasions in eastern and southern Africa	Africa
Vossia cuspidata	Poaceae	Tropical Africa	Tropical Africa	
Typha spp	Typhaceae	Tropical and temperate Africa	Tropical and temperate Africa	
Aeschynomene elaphroxylon	Leguminosae Papilionoidea	Africa	West Africa	Africa
Ipomoea aquatica	Convolvulaceae	Pan-tropical	Africa-wide	Central, West, southern and eastern Africa
Submerged plants				
Najas horrida	Najadaceae	Africa-wide	(Uganda)	Southern and eastern Africa

(Lemnaceae). He also listed the four infamous pan-tropical invasive species [*Eichhornia crassipes* (Mart.) Solms, *Pistia stratiotes* Linnaeus, *Salvinia molesta* D. S. Mitch., and *Azolla filiculoides* Lam., see Table 1], which all probably originated in Central/South America. One of these, *P. stratiotes*, has been known in Africa for centuries and has local names and traditional uses in some parts of the continent (hence its occasionally-used common name of "Nile cabbage"). Nevertheless, *P. stratiotes* is regarded as an NIS in Africa and has spread across the watery ecosystems of the continent as aggressively as the other three. All four of these pan-tropical floating invasive plants are present in most of the larger (and many of the smaller) river systems and lakes of Africa. However, some water bodies are still free of all or some of these (e.g. the Okavango delta in Botswana and, until very recently, Lake Tanganyika), so there is a need for prevention of further spread of these four species. For many years, there was a search for "who introduced water hyacinth", for example, to a particular system such as the Congo River or Lake Victoria. Latterly, it is becoming clear that most of these species have been on the continent for many decades and that their spread can no longer be attributed to a person, a process, or a date. Low density and isolated occurrences of these species have been noticed for many years in the low nutrient waters of some wild rivers, crater lakes, and rift valley lakes where they could not be regarded as invasive. But in the last few years, with increased nutrient input from feeder streams and surface run-off, they have become more obvious as they increase in density and impact – becoming potentially invasive. Further, the growth of all of these species near human habitation, towns, and cities has been enhanced by the entry of untreated (or partly treated) sewage, industrial wastes, and urban run-off after storms which have made the four infamous floating invasive species more noticeable.

For many years, the passage of floating *E. crassipes* and *P. stratiotes* plants down the Zambezi River upstream of the Victoria Falls was part of the charm of the scenery. The same is true of other African rivers, such as the Awash River in Ethiopia and the Pangani and Rufiji Rivers in Tanzania. Now, these are seen as sources of invasions downstream and of local infestations in backwaters and slow-flowing tributaries which are "fuelled" by agricultural, urban, and industrial wastes. Very few of the main river and lake systems in continental Africa and its island States are free of all four of these species – and the same is true of the larger wetlands. For example, in Somalia, *P. stratiotes* is present in the two large rivers that flow from Ethiopia – the Shebelle and the Jubba, but *E. crassipes* is not recorded in the most recent literature from that country (Thulin 2006). However, *E. crassipes* is present in several lakes and wetlands of Ethiopia and it is likely that it will be found in Somalia in the near future – as it has been in so many other freshwater systems across Africa.

Indigenous floating plant species in Africa are few (although there are many species of Lemnaceae in Africa, some of them pan-tropical) and most are small compared to the size that can be attained by the four major "alien invasive species" when conditions of temperature, humidity, and nutrients are in their favour. For instance, G. Howard has recorded water hyacinth plants in a nutrient-rich bay of Lake Victoria that were 2 m above water and 1 m below; he also recorded water lettuce plants 60 cm in diameter in a nutrient-rich bay of Lake Baringo in Kenya. Thus, in a sense, the floating water plant habitat/ niche is empty of large indigenous species and so most easily exploited by non-indigenous plants – like the four mentioned above.

The impacts of the four main non-indigenous floating species are well-known (e.g. Mitchell 1985a, Howard and Harley 1998, Howard and Matindi 2003) as they affect previously open waters, often covering the surfaces of lakes, ponds, dams, swamps, and even (slow-flowing) rivers with disastrous effects on biodiversity, water use, and management and people's access – as well as human health and economies. The smaller, native floating plants can become weeds under certain conditions (e.g. farm dams and irrigation run-off canals) and have the potential to become invasive – but they have rarely done so to date (Table 2).

EMERGENT PLANTS - INCLUDING THOSE ON SEASONALLY-FLOODED FLOOD PLAINS

This group of plants is very diverse in life form and size in Africa, ranging from very small sedges (Cyperaceae) to very tall reeds, shrubs, and trees. All are rooted in soil which is inundated, either above or below ground, at some time in a year. As the name implies, emergent species have aerial parts that extend beyond the water surface (Mesléard and Perennou 1996) (those that have leaves that float on the water surface are considered in the next section). They include herbs, grasses and sedges, reeds, shrubs, and trees, and occupy the margins of lakes and rivers as well as forming swamps and marshes in many wetland types. This group also includes those plants that are tolerant of seasonal flooding and is made up of representatives of many different plant families. A description of the range of plants in this group in Africa is given by Thompson (1985) and the main NIS listed in Table 1, while those that are indigenous are in Table 2.

In this group, there is less impact and perhaps less severity of invasion compared to the floating plants as the indigenous vegetation leaves little opportunity for the expansion of new species. However, the spread of catchment mismanagement and other forms of habitat disturbance (water level and flow changes, water quality changes, and wetland conversion) have led to several indigenous and NIS becoming serious threats to biodiversity and development.

One non-indigenous flood plain and riparian shrub species is worthy of mention as it is steadily spreading and creating serious problems in many wetlands in Africa. This is *Mimosa pigra* Linnaeus (the "giant sensitive plant") which originates from central and southern America but which has been known in Africa for at least two centuries (and has local names and associated beliefs in some areas). *Mimosa pigra* has been relatively widespread although in low density in the past, but is now appearing as a serious invader of flood plains and riverine wetlands in many parts of the continent. It develops dense thickets that can cover both permanent and seasonal wetlands, as it has done in northern Australia and South-east Asia (Julien *et al.* 2004, Triet *et al.* 2004), and in doing so it excludes many large vertebrates (including livestock) and most other wetland plants and completely impedes access and passage in former open waters and plains.

The (four) indigenous species of *Typha* Linnaeus (Typhaceae, the Cattail family, the bulrushes, or reed maces) are notable as they are also spreading with disturbance and the development of water systems (water supply, irrigation, and hydropower development) in previously dry areas. These reeds are essentially swamp plants that are only otherwise found on the edges of lakes and slow-flowing streams. However, they have the capacity to produce millions of wind-dispersed seeds which can germinate and survive in even small sources of water and then develop into one-species reed swamps as long as water is available. Unlike most European reeds, bulrushes in tropical Africa grow throughout the year and can have a great impact on other wetland vegetation. *Typha domingensis* Pers. is capable of aggressive competition with other reeds (such as papyrus, *Cyperus papyrus* Linnaeus, and *Phragmites* Adans. spp.) and can dominate previously stable swamp communities as a result of water level or water quality changes. It is very tolerant of increases in salinity and other dissolved salts in freshwater as well as being able to withstand changes in water levels – even to the point of becoming dried out for several months. Other local species of reeds and other swamp and lake vegetation cannot compete in such situations and so *Typha* tends to "take over", to the detriment of both larger and smaller wetland plants as well as the fauna that depends on them.

Several species of non-indigenous stream-side and lake-side plants are able to withstand flooding and so can become invasive as they can occupy a catena of inundation from almost dry to almost completely submerged. Some of these are often not seen as primarily wetland species but can nevertheless take on that role under some circumstances and dominate other wetland and wetland-edge vegetation. Examples are the non-indigenous wild canna (*Canna indica* Linnaeus) and *Sesbania punicea* (Cav.) Benth., and the indigenous *Aeschynomene elaphroxylon* (Guill. and Perr.) Taub. One tropical indigenous (and pan-tropical) scrambling plant that can become invasive is *Ipomoea aquatica* Forssk. This species is rooted in the soil and scrambles on other aquatic vegetation as well as floating on and above the water surface. While it is seen as invasive in some situations, it is prized in others as a valuable green vegetable.

SUBMERGED AND FLOATING-LEAVED PLANTS

Denny (1985) refers to this group of wetland plants as "euhydropytes" because of their complete reliance on water and gives examples of the life forms and taxa found in Africa. All are rooted or anchored in the substrate of a wetland, lake, or river and may have leaves floating below or upon the water surface. They are typified by the "pond weeds" and water lilies, although there are many species in Africa from the algae, liverworts, mosses, ferns, and angiosperms. Many are local weeds, especially in agricultural and water supply situations, but few are invasive.

The NIS of *Hydrilla* Rich. and *Ceratophyllum* Linnaeus (see Table 1) are now well-established in African waters; they can cause problems in both still and running waters and can be regarded as invasive. A potential invasive species is *Hydrocleys nymphoides* (Willd.) Buchen. (from South America), which has been established as a horticultural decorative plant in some tropical water gardens and which is occasionally used in constructed wetlands for wastewater treatment. The local species of water lilies (*Nymphaea* Linnaeus spp., *Nymphoides* Hill spp.) can cause local weedy problems but are not considered as invasive and are valuable as both food and shelter for a wide range of wetland animals. A recent report of the indigenous, submerged *Najas horrida* A. Braun and Rendle (Table 2) becoming invasive in a lake and riverine situation in eastern Africa may require further attention. A concern with the truly submerged plants is that they are not noticed as much as others and often become problems that could have been addressed if they had been identified sooner.

IMPACTS OF WETLAND/WATER INVASIONS BY PLANTS

Invasive species in freshwater ecosystems have negative impacts on biodiversity as well as on human development and human well-being, including health. The relationship between invasive plants in Africa and biodiversity in freshwater ecosystems is quite complex, rarely researched, and so not well-understood. Most available information is about the effects of aquatic "weeds" (e.g. Mitchell 1985a, Pieterse and Murphy 1990) on water and wetland systems, while accounts of the impacts of invasive species on biodiversity are often anecdotal. Floating species are accused of reducing light below them, reducing oxygen in water, competing with other water plants for space, increasing evapotranspiration from water bodies, altering water flows, and causing changes in wetland vegetation communities (Howard and Harley 1998, Navarro and Phiri 2000, Howard and Matindi 2003). All types are capable of changing the structure of aquatic vegetation through competition above and below water, often to the detriment of species diversity both of plants and the animals that depend on them for food, shelter, and breeding sites.

Specific examples of changes include the alteration of vegetation patterns following some invasions of water hyacinth and the other non-indigenous floating species. There are situations where these invasive floating water plants provide platforms for the growth of other plants that would otherwise not be able to survive on the surface of open waters. Clumps of water hyacinth can be bound together by semiaquatic sedges and grasses to form larger floating mats upon which even terrestrial species can survive and spread as the mats move with water currents and wind. Adams *et al.* (2002) described the formation of mats involving *S. molesta* and *E. crassipes* on Lake Naivasha in Kenya and their association with other organisms. Similar mat formation is sometimes seen with *S. molesta* and *P. stratiotes* and, in some situations, all three combine to together to form multiple species floating mats.

There is a special relationship between water hyacinth and "hippo grass" [*Vossia cuspidata* (Roxb.) Griff], itself an occasional (indigenous) invasive species. *Vossia cuspidata* is a widespread wetland species that is rooted at the edges of lakes, rivers, and wetlands and which has stems that stretch out across the water surface – floating as far as their buoyancy will allow. While *V. cuspidata* is usually restricted in the distance over which its stems can reach, this changes when *E. crassipes* forms mats that become "anchored" by the hippo grass, which then uses their buoyancy to stretch further out into open water. This has been seen in many situations, particularly at river edges, and can result in a complete coverage of the water surface from one river bank to the other with a mat that is anchored by the grass roots. This can then slow or block the river flow as well as encouraging other plants to establish over the water surface and change the flowing river to a slow-moving swamp or sudd. Waterbird, fish, and aquatic invertebrate assemblages may be altered as a result and fisheries, transport, and water use affected. Similar situations are possible with *I. aquatica* and the floating water plants as well as with other emergent species. Often the submerged plants are placed at a greater disadvantage by these combinations than they are with one or the other invasive types. Changes in the vegetation patterns of submerged species are even less well-documented (partly because they are "out of sight"). These can be altered by the floating invasive plants, by the emergent plants, and by other submerged species. Changes in the submerged and emergent vegetation of Lake Naivasha in Kenya over several decades were described by Harper *et al.* (1990), who ascribed these changes to (non-indigenous) invasive freshwater crayfish [*Procambarus clarkii* (Girard)] as well as to an infestation of *S. molesta* and changes in climate. Together with several species of introduced (non-indigenous) fish and coypu [*Myocastor coypus* (Molina)], the crayfish and floating plants (including the more recent infestations of water hyacinth) have wrought significant changes to the ecology and utility of Naivasha as a centre of tourism, commercial fisheries, irrigated agriculture, and conservation.

Other impacts of the range of invasive water plants on African waters and wetlands are listed by, e.g. Howard and Harley (1998), Navarro and Phiri (2000), Howard and Matindi (2003), and include:

- Direct effects on water flows and availability – affecting water supply, hydropower generation, and irrigation
- Changes in water quality with impacts on household, agricultural, and industry water supply

- Blockage of water supply and drainage channels, and alterations to water storage facilities
- Alteration of access of wildlife and livestock to water and water pastures
- Blockage of peoples' access to water and transport on water
- Alteration of fisheries in diversity and yields
- Encouragement of intermediate hosts of human diseases such as malaria and schistosomiasis
- Provision of habitats for venomous animals and predators such a crocodiles

In some cases, there are benefits to be gained from the presence of invasive water plants in Africa (for human food, livestock food, mulch, biogas, fibre, fuel, and, occasionally, enhanced fisheries). However, in most cases, the negative impacts outweigh the benefits unless the introduced species can be proactively managed to ensure the opposite.

There is a dearth of information and published accounts on the economic impacts of invasive plants on the aquatic systems of Africa – apart from those associated with water hyacinth (e.g. Goodland 1995). This arises partly from a lack of awareness of the threats posed by invasive plants and partly from available expertise and experience to make the necessary economic assessments. Absence of this important information further holds back awareness about the threat of invasive species by those decision-makers and policy developers who are likely to be influenced by economic arguments.

ADDRESSING THE IMPACTS AND THREATS OF INVASIVE WATER PLANTS

Across Africa, there is a general lack of available information and awareness products about invasive species in general. Water hyacinth is quite widely known as are its threats and benefits, but other invasive water plants are hardly known in most countries. Several countries have begun to address this at national level, while there are emerging responses at regional levels. The New Partnership for Africa's Development (NEPAD) has developed a continent-wide strategy to address environmental problems (NEPAD 2003), which includes a sector specific to "alien invasive species" and their management (UNEP 2003). This strategy is now being further expanded at sub-regional level with the development of NEPAD Sub-regional Environment Actions Plans – each of which in the five sub-regions of Africa has a section on invasive species. The African Convention on the Conservation of Nature and Natural Resources (see IUCN 2004) is a recent revision of a 1968 agreement between African States to conserve biodiversity: it has a specific paragraph on the strict control of introduction of NIS and promotes the eradication of those that have become established. International agencies, such as the Global Invasive Species Programme, United Nations Environment Programme (UNEP), CABI Africa, and IUCN are developing and distributing information and awareness materials, but it will be some time before the whole continent is aware of the nature and extent of the invasive species problem in fresh waters and has the information to address it.

There are techniques available for the management of some of the invasive plants in freshwater systems and this is being addressed in some countries at national level. However, the distribution of large rivers, lakes, and wetlands, as mentioned above, crosses many national boundaries. Thus there is need for sub-regional and international cooperation in this endeavour and a network of experts and expertise that can eradicate, control, or manage invasive plants in these vast international ecosystems (the Nile, for example, is the longest river in the world and its drainage basin covers ten countries). Here it is institutions like the Africa Union and NEPAD, which can play a part as well as the five sub-regions and their economic and development commissions. Basic to these solutions, however, is the concept of the "ecosystem approach" to address the threats of invasive species. This is beginning to take effect as proposed by the Convention on Biological Diversity. The ecosystem approach is especially helpful when deciding upon management actions across borders and in ecosystems where the objectives for that management are the restoration of invaded ecosystems rather than solely the eradication of the invasive species. Eradication of invasive water plants is especially difficult (if not impossible) as all methods – apart from specific biological control – have impact on other organisms and uses of the waters. Eradication is certainly not attainable where invasive plants have become established in complex ecosystems that are part of, or connected to, other similar habitats as in the water systems of Africa. In these circumstances, only management (to reduce populations of invaders and their impacts) is possible or feasible. Biocontrol is a useful tool (where available) to manage invasions but it cannot be expected to eradicate invasive species in the waters and wetlands of Africa or anywhere else. Mechanical control is feasible in some situations as an option for management of invasive plants, but the rate of growth and spread of many species in the tropics is so fast that it is not usually a sustainable solution. Chemical control (using "weedicides") is possible, but most often it has negative effects on non-target species and is ineffective in large and moving water bodies because of dilution effects.

The spread of invasive species in water systems and wetlands in Africa can be addressed by regional cooperation if there is agreement to try to prevent "out-of-basin transfers" of both water and species. Natural spread of invasive species by water currents and winds within lakes and river systems is inevitable, so different types of cooperation between States are necessary when they share a river basin or a lake basin. Joint programmes for management of shared ecosystems are becoming more possible as regionality is explored, so that cooperation in addressing the management of invasive species in Africa's waters and wetlands should become more likely in the near future. At the same time, research at government institutions and universities is now including the dynamics and impacts of plant invasions – so there is hope for the future that the knowledge gap will be narrowed, especially if support from such organisations as the European Commission becomes available.

CONCLUSIONS

Invasions by plants in the inland waters of Africa are a significant threat to biodiversity and human development. Both non-indigenous and indigenous species are involved in all aspects of the freshwater environment – both above, upon, and below the water surface. Detailed information and research is lacking in many cases and needs to be supported in the future so that the species and their impacts are better known and so that the ecosystems affected can be restored. Economic analyses of the impacts of invasive plants on biodiversity and development in Africa need to be developed, published, and distributed, so that these threats can be appreciated by those who develop policy and make national and regional decisions. Any such information needs to be disseminated across Africa, so that international and drainage basin cooperation can be enhanced in the management of invasive species – within and between countries. The ecosystem approach to understanding the threats, impacts, and solutions to freshwater invasions is likely to be the most fruitful and can take into account the cross-border nature of many of the waters and wetlands of the continent and their catchments. This must be enhanced, however, by cooperation between the national, sub-regional, and continental agencies of government and civil society to ensure that the spread and management of invasive water plants is effective. Europe has developed an invasive species strategy and has many experts and much technical expertise: perhaps there are grounds for more cooperation between Africa and Europe in this arena of conservation and development. African institutions are trying to address this problem and awareness is growing about freshwater invasions – but it may take some time before continent-wide solutions are in place.

ACKNOWLEDGEMENTS

This review would not have been possible without the support of institutions that allowed the authors to travel widely in Africa to see and understand invasive plants in the waters and wetlands across the continent and in some of its island States. These institutions include the University of Zambia in Lusaka and the World Conservation Union (IUCN). The generous support of the organizers and sponsors of the International Workshop on Biological Invasions in Inland Waters, especially Dr. Francesca Gherardi, are gratefully acknowledged for encouraging these ideas to be presented in Firenze and subsequently published here. The Invasive Species Specialist Group of the IUCN Species Survival Commission had been especially helpful as have our colleagues in CABI Africa.

REFERENCES

Adams, C. S., R. R. Boar, D. S. Hubble, M. Gikungu, D. M. Harper, P. Hickley, and N. Tarras-Wahlberg. 2002. The dynamics and ecology of exotic tropical species in floating plant mats: Lake Naivasha, Kenya. Hydrobiolgia **488**, 115–2002.

Cronk, Q. C. B. and J. L. Fuller. 2001. Plant invaders. The threat to natural ecosystems. Earthscan, London, UK.

Denny, P. 1985. The structure and functioning of African euhydrophyte communities. The floating-leaved and submerged vegetation. Pages 125–151 in P. Denny, editor. The ecology and management of African wetland vegetation. A botanical account of African swamps and shallow waterbodies. Dr W. Junk Publishers, Dordrecht, The Netherlands.

Goodland, R. 1995. Uganda: owen falls hydroproject: an environmental reconnaissance of the water hyacinth problem. Environment Department Report, World Bank, Washington, DC, 23 pp.

Harper, D. M., K. M. Mavuti, and S. M. Muchiri. 1990. Ecology and management of Lake Naivasha, Kenya, in relation to Climate Change, Alien Species Introductions, and Agricultural Development. Environmental Conservation **17**, 329–335.

Henderson, L. and C. J. Cilliers. 2002. Invasive aquatic plants. A guide to the identification of the most important and potentially dangerous invasive aquatic and wetland plants in South Africa. Plant Protection Research Institute Handbook No.16, Agricultural Research Council, Pretoria, South Africa.

Howard, G. W. 1985. The Kafue Flats of Zambia – a wetland ecosystem comparable with floodplain areas of northern Australia. Proceedings of the Ecological. Society of Australia **13**, 293–306.

Howard, G. W. 2004. Invasive species in water-dependent ecosystems. Pages 22–26 in M. V. Gupta, D. M. Bartley and B. O. Acosta, editors. Use of genetically improved and alien species for aquaculture and conservation of aquatic biodiversity in Africa. World Fish Centre Conference Proceedings, 68 pp.

Howard, G. W. 2007. Wetlands of the Nile River Basin. Hydrobiologia, in press.

Howard, G. W. and K. L. S. Harley. 1998. How do floating aquatic weeds affect wetland conservation and development? How can these effects be minimized? Wetlands Ecology and Management **5**, 215–225.

Howard, G.W. and S. W. Matindi. 2003. Alien invasive species in Africa's wetlands. Some threats and solutions. IUCN Eastern Africa Regional Programme, Nairobi, Kenya. (also in French as Les Espèces étrangères envahissantes dans les zones humides de l'Afrique).

IUCN. 2004. An introduction to the African Convention on the conservation of nature and natural resources. IUCN, Gland, Switzerland and Cambridge, UK.

Julien, M., G. Flanagan, T. Heard, B. Hennecke, Q. Paynter, and C. Wilson. 2004. Research and management of *Mimosa pigra*. CSIRO Entomology, Canberra, Australia.

Mesléard, F. and C. Perennou. 1996. Aquatic emergent vegetation. Ecology and management. Conservation of Mediterranean wetlands, No. 6, Tour de Valat, Arles, France, 86 pp.

Mitchell, D. S. 1985a. African aquatic weeds and their management. Pages 177–202 in P. Denny, editor. The ecology and management of African wetland vegetation. A botanical account of African swamps and shallow waterbodies. Dr W. Junk Publishers, Dordrecht, The Netherlands.

Mitchell, D. S. 1985b. Surface-floating aquatic macrophytes. Pages 109–124 in P. Denny, editor. The ecology and management of African wetland vegetation. A botanical account of African swamps and shallow waterbodies. Dr W. Junk Publishers, Dordrecht, The Netherlands.

Navarro, L. and G. Phiri. 2000. Water Hyacinth in Africa and the Middle East. A Survey of Problems and Solutions. IDRC, Ottawa, Canada.

NEPAD, 2003. Action plan for the environment initiative of the New Partnership for Africa's Development (NEPAD). African Union, NEPAD, UNEP and GEF.

Pieterse, A. H. and K. J. Murphy. 1990. Aquatic weeds. The ecology and management of nuisance aquatic vegetation. Oxford University Press, Oxford, UK.

Thompson, K. 1985. Emergent plants of permanent and seasonally-flooded wetlands. Pages 43–107 in P. Denny, editor. The ecology and management of African wetland vegetation. A botanical account of African swamps and shallow waterbodies. Dr W. Junk Publishers, Dordrecht, The Netherlands.

Thulin, M. 2006. Flora of Somalia. volume 3. Royal Botanic Gardens, Kew, Richmond, UK.

Triet, T., L. C. Man, and N. P. Nga. 2004. Impacts of *Mimosa pigra* on native plants and soil insect communities in Tram Chim National Park, Vietnam. Pages 22–27 in M. Julien, G. Flanagan, T. Heard, B. Hennecke, Q. Paynter, and C. Wilson C, editors. Research and management of *Mimosa pigra*. CSIRO Entomology, Canberra, Australia.

UNEP. 2003. Development of an action plan for the environment initiative of the New Partnership for Africa's Development (NEPAD). Programme Area 3. Prevention, Control and Management of Invasive Alien Species. UNEP, Nairobi, Kenya.

UN/Water Africa. 2006. African Water Development Report, 2006. UNECA, Addis Ababa, Ethiopia. 370 pp. Compact Disk.

Weber, E. 2003. Invasive plant species of the world. A reference guide to environmental weeds. CAB International, Wallingford, UK.

Databases:

Aquatic, Wetland and Invasive Plant Database. www.plants.ifas.ufl.edu

Global Invasive Species Database. www.issg.org/database

Global Invasive Species Information Network. www.gisinetwork.org

Smithsonian Environment Research Center, Aquatic Invasions Research Directory. http://invasions.si.edu/aird.htm

Chapter eleven

Non-indigenous aquatic and semiaquatic plant species in France

Gabrielle Thiébaut

INTRODUCTION

The invasion of natural communities by introduced plants constitutes one of the most serious threats to biodiversity (Heywood 1989). What is the current situation in France? What do we know about these invasions and their consequences? What measures have been implemented to manage non-indigenous plant species populations? To respond to these questions, the French Ministry for Ecology and Sustainable Development has supported various biological invasion research projects ("INVABIO") and the National Museum of Natural History entrusted Muller *et al.* (2004) to evaluate plant species invasiveness in metropolitan France.

While the invasibility of riparian plants communities, patterns, and causes of river corridor invasion by non-indigenous plant species have been studied (Planty-Tabacchi *et al.* 1995, 2001, Tabacchi and Planty-Tabacchi 2002), significantly less attention has been paid to the introduced aquatic and semi-aquatic plants. Biological invasion research in freshwater systems has focused on a few plants such as *Fallopia* taxa (Schnitzler and Muller 1998, Bailey and Schnitzler 2003), *Elodea* species (Thiébaut *et al.* 1997, Barrat-Segretain 2001, 2004, 2005, Barrat-Segretain *et al.* 2002, Greulich and Trémolières 2002) or *Ludwigia* sp. (Dutartre and Oyarzabal 1993, Dutartre *et al.* 1997, 1999, 2002, Cazaubon *et al.* 2002, Cornier *et al.* 2002, Dandelot *et al.* 2005). To elucidate the reasons for the success of non-indigenous species (NIS), studies have gathered

Francesca Gherardi, Biological invaders in inland waters: Profiles, distribution, and threats, 209–229.

data on the biology and ecology of emblematic, non-indigenous macrophytes. Numerous studies and reports have focused on the management of invaders such as *Ludwigia* sp. (Dutartre and Oyarzabal 1993, Damien 2002, Fournier and Oyarzabal 2002, Pipet 2002, Rebillard *et al.* 2002, Grillas 2004) and, to a lesser extent, *Elodea* species (Di Nino *et al.* 2005).

NON-INDIGENOUS AQUATIC AND SEMIAQUATIC MACROPHYTES IN FRANCE

Compilation of a species list

The term "aquatic macrophyte" is commonly used for all macroscopic forms of aquatic vegetation including algae, bryophytes, some pteridophytes, and many flowering plants (angiosperms). This assemblage contains extremely heterogeneous species which survive in similar habitats but result from fundamentally different evolutionary pathways. Non-indigenous aquatic plants do not belong to one distinct taxonomic group, but rather form a collection of many plant taxa.

This chapter does not claim to be an exhaustive review of introduced plants in metropolitan France, but rather an overview of the present situation. This review covers aquatic macrophytes in inland waters, excluding marine algae, such as *Caulerpa taxifolia* (Vahl.) C. Agardh, as well as woody species (except for *Hibiscus roseus*). Some taxa, regarded as non-indigenous by certain authors (although with a wide margin of doubt), have been included here: *Acorus calamus*, *H. roseus* (three European populations; E. Tabacchi 2005, personal communication), and *Azolla mexicana* (formerly known as *Azolla caroliniana* Willd). The status of several species from southern Europe varies according to the author: *Vallisneria spiralis*, *Stratiotes aloides*, *Scirpus mucronatus*, and *Scirpus pungens* are protected locally, whereas Dutartre *et al.* (1997) considered these species non-indigenous. In addition, van der Velde *et al.* (2002) considered *Salvinia natans* and *Octodiceras fontanum* as non-indigenous in the Netherlands, while they are protected in France. Some NIS are well-established, whereas others are found only occasionally in aquatic environments.

For aquatic macrophytes, several regional floras (Abbayes *et al.* 1971, Corillion 1982, Bournerias 1984, Lambinon *et al.* 1992) and local scientific journals (Bulletin de la Société Scientifique de Bretagne, Bulletin de la Société d'Histoire Naturelle de Moselle, Bulletin de la Société Linnéenne de Normandie, Bulletin de l'Association Philomathique d'Alsace Lorraine, Bulletin de la Société des Sciences de Nancy, le Monde des Plantes) were consulted. They provided information about plant histories in the geographical area, specifically on whether it is an indigenous or an NIS. Several sources containing information on NIS were also used (Dutartre *et al.* 1997, Aboucaya 1999, Muller *et al.* 2004).

NIS taxonomic groups and eco-morphological types

My resulting list of NIS, representing 24 families and 58 species in metropolitan France (Table 1), includes species found occasionally as well some hybrids. Some aquatic families are well represented, such as *Lemnaceae* and *Hydrocharitaceae*. Furthermore, numerous non-indigenous semiaquatic species were also present, such as *Poaceae* and *Cyperaceae*. As compared to Wallentinus's list (Wallentinus 2002), 10 additional species, including two bryophytes, have been identified as NIS in France.

Table 1 List of non-indigenous aquatic and semiaquatic plant species in France and their vectors of introduction: (a) escaped from aquaria, (b) arrived on seagoing vessels (ballast water, timber trade), (c) introduced intentionally (ornamentals, pond gardens, medicinal plant), (d) seed or grain contaminant, (e) wool industry, (f) natural expansion, (?) unknown source.

	Family	Vectors
Alga		
Hydrodictyon reticulatum (L.) Lagerh.	Hydrodictyaceae	?
Bryophyta		
Dumortiera hirsuta (Sw.) Nees	Marchantiaceae	f
Octodiceras fontanum (Bach. Pyl) Lindb.	Fissidentaceae	f
Pteridophyta		
Azolla filiculoides Lam	Azollaceae	a
Azolla mexicana C. Presl	Azollaceae	a
Salvinia natans (L.) All.	Salviniaceae	a?
Vascular plants (monocots and dicots)		
Acorus calamus L.	Araceae	c
Althernanthera philoxeroides (Martius) Griseb	Amaranthacea	a
Aponogeton distachyos Thunb.	Aponogetonaceae	c
Callitriche peploides Nutt.	Callitrichaceae	?
Callitriche terrestris Rafin	Callitrichaceae	?
Cortadaria selloana (Schultes and Schultes fil.) Ascherton and Graebner	Poaceae	c
Cotula coronopifolia L.	Asteraceae	c?
Cyperus difformis L.	Cyperacea	?
Cyperus eragrostis Lam.	Cyperacea	c
Cyperus esculentus L.	Cyperacea	e
Cyperus reflexus Vahl	Cyperacea	e
Egeria densa Planchon	Hydrocharitaceae	a
Eichhornia crassipes (Mart.) Solms	Pontederiaceae	c
Eleocharis bonariensis Nees	Cyperaceae	?

Table 1 Continued.

	Family	Vectors
Elodea canadensis Michaux	Hydrocharitaceae	a/b
Elodea ernstiae H. St. John	Hydrocharitaceae	a
Elodea nuttallii (Planchon) H. St. John	Hydrocharitaceae	a
Fallopia japonica (Houtt.) Ronse Decraene	Polygonaceae	c
Fallopia sachalinensis (F. Schmidt Petrop.) Ronse Decraene	Polygonaceae	c
Fallopia x bohemica Chrtek and Chrtkova	Polygonaceae	f
Glyceria striata (Lam.) A. S. Hitchc	Poaceae	d
Heracleum mantegezzianum Sommier and Lev	Apiaceae	c
Hibiscus roseus Thore	Malvaceae	c
Hydrilla verticillata (L.f.) Royle	Hydrocharitaceae	a/c
Hydrocotyle ranunculoides L.fil	Apiacae	a/c
Impatiens balfouri Hooker fil.	Balsaminaceae	c
Impatiens glandulifera Royle	Balsaminaceae	c
Juncus tenuis Willd.	Juncaceae	?
Lagarosiphon major (Ridley) Moss	Hydrocharitaceae	a
Lemna aequinoctialis Welw.	Lemnacaeae	f?
Lemna minuta H.B.K.	Lemnacaeae	f?
Lemna perpusilla Torrey	Lemnacaeae	f?
Lemna turionifera Landolt	Lemnacaeae	f?
Lindernia dubia (L.) Pennel	Scrophulariaceae	f
Ludwigia grandiflora subsp. hexapetala (Hook. and Arn.) Nesom and Kartesz	Onagraceae	c
Ludwigia peploides subsp. montevidensis (Spreng.) Raven	Onagraceae	c
Myriophyllum aquaticum (Velloso) Verdcourt	Haloragaceae	a
Myriophyllum heterophyllum Michaux	Haloragaceae	a/c
Najas graminea Delile	Najadaceae	f?
Paspalum dilatatum Poiret	Poaceae	e
Paspalum distichum L	Poaceae	f?
Pistia stratiotes L.	Araceae	a/c
Pontederia cordata L.	Pontederiaceae	c
Sagittaria latifolia Willd.	Alismataceae	a/c
Schoenoplectus prolifer Rottb.	Cyperaceae	?
Scirpus mucronatus L.	Cyperaceae	?
Scirpus pungens Valh.	Cyperaceae	?
Spartina alterniflora Loisel	Poaceae	b/c
Spartina x townsendii H. and J. Groves	Poaceae	f
Spirodela oligorhiza (Kurz) Hegelm.	Lemnacaeae	f/a
Stratiotes aloides L.	Hydrocharitaceae	c
Vallisneria spiralis L.	Hydrocharitaceae	a

The composition of the non-indigenous aquatic flora is illustrated in Fig. 1A, with one alga (hydrodictyaceae), two bryophytes (Marchantiaceae, Fissidentaceae), three pteridophytes (Azollaceae and Salviniaceae), and 52 vascular plants (34 monocotyledons and 18 dicotyledons). More NIS are monocots than dicots proportionally, perhaps due to the monocots' high incidence of rhizomatous growth.

Of these non-indigenous plants, helophytes, amphiphytes, floating, and submersed species represented 20, 20, 19, and 16% of the introduced macrophytes, respectively (Fig. 1B). Helophytes are emergent plants which occupy permanent, standing water, or wet soil. Floating macrophytes are not rooted in sediment, but live unattached in the water. The life forms within this group included very small floating or submersed plants with few or no roots (*Lemna minuta* and the water fern *Azolla* sp.). Submersed macrophytes include many flowering plants, for example *Elodea nuttallii* and *Egeria densa* which complete their life cycle under water.

Invasion histories and pathways

Many aquatic macrophytes were introduced more than a hundred years ago, while others are more recent arrivals. The majority of introductions took place at the end of the 19th and at the beginning of the 20th centuries (Table 2). Of all the introduced species listed here, more than 50% came from America and almost 40% came from Asia and/or Africa (Table 2).

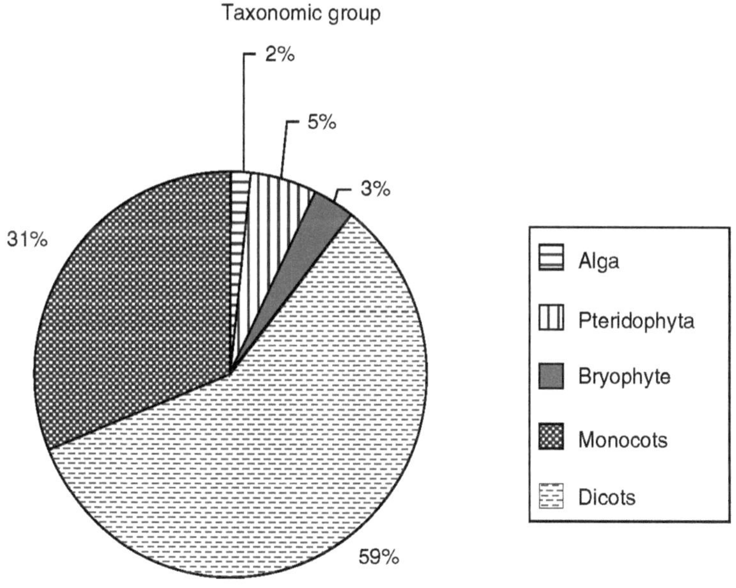

Fig. 1A Taxonomic groups of non-indigenous aquatic and semiaquatic plants in France.

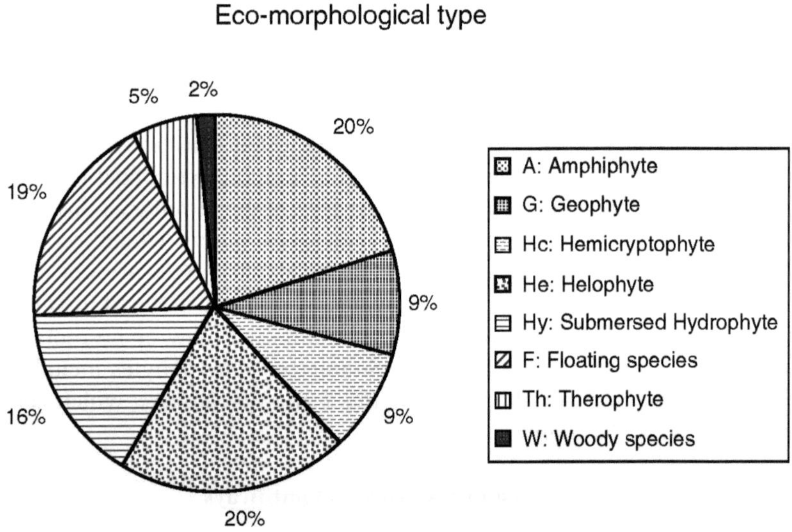

Fig. 1B Eco-morphological types of non-indigenous aquatic and semiaquatic plants in France.

Some NIS are only reported occasionally, like *Eichhornia crassipes* (Planty-Tabacchi 1993) or *Althernanthera philoxeroides* in south-west France (Dupont 1989, Georges 2004) or *Pistia stratiotes*; *Pontederia cordata* and *Hydrilla verticillata* were observed only once in an aquatic environment.

The time-lag phenomenon, during which a given population remains small and geographically restricted, is well documented for riparian species (e.g. *Impatiens glandulifera, Fallopia japonica, Fallopia sachalinensis*) or amphiphytes, such as *Ludwigia grandiflora* subsp. *hexapetala* and *Ludwigia peploides* subsp. *montevidensis* (Dutartre and Oyarzabal 1993). The fact that certain introduced species became aggressive after a lag phase is ecologically significant.

The invasion histories of some NIS are well known:

- The hybrid *Spartina x townsendii* was first observed in France in 1906 along the Atlantic Coast. The first sighting in south-west France occurred in 1985 and then spread quickly via mud flats, resulting in hundreds of hectares being colonised by this species. Moreover, a second North American species, *Spartina versicolor* Fabre, not included in our list, has been reported occasionally. *Spartina alterniflora* spread, but only around the Bay of Brest (Goulletquer *et al.* 2002).
- The two Japanese Knotweed taxa (*F. japonica* and *F. sachalinensis*) have been expanding throughout Europe ever since their deliberate introduction in the 19th century. The spread has increased dramatically since the 1980s: *F. japonica* and *F. sachalinensis* and hybrids have developed in large

Table 2 NIS origin and introduction date in France. (? = unknown).

	Origin area	First found
Alga		
Hydrodictyon reticulatum (L.) Lagerh.	Cosm. subtropic.	1989
Bryophyta		
Dumortiera hirsuta (Sw.) Nees	N. America	before 1997
Octodiceras fontanum (Bach. Pyl) Lindb.	S-Europe	?
Pteridophyta		
Azolla filiculoides Lam.	N. America, S. America, Australia	1880
Azolla mexicana C. Presl.	N. America	1901
Salvinia natans (L.) All.	S. Europe, India, Japan.	before 1997
Vascular plants (monocots and dicots)		
Acorus calamus L.	Asia, N. America, India	XIV
Althernanthera philoxeroides (Martius) Griseb	S. America	1971
Aponogeton distachyos Thunb.	S. Africa	1830
Callitriche peploides Nutt.	N. America	before 1997
Callitriche terrestris Rafin	N. America	before 1997
Cortadaria selloana (Schultes and Schultes fil.) Ascherton and Graebner	S. America	before 1977
Cotula coronopifolia L.A	S. Africa	before 1980
Cyperus difformis L.	Pan-tropical	1850
Cyperus eragrostis Lam.	S. America	XIX
Cyperus esculentus L.	Tropical Asia and Africa	1952
Cyperus reflexus Vahl	Tropical America	2003
Egeria densa Planchon	S. America	1961
Eichhornia crassispes (Mart.) Solms	S. America	before 1993
Eleocharis bonariensis Nees	S. America	1750
Elodea canadensis Michaux	N. America	1845
Elodea ernstiae H. St. John	S. America	1959
Elodea nuttallii (Planchon) H. St. John	N. America	1959
Fallopia japonica (Houtt.) Ronse Decraene	Asia	1825
Fallopia sachalinensis (F. Schmidt Petrop.) Ronse Decraene	Japan	1869
Fallopia x bohemica Chrtek and Chrtkova	Hybrid	before 2003
Glyceria striata (Lam.)A. S. Hitchc	N. America, Central America	1906
Heracleum mantegezzianum Sommier and Lev	Caucasus	1993
Hibiscus roseus Thore	Europe ?	before 1995
Hydrilla verticillata (L.f.) Royle	Australia, Asia, Africa	before 1997
Hydrocotyle ranunculoides L.fil	N. and S. America	1820
Impatiens balfouri Hooker fil.	Himalayas	1943

Table 2 Continued.

	Origin area	First found
Impatiens glandulifera Royle	Himalayas	1842
Juncus tenuis Willd	N. America	XIX
Lagarosiphon major (Ridley) Moss	S. Africa	1960
Lemna aequinoctialis Welw.	S. America	?
Lemna minuta H.B.K.	N. and S. America	1965
Lemna perpusilla Torrey	Asia, N. Africa and S. America	before 1997
Lemna turionifera Landolt	N. America	1992
Lindernia dubia (L.) Pennel	N. America	before 1997
Ludwigia grandiflora subsp. hexapetala (Hook. and Arn.) Nesom and Kartesz	S. America	1820–1830
Ludwigia peploides subsp. montevidensis (Spreng.) Raven	S. America	1820–1830
Myriophyllum aquaticum (Velloso) Verdcourt	S. America	1880
Myriophyllum heterophyllum Michaux	N. America	before 1997
Najas graminea Delile	S. Europe to E. Asia ?	before 1997
Paspalum dilatatum Poiret	S. America	1937
Paspalum distichum L.	Trop. America	1965
Pistia stratiotes L.	S. America	?
Pontederia cordata L.	N. America	?
Sagittaria latifolia Willd.	N. America	1936
Schoenoplectus prolifer Rottb.	S. Africa, Australia?	1920
Scirpus mucronatus L.	Paleo Subtrop.	1859
Scirpus pungens Valh.	S. America	1849
Spartina alterniflora Loisel	N. America	1906?
Spartina x townsendii H. and J. Groves	hybrid	1906
Spirodela oligorhiza (Kurz) Hegelm.	Asia, Australia	?
Stratiotes aloides L.	S. Europe, Asia	1834
Vallisneria spiralis L.	S. Europe, N. Africa, Asia	1787

patches along many riparian and man-made habitats, often far from the original introduction point (Bailey and Schnitzler 2003).

- Coming from South America, *Ludwigia* sp. (*L. grandiflora* and *L. peploides*) were introduced by accident in southern France in the 1820s. Long restricted to the southern part of the country, from Camargue to Aquitaine, they have been migrating north for nearly 30 years. Today, *L. peploides* has reached the Belgian border. Others have been observed at sites in Belgium and in the Netherlands (Dandelot 2004).
- In a similar pattern, *E. densa*, a species first observed in France in 1960 (Feuillade 1961a, b), has spread along the entire Atlantic coast (Dutartre *et al.* 1999).

- An indigenous of North America, *Elodea canadensis*, first recorded in the early 19th century in the British Isles (Simpson 1984, 1990), is now naturalized and widespread in Europe. *Elodea canadensis* became a persistent weed following its naturalization, choking waterways before declining to its present, less-abundant (but still common) level (Thiébaut *et al.* 1997, Barrat-Segretain 2001).
- Another species from North America, *E. nuttallii* was first found in Belgium in 1939 and had spread into northern France by the end of the 1950s (Sell 1959). For the past 30 years it has been colonizing numerous ponds and streams in metropolitan France, except in the south-east (G. Thiébaut 2006, unpublished data). *Elodea nuttallii* is replacing *E. canadensis* at many sites (Mériaux 1979a, b, Thiebaut *et al.* 1997, Barrat-Segretain 2001). Although *E. canadensis* and *E. nuttallii* have been spreading for several years in eastern France, this species is relatively more problematic in other European countries. For example, *E. nuttallii* was classified as one of the "top ten" invasive species in Germany (F. Klingenstein 2005, personal communication); colonies have been expanding in Lake Leman in Switzerland since 1993 (Demierre and Perfetta 2002), in numerous ponds, reservoirs, and streams in Brittany (Simpson 1990), in Belgium (G. Verniers 2004, personal communication), as well as in Sweden (D. Larson 2006, personal communication).

I have come to the conclusion that most invasive plant species arrived in France as a result of human intervention (aquarium plants, ornamental use). Of all the plants, 38% are ornamentals, by far the dominant vector for introduced plants (Table 1). Another 17 of the introduced plant species are sold for use in freshwater aquaria (29% of the aquatic and semiaquatic plants; Table 1). Among the well-known examples of aquarium plants are certain hydrocharitaceae species (*E. densa, Elodea sp., Lagarosiphon major, H. verticillata*) and some other taxa (*Myriophyllum aquaria*). Plants which escaped from aquaria (Hydrocharitaceae, *Ludwigia* sp., *Myriophyllum* sp.) easily colonized freshwater environments. Highly invasive aquatic and semiaquatic species, including *Ludwigia* sp. and *Fallopia* taxa, have resulted to a large extent from either intentional introduction for ornamental use (outdoor ponds) or use as ornamentals after the initial introduction (Table 1). A few species are medicinal plants (e.g. *A. calamus*). *Spartina alterniflora* from the East coast of North America was introduced accidentally in ship ballast at the end of the 19th century in southern England, but it was also intentionally introduced into France to stabilize sand dunes, given its ability to increase sediment accumulation.

NIS distribution in France

Plant distribution varies according to the different climate in each of three biogeographical zones (Atlantic A, Continental C, Mediterranean M). For example, the diploid *L. peploides* colonises mainly the Mediterranean region of France (except for the south-eastern part), while the polyploid *L. grandiflora* predominates in all the other regions (Dandelot 2004). These data are taken primarily from Aboucaya (1999), from other publications (Felzines and Loiseau 2003, Muller *et al.* 2004, Felzines 2004), and from personal data (Table 3).

Of the 11 taxa present in the three biogeographical areas, there are five widespread invasive and one potentially invasive plants: three riparian species (*F. japonica, F. sachalinensis, I. glandulifera*), two floating species (*Azolla filiculoides, L. minuta*), and a single amphibious species (*L. grandiflora*). *Ludwigia* species are considered to be the most invasive aquatic plants in France. For

Table 3 Classification of NIS according to their invasiveness in France. Data taken primarily from Aboucaya (1999) and from: Felzines and Loiseau (2003), Muller *et al.* (2004), Felzines (2004), and personal data.
M: Mediterranean area, A: Atlantic area, C: Continental zone.

	Invasive		Potentially invasive	
	widespread	restricted	widespread	restricted
Acorus calamus L.				AC
Althernanthera philoxeroides (Martius) Griseb				M?
Aponogeton distachyos Thunb.				A
Azolla filiculoides Lam			MAC	
Azolla mexicana C. Presl				AC
Callitriche peploides Nutt.				?
Callitriche terrestris Rafin				?
Cortadaria selloana (Schultes and Schultes fil.) Ascherton and Graebner		M		A
Cotula coronopifolia L. A		M		A
Cyperus difformis L.				M
Cyperus eragrostis Lam.				C
Cyperus esculentus L.				A
Cyperus reflexus Vahl				A
Dumortiera hirsuta (Sw.) Nees				A?
Egeria densa Planchon		A		
Eichhornia crassispes (Mart.) Solms				M
Eleocharis bonariensis Nees				AC
Elodea canadensis Michaux		C		MA
Elodea ernstiae H. St. John				C
Elodea nuttallii (Planchon) H. St. John		C		A
Fallopia japonica (Houtt.) Ronse Decraene	MAC			
Fallopia sachalinensis (F. Schmidt Petrop.) Ronse Decraene	MAC			

Table 3 Continued.

	Invasive		Potentially invasive	
	widespread	restricted	widespread	restricted
Fallopia x bohemica Chrtek and Chrtkova			MAC	
Glyceria striata (Lam.)A. S.Hitchc				AC
Heracleum mantegezzianum Sommier and Lev		AC		
Hibiscus roseus Thore				A
Hydrilla verticillata (L.f.) Royle				MA?
Hydrocotyle ranunculoides L.fil				AC
Hydrodictyon reticulatum (L.) Lagerh.			MAC?	
Impatiens balfouri Hooker fil.				AC
Impatiens glandulifera Royle	MAC			
Juncus tenuis Willd		AC		
Lagarosiphon major (Ridley) Moss		A		
Lemna aequinoctialis Welw.				M
Lemna minuta H. B. K.	MAC			
Lemna perpusilla Torrey				M
Lemna turionifera Landolt		C		
Lindernia dubia (L.) Pennel		AC		M
Ludwigia grandiflora subsp. hexapetala (Hook. and Arn.) Nesom and Kartesz	MAC			
Ludwigia peploides subsp. montevidensis (Spreng.) Raven				C
Myriophyllum aquaticum (Velloso) Verdcourt		A		M
Myriophyllum heterophyllum Michaux				?
Najas graminea Delile				?
Octodiceras fontanum (Bach. Pyl) Lindb.				?
Paspalum dilatatum Poiret		MA		C
Paspalum distichum L.		MA		C
Pistia stratiotes L.				M
Pontederia cordata L.			MAC?	
Sagittaria latifolia Willd.				AC
Salvinia natans (L.) All.				MA?
Schoenoplectus prolifer Rottb.				A
Scirpus mucronatus L.			MAC	
Scirpus pungens Valh.				AC
Spartina alterniflora Loisel				A
Spartina x townsendii H.and J. Groves				A?
Spirodela oligorhiza (Kurz) Hegelm.				A
Stratiotes aloides L.				AC
Vallisneria spiralis L.			MAC	

the 567 sites investigated in France by Dutartre (2004), *Ludwigia* sp. were found in rivers with low water velocity in summer (29%), in shallow wetlands (20%), in ditches and channels (20%), in ponds and on lake shores (13%), in oxbows (9%), and in wet meadows (4%).

Many of the NIS listed have a restricted invasion range in France: for example the hybrid *S. x townsendii* along the Atlantic Coast or *S. alterniflora* in Bay of Brest (Goulletquer *et al.* 2002). Fifteen species are restricted to Atlantic sites only, 10 to the Mediterranean zone, and eight to the Continental area. Twelve species have been observed in both Atlantic and Continental areas, whereas only five taxa were listed for both Atlantic and Mediterranean areas (Table 3). Plants normally found growing in rice-fields were often limited in range to southern France, since the warm climate they need is not found further north. Invasibility potential is highest in the Atlantic area. In many cases, tropical–subtropical species thrived in Mediterranean and Atlantic areas but were absent in colder, northern France.

Numerous NIS (e.g. *A. philoxeroides, E. crassipes, P. stratiotes)* have spread worldwide, but are restricted to the Mediterranean zone in France. These species, sold in the aquarium trade, are potentially invasive (Table 3). Other species, such as *Cotula coronopifolia*, are invasive in some habitats (salt marshes, estuaries) but their low actual invasiveness allows them to be considered as potentially invasive in the Atlantic area. The number of sites at which a species occurs is a misleading indicator for degree of invasiveness. Some species, such as the non-indigenous *Lemna* sp. or water fern, which are found at a much more restricted number of sites than *Elodea* species, for example, are in fact highly invasive and spreading at those sites.

NIS MANAGEMENT IN FRANCE

Some invasive species are considered to cause "nuisance growth", where the degree of nuisance is judged in relation to the water body management aim (for transportation, recreation, fishery management, or conservation). The ultimate goal is to prevent the establishment of new invasive species proactively, while setting control priorities for established plants. The action plan takes into account the plants' actual and potential impact on ecosystem functioning, as well as the indigenous species and communities present, particularly if rare and/ or ecologically important species are targeted for conservation. Action is recommended only after careful analysis indicates that leaving the spreading species unchecked will result in greater damage than that caused by control efforts.

In general, the biological invasion control priority is to prevent new infestations from taking hold, especially for the fastest growing and most disruptive species. NIS that are not rapidly increasing in numbers, proliferating in undisturbed habitats, or interfering in areas recovering from disturbance have a lower priority for control. Large infestations of plants which cause considerable environmental impact, such as *Ludwigia* spp. or Hydrocharitaceae (*L. major*, *E. densa*) have the highest priority for control.

Each site has its own management plan based on individual characteristics. Hand-pulling has been tested to limit unwanted proliferation of *E. nuttallii* in a small stream (Di Nino *et al.* 2005). In order to fight *Ludwigia* sp., various solutions adapted to individual sites were tested: manual removal and/or treating with herbicides (Dutartre and Oyarzabal 1993, Damien 2002, Fournier and Oyarzabal 2002, Pipet 2002, Rebillard *et al.* 2002). The removal operation was manual at the beginning of *Ludwigia* sp. colonisation. When it became well-established, mechanization was necessary (Dutartre and Oyarzabal 1993, Dutartre *et al.* 1999). Although chemical treatment can replace or enhance manual removal operations, it has been used only as a last resort, where water use and environmental considerations made it possible. In some wetlands in southern France, salt water has been used to eradicate salt-sensitive *L. peploides* (Grillas 2004).

Management plans established early on were the first steps towards sustainable management of aquatic environments. However, these efforts are compromised as long as invasive aquatic plant species continue to be sold to individuals. Stronger enforcement of existing laws, coupled with an intensive public education campaign, is needed to prevent further NIS introduction.

INVASIBILITY AND INVASIVENESS

Habitat invasibility

Invasibility is an emergent property of an environment, the outcome of several factors including the region's climate, the environmental disturbance regime, and the competitiveness of the resident species (Lonsdale 1999). The actual invasion of an environment by a new species is influenced by three additional factors: the number of propagules entering the new environment, the characteristics of the new species, and the susceptibility of the environment to invasion (Lonsdale 1999).

For example, whether or not cut-off channels are connected to the main river is probably the essential parameter of colonization by *E. canadensis* and *E. nuttallii* in the Rhône River flood plain. These two species colonize new areas most often by vegetative fragments transported by water currents (Barrat-Segretain 2001). Flood disturbances can, in particular, damage or destroy some resident vegetation and allow for the introduction of *Fallopia* taxa. In some cases, restoration work or river management efforts may be considered as disturbances that facilitated NIS invasion (Schnitzler and Mulller 1998).

Fluctuation in resource availability is identified as another key factor controlling habitat invasibility (Davis *et al.* 2000). In a previous study, I established that the eutrophication process increases the invasibility of *Elodea* species while inducing competition between *Elodea* species and indigenous macrophyte species. *Elodea nuttallii* and *E. canadensis* take advantage of eutrophication because they are adapted for the quick nutrient uptake necessary for growth and can avoid turbidity by covering the water's surface (Thiébaut 2005). However, when increased levels of eutrophication induced the disappearance of submersed macrophytes as a result of phytolankton blooms and increased turbidity, they were replaced by free-floating plants such as duckweed. This type of vegetation allowed *Azolla* species to invade, particularly *A. mexicana* and *A. filiculoides. Azolla* species live in symbiosis with the nitrogen-fixing cyanobacterium *Anabaena azollae* Strass and are therefore efficient phosphorus removers in the absence of nitrate.

Life history traits of invasive plants

Many studies have focused on identifying plant traits that define invasiveness (e.g. Goodwin *et al.* 1999, Vaźquez 2005). There have been many attempts in invasion biology to predict outcomes by focusing on the traits of potential invaders and of the invaded community. Unfortunately, most of these attempts have been unsuccessful (Vásquez 2005).

Below is my review of the studies that have evaluated the relationship between the traits of some emblematic species in France and their invasion success (Barrat-Segretain *et al.* 2002, 2004, Barrat-Segretain 2004, 2005, Barrat-Segretain and Elger 2004, Dandelot 2004, Petit 2004, Thiébaut 2006). However, less is known about the invasiveness of the majority of NIS and the invasibility of aquatic habitats.

Biological attributes as key factors for invasion

Ploidy level: Invasion outcomes might be influenced by variations in clonal architecture and ecological attributes of emergent hybrids, as well as differentiated abilities for sexual reproduction. Effective hybridization is known to increase a species' invasive potential in its secondary distribution area, if the parents themselves are invasive (Bailey 2003). For example, hybridization with local *S. maritima* (Curtis) Fernald resulted in a sterile hybrid, *S. alterniflora* being the seed parent in the cross. Chromosome doubling in this hybrid gave rise to a new fertile allopolyploid species, *S. x townsendii*. This new species, genetically isolated from its parents, is very aggressive (Petit 2004). Similarly, the ploidy level of *Fallopia* taxa was studied in north-eastern France. All plants were hybrid *Fallopia x bohemica* and male fertile. The population analysed was a mixture of hexaploids, octoploids, and aneuploids. The seedlings found were octoploids, indicating the ability of octoploid plants to produce seeds (A. Schnitzler 2006, personal communication). A possible outcome of hybridization is heterosis or "hybrid vigour". Although dissolution of heterosis can occur in hybrid populations that retain sexual reproduction, vegetatively reproducing aquatic plants can propagate hybrid genotypes indefinitely. Molecular data demonstrate clearly that invasive water milfoil populations in North America have resulted from hybrization between NIS and indigenous species. These observations suggest that invasiveness in these aggressive NIS may be linked to heterosis maintained by vegetative propagation (Moody and Les 2002). In France, no plant morphologically intermediate between the indigenous *M. spicatum* Linnaeus and the non-indigenous *Myriophyllum heterophyllum* has been discovered yet, but the potential exists.

Reproductive biology: The reproductive biology of numerous aquatic NIS, especially in their foreign ranges, is relatively poorly understood. Levels of inbreeding and other mating-system parameters have been measured in several emergent species but are lacking for free-floating or submerged taxa. Invasion capacity may be influenced by the balance between sexual versus clonal reproduction. The relative importance of sexual versus clonal recruitment may vary among populations of clonal plants because reproduction allows populations to persist in habitats or regions where sexual reproduction cannot occur. For example, the spread of dioecious *E. canadensis* or *E. nuttallii* across Europe involved only female plants, and male and female *S. aloides* plants tend to be confined to different parts of the species' European range so that sexual reproduction is not possible in most populations.

Sexual reproduction generates genotypic diversity which may increase the adaptive evolution rate during expansion into new habitats. The seeds produced by sexual reproduction are also more likely to participate in long-distance colonization than vegetative clonal propagules which are often larger, more vulnerable to desiccation, lack dispersal and dormancy mechanisms, and therefore have less capacity for dispersal (Eckert 2002). Asexual reproduction includes both seed production without fertilization and vegetative reproduction (rhizomes, turions, tubers, and stolons). Asexual reproduction is important in the establishment, growth, and maintenance of NIS. Each aquatic species has followed a unique evolutionary path representing a complex balance between sexual and asexual reproduction, levels of genetic variation in offspring, and the ability to maximize survival. Because of the highly diverse evolutionary histories of aquatic plants, it is difficult to identify general evolutionary models.

Dispersal of propagules: Gene flow in aquatic plants may be greatly affected by the discrete and patchy nature of many aquatic habitats and the directional transport of propagules in running water. Transport of vegetative fragments may lead more frequently to successful gene establishment than seed dispersal and may, in part, explain the extensive geographical ranges of many clonal aquatic species (Barret *et al.* 1993). Semiaquatic invaders differ from many aquatic invaders in that seeds are often dispersed *via* water, whereas aquatic plants and plant fragments can be dispersed *via* flotation. In aquatic species, reproduction occurs primarily from asexually rooting plant fragments. *Ludwigia grandiflora* produces viable seeds and plantlets in the south of France (Dutartre *et al.* 1997, Dandelot 2004). Stem fragmentation is the main dispersal mode for *Ludwigia* spp., *Elodea* sp. and *E. densa*. After establishing themselves in the bank or channel bottom, prostrate stems grow laterally, rooting adventitiously at nodes. Water plants excel in this capacity with a variety of vegetative structures that are highly specialized to function efficiently as propagules, some being even capable of long-distance dispersal (Dutartre *et al.* 1997, Thiébaut *et al.* 1997, Dandelot 2004). Life history traits, regeneration (regrowth into viable plants) and colonization (establishment in the sediment) of vegetative plant fragments, and resistance to water current were compared in two invasive macrophyte species, *E. canadensis* and *E. nuttallii* (Barrat-Segretain *et al.* 2002). Both species showed similar resistance to currents, while fragment regeneration and colonization were only slightly higher in *E. nuttallii* than in *E. canadensis.*

Physiological traits as key factors for invasion

Allelopathy: NIS are considered less vulnerable than indigenous species to phytophagous animals, due to a lack of natural herbivores in their introduced range or efficient defence mechanisms. For example, *Ludwigia* spp. are consumed less by herbivores probably due to their high content of saponins and calcium oxalate (Dandelot 2004). In the same way, a slightly higher palatability was established for *E. nuttallii* than for *E. canadensis* (Barrat-Segretain *et al.* 2002, Barrat-Segretain and Elger 2004). The difference in palatability between the two *Elodea* species was also partly related to the smaller dry matter content of *E. nuttallii*. At an intraspecific level, the effect of time of year is also fully explained by the temporal variability in dry matter content for the *Elodea* species (Elger and Wilby 2005). Palatability is a multi-factorial feature of plants, resulting from chemical (e.g. nutrient content and amount of secondary compounds) and physical (e.g. toughness and hairiness) tissue characteristics. Despite these studies, there is no comprehensive view of biotic interactions occurring in fresh waters.

Competition: The success of invasive species has also been attributed to their ability to displace other species by direct competition. The formation of an *E. nuttallii* canopy which shades *E. canadensis* is a key factor in explaining the success of *E. nuttallii*, particularly under eutrophic conditions (Barrat-Segretain and Elger 2004). Later, Barrat-Segretain (2005) established that both spatial pattern and development stage of *E. canadensis* may influence the outcome of competition with *E. nuttallii*. The coexistence of the two *Elodea* species is enhanced by river disturbances (Barrat-Segretain 2001), whereas *E. nuttallii* dominates in less-disturbed waters as a result of its higher growth rate.

Phenotypic plasticity: NIS have been shown to modify resource allocation through changes in their morphology and physiology. Plant plasticity when facing fluctuating resources is one characteristic that contributes to competitiveness and invasibility. By changing leaf area, *E. nuttallii* individuals can maximize growth and reproduction under a variety of environmental conditions (F. Di Nino 2006, unpublished data). Phenotypic plasticity may play a key role in the adaptation of organisms to changing environmental conditions. This trait is especially important for aquatic plant species which often spread asexually and thus lack genetic variation.

A broad ecological tolerance: Wide ecological amplitude seems necessary because any changes in the water potentially influence all plants in contact with it; for example, *Ludwigia* sp. has rather good resistance to frost in Europe. The growth of *E. canadensis* is affected by reduced light intensity, contrary to that of *E. nuttallii*. Increasing water phosphate levels increased the growth rate of *E. nuttallii* (Barrat-Segretain 2004). *Elodea nuttallii* and *E. canadensis* have wide amplitude in nutrient levels (Dendène *et al.* 1993, Robach *et al.* 1995, Rolland *et al.* 1999, Thiébaut and Muller 2003, Thiébaut 2005). Adaptation to dynamic water conditions is apparent in widespread aquatic species such as *Lemna aequinoctialis* and *L. turionifera* which can tolerate extreme ranges in pH from 3.2 to more than 9.0 (Landolt 1986).

CONCLUDING REMARKS

This review shows that a total of 58 plant species has been introduced into aquatic environments in France over the last three centuries. Most NIS arrived in France through human intervention (aquarium or ornamental plant use). Plant distribution varies according to the different climates in three biogeographical zones.

Despite several recent contributions concerning biology and ecology in the rapidly developing field of invasion biology, less is known about the invasiveness characteristics of aquatic or semiaquatic species and the specific features associated with habitat invasibility. This synthesis highlights the gaps in our understanding and contributes to identifying areas for further research which should be encouraged in order to prevent biological invasions of aquatic and semiaquatic species in France, other parts of Europe, and the world.

ACKNOWLEDGEMENTS

I am grateful to D. Larson, F. Klingenstein, and G. Verniers for information about *Elodea* species. Many thanks to Dr. E. Tabacchi (University Paul Sabatier of Toulouse) and Dr. A. Schnitzler (University Paul Verlaine of Metz) for their assistance about invasive taxa in France and to the two referees for their helpful comments.

REFERENCES

Abbayes des, H. G. Claustres, R. Corillon, and P. Dupont. 1971. Flore et végétation du massif armoricain. Presses universitaires de Bretagne, Saint Brieuc, France.

Aboucaya, A. 1999. Premier bilan d'une enquête nationale destinée à identifier les xénophytes invasifs sur le territoire métropolitain français (Corse comprise). Bulletin de la Société Botanique du Centre-Ouest **19**, 463–482.

Bailey, J. 2003. Japanese knotweed s.1 at home and abroad. Pages 183–196 in L. E. Child, J. H. Brock, G. Brandu, D. Prach, P. Pysek, P. M. Wade, and M. Williamson, editors. Plant Invasions: ecological threats and management solutions. Backhuys Publishers, Leiden, The Netherlands.

Bailey, J. and A. Schnitzler. 2003. La Renouée du Japon: une dangereuse conquête de l'Ouest. La Recherche **364,** 50–54.

Barrat-Segretain, M.-H. 2001. Invasive species in the Rhône River floodplain (France): replacement of *Elodea canadensis* Michaux by *E. nuttallii* St. John in two former river channels. Archiv für Hydrobiologie **152**, 237–251.

Barrat-Segretain, M.-H. 2004. Growth of *Elodea canadensis* and *Elodea nuttallii* in monocultures and mixture under different light and nutrient conditions. Archiv für Hydrobiologie **161**, 133–144.

Barrat-Segretain, M.-H. 2005. Competition between invasive and indigenous species: impact of spatial pattern and developmental stage. Plant Ecology **180**, 153–160.

Barrat-Segretain, M.-H. and A. Elger. 2004. Experiments on growth interactions between two invasive macrophyte species. Journal of Vegetation Sciences **15**, 109–114.

Barrat-Segretain, M.-H., A. Elger, P. Sagnes, S., and Puijalon. 2002. Comparative life-history traits of two invasive macrophyte species, *Elodea canadensis* Michaux and *Elodea nuttallii* (Planchon) H. St John. Aquatic Botany **74**, 299–313.

Barret, S. C. H., C. G. Eckert, and B. C. Husband. 1993. Evolutionary processes in aquatic plant population. Aquatic Botany **44**, 105–145.

Bournerias, M. 1984. Guide des groupements végétaux de la région parisienne. SEDES, Paris, France.

Cazaubon, A., S. Dandelot, C. Bertrand, and S. Fayolle. 2002. Ecological study on *Ludwigia* spp., invasive plant, on aquatic ecosystems (artificial and natural areas) of south eastern France (Mediterranean). Pages 11–13 in Proceedings of the 11th EWRS International Symposium on Aquatic Weeds, Moliets Maâ, France.

Corillion, R. 1982. Flore et végétation de la Vallée de la Loire (cours occidental: de l'Orléanais à l'estuaire). Jouve, Paris, France.

Cornier, T., D. Lejas, E. Lambert, A. Dutartre, J. P. Berton, and J. Haury. 2002. Preliminary data on distribution, autoecology and effects of *Ludwigia peploides* (Kunth) P. H. Raven (Jussie) and *L. grandiflora* (Michaux) Greuter and Burdet communities in the river Loire (France). Pages 19–22 in Proceedings of the 11th EWRS International Symposium on Aquatic Weeds, Moliets Maâ, France.

Damien, J. P. 2002. Control of *Ludwigia grandiflora* communities in the Brière Regional Nature Park. Pages 341–344 in Proceedings of the 11th EWRS International Symposium on Aquatic Weeds, Moliets Maâ, France.

Dandelot, S. 2004. Les *Ludwigia* spp. invasives du Sud de la France: Historique, Biosystématique, Biologie, Ecologie. Thesis, Université d'Aix-Marseille III, Marseille, France.

Dandelot, S., R. Verlaque, A. Dutartre, and A. Cazaubon. 2005. Ecological, dynamic and taxonomic problems due to *Ludwigia* (Onagraceae) in France. Hydrobiologia **551**, 131–136.

Davis, M. A., J. P. Grime, and K. Thompson. 2000. Fluctuating resources in plant communities: a general theory of invasibility. Journal of Ecology **88**, 528–534.

Demierre, A. and J. Perfetta. 2002. Gestion du faucardage des macrophytes sur les rives genevoises du Léman (Suisse). Pages 345–347 in Proceedings of the 11th EWRS International Symposium on Aquatic Weeds, Moliets Maâ, France.

Dendène, M.-A., T. Rolland, M. Trémolières, and R. Carbiener. 1993. Effect of ammonium ions on the net photosynthesis of three species of *Elodea*. Aquatic Botany **46**, 301–315.

Di Nino, F., G. Thiébaut, and S. Muller. 2005. Response of *Elodea nuttallii* (Planch.) H. St. John to manual harvesting in the north-eastern France. Hydrobiologia **551**, 147–157.

Dupont, P. 1989. Une découverte inédite d'Emile Contre: *Althernanthera philoxeroides* dans la vallée de la Garonne. Bulletin de la Société Botanique du Centre-Ouest **20**, 27–28.

Dutartre, A. 2004. Invasion and management of the water primrose (*Ludwigia* spp.) in France: a panorama. In 13th International Conference on Aquatic Invasive Species, Ennis, County Clare, Ireland, September 2004.

Dutartre, A., C. Charbonnier, V. Dosda, A. Fare, C. Lebougre, I. Saint Macary, and O. Touzot. 2002. Primary production of *Ludwigia* spp. in the South West of France. Pages 23–26 in Proceedings of the 11th EWRS International Symposium on Aquatic Weeds, Moliets Maâ, France.

Dutartre, A., J. Haury, and A. Jicorel. 1999. Succession of *Egeria densa* in a drinking water reservoir in Morbihan (France). Hydrobiologia **415**, 243–247.

Dutartre, A., J. Haury, and A.-M. Planty-Tabacchi. 1997. Introductions of aquatic and riparian macrophytes into continental French hydrosystems: an attempt at evaluation. Bulletin Français de la Pêche et de la Pisciculture **344–345**, 407–426.

Dutartre, A. and J. Oyarzabal. 1993. Gestion des plantes aquatiques dans les lacs et les étangs landais. Hydroécologie Appliquée **5**, 43–60.

Eckert, C. G. 2002. The loss of sex in clonal plants. Evolutionary Ecology **15**, 501–520.

Elger, A. and N. J. Wilby. 2005. Leaf dry matter content as an integrative expression of plant palatability: the case of freshwater macrophytes. Functional Ecology **17**, 58–65.

Feuillade, J. 1961a. Une plante aquatique nouvelle pour la France *Elodea densa* (Planch.) Casp. Bulletin de la Société Linnéenne de Normandie **10**, 47–51.

Feuillade, J. 1961b. Note complémentaire sur *Elodea densa* (Planch.) Casp. Bulletin de la Société Linnéenne de Normandie **10**, 185–188.

Felzines, J.-C. 2004. Introduction et naturalisation d'espèces dans les groupements végétaux aquatiques et alluviaux de la Dordogne quercynoise: situation actuelle et modifications au cours du XXème siècle. Le Monde des Plantes **484**, 21–24.

Felzines, J.-C. and J.-E. Loiseau. 2003. *Cyperus reflexus* Vahl et *Glyceria striata* (Lam.) A. S. Hitch., deux adventices en cours de naturalisation dans la vallée de la Dordogne moyenne. Le Monde des Plantes **478**, 9–11.

Fournier, L. and J. Oyarzabal. 2002. Management of aquatic invasive plants in lakes and ponds (Landes, France). Pages 287–290 in Proceedings of the 11th EWRS International Symposium on Aquatic Weeds, Moliets Maâ, France.

Georges, N. 2004. L'herbe à l'alligator (*Althernanthera philoxeroides* (Martius) Grisebach) atteint le département du Tarn-et-Garonne. Le Monde des Plantes **484**, 1–3.

Goodwin, B. J., A. J. McAllister, and L. Fahrig. 1999. Predicting invasiveness of plant species based on biological information. Conservation Biology **13**, 422–426.

Goulletquer, P., G. Bachelet, P. G., Sauriau, and P. Noel. 2002. Open Atlantic coast of Europe. A century of introduced species into French waters. Pages 276–290 in E. Leppäkoski, S. Gollasch, and S. Olenin, editors. Invasive Aquatic Species of Europe:

Distribution, Impacts and Management. Kluwer Academic Publishers, Dordrecht, The Netherlands.
Greulich, S. and M. Trémolières. 2002. Distribution and expansion of *Elodea nuttallii* in the Alsatian Upper Rhine Floodplain (France). Pages 427–430 in Proceedings of the 11th EWRS International Symposium on Aquatic Weeds, Moliets Maâ, France.
Grillas, P. 2004. Bilan des actions de gestion de *Ludwigia grandiflora* et *L. peploides* (jussies) dans les espaces protégés du Languedoc-Roussillon. Pages 148–152 in Patrimoines naturels, editors. Plantes invasives en France – Etat des connaissances et propositions d'actions. Museum National d'Histoire Naturelle, Paris, France.
Heywood, V. H. 1989. Patterns, extends and modes of invasions by terrestrial plants. Pages 31–60 in J.A. Drake, H. A. Mooney, F. di Castri, R. H. Groves, F. J. Kruger, M. Rejmanek, and M. Williamson, editors. Biological Invasions: A Global Perspective. Wiley, Chichester, UK.
Lambinon, J., J. E. De Langhe, L. Delvosalle, J. Duvigneaud, and C. Vanden Berghen. 1992. Nouvelle Flore de la Belgique, du Grand-Duché de Luxembourg, du Nord de la France des Régions Voisines (Ptéridophytes Spermaphytes). Patrimoine du Jardin botanique national de Belgique editors. Meise, Belgium.
Landolt, E. 1986. The family of Lemnaceae – a monographic study. Veröffentlichungen des Geobotanischen Institutes der Eidgenoessische Technische Hochschule Stiftung Ruebel in Zuerich **71**.
Lonsdale, W. M. 1999. Global patterns of plant invasions and the concept of invasibility. Ecology **80**, 1522–1536.
Mériaux, J.-L. 1979a. *Elodea nuttallii* St. John, espèce nouvelle pour le Nord de la France. Bulletin de la Société Botaique Nord France **32**, 30–32.
Mériaux, J.-L. 1979b. *Elodea ernstae* St. John, espèce nouvelle pour la flore régionale (Nord de la France). Documentation Floristique **2**, 55–60.
Moody, M. L. and D. H. Les. 2002. Evidence of hybridity in invasive water milfoil (*Myriophyllum*) populations. Proceedings of the National Academy of Sciences USA **99**, 14867–14871.
Muller, S., A. Aboucaya, L. Affre, S. Cassan, G. Decocq, F. Dinger, A. Dutartre, L. Fournier, L. Gavory, P. Grillas, G. Largier, J. Maillet, F. Médail, L. Moiroud, C. Mony, J. Oyarzabal, L. Pénélon, D. Pénin, P. Quertier, S. Saïd, J-M. Sinnassamy, C. Suehs, E. Tabacchi, G. Thiébaut, B. Toussaint, and N. Yavercovski. 2004. Plantes invasives en France. Etat des connaissances et propositions d'actions. In Patrimoines naturels, editors. Plantes invasives en France – Etat des connaissances et propositions d'actions. Museum National d'Histoire Naturelle, Paris, France.
Petit, R. J. 2004. Biological invasions at the gene level. Diversity and Distribution **10**, 159–165.
Pipet, N. 2002. Management of *Ludwigia spp* in Marais Poitevin (West of France). Pages 389–392 *in* Proceedings of the 11th EWRS International Symposium on Aquatic Weeds, Moliets Maâ, France.
Planty-Tabacchi, A.-M. 1993. Invasion des corridors riverains fluviaux par des espèces végétales d'origine étrangère. Thesis, Université Toulouse III, Toulouse, France.
Planty-Tabacchi, A.-M., E. Tabacchi, R. Naiman, C. DeFerrari, and H. Décamps. 1995. Invasibility of species-rich communities in riparian zones. Conservation Biology **10**, 598–607.
Planty-Tabacchi, A.-M., E. Tabacchi, and M. J. Salinas-Bonillo. 2001. Invasions of river corridors by exotic plant species: patterns and causes. Pages 221–234 in Brundu,

G., J. Brock, I. Camarda, L. Child L., and M. Wade, editors. Plant Invasions: Species Ecology and Ecosystem Management. Backhuys Publisher, Leiden, The Netherlands.

Rebillard, J.-P., A. Dutartre, A. Far, and J.-M. Ferroni. 2002. Management of the development of aquatic plants in the Adour-Garonne river basin (South West of France). Pages 307–310 in Proceedings of the 11th EWRS International Symposium on Aquatic Weeds, Moliets Maâ, France.

Robach, F., I. Hajnssek, I. Eglin, and M. Trémolieres. 1995. Phosphorus sources for aquatic macrophytes: water or sediment? Acta Botanica Gallica **142**, 719–732.

Rolland, T., G. Thiébaut, H. Daniel, M. Trémolières, J. Haury, and S. Muller. 1999. Response of nitrate reductase activity to ammonium in three populations of *Elodea canadensis* Michx. Ecologie **30**, 119–124.

Schnitzler, A. and S. Muller. 1998. Ecologie et biogéographie de plantes hautement invasives en Europe: les renouées géantes du Japon (*Fallopia japonica* et *F. sachalinensis*). Terre et Vie **53**, 3–38.

Sell, Y. 1959. Etude comparative de quelques espèces du genre *Elode*a à propos de l'apparition à Strasbourg et de ses environs d'une espèce nouvelle pour l'Europe continentale. Bulletin de l'Association Philomathique d'Alsace Lorraine **10**, 121–133.

Simpson, D. A. 1984. A short history of the introduction and spread of *Elodea* Michx. in the British Isles. Watsonia **15**, 1–9.

Simpson, D. A. 1990. Displacement of *Elodea canadensis* Michx. by *Elodea nuttallii* (Planch.) St. John in the British Isles. Watsonia **18**, 173–177.

Tabacchi, E. and A.-M. Planty-Tabacchi. 2002. Changes in Alien and Native Weedy Vegetation along Rivers: What's news? Pages 439–442 in Proceedings of the 11th EWRS International Symposium on Aquatic Weeds, Moliets Maâ, France.

Thiébaut, G. 2005. Does competition for phosphate supply explain the invasion pattern of *Elodea* species? Water Research **39**, 3385–3393.

Thiébaut, G. 2006. Invasion success of exotic aquatic plants in their native and introduced ranges. A comparison between their invasiveness in North America and in France. Biological Invasions **9**, 1–12.

Thiébaut, G. and S. Muller. 2003. Linking phosphorus pools of water, sediment and macrophytes in running waters. International Journal of Limnology **39**, 307–316.

Thiébaut, G., T. Rolland, F. Robach, M. Trémolières, and S. Muller. 1997. Quelques conséquences de l'introduction de deux espèces de macrophytes, *Elodea canadensis* Michaux *Elodea nuttallii* (Planch.) St. John dans les écosystèmes aquatiques continentaux: exemples de la plaine d'Alsace et des Vosges du Nord (Nord-Est de la France). Bulletin Français de la Pêche et de la Pisciculture **344–345**, 441–452.

Vasquez, D.P. 2005. Exploring the relationships between niche breath and invasion success. Pages 317–332 in M. W. Cadotte, S. M. McMahon, and T. Fukami, editors. Conceptual Ecology and Invasions Biology. Springer, Dordrecht, The Netherlands.

Van der Velde, G., I. Nagelkerken, S. Rajagopal, and A. Bij de Vaate. 2002. Invasions by aliens species in inland freshwater bodies in Western Europe: the Rhine delta. Pages 360–372 in E. Leppäkoski, S. Gollasch, and S. Olenin, editors. Invasive Aquatic Species of Europe: Distribution, Impacts and Management. Kluwer Academic Publishers, Dordrecht, The Netherlands.

Wallentinus, I. 2002. Introduced marine algae and vascular plants in European aquatic environments. Pages 27–52 in E. Leppäkoski, S. Gollasch, and S. Olenin, editors. Invasive Aquatic Species of Europe: Distribution, Impacts and Management. Kluwer Academic Publishers, Dordrecht, The Netherlands.

Profiles of invaders

Learning to identify invaders in advance would tell us a great deal about how life history traits evolve and how biotic communities are assembled. In practical terms, it could reveal the most effective means to prevent future invasions.

Richard Mack *et al.* (2000)

Chapter twelve

Ecological traits of aquatic NIS invading Austrian fresh waters

Leopold Füreder and Manfred Pöckl

HABITATS, FAUNA, AND NIS IN AUSTRIAN FRESH WATERS

The modern geographical distribution of aquatic organisms in Austria has been highly influenced by glaciation. Accordingly, most of the today temperate freshwater fauna in this area has become established within the last 10,000–15,000 years, a process that has not yet stopped. Although prehistoric man had already favoured the distribution of various plant and animal species, a considerable change in species distributions started with the onset of the great expeditions and continued with the subsequent alterations of the cultural landscape. Predominantly, plant species were transported to Europe and with them various species of animals and fungi arrived. In addition to the intentional introductions, many species were transferred to European fresh waters accidentally. The general decline of species diversity in many types of ecosystems has been a result of all these activities in central Europe. This is expected to be intensified, leading to a drastic increase of non-indigenous species (NIS).

Freshwater systems and their faunas were subject to dramatic changes due to the human impact (Strayer 2006). Similarly, organic pollution, river engineering, impoundments, hydropower development, tourism, and fisheries have considerably altered the structure and function of various lentic and lotic systems. In some extreme cases, stenotopic organisms (i.e. species occupying a narrow range of environmental gradients) have declined or disappeared, while eurytopic organisms (species occurring across a wide range of environmental

Francesca Gherardi, Biological invaders in inland waters: Profiles, distribution, and threats, 233–257.

gradients) have survived within generally species-poorer assemblages. As a consequence, NIS may have used the opportunity to compensate the obvious deficit in indigenous species.

Austria has about 6,000 natural lakes and about 100,000 km of rivers, with an estimated surface area of 1,300 km^2, which is about 1.6% of the area of the country. Its streams and rivers drain to three different systems, which are the Danube River system; its Austrian catchment covers well above 90% of the country, the Rhine River system in the extreme west, and the Elbe River system with the River Lainsitz or Lusnice and tributaries in the extreme north.

In a recently published book on the "neobiota" in Austria, Essl and Rabitsch (2002) recorded a still relatively small number of 89 aquatic taxa compared to 411 terrestrial NIS, which together – although based on a preliminary data set – represent an estimated 1.1% of the Austrian fauna (Rabitsch and Essl 2006). The Austrian freshwater NIS (Table 1) are predominantly found in river systems recently connected by shipping canals that facilitated the active migration and the passive transport of organisms. Major donor centres for freshwater NIS are the Ponto–Caspian Basin (Black Sea, Caspian Sea), North America, and South-east Asia. Due to the mainly passive transport, there is no preference for any specific organism; therefore, NIS can be found in a variety of taxonomic groups, e.g. fish, flatworms, isopods, molluscs, microcrustaceans, and crayfish.

Considerable anthropogenic alterations of river systems have expanded the available habitat especially for ecologically indifferent NIS (Spindler 1997, Pöckl 2002, Mikschi 2005). The species inventory of the Danube has been notably altered since 1992, when the Rhine–Main–Danube Canal was established. A well-documented example is the migration and succession of gammarid species in the Danube (Pöckl 2007). In the headwaters of river systems, few NIS have been found to date, although in these freshwater habitats noticeable examples of ecological relevance exist, e.g. the North American signal crayfish *Pacifastacus leniusculus* (Dana) and the New Zealand mudsnail *Potamopyrgus antipodarum* (Gray).

In isolated ponds, lakes, and marshes, which are generally less reachable and consequently less affected by accidental introductions (although exceptions may occur due to aquatic birds, boats, fishermen, and the aquarium trade), active release is more likely. Several examples can be found in fish, snails, crayfish, and turtles. For example, among the seven fish species of relevance for conservation, the cyprinids *Ctenopharyngodon idella* (Valenciennes), *Hypophthalmichthys molitrix* (Valenciennes), *Hypophthalmichthys nobilis* (Richardson), and the centrarchid *Lepomis gibbosus* (Linnaeus) in lentic waters, the salmonid *Oncorhynchus mykiss* (Walbaum) and the gobiid *Neogobius kessleri* (Günther) in lotic waters, and the cyprinid *Pseudoraspora parva* in both types of waters, the first five fish species were deliberately introduced (Mikschi 2002). The non-indigenous crayfish species are another example.

The causes of the occurrence of NIS in Austria and their vectors of propagation are diverse, being the result of deliberate and non-deliberate introductions

Table 1 Overview of 92 non-indigenous species recorded in Austrian fresh waters among bivalves, snails, crustaceans and other invertebrates, amphibians, reptiles, and fish. Information on their habitat, distribution, and potential impact is shown. For the remaining species, no adequate information is available. Information compiled from a draft data base of W. Rabitsch from the Federal Environmental Agency Ltd, Essl, Rabitsch (2002), and several specialists.

Taxa	No. of species	Habitat	Distribution				Impact		
			Single/local	Rare	Common	Abundant	No	Some	Strong
Various invertebrates	19	Lentic, lotic, benthic, host	6	9			7		
Bivalvia	3	Lotic, lentic benthic	2		1		2	1	
Gastropoda	10	Lotic, lentic benthic, thermal springs							
Crustacea	28	Lentic pelagic and benthic, semi-terrestrial, thermal springs, host	9	8	1	6	12	5	3
Amphibia	2	Cave, lotic		2			2		
Reptilia	2	Lentic pelagic	1	1			2		
Pisces	28	Lentic pelagic, lotic, thermal springs	7	4	6	3	19		

and displacements, active immigration (following displacement), and passive transportations and translocations by ships, most often due to the construction of canals that connect formerly separated faunistic provinces.

DELIBERATE INTRODUCTIONS WITH AN ECONOMIC BACKGROUND

Several species of fish and crayfish were originally imported and deliberately released in streams, rivers, and other water bodies (Spindler 1997, Füreder and Machino 1998, 1999, Eder 2002, Machino *et al.* 2004). Although ecological assessment studies had not been undertaken decades ago, some of these introductions were welcomed by the responsible authorities because of the expected economic value. Recreational fisheries are worth a lot of money; therefore, any supposed means of securing this income is welcome. For example, in Austria its value is estimated at approximately 75 million Euro per year, and there are officially at least 200,000 licensees (Spindler 1997).

Meanwhile, species like rainbow trout *O. mykiss* and American brook char *Salvelinus fontinalis* (Mitchell), introduced during the 1880s from North America, are well established and, despite their potential impact on indigenous fish species, were not considered as NIS according to the Fisheries Acts of the Austrian Federal States. Similarly, the European eel *Anguilla anguilla* (Linnaeus), non-indigenous to the Danube river drainage system, has been intensively stocked in many waters. As an aggressive predator it preferably feeds on indigenous crayfish and on small, endangered fish species.

Presumably, the oldest non-indigenous fish is the goldfish *Carassius auratus* (Linnaeus) from East Asia, which was imported during the 17th century for ornamental reasons in artificial park ponds by the aristocracy. Other cyprinid fish from the Amur drainage system in South-east Asia, such as *C. idella*, *H. molitrix*, *H. nobilis*, and *Mylopharyngodon piceus* (Richardson) (also known as grass carps), have been introduced and stocked since the 1970s. The common grass carp *(C. idella)* was especially used to consume and reduce aquatic plants in Austrian water bodies.

The notorious history of the non-indigenous crayfish invasion in Europe started with the introduction of the spiny-cheek crayfish *Orconectes limosus* (Rafinesque) in 1890, when 100 specimens from the USA were released into a 0.1 ha fish farm pond north-east of Berlin, Germany. Many introductions followed throughout Europe, as this species was expected to replace the indigenous noble crayfish *Astacus astacus* (Linnaeus), whose populations were decreasing due to the crayfish plague (Chapters 2 and 28). *Orconectes limosus* is today the most widely spread non-indigenous crayfish species in European inland waters and is responsible for the spreading of *Aphanomyces astaci* Schikora (the oomycete fungus that causes the crayfish plague). In 1970, about 2,000 specimens of the signal crayfish (*P. leniusculus*) were imported directly from California into Austria and the water bodies of several of its provinces were stocked. More specimens for further stockings came from Sweden. Within the last decades, the crayfish plague developed to become one of the major threats to indigenous crayfish in Europe. Although impact studies have not been performed up to now in Austria, we have learnt from other countries that indigenous crayfish species are highly threatened by the crayfish plague for which the vectors are predominantly non-indigenous crayfish, as well as by direct competition (Söderbäck 1991, 1995, Maiwald *et al.* 2006).

NON-DELIBERATE AND OTHER DELIBERATE INTRODUCTIONS

While sticklebacks, such as *Gasterosteus aculeatus* Linnaeus and *Pungitius pungitius* Linnaeus, have been introduced deliberately from North Europe and North America; the cyprinid *Pseudorasbora parva* (Temminck and Schlegel), which is now regarded as a pest in pond systems, has been introduced accidentally from South-east Asia together with the grass carp mentioned above. Together with introduced fish, parasitic worms (Monogenea, Cestoda, Nematoda) were also imported (Pöckl and Rabitsch 2002). Already in the 1870s, the giant trematode *Fascioloides magna* (Bassi), which has aquatic juvenile stages, was imported with Wapitis from North America. The parasite had severe impacts on the indigenous red deer *Cervus elaphus* Linnaeus and roe deer *Capreolus capreolus* (Linnaeus), as well as on the non-indigenous fallow deer *Dama dama* (Linnaeus) and on other related ruminants.

Aquarists are well known for having released a number of specimens in natural waters. Cichlids from Africa such as *Hemichromis letourneauxi* (Sauvage) and *H. fasciatus* Peters, as well as poecilids from Central and South America such as *Poecilia reticulata* Peters, *Xiphophorus maculatus* (Günther), and *X. helleri* Heckel, have been reported to be free-living in thermal waters near Villach in Carinthia. This is also the case for some specific warm-water gastropods; recently, the American red swamp crayfish *Procambarus clarkii* (Girard) was recorded from there (J. Petutschnig 2005, personal communication).

In Austria, several invertebrate and vertebrate NIS have been introduced accidentally. The Chinese mitten crab (*Eriocheir sinensis* H. Milne Edwards) was caught in the Danube downstream from Vienna (at Fischamend) by lift-net fishing in 2002 (Rabitsch and Schiemer 2003). The direct vector for these crabs is unknown, although they were most probably displaced by ships. They are, however, not expected to reproduce in the Danube, as the pelagic larvae of this species need brackish water for their development. A subspecies of the smooth newt, *Triturus vulgaris graecus* (Wolterstorff), from the Greek Balkan region was found in ponds near Vienna. The animals must have been set free by aquarists, as there are no other plausible reasons for their existence (Cabela and Grillitsch 2005). A recent record of young crocodiles, *Caiman crocodylus* (Linnaeus), in the Danube Canal in Vienna underlines this potential threat of NIS.

THE BUILDING OF CANALS AND SHIP TRAFFIC

Canals connect former separate drainage systems and consequently different faunistic provinces. Aquatic organisms are therefore presented with new opportunities to spread. Examples for the successful invasion of NIS are numerous. In most cases, the active upstream or downstream migration, drift, and passive transportation in the ballast waters of ships overlap. Biofouling, i.e. the development of biofilm or sessile biocoenosis on hard substrates, including natural (especially drifting wood) and artificial ones (the metal surfaces of vessels, boats, chains, and ropes), is responsible for species displacement from one water body to another. The migration of small non-indigenous organisms can further be facilitated by natural vectors: the feathers of waterfowl and the fur of water mammals, such as otters, minks, and voles.

The construction of the Rhine–Main–Danube Canal (25 September 1992) has opened an important migration and passive transportation route for aquatic organisms in Austria. With a total length of approximately 3,550 km it is one of the longest waterways in Europe, allowing the direct shipping of goods from the North Sea to the Black Sea. Since its opening, many species from the Ponto–Caspian Basin have successfully moved up- and downstream in the Danube, the Canal, and the Rhine in Germany and in the Netherlands. Many of them are now widespread, well established, and reach considerable densities and productivity. While the gudgeons *Neogobius kessleri* (Günther), *N. melanostomus* (Pallas), and *N. gymnotrachelus* (Kessler) from the Ponto–Caspian Basin are reported to be displaced in the embryonic stage by ships, some species from the Black and Caspian seas also reached Austria from the west in the downstream direction. An example is the mysid shrimp *Hemimysis anomala* G. O. Sars. Although this species originates from the Ponto–Caspian Basin, it was first displaced to the Baltic Sea, and reached the Danube via the Rhine and the Canal. Other examples for new species having invaded Austria from the west via the Rhine–Main–Danube Canal are *Atyaephyra desmaresti* (Millet) and *Echinogammarus trichiatus* Martynov (Pöckl and Rabitsch 2002).

STRATEGIES OF SUCCESSFUL INVADERS

In spite of increased opportunities for the immigration of NIS, not all of them are successful. Williamson (1996) formulated the "tens rule", to explain that only 10% of the established immigrant species can turn into pests. Species that were successful during a previous invasion on one continent have also proved successful as they entered new territories on another continent. This means that such species may become cosmopolitans, making ecosystems all over the world more and more similar with respect to species composition.

Successful invasions by new species usually proceed via a number of stages. The initial introduction must occur across a natural barrier, which is circumvented most often with the help of human activities. Enormous spatial leaps may be made, e.g. from one continent to another. Subsequently, the invasive species must be able to settle or adapt to the habitat in a new area that is conducive to its survival. There, it must be able to reproduce and complete its life cycle. The species will then spread over the new range successfully, usually by gradual local dispersal but also by jumps aided by transport action. It will show exponential population growth, reaching densities sometimes higher than ever recorded before, followed by a somewhat lower steady-state fluctuation.

Attempts have been made to define the possible factors likely to be important for successful invasions (Bij de Vaate *et al.* 2002). These include (a) invader properties, adaptations, genetic characteristics, mode of reproduction, growth rate, and dispersive capability; (b) competition, predation, diseases, and parasitism; and (c) climate match, habitat modification, vacant niche, and ancestral habitat.

The qualities of successful invaders are generally those of opportunistic species, often termed *r*-strategists. Morton (1997) listed several ecological/biological traits for molluscan invaders, such as a short lifespan and generation time, rapid growth with early sexual maturity, high fecundity, larger size than phylogenetically close species, habitat generalism with wide physiological tolerances, wide genetic variability and phenotypic plasticity, omnivorous and/or suspension feeding, and others. The success of invaders is ultimately also dependent on a variety of other factors, such as the lack of indigenous competitors, predators, diseases, and parasites, allowing uncontrolled population expansion.

Below, we will provide some well-studied examples of invasive species recorded in Austria and will highlight their success in relation to their life history traits and their potential threat to the indigenous fauna. These include (a) Asian clams and zebra mussels; (b) the amphipod *Dikerogammarus villosus* (Sovinski); and (c) the non-indigenous astacid and cambarid crayfish species.

Example 1: Asian clams, zebra and unionoidean mussels

The bivalve fauna of European rivers and lakes consists of species of the superfamily Unionoidea. In riverine habitats, they inhabit stable substrates of course and sand–gravel mixtures subject to moderate current velocities. Although there is interspecific variation, members of the Unionoidea display most of the K-selected characteristics expected of species adapted to the intense competition associated with stable habitats (Table 2). They have long lifespans and delayed maturity, are gonochoristic (i.e. species with sexes separate, the male and female reproductive organs being in different individuals), grow rapidly to maturity, and, thereafter, grow slowly, have extremely low juvenile survivorship but high adult survivorship, are highly iteroparous (i.e. have many reproductive periods within their extended lifespans) with

Table 2 Summary of the life history characteristics of indigenous, European freshwater bivalves (Unionoidea) and of the non-indigenous *Corbicula fluminea* and *Dreissena polymorpha* (after McMahon 2002).

	Unionoidea	*Corbicula fluminea*	*Dreissena polymorpha*
Origin	Indigenous, partly endangered	Asia, Australia, Africa	Caspian Sea, Ural River
Habitat stability	Generally stable	Stable to unstable	Stable to moderately unstable (can inhabit high-flow habitats)
Life history trait			
Lifespan (years)	<6 to >100, depending on species	1–4	4–7
Age at maturity (years)	6–12	0.25–0.75	0.5–2
Reproductive mode	Gonochoristic (few hermaphroditic species)	Hermaphroditic (self-fertilizing)	Gonocharistic
Growth rate	Rapid before maturity, much slower thereafter	Rapid throughout life	Relatively rapid throughout life
Fecundity (no. young per adult per breeding season)	200,000 – 17,000,000 per female depending on species	35,000 per hermaphroditic individual	30,000 – 1,000,000 per female
Juvenile size at release	Very small, 50–450 μm depending on species	Small, 250 μm	Extremely small, 40 μm
Relative juvenile survival	Extremely low	Extremely low	Extremely low
Relative adult survival	High	Low, 2–41% per year	Intermediate, 26–88% per year
Degree of iteroparity	Highly iteroparous, generally > 10 reproductive periods	Moderately iteroparous, 1–7 reproductive periods	Moderately iteroparous, 1–4 reproductive periods
Reproductive efforts per year	One	Two (spring to autumn)	One (2–8 months long)
Assimilated energy respired (%)	–	11–42%	> 90% from May to October
Non-respired energy allocated to growth (%)	85.2–97.5% depending on species	58–71% depending on cohort and season	10–26% depending on temperature and ration
Non-respired energy allocated to reproduction (%)	2.8–14.8% depending on species	5–15% depending on cohort and season	74–90% depending on temperature and ration
Turnover time in days (= mean standing crop biomass : biomass produced per day)	1,790–2,849 depending on species	73–91 depending on cohort	53–869 depending on habitat

one reproductive period per year, and tend to allocate high proportions of non-respired assimilated energy to growth and low proportions to reproduction (McMahon and Bogan 2001). Low juvenile survival and slow adult growth rates lead to low population productivity, reflected in extended turnover times (i.e. time in days for population to produce the equivalent of their mean population standing crop biomass) up to 2,849 days (Table 2). High adult survival, long life spans, and low juvenile survival result in domination of unionoidean populations by adults relative to juveniles, a trait characteristic of species adapted to stable habitats (Sibly and Calow 1986).

Unionoids deviate from the life history traits expected for species adapted to stable habitats in that females produce very large numbers of small young individuals (Table 2). Females retain eggs in marsupial chambers within the exhalant water channels of their outer gills where they are fertilized by sperm carried to them on inhalant currents. After fertilization, eggs develop into a small, externally released, bivalved larva called glochidium (McMahon and Bogan 2001). The glochidium is parasitic on specific fish hosts, encysting in their fins or gills for periods of less than 200 days to more than 1,000 days depending on species, allowing dispersal and growth to a more competitive size before excystment as a free-living juvenile (Bauer 1994). Because of the high risk within this parasitic stage, the effective fecundity of unionoidean species is quite low, leading to the production of a few, large, well-developed offspring (i.e. excysted juveniles), a characteristic of K-selected species from stable habitats (Sibly and Calow 1986).

Extended lifespans, delayed maturity, low effective fecundities, reduced powers of dispersal, high habitat selectivity, poor juvenile survival, and long turnover times (Table 2) make unionoidean populations highly susceptible to human perturbations (Strayer *et al.* 1999, McMahon and Bogan 2001). These unionoidean life history traits (particularly long life spans and low effective fecundities) slow population recovery from human- or naturally mediated habitat disturbances (Strayer *et al.* 1999, McMahon and Bogan 2001).

The introduced freshwater clam, *Corbicula fluminea* (O. F. Müller), unlike unionoideans, displays a majority of the life history traits (Table 2) that adapt it for life in unstable, unpredictable habitats (McMahon 1999). Arguably, it is the most invasive of all freshwater bivalves, occurring in southern Asia, Australia, and Africa and having been introduced into Europe, North America, and South America within the last 100 years (McMahon 1999). *Corbicula fluminea* grows rapidly, in part because it has higher filtration and assimilation rates than other freshwater bivalve species. Only a relatively small proportion of its assimilated energy is devoted to respiration, the majority being allocated to growth and reproduction. This species allocates a high proportion of non-respired assimilation to growth, allowing individuals to reach 15–30 mm in shell length in the first year of life and 35–50 mm in the terminal 3rd–4th year (McMahon 1999). Thus, *C. fluminea* has the highest net production efficiencies recorded for any freshwater bivalve, reflected by short turnover times (Table 2).

Newly released juveniles of *C. fluminea* are small but completely formed, with a well-developed bivalve shell, adductor muscles, foot, statocysts, gills, and digestive systems. They anchor to sediments or hard surfaces with a mucilaginous byssal thread but can resuspend in turbulent flows to be dispersed long distances downstream (McMahon 1999). A relatively low percentage of non-respired assimilation in *C. fluminea* is allocated to reproduction (equivalent to that expended by unionoideans); however, its elevated assimilation rates allow higher absolute energy allocation to reproduction than in other freshwater bivalves. Fecundity is high, estimated at 68,678 juveniles per adult per year (Aldridge and McMahon 1978). Juvenile survival, while higher than that of unionoideans, is still low, and unlike unionoideans, mortality rates remain high throughout adult life (74–98% in the first year, 59–69% in the 2nd year, and 93–97% in the 3rd year of life). Low adult survival leads to populations dominated by juveniles and immature individuals (McMahon 1999), a characteristic of species adapted to unstable habitats (Sibly and Calow 1986). Most European *C. fluminea* populations have two reproductive periods per year (i.e. spring through early summer and late summer through early autumn). *Corbicula fluminea* is hermaphroditic and self-fertilizing (Kraemer *et al.* 1986), allowing single individuals to found new populations. Maturation occurs within 3–6 months at a shell length of 6–10 mm, thus spring-born juveniles can participate in autumn reproduction. Maximum lifespan is highly variable, ranging from 1 to 4 years, within which early maturity and bivoltine reproduction allows individuals to participate in one to seven reproductive efforts (McMahon 1999).

The relatively short lifespan, early maturity, high fecundity, bivoltine juvenile release patterns, high growth rates, small juvenile size, and capacity for downstream dispersal of *C. fluminea* make it highly invasive and adapted for life in unstable lotic habitats subject to unpredictable catastrophic environmental disturbance. Its high reproductive potential and growth rate allow it to achieve high densities after having invaded a new habitat or re-establish dense populations soon after experiencing catastrophic population declines. Its high metabolic rates allow for rapid burrowing, being the species tolerant of suspended silt (McMahon 1999). Thus, it is highly successful in European and American drainage systems, subject to periodic anthropogenic interference, such as channelization, navigational dredging, "pearling" acitivities in unionoidean beds, sand and gravel dredging, commercial and recreational boating, and organic and chemical pollution, compared with the less resilient indigenous unionoidean species (McMahon and Bogan 2001).

Unlike unionoideans in which dispersal generally occurs within a population's drainage by host-fish glochidial transport, *C. fluminea* is adapted for transport between isolated drainages. The juvenile's mucilaginous byssal thread or the filamentous algae on which it settles entangles in the feet or feathers of shore birds and waterfowl, which then transport them between drainages (McMahon 1999).

The zebra mussel, *Dreissena polymorpha* (Pallas), was introduced into Austria long before *Corbicula* and *Sinanodonta*. Like *C. fluminea*, many of its life history characteristics (reviewed in Mackie and Schloesser 1996, McMahon 1996, Nichols 1996) make it highly successful as an invader (Table 2). Unlike all other European bivalve species, it has external fertilization that results in the development of a free-swimming, planktonic Veliger larva. The Veliger remains in the water column for 8–10 days before settlement, allowing long-distance downstream dispersal (Nichols 1996) or even displacement attached to small ships or fishermen's equipment. Veligers released into the Illinois River in North America were estimated to travel > 306 km downstream before settlement with total annual Veliger flux ranging approximately from 1.94×10^{14} to 2.13×10^{14} Veligers (Stoeckel *et al.* 1997). Adults byssally attached to floating objects can be transported long distances downstream, and juveniles attached to macrophytic vegetation can be carried between drainages by waterfowl. Thus, zebra mussels may be dispersed by a number of passive and active mechanisms (Carlton 1993).

Zebra mussels sustain relatively high growth rates throughout their lifespan, becoming sexually mature in the first year of life and reaching a terminal adult shell length of 3.5–5.0 cm. *Dreissena polymorpha* is gonochoristic, iteroparous, and univoltine, with individuals participating in three to four annual reproductive periods within their life span (Mackie and Schloesser 1996). The egg and freshly hatched Veliger larvae are small. *Dreissena polymorpha* allocates a lower percentage than *C. fluminea* of non-respired assimilation to somatic growth (Stoeckmann and Garton 2001; Table 2). Allocation of a relatively large proportion of non-respired assimilation to growth allows individuals to rapidly increase in size, making them more competitive and less vulnerable to predation (Sibly and Calow 1986). The elevated levels of energy devoted to reproduction by *D. polymorpha* relative to *C. fluminea* and its very small egg size allow for high fecundity (Table 2; Sprung 1991, Mackie and Schlosser 1996).

Dreissena polymorpha population densities can range from 7,000 to 114,000 individuals m^{-2}, and standing crop biomasses from 0.05 to 15 kg m^{-2} (Claudi and Mackie 1994, Mackie and Schloesser 1996). High standing crop biomass results from juveniles settling on the shells of adults and substrates inhabited by adults, forming dense mats. High individual growth rates and population densities lead to high population productivity, estimated to be 0.05–15 g C m^{-2} $year^{-1}$ in European populations and approximately 75 g C m^{-2} $year^{-1}$ in North American Great Lakes populations when converted from dry tissue mass productivity values (Mackie and Schloesser 1996). These productivity values, although higher than those of unionoideans, are still relatively low compared with those of 1,000–4,500 g C m^{-2} $year^{-1}$ estimated for dense North American *C. flumnea* populations (McMahon and Bogan 2001). Population growth and productivity are habitat-dependent in *D. polymorpha*, yielding variable turnover times, ranging from 53 days (high productivity) to 869 days (low productivity) (McMahon and Bogan 2001).

As for *C. fluminea*, the life history traits of *D. polymorpha* cause its populations to be dominated by juvenile and immature individuals (Smit *et al.* 1993), characteristic of a species adapted to unstable habitats (Sibly and Calow 1986).

The *r*-selected life history traits, high growth rate throughout life, elevated fecundity, short life spans, and long-distance downstream dispersal of adult and larval stages make *D. polymorpha* a successful invasive species. However, unlike *C. fluminea*, *D. polymorpha* populations tend to be characterized by a restriction to more stable habitats in medium- to large-sized lakes and rivers, reflected by its original distribution in the Caspian Sea and Ural River, avoidance of shallow, near-shore, lentic habitats and small, variable-flow, lotic habitats, relatively long age to maturity (generally at least 1 year), iteroparity, gonochorism, and relatively high adult survival (Mackie and Schloesser 1996). Thus, *D. polymorpha*, while adapted to rapidly recover from catastrophic reductions in population densities, appears less *r*-selected than *C. fluminea* but it is more *r*-selected than members of the Unionoidea.

Example 2: *Dikerogammarus villosus*

Dikerogammarus villosus was not found before 1989 in the Austrian stretch of the River Danube (Nesemann *et al.* 1995). It is reported to be a successful invader by competition and predation in the Rhine system (Müller *et al.* 2002) and in the large rivers in northern Germany (Grabow *et al.* 1998), as well as in the Moselle and in other French streams, rivers, and canals (Devin *et al.* 2001). In the Netherlands, Dick and Platvoet (2000) have found that *D. villosus* is having a marked impact on the indigenous *Gammarus duebeni* Lilljeborg, as well as on the non-indigenous *G. tigrinus* Sexton, until now a successful invader from North America, and they predict that it will further reduce amphipod diversity in a range of freshwater habitats in Europe. Besides its occurrences in the Danube, *Dikerogammarus villosus* has also been found to occur in several lakes, e.g. Traunsee and its outflow, the River Traun, Austria (O. Moog 2003, personal communication), Lake Constance, Germany (K. O. Rotthaupt 2003, personal communication), Lake Garda, Italy (Casellato *et al.* 2006), where it is observed to partially replace the indigenous *Echinogammarus stammeri* (S. Karamann).

Although in field studies and laboratory experiments the exceptional predatory capabilities of *D. villosus* on other amphipod species, *Asellus aquaticus* (Linnaeus), insect larvae, including the hard-shelled water boatman, and fish eggs have been observed – even small fish are reported to be attacked (Dick and Platvoet 2000, 2001, Dick *et al.* 2002, Chapter 27), Platvoet (2005) showed that the species is able to nourish itself by a wide range of feeding methods. Feeding habits included shredding, grazing, collecting micro-algae, macro-algae, coprophagy, and carnivory, and were dependent on the water temperature and the (micro)-distribution of prey organisms. Finally, van Riel *et al.* (2005) showed that *D. villosus* was less predatory when a population was well established in a respective habitat compared to the phase when a population was rapidly increasing its number in a recently invaded habitat. In Table 3, life history and ecological traits are compared between *D. villosus* and the indigenous gammarid species *Gammarus fossarum* Koch and *G. roeseli* Gervais.

Currently, the high reproductive potential of *D. villosus* in the Austrian Danube is greater than that reported for other Danubian gammarids with larger females carrying more than 100 developing embryos and juveniles in the brood pouch. Just a couple of animals is probably sufficient to found a new population. Moreover, the number of offspring produced by a female *D. villosus* in a single clutch is much higher than the total numbers that most of the indigenous freshwater gammarids, such as *G. fossarum*, *G. roeseli*, or *G. pulex*, can produce during their entire lifespan of ca. 1.5–2 years in 6–8 successive broods (Pöckl 1993, 2007, Pöckl *et al.* 2003). In Fig. 1, the frequency distribution of classes of egg numbers are plotted for *D. villosus* in comparison with *G. roeseli* and *G. fossarum*.

Also, the time taken to reach sexual maturity is considerably shorter at similar water temperatures, compared with other gammarid species (Piscart *et al.* 2003). Moreover, in the Moselle River, where the temperature range is similar to that of the Danube, *D. villosus* has a short generation time and rapid growth rates, particularly in younger specimens (Piscart *et al.* 2003). Altogether, some six life history parameters – early sexual maturity, short generation time, high growth rates, large reproductive capacity, comparatively small eggs, and optimal timing to release the maximum number of neonates per female in April/May (maximizing rapid growth at the period of high summer temperatures and plentiful food) – are important biological traits in the invasion process (Mordukhai-Boltovskoi 1949, Ricciardi and Rasmussen 1998, Kolar and Lodge 2001). Bruijs *et al.* (2001) highlighted the tolerance of *D. villosus* to a wide range of environmental conditions and several workers described the exceptional predatory capabilities of the species (Dick and Platvoet 2000, 2001, Dick *et al.* 2002, Chapter 27).

In contrast to the disturbed ecosystem of the River Rhine, where *D. villosus* is reported to have eliminated indigenous species, several other gammarid species cohabit with *D. villosus* at the sampling site of the Danube: *D. haemobaphes*, *D. bispinosus*, *Echinogammarus ischnus* Stebbing, *Obesogammarus obesus* (G. O. Sars), *Chelicorophium curvispinum* (G. O. Sars), and *G. roeseli*.

Bruijs *et al.* (2001), who did laboratory experiments on the oxygen consumption, temperature, and salinity tolerance of *D. villosus*, speculated that the species might survive (incomplete) ballast water exchange and subsequently be dispersed over intercontinental distances. It has thus the potential to develop large populations in temperate biomes all over the world and could develop into a cosmopolitan species. In all life history variables, *D. villosus* is competitively superior to many of the indigenous species in north-western Europe and elsewhere. If it will ever spread into North America, its impact on freshwater and brackish water ecosystems could be disastrous.

Table 3 Summary of the life history characteristics of the non-indigenous gammarid *Dikerogammarus villosus* compared to the two indigenous *Gammarus roeseli* and *G. fossarum*.

	D. villosus	*G. roeseli*	*G. fossarum*
Origin	Ponto–Caspian Basin	Indigenous	Indigenous
Habitat	R. Danube	warm rivers (high order streams)	small brooks (low order streams)
Life history trait			
Life span (years)	1 ??	1–2	1–2
Age at maturity (months)	3	4	5
Growth rate	Rapid throughout life	Rapid until sexual maturity	Slow throughout life
Fecundity per clutch (No. of eggs)	43 (5–194)	32 (5–87)	15 (5–59)
Fecundity in total life span (No. of eggs)	350 ??	120–169	127–208
Egg size (mm^3)	Small (0.05)	Large (0.075)	Large (0.08)
Juvenile size at hatching	1.5 mm	1.5 mm	1.5 mm
Relative juvenile survival	Extremely low	Low	Intermediate
Size at maturity (mm)	7.5	8.5	6.5
Brood development time	Rapid	Slow at cold temperatures	Intermediate
Maximum size (mm, M/F)	22/18	19/16.5	17.5/12
Embryonic survival	High	Intermediate	High
Relative adult survival	Low	Intermediate	High
Degree of iteroparity	Low (reproductive peaks with cohorts)	Intermediate	Low (+ continuous, no cohorts)
Reproductive resting stage	October–December (3 months)	October–March (5 months)	October–November (2 months)

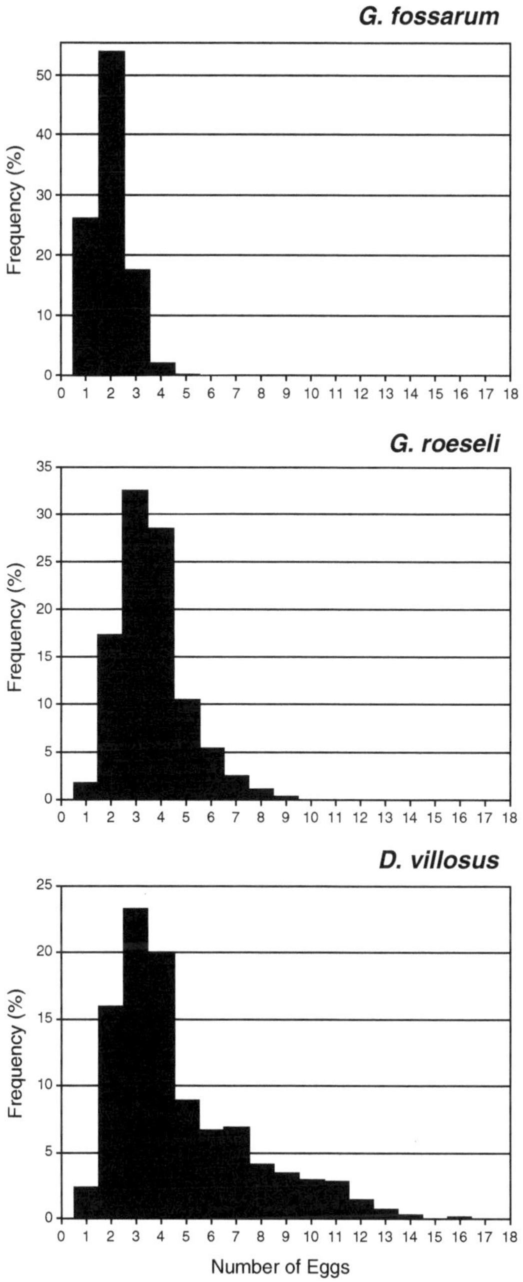

Fig. 1 Frequency distribution of egg numbers per clutch in classes (1 is up to 10 eggs, 2 is 11–20 eggs, 3 is 21–30 eggs, etc.). Amphipods are from Austrian locations: *Gammarus fossarum* from six Austrian streams, *G. roeseli* from two Austrian streams, *Dikerogammarus villosus* from the Danube.

Example 3: Indigenous and non-indigenous freshwater crayfish

In Austria, the two astacids *Astacus astacus* and *Austropotamobius torrentium* (Schrank) are considered indigenous, whereas the position of the two astacids *A. leptodactylus* and *Austropotamobius pallipes* (Lereboullet) is still debated. The astacid *P. leniusculus* and the cambarids *O. limosus* and *P. clarkii* are of North American origin (Füreder *et al.* 2006, Chapter 28). The non-indigenous crayfish considerably increased their distribution within the last five years: the available data show an increase in records of the spiny-cheek crayfish from 16 to 25 and of the signal crayfish from 126 to more than 320 since 1998 (L. Füreder 2006, unpublished data). Meanwhile, *P. clarkii* has also been found very recently near Villach (J. Petutschnig 2005, personal communication).

Neveu (2006) has classified six crayfish species according to *r* and K strategies (Table 4), the European species *A. astacus*, *A. pallipes*, and *A. leptodactylus*, and the North American species *P. leniusculus*, *O. limosus*, and *P. clarkii*. According to their biological and life history traits, these species can clearly be separated, which demonstrates these species pose increasing threats to the indigenous species.

Annual life cycle and activity patterns differ between the crayfish species. The indigenous *A. astacus*, the two *Austropotamobius* species, and the North American *P. leniusculus* mate and spawn in October and November, and carry the eggs during the cold season of the year. The mating and spawning of *A. leptodactylus* occurs about one month later in December. The juveniles of these five species hatch in May and June. The offspring leave the mother ~2 weeks later in June and in the beginning of July. The incubation time of the eggs of *A. torrentium*, *A. pallipes*, *A. astacus*, *A. leptodactylus*, and *P. leniusculus* ranges from 158 to 214 days.

Orconectes limosus and *P. clarkii* spawn at other periods of the year. Mating in *O. limosus* occurs from the end of August until the beginning of April, but egg-carrying females are not found until mid-April. The juveniles hatch at the end of May and in June, as in the European species and in *P. leniusculus*. The incubation time of the eggs of *O. limosus* is therefore much shorter (49–55 days). Females of *P. clarkii* spawn at different times during the year. The incubation time for the eggs is ~48–55 days. Females bearing eggs at different stages of development and females carrying embryos or juveniles are found throughout the year. The hatching and incubation time for the eggs carried by females spawning in late autumn is not known. Most probably, they are carried by the female during winter, and hatching is in spring as for other species in moderate climates.

Females of *A. torrentium* and *A. pallipes* become sexually mature at a size of between 23 and 26 mm in cephalothorax length (CL). Single crayfish reach this size at the end of the 2nd year, but the majority of crayfish in both species become sexually mature in the 3rd year. Females of *A. astacus* reach sexual maturity at a CL of 41–44 mm in the 3rd–4th year. *Astacus leptodactylus*

Table 4 Classification of crayfish species from Brittany (France) according to the two strategies *r* and K in relation to their life history characteristics. Code: ***: strong test; **: medium test; * low test; total score: stars sum for each species (AUP *Austropotamobius pallipes*, ASA *Astacus astacus*, ASL *Astacus leptodactylus*, PAL *Pacifastacus leniusculus*, ORL *Orconectes limosus*, PRC *Procambarus clarkii*). After Neveu (2006).

	AUP	ASA	ASL	PAL	ORL	PRC
***r*-selection?**						
Early sexual maturity			*		***	***
High fecundity				*	**	***
Small egg size				*	***	***
Rapid egg development					**	***
Long egg-laying duration						***
Short hatching duration						***
Fast growth 1st summer			***		***	*
Fast growth 2nd summer		**	***	*	**	***
Small body 2nd summer	**			*	***	
Variable size 1st summer			*	*	*	***
Short life span					**	***
Susceptible to plague				***	***	***
Score total	**2**	**2**	**8**	**8**	**24**	**31**
K-selection?						
Extended sexual maturity	***	**		**		
Low fecundity	***	***	**			
Large egg size	***	**	*			
Slow egg development	***	**	*	*		
Short egg-laying duration	***	*	*	***	***	
Short hatching duration	***	*	*	***	***	
Slow growth 1st summer	***					
Slow growth 2nd summer	***					
Larger size 2nd summer		**	***	**		***
Same size 1st summer	***	***	*			
Long life span	**	***	***	*		
Susceptible to plague	***	***	***			
Score total	**32**	**22**	**16**	**12**	**6**	**3**

females reach maturity at a similar size (39–40 mm CL) as *A. astacus* but, due to faster growth, some individuals reach that size during the 2nd year. In contrast to the European species, the females of all the American species, *O. limosus*, *P. leniusculus*, and *P. clarkii*, become mature in their 2nd year. Also the males of *P. leniusculus* and *P. clarkii* reach maturity in the 2nd year.

The American crayfish species display a high capacity for recruiting offspring. The three species reach sexual maturity in their 2nd year. Young females are already able to spawn high numbers of eggs, and the average egg number per female is high. Females of *P. clarkii* are reported to spawn up to 700 eggs, and the average number of eggs can reach up to 300 per female in populations in Portugal and Louisiana (Correia 1995, Avery and Lorio 1996, Ilhéu and Bernardo 1997). In *P. leniusculus*, average numbers of 160–300 eggs per female are described for Scandinavia, Germany, France, and the western USA (Abrahamsson 1972, Müller 1978, Shimizu and Goldman 1983, Westman *et al.* 1995, Mankampa and Chaisemartin 1996). In contrast, *A. torrentium* and *A. pallipes* spawn a lower number of eggs, averaging 39–75 eggs. The average egg diameter is between 2.52 and 3.06 mm in *A. astacus*, *A. leptodactylus*, *A. torrentium*, *A. pallipes*, and *P. leniusculus*. The eggs of *O. limosus* and *P. clarkii* are smaller, with average diameters of 1.76–2.00 mm.

NIS display several characteristics that differ from the life history traits of the indigenous species but also show a different behavioural pattern. *Astacus leptodactylus*, *O. limosus*, and *P. clarkii* can be active during both nocturnal and diurnal hours and their activity period during the year is longer. The flexibility in the diurnal activity pattern enables these crayfish populations to extend their feeding activity during times of favourable conditions and to avoid predators. According to Cukerzis (1988), the better adaptability of *A. leptodactylus* to different light regimes and the better utilization of the food supply are reasons for its success in competition with *A. astacus*.

Orconectes limosus and *P. clarkii* show not only a more flexible activity pattern, but also a different life cycle and a more flexible reproductive pattern than the other species. As cambarids, females are able to store spermatophores in their *annulus ventralis*, which enables a temporal separation between mating and spawning. The European species and *P. leniusculus* lack this morphological feature, and spawn one or two weeks after mating. As female *O. limosus* and *P. clarkii* carry eggs in spring and summer, the temperature dependent embryonic development is much faster. Moreover, females of *P. clarkii* are able to spawn at different times of the year, and there are always crayfish at different stages of development in the population. Therefore, a population is able to respond very quickly to changing environmental conditions and, due to this ability to store spermatophores, times of unfavourable conditions can be overcome. This strategy which may have evolved as an adaptation to temporary waters in their original distribution area is maintained also in the permanent water bodies in Europe.

In summary, *O. limosus* and *P. clarkii* show several characteristics of *r*-selected organisms. Both species have an early sexual maturity, are fast growing, spawn many small eggs, and have a relatively short lifespan. On the contrary, the life history of the indigenous species shows characteristics of more K-selected organisms. Sexual maturity is later, growth is slower, the females spawn fewer but larger eggs and the lifespan is longer. *Astacus astacus* and *P. leniusculus* show characteristics between these two extreme positions. *Pacifastacus leniusculus* grows fast and has a high fecundity, but the crayfish reaches a large body size and has a long lifespan. Growth is also fast in *A. leptodactylus*; females spawn high quantities of large eggs and their life is long.

The comparison of life history and ecological traits demonstrates that non-indigenous crayfish are clearly able to exploit European water bodies. Due to their high fecundity, early maturity, fast growth, and flexibility in life cycle and activity, the introduced species show characteristics of successful dispersal and are superior in the competition with indigenous species (Stucki 2000, Holdich *et al.* 2006). Additionally, populations of the indigenous European species, particularly *A. astacus*, were badly affected by the introduction of the crayfish plague.

CONCLUSIONS

Habitat destruction and degradation, pollution, introductions of NIS, direct harvest, and global climate change have been considered as the major threats to the freshwater fauna (Strayer 2006). Introductions of non-indigenous species are now widespread around the world (e.g. Cox 1999). Also in Austria, some of the introductions were deliberate, but most were unintentional (Essl and Rabitsch 2002, Rabitsch and Essl 2006). Non-indigenous species often have strong, long-lasting, and irreversible ecological effects. Our examples of successful invaders of Austrian waters demonstrate their potential threats for ecologically similar indigenous species. In some cases, alterations of the whole ecosystem have to be expected. The two bivalve species, *D. polymorpha* and *C. fluminea* are highly competitive: it was demonstrated at other locations that they can radically change the physical, chemical, and biological characteristics of lakes and rivers (e.g. Strayer *et al.* 1999). Competition with zebra mussels led to the extirpation of many populations of indigenous unionid mussels in North America (Ricciardi *et al.* 1998, Strayer 1999); Ricciardi *et al.* (1998) even predicted that the zebra mussel invasion will drive unionids into global extinction. For Austria, the time of monitoring is still too short or the effort even not existent; therefore, drastic changes are hardly reported and may occur unnoticed. However, given the examples from other countries, worse scenarios have to be expected.

The life history and ecological traits of the non-indigenous molluscan and crustacean species show that most of them have the potential to become highly invasive. Similarly to *D. villosus*, these are considerably different from those of the indigenous species, giving them the ability to potentially change the biological characteristics of the invaded habitats. The same is true for the American crayfish species and also for some non-indigenous fishes. Today, the rainbow trout is distributed throughout Austria's epirhithral zone and represents a strong but often neglected competitor for space and food with the indigenous fish species.

Many other non-indigenous plants, microbes, fungi, invertebrates, fishes and other vertebrates, and diseases have been moved around the world and have had strong ecological effects (e.g. Cox 1999, Mack *et al.* 2000). The repeated outbreak of the crayfish plague is the unmistakable evidence of the threats that NIS introductions bring along. The introduction of NIS continues to be a difficult and growing problem in freshwater biodiversity conservation because the number of new invasions is expected to rise also in Austria and the effects of established NIS tend to be cumulative and difficult or impossible to reverse.

Our examples also demonstrate that ecosystems can easily be colonized by invasive NIS and, due to their dominance over indigenous species, endemic, unique, rare, and stenotopic species may be lost. Due to the impacts on freshwater ecosystems all over the world and the effects of global warming, the establishment of NIS will make the composition of the aquatic fauna more and more similar and dominated by globally distributed, formerly non-indigenous, invaders. The "homogenization of the earth's biota" (Rahel 2002) is acknowledged as a major threat to global biodiversity. In Austria, this is expected to be especially the case in larger rivers, but it may also affect lakes where human activities threaten the existence of NIS. Endemic, unique, and endangered indigenous species may have a better chance to survive in woodland brooks, headwaters, and isolated water bodies at higher altitudes where the pressure of human activities is less severe.

Once a population of NIS is well established, it is impossible to eradicate it under normal conditions, and even management methods for an effective population control will be intensive in cost, labour, and time. Hence, it is the duty of scientists and authorities to raise public awareness. As in other European countries, there is an Austrian Action Plan on invasive aquatic species (Essl and Rabitsch 2004), where the spread and impact of NIS should be prevented through concerted national actions and international cooperation. Appropriate adaptations in customs acts (legislation) for prohibiting the import of NIS and enforced control were considered necessary (Holdich and Pöckl 2005). However, when we approached the topic of the ecological impact of NIS in Austria, we became rapidly aware that there exists a great lack of knowledge and data. Ecological impact studies are very necessary, compulsory, and must be seriously conducted before an NIS is set free. The area has to be extremely limited, and the performance of the non-indigenous population and the reaction of indigenous species studied in great detail. Given today's knowledge, well accustomed practices in the introduction of NIS have to be questioned in terms of necessity, economy, effects, and (better) alternatives.

ACKNOWLEDGEMENTS

We thank Wolfgang Rabitsch who provided the database of non-indigenous freshwater animal species which was prepared within the EU-STREP-Project DAISIE (Delivering Alien Invasive Inventories for Europe). We are grateful for valuable suggestions by Francesca Gherardi and Wolfgang Rabitsch as well as two anonymous reviewers.

REFERENCES

Abrahamsson, S. A. A. 1972. The crayfish *Astacus astacus* in Sweden and the introduction of the American crayfish *Pacifastacus leniusculus*. Freshwater Crayfish **1**, 27–40.

Aldridge, D. W. and R. F. McMahon. 1978. Growth, fecundity, and bioenergetics in a natural population of the Asiatic freshwater clam, *Corbicula fluminea* Phillippi, from north central Texas. Journal Molluscan Studies **44**, 49–70.

Avery, J. and W. Lorio. 1996. Crawfish Production Manual. Louisiana Cooperative Extension Service Publication 2637. Louisiana State University, Baton Rouge, LA.

Bauer, G. 1994. The adaptive value of offspring size among freshwater mussels (Bivalvia; Unionidae). Journal of Animal Ecology **63**, 933–944.

Bij de Vaate, A., K. Jaźdźewski, H. A. M. Ketelaars, S. Gollasch, and G. Van der Velde. 2002. Geographical patterns in range extension of Ponto–Caspian macroinvertebrate species in Europe. Canadian Journal of Fisheries and Aquatic Sciences **59**, 1159–1174.

Bruijs, M. C. M., B. Kelleher, G. Van der Velde, and A. Bij de Vaate. 2001. Oxygen consumption, temperature and salinity tolerance of the invasive amphipod *Dikerogammarus villosus*: indicator of further dispersal via ballast water transport. Archiv für Hydrobiologie **152**, 633–646.

Cabela, A. and H. Grillitsch. 2005. On the presence of a southeastern European Smooth Newt near Vienna (Austria). Short notice. Herpetozoa **18**, 84–87.

Carlton, J. T. 1993. Dispersal mechanisms of the zebra mussel, *Dreissena polymorpha*. Pages 677–697 in T. F. Nalepa and D. W. Schloesser, editors. Zebra mussels: biology, impacts and control. Lewis Publishers, Boca Raton, FL.

Casellato, S., G. La Piana, L. Latella, and S. Ruffo. 2006. Dikerogammarus villosus (Sowinsky, 1894) (Crustacea, Amphipoda, Gammaridae) for the first time in Italy. Italian Journal of Zoology **73**, 97–104.

Claudi, R. and G. L. Mackie. 1994. Practical manual for zebra mussel monitoring and control. Lewis Publishers, Boca Raton, FL.

Correia, A. M. 1995. Population dynamics of *Procambarus clarkii* (Crustacea: Decapoda) in Portugal. Freshwater Crayfish **8**, 287–290.

Cox, G. W. 1999. Alien species in North America and Hawaii: impacts on natural ecosystems. Island Press, Washington, DC.

Cukerzis, J. M. 1988. *Astacus astacus* in Europe. Pages 309–340 in D. M. Holdich and R. S. Lowery, editors. Freshwater Crayfish – Biology, Management and Exploitation. Timber Press, Portland, OR.

Devin, S., J.-N. Beisel, V. Bachmann, and J.-C. Moreteau. 2001. *Dikerogammarus villosus* (Amphipoda: Gammaridae): another invasive species newly established in the Moselle river and French hydrosystems. Annales de Limnologie **37**, 21–28.

Dick, J. T. A. and D. Platvoet. 2000. Invading predatory crustacean *Dikerogammarus villosus* eliminates both native and exotic species. Proceedings of the Royal Society of London B **267**, 977–983.

Dick, J. T. A. and D. Platvoet. 2001. Predicting future aquatic invaders; the case of *Dikerogammarus villosus*. Aquatic Nuisance Species **4**, 25–27.

Dick, J. T. A., D. Platvoet, and D. W. Kelly. 2002. Predatory impact of the freshwater invader *Dikerogammarus villosus* (Crustacea: Amphipoda). Canadian Journal of Fisheries and Aquatic Sciences **59**, 1078–1084.

Eder, E. 2002. Krebstiere: Flusskrebse (Crustacea: Decapoda: Astacidae und Cambaridae). Pages 287–291 in F. Essl and W. Rabitsch, editors. Neobiota in Österreich. Umweltbundesamt, Vienna, Austria.

Essl, F. and W. Rabitsch. 2002. Neobiota in Österreich. Umweltbundesamt, Vienna, Austria.

Füreder, L., D. Holdich, R. Schulz, H. Schulz, Y. Machino, D. Sint, J. Reynolds, M. C. Trouilhé, P. Kozák, T. Taugbøl, L. Edsman, B. Renai, and M. Pöckl. 2006. Chapter 2: Indigenous crayfish – habitat and threats. Pages 26–47 in C. Souty-Grosset, D. M. Holdich, P. Y. Noël, J. R. Reynolds, and P. Haffner, editors. Atlas of crayfish in Europe. Patrimoines Naturels 63, Muséum national d'Histoire naturelle, Paris, France.

Füreder, L. and Y. Machino. 1998. Verbreitung von Flußkrebsen in Tirol, Südtirol und Vorarlberg. Stapfia **58**, 77–88.

Füreder, L. and Y. Machino. 1999. Past and present crayfish situations in Tyrol (Austria). Freshwater Crayfish **12**, 751–764.

Grabow, K., T.-O. Eggers, and A. Martens. 1998. *Dikerogammarus villosus* Sowinsky (Crustacea: Amphipoda) in navigable canals and rivers in northern Germany. Lauterbornia **33**, 103–107.

Holdich, D. M., P. Haffner, P. Noël, J. Carral, L. Füreder, F. Gherardi, Y. Machino, J. Madec, M. Pöckl, P. Smietana, T. Taugbøl, and E. Vigneux. 2006. Species files. Chapter 3. Pages 50–129 in C. Souty-Grosset, D. M. Holdich, P. Y. Noël, J. D. Reynolds, and P. Haffner, editors. Atlas of crayfish in Europe. Patrimoines Naturels 63, Muséum national d'Histoire naturelle, Paris, France.

Holdich, D. M. and M. Pöckl. 2005. Does legislation work in protecting vulnerable species? Pages 809–827 in L. Füreder and C. Souty-Grosset, editors. European native crayfish in relation to land-use and habitat deterioration with a special focus on *Austropotamobius torrentium*. CRAYNET, vol. **3**, Bulletin Francais de la Pêche et de la Pisciculture **376–377**, 809–827.

Ilhéu, M. and J. M. Bernardo. 1997. Life history and population biology of the red swamp crawfish, *Procambarus clarkii*, in a Mediterranean reservoir. Freshwater Crayfish **11**, 54–59.

Kolar, C. S. and D. M. Lodge. 2001. Progress in invasive biology: predicting invaders. Trends in Ecology and Evolution **16**, 199–204.

Kraemer, L. R., C. Swanson, M. Galloway, and R. Kraemer. 1986. Biological basis of behavior in *Corbicula fluminea*. II. Functional morphology of reproduction and development and review of evidence for self-fertilization. American Malacological Bulletin, Special Edition **2**, 193–201.

Machino, Y., L. Füreder, P. J. Laurent, and J. Petutschnig. 2004. Introductions of white-clawed crayfish *Austropotamobius pallipes* in Europe. Berichte naturwissenschaftlich-medizinischer Verein Innsbruck **91**, 187–212.

Mack, R. N., D. Simberloff, W. M. Lonsdale, H. Evans, M. Clout, and F. A. Bazzaz. 2000. Biotic invasions: causes, epidemiology, global consequences, and control. Ecological Applications **10**, 689–710.

Mackie, G. L. and D. W. Schloesser. 1996. Comparative biology of zebra mussels in Europe and North America: an overview. American Zoologist **36**, 244–258.

Maiwald, T., H. K. Schulz, P. Smietana, and R. Schulz. 2006. Aggressive interactions and interspecific competition between the indigenous crayfish *Astacus astacus* (Linne) and the non-indigenous *Orconectes limosus* (Rafinesque). Freshwater Crayfish **15**, 203–211.

Mankampa, M. and C. Chaisemartin. 1996. Etude comparative de cinq espèces d'écrevisses, définis par leurs populations naturelles du même secteur géographique. Vie et Milieu **46**, 79–85.

McMahon, R. F. 1996. The physiological ecology of the zebra mussel, *Dreissena polymorpha*, in North America. American Zoologist **36**, 339–363.

McMahon, R. F. 1999. Invasive characteristics of the freshwater bivalve, *Corbicula fluminea*. Pages 315–343 in R. Claudi and J. H. Leach, editors. Nonindigenous freshwater organisms: vectors, biology and impacts. Lewis Publishers, Boca Raton, FL.

McMahon, R. F. 2002. Evolutionary and physiological adaptations of aquatic invasive animals: *r*-selection versus resistance. Canadian Journal of Fisheries and Aquatic Sciences **59**, 1235–1244.

McMahon, R. F. and A. E. Bogan. 2001. Mollusca: Bivalvia. Pages 331–429 in J. H. Thorp and P. Covich, editors. Ecology and classification of North American freshwater invertebrates. 2nd edition. Academic Press, San Diego, CA.

Mikschi, E. 2002. Fische (Pisces). Pages 197–204 in F. Essl and W. Rabitsch, editors. Neobiota in Österreich. Umweltbundesamt, Vienna, Austria.

Mikschi, E. 2005. Fische. Pages 133–147 in Bundesministerium für Land- und Forstwirtschaft, Umwelt und Wasserwirtschaft, editor. Aliens. Neobiota in Österreich. Grüne Reihe des Lebensministeriums, 15, Vienna, Austria.

Mordukhai-Boltovskoi, D. 1949. Life cycle of some Caspian gammarids. Dokl. Akad. Nauk SSSR **46**, 997–999.

Morton, B. S. 1997. The aquatic nuisance species problem: a global perspective and review. Pages 1–54 in F. M. D'Itri, editor. Zebra mussels and aquatic nuisance species. Ann Arbor Press, Chelsea, MI.

Müller, G. 1978. Studie zur Neueinbürgerung des gegen die Krebspest resistenten Signalkrebses (*Pacifastacus leniusculus* Dana). Freshwater Crayfish **4**, 381–383.

Müller, J. C., S. Schramm, and A. Seitz. 2002. Genetic and morphological differentiation of *Dikerogammarus* invaders and their invasion history in Central Europe. Freshwater Biology **47**, 2039–2048.

Nesemann, H., M. Pöckl, and K. J. Wittmann. 1995. Distribution of epigean Malacostraca in the middle and upper Danube (Hungary, Austria, Germany). Miscellanea Zoologica Hungarica **10**, 49–68.

Neveu, A. 2006. Les écrevisses étrangères sont elles invasives? Quelles cractéristique expliquent leur dévelopment? L'Astacuculteur de France **86**, 2–11.

Nichols, S. J. 1996. Variation in the reproductive cycle of *Dreissena polymorpha* in Europe, Russia, and North America. American Zoologist **36**, 311–325.

Piscart, C., S. Devin, J.-N. Beisel, and J.-C. Moreteau. 2003. Growth-related life-history traits of an invasive gammarid species: evaluation with a Laird-Gompertz model. Canadian Journal of Zoology **81**, 2006–2014.

Platvoet, D. 2005. Temperature dependent feeding in *Dikerogammarus villosus* Sowinsky. Page 60 in abstracts for Biological Invasions in Inland Waters (INWAT) workshop, Firenze, Italy, May 5–7, 2005.

Pöckl, M. 1993. Reproductive potential and lifetime potential fecundity of the freshwater amphipods *Gammarus fossarum* and *G. roeseli* in Austrian streams and rivers. Freshwater Biology **30**, 73–91.

Pöckl, M. 2002. Krebstiere: Flohkrebse (Crustacea: Amphipoda: Corophiidae und Gammaridae). Pages 273–284 in F. Essl and W. Rabitsch, editors. Neobiota in Österreich. Umweltbundesamt, Vienna, Austria.

Pöckl, M. 2007. Strategies of a successful new invader to European fresh waters: fecundity and reproductive potential of the Ponto–Caspian amphipod *Dikerogammarus villosus* in the Austrian Danube, compared with the indigenous *Gammarus fossarum* and *G. roeseli*. Freshwater Biology **52**, 50–63.

Pöckl, M. and W. Rabitsch. 2002. Ausgewählte aquatische Neozoen (Cnidaria, Plathelminthes, Kamptozoa, Annelida, Isopoda). Pages 228–239 in F. Essl and W. Rabitsch, editors. Neobiota in Österreich. Umweltbundesamt, Vienna, Austria.

Pöckl, M., B. W. Webb, and D. W. Sutcliffe. 2003. Life history and reproductive capacity of *Gammarus fossarum* and *G. roeseli* (Custacea: Amphipoda) under naturally fluctuating water temperatures: a simulation study. Freshwater Biology **48**, 53–66.

Rabitsch, W. and F. Essl. 2006. Biological invasions in Austria: patterns and case studies. Biological Invasions **8**, 295–308.

Rabitsch, W. and F. Schiemer. 2003. Chinesische Wollhandkrabbe (Erichoir sinensis) in der Österreichischen Donau festgestellt. Österreichs Fischerei **56**, 61–65.

Rahel, F. J. 2002. Homogenization of freshwater faunas. Annual Review of Ecology and Systematics **33**, 291–315.

Ricciardi, A. and J. B. Rasmussen. 1998. Predicting the identity and impact of future biological invaders: a priority for aquatic resource management. Canadian Journal of Fisheries and Aquatic Sciences **55**, 1759–1765.

Ricciardi, A., R. J. Neves, and J. B. Rasmussen. 1998. Impending extinctions of North American freshwater mussels (Unionoida) following the zebra mussel (*Dreissena polymorpha*) invasion. Journal of Animal Ecology **67**, 613–619.

Shimizu, S. and C. Goldmann. 1983. *Pacifastacus leniusculus* production in the Sacramento River. Freshwater Crayfish **5**, 210–228.

Sibly, R. M. and P. Calow. 1986. Physiological ecology of animals: an evolutionary approach. Blackwell, London, UK.

Smit, H., A. Bij de Vaate, H. H. Reeders, E. H. Van Nes, and R. Noordhuis. 1993. Colonization, ecology, and positive aspects of zebra mussels: biology, impacts and control. Pages 55–77 in T. F. Nalepa and D. W. Schloesser, editors. Zebra mussels: biology, impacts and control. Lewis Publishers, Boca Raton, FL.

Söderbäck, B. 1991. Interspecific dominance relationship and aggressive interactions in the freshwater crayfishes *Astacus astacus* (L.) and *Pacifastacus leniusculus* (Dana). Canadian Journal of Zoology **69**, 1321–1325.

Söderbäck, B. 1995. Replacement of the native crayfish *Astacus astacus* by the introduced species *Pacifastacus leniusculus* in a Swedish lake: possible causes and mechanisms. Freshwater Biology **33**, 291–304.

Spindler, T. 1997. Fischfauna in Österreich. Ökologie-Gefährdung-Bioindikation-Fischerei-Gesetzgebung. Umweltbundesamt, Monographien Band 53, Bundesministerium für Umwelt, Vienna, Austria.

Sprung, M. 1991. Costs of reproduction: a study on the metabolic requirements of the gonads and fecundity of the bivalve *Dreissena polymorpha*. Malacologia **33**, 63–70.

Stoeckel, J. A., D. W. Schneider, L. A. Soeken, K. D. Blodgett, and R. E. Sparks. 1997. Larval dynamics of a riverine metapopulation: implications for zebra mussel recruitment, dispersal, and control in large river systems. Journal of the North American Benthological Society **16**, 586–601.

Stoeckmann, A. M. and D. W. Garton. 2001. Flexible energy allocation in zebra mussels (*Dreissena polymorpha*) in response to different environmental conditions. Journal of the North American Benthological Society **20**, 486–500.

Strayer, D. L. 1999. Effects of alien species on freshwater molluscs in North America. Journal of the North American Benthological Society **18**, 74–98.

Strayer, D. L. 2006. Challenges for freshwater invertebrate conservation. Journal of the North American Benthological Society **25**, 271–287.

Strayer, D. L., N. F. Caraco, J. J. Cole, S. Findlay, and M. L. Pace. 1999. Transformation of freshwater ecosystems by bivalves: a case study of zebra mussels in the Hudson River. BioScience **49**, 19–27.

Stucki, T. P. 2000. Differences in life history of native and introduced crayfish species in Switzerland. Freshwater Crayfish **13**, 463–476.

van Riel, M. C., G. van der Velde, and A. Bij de Vaate. 2005. Trophic relationship in the Lower Rhine food web during invasion and after establishment of *Dikerogammarus villosus*. Page 67 in abstracts for Biological Invasions in Inland Waters (INWAT) workshop, Firenze, Italy, May 5–7, 2005.

Westmann, K., R. Savolainen, and M. Pursiainen. 1995. Development of European noble crayfish *Astacus astacus* (L.) and American signal crayfish *Pacifastacus leniusculus* (Dana) populations in a small Finnish lake – a 20-year study. Freshwater Crayfish **8**, 235–248.

Williamson, M. 1996. Biological invasions. Chapman & Hall, London, UK.

Chapter thirteen

Growth and reproduction of the goldfish Carassius auratus*: a case study from Italy*

Massimo Lorenzoni, Massimiliano Corboli, Lucia Ghetti, Giovanni Pedicillo, and Antonella Carosi

INTRODUCTION

The goldfish *Carassius auratus* (Linnaeus) is a scaly, high-bodied, laterally compressed fish; its mouth is small and terminal, without barbels. The dorsal fin is long with a slightly serrated third spine. This species is very similar to the crucian carp *Carassius carassius* (Linneaus), but is more elongated and has a slightly concave dorsal fin and slightly larger scales (Lelek 1987).

There has been considerable confusion concerning the taxonomic status of *C. auratus*. Many authors have recognized two subspecies in its native range: *C. a. auratus* (goldfish, Chinese goldfish, or Asian goldfish) from Asia, and *C. a. gibelio* Bloch (Prussian carp, gibele carp, or European goldfish) from eastern Europe (Hanfling *et al.* 2005). Howells (1992, in Nico and Schofield 2006) reported that goldfish typically observed in the US waters are crucian carp × goldfish hybrids. Goldfish commonly hybridises with the carp *Cyprinus carpio* Linnaeus, giving rise to individuals that are intermediate in morphology between the two parent species. Recent studies have indicated that European

Francesca Gherardi, Biological invaders in inland waters: Profiles, distribution, and threats, 259–273.

goldfish populations represent probably an assemblage of lineages of different origins (various clonal lineages as well as hybrids between goldfish and crucian carp) which might have contributed to the taxonomic confusion in the genus *Carassius* Jarocki (Hanfling *et al.* 2005).

Goldfish may grow to 45 cm total length (TL) and 3 kg; however, they generally reach only 20 cm TL and weigh 100–300 g (Muus and Dahlström 1967). Their lifespan is typically 6–7 years, but it has been reported to be as long as 30 years (Menassè 1974). There is no parental care of the eggs or larvae.

Typical habitat includes weedy ponds, shallow lakes, and slow-flowing rivers, especially those with submerged aquatic vegetation (Lelek 1987, Maitland 2004). Many different varieties of goldfish have been produced by man through selective breeding to create a wide range of colours and fin shapes. When released from captivity, these fishes usually revert to their natural olive-bronze colour and normal fin shapes.

DISTRIBUTION

There is still considerable uncertainty regarding the distribution of goldfish. This species is indigenous to eastern Asia (Lelek 1987), including China and neighbouring countries, and, if *C. a. gibelio* is a valid subspecies and not just a feral introduction (Raicu *et al.* 1981), also to some parts of central-eastern Europe. However, the distribution of goldfish in Europe today extends from the Iberian Peninsula to the Black Sea area, with the exception of northern regions (Ireland, Scotland, and part of the Scandinavian Peninsula) (Lelek 1987, Maitland 2004). Wild populations have often been established by released pet goldfish, but many have been introduced unintentionally through restocking with young carp, from which goldfish are difficult to distinguish (Halacka *et al.* 2003); goldfish have been also introduced as bait fish (Nico and Schofield 2006). The species can also spread spontaneously using the connections of hydrological networks. The range of this species in Europe is currently expanding (Lelek 1987).

The goldfish was probably the first foreign fish species to be introduced into North America, arriving in the late 1600s (Jenkins and Burkhead 1994). Today this species is established or reported in all the American States except Alaska (Nico and Schofield 2006). The species can also be found in South America, where it was introduced at the beginning of the 1900s by European immigrants (Gomez *et al.* 1997). The goldfish was first taken to New Zealand in the late 1860s and is now widespread and well established in the country. A large number of *C. auratus* has been reported in many lakes, dams, and rivers in Australia since the late 1870s (Department of Fisheries of Western Australia 2005).

ECOLOGICAL AND BIOLOGICAL CHARACTERISTICS

The goldfish has the potential to be invasive on account of some of its ecological and biological characteristics: high tolerance of water pollution, high fecundity, and wide-ranging diet. Its omnivorous diet includes planktonic crustaceans, phytoplankton, insect larvae, fish eggs and fry, benthic vegetation, and detritus (Muus and Dahlström 1967, Scott and Crossman 1973, Maitland 2004, Nico and Schofield 2006). Its populations grow rapidly, as the species can reproduce through gynogenesis (Abramenko *et al.* 1997, Xie *et al.* 2001, Kuznetsov 2004).

Goldfish are extremely tolerant of environmental stress (Abramenko *et al.* 1997), including high levels of turbidity and fluctuations in pH and temperature (Spotila *et al.* 1979). Laboratory tests have revealed pH tolerance levels between 4.5 and 10.5, and a preference for pH levels between 5.5 and 7.0 (Szczerbowski 2001). Goldfish have been captured in waters with salinity levels as high as 17 ppt, and adults can survive water temperatures between 0 °C and 41 °C (Nico and Schofield 2006). Moreover, the species is highly tolerant of water pollution (Abramenko *et al.* 1997) and can cope with low levels of dissolved oxygen and even prolonged periods (several months at 2 °C) of total anoxia (Walker and Johansen 1977, Van den Thillart *et al.* 1983). This ability requires metabolic adaptations: below critical oxygen content in the water, the fish is able to exploit an anaerobic, or mixed aerobic-anaerobic, metabolism (Holopainen and Hyvarinen 1985, Nilsson 2001). This ability allows them to colonize a wide variety of habitats, including small ponds. In shallow pond conditions in Finland, the crucian carp *C. carassius* abounds and dominates the ecosystem (Holopainen and Pitkanen 1985, Holopainen *et al.* 1991).

Goldfish are considered to be vulnerable to competition (Piironen and Holopainen 1988, Paszowski *et al.* 1990) and to predation (Tonn *et al.* 1991); however, the rapidity of their growth limits their vulnerability as prey for ichthyophagous fish (Nico and Schofield 2006).

Concerns have been raised about the impact that goldfish have on the aquatic community, including increasing turbidity (Cowx 1997) and competition with indigenous fish (Scheffer *et al.* 1993). Indeed, declines in invertebrate numbers have been attributed to the establishment of this species (Richardson and Whoriskey 1992) and local eradication of aquatic macrophytes through direct consumption and uprooting has also been documented (Richardson *et al.* 1995). The bottom-sucking feeding methods of goldfish can also contribute to algal blooms by re-suspending nutrients, which makes them available to phytoplankton (Richardson *et al.* 1995). Furthermore, recent studies have demonstrated that growth of cyanobacteria is stimulated by the passage through goldfish intestines (Kolmakov and Gladyshev 2003). The primary threat to indigenous fish species is probably competition for food and other resources (Moyle 1976). Goldfish have also been known to prey upon eggs, larvae, and adults of indigenous fishes (Scott and Crossman 1973). Other threats may include the introduction and persistence of parasites (such as *Lernea* sp.) that commonly live on goldfish. In the US, the introduction of goldfish was believed to be a major cause of the decline of populations of *Empetrichthys latos* Miller during the early 1960s (Deacon *et al.* 1964); it seems that also the Sacramento sucker *Catostomus occidentalis* Ayres suffers in the presence of goldfish (Moyle 1976). In Europe it has been reported that in some habitats the introduced goldfish affects resident fish, such as crucian carp and tench *Tinca tinca* (Linnaeus) (Halacka *et al.* 2003); in addition, declines in pike abundance (*Esox lucius* Linnaeus) can occur as a result of increased water turbidity (Cowx 1997).

CASE STUDY

Lake Trasimeno is a lake of tectonic origin situated in central Italy (43°9′11″ N and 12°15′ E) between the Tiber and Arno rivers basins. It is the fourth largest lake in Italy (124.3 km^2) and the most extensive of the Italian peninsula. Its shallowness (average depth: 4.72 m; maximum depth: 6.3 m) makes Lake Trasimeno the largest laminar lake in Italy. The catchment basin is made up of lands with low permeability and covers an area of 357.98 km^2, about three times greater than the lake surface (Mearelli *et al.* 1990). The water is supplied by short intermittent streams which have little or no water in the summer. Owing to the morphologic characteristics of Lake Trasimeno, the water temperature is almost the same as the air temperature, exceeding 30 °C in the summer; thermal stratification being usually absent (Lorenzoni *et al.* 1993). Lake Trasimeno is classified as mesotrophic (Mearelli *et al.* 1990).

The fish community, composed of 19 species (Mearelli *et al.* 1990), is dominated by cyprinids. Fishing is still one of the main commercial activities of the local population and, although it has declined in recent years, the number of professional fishermen is the highest in Italy with regard to inland lakes (Lorenzoni *et al.* 2002). Goldfish have been found in Lake Trasimeno since the end of the 1990s (Mearelli *et al.* 1990) and, owing to the absence of predators, man included, their numbers are currently high. This probably exerts a negative impact on fish communities owing to interspecific competition.

Little information is available on the biological characteristics of goldfish populations in Italy and in western Europe in general. A study was conducted to collect information on the growth and reproductive biology of goldfish, in order to investigate the causes of their rapid expansion in Lake Trasimeno and to gather data on which to design a plan for the control of these unwanted populations. Sampling was conducted monthly, from February 2003 to January 2004; individuals were caught by means of electrofishing and multi-mesh gill-nets at 6 sampling stations along the perimeter of the lake. Two types of net were used: fyke nets and gill-nets. The gill-nets were assembled using panels with differently sized mesh (22, 25, 28, 35, 40, 50, and 70 mm), to allow more efficient and representative sampling (Craig *et al.* 1986, Degerman *et al.* 1988). The panels, each of which was 1 m high and 50 m long, were positioned for one night near the bottom, perpendicular to and about 1,000 m from the shore. The fyke nets were positioned for one night in the vicinity of the gill-nets.

Electrofishing was conducted monthly, except in April, when it was conducted weekly. Sampling was carried out from boats by means of 4.5 kW electric stunning devices; these devices supplied continuous pulsating current. Electrofishing has been used to study fish populations in lotic wadable waters for some considerable time, but is seldom used in lentic systems, where it is effectively restricted to the littoral area (Eloranta 1990, Reynolds 1996). In Lake Trasimeno, however, this technique is more efficient, in that the water is shallow in most of the lake, as pointed out by a previous research (Mearelli *et al.* 2004). During each sampling at each of the six stations, a variable number of transects of varying lengths were examined. These transects were chosen on the basis of their different environmental conditions (in terms of substrate, vegetation, depth, and transparency) in order to determine in which conditions catches would be optimised.

The fish caught (expressed as biomass) were standardized with regard to the "fishing effort" (CPUE = catch per unit effort) (Degerman *et al.* 1988, Wilderbuer and Kappenman 1998). For fyke nets, fishing effort was defined as the time of sampling, and CPUEs are expressed as g h^{-1}; for gill-nets, fishing effort was the area of nets (CPUEs = g 10^{-2} m^{-2}); for electrofishing, fishing effort was the time of sampling (CPUEs = g min^{-1}). The lengths of the sampling areas were measured by a GPS meter.

Laboratory analysis and data elaboration

All specimens were measured in terms of total length (TL) and standard length (SL) with an accuracy of 1 mm, and weighed (W) with an accuracy of 1 g (Anderson and Neumann 1996). Sex was determined by macroscopic examination of the gonads (Bagenal 1978) and gonads were weighed (W_g) with an accuracy of 0.1 g. Age was evaluated in the laboratory by a microscopic scalimetric method (Bagenal 1978, Britton *et al.* 2004): the scales were removed from the left side of the fish, above the lateral line, near the dorsal fin (De Vries and Frie 1996) and stored in ethanol (33%). The TL-SL relationship ($TL = a + b\ SL$) and TL-weight relationship ($W = a\ TL^b$) were calculated separately for the two sexes, using a least-squares method (Ricker 1975). The relationships between the sexes were compared by analysis of covariance (ANCOVA).

The theoretical growth in length was described by the Von Bertalanffy growth equation (1938): $L_t = L_\infty(1\text{-}\exp^{(-K(t-t_0))})$, where L_t is the theoretical total length (in cm) at age t, L_∞ the asymptotic length, K the coefficient of growth, t_0 the theoretical age (in years) at length = 0 (Bagenal 1978). The analysis was conducted using the values of total length and age of the single individuals. Because no difference emerged in the TL-SL and TL-weight relationships, the theoretical growth in length was analysed without distinction between sexes.

Gonadosomatic index (GSI) was evaluated by the following formula (Ricker 1975): GSI $= (100\ W_g)/W$, where W_g is gonad weight (in g) and W is total weight (in g). The ovaries of 92 females were excised, weighed, and fixed immediately in 10% buffered formalin. Some cross sections of ovaries from each fish were weighed and microscopically examined, and the oocytes were counted. Ten oocytes were selected for each female and the diameter was measured by means of a computerized system of image analysis (IAS2000) connected to the microscope. The relationship between TL and number of eggs ($N = a\ TL^b$) was calculated using a least-squares method (Ricker 1975).

Age structure and sex ratio

The sample was composed of 3,111 specimens. The TL, weight, and age of the specimens analysed ranged between 4.30 and 40.60 cm, 1 and 1137 g, 0.2 and 7.9 years, respectively. Eight age-classes were found, with most specimens in the 2+ age-class; the number of captured individuals decreased progressively as the age increased. The 0+ age group is not well represented in the sample, probably because of the selectivity of the capture nets. Females were grouped into 7 age-classes, while in the male subsample, 8 age-classes were found. Results showed that the population was composed mostly of females (males = 102, females = 1953, sex ratio: 1:19). The sex ratio also seems to be unbalanced in May, when sampling was carried out among the groups during reproduction (males = 20, females = 575, sex ratio: 1:29).

In many European populations of *C. auratus* a similar imbalance in the sex ratio has been observed, which is probably due to the reproductive system of the population (Abramenko *et al.* 1997, Xie *et al.* 2001, Kuznetsov 2004). Indeed, in Europe many populations are made up exclusively of females that reproduce by gynogenesis through mitotic divisions of eggs due to heterologous species of sperm (Muus and Dahlstrom 1967, Sani *et al.* 1999). By contrast, in Asia the sex ratio is around 1:1 (Muus and Dahlstrom 1967, Abramenko *et al.* 1997, Kuznetsov 2004).

Growth

The TL-SL relationship estimated for the whole sample was $TL = 0.0822 + 1.2155\ SL$ ($R^2 = 0.992$, $P = 0.000$). On covariance analysis, the difference between the two sexes was not statistically significant ($F = 3.700$, $P = 0.054$). The weight-length relationship estimated for the whole sample was: $W = 0.0147\ TL^{3.062}$ ($R^2 = 0.990$, $P = 0.000$), without any significant difference between sexes ($F = 3.124$, $P = 0.077$). The results show that in Lake Trasimeno the species displays allometric growth ($b > 3$ in both sexes). The

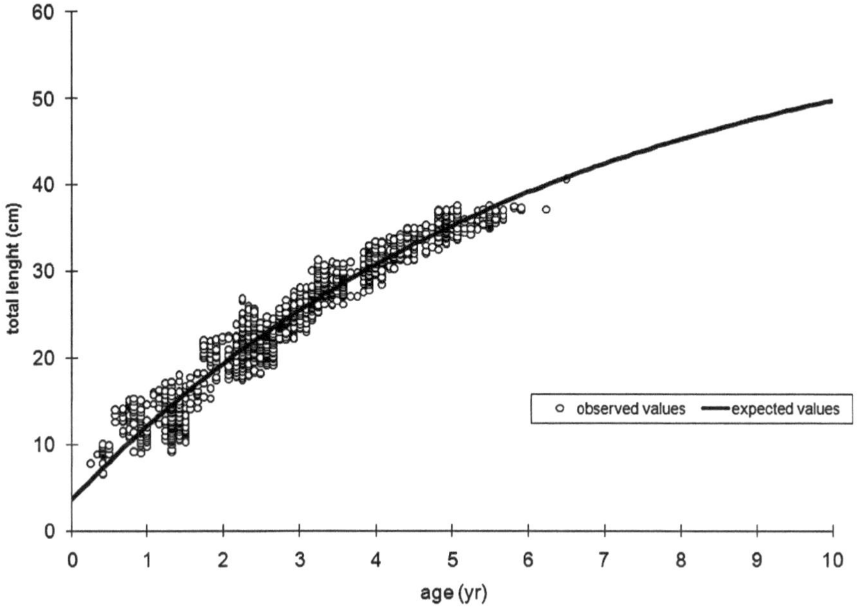

Fig. 1 Goldfish: curve of theoretical growth in length.

regression coefficient values for the population of Lake Trasimeno are generally higher than those of other populations investigated (Froese and Pauly 1998).

The curve of the theoretical growth in length was TL = 46.967 $\{1\text{-}e^{(-0.154(t+1.048))}\}$ ($R^2 = 0.982$) for the total sample (Fig. 1). Froese and Pauly (1998) and Kuznetsov (2004) report data on numerous other European and Asian populations of goldfish.

Reproductive biology

The GSI was calculated in both sexes. The average GSI value was 6.25 in females, varying between a minimum of 0.18 and a maximum of 46.51, while in males it was 1.69, varying between 0.19 and 11.19. The differences between the two sexes were highly significant on *t*-test ($t = 7.36$, $P = 0.000$). The monthly trend in GSI for the female sample (Fig. 2) showed that maturation of ovarian eggs reached a maximum in April, while in August reproduction ended and gonads were in a resting condition. The GSI value began to rise in autumn; during the winter, when fish metabolism is slowed and food supply is scarce, the GSI value tended to decrease slightly and then increased rapidly from February onwards. In Lake Trasimeno, the female reproductive investment was high, reaching almost 50% of the entire body mass at its maximum peak. Analyses of the GSI showed that the reproductive period of the population

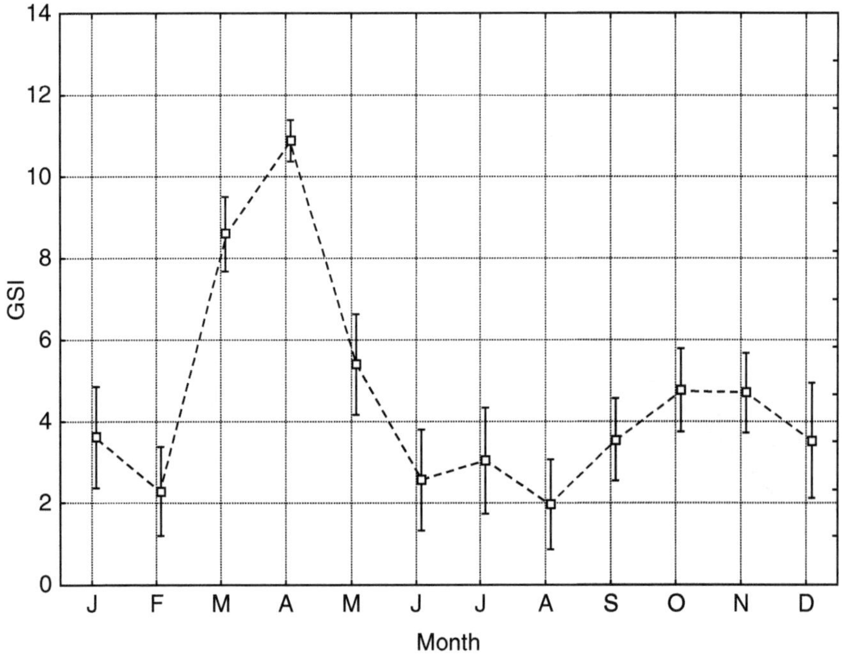

Fig. 2 Goldfish: monthly average values (with confidence limits) of gonadosomatic index (GSI).

extends over an ample time period, from the end of March until June. The water temperature recorded at the beginning of reproduction was about 13 °C.

Most of the females (85% of the population sampled) attained sexual maturity in the second year, while in the third and subsequent years this percentage rose to 100%; however, some (7.55%) 1-year-old females were able to spawn. The smallest sexually mature female was 12.20 cm TL. The relationship between TL and the number of eggs was $N = 0.0198\ TL^{4.339}$ ($R^2 = 0.743$, $P = 0.000$). Fecundity varied from 286 to 219,104 eggs, averaging (± SE) 46,253 ± 3,921 eggs. The diameter of mature eggs in the spawning season ranged from 0.74 to 1.71 mm, with a 1.27 ± 0.01 mm average. Average relative fecundity was 103 ± 5 eggs g^{-1}. The relative fecundity and the diameter of mature eggs of the population sampled were positively correlated with TL (cm) (fecundity: $r = 0.315$, $P = 0.002$; egg diameter: $r = 0.561$, $P = 0.000$) and body weight (fecundity: $r = 0.216$, $P = 0.012$; egg diameter: $r = 0.511$, $P = 0.000$).

Catch per unit effort

In Lake Trasimeno the statistical records kept by commercial fishermen do not include catches of goldfish, as this species is not marketed. In the 1980s, the total commercial yield of the lake was 0.048 t ha^{-1} y^{-1}, while recently (2002–2004) it has fallen to 0.023 ha^{-1} y^{-1} (unpublished data). At the end of the 1980s, tench (31.05% of total yield), European perch *Perca fluviatilis* (Linnaeus) (21.28%), sandsmelt *Atherina boyeri* (Risso) (20.59%), and eel *Anguilla anguilla* (Linnaeus) (13.46%) were the species most caught in the lake. By 2002–2004, the situation had changed markedly: sandsmelt (29.76%) was the most caught species, followed in decreasing order by tench (21.95%), largemouth bass *Micropterus salmoides* (Lacépède) (10.39%), eel (10.28%), carp (10.16%), and black bullhead *Ictalurus melas* (Rafinesque) (8.53%); catches of European perch, the most lucrative fish in the lake, had plummeted to 5.07% of the total yield.

Our research confirmed concerns over the abundance of goldfish in Lake Trasimeno: in the last few years the population has noticeably increased. In our samples, it was by far the most abundant species, representing 58.08% of the individuals and 73.23% of the whole biomass caught with nets, and 48.24% of individuals and 62.63% of the biomass captured by means of electrofishing. The average value of the CPUEs of goldfish caught by gill-nets was 10,175.48 g 10^{-2} m^{-2}, while with fyke nets the average value was 246.97 g h^{-1} (Table 1). With regard to electrofishing, the number of transects was 97, while the average fishing effort applied was 10.72 min, for a length of 387.93 m. Goldfish were also the species most captured by electrofishing: the average value of the CPUEs was 606.50 g min^{-1}. A similar monitoring campaign conducted in 1993 by means of electrofishing turned up on *C. auratus*; comparison with the present data points up the changes that have occurred in the fish populations (Table 1).

Figure 3 shows monthly average values and pertinent confidence limits in the CPUEs of goldfish. The efficiency of electrofishing was not equal in all periods of the year: no fish were caught in winter; catches increased in spring, reaching a maximum in May; from June to August they declined and then increased again in autumn. The Kruskall-Wallis non-parametric test showed that the differences in CPUEs among monthly median values were highly significant ($\chi^2 = 26.05$, $P = 0.006$). The variability in catches was partly due to the different environmental characteristics of the sampling sites; however, statistically significant differences among the CPUEs emerged only with regard to vegetation. Indeed, goldfish were far more abundant near submerged vegetation (mean CPUEs $\pm$ S.E. = 758.46 $\pm$ 123.51 g min^{-1}) than in areas lacking vegetation (212.51 $\pm$ 74.32 g min^{-1}); these differences were significant (Kruskall-Wallis non-parametric test: $\chi^2 = 5.90$, $P = 0.015$). When the sample was subdivided on the basis of the sampling season, the average yields were higher in all seasons in the areas with submerged vegetation, but the CPUE was particularly elevated in spring, when the population was reproducing (areas with vegetation: 1,268.52 $\pm$ 291.60 g min^{-1}; areas without vegetation: 169.52 $\pm$ 181.90 g min^{-1}).

Gill-nets yielded abundant catches of goldfish at all times, without marked differences from one month to another (Fig. 3); the average CPUEs reached their

Table 1 Descriptive statistics of the CPUEs yielded by nets and by electrofishing.

	Gill-nets (g 10^{-2} m^{-2})			Fyke nets (g h^{-1})		
	Sample size	Mean	Standard deviation	Sample size	Mean	Standard deviation
Goldfish	84	10,175.48	9,733.86	12	246.97	202.88
European perch	84	275.25	1,485.69	12	1.99	4.90
Pumpkinseed[1]	84	214.39	1,523.93	12	5.87	9.05
Rudd	84	2,018.21	4,319.34	12	1.39	4.81
Tench	84	1,388.68	2,899.12	12	13.54	20.50
Largemouth bass	84	520.32	1,304.61	12	133.33	456.66
Black bullhead	84	363.76	1,205.45	12	16.04	15.83
Pike	84	80.05	648.55	12	0.00	0.00
Eel	84	0.00	0.00	12	3.47	6.61

	2003–2004			1993		
Electrofishing (g min^{-1})	Sample size	Mean	Standard deviation	Sample size	Mean	Standard deviation
Effort (min)	97	10.72	6.17	14	56.07	33.75
Length (m)	97	387.93	233.31			
Goldfish	97	606.50	931.89	14	0.00	0.00
Tench	97	45.62	89.35	14	199.40	95.35
Rudd	97	17.13	51.20	14	17.65	32.91
Topmouth gudgeon[2]	97	0.51	1.90	14	0.00	0.00
Black bullhead	97	3.10	15.32	14	13.05	12.42
Largemouth bass	97	39.84	82.74	14	20.65	24.09
Pumpkinseed[1]	97	0.98	4.51	14	9.29	7.10
European perch	97	0.65	2.17	14	1.94	1.69
Pike	97	3.98	17.24	14	47.06	35.59
Sandsmelt	97	1.26	4.41	14	0.00	0.00
Carp	97	253.63	709.63	14	0.00	0.00
Eel	97	6.64	25.94	14	2.22	3.08
Bleak[3]	97	0.10	0.63	14	0.00	0.00

[1]*Lepomis gibbosus* (Linnaeus), [2]*Pseudorasbora parva* (Schlegel), [3]*Alburnus alburnus alborella* (De Filippi)

highest values in the period preceding reproduction and during the reproduction period. However, the nets also showed good sampling efficiency in winter, when goldfish reduce their activity and move offshore. The Kruskall-Wallis non-parametric test showed that the differences in CPUEs among monthly median values were not significant ($\chi^2 = 10.70$, $P = 0.469$).

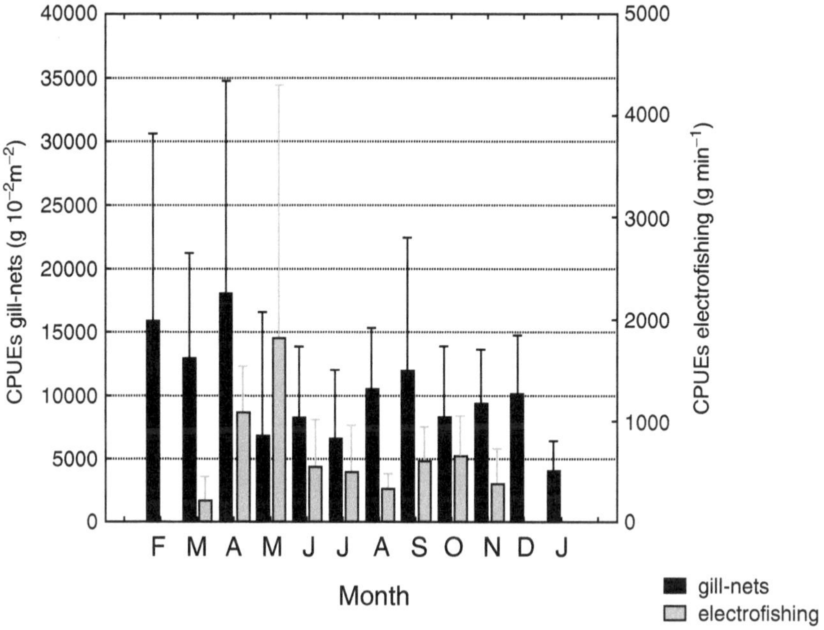

Fig. 3 Goldfish: monthly average values (with upper confidence limits) of the CPUEs yielded by electrofishing and by gill-nets.

CONCLUSIONS

The main characteristic that determines the high invasive potential of the goldfish is its great adaptability and its ability to tolerate extreme environmental conditions. Another important factor in its success is its growth capability, which enables this species to rapidly reach a size that makes it safe from predators in Lake Trasimeno. Predation on *C. auratus*, as on *C. carassius*, is size-dependent (Piironen and Holopainen 1988, Bronmark *et al.* 1995), predators preferring individuals of small size (Holopainen *et al.* 1991, Tonn *et al.* 1991). Its reproductive biology – precocious maturity, ability to reproduce by gynogenesis, high fecundity, and reproduction several times per year – is also a prerequisite to the invasive potential of this species.

In dealing with invasive species, eradication is obviously the favoured strategy and several studies have demonstrated its success (Chapter 34). However, success has been limited to small, isolated biotopes, on a local scale and in the first stages of invasion (Zavaleta *et al.* 2001). Efforts fail when eradication is not complete and if re-invasion is likely; in such cases, an *r*-strategist like the goldfish can rapidly increase. For this species, therefore, it seems preferable to adopt a control programme aimed at reducing the density of the unwanted populations and at maintaining it below an impact threshold (Mueller 2005).

The adoption of a particular strategy depends both on the assessment of its costs/benefits and on its potential to be successful (Myers *et al.* 2000). Generalization is difficult, as the choices vary according to several aspects, which also regard the economic and social context. In lakes, netting probably remains the most common and effective method of keeping down an invasive fish population, although a substantial effort was required: various mesh size were necessary to target the full range of the size classes present. In the Rotopiko Lake (New Zealand) the eradication of rudd made with nets alone was unlikely and additional control techniques were required (Barnes *et al.* 2003). Where commercial fishermen operate, a good strategy of control may be to encourage the harvesting of invasive populations, for example by offering financial incentives or encouraging the trade in fish or fish parts (e.g. eggs).

Electrofishing is a specific sampling technique for shallow water, and habitat preference among species or life stages affects their vulnerability to it (Reynolds 1996). In favourable situations, this sampling technique can be effectively combined with the use of nets in the containment of some invasive fish populations. These results have application for managing goldfish because they should be easier to remove when they are aggregated: in a goldfish removal project conducted at Medical Lake (Washington, USA) in 1983, 17,837 goldfish were harvested by electrofishing over a period of seven days. The efficacy of the procedure was quantified by comparing goldfish county and gill-net CPUEs: an estimated 95–99% of goldfish was removed (Scholz 1984). Electrofishing has some advantages over gill-nets: it causes little injury to the fish captured and it exerts modest selectivity regarding the size of the specimens. In spring, when goldfish and other cyprinids congregate in shallow, vegetated areas to reproduce, it can be effectively used without greatly affecting other species.

REFERENCES

Abramenko, M. I., O. V. Kravchenko, and A. E. Velikoivanenko.1997. Population genetic structure of the goldfish *Carassius auratus gibelio* diploid-triploid complex from the Don River Basin. Journal of Ichthyology **37**, 56–65.

Anderson, R. O. and R. M. Neumann.1996. Length, weight and associated structural indices. Pages 447–482 in B.R. Murphy and D.W. Willis, editors. Fisheries techniques. American Fisheries Society, Bethesda, MD.

Bagenal, T. B. 1978. Fish production in fresh waters. Blackwell, London, UK.

Barnes, G., D. Speirs, K. Neilson, and R. Kelleher. 2003. The use of fine mesh monofilament mist nets to control rudd (*Scardinius erythrophthalmus*) from a small lake complex, in the Waikato region, New Zealand. ASFB Conference, Wellington, New Zealand.

Britton, J. R., I. G. Cowx, and G. Peirson. 2004. Sources of error in the ageing of stocked cyprinids. Fisheries Management and Ecology **11**, 415–417.

Bronmark, C., C. A. Paszkowski, W. M. Tonn, and A. Hargeby. 1995. Predation as a determinant of size structure in populations of crucian carp (*Carassius carassius*) and tench (*Tinca tinca*). Ecology of Freshwater Fish **4**, 85–92.

Cowx, I. G. 1997. Introduction of fish species into European fresh waters: economic successes or ecological disasters? Bulletin Français de la Pêche et de la Pisciculture **344–345**, 57–77.

Craig, J. F., A. Sharma, and K. Smiley. 1986. The variability in catches from multi-mesh gillnets fished in three Canadian Lakes. Journal of Fish Biology **28**, 671–678.

Deacon, J. E., C. Hubbs, and B. J. Zahuranec. 1964. Some effects of introduced fishes on the native fish fauna of southern Nevada. Copeia **1964**, 384–388.

Degerman, E., P. Nyberg, and M. Appelberg. 1988. Estimating the number of species and relative abundance of fish in oligotrophic Swedish lakes using multi-mesh gillnets. Nordic Journal of Freshwater Research **64**, 91–100.

Department of Fisheries of Western Australia. 2005. http://www.fish.wa.gov.au Published: May 2005.

De Vries, D. R., and R. V. Frie. 1996. Determination of age and growth. Pages 483–512 in B.R. Murphy and D.W. Willis, editors. Fisheries techniques. American Fisheries Society, Bethesda, MD.

Eloranta, A. 1990. Electric fishing in the stony littoral zone of lakes. Pages 91–95 in I.G. Cowx, editor. Developments in electric fishing. Fishing News Book, Cambridge, UK.

Froese, R and D. Pauly. 1998. Fishbase 98: concepts, design and data sources. Iclarm, Manila, Philippines.

Gomez, S. E., H. Ferré, H. Cassará, and S. Bordone. 1997. Cultivo de peces ornamentales (*Carassius auratus* y *Cyprinus carpio*) en sistemas semiintensivos en la Argentina. Aquatec **4**, 1–13.

Halacka, K., V. Luskova, and S. Lusk. 2003. *Carassius gibelio* in fish communities of the Czech Republic. Ecohydrology and Hydrobiology **3**, 133–138.

Hanfling, B., P. Bolton, M. Harley, and G. R. Carvalho. 2005. A molecular approach to detect hybridisation between crucian carp (*Carassius carassius*) and non-indigenous carp species (*Carassius spp.* and *Cyprinus carpio*). Freshwater Biology **50**, 403–417.

Holopainen, I. J. and H. Hyvarinen.1985. Ecology and physiology of crucian carp (*Carassius carassius* (L.)) in small Finnish pounds with anoxic conditions in winter. Verhandlungen der Internationale Vereinigung Limnologie **22**, 2566–2570.

Holopainen, I. J. and A. K. Pitkanen. 1985. Population size and structure of crucian carp (*Carassius carassius* (L.)) in two small, natural ponds in Eastern Finland. Annales Zoologici Fennici **22**, 397–406.

Holopainen, I. J., W. M. Tonn, and C. A. Paszkowski. 1991. Ecological responses of crucian carp populations to predation by perch in a manipulated pond. Verhandlungen der Internationale Vereinigung Limnologie **14**, 2412–2417.

Jenkins, R. E. and N .M. Burkhead. 1994. Freshwater fishes of Virginia. American Fisheries Society, Bethesda, MD.

Kolmakov, V. I. and M. I. Gladyshev. 2003. Growth and potential photosynthesis of cyanobacteria are stimulated by viable gut passage in crucian carp. Aquatic Ecology **37**, 237–242.

Kuznetsov, V. A. 2004. Changes in the population structure and biological indices of the goldfish *Carassius auratus gibelio* in the Volga Stretch of the Kuibyshev Reservoir under conditions of intense anthropogenic load on the ecosystem. Journal of Ichthyology **44**, 167–174.

Lelek, A. 1987. The freshwater fishes of Europe. Threatened fishes of Europe. Aula-Verlag, Wiesbaden, Germany.

Lorenzoni, M., A. J. M. Dorr, R. Erra, G. Giovinazzo, M. Mearelli, and S. Selvi. 2002. Growth and reproduction of largemouth bass (*Micropterus salmoides* Lacépède, 1802) in Lake Trasimeno (Umbria, Italy). Fisheries Research **56**, 89–95.

Lorenzoni, M., G. Giovinazzo, M. Mearelli, and M. Natali. 1993. Growth and biology of perch (*Perca fluviatilis* L.) in Lake Trasimeno (Umbria, Italy). Polskie Archiwum Hydrobiologii **40**, 313–328.

Maitland, P. S. 2004. Keys to the freshwater fish of Britain and Ireland, with notes on their distribution and ecology. Freshwater Biological Association, Amblesid, UK.

Mearelli, M., A. Carosi, A. J. M. Dorr, G. Giovinazzo, M. Natali, G. La Porta, and M. Lorenzoni. 2004. Primi risultati relativi alla messa a punto di un protocollo operativo per l'uso della pesca elettrica nel lago Trasimeno. Biologia Ambientale **18**, 201–206.

Mearelli, M., M. Lorenzoni, and L. Mantilacci. 1990. Il lago Trasimeno. Rivista di Idrobiologia **29**, 353–389.

Menassè, V. 1974. Pesci rossi o carassi. Edagricole, Bologna, Italy.

Moyle, P. B. 1976. Inland fishes of California. University of California Press, Berkeley, CA.

Mueller, G. A. 2005. Predatory fish removal and native fish recovery in the Colorado River mainstream: what have we leaned? Fisheries **30**, 10–19.

Muus, B. J. and P. Dahlström. 1967. Guide des Poissons d'eau douce et Pêche. Delachaux & Niestlè, Neuchatel, Switzerland.

Myers, J. H., D. Simberloff, A. M. Kuris, and J. R. Carey. 2000. Eradication revisited: dealing with exotic species. Trends in Ecology and Evolution **15**, 316–320.

Nico, L. and P. J. Schofield. 2006. *Carassius auratus*. USGS Non-indigenous Aquatic Species Database, Gainesville, FL.

Nilsson, G. E. 2001. Surviving anoxia with the brain turned on. News in Physiological Sciences **16**, 217–221.

Paszkowski, C. A., W. M. Tonn, J. Piironen, and I. J. Holopainen. 1990. Behavioural and population-level aspects of intraspecific competition in crucian carp. Annales Zoologici Fennici **27**, 77–85.

Piironen, J. and I. J. Holopainen. 1988. Length structure and reproductive potential of crucian carp (*Carassius carassius* L.) populations in some small forests ponds. Annales Zoologici Fennici **25**, 203–208.

Raicu, P., E. Taisescu, and P. Banarescu. 1981. *Carassius carassius* and *C. auratus*, a pair of diploid and tetraploid representative species (*Pisces, Cyprinidae*). Cytologia **46**, 233–240.

Reynolds, J. B. 1996. Electrofishing. Pages 221–253 in B.R. Murphy and D.W. Willis, editors. Fisheries techniques. American Fisheries Society, Bethesda. MD.

Richardson, M. J. and F.G. Whoriskey. 1992. Factors influencing the production of turbidity by goldfish. Canadian Journal of Zoology **70**, 1585–1589.

Richardson, M. J., F.G. Whoriskey, and H. Roy. 1995. Turbidity generation and biological impacts of an exotic *Carassius auratus*, introduced into shallow seasonally anoxic pounds. Journal of Fish Biology **47**, 576–585.

Ricker, W. E. 1975. Computation and interpretation of biological statistics of fish population. Bulletin of the Fisheries Research Board of Canada **191**, 1–382.

Sani, L., A. Rongoni, and G. Alessio. 1999. Biologia riproduttiva delle principali specie ittiche dulcicole di un ecosistema eutrofizzato (lago Massaciuccoli, Toscana). Quaderni ETP **28**, 191–203.

Scheffer, M., S. H. Hosper, M. L. Meijer, B. Moss, and E. Jeppesen. 1993. Alternative equilibria in shallow lakes. Trends in Ecology and Evolution **8**, 275–279.

Scholz, A. T. 1984. The seasonal distribution and aggregation behaviour of goldfish (*Carassius auratus* L.) in eastern Washington lakes: new technology for control of goldfish populations based on their behavioural ecology. Final report to Office of Water Policy, Washington, DC.

Scott, W. C. and E. J. Crossman. 1973. Freshwater fishes of Canada. Bulletin of the Fisheries Research Board of Canada **184**, 1–966.

Spotila, J. R., K. M. Terpin, R. R. Koons, and R. L. Bonati. 1979. Temperature requirements of fishes from eastern Lake Erie and upper Niagara River. Environmental Biology of Fishes **4**, 281–307.

Szczerbowski, J. A. 2001. *Carassius auratus* (Linneaus, 1758). Pages 5–41 in P. M. Banarescu and H. J. Paepke, editors. The Freshwater Fishes of Europe, vol. **5**/III; *Cyprinidae* 2/III and *Gasterosteidae*. AULA-Verlag, Wiebelsheim, Germany.

Tonn, W. M., C. A. Paszkowski, and I. J. Holopainen. 1991. Selective piscivory by perch: effects of predator size, prey size, and prey species. Verhandlungen der Internationale Vereinigung Limnologie **24**, 2406–2411.

Van den Thillart, G., M. Van Berge Henegounen, and F. Kesbete. 1983. Anaerobic metabolism of goldfish, *Carassius auratus*: ethanol and CO_2 excretion rates and anoxic tolerance at 20, 10, and 5 degrees C. Comparative Biochemistry and Physiology **76**, 295–300.

Von Bertalanffy, L. 1938. A quantitative theory of organic growth. Human Biology **10**, 11–243.

Walker, R. and P. Johansen. 1977. Anaerobic metabolism in goldfish, *Carassius auratus*. Canadian Journal of Zoology **55**, 304–311.

Wilderbuer, T. K. and R. F. Kappenman. 1998. Analysis of Fishing Power Correction Factor Estimates from a Trawl Comparison Experimental. North American Journal of Fisheries Management **18**, 11–18.

Xie, J., J. Wen, B. Chen, and J. F. Gui. 2001. Differential gene expression in fully-grown oocytes between gynogenetic and gonochoristic crucian carp. Gene **272**, 109–116.

Zavaleta, E. S., R. J. Hobbs, and H. A. Mooney. 2001. Viewing invasive species removal in a whole-ecosystem context. Trends in Ecology and Evolution **16**, 454–459.

Chapter fourteen

Epigenetic context in the life history traits of the round goby, Neogobius melanostomus

Mária Balážová-L'avrinčíková and Vladimír Kováč

INTRODUCTION

The round goby, *Neogobius melanostomus* (Pallas), is a typical representative of gobiids, a distinct family within the order Perciformes (Nelson 1994). Most species in this family have their pelvic fins fused into a suction disk with weak adhesion, the external lateral line being usually absent. The subfamily Gobionellinae, which includes the round goby, is mostly confined to the Caspian, Black, and Azov seas (Nelson 1994). Members of this subfamily are primarily benthic and therefore do not have a swim bladder (Nikolsky 1954). *Neogobius melanostomus* is a small, soft-bodied fish. Its head width is about equal to its depth; the snout is oblique and convex in profile. The anterior and middle nape, upper opercule, breast, pectoral fin lobes, and part of abdomen are covered with cycloid scales, the rest of body with weak ctenoid scales. The dorsal fins are contiguous. The second dorsal fin is uniform in height, while the first dorsal fin has a prominent posterior black spot (Berg 1949, Miller 1986). Both sexes have an erectile urogenital papilla between the anus and the base of the anal fin, a subject of sexual dimorphism (Miller 1984).

Its native range covers a wide area of Ponto–Caspian basin. In the Black Sea, round gobies were reported from all shore areas (Borcea 1934, Bănărescu

Francesca Gherardi, Biological invaders in inland waters: Profiles, distribution, and threats, 275–287.

1964), including estuaries and ascended tributaries, the River Dniester as far as the city of Kamianets–Podolsky, the River Southern Bug to the city of Ladyzhino, the River Dnieper up to the city of Dnepropetrovsk, and the River Don up to the city of Rostov (Berg 1949, Pinchuk *et al.* 2003). Recently, an upstream extension of the species' range has been recorded in these rivers. At the southern shores of the Black Sea, round gobies have been found at Sinope and Samsun, in the Bosporus and throughout the Sea of Marmara (Pinchuk *et al.* 2003). It is also native to the lakes and rivers of the Caucasian west coast. The round goby has also been found throughout the Sea of Azov, though less abundant in the central area (Kostyuchenko 1969). In the Caspian Sea, this species occurs along almost all seacoast, as well as in southern parts of the rivers Volga and Ural (Pinchuk *et al.* 2003). Concerning the River Danube, *N. melanostomus* was known earlier as far as Vidin, and along the River Vit up to Pleven (Berg 1949, Drensky 1951).

As the second longest river in Europe, the Danube serves as an important waterway across the continent. Such a geographical predisposition, together with intensive shipping, makes the Danube an ideal route for biological invasions. Indeed, over the last decade, four Ponto–Caspian species of the genus *Neogobius* have been observed to invade the middle stretches (as much as 2,000 km upstream) of the river. *Neogobius melanostomus* has become a new member of the fish fauna in the Slovak stretch of the Danube since 2003 (Stráňai and Andreji 2004). In this chapter, some biological traits and aspects of the species' invasion biology are briefly reviewed, which is followed by data on life history traits of the Danubian population of the round goby, and finally, a hypothesis about the role of epigenetic mechanisms in helping this species to become a successful invader is developed.

INTRODUCTION, EXPANSION, AND INVASION

In the late 1950s, the round goby was introduced into the Aral Sea, together with the grey mulet (Moskalkova 1996). Despite increased salinity in the sea, this species survived at least until late 1980s, although it is now reported extinct (Skóra 1996). In 1968, the round goby was found in the Kuybyshev Reservoir on the River Volga, from where it has been spreading continuously (Tsyplakov 1974). In the River Moscow, the species was found in the 1980s, probably as a consequence of the accidental introduction of eggs attached to the hulls of barges (Sokolov and Tsepkin 1992).

In 1990, *N. melanostomus* was found outside the Ponto–Caspian basin – in the Gulf of Gdansk in the Baltic Sea. The species was most likely transported to the Baltic with the ballast water of vessels sailing from the Black and/or Caspian seas, or directly through the riverine routes going to the north. In general, there are three possible routes of introduction (Fig. 1). The first route starts in the Black Sea and continues via the rivers Dnieper, Pripet, Pina, Kanal Krolewski,

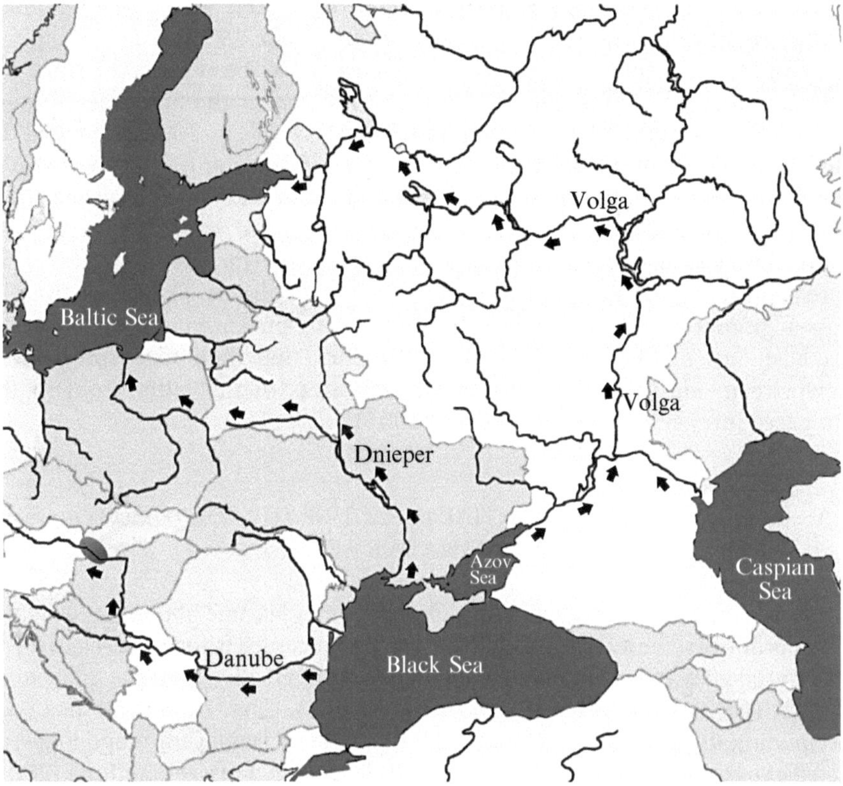

Fig. 1 Major spreading routes of *Neogobius melanostomus* in Europe. Note that there is a network of canals in the river systems, which facilitates fast spreading of the species (not illustrated). The native range of the round goby covers the Black Sea, Azov Sea, Caspian Sea, and the surrounding waters. The oval indicates the Slovak stretch of the Danube, the area of this study.

Bug, and Vistula; the second route is that from the Azov Sea, which runs via the rivers Don to Volga, and then joins the third possible route, which originates in the Caspian Sea and continues via the River Volga, the Rybinskoe Reservoir, lakes Onega and Ladoga to the Gulf of Finland. The first specimen caught was 3–4 years old; therefore, the initial introduction could have occurred 3 years earlier (Skóra 1996). In summer 1999, the round goby was also reported from Rugia Island, Germany, and in 2003, from several locations along the north coast of Germany (Corkum *et al.* 2004).

The first record of *N. melanostomus* in North America comes from summer 1990, when the species was found in the River St. Clair (Jude *et al.* 1992). It is thought that it was initially transported to the Great Lakes in the ballast water of transoceanic ships. The round goby needed only 5 years to spread into all five

Great Lakes. It has been termed a "cyberfish" able to disperse rapidly, using Great Lakes freighters as transport vectors (Jude 1997).

In the River Danube, its upstream expansion has been reported since the 1960s (Bănărescu 1964). In 1997, the species was for the first time found in the Serbian part of the Danube (Simonović *et al.* 1998), in 2000 it reached the stretch near Vienna (Wiesner *et al.* 2000), and 3 years later it was found also in the Slovak stretch of the Danube (Stráňai and Andreji 2004). Since the expansion of the round goby over the Danube appears to have been very fast, especially close to industrial areas and large towns (Jurajda *et al.* 2005), its spreading is likely to have been facilitated by freight vessels. The idea is that the vessels primarily help the gobies to move upstream for longer distances, to establish foci of their populations in harbours, and then they spread both downstream and upstream (for shorter distances in the latter case) in the river (see Jurajda *et al.* 2005, Wiesner 2005).

EXISTING AND POTENTIAL IMPACT OF THE ROUND GOBY IN NON-NATIVE AREAS

In North America, in areas where *N. melanostomus* became abundant, a significant decline of some indigenous species has been recorded (Crossman *et al.* 1992). The round goby is larger, more fecund and more aggressive than the indigenous benthic species (Jude *et al.* 1995). In the River St. Clair, the species has been decimating the populations of mottled sculpin *Cottus bairdi* Girard and logperch *Percina caprodes* Rafinesque (Jude *et al.* 1995). Indeed, Janssen and Jude (2001) found that the round goby potentially affects the mottled sculpin through competition for food resources in small specimens (<60 mm), competition for space in fish of intermediate sizes (60–100 mm), and competition for spawning substrates at larger sizes (>100 mm). Apart from that, *N. melanostomus* also benefits from a longer reactive distance to prey, which suggests that they may possess a competitive advantage in acquiring prey at night. The decline in logperch populations may be due to the predation of the round goby on logperch eggs (Jude *et al.* 1995). Other species impacted by *N. melanostomus* predation on their eggs and larvae are the lake trout, *Salvelinus namaycush* Walbaum (Chotkowski and Marsden 1999), and the lake sturgeon, *Acipenser fulvescens* Rafinesque (Corkum *et al.* 2004). Another local species – the deepwater sculpin, *Myoxocephalus thompsoni* Girard – may be also at risk because, in winter, its normal bathymetric range may be penetrated by the round goby (Jude *et al.* 1995).

In the Baltic Sea, *N. melanostomus* appears to have affected the populations of the flounder, *Platithys flesus* Linnaeus, a mollusc feeder. Some species of Gobiidae (the sand goby, *Pomatoschistus minutus* Pallas, the common goby, *P. microps* Krøyer, and the black goby, *Gobius niger* Linnaeus), as well as the eelpout, *Zoarces viviparus* Linnaeus, have been affected due interference with their similar life space (Sapota 2001a, b).

The appearance of the round goby in new areas may also affect the food webs. In the Gulf of Gdansk, cormorants have directed their diet at the round goby, which resulted in increased density of eel and planktivorous sprat (original main diet components of cormorants); this in turn caused a reduction of zooplankton and a corresponding increase in the algal biomass (Bzoma and Stempniewicz 2001). *Neogobius melanostomus* predation on grazing invertebrates enhanced the algal biomass, as apparent from an increased chlorophyll *a* concentration (Kuhns and Berg 1999). Moreover, piscivorous fish consume round gobies, which have consumed zebra mussels. Thus, the round goby introduces the otherwise inaccessible zebra mussel into the food webs, contributing indirectly to increase populations of piscivores (Charlebois *et al.* 1997).

The round goby may also indirectly release detrimental material from the bottom, because benthic prey organisms are exposed to contaminated sediments. In turn, *N. melanostomus* is preyed upon by various sport and commercial fish and thus toxic substances can then be passed to humans who consume piscivorous fish (Charlebois *et al.* 1997, Corkum *et al.* 2004). Researchers have also suggested a possible link between this species and botulism, *Clostridium botulinum* van Ermengem type E, a mortal disease of wild migratory birds. The infected birds had a higher incidence of gobies in their guts than did uninfected birds (Corkum *et al.* 2004).

Recently, the first possible current impact of the high frequency and abundance of invasive *Neogobius* species in the Danube has been observed: a progressive decline in indigenous benthic fishes, such as the bullhead (*Cottus gobio* Linnaeus), the stone loach (*Barbatula barbatula* Linnaeus), and the white-finned gudgeon (*Gobio albipinnatus* Lukasch) (Jurajda *et al.* 2005).

WHY IS IT SO SUCCESSFUL?

The main advantages that *N. melanostomus* possesses for the occupation of new areas appear to be its tolerance for a wide range of environmental conditions, a broad diet, aggressive behaviour, parental care, and larger size compared to species of a similar benthic lifestyle (Charlebois *et al.* 1997). Adult round gobies are euryhaline and thus able to colonize various habitats from the freshwater of rivers or lakes to brackish waters with polyhaline (up to 37‰) salinities (Smirnov 1986). Development inside their eggs can also take place at a wide range of salinities and the species can even reproduce in freshwater (Smirnov 1986, Moskalkova 1996). The round goby also tolerates big contrasts in water depth. During the breeding seasons it migrates close inshore to depths 0.2–0.5 m and the mature specimens concentrate in shallow areas where they spawn (Kostyuchenko 1969). Young fish, a few days after hatching, gather in sunlit shallows in pursuit of plankton (Moskalkova 1996). In winter, the round goby moves back down to depths of 10–15 m (Kostyuchenko 1969) or even 60–70 m (Ragimov 1991).

Neogobius melanostomus can be found in various habitats, for example, on coarse gravel, shell sands, and even macrophytes (Smirnov 1986). In the Gulf of Gdansk, the gobies are associated with stony and sandy areas, mussel beds, marine structures and/or sunken objects (Skóra and Rzezník 2001). However, in Puck Bay, they occupy primarily a sandy-sludgy bottom covered with benthic flora (Skóra 1996). In the middle section of the Danube (Hungary, Slovakia, Austria) the round goby occupy most of the available shorelines, being found in highest relative densities along artificial rip-rap habitats and in relative low densities in sandy areas (Erős *et al.* 2005, Jurajda *et al.* 2005, Wiesner 2005). They can tolerate a flow rate up to $0.34\,\mathrm{m\ s^{-1}}$ for 3–4 min. At higher current velocities, they remain close to the bottom, presumably using their pelvic fins to brace against the current (Skazkina 1972).

Water transparency does not seem to be a problem for the round goby. In the Azov Sea, the species uses acoustic signalling in the areas with decreased visibility (Kostyuchenko 1960). It is a eurythermic species, tolerating temperatures from -1 to 30 °C (Kalinina 1976), being well resistant to temperature stress. Round gobies prefer at least 60% oxygen saturation (Skazkina 1966), though their threshold level is relatively low, depending on their body weight, from 0.3 to $0.9\,\mathrm{mgl^{-1}}$.

Neogobius melanostomus is essentially a benthic feeder, preying mainly, though not exclusively, on molluscs, followed by crustaceans (Lus 1963). For example, in the Azov Sea, molluscs make 93.5% of total weight of food consumed (Lus 1963), whereas in the Black Sea the ratio of molluscs comprise only 60% (Smirnov 1986) and in the Caspian Sea only 43.6% of total food weight, together with crustaceans at about 40% in the latter (Lus 1963). Even in non-native areas of the distribution, *N. melanostomus* retains its diet preference for molluscs. In the River St. Clair, their dominant diet item is the zebra mussel (another non-indigenous species accidentally introduced into the Great Lakes). Round gobies even prefer them to indigenous molluscs (Ghedotti *et al.* 1995). In the Baltic Sea, their diet composition is similar to that in the Ponto–Caspian area (Skóra and Rzezník 2001). At the onset of the round goby invasion into the Baltic Sea, a prominent increase in bivalve quantities was noticed, that might have facilitated their success (Sapota 2004). In the Serbian part of the Danube, molluscs represent 70.59% of the round goby diet, with zebra mussel predominating (Simonović *et al.* 1998).

A parasite release may be important in allowing an introduced species to become invasive. In the Slovak stretch of the Danube, the number of parasite species found in round gobies was similar to those in their native range, and no parasites specific to the genus *Neogobius* in their indigenous populations were found. Moreover, reduced parasite abundance combined with strong host reaction to many parasite species may be in favour of the round goby in new habitats of the Middle Danube (Ondračková *et al.* 2005).

Since a wide ecological tolerance itself may not be sufficient to turn a species into a successful invader, we have also evaluated the round goby in terms of possible epigenetic relations among the ontogenies of its indigenous and invasive populations.

LIFE HISTORY TRAITS OF THE ROUND GOBY WITHIN AN EPIGENETIC CONTEXT

The phenotypic plasticity of a species can be a function of epigenetic mechanisms usually expressed in creating both altricial and precocial forms within and/or among populations (Balon 2001, 2002). These forms can be best evaluated by a thorough examination of their ontogenies, especially early development, associated with studies of life history traits such as fecundity, number of spawning acts per season, parental care, egg size, age at maturation, as well as both between- and within-population ontogenetic variability in external morphology. Thus, the ontogenetic patterns of the round goby from the Slovak stretch of the Danube, as related to external morphology, have been examined using both distance-based and geometrical (Bookstein coordinates-based) measurements (L'avrinčíková *et al.* 2005). The overall development of the Danubian round goby (17–153 mm standard length, SL) appeared to be rather fuzzy and no remarkable change in external shape during their ontogeny was observed (L'avrinčíková *et al.* 2005). This means that all substantive morphological developments must have occurred very early. Indeed, the eggs of the round goby are relatively large and high in protein and fat (Moiseyeva 1983), and studies of early development define the round goby a species with direct development (Moskalkova 1996). Such a direct development represents a strongly precocial (specialized) life history (L'avrinčíková *et al.* 2005). To summarize, if the life history traits of *N. melanostomus* (indigenous populations) are evaluated within the epigenetic concept (Balon 2004), it is a typical precocial species: it has large eggs with a dense yolk, a small egg number, the larval period is absent, it has a large size at first exogenous feeding, high parental investment in the young, advanced developmental state of the young, and implicitly a longer juvenile period.

Successful colonists typically have a short generation time and mature earlier (Lodge 1993). In their non-native areas, such as brackish waters and lakes, as well in the Detroit River, newly arrived round gobies really mature earlier compared to those from native marine habitats (MacInnis and Corkum 2000b, Corkum *et al.* 2004). Similarly, round goby females (n = 109) from the Slovak stretch of the Danube can be mature at only 45 mm SL (minimum size), and thus they reach reproductive maturation at smaller size than in most indigenous populations, where maturation occurs mostly in females longer than 55 mm SL (Table 1; L'avrinčíková and Kováč 2007).

Building a phenotype of any multicellular organism requires two sources of information: programmatic (coming from the genotype) and developmental (coming from the environment). At the same time, both of these sources of

Table 1 Life history traits of round goby from native and non-native (rivers Danube and Detroit) areas of distribution. Note that the invasive populations reach reproductive maturation at smaller size than in most indigenous populations. Data for the Danube from L'avrinčíková and Kováč (2007).

	Black Sea	Azov Sea	Caspian Sea[7] north	Caspian Sea[7] middle	River Danube (Slovakia)	River Detroit[8] (North America)
Minimum size at maturity	55 mm[1]	55 mm[4]	32 mm	50 mm	45 mm	42 mm
Fecundity at the same SL						
42.4–92 mm	–	–	–	–	80–600	84–606
53–120 mm	200–1500[2]	–	–	–	206–1402	–
70–130 mm	–	578–2693[5]	–	–	443–1640	–
Diameter of mature group of eggs	1.0–(1.9–2.5) mm[3]	1.3–2.5 mm[6]	–	–	0.8–2.3 mm	–

References: [1]Pavlov (1964)*; [2]Burnashev *et al.* (1955)*; [3]Vinogradov and Tkacheva (1950)*; [4]Trifonov (1955)*; [5]Kovtun (1977); [6]Mikhman (1963)*; [7]Azizova (1962) **; [8]MacInnis and Corkum (2000b)
*after Smirnov (1986)
**after Pinchuk *et al.* (2003)

information are also sources of variation, which enters the process at each transition from one developmental step to another (this is in terms of the theory of saltatory ontogeny, and the theory of synchrony and heterochrony in ontogeny; see Balon 1990 and Kováč 2002 for details). As a result, the life history of each population and/or species can slide over the trajectory between the most altricial extreme to the most precocial extreme, back and forth, from generation to generation. Of course, such intraspecific differences between altricial and precocial forms in ontogeny (i.e. the shifts over the trajectory) are usually very small. "Epigenesis, the mechanism of ontogenies, creates in every generation alternative variations in a saltatory way that enable the organisms to survive in the changing environments as either altricial or precocial forms" (Balon 2004). In the light of such a notion, earlier maturation (at smaller size, Table 1) of non-indigenous round goby populations, which is associated with shortening the juvenile period of life, suggests a shift back from the strongly precocial towards a more altricial (more generalized) life history (L'avrinčíková *et al.* 2005, L'avrinčíková and Kováč 2007). On the other hand, the absolute fecundity of the smaller females from the Danube did not decline significantly compared to those of indigenous round gobies (Table 1; L'avrinčíková and Kováč 2007). Thus, the shift from the highly precocial life history of the indigenous round goby back towards a more altricial life history in the invasive populations (see also MacInnis and Corkum 2000a) also appears to have occurred in the Danubian population. It has been hypothesized that such a combination of altricial-precocial epigenetic trajectories may be one of the key factors for successful colonization of new environments.

CONCLUSIONS

Neogobius melanostomus has several attributes typical of successful invaders, allowing it to establish rapidly in novel environments: increased phenotypic plasticity, tolerance of a wide range of environmental conditions, broad diet, aggressive behaviour, high reproductive capacity, nest guarding by males, and a large size compared to species of a similar benthic lifestyle. Nevertheless, a wide ecological tolerance itself may not be enough to explain how a species can turn into a successful invader. Complementary insights can be obtained from examination of epigenetic interactions among the overall ontogenies (i.e. including life history traits) of its indigenous and invasive populations. In the round goby, it appears that a similar shift from the highly precocial life history of indigenous populations back towards a more altricial life history in its invasive populations (earlier maturation at smaller size, i.e. a shortened juvenile period) occurred at least in North American and Danubian populations. This similarity in patterns that occurred independently in two continents has led us to the hypothesis that it is the combination of altricial-precocial epigenetic trajectories that may be one of the key factors for successful colonization of new environments.

ACKNOWLEDGEMENTS

Jozef Tomeček, Michal Baláž, and Daniel Grul'a helped us in the field and Andrea Novomeská assisted in the laboratory. We thank them all. This study was funded by the Slovak VEGA Grant Agency, projects No. 2/3126/23 and 1/2341/05.

REFERENCES

Azizova, N. A. 1962. The possibility of a goby fishery in the Caspian. Rybnoye Khozaystvo **3**, 14. (In Russian)

Balon, E. K. 1990. Epigenesis of an epigeneticist: the development of some alternative concepts on the early ontogeny and evolution of fishes. Guelph Ichthyology Review **1**, 1–48.

Balon, E. K. 2001. Saltatory ontogeny and life-history model: neglected processes and patterns of evolution. Journal of Bioeconomics **3**, 1–26.

Balon, E. K. 2002. Epigenetic processes, when natura non facit saltum becomes a myth, and alternative ontogenies a mechanism of evolution. Environmental Biology of Fishes **65**, 1–35.

Balon, E. K. 2004. Evolution by epigenesis: Farewell to Darwinism, neo- and otherwise. Rivista di Biologia/Biology Forum **97**, 269–312.

Bănărescu, P. 1964. Pisces – Osteichthyes. Fauna Republicii populare Romîne, vol. XIII. Ed. Acad. RPR, Bucharest, Romania.

Berg, L. S. 1949. Freshwater fishes of USSR and adjacent countries, vol. 3. Izvestiya Akademii Nauk, Moscow, Russia. (In Russian)

Borcea, I. 1934. Révision systématique et distribution géographique des gobiidés de la Mer Noire et particulièrement des eaux Roumaines. Annales scientifiques de l'Université de Jassy, Jassy, Romania.

Burnashev, M. S., V. S. Chepurnov, and N. P. Rakitina. 1955. Fishes of the Dubossarskoe Reservoir and questions of the development of fishery industry in it. Uchen. Zap. Kishinev **20**, 3–31. (In Russian)

Bzoma, Sz. and L. Stempniewicz. 2001. Great Cormorants (*Phalacrocorax carbo*) diet in the Gulf of Gdańsk in 1998 and 1999. Third International Symposium on Functioning of Coastal Ecosystems in Various Geographical Regions, Institute of Oceanography, University of Gdańsk, Poland.

Charlebois, P. M., J. E. Marsden, R. G. Goettel, R. K. Wolfe, D. J. Jude, and S. Rudnika. 1997 The round goby *Neogobius melanostomus* (Pallas). A review of European and North American Literature. Illinois-Indiana Sea Grant Program and Illinois Natural History Survey. INHS Special publications **20**.

Chotkowski, M. A. and J. E. Marsden. 1999. Round goby and mottled sculpin predation on trout eggs and fry: field prediction from laboratory experiments. Journal of Great Lakes Research **25**, 26–35.

Corkum, L. D., M. R. Sapota, and K. E. Skóra. 2004. The round goby, *Neogobius melanostomus*, a fish invader on both sides of the Atlantic Ocean. Biological Invasions **6**, 173–181.

Crossman, E. J., E. Holm, R. Cholmondeley, and K. Tuininga. 1992. First record for Canada of the rudd, *Scardinius erythropthalmus*, and notes on the introduced round goby, *Neogobius melanostomus*. Canadian Field-Naturalist **106**, 206–209.

Drensky, P. 1951. Fishes of Bulgaria. Izvestiya Akademii Nauk, Sofia, Bulgaria. (In Bulgarian)

Erős, T., A. Sevcsik, and B. Tóth. 2005. Abundance and night-time habitat use patterns of Ponto–Caspian gobiid species (Pisces, Gobiidae) in the littoral zone of the River Danube, Hungary. Journal of Applied Ichthyology **21**, 350–357.

Ghedotti, M., J. Smihula, and J. Smith. 1995. Zebra mussel predation by round gobies in the laboratory. Journal of Great Lakes Research **21**, 665–669.

Janssen, J. and D. J. Jude. 2001. Recruitment failure of mottled sculpin, *Cottus bairdi*, in Calumet Harbor, Southern Lake Michigan, induced by the newly introduced round goby *Neogobius melanostomus*. Journal of Great Lakes Research **27**, 319–328.

Jude, D. J. 1997. Round gobies: cyberfish of the third millennium. Great Lakes Research Review **3**, 27–34.

Jude, D. J., J. Janssen, and G. Crawford. 1995. Ecology, distribution, and impact of the newly introduced round and tubenose gobies on the biota of the St. Clair and Detroit Rivers. Pages 447–460 in M. Munawar, T. Edsall, and J. Leach, editors. The Lake Huron ecosystem: ecology, fisheries, and management. Ecovision World Monograph Series. S.P.B. Academic Publishing, Amsterdam, The Netherlands.

Jude, D. J., R. H. Reider, and G. R. Smith. 1992. Establishment of Gobiidae in the Great Lakes basin. Canadian Journal of Fisheries and Aquatic Sciences **49**, 416–421.

Jurajda, P., J. Černý, M. Polačik, Z. Valová, M. Janáč, R. Blažek, and M. Ondračková. 2005. The recent distribution and abundance of non-native *Neogobius* fishes in the Slovak section of the River Danube. Journal of Applied Ichthyology **21**, 319–323.

Kalinina, E. M. 1976. Spawning and development of the Azov and Black Sea gobies. Naukova Dumka, Kiev, Russia. (In Russian)

Kostyuchenko, A. A. 1960. Feeding of the round goby and utilization of its food supply in the Sea of Azov. Trudy AZNIIRCH **1**, 341–360. (In Russian)

Kostyuchenko, A. A. 1969. Regularity of distribution and migrations of the round goby in the Sea of Azov. Trudy AZNIIRCH **26**, 14–19. (In Russian)

Kováč, V. 2002. Synchrony and heterochrony in ontogeny (of fish). Journal of Theoretical Biology **217**, 499–507.

Kovtun, I. F. 1977. On the fecundity of the round goby, *Gobius melanostomus*, from the Sea of Azov. Voprosy Ikhtiologii **17**, 642–644.

Kuhns, L. A. and M. B. Berg. 1999. Benthic invertebrate community responses to round goby (*Neogobius melanostomus*) and zebra mussel (*Dreissena polymorpha*) invasion in Southern Lake Michigan. Journal of Great Lakes Research **25**, 910–917.

L'avrinčíková, M. and V. Kováč. 2007. Invasive round goby *Neogobius melanostomus* from the Danube mature at small size. Journal of Applied Ichthyology, in press.

L'avrinčíková, M., V. Kováč, and S. Katina. 2005. Ontogenetic variability in external morphology of round goby *Neogobius melanostomus* from Middle Danube, Slovakia. Journal of Applied Ichthyology **21**, 328–334.

Lodge, D. M. 1993. Biological invasions: lessons for ecology. Trends in Ecology and Evolution **8**, 133–137.

Lus, V. Ja. 1963. The diet of Azov Sea gobies. Trudy Instituta Okeanologii **62**, 96–127. (In Russian)

MacInnis, A. J. and L. D. Corkum. 2000a. Age and growth of round goby *Neogobius melanostomus* in the Upper Detroit River. Transactions of American Fisheries Society **129**, 852–858.

MacInnis, A. J. and L. D. Corkum. 2000b. Fecundity and reproductive seasons of the round goby *Neogobius melanostomus* in the upper Detroit River. Transactions of the American Fisheries Society **129**, 136–144.

Mikhman, A. S. 1963. On fecundity of Azovian gobies – round goby and syrman. Trudy AZNIIRCH **6**, 105–109. (In Russian)

Miller, P. J. 1984. The tokology of gobioid fishes. Pages 119–153 in G. W. Potts and R. J. Wootton, editors. Fish reproduction strategies and tactics. Academic Press, London, UK.

Miller, P. J. 1986. Gobiidae. Pages 1019–1095 in P. J. P. Whitehead, M. L. Bauchot, J. C. Hureau, J. Nielsem, and E. Tortonese, editors. Fishes of the northeast Atlantic and Mediterranean. UNESCO, Paris, France.

Moiseyeva, E. B. 1983. The development of the gonads of the round goby, *Neogobius melanostomus* (Gobiidae), during the embryonic period. Voprosy Ikhtiologii **23**, 64–74. (In Russian)

Moskalkova, K. I. 1996. Ecological and morphophysiological prerequisites to range extension in the round goby *Neogobius melanostomus* under conditions of anthropogenic pollution. Voprosy Ikhtiologii **36**, 584–590. (In Russian)

Nelson, J. S. 1994. Fishes of the world. Wiley, New York, NY.

Nikolsky, G. V. 1954. Special ichthyology. Sovetshkaya nauka, Moscow, Russia.

Ondračková, M., M. Dávidová, M. Pečínková, R. Blažek, M. Gelnar, Z. Valová, J. Černý, and P. Jurajda. 2005. Metazoan parasites of *Neogobius* fishes in the Slovak section of the River Danube: a preliminary study. Journal of Applied Ichthyology **21**, 345–349.

Pavlov, P. I. 1964. Current status of resources of industry fishes in lower Dnepr and Dneprovsko-Bugskoy liman and their conservation. (Unpublished manuscript deposited in VINITI, in Russian)

Pinchuk, V. I., E. D. Vasil'eva, V. P. Vasil'ev, and P. J. Miller. 2003. *Neogobius melanostomus* (Pallas, 1814). Pages 293–345 in P. J. Miller, editor. The freshwater fishes of Europe, vol. 8/I, Aula Wiebelsheim, Germany.

Ragimov, D. B. 1991. Gobies of the Caspian Sea (Systematics, Ecology, Commercial Importance). Avtoref. Doctoral dissertation, St. Petersburg, Russia. (In Russian)

Sapota, M. R. 2001a. The round goby (*Neogobius melanostomus*) dominating fish species in the shallow zone of the Puck Bay-sequence of invasion and population features. International round goby workshop, June 18–19, Hel Marine Station, Hel, Poland.

Sapota, M. R. 2001b. Tne three-spined stickleback and the round goby dominating fish species in the shallow water zone of the Puck Bay. Third International Symposium on functioning of coastal ecosystems in various geographical regions, June 19–22, Institute of Oceanography, University of Gdańsk, Gdańsk, Poland.

Sapota, M. R. 2004. The round goby (*Neogobius melanostomus*) in the Gulf of Gdansk – a species introduction into the Baltic Sea. Hydrobiologia **514**, 219–224.

Simonović, P., B. Valković, and M. Paunović. 1998. Round goby *Neogobius melanostomus*, a new Ponto–Caspian element for Yugoslavia. Folia Zoologica **47**, 305–312.

Skazkina, E. P. 1966. On the problem of ecological differences in Azov gobies (round goby and syrman goby). Trudy AzerNIRO **24**, 35–38. (In Russian)

Skazkina, E. P. 1972. On active metabolism of gobies from the Azov Sea. Trudy VNIRO **85**, 138–143. (In Russian)

Skóra, K. E. 1996. Nowe I rzadkie gatunki ryb z rajonu Zatoki Gdanskiej. Zoologica Poloniae **41**, 113–130.

Skóra, K. E. and J. Rzezník 2001. Observations on the diet composition of *Neogobius melanostomus* Pallas 1811 (Gobiidae, Pisces) in the Gulf of Gdansk (Baltic Sea). Journal of Great Lakes Research **27**, 290–299.

Smirnov, A. I. 1986. *Perciformes (Gobioidei), Scorpaeniformes, Pleuronectiformes, Lophiiformes*. Fauna Ukrainy 8, Kiev, Russia. (In Russian)

Sokolov, L. I. and E. A. Tsepkin. 1992. The changes of ichthyofauna in the river systems of the Russian central region as the result of antropogenic factors. Vestnik Moscow University, Series **16**. (In Russian)

Stráňai, I. and J. Andreji. 2004. The first report of round goby, *Neogobius melanostomus* (Pisces, Gobiidae) in the waters of Slovakia. Folia Zoologica **53**, 335–338.

Trifonov, G. P. 1955. To a study of reproductive biology of Azovian gobies. Trudy Karadagskoy Biologicheskoy Stantsii **13**, 5–46. (In Russian)

Tsyplakov, E. P. 1974. Expansion of the ranges of some fishes in connection with hydraulic engineering works on the Volga and aclimatization operations. Journal of Ichthyology **14**, 343–351. (In Russian)

Vinogradov, K. A and K. S. Tkacheva. 1950. Materials on fecundity of fishes from the Black Sea. Trudy Karadagskoy Biologicheskoy Stantsii **9**, 9–63. (In Russian)

Wiesner, C. 2005. New records of non-indigenous gobies (*Neogobius* spp.) in the Austrian Danube. Journal of Applied Ichthyology **21**, 324–327.

Wiesner, C., R. Spolwind, H. Waidbacher, S. Guttmann, and A. Doblinger. 2000. Erstnachweis der Schwarzmundgrundel *Neogobius melanostomus* (Pallas, 1814) in Österreich. Österreichs Fischerei **53**, 330–331.

Chapter fifteen

Growth and life history traits of introduced pumpkinseed (Lepomis gibbosus) *in Europe, and the relevance to its potential invasiveness*

Gordon H. Copp and Michael G. Fox

INTRODUCTION

The North American centrarchid, pumpkinseed *Lepomis gibbosus* (Linnaeus), is a good example of how the life history traits and biogeographical patterns of growth can vary in populations of introduced species, facilitating the establishment of new populations in novel environments. Introduced into Europe from North America in the late 19th century (Maes 1898), the pumpkinseed is now established in at least 28 countries of Europe and Asia Minor (Fig. 1). Scientific interest in the species in Europe began early (e.g. Künstler 1908, Roule 1931), but studies of the species (i.e. papers with either the Latin or common names, pumpkinseed, or sunfish, in the title) only really began in the mid-20th century (Fig. 2). Scientific output rose to about one publication per year in the 1970s, though half of these were laboratory experiments on

Francesca Gherardi, Biological invaders in inland waters: Profiles, distribution, and threats, 289–306.

Fig. 1 Countries in Europe in which introduced pumpkinseed *Lepomis gibbosus* are known to have established populations.

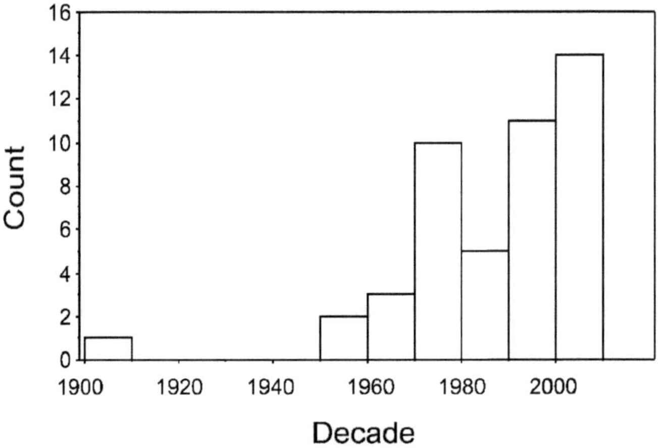

Fig. 2 Number per decade of publications in Europe specifically on pumpkinseed (i.e. title contains one or more of the following terms: pumpkinseed, *Lepomis gibbosus*, sunfish), based on the authors' bibliographic collections and a review of listings in ASFA (Aquatic Sciences and Fisheries Abstracts) undertaken in April 2006 using these search terms.

pumpkinseed physiology (Kramer 1971, 1972, 1973, Krapp 1975, Piavaux 1977). With a sharp rise in papers during the 1990s and 2000s, approximately half of the 12 papers on growth and life history of European pumpkinseed appeared in the last 5 years (Gutiérrez-Estrada *et al.* 2000a, b, Fox and Crivelli 2001, Copp *et al.* 2002, 2004, Villeneuve *et al.* 2005). Virtually, all of the studies on pumpkinseed growth and life history included some form of limited comparison with populations from the species' indigenous and/or introduced ranges, and a common finding in most studies was the absence of sexual dimorphism in growth. However, Fox and Crivelli (2001) were the first to examine the biological traits of pumpkinseed populations introduced into Europe, in contrast with indigenous populations, within the context of ecological theory. They tested whether ectotherms living in warm thermal environments should mature early because of fast juvenile growth (Atkinson 1994), restricting their analyses to female pumpkinseed because male reproductive indices are poor indicators of population responses to environmental variability (see Danylchuk and Fox 1994). This led to a critical review of the available data on the biological traits of European pumpkinseed (Copp *et al.* 2002), which examined the influence of latitude on female growth rate and life history traits, as well as the interrelationships among life history traits.

Building on empirical relationships established in studies of North American female pumpkinseed (Deacon and Keast 1987, Fox 1994, Fox and Crivelli 2001), the life history theory (e.g. Gadgil and Bossert 1970) that underpinned subsequent European studies (e.g. Copp *et al.* 2002, Villeneuve *et al.* 2005) predicts that age at maturity and reproductive allocation in females would be influenced by prematurational growth rates (which determines size at maturity), and that these reproductive variables would, in turn, influence postmaturational growth rates. The aim of the present chapter is to review the available data on growth and life history traits of pumpkinseed in Europe and consider whether any of these indices may be used to assess the species invasiveness [Copp *et al.* (2005a) define an invasive organism as a "indigenous or non-indigenous species, NIS, that spreads, with or without the aid of humans, in natural or semi-natural habitats, producing a significant change in composition, structure, or ecosystem processes, or cause severe economic losses to human activities"]. In the present paper, specific reference is made to the following geographical and life history predictions for female pumpkinseed originally hypothesized by Copp *et al.* (2002):

(1) Juvenile growth rate decreases with latitude.
(2) Age at maturity increases with latitude.
(3) Age and size at maturity are inversely related to juvenile growth rate.
(4) Reproductive effort increases with size at maturity and decreases with temperature.

SOMATIC GROWTH PATTERNS IN EUROPEAN PUMPKINSEED

The growth trajectories of freshwater fishes, often attributed to the influence of environmental conditions, are usually quite variable (Mann 1991). This is apparent in the pumpkinseed populations introduced into Europe, which generally show much slower growth in the adult stage than their indigenous North American counterparts (Copp *et al.* 2004). As is the case in North America (Fox 1994), growth trajectories in introduced European populations show a high degree of variation, with body length at a given age typically varying more than twofold from age 2 and above (Table 1).

In their initial examination of growth in European populations, Copp *et al.* (2002) hypothesized that mean juvenile growth decreases significantly with latitude, but this relationship disappeared when a pumpkinseed population from the River Odra (Poland) was included (Copp *et al.* 2004). This data point is spurious, however, as the Odra population inhabits the plume of a thermal discharge from a power plant (Piasecki and Falandysz 1994). Using all available data for European pumpkinseed populations existing under natural thermal conditions, juvenile growth rate (length at age 2) appears to decrease significantly with increasing latitude, although at any given latitude there is considerable variation (Fig. 3). This latitudinal cline in juvenile growth rate is consistent with other studies in which juvenile growth of centrarchids is expected to be faster in warmer waters (McCauley and Kilgour 1990, Fox and Crivelli 2001). Juvenile growth rates in more northerly latitudes (i.e. English populations) do appear to be lower than those in the rest of Europe (Fig. 4); these differences in growth rate between English populations and those of more southerly latitudes extend into the adult stage, and adult body size in English pumpkinseed is generally stunted despite a relatively long lifespan. At present, growth data are not available for pumpkinseed populations (under natural thermal conditions) in other parts of northern Europe, such as Belgium (Declerck *et al.* 2002, H. Verreycken 2006, personal communication), the Netherlands (Vooren 1971, Aarts and Nienhuis 2003, Denoel *et al.* 2005, H. van Kleef 2006, personal communication), and Norway (E. Sterud 2006, personal communication). Data on these northern populations would be useful for providing a broader-based determination of latitudinal patterns, and thus determine whether English pumpkinseed populations are unusual or representative of northern European populations.

Slow growth in European pumpkinseed, relative to indigenous North American populations, was initially attributed to the selective importation of small-bodied specimens for the aquarium trade (Crivelli and Mestre 1988). However, bibliographic sources suggest that pumpkinseed were introduced for angling as well as for ornamental purposes (pumpkinseed introductions reviewed by Copp *et al.* 2005a). Furthermore, comparative studies of the longitudinal growth of introduced European and indigenous North American pumpkinseed populations have revealed overall differences in adult, but not in juvenile growth patterns (Copp *et al.* 2004). Although there are a few European populations that produce

Table 1 Latitude (Lat) in °N, back-calculated total lengths (TL) at age, reproductive effort (E_R), age at maturity (AaM) in years, and TL at maturity (LaM) in mm for all ages and for ages 3–5 only of pumpkinseed populations in Europe (using the internationally recognized country codes except for England: Eng). Continental European data and those on Cottesmore School Pond were taken from Copp *et al.* (2002a, 2004), with back calculations using a common body-scale intercept of 24 mm TL and with standard length to TL conversions from the original data source or when unavailable, by using TL = 1.27SL + 0.981 as per Copp *et al.* (2004). Underlined E_R values are populations believed to have been established less than 50 years prior to 2006, which are significantly higher (Students' unpaired $t = 2.66$, df = 6, $P = 0.038$) than those for populations older than 50 years (all others).

		Mean back-calculated TL at age (mm)												
Site	Lat	1	2	3	4	5	6	7	8	9	AaM	LaM	E_R	Bibliographic source
1 Spain (Guadiato River)	38	35	93	109	114						1.4		0.306	Gutiérrez-Estrada *et al.* (2000a)[1]
2 Tapada Pequena Reservoir (P)	38	55	73	81	90	92	*115*							Godinho and Ferreira (1996)[2]
3 Tapada Grande Reservoir (P)	38	48	65	75	82	96								Godinho (2004)
4 Monte Novo Reservoir (P)	39	55	84	104	113	118	123	126						Godinho (2004)
5 Divor Reservoir (P)	39	58	77	90	103	109	109	109	115					Brabrand and Saltveit (1989)
6 Tavropos Reservoir (GR)	39	88	100	107										Bobori *et al.* (2006)
7 Kerkini Reservoir (GR)	41	74	88	101	113	122	132				2.0			Neophitou and Giapis (1994)
8 Lake Banyoles (E)	42	53	98	129	145	152	155	162			2.0	61.4		Copp *et al.* (2004)
9 Fumemorte Canal (F)	43	40	72	89	105	116	126	128			1.8	70.0	0.142	Fox and Crivelli (2001)[3]
10 Ligagneau Canal (F)	43	43	70	94	114						1.3	66.1		Fox and Crivelli (2001)
11 Sollac Marsh (F)	43	39	73	105	125	136	142	153			2.3	75.9	0.156	Fox and Crivelli (2001)
12 Fundata Lake (R)[4]	44	55	94	130							3.0	83.5		Constantinescu (1981)
13 Danube delta (R)	44										2.0	96.2		Papadopol and Ignat (1967)
14 Dabas Pond (HU)	47	41	64	84	97	102						75.9		Tandon (1977)
15 Danube (SK)	47	51	86	111	132	150	164							Krupka (1973)
16 Boringwheel Lake (Eng)	51	35	53	68	83	93	109				3.2	78.9	0.321	Villeneuve *et al.* (2005)
17 Batts Bridge upper Stream (Eng)	51	36	57	75	87	98	108				3.3	84.6		Villeneuve *et al.* (2005)
18 Powdermill Upper Lake (Eng)	51	41	77	96	107	122	131	141			2.0	70.0	0.167	Villeneuve *et al.* (2005)

Table 1 Continued.

Site	Lat	Mean back-calculated TL at age (mm) 1	2	3	4	5	6	7	8	9	AaM	LaM	E_R	Bibliographic source
19 Powdermill Side Pond (Eng)	51	41	78	97	102						2.2	81.7		Villeneuve *et al.* (2005)
20 Powdermill Lower Lake (Eng)	51	40	72	104	118	136					2.1	83.8		Villeneuve *et al.* (2005)
21 Tanyards Lake 1 (Eng)	51	42	69	85										Villeneuve *et al.* (2005)
22 Tanyards Lake 2 (Eng)	51	40	70	92	103						2.1	75.5		Villeneuve *et al.* (2005)
23 Tanyards Lake 3 (Eng)	51	38	58	68	78	90	97	100			2.7	72.1		Villeneuve *et al.* (2005)
24 Tanyards Lake 4 (Eng)	51	36	58	70	82	93					3.4	78.3		Villeneuve *et al.* (2005)
25 Tanyards Lake 5 (Eng)	51	38	59	73	88	105					2.8	75.7		Villeneuve *et al.* (2005)
26 Tanyards Lake 6 (Eng)	51	40	68	84	105						2.0	65.0	0.306	Villeneuve *et al.* (2005)
27 Tanyards Lake 7 (Eng)	51	36	59	71	86	97					2.6	66.0	0.248	Villeneuve *et al.* (2005)
28 Milton Mount Pond (Eng)	51	40	71	89	99	109					2.1	80.0		Villeneuve *et al.* (2005)
29 Cottesmore School Pond (Eng)	51	36	48	63	72	81					3.9	78.6	0.194	Copp *et al.* (2002a)
30 Cottesmore Middle Pond (Eng)	51	37	55	66	76	91	104	107			3.1	80.8		Villeneuve *et al.* (2005)
31 Cottesmore Lower Pond (Eng)	51	36	53	65	78	93	104	113	126		3.1	69.1		Villeneuve *et al.* (2005)
32 Island Pond (Eng)	51	34	54	78	92	107	120				3.0	75.0		Villeneuve *et al.* (2005)
33 Douster Pond (Eng)	51	35	49	61	72	82	90	102	110	114	2.8	66.9		Villeneuve *et al.* (2005)
34 Dunwear Pond (Eng)	51	41	67	98	107	116	129				2.5	83.3		Villeneuve *et al.* (2005)
35 Mirgenbach Reservoir (F)	49										1.0	76.6		Dembski *et al.* (2006)[5]
36 Odra River (PL)	53	81	109	136	155	173	178	181						T. Heese and M. Przybylski (2004)[6]

[1]Back-calculated data from Table 1 of that paper (fish from 1993 only, i.e. up to age 4 only), E_R values from Table 2 of that paper, which includes fish (up to age 5) from 1994; [2]Back-calculated value not given for age 6, value given here is the estimated mean of the observed TL values presented in Fig. 2 of the cited paper; [3]E_R values from Copp *et al.* (2002a): [4]The age at maturity value used for the Lake Fundata population was 2.0 years, as per Copp *et al.* (2002a) rather than that (3.0 years) given in Villeneuve *et al.* 2005), a presumed transcription error. [5]Published data for a thermally influenced population. [6]Unpublished data for a thermally influenced population cited in Copp *et al.* (2004). [6]2004, personal communication.

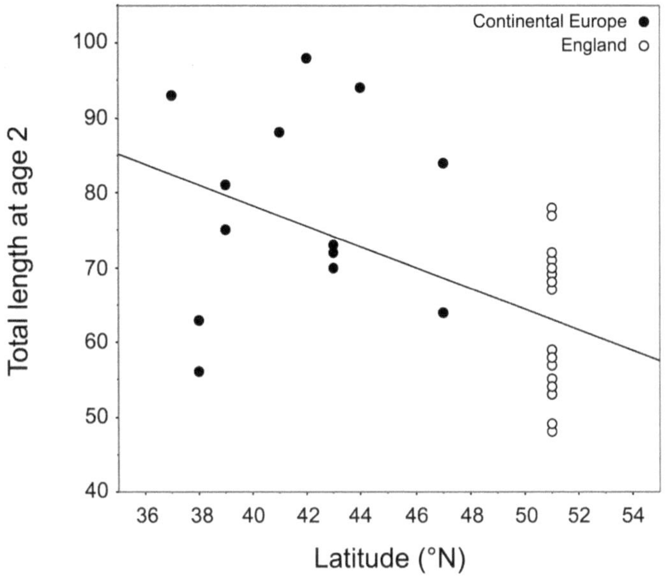

Fig. 3 Juvenile growth (indicated by mean TL at age 2) of pumpkinseed populations as a function of latitude, with the best-fit least squares relationship ($r^2 = 0.271$, $P < 0.005$) shown as a solid line. Population data are from Table 1.

adults comparable in size to those of typical North American populations (e.g. Lake Banyoles, Spain, and the thermally influenced Odra River population in Poland), these are the exception rather than the rule. The processes that determine the differential growth between the pumpkinseed populations in England and in southern Europe remain unclear, but differences in the length of the vegetative season and variations in diet may play a role, e.g. limited food resources due to elevated population density (Copp *et al.* 2002, Klaar *et al.* 2004) and/or the greater energetic value of mollusc (Godinho *et al.* 1997, García-Berthou and Moreno-Amich 2000) and non-mollusc dominated diets (Copp *et al.* 2002, Declerck *et al.* 2002).

LIFE HISTORY TRAITS OF EUROPEAN FEMALE PUMPKINSEED

Female pumpkinseed in their indigenous range show a high degree of variation in mean age (α_A) and size at maturity (α_{TL}), where α is calculated from the females collected in just prior to or during the height of spawning, using the formula from DeMaster (1978) as adapted by Fox (1994) and Fox and Crivelli (2001):

$$\alpha = \sum_{x=0}^{w} (x)\ [f(x) - f(x-1)]$$

Fig. 4 Mean growth trajectories (total length at age) from Villeneuve *et al.* (2005) for populations of pumpkinseed in the northern (cool-water: England) and the southern (warm-water: France, Greece, Portugal, Romania, Slovakia, Spain), parts of its introduced European range. Data points in the graph are means from the individual populations, listed in Table 1.

This high degree of variation in age and length at maturity is also true of the European populations, with the available data (Table 1) demonstrating that mean age at maturity among European populations is almost three times more variable (coefficient of variation = 26.3) than the mean length at maturity (coefficient of variation = 9.06). Nonetheless, the latter extends over a range of 23.2 mm total length.

The third hypothesis proposed by Copp *et al.* (2002) was that the age at maturity of European pumpkinseed would increase with latitude. This hypothesis was supported with the addition of new data for England (Villeneuve *et al.* 2005) to the existing data for continental Europe (Copp *et al.* 2002), which resulted in a significant relationship between age at maturity and latitude (Fig. 5). Again, however, there was considerable variability in the mean age at maturity of populations within any given degree of latitude, especially in England.

Age and length at maturity were found to be inversely related to the juvenile growth rate in European pumpkinseed populations (Copp *et al.* 2002), although the relationship with length at maturity was not quite significant ($P = 0.07$). With the inclusion of new data (Villeneuve *et al.* 2005), the relationship

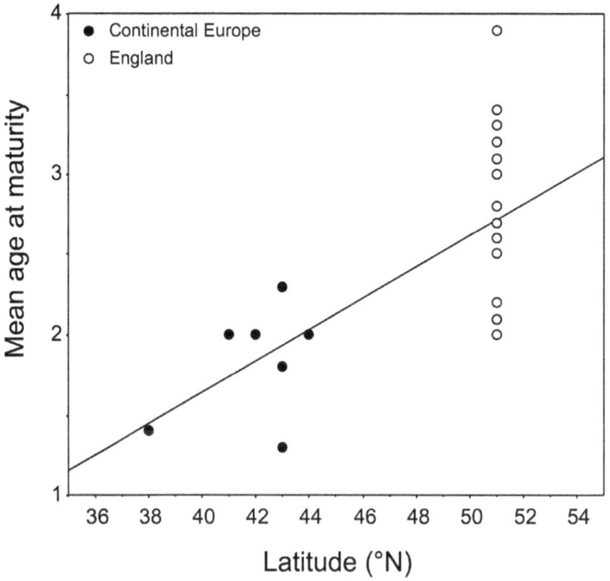

Fig. 5 Age at maturity *vs*. latitude for European pumpkinseed populations. Solid line is the best-fit least-squares regression line ($r^2 = 0.41$, $P < 0.001$). Data are listed in Table 1.

between the juvenile growth rate and age at maturity was found to be highly significant ($r = 0.78$, $P < 0.001$, Fig. 6), whereas no significant relationship was found between the juvenile growth rate and length at maturity ($r = 0.08$, $P = 0.71$). The relationship between fast growth and early maturation is predicted from life history models (e.g. Stearns and Koella 1986) and has been found at the population level in many fish species (reviewed in Fox 1994). This relationship appears to explain some of the interpopulation variation in age at maturity, particularly in England, that cannot be explained by latitude alone. This relationship may provide a useful means for assessing the potential invasiveness of pumpkinseed populations in Europe (Fig. 6), whereby the transition phase between non-invasive and invasive pumpkinseed populations is defined at its lower extent by the minimum age at maturity (age 1; Fox 1994) and at its upper extent by the end of juvenile growth (which for many pumpkinseed populations is age 2; Fox 1994, Fox and Crivelli 2001, Copp *et al.* 2004).

Virtually all of the populations from Southern Europe, where the species is considered invasive, are in the lower right of the graph (Fig. 6), with the apparent non-invasive sites to the upper left. Populations in the lower right quadrat of the graph would be those with the highest potential for population expansion by virtue of their short generation time and their high fecundity early in life because of fast juvenile growth. This of course presumes that the fitness

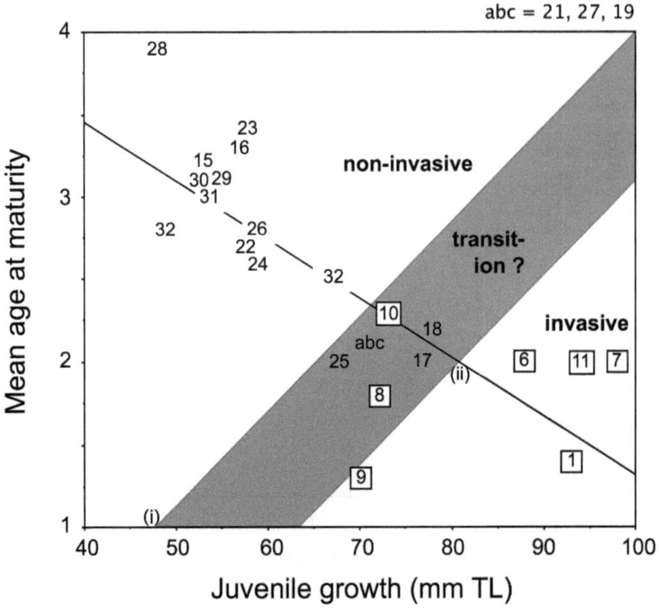

Fig. 6 Mean age at maturity (in years) as a function of mean juvenile growth (TL at age 2) for European pumpkinseed populations (solid: $r^2 = 0.61$, $P < 0.001$), with boxed numbers being those in regions where the pumpkinseed is considered invasive. Sites are numbered as in Table 1. The proposed physiological transition phase between non-invasive and invasive pumpkinseed populations is hypothesized as extending from the minimum age at maturity [the 45° line that traces from the intercept, at (i)] and the end of juvenile growth [which for many pumpkinseed populations is age 2; the 45° line that traces through the age 2 intercept with the regression slope, at (ii)].

benefit of early maturity and high reproductive output is not offset by high rates of juvenile and adult mortality. Although southern European populations with these "*r*-selected" traits do appear to have shorter lifespans than populations with slower juvenile growth rates and later maturity (see Table 2 in Villeneuve *et al.* 2005), these differences may not be reflected in juvenile mortality rates (see Bertschy and Fox 1999), and the rapid expansion of southern populations, particularly in the Iberian Peninsula, would seem to support this explanation. Unfortunately, there are no data on juvenile mortality in European pumpkinseed populations that can be used to test this hypothesis in a fitness model. Such data would require more intensive sampling than has typically been collected for European pumpkinseed populations (see Bertschy and Fox 1999). Of particular interest in the relationship between juvenile growth and age at maturity are the sites in the proposed transition phase between invasive and non-invasive (i.e. end of lag phase). All but one of the English sites in this zone are in the Sussex Ouse River catchment. The two continental sites in this intermediate/potentially invasive phase are a canal and a marsh/canal in southern France, an area where migration is much less difficult, and where pumpkinseed are widely distributed (Keith and Allardi 2001).

The last of the hypotheses proposed by Copp *et al.* (2002) for introduced pumpkinseed in Europe was the positive (but significant at 10% only) relationship between reproductive effort (E_R) and mean length at maturity. Similar to reproductive allocation, which is commonly measured using the gonado-somatic index (GSI = 100 × ovary weight ÷ total body weight), reproductive effort is the proportion of available resources used for reproduction, representing the relative annual energetic investment in gonadal growth (for a review, see Wootton 1979). Biological theory predicts that species in unpredictable (i.e. novel) environments will exhibit fast juvenile growth and high reproductive investment at an early age (Stearns 1976). Within the context of invasion biology theory, these same attributes are expected to be characteristic of recently introduced species as they strive to establish a self-sustaining population in a novel environment where intraspecific competition would be absent, and interspecific competition might be reduced due to the absence of indigenous competitors. However, after an extended period of establishment in a given water body, the introduced populations are expected to shift their allocation of resources gradually to somatic growth, resulting in lower E_R values, as intraspecific competition becomes a factor and the population shifts from a more "*r*"-selected to a "K"-selected life history strategy.

To examine reproductive effort, we used the Mills and Eloranta (1985) version of the formula described by Wootton (1979): $E_R = W_G \div (W_G + W_I)$, where W_G is the mean weight (g) of the gonad for a given age class and W_I is the somatic weight increment (g) between consecutive year classes. E_R will therefore be high when allocation to somatic growth (i.e. increment) is low, and E_R will be low when somatic growth increment is elevated. E_R is difficult to obtain from bibliographic sources, as it requires the age-specific mean values of gonad weight and of the incremental change in somatic weight (W_I). Additionally, potential use of the index can be hampered when the number of female fish obtained per age class is low, which can result in negative W_I values (i.e. a few specimens from a given age that are small for their age will render a mean size below the mean of the next younger year class). E_R tends to increase with age (Gutiérrez-Estrada *et al.* 2000a) because the allocation to somatic growth is disproportionately high in immature fish (Wootton 1979). Therefore, we calculated E_R for ages 3–5 only, to reduce the impact of size-related growth, but this was possible for only five of the 19 populations examined by Villeneuve *et al.* (2005).

In the limited data for European pumpkinseed populations (Table 1), E_R is not correlated with age at maturity (Pearson $r = 0.16$, $P = 0.70$, $n = 7$), and no latitudinal pattern in E_R is evident in reproductive effort ($r = 0.08$, $P = 0.85$, $n = 7$). We also contrasted E_R in populations from southern Europe with those in England, in order to compare locations with warmer water and more widespread pumpkinseed distribution with those of cooler water and more limited distribution. Both of these factors would favour higher reproductive effort in English pumpkinseed populations. Although E_R was greater, on average, in England (mean $E_R = 0.247$, SE = 0.03, *vs.* mean = 0.201, SE = 0.05 in southern European populations), this difference was not significant ($t = 0.83$, df = 6, $P = 0.44$). However, pumpkinseed populations that are thought to have existed for less than 50 years had a significantly higher (Students' unpaired $t = 2.66$, df = 6, $P < 0.05$) mean reproductive effort (mean $E_R = 0.27$, SE = 0.029) than those established prior to 1950 (mean $E_R = 0.164$, S.E. = 0.016). This apparent difference is provisional in nature and requires further study for verification.

DISCUSSION

The inverse relationship between the juvenile growth rate of a pumpkinseed population and its mean age at maturity may provide a useful means for assessing the potential invasiveness of pumpkinseed populations in Europe. The parts of Europe where this species is considered invasive (rapid spread, producing significant ecological change, or severe economic losses; Copp *et al.* 2005a) are in southern Europe (Neophitous and Giapis 1994, Godinho and Ferreira 1996, García-Berthou and Moreno-Amich 2000, Keith and Allardi 2001) and demonstrate characteristics (Fig. 6) that generate high potential for population expansion by virtue of short generation time and high fecundity early in life. There are six English pumpkinseed populations that show juvenile growth and age at maturity patterns similar to those of the southern France populations. Yet, pumpkinseed in England have demonstrated minimal expansion since the species' introduction about one hundred years ago (Wheeler and Maitland 1973, Lever 1977). In the English river catchments where pumpkinseed occurs, the species has remained confined mainly to still waters. Few individuals are found in English watercourses except in the Sussex Ouse catchment (Klaar *et al.* 2004), where ripe male and female pumpkinseed have been captured, but no evidence of in-stream reproduction (e.g. no nests or larvae) has been found (Copp *et al.* 2004, Klaar *et al.* 2004, Villeneuve *et al.* 2005), with only one specimen <40 mm TL observed (S. Stakėnas and F. Villeneuve, unpublished data). Therefore, the small specimens of pumpkinseed found in English streams are most likely to be escapees from upstream online water bodies, which act as "drip-feeds" of small pumpkinseed into the receiving streams. The intensity of this drip-feeding is determined by the type of outlet from still waters to the adjacent streams.

Slow expansion of pumpkinseed in England does not appear to be due to elevated water velocities in the streams. The English water courses inhabited by pumpkinseed are not particularly fast flowing, and telemetry data for pumpkinseed in Batts Bridge Stream (Sussex Ouse catchment) indicate that the species is able to move through rapids at least as fast as $0.7\,\mathrm{m\,s^{-1}}$ (S. Stakėnas 2005, personal communication). However, most water courses in southern England (as in much of Europe) are interrupted by weirs and small reservoir dams, which can impede upstream movement. The environmental conditions available to pumpkinseed in the receiving water courses appear to be favourable to the establishment of pumpkinseed populations, as there are lentic areas similar to those found in French watercourses where pumpkinseed have established populations (e.g. Copp and Cellot 1988). However, the species is rarely recorded in water courses outside the Sussex Ouse catchment (Copp *et al.* 2006). The reason(s) for the absence of in-stream reproduction may be related to an establishment lag-phase, as suggested by Klaar *et al.* (2004). Lag phases are common in NIS (Crooks and Soule 1999) and represent a post-establishment period during which the introduced species adapts its reproduction to local conditions.

As previously mentioned, recently-established populations are expected to express a more *r*-selected life history strategy, investing more effort into reproduction, so as to ensure continuation of the species. The available data on E_R do suggest that recently-established populations (i.e. post-1950s) are putting more into reproduction than longer-established populations (i.e. pre-1950s). One possible explanation for this is compensatory reproductive effort in response to cannibalism. Introduced pumpkinseed are known to be cannibalistic (Copp *et al.* 2002) and during establishment the young-of-the-year pumpkinseed represent a "familiar" prey type that could smooth the transition to novel prey types of the new environment. Elevated reproductive effort would compensate for the high juvenile mortality (e.g. Copp *et al.* 2002), which can express itself in longer lifespans or a greater number of spawning events, such as has been suggested for other nest-guarding fish species (Copp *et al.* 2002b).

However, there are problems with both GSI and E_R in that they assume that the number of spawning events is equal among populations. When E_R is calculated from "one off" assessments (i.e. just before spawning), it may misrepresent the real allocation to reproduction. Previous studies have found pumpkinseed to spawn over shorter periods and with fewer batches, in colder water (Fox and Crivelli 2001). Indeed, only four population data points (7, 8, 9, and 27 in Fig. 6), from France and England, are based on collections made during several periods of the spawning season. Therefore, the difference observed in E_R between recently established and longer-established populations European pumpkinseed populations may be spurious and should be viewed as suggestive, pending further study.

Of particular interest to environmental managers is whether the invasiveness of pumpkinseed, in its current and future introduced range (i.e. under conditions of climate change), can be predicted by the simple relationship of juvenile growth and mean age at maturity (Fig. 6). This relationship is influenced by a suite of environmental factors, not the least of which are temperature, intraspecific competition, and predation. It is interesting to note that pumpkinseed and its indigenous predator, the largemouth bass *Micropterus salmoides* (Lacépèdé), have generally been introduced into the same European countries (i.e. 22 of 28 countries; see Elvira 2001). And largemouth bass introductions have generally been successful in southern European water bodies where pumpkinseed is considered invasive (Elvira 1998), but rarely in other parts of Europe where pumpkinseed is not invasive (Copp *et al.* 2004). In Iberia, where freshwater fishes, and in particular its endemic species, evolved in the absence of piscivorous fishes, largemouth bass could be facilitating the invasion of pumpkinseed by reducing (or removing) potential competitors (see Ricciardi 2001) rather than impeding its establishment through predation pressure. This would be particularly true for Iberian river systems that have been rendered largely lentic by the construction of reservoirs, which provide ideal source populations for pumpkinseed establishment and dispersal. Human assistance in the spread of pumpkinseed (e.g. by anglers) may be more common in southern Europe than elsewhere. Pumpkinseed were introduced into France for angling (Künstler 1908) and in Iberia as prey for largemouth bass in order to keep them from eating indigenous forage (Elvira 2001). But pumpkinseed are rarely sought after in Europe as an angling amenity, and indeed the species has been considered to be a pest throughout most of Europe for some time (e.g. Roule 1931, Vooren 1972, Belpaire *et al.* 2000).

In conclusion, the initial evaluation of the available data on European pumpkinseed populations, Copp *et al.* (2002) suggested that the later maturation and slower growth of a pumpkinseed population in southern England resulted from the combined effects of thermal regime, limited food resources, and relatively high juvenile survivorship. Subsequent study of other English pumpkinseed populations (Villeneuve *et al.* 2005) revealed the initial data point for northern Europe to be representative of the upper extent of the range of observable values for age and length at maturity (Fig. 6) as well as age-specific growth (Table 1). From the information currently available, the degree of invasiveness of pumpkinseed in Europe appears to be a function of three factors: (1) life history traits (Fig. 6), with particular reference to age at maturity and juvenile growth rate of which both depend to a degree on temperature; (2) characteristics of the drainage basin/landscape into which the species was first introduced, with dispersal facilitated by watercourse and water body connectivity; and (3) human-assisted dispersal such as unauthorized fish movements and releases for angling amenity or religious reasons (Copp *et al.* 2005b). To verify the importance of these factors, and provide advice on the management needs as regards non-indigenous fish species, further study is needed of pumpkinseed populations in northern Europe as well as a detailed comparison between indigenous and introduced populations of this species.

ACKNOWLEDGEMENTS

The authors' research is funded by UK Defra and the National Science and Engineering Research Council of Canada, respectively.

REFERENCES

Aarts, B. G. W. and P. H. Nienhuis. 2003. Fish zonations and guilds as the basis for assessment of ecological integrity of large rivers. Hydrobiologia **500**, 157–178.

Atkinson, D. 1994. Temperature and organism size a biological law for ectotherms. Advances in Ecological Research **25**, 1–58.

Belpaire, C., R. Smolders, I. vanden Auweele, D. Ercken, J. Breine, G. van Thuyne, and F. Ollevier. 2000. An index of biotic integrity characterizing fish populations and the ecological quality of Flandrian water bodies. Hydrobiologia **434**, 17–33.

Bertschy, K. A. and M. G. Fox. 1999. The influences of age-specific survivorship on pumpkinseed sunfish life histories. Ecology **80**, 2299–2313.

Bobori, D. C., A. C. Tsikliras, and N. I. Economidis. 2006. Some morphological and biological characteristics of fishes from Tavropos Reservoir (western Greece). Folia Zoologica **55**, 199–210.

Brabrand, A. and S. J. Saltveit. 1989. Ecological aspects of the fish fauna in three Portuguese reservoirs. Archiv für Hydrobiologie **114**, 575–589.

Constantinescu, V. 1981. Relationship of total length, body length, weight with standard length in *Lepomis gibbosus* from Fundata Lake (Romania). Travaux du Museum d'Histoire Naturelle 'Grigore Antipa' **23**, 213–223.

Copp, G. H., P. G. Bianco, N. Bogutskaya, T. Erős, I. Falka, M. T. Ferreira, M. G. Fox, J. Freyhof, R. E. Gozlan, J. Grabowska, V. Kováč, R. Moreno-Amich, A. M. Naseka, M. Peňáz, M. Povž, M. Przybylski, M. Robillard, I. C. Russell, S. Stakėnas, S. Šumer, A. Vila-Gispert, and C. Wiesner. 2005a. To be, or not to be, a non-native freshwater fish? Journal of Applied Ichthyology **21**, 242–262.

Copp, G. H. and B. Cellot. 1988. Drift of embryonic and larval fishes, especially *Lepomis gibbosus* (L.), in the Upper Rhône River. Journal of Freshwater Ecology **4**, 419–424.

Copp, G. H., M. G. Fox, and V. Kováč. 2002. Growth, morphology and life history traits of a coolwater European population of pumpkinseed *Lepomis gibbosus*. Archiv für Hydrobiologie **155**, 585–614.

Copp, G. H., M. G. Fox, M. Przybylski, F. Godinho, and A. Vila-Gispert. 2004. Life-time growth patterns of pumpkinseed *Lepomis gibbosus* introduced to Europe relative to native North American populations. Folia Zoologica **53**, 237–254.

Copp, G. H., V. Kováč, and F. Blacker. 2002b. Differential reproductive allocation in sympatric stream-dwelling sticklebacks *Gasterosteus aculeatus* and *Pungitius pungitius*. Folia Zoologica **51**, 337–351.

Copp, G. H., S. Stakėnas, and P. Davison. 2006. The incidence of non-native fishes in water courses: example of the United Kingdom. Aquatic Invasions **1**, 72–75.

Copp, G. H., K. J. Wesley, and L. Vilizzi. 2005b. Pathways of ornamental and aquarium fish introductions into urban ponds of Epping Forest (London, England): the human vector. Journal of Applied Ichthyology **21**, 263–274.

Crivelli, A. J. and D. Mestre. 1988. Life history traits of pumpkinseed *Lepomis gibbosus* introduced into the Camargue, a Mediterranean wetland. Archiv für Hydrobiologie **111**, 449–466.

Crooks, J. A. and M. E. Soule. 1999. Lag times in population explosions of invasive species: causes and implications. Pages 103–125 in O. T. Sandlund, P. J. Schei, and A. Viken, editors. Invasive species and biodiversity management. Kluwer Academic Publishers, Dordrecht, The Netherlands.

Danylchuk, A. and M. G. Fox. 1994. Age and size dependent variation in the seasonal timing and probability of reproduction among mature female pumpkinseed. Environmental Biology of Fishes **39**, 119–127.

Deacon, L. I. and J. A. Keast. 1987. Patterns of reproduction in two populations of pumpkinseed sunfish, *Lepomis gibbosus*, with differing food resources. Environmental Biology of Fishes **19**, 281–296.

Declerck, S., G. Louette, T. de Bie, and L. de Meester. 2002. Patterns of diet overlap between populations of non-indigenous and native fishes in shallow ponds. Journal of Fish Biology **61**, 1182–1197.

DeMaster, D. P. 1978. Calculation of the average age of sexual maturity in marine mammals. Journal of the Fisheries Research Board of Canada **35**, 912–915.

Dembski, S., G. Masson, D. Monnier, P. Wagner, and J. C. Pihan. 2006. Consequences of elevated temperatures on life-history traits of an introduced fish, pumpkinseed *Lepomis gibbosus*. Journal of Fish Biology **69**, 331–346.

Denoel, M., G. Dzukic, and M. L. Kalezic. 2005 Effects of widespread fish introductions on paedomorphic newts in Europe. Conservation Biology **19**, 162–170.

Elvira, B. 1998. Impact of introduced fish on the native freshwater fish fauna of Spain. Pages 186–190 in I. G. Cowx, editor. Stocking and introduction of fish. Fishing News Books, Blackwell Science, Oxford, UK.

Elvira, B. 2001. Identification of non-native freshwater fishes established in Europe and assessment of their potential threats to the biological diversity. Convention on the Conservation of European Wildlife and Natural Habitats, Strasbourg, France. (Bern\T-PVS 2001\tpvs06e_2001)

Fox, M. G. 1994. Growth, density, and interspecific influences on pumpkinseed sunfish life histories. Ecology **75**, 1157–1171.

Fox, M. G. and A. J. Crivelli. 2001. Life history traits of pumpkinseed (*Lepomis gibbosus*) populations introduced into warm thermal environments. Archiv für Hydrobiologie **150**, 561–580.

Gadgil, M. and W. H. Bossert. 1970. Life historical consequences of natural selection. American Naturalist **104**, 1–24.

García-Berthou, E. and R. Moreno-Amich. 2000. Food of introduced pumpkinseed sunfish: ontogenetic diet shift and seasonal variation. Journal of Fish Biology **57**, 29–40.

Godinho, F. N. 2004. The ecology of largemouth bass, *Micropterus salmoides*, and pumpkinseed sunfish, *Lepomis gibbosus*, in the lower Guadiana basin: the environmental mediation of biotic interactions. Universidad Téchica de Lisboa, Portugal.

Godinho, F. N. and M. T. Ferreira. 1996. The application of size-structure indices to *Micropterus salmoides* (Lacépède, 1802) and *Lepomis gibbosus* (L., 1758) populations as a management tool for southern Iberian reservoirs. Publicaciones Especiales, Instituto Español de Oceanografía **21**, 275–281.

Godinho, F. N., M. T. Ferreira, and R. V. Cortes. 1997. The environmental basis of diet variation in pumpkinseed sunfish, *Lepomis gibbosus*, and largemouth bass, *Micropterus salmoides*, along an Iberian river basin. Environmental Biology of Fishes **50**, 105–115.

Gutiérrez-Estrada, J. C., I. Pulido-Calvo, and C. G. Fernández-Delgado. 2000a. Age-structure, growth and reproduction of the introduced pumpkinseed (*Lepomis gibbosus*, L. 1758) in a tributary of the Guadalquivir River (Southern Spain). Limnetica **19**, 21–29.

Gutiérrez-Estrada, J. C., I. Pulido-Calvo, and J. Prenda. 2000b. Gonadosomatic index estimates of an introduced pumpkinseed (*Lepomis gibbosus*) population in a Mediterranean stream, using computational neural networks. Aquatic Sciences **62**, 350–363.

Keith, P. and J. Allardi. 2001. *Atlas des poissons d'eau douce de France*. Patrimoines Naturels no. 47. Muséum National d'Histoire Naturelle, Paris, France, 387 pp.

Klaar, M., G. H. Copp, and R. Horsfield. 2004. Autumnal habitat use of non-native pumpkinseed *Lepomis gibbosus* and associations with native fish species in small English streams. Folia Zoologica **53**, 189–202.

Kramer, B. 1971. Zur hormonalen Steuerung von Verhaltensweisen der Fortpflanzung beim Sonnenbarsch *Lepomis gibbosus* (L.) (Centrarchidae, Teleostei). Zeitschrift für Tierpsychologie **28**, 351–386.

Kramer, B. 1972. Behavioural effects of an antigonadotropin, of sexual hormones, and of psychopharmaka in the pumpkinseed sunfish, *Lepomis gibbosus* (Centrarchidae). Experientia **28**, 1195–1196.

Kramer, B. 1973. Chemische Wirkstoffe im Nestbau-Sexual- und Kampfverhalten des Sonnenbarsches *Lepomis gibbosus* (L.) (Centrarchidae, Teleostei). Zeitschrift für Tierpsychologie **32**, 353–373.

Krapp, C. 1975. Bau und Vaskularisation des Ovals in der Schwimmblase des Sonnenbarsches *Lepomis gibbosus* (Teleostei, Fam. Centrarchidae). Anatomischer Anzeiger **137**, 486–491.

Krupka, I. 1973. Ďalšie poznatky o veku a raste rýb [Further knowledge on age and growth of fishes]. Pol'ovníctvo a rybárstvo **25**, 28. (In Slovak)

Künstler, J. 1908. *Amiurus nebulosus* et *Eupomotis gibbosus*. Pages 238–244 in Bulletin de la Societé d'Acclimatation.

Lever, C. 1977. The naturalised animals of the British Isles. Hutchinson, London, 600 pp.

Maes, L. 1898. Notes sur la pêche fluviale et maritime en Belgique. Administration des Easux et Forêts, Imprimerie Scientifique, Bruxelles, Belgium, 295 pp.

McCauley, R. W. and D. M. Kilgour. 1990. Effect of air temperature on growth of largemouth bass in North America. Transactions of the American Fisheries Society **119**, 276–281.

Mann, R. H. K. 1991. Growth and production. Pages 446–481 in I. J. Winfield and J. S. Nelson, editors. Cyprinid Fishes. Systematics, Biology and Exploitation. Chapman & Hall, London, UK.

Mills, C. A. and A. Eloranta. 1985. Reproductive strategies in the stone loach *Noemacheilus barbatulus*. Oikos **44**, 341–349.

Neophitous, C. and A. J. Giapis. 1994. A study of the biology of pumpkinseed (*Lepomis gibbosus* (L.)) in Lake Kerkini (Greece). Journal of Applied Ichthyology **10**, 123–133.

Papadopol, M. and G. H. Ignat. 1967. Contribution to the study of the reproduction biology and growth of the American sun-fish (*Lepomis gibbosus* (L.)) in the Lower Danube (flooded zone). Buletinul Instutului de Cercetări i Projectări Piscicole **26**, 55–68.

Piasecki, W. and M. Falandysz. 1994. Preliminary survey on parasite fauna of pumpkinseed sunfish *Lepomis gibbosus* (Linnaeus, 1758) (Pisces, teleostei, centrarchidae) from warm-water discharge canal of the 'Pomorzany' power plant in Szczecin, Poland. Acta Ichthyologica et Piscatoria **24**, 87–100.

Piavaux, A. 1977. L'influence du cadmium sur l'activite de la phosphatase alcaline dans divers tissus de *Lepomis gibbosus* L. (Pisces; Centrarchidae). Hydrobiologia **55**, 251–256.

Ricciardi, A. 2001. Facilitative interactions among aquatic invaders: is an "invasional meltdown" occurring in the Great Lakes? Canadian Journal of Fisheries and Aquatic Sciences **54**, 2513–2525.

Roule, L. 1931. Les poissons et le monde vivant des eaux. Tome II. Les oeufs et les nids. Delagrave, Paris, France.

Stearns, S. C. 1976. Life-history tactics: a review of the ideas Quarterly Reviews in Biology **51**, 3–47.

Stearns, S. C. and J. C. Koella. 1986. The evolution of phenotypic plasticity in life-history traits: predictions of reaction norms for age and size at maturity. Evolution **40**, 893–913.

Tandon, K. K. 1977. Age and growth of pumpkinseed, *Lepomis gibbosus*, (Perciformes, Centrarchidae) from Hungaria. Věstník Československé společnosti zoologické **41**, 74–79.

Villeneuve, F., G. H. Copp, M. G. Fox, and S. Stakėnas. 2005. Interpopulaton variation in the growth and life history traits of the introduced sunfish, pumpkinseed *Lepomis gibbosus*, in Southern England. Journal of Applied Ichthyology **21**, 275–281.

Vooren, C. 1971. Ecological aspects of the introduction of fish species into natural habitats in Europe, with special reference to the Netherlands. Journal of Fish Biology **4**, 565–583.

Wheeler, A. and P. Maitland. 1973. The scarcer freshwater fishes of the British Isles. Journal of Fish Biology **5**, 49–68.

Wootton, R. J. 1979. Energy costs of egg production and environmental determinants of fecundity in teleost fishes. Symposia of the Zoological Society of London **44**, 133–159.

Chapter sixteen

The biological flexibility of the pumpkinseed: a successful colonizer throughout Europe

Jozef Tomeček, Vladimír Kováč, and Stanislav Katina

INTRODUCTION

Among the introduced species of fishes, the pumpkinseed *Lepomis gibbosus* (Linnaeus) is one of the most successful. During the last century, the pumpkinseed became established in most European countries, not merely by introductions but also by natural spreading into adjacent water bodies, and in many cases they attained high population densities in the receiving waters. Impacts on local ichthyofauna were first reported in Portugal and Spain, where pumpkinseed spread invasively (e.g. Zapata and Granado-Lorencio 1993, Godinho *et al.* 1997a, Godinho *et al.* 1998, Blanco *et al.* 2003). So, what makes this fish such a successful colonizer of new environments? It appears that it is the flexibility and plasticity of the pumpkinseed, so obvious in several aspects of its life history. This chapter reviews some of the most important literary sources on various aspects of the life history of the pumpkinseed, emphasising its flexibility. Where possible, the ecology of indigenous North American and introduced European populations is compared. The review is combined with our own original data that focus on the phenotypic plasticity of the species, thus addressing the high potential of pumpkinseed to colonize new environments.

Francesca Gherardi, Biological invaders in inland waters: Profiles, distribution, and threats, 307–336.

DISTRIBUTION

Native area

The native distribution of the pumpkinseed is restricted to the fresh waters of eastern North America, from New Brunswick, Canada, to north-eastern Georgia, USA (Scott and Crossman 1973). In the post-glacial era, pumpkinseed recolonized the Great Lakes region, from Mississippi–Missouri river system or through the Hudson River outlet (Fox *et al.* 1997), as St. Lawrence River probably provided them a refuge during glaciation (Arendt and Wilson 1999). Over the past centuries, the pumpkinseed was introduced into other parts of the USA and Canada (Scott and Crossman 1973, Holčík 1995), and now it is common in both lakes and flowing waters in north-east of North America, from North Carolina to South Ontario (Laughlin and Werner 1980, Hanson and Leggett 1985, Mittelbach 1986).

Non-native area

Around 1880, the pumpkinseed was first introduced into Europe (García-Berthou and Moreno-Amich 2000a), as one of six centrarchids introduced to Europe by the end of the 19th century (Holčík 1991). Since 1887, the pumpkinseed has acclimatized in France and since 1890 also in South England (Copp *et al.* 2002). During the 19th century, the pumpkinseed was introduced into nearly all European and South American countries (Welcomme 1981), as a sport fish (e.g. in France) or as an ornamental fish (e.g. in England, Copp *et al.* 2002, 2004; in Slovenia, Povž and Šumer 2005). In 1910–1913, the pumpkinseed was also introduced into Lake Banyoles (Spain), where it has become a dominant species in the littoral zone (García-Berthou and Moreno-Amich 2000b). Being used by anglers as a bait-fish, the pumpkinseed has spread over many catchment areas of the Iberian Peninsula (Zapata and Granado-Lorencio 1993), including Portugal, where it was recorded in 1977 (Godinho *et al.* 1998). Thereafter, the pumpkinseed continued to spread, and is now found in most Portuguese river basins (Godinho and Ferreira 1998a, Godinho *et al.* 1998). The success story of this species is similar in other parts of Europe. Therefore, it is considered to be one of the most successfully introduced fish species (Holčík 1991), especially in water reservoirs and coastal wetlands (Crivelli and Mestre 1988). For example, in Slovakia, pumpkinseeds have increased their densities and distribution along the Danube and adjacent canals over the last decade so notably that the species was included into the list of invasive species of fish in Slovakia (J. Černý, V. Kováč and J. Kautman 2007, unpublished data). Currently, it occurs in Austria, Belgium, Bulgaria, Czech Republic, France, Germany, Greece, Hungary, Italy, the Netherlands, Poland, Portugal, Romania, Slovakia, Spain, Switzerland, Ukraine, former Yugoslavia (Serbia and Montenegro, Slovenia), and Great Britain (Băcesku 1942, Blanc *et al.* 1971,

Holčík 1976, 1991, Witkowski 1979, Reshetnikov *et al.* 1997, Copp *et al.* 2002, Povž and Šumer 2005, Šumer *et al.* 2005).

FLEXIBLE IN HABITAT REQUIREMENTS

The pumpkinseed inhabits lentic, clear, and transparent waters of lakes, river inlets, side arms, and channels (Holčík 1995), but individuals can also be found in river beds, and periodically even in the turbid waters of the main channel of large rivers and flow-through side channels (Balon 1959a, b, 1966). Although pumpkinseeds require a lentic habitat for reproduction, they can be occasionally found in moderately fast water (Klaar *et al.* 2004). However, in rivers, pumpkinseeds prefer backwaters and calm sites to avoid direct water flow (Keenleyside 1978), since at current velocities >0.12 m sec^{-1}, they have to turn their swimming behaviour from sculling with pectoral fins to propulsion with the caudal fin, and to move their body to maintain position (Brett and Sutherland 1965). In streams, young pumpkinseeds migrate mostly downstream, whereas older, large individuals do more upstream migration (Hall 1972).

Concerning the substrate, the pumpkinseed is ubiquitous, occurring in all lake types with various organic content in the sediment, though more abundant in lakes with more organic sediment (Laughlin and Werner 1980). They seem to prefer mud-bottom sites to hard-bottom sites (Nash 1950). Nevertheless, their presence is more dependent on aquatic macrophytes, especially when young, as smaller specimens, prefer heavy vegetation cover. Indeed, in the lakes with sparse vegetation, pumpkinseeds concentrate in areas with vegetation cover (Laughlin and Werner 1980, García-Berthou and Moreno-Amich 2002). The height of vegetation also plays an important role in habitat preferences of smaller-sized classes (Laughlin and Werner 1980). After hatching, larvae move to the limnetic zone to feed on zooplankton; at a total length (TL) >14 mm they move back to the littoral zone (Vila-Gispert and Moreno-Amich 1998, Hall and Rudstam 1999). Larvae can also undertake diurnal vertical migrations. During the day, they swim in 3.5–4.5 m depth compared to 1 m at night (Copp and Cellot 1988, Vila-Gispert and Moreno-Amich 1998). Small juveniles prefer vegetation in the shallow littoral zone, swimming near the surface in loose schools. Adults swim mostly in pairs or in small aggregations of 3–4 individuals, only rarely schooled (Emery 1973). Larger specimens also exhibit closer association with offshore vegetation in deeper water; however, some individuals occur over bare sediments (Laughlin and Werner 1980). In the lakes with thermal stratification, pumpkinseeds do not move to feed in cold water below the thermocline, whereas in the lakes without a thermocline they can feed in deep waters (Hartleb and Haney 1998).

The pumpkinseed is better adapted to hypoxic conditions than its congener bluegill (*Lepomis macrochirus* Rafinesque), which often dominates pumpkinseeds in nearby lakes but is absent in water bodies with winter hypoxia (Fox and

Keast 1990, Keast and Fox 1990, Osenberg *et al.* 1992). Furthermore, pumpkinseeds are also able to reduce metabolic rates (oxygen consumption) and maintenance costs during winter, reducing the risk of overwinter starvation (Evans 1984). On the other hand, pumpkinseeds are eurythermic, preferring 22–31 °C in summer (Müller and Fry 1976), and thus tolerating easily temperatures over 30 °C (Carlander 1977, Houston 1980). In non-native areas, pumpkinseeds are able to survive at 36 °C (e.g. in Arrocampo Reservoir, Spain; Zapata and Granado-Lorencio 1993), or even at 36.6 °C (in River Guadiato, Spain; Gutierréz-Estrada *et al.* 2000).

In the lakes of Ontario, pumpkinseeds tolerate a pH-value of 5.2–8.4 (Wren and MacCrimmon 1983, Scheuhammer and Graham 1999), though a low pH during winter probably affects their survival, as they need more oxygen under low pH conditions (Rahel 1984). As in other fishes, larger individuals have lower oxygen consumption per gram of body mass but as the temperature increases, they consume more oxygen, probably because they are less adaptable than smaller ones (O'Hara 1968). Pumpkinseeds also show a high tolerance to nitrites (Brunson and Morris 2000).

FLEXIBLE IN DIET

Diet composition

The pumpkinseed is classified as a secondary–tertiary consumer (Vander Zanden *et al.* 1997) and is considered an opportunistic feeder. Both the diversity and size of food items increase with age (Guti *et al.* 1991, Zapata and Granado-Lorencio 1993). Important food items of young pumpkinseeds are small crustaceans (zooplankton), but with increasing age, their proportion decreases in favour of larger prey. This transition is believed to represent an adaptation that minimalises the energetic costs of capturing and processing the prey (Keast 1978, Vinyard 1980, Hanson and Legett 1986, Zapata and Granado-Lorencio 1993). The diet of adults consists mainly of dipteran larvae and pupae (mostly of the family Chironomidae), molluscs, caddisflies, dragonflies, isopods, and amphipods.

The diet composition of pumpkinseeds varies seasonally and it is highly correlated with the abundance of local macroinvertebrates (e.g. Deacon and Keast 1987, Fox and Keast 1990, Mittelbach *et al.* 1992, Zapata and Granado-Lorencio 1993, Sutter and Newman 1997, Wolfram-Wais *et al.* 1999, Copp *et al.* 2002). If the abundance of invertebrates is low, they also prey on eggs and fish, including those of their own species (e.g. García-Berthou and Moreno-Amich 2000a, Copp *et al.* 2002).

The pumpkinseed can consume large quantities of molluscs, a prey type that is ignored by many other species of fish (Pearse 1924). They are also able to benefit from feeding on the invasive zebra mussel [*Dreissena polymorpha*

(Pallas)]. In fact, it has been shown that their growth rates can increase following zebra mussel invasions (Mercer *et al.* 2000). However, pumpkinseeds do not hesitate to consume other prey, such as dragonfly nymphs or *Simocephalus* sp. (Reid 1930, Werner and Hall 1976, 1979); in some locations, molluscs may be just a competitive substitute, especially for American populations living in sympatry with the bluegill (see Keast 1978, Werner and Hall 1979, Mittelbach 1984). Nevertheless, zebra mussels were also observed to be important prey for non-indigenous pumpkinseeds from the lower Danube inundation area (in Romania; Spătaru 1967).

Foraging patterns

It seems that pumpkinseeds preferentially search for prey on macrophytes; however, if vegetation density is low, they are also able to search for prey in sediments (Laughlin and Werner 1980). When competing with the bluegill and green sunfish (*Lepomis cyanellus* Rafinesque), they are able to switch from prey on vegetation to prey in the sediments, as the green sunfish is more efficient in catching prey on vegetation. However, if the vegetation cover is removed, the growth of pumpkinseed is minimally affected, as they are more efficient in searching for prey in sediments (Werner and Hall 1976, 1979).

The foraging rate of pumpkinseeds can significantly differ within or between lakes, depending on the composition of littoral vegetation, as the physical structure of the vegetation can affect the ability of pumpkinseeds to localize prey, and increasing habitat complexity through plant growth reduces fish-foraging rate (without increasing prey abundance). Prey capture rates of pumpkinseeds foraging for cladoceran *Sida crystallina* (O. F. Müller) and larval damselfly (Coenagrionidae) were 53% and 365% higher, respectively, among *Scirpus validus* Vahl with cylindrical stems than for *Potamogeton amplifolius* Tuckerman that has leafy stems. The antipredation behaviour (hiding) of damselfly larvae, especially at their higher abundance, may also reduce capture rate by reducing encounter rate (Dionne and Folt 1991).

Mean gut contents, which ranges from 1.09% to 1.96% of body weight, does not seem to vary significantly at different pumpkinseed densities (6.5, 13, and 26 g m^{-2}), nor with the presence of yellow perch [*Perca flavescens* (Mitchill)] (Hanson and Legett 1986). The amount of food eaten by pumpkinseeds during 24 h can be up to 16% of their own weight, and young specimens can consume up to 1.5 times more food per weight than larger fish (Pearse 1924). At high densities (26 g m^{-2}), macrophyte fragments may appear in their guts (19.5–31.6% of total weight of intestine content). However, the fragments recovered from faeces have provided no evidence of digestion (Hanson and Legett 1986). Macrophyte fragments also appeared in the diet of introduced pumpkinseeds from Lake Banyoles (García-Berthou and Moreno Amich 2000a); sediment, algae, and detritus were found in their guts from Lake Albufera (Blanco *et al.* 2003).

Pumpkinseeds are diurnal feeders using visual orientation for foraging (Shao 1997a), and UV light reception enhances their prey search, probably by improving target contrast (Browman *et al.* 1994). They may begin to feed from midnight to 8:00 am (Hanson and Legett 1986), being most active at dawn, and during the first hours of sunlight (Hanson and Legett 1985, Dionne and Folt 1991). In Lakes Memphremagog (Québec-Vermont, Canada) and Lawrence (Michigan, USA), their stomach attains maximum fullness between 8:30 and 11:30 (Mittelbach 1984, Hanson and Legett 1986). The foraging activity of pumpkinseeds is influenced by reproduction, mainly in nesting males. At dawn, non-nesting fish had significantly fuller stomachs, and a similar, though not significant, trend has been observed at dusk. Nesting males ate less chironomid pupae, cladocerans, and isopods at dawn, and less odonates and isopods at dusk. During the daytime, only one occasional prey attack was observed. At night, nesting males probably leave their nest for short periods to forage in the vicinity of their nests (Thorp *et al.* 1989).

Plasticity in feeding mechanisms

Pumpkinseeds use suction feeding (Lauder 1983, Wainwright 1996) but their considerable plasticity can be illustrated by their variability in feeding mechanisms. Snails are captured by the jaws, drawn in the buccal cavity, and then passed to the pharyngeal jaw apparatus. Pumpkinseeds possess strong, molariform teeth, which allow them to crush mollusc shells quickly and effectively and to extract soft body parts (Mittelbach 1984). This requires a special muscular activity that is absent in other species of the genus, who are unable to crush shells (Wainwright 1996). In crushing, all the pharyngeal jaw muscles act simultaneously in an intense burst, including antagonistic muscles, probably stabilizing the pharyngeal jaw apparatus (Wainwright *et al.* 1991a). The *musculus pharyngocleithralis externus* exhibits short bursts of activity before crushing, presumably helping to position the prey in the pharyngeal jaw apparatus. However, this type of muscular motor pattern has not been observed in all pumpkinseed populations. Pumpkinseeds are able to adjust their motor pattern to novel prey over a period of several weeks (Wainwright 1996). The force necessary to crush hard shells is created mainly by the *musculus levator posterior*. Depending on the type of prey, this muscle shows a high degree of phenotypic plasticity – in pumpkinseeds feeding mostly on molluscs this muscle is up to 2.34 times heavier than in populations feeding on soft bodied invertebrates (Mittelbach 1984, Wainwright *et al.* 1991b). This also applies for other muscles, as well as for the bones of crushing apparatus, which are more robust (Wainwright *et al.* 1991b). Furthermore, the size of the molars and gill rakers is also a subject of variability, both within and between populations (Robinson *et al.* 1993, 2000, Gillespie and Fox 2003). Such a high degree of phenotypic plasticity in feeding mechanics and the robustness of the crushing apparatus have also been confirmed experimentally (Mittelbach *et al.* 1999). In correlation with the inter-individual differences in preferred food, observed both in indigenous as well as non-indigenous populations (Mittelbach *et al.* 1999, García-Berthou and Moreno-Amich 2000a), a great variability in the morphology of feeding apparatus exists even within populations (Mittelbach *et al.* 1999).

GROWTH, AGE, AND LONGEVITY

Pumpkinseed growth exhibits a high degree of variation in both indigenous North American and non-indigenous European populations. In general, growth-rate differences (especially in the juveniles) may be due to environmental variation rather than genetic diversity, as the growth of fish originating from stunted and normal populations did not differ under the same laboratory conditions (Heath and Roff 1987). Adult growth rates and ultimate lengths in non-indigenous European populations are generally lower than in indigenous North American populations, which may result from a higher reproductive effort associated with the colonization of new ecosystems (Copp *et al.* 2004). Latitudinal clines in growth rates and body sizes were significant only in adults, but in a reverse correlation for each of the two continents: with increasing latitude, the adult body size decreased in North America but increased in Europe (Copp *et al.* 2004). The main factors affecting growth in pumpkinseeds appear to be temperature, pH, density, and interspecific competition. Growth can be also affected by a behavioural influence on feeding, as dominant specimens acquire more food than subordinates (Blanckenhorn 1992).

TEMPERATURE AND pH

The growth rate of the pumpkinseed is positively correlated with temperature (Pessah and Powles 1974). For instance, temperatures over 15 °C induced accelerated growth for 4–7 weeks, though then the growth rate declined. At 10 °C, growth was very slow, and at 5 °C, no growth was recorded. When the temperature increased to 15 °C, and then (after 16 weeks) to 25 °C, "overcompensatory" growth was observed in both cases, followed by a growth-rate decline after 6 weeks. It seems that an increasing temperature can be a signal to begin the spring growth cycle (Pessah and Powles 1974).

High temperatures, together with an extended growth period, have been attributed as the cause of the high growth rate seen in young pumpkinseeds in the warm Arrocampo Reservoir (Spain; Zapata and Granado-Lorencio 1993). Similarly, early warming in shallow beaver ponds in spring promotes the growth of pumpkinseeds (Fox and Keast 1990, 1991). In contrast, short summers and low temperatures result in poor growth and high mortality, especially in 0+ specimens (Jackson and Harvey 1989). Slow growth is also characteristic for the populations from England with mild winters but low temperatures and limited food sources during the growth period (Copp *et al.* 2002). However, a water temperature that is too high also results in slower growth (Crivelli and Mestre 1988).

Growth of the pumpkinseed, and its coefficient of condition as well, can also be influenced by pH. At low pH-values, growth is reduced and, in addition to acid stress, fish from acidic lakes may be exposed to elevated metal levels, e.g. in form of methylmercury (Wren and MacCrimmon 1983, Scheuhammer and Graham 1999).

GROWTH RATE, POPULATION DENSITY, AND INTERSPECIFIC COMPETITION

Pumpkinseed growth may be negatively correlated with its population density, as well as with interspecific competition, though in experiments with yellow perch the growth of the pumpkinseeds was not affected by the presence of this competitor (Hanson and Leggett 1985, 1986). Because no significant difference in composition and quantity of food eaten was observed when compared to low densities, fish probably expend more energy to obtain the same amount of food at higher densities. However, the growth of the pumpkinseed appeared to be less affected by interspecific competition, compared to *Lepomis megalotis* (Rafinesque) and *L. cyanellus*, when all three species were reared together (Werner and Hall 1976). Density mostly affects the growth of smaller specimens, especially in the case of competition between young *L. macrochirus* and *L. gibbosus* (Osenberg *et al.* 1988). In Lake Lawrence and Three Lakes II (USA), the growth rate of pumpkinseed attained only 13–45% of that exhibited in experimental ponds with no other fish species (Mittelbach 1986).

When a part of each population of sunfish (*L. gibbosus*, *L. macrochirus*, and their hybrids) was removed from Lake Flora (Wisconsin, USA), the growth of older age classes improved considerably (Parker 1958). In Lake Alanconie (Pennsylvania, USA), partial removal of a fish population positively affected growth of the remaining sunfish, growing up to 221 mm TL (Cooper *et al.* 1971). When only the bluegill population was removed, the improved growth and the higher survival rate of the remaining juveniles reduced growth in older pumpkinseeds (Osenberg *et al.* 1992). However, competition between bluegill and pumpkinseed juveniles does not need to affect the growth rate of pumpkinseeds only negatively. If predators are present, juveniles of both species are confined to the shallow littoral areas, where they compete for food. At 70 mm TL, pumpkinseeds were found to be able to crush mollusc shells effectively, and also to escape from the danger of predators limited by their gape. Therefore, it can be an advantage to grow faster and to escape from competition, even if juvenile resources are limited (Arendt and Wilson 1997, 1999, Arendt *et al.* 2001). Of course, in such a case, there is a trade-off between growth rate and skeletal development (e.g. cranial ossification; Arendt and Wilson 2000), regeneration of damaged fins (Arendt and Wilson 1999), and/or scale strength in the faster growing specimens. Such a compromise in turn results in inferior feeding and swimming abilities, and in weaker defence against predators.

LONGEVITY AND MAXIMUM SIZE

The pumpkinseed appears to be a medium-lived fish with a lifespan not exceeding 10 years in nature and 12 years in captivity (Holčík 1995). It seems that indigenous populations not only grow faster but also live longer compared to the European pumpkinseeds. For example, Scott and Crossmann (1973) have reported pumpkinseeds of age 9 years, attaining 241 mm TL. In the Lower and Upper Beverley Lakes (Canada), males are usually larger than females of the same age, though this becomes significant only at the age classes 5 and 6 (Deacon and Keast 1987). In Europe, the maximum age recorded in most pumpkinseed populations has been 6 years (e.g. Lake Banyoles, Spain, García-Berthou and Moreno-Amich 2000a; backwaters of the rivers Danube and Latorica, Slovakia, Krupka 1973, Koščo *et al.* 2001). However, some populations in England have specimens aged up to 9 years (Villeneuve *et al.* 2005). The maximum size of pumpkinseeds from non-indigenous European habitats also appears to be smaller than in indigenous populations. The largest specimens have been recorded in Spain (176 mm Standard length [SL], García-Berthou and Moreno-Amich 2000a) but those from the Slovakian habitats did not exceed 112–128 mm SL (Krupka 1973, Koščo 1997, Koščo *et al.* 2001), which is similar to pumpkinseeds from the lower Danube in Romania (maximum 113.7 mm SL; Constantinescu 1981). In England, the growth of pumpkinseeds is stunted; for example, 9 years old specimens from the Douster pond attained only 114 mm TL, and the mean TL of 5 to 9 years old specimens ranged from 101 to 114 mm (Villenueve *et al.* 2005).

MORPHOLOGICAL PLASTICITY

Phenotype *vs.* genotype

A very important source of variability in the body shape of pumpkinseeds appears to be their phenotypic plasticity. For example, in a study of limnetic and littoral pumpkinseeds from Paradox Lake, phenotypic plasticity was estimated to account for 53% of total variation in body shape, whereas heritable genetic variation accounted for only 14% of this variation (Robinson *et al.* 2000). In general, pumpkinseeds adapt their body shape according to the habitat occupied. Riverine populations tend to be more slender-bodied than lacustrine pumpkinseeds, though no two discrete morphs can be distinguished. Riverine specimens also have longer pectoral fins, however the other fins are shorter. This is probably due to the necessity to produce higher drag and to minimize resistance when the fish is not oriented precisely in an upstream direction (Brinsmead and Fox 2002).

Littoral and pelagic morphology

In North American lakes, adult pumpkinseeds usually inhabit the littoral zone, whereas the pelagic zone is typically occupied by congeneric adult bluegill where these species co-occur. But if bluegills are absent, then the pumpkinseeds may occupy bluegill's pelagic niche, if offshore reefs are present. Littoral and pelagic pumpkinseed are segregated most of their lifespan, as indicated by the analysis of their parasite load (Robinson *et al.* 2000, Gillespie and Fox 2003). A high degree of reproductive isolation is also probable (Gillespie 2000). Pelagic pumpkinseed have a more slender shaped body, which is thought to be energetically efficient for cruising open waters, shorter heads, increased length and depth of the caudal peduncle, and reduced paired fins (Robinson *et al.* 1996, 2000, Gillespie and Fox 2003). Littoral pumpkinseeds from Monck Lake (Ontario, Canada) were found to have longer pectoral fins, as expected in a habitat requiring more precise manoeuvring. They also tended to have longer heads, and more anteriorly placed dorsal and pectoral fins (Gillespie and Fox 2003). Pelagic pumpkinseeds had thicker, more closely (up to 38% closer) spaced gill rakers, thus improving their efficiency to capture zooplankton (Robinson *et al.* 1993, Gillespie and Fox 2003). Littoral forms may tend to have larger molariform teeth of the pharyngeal jaw apparatus and more distant gill rakers (significantly in three of five lakes; Gillespie and Fox 2003), though this is not necessarily always the case (Robinson *et al.* 1993).

The degree of differentiation between littoral and pelagic morphs varies from lake to lake. For example, in Lake Paradox (New York), pumpkinseeds from both habitats are only moderately differentiated, whereas in Round Lake (New York), the differentiation is bimodal, with a reduced number of intermediate phenotypes. Specialised phenotypes have a significantly better condition factor than generalist phenotypes (Robinson *et al.* 1996).

Indigenous *vs.* non-indigenous populations

Detailed comparison of ontogenetic patterns in external morphology among two indigenous Canadian populations of pumpkinseeds and several non-indigenous populations from Slovakia, Slovenia, and England (see Figs. 1 and 2 for details) revealed that each population had its own developmental pattern (Tomeček *et al.* 2005). The differences increased with the size of the fishes, i.e. during their ontogeny. In general, smaller specimens showed no significant differences among populations, except for pumpkinseeds from the thermal Topla struga oxbow (Slovenia), whereas in pumpkinseeds >60 mm SL populations differed from each other in numerous characters (Tomeček *et al.* 2005).

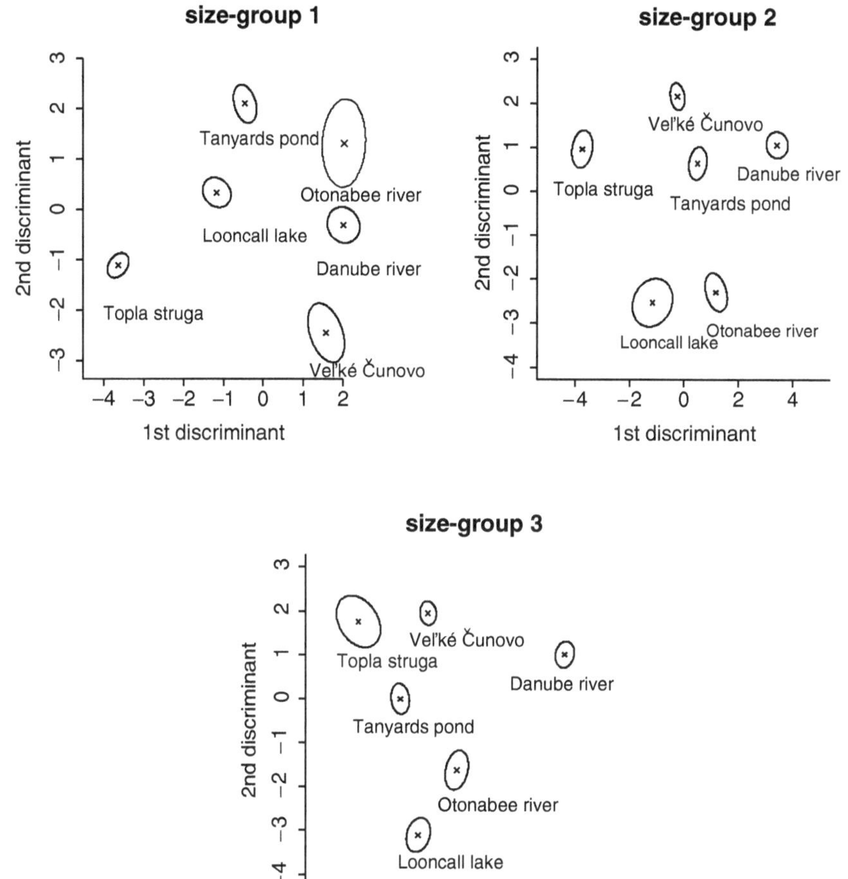

Fig. 1 Graphical representations of comparisons among pumpkinseed populations (Slovakia: Vel'ké Čunovo gravel pit and River Danube; Slovenia: Topla struga; England: Tanyards pond; Canada: Looncall Lake and River Otonabee) in geometrical discriminant analysis, where the x and y axes correspond to the two most important discriminants. Ellipses cover the interval where the mean is located with 95% probability. Size groups 1, 2, and 3 approximately correspond to specimens < 40 mm SL, 40–60 mm SL, and > 60 mm SL, respectively (each group was identified by searching for lowest within group variability, see Katina 2003 for the algorithm used).

Importantly, geographical proximity (and presumably also genetic similarity) seemed to be less important in this respect than local environmental conditions, which supports the above statement that most of the plasticity in pumpkinseeds is of phenotypic rather than of genotypic origin. Detailed analysis revealed that each population differed from every other population – and although the

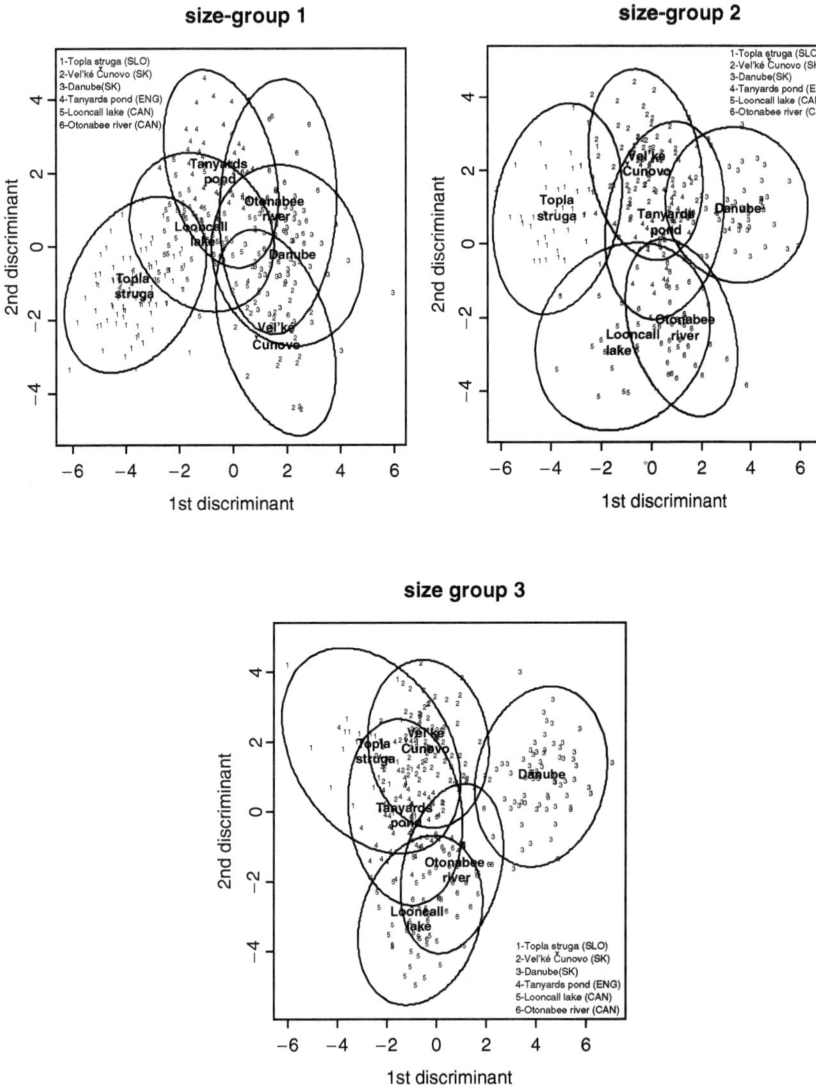

Fig. 2 Graphical representations of comparisons among the pumpkinseed populations. Ellipses cover the interval, where 95% of fish from respective population should be predictively classified. All details as in Fig. 1.

differences were not great, they were significant. Indeed, a model provided by geometrical-discriminant analysis permitted re-classification of the specimens studied with >95% success. Populations from the River Danube, Looncall Lake, and the Topla struga oxbow appeared to be the most distinct (Figs. 1 and 2), the latter being also most extreme, probably due to the extreme thermal conditions in this habitat (Šumer *et al.* 2005).

REPRODUCTION AND VARIABILITY IN LIFE-HISTORY

Among all aspects of its biology, the reproductive characteristics of the pumpkinseed are subject to the most intense variations depending on local environmental conditions. It is an iteroparous species (Deacon and Keast 1987, Fox and Keast 1991), classified to the guild of egg-protecting indifferent species (C2; Balon 1966).

Nesting

Pumpkinseed males build nests prior to spawning, and actively protect eggs and embryos (e.g. Balon 1959a,b, 1966), which can be one of the key factors of pumpkinseed success. Nests are built in shallow littoral areas near the shores of lakes; in rivers, nests are built only in calm backwaters (Keenleyside 1978), usually in depths around 40 cm (Shao 1997a) to 110 cm (Thorp *et al.* 1989), rarely 25 cm (Forbes 1989) and very rarely up to 200 cm (Thorp *et al.* 1989). The manner of nesting and the morphology of the embryonic respiratory system (*aorta dorsalis, vena caudalis profunda, vena subintestinalis*, and elongated *ducti Cuvieri*) presuppose that pumpkinseeds reproduce preferably in clean waters with gravel or clay bottom and good oxygen conditions (Balon 1959a,b). However, the substrate for the nest depends on local conditions. It can be gravel or coarse sand (Balon 1959a,b, 1966), or clay with varying amounts of gravel (Shao 1997a). If such substratum is not available, the pumpkinseed can also use branchlets and vegetation remnants to build its nests (Balon 1966, Colgan and Ealey 1973). The presence of micro-organisms around the nest is also important, as they play a vital role for the pumpkinseeds during the transition of the embryos to exogenous feeding (Balon 1959a,b, 1966).

Nests are often grouped (Balon 1959a,b, 1966, Keenleyside 1978) as pumpkinseeds tend to aggregate at preferred sites (Gross 1979); though, in contrast to the bluegill, pumpkinseed males also build solitary nests, not only colonies (Fox and Crivelli 1998, Garvey *et al.* 2002). Nests on a muddy substrate hidden in vegetation are more dispersed (>1 m in diameter) than those on a gravel bottom (0.5 m in diameter; Shao 1997b). The area covered by nests range from 1150–1700 cm^2 (Shao 1997a), with a diameter 40–125 cm (Balon 1959a,b; Thorp *et al.* 1989).

Spawning behaviour

Reproduction usually takes 10 days (8–15 days), and can be divided into 4 stages:

(1) Nesting – males leave deep water and take up territories in shallow water. Establishment of a territory is accompanied by nest building. Nests and territories are co-extensive. Typical activity of males during this period is sweeping with the caudal fin across the substrate and removing debris to form a nest depression.

This usually lasts two days (Colgan and Gross 1977). (2) Spawning – females enter the colonies, courtship begins. It consists of a series of grunts; males pursue females around the nest (Balon 1959a,b, Morris and Mischke 2000). Successful courtship leads to spawning into the nest. Eggs are laid into the centre of the nest or at its periphery (Shao 1997a). A spent female leaves the nest (Morris and Mischke 2000); if not, the male drives her away and exhibits fanning behaviour (Colgan and Gross 1977, Thorp *et al.* 1989), which lasts three days. Fanning behaviour is characteristic only for this period, and has never been observed beyond the third day (Colgan and Gross 1977). (3) Brooding – this period starts with the hatching of the embryos and ends approximately four days later when the larvae leave the nest. (4) Vacating – reproductive activities terminate; the male leaves its nest (Colgan and Gross 1977).

In pumpkinseeds, an alternative reproductive strategy is also known in young non-nesting males. Intruders (sneakers) usually hide behind physical objects (plants, rocks, woody debris), to avoid the parental male and when a female releases eggs, they enter the nest and release sperm. As nesting males react very aggressively, the intruders hide behind the female and can stay in the nest for several egg releases (Gross 1979).

Spawning period

The beginning of the spawning period coincides with changes in physico-chemical characteristics of the environment, such as water temperature, pH, and water level (Neophitou and Giapis 1994). The breeding season usually begins at the end of May or early June (Bertschy and Fox 1999), which is true not only for indigenous populations, e.g. those from the lakes of Creston valley in British Columbia, Canada (water temperatures 16–26 °C; Forbes 1989) or Lake Opinicon, Ontario, Canada (19 °C; Garvey *et al.* 2002) but also for non-indigenous populations (Lion side-arm of the Danube, Slovakia, Balon 1966; Lake Banyoles, Spain, Vila-Gispert and Moreno-Amich 1998). The spawning season extends, depending on climate conditions, to late July (Balon 1966, Garvey *et al.* 2002), early August (Forbes 1989), or even late August (Vila-Gispert and Moreno-Amich 1998). Larger individuals reproduce earlier in the season than smaller ones; thus, the size structure of a population can markedly influence offspring survival (Danylchuk and Fox 1994). In Little Round Lake (Ontario, Canada), the proportion of large males decreased with the progression of the season; later in the season > 70% of nesting males were the small ones (Danylchuk and Fox 1996).

In pumpkinseeds, reproductive success can be enhanced with repeated spawning (Deacon and Keast 1987). One female can lay eggs in more than one nest, and one male can breed with more than one female in a single nest (Morris and Mischke 2000). Certain males repeat spawning, once the previous brood has left the nest. Some males can fertilize up to four broods, although most males fertilize only one brood. The number of broods appears to be correlated positively with the SL of males and with the duration of nesting. However, the brood size may not be correlated with the size of the nesting males (Shao 1997a). Pumpkinseeds from the delta of the River Rhône (France) repeated their spawning 3 times per season on average (1–6 times) at Fumemorte, and 3.1 times (2–5 times) at Sollac. The number of broods was not correlated with the size of females. In the Lake Opinicon, in Canada, pumpkinseeds breed 2.1 times per season on average (1–4 times); again, no significant correlation between the number of spawning periods and the body size of the female was found (Fox and Crivelli 1998).

Maturation of gonads and gonadosomatic index (GSI)

Maturation of gonads in males and females (also observed as changes in GSI) coincide with each other. This can be illustrated with populations from ponds in Massachusetts (USA). From January to early May, the gonads of both sexes remained small. However, in late May and early June, both the testes and ovaries in adult pumpkinseeds began to increase, and reached their maximum size in late June and/or early July. Then, in mid-August, the size of gonads decreased and spawning stopped. During autumn and winter, GSI in both sexes remains low. Changes in GSI are correlated with changes in the histology of the gonads. Thus, histologically, the annual cycle of the testes in pumpkinseeds contains seven developmental stages, and that of the ovaries, four developmental stages (Burns 1976; see Table 1). Gonads of mature and immature fish can be distinguished visually. For example, mature males have large testes, usually opaque creamy-white, whereas the immature testes are translucent and threadlike (Deacon and Keast 1987).

Mean values of GSI vary widely among populations, reaching 3–9.3% in females (Deacon and Keast 1987, Bertschy and Fox 1999, Copp *et al.* 2002, Gillespie and Fox 2003) and 0.25–1.16% in males (Deacon and Keast 1987, Neophitou and Giapis 1994; Tables 1 and 2). On the other hand, differences in GSI values between littoral and pelagic forms are rare, and have been recorded only in Shadow Lake (Canada), where females from the littoral zone had higher GSI than those from pelagic (Gillespie and Fox 2003).

Fecundity

Absolute fecundity (Table 2) increases with age or SL; relative fecundity remains more or less unchanged, being lower only in the youngest females. Batch fecundity is highly correlated with SL (Fox and Crivelli 1998). During the spawning season, the diameter of eggs in the gonads ranges from 0.48 mm up to 1.3 mm, depending on the stage of their maturity (Deacon and Keast 1987, Neophitou and Giapis 1994, Holčík 1995, Copp *et al.* 2002).

Table 1 Seasonal histological changes in testes and ovaries of pumpkinseeds (from Burns 1976).

Stage	Description	Period
Testes		
1	Collapsed seminiferous lobules, secondary spermatogonia	February–early May
2	Seminiferous lobules collapsed, active spermatogenesis, spermatocytes present, but no free spermatozoa	end of May
3	Free spermatozoa present in lobules, all stages of spermatogenesis present	early June
4	All stages of spermatogenesis present, expansion of lobules with free spermatozoa	mid-June–early July
5	Lobules totally distended and filled with free spermatozoa	late July
6	Lobules collapsing, spermatozoa still present in many of them, lobule walls lined with primary and secondary spermatogonia	late August
7	Only occasional residuals of spermatozoa present, lobules collapsed, secondary spermatogonia increased against primary spermatogonia	September–February
Ovaries		
1	Oocytes diameter not exceeding 0.25 mm, very little yolk present	February–early May
2	Active vitellogenesis, largest oocytes 0.35 mm in diameter	late May–early June
3	Oocytes reach largest size, yolk is extensive	mid-June–late July
4	Similar to stage 1	late August–January

Age and size at maturity

Age and length at maturity is also subject to variation between indigenous Canadian and non-indigenous European populations. The Canadian pumpkinseeds usually mature later than the European ones (Tables 3 and 4; Fox 1994, Copp *et al.* 2002), though mean age at maturity varies within adjacent waterbodies (Fox 1994). However, within a single population, both sexes reach maturity at a similar age and size. No differences in maturation were found between littoral and pelagic forms, except at Shadow Lake (Canada), where littoral forms matured at 2.6 years and 81 mm TL, compared with 3.6 years and 108 mm TL for the pelagic ones (Gillespie and Fox 2003).

Age and size at maturity have not been found to be consistent with the genetic proximity of populations, but rather they are influenced by a combination of ecological factors (biotic or abiotic), e.g. temperature/survival ratio of adults to juveniles, adult to juvenile growth ratio, presence/absence of sympatric bluegill, occurrence of winterkills, predator presence, and/or history of

Table 2 Absolute and relative fecundity of some pumpkinseed populations. For some locations, regression equations are available. F = fecundity, FL = Fork length, TL = Total length.

Location	Absolute fecundity	Relative fecundity	Source
Kerkini (Greece)	7,169 (1,122–12,293)	143.5 (82–165)	Neophitou and Giapis 1994
Upper Beverley Lake (Canada)	1,844–10,632		Deacon and Keast 1987
Lower Beverley Lake (Canada)	2,451–5,387		Deacon and Keast 1987
Banyoles Lake (Spain)	4,485 (log F = −0,59 + 2,16 log FL)		Vila-Gispert and Moreno-Amich 2000
Camargue (France)	7,613 (F = 276, 794TL–18,129)		Crivelli and Mestre 1988
Lower Danube (Romania)	3,200–29,172	158–588	Holčík 1995

Table 3 Mean age, length at maturity and female GSI (gonadosomatic index) values of some European pumpkinseed populations. SL = Standard length (from Copp *et al.* 2002).

Locality	Age at maturity (years)	Length at maturity (mm SL)	GSI (%)
Ligagneau (France)	1.3	51.6	5.4
Fumemorte (France)	1.8	54.6	6.5–10.5
Sollac (France)	2.3	59.2	6.9
Guadalquivir (Spain)	1.4	91.2	6.6
Cottesmore pond (England)	3.9	61.1	6.1
Kerkini (Greece)	2.0	–	9.0
Danube (Romania)	2.0	75	9.0
Fundata Lake (Romania)	2.0	65	–

post-glacial recolonization (Fox and Keast 1991, Fox 1994, Fox *et al.* 1997, Bertschy and Fox 1999, Fox and Crivelli 2001).

Patterns in early development

Pumpkinseed eggs have a single large oil globule and are adhesive (Balon 1966, Shao 1997b). As parental males do not remove ailing eggs (Shao 1997b), successful hatching is limited due to fungal infections (Shao 1997a). Hatching

Table 4 Male and female mean age, length at maturity, length at age 2, and female GSI values of some native Canadian populations; TL = Total length, FL = Fork length (from Deacon and Keast 1987, Fox *et al.* 1997, and Bertschy and Fox 1999).

	Males			Females			
Locality	Age at maturity (years)	Length at maturity (mm TL)	Size at age 2 (mm TL)	Age at maturity (years)	Length at maturity (mm TL)	GSI (%)	Size at age 2 (mm TL)
Otonabee River basin							
Balsam Lake	3.4	125	85	3.6	129	4	85
Buckhorn Lake	3.5	104	75	3.1	91	5.07	77
Rice Lake	3.6	112	72	2.7	82	3.67	72
Scugog Lake	2.7	84	74	2.9	88	5.59	75
Crowe River basin							
Beloporine Lake	3.2	102	73	3.1	92	7.01	71
Belmont Lake	2.75	81	68	3.36	89	6.72	62
Crowe Lake	3.06	91	74	3.49	105	4.67	73
Round Lake	3.6	107	72	3.2	89	5.52	71
Cataraqui River basin							
Loughborough Lake	4.2	114	66	4.5	115	3.44	65
Opinicon Lake	4.3	116	70	3.9	108	5.9	71
Smiths B Lake	4.2	115	69	4.3	115	3	70
Upper Rock Lake	4	122	67	3.7	102	3.88	66
Rideau River basin							
Bass Lake	4	117	73	4.2	130	5.45	67
Long Lake	4.6	125	65	4.6	131	7.42	65
Otter Lake	3	86	68	3.52	98	6.58	67
Little Round Lake	–	–	–	2.4	65	7.7	–
Warrens Lake	–	–	–	2.9	74	9.3	–
Black Lake	–	–	–	3.1	84	6.9	–
Vance Lake	–	–	–	3.4	95	7.5	–
Upper Beverley Lake	5,76	147 (FL)	–	5.52	138 (FL)	3.13	–
Lower Beverley Lake	5	117 (FL)	–	5	110 (FL)	3.36	–

normally occurs in 2–3 days (Shao 1997a,b). Hatched embryos have a round yolk sac with a large oil globule (Shao 1997a) and are very small, about 3.1 mm TL (Balon 1966, Vila-Gispert and Moreno-Amich 1998). The embryonic respiratory system is quite weak (Balon 1959a,b), but instantly after hatching the embryos vehemently move off the above bottom and disperse around the nest, where they have better oxygen conditions and protection from predators (Balon 1966, Shao 1997a). During the next 4–5 days, the embryos rest on the bottom and swim from time to time towards the surface to enhance respiration (Balon 1959a,b).

With the onset of exogenous feeding, at TL about 5.3 mm, the actively swimming embryos become larvae (Brown and Colgan 1984). The mouth and opercular aperture are already opened at this time. Such early mobility of the jaws and opercles indicates not only the early beginning of gill respiration, but also the ability to feed from the first day of free swimming. The yolk sac with its oil globule serves as a reserve source of energy during the next two days (Balon 1959a,b, Brown and Colgan 1984). The relative size of the oil globule in pumpkinseed larvae is the lowest of all Percoidei, and it is not large enough to serve as hydrostatic organ only (Balon 1959a,b). However, unlike other percids, swim bladder fills in on the 6th day (i.e. the first day of swimming) and so larvae are fully prepared to swim and seek for food. At transition to exogenous feeding, larvae must have small micro-organisms at their disposal (Balon 1966), as they are probably selective feeders (Hart and Werner 1987) and suffer high mortality rates as they are more dependent on the availability of suitable prey at transition to exogenous feeding than fish with larger yolk reserves. However, the strategy of the pumpkinseed is based on a continual production of offspring, so that the chance that some of the batches will encounter good prey availability is higher (Hart and Werner 1987). After hatching, larvae become pelagic for 2–3 weeks and are obligate zooplanktivorous (Vila-Gispert and Moreno-Amich 1998). From day 19 to 60, caudal, dorsal, pectoral, and ventral fins form completely. Calcification of cranial structures with a clearly given order varies following epigenetical interactions (Arendt and Wilson 1997, 2000, Arendt *et al.* 2001). The first scales appear at 15–19 mm SL (Tandon 1976, 1977a,b Crivelli and Mestre 1988, Koščo *et al.* 2001). At 20 mm TL, the larvae return back to the littoral area (Garvey *et al.* 2002).

PUMPKINSEED AS A PREY SPECIES

Theoretically, strong predation on pumpkinseeds might be one of the ways to control its population density, though such a control in non-native areas would require the introduction of another non-indigenous species, which itself implies controversy. In native areas, the largemouth bass *Micropterus salmoides* (Lacepède) is an important predator of pumpkinseeds (e.g. Godinho *et al.* 1997b). Its presence affects not only habitat use, but also the feeding resources of young pumpkinseeds, as these have to keep hidden in vegetation up to 50–100 mm SL (Wainwright 1996). However, in the Spanish Lake Banyoles, non-indigenous pumpkinseeds were not found in the diet of the largemouth bass, which apparently prefer to prey on indigenous cyprinids – presumably for lower costs of capture (García-Berthou 2002). Thus, the largemouth bass does not appear to be a species that could control non-indigenous pumpkinseed populations. Of course, pumpkinseeds can also be a prey of indigenous species of fish, such as wels *Silurus glanis* Linnaeus, sander *Stizostedion lucioperca* (Linnaeus), and/or Eurasian perch *Perca fluviatilis* Linnaeus, as observed in the Danube inundation (Romania; Spătaru 1967). According to Scott and Crossman (1973), small pumpkinseeds form part of diet for almost all predatory fish, including northern pike *Esox lucius* Linnaeus. Guti *et al.* (1991) found that the pumpkinseed was the second most abundant fish prey in pike from moorland water bodies in Hungary. However, in some places pike seem to ignore pumpkinseeds as a potential prey (Slovakia, Sedlár 1957; Camargue, France, Crivelli and Mestre 1988), and it appears that in general, indigenous predators may only have a moderate impact on non-indigenous populations of pumpkinseeds.

NEGATIVE IMPACTS OF PUMPKINSEED INTRODUCTIONS AND INVASIONS

The pumpkinseed, as a flexible generalist, can, by competition, limit cyprinid communities (Rahel 1984). They can survive in localities with fluctuating hydrological conditions and may be able to invade new environments (Poff and Allan 1995). As they are considerably less frequent in riverine environment, the damming of rivers can help their successful establishment, so river reservoirs have been identified as the most probable source for pumpkinseed inoculation in Portugal and in England (Godinho *et al.* 1998, Klaar *et al.* 2004). The introduction of pumpkinseeds into some water bodies has led to a suppression and density decline of indigenous fish species (Holčík 1991), mainly due to feeding on eggs and young fish, and to feeding competition (Sedlár 1957, 1965, García-Berthou and Moreno-Amich 2000a).

Indeed, the spread of the pumpkinseed and largemouth bass is considered a major cause of indigenous fish species decline in Portugal and Spain (Godinho and Ferreira 1998b). For example, in Lake Banyoles, indigenous fish decline seems to be entirely an effect of pumpkinseed, largemouth bass, and roach introduction (García-Berthou and Moreno-Amich 2000b). In Lake Albufera, *Mugil cephalus* Linnaeus was a dominant species until 2000, but in 2002 this species was no longer recorded, and the density of the pumpkinseed doubled (Blanco *et al.* 2003). Similarly, in the lower Guadiana basin, indigenous *Anaecypris hispanica* (Steindachner) disappeared at the same time as a notable predominance of pumpkinseeds was recorded (Godinho *et al.* 1997a). In three reservoirs located in the Madrid Community (Spain), overpopulation with pumpkinseeds has been attributed as the cause for lack of spawning of other fish species (García de Jalón *et al.* 1993). The pumpkinseed, together with the largemouth bass and mosquitofish, are also responsible for the decline of indigenous fish fauna in the Portuguese River Raia (Godinho *et al.* 1998, Godinho and Ferreira 2000). Furthermore, in the Portuguese reservoirs Divor and Montargil, it was the only species found in open water, feeding intensively on zooplankton, probably because water level fluctuations reduced the littoral fauna. This predation pressure resulted in changes in local zooplankton communities (Braband and Saltveit 1989). Finally, in the Portuguese reservoir Tapada Pequena, only non-indigenous species can now be found, the pumpkinseed dominance reaching 93.9% (Godinho and Ferreira 1996).

Another example of competition for food comes from the Lake Neusiedler (Austria), where a very abundant population of pumpkinseeds competes with white bream [*Blicca bjoerkna* (Linnaeus)], eel [*Anguilla anguilla* (Linnaeus)], and ruffe [*Gymnocephalus cernuus* (Linnaeus)]. Surprisingly, the dietary overlap with another non-indigenous fish, topmouth gudgeon [*Pseudorasbora parva* (Temminck and Schlegel)] was found to be much less. Although both species feed on chironomid larvae, their feeding microhabitat is clearly differentiated, as pumpkinseeds feed here mainly on sediment dwelling chironomids, whereas the topmouth gudgeon prefers epiphytic species (Wolfram-Wais *et al.* 1999). Guti *et al.* (1991) also recorded a high dietary overlap between the pumpkinseed and the crucian carp, with both species consuming mainly chironomids, ceratopogonid larvae, and gastropods.

The presence of non-indigenous pumpkinseeds, especially in high densities, also leads to a strong decline in mollusc abundance (Osenberg *et al.* 1992). Moreover, predation on molluscs can induce morphological and life-history changes in molluscs, and adversely affects endemic species (García-Berthou and Moreno-Amich 2000a). Molluscs, like *Physa acuta* (Draparnaud), use chemical cues (from the molluscs eaten) to detect a predator and in response change their morphology, life history, and behaviour. Stationary pumpkinseeds can influence prey behaviour over an area of $3.1\,m^2$, but, as pumpkinseed movement rate is estimated to achieve $100\,m\,h^{-1}$, each fish can change snail behaviour over an area of $8{,}000\,m^2$. Snails remain in refuges and their growth is negligible (Turner and Montgomery 2003). Direct predation of pumpkinseeds in experimental enclosures led to a dramatic decline in snail density and modified snail assemblages (Brönmark *et al.* 1992). The presence of pumpkinseeds can also reduce growth, rate of development (in males), and fecundity in the chironomid *Chironomus tentans* (Fabricius) (Ball and Baker 1996). High pumpkinseed density can reduce the density of the aquatic weevil [*Euhrychiopsis lecontei* (Dietz), Coleoptera: Curculionidae], a potential control agent for Eurasian watermilfoil (*Myriophyllum spicatum* Linnaeus) introduced to North America in the mid-1900 (Sutter and Newman 1997). Finally, pumpkinseeds, similar to the common carp, were found to be able to contribute considerably to the increase of turbidity, as well as chlorophyll *a*, total phosphorus, and total nitrogen levels (Angeler *et al.* 2002).

CONCLUSIONS

Pumpkinseeds have been in Europe for more than 120 years, and they have managed to establish their populations in various types of water, in most catchments of rivers flowing through a considerable part of the continent. The pumpkinseed appears to be a typical generalist species with flexible habitat requirements, diet, growth rates, and overall life history. However, being a generalist species does not necessarily mean a successful invader. Indeed, pumpkinseeds are not only flexible generalists but they also have a great capacity for phenotypic plasticity, both in their native and non-native areas of distribution. The review of various studies, as well as our own research, shows that this plasticity can be expressed in many aspects of this species' life, for example in its external and/or internal morphology, early development or feeding mechanisms, thus affecting its overall life history in general. So, it is probably this high degree of phenotypic plasticity, based on the generalist genotypic attributes of the species, that makes this species such a successful colonizer – and in some regions even an invader – throughout Europe.

ACKNOWLEDGEMENTS

We thank G. H. Copp for his bibliographic assistance. This study was funded by the Slovak VEGA Grant Agency projects No 2/3126/23 and 1/2341/05.

REFERENCES

Angeler, D. G., M. Álvarez-Cobelas, S. Sánchez-Carillo, and M. A. Rodrigo. 2002. Assesment of exotic fish impacts on water quality in a degraded semi-arid floodplain wetland. Aquatic Science **64**, 76–86.

Arendt, J. D. and D. S. Wilson. 1997. Optimistic growth: competition and an ontogenetic niche-shift select for rapid growth in pumpkinseed sunfish (*Lepomis gibbosus*). Evolution **51**, 1946–1954.

Arendt, J. D. and D. S. Wilson. 1999. Countergradient selection for rapid growth in pumpkinseed sunfish: disentangling ecological and evolutionary effects. Ecology **80**, 2793–2798.

Arendt, J. D. and D. S. Wilson. 2000. Population differences in the onset of cranial ossification in pumpkinseed (*Lepomis gibbosus*), a potential cost of rapid growth. Canadian Journal of Fisheries and Aquatic Sciences **57**, 351–356.

Arendt, J. D., D. S. Wilson, and E. Stark. 2001. Scale strength as a cost of rapid growth in sunfish. Oikos **93**, 95–100.

Băcescu, M. 1942. *Eupomotis gibbosus* (Lin.) studiu etnozoologic, zoogeografic şi morphologic. Analele Academiei Române, Memoriile Secţiunii Ştiinţifice, Seria III, Bucharest, Romania **17**, 547–560.

Ball, S. L. and R. L. Baker. 1996. Predator – induced life history changes: facultative life history shifts? Ecology **77**, 1116–1124.

Balon, E. K. 1959a. Die Entwicklung des akklimatisierten *Lepomis gibbosus* (Linné 1758) wahrend der embryonalen Periode in den Donauseitenwassern. Zeitschrift für Fischerei **VIII**, 1–27.

Balon, E. K. 1959b. Neres slnečnice pestrej (*Lepomis gibbosus* Linné 1758) aklimatizovanej v bočných vodách československého úseku Dunaja. Věstník československé společnosti zoologické **23**, 1–22.

Balon, E. 1966. In V. Mucha and O. Dub, editors. Limnológia československého úseku Dunaja, Vydavatel'stvo SAV, Bratislava, Slovakia, 328 pp.

Bertschy, K. A. and M. G. Fox. 1999. The influence of age-specific survivorship on pumpkinseed sunfish life histories. Ecology **80**, 2299–2313.

Blanc, M., J. L. Gaudet, P. Bănărescu, and J. C. Hureau. 1971. European inland water fish: a multilingual catalogue. Fishing News (Books), London, UK, 683 pp.

Blanckenhorn, W. U. 1992. Group size and the cost of agonistic behavior in pumpkinseed sunfish. Ethology, Ecology and Evolution **4**, 255–271.

Blanco, S., S. Romo, M. J. Villena, and S. Martinéz. 2003. Fish communities and food web interactions in some shallow Mediterranean lakes. Hydrobiologia **506–509**, 473–480.

Braband, Å. and S. J. Saltveit. 1989. Ecological aspects of the fish fauna in three Portuguese reservoirs. Archiv für Hydrobiologie **114**, 575–589.

Brett, J. R. and D. B. Sutherland. 1965. Respiratory metabolism of pumpkinseed (*Lepomis gibbosus*) in relation to swimming speed. Journal of the Fisheries Research Board of Canada **22**, 405–409.

Brinsmead, J. and M. G. Fox. 2002. Morphological variation between lake- and stream-dwelling rock bass and pumpkinseed populations. Journal of Fish Biology **61**, 1619–1638.

Brönmark, Ch., S. P. Klosiewski, and R. A. Stein. 1992. Indirect effects of predation in a freshwater, benthic food chain. Ecology **73**, 1662–1674.

Browman, H. I., I. Novales-Flamarique, and C. W. Hawryshyn. 1994. Ultraviolet photoreception contributes to prey search behaviour in two species of zooplanktivorous fishes. Journal of Experimental Biology **186**, 187–198.

Brown, J. A and P. W. Colgan. 1984. The ontogeny of feeding behaviour in four species of centrarchid fish. Behavioural Processes **9**, 395–411.

Brunson, M. W. and J. E. Morris. 2000. Species profile: Sunfish. Southern Regional Aquaculture Center Publication No. 724. Stoneville, Mississippi, MS, 6 pp.

Burns, J. R. 1976. The reproductive cycle and its environmental control in the pumpkinseed, *Lepomis gibbosus* (Pisces: Centrarchidae). Copeia **3**, 449–455.

Carlander, K. D. 1977. Handbook of freshwater fishery biology. Life history data on Centrarchid fishes of the United States and Canada, Vol 2. Iowa State University Press, Ames, IA.

Colgan, P. and D. Ealey. 1973. Role of woody debris in nest site selection by pumpkinseed sunfish, *Lepomis gibbosus*. Journal of the Fisheries Research Board of Canada **30**, 853–856.

Colgan, P. W. and M. R. Gross. 1977. Dynamics of agression in male pumpkinseed sunfish (*Lepomis gibbosus*) over the reproductive phase. Zeitschrift für Tierpsychologie **43**, 139–151.

Constantinescu, V. 1981. Relationship of total length, body depth, weight with standard length in *Lepomis gibbosus* (Pisces, Perciformes, Centrarchidae) from Fundata Lake (Romania). Travaux du Museum d'Histoire Naturelle "Grigore Antipa" **XXIII**, 213–223.

Cooper, E. L., C. C. Wagner, and G. E. Krantz. 1971. Bluegills dominate production in a mixed population of fishes. Ecology **52**, 280–290.

Copp, G. H. and B. Cellot. 1988. Drift of embryonic and larval fishes, especially *Lepomis gibbosus* (L.), in the upper Rhône River. Journal of Freshwater Ecology **4**, 419–424.

Copp, G. H., M. G. Fox, and V. Kováč. 2002. Growth, morphology and life history traits of a cool-water European population of pumpkinseed *Lepomis gibbosus*. Archiv für Hydrobiologie **155**, 585–614.

Copp, G. H., M. G. Fox, M. Przybylski, F. N. Godinho, and A. Vila-Gispert. 2004. Lifetime growth patterns of pumpkinseed *Lepomis gibbosus* introduced to Europe, relative to native North American populations. Folia Zoologica **53**, 237–254.

Crivelli, A. J. and D. Mestre. 1988. Life history traits of pumpkinseed, *Lepomis gibbosus*, introduced into the Camargue, a Mediterranean wetland. Achiv für Hydrobiolgie **111**, 449–466.

Danylchuk, A. J. and M. G. Fox. 1994. Age and size-dependent variation in the seasonal timing and probability of reproduction among mature female pumpkinseed, *Lepomis gibbosus*. Environmental Biology of Fishes **39**, 119–127.

Danylchuk, A. J. and M. G. Fox. 1996. Size- and age-related variation in the seasonal timing of nesting activity, nest characteristics, and female choice of parental male pumpkinseed sunfish (*Lepomis gibbosus*). Canadian Journal of Zoology **74**, 1834–1840.

Deacon, L. I. and J. A. Keast. 1987. Patterns of reproduction in two populations of pumpkinseed sunfish, *Lepomis gibbosus*, with differing food resources. Environmental Biology of Fishes **19**, 281–296.

Dionne, M., C. and L. Folt. 1991. An experimental analysis of macrophyte growth forms as fish foraging habitat. Canadian Journal of Fisheries and Aquatic Sciences **48**, 123–131.

Emery, A. R. 1973. Preliminary comparisons of day and night habits of freshwater fish in Ontario lakes. Journal of the Fisheries Research Board of Canada **30**, 761–774.

Evans, D. O. 1984. Temperature independence of the annual cycle of standard metabolism in the pumpkinseed. Transactions of the American Fisheries Society **113**, 494–512.

Forbes, L. S. 1989. Spawning, growth, and mortality of three introduced fishes at Creston, British Columbia. Canadian Field-Naturalist **103**, 520–523.

Fox, M. G. 1994. Growth, density, and interspecific influences on pumpkinseed sunfish life histories. Ecology **75**, 1157–1171.

Fox, M. G., J. E. Claussen, and D. P. Philipp. 1997. Geographic patterns in genetic and life history variation in pumpkinseed populations from four east-central Ontario watersheds. North American Journal of Fisheries Management **17**, 543–556.

Fox, M. G. and A. J. Crivelli. 1998. Body size and reproductive allocation in a multiple spawning centrarchid. Canadian Journal of Fisheries and Aquatic Sciences **55**, 737–748.

Fox, M. G. and A. J. Crivelli. 2001. Life history traits of pumpkinseed (*Lepomis gibbosus*) populations introduced into warm thermal environments. Archiv für Hydrobiologie **150**, 561–580.

Fox, M. G. and A. Keast. 1990. Effects of winterkill on population structure, body size, and prey consumption patterns of pumpkinseed in isolated beaver ponds. Canadian Journal of Zoology **68**, 2489–2498.

Fox, M. G. and A. Keast. 1991. Effect of overwinter mortality on reproductive life history characteristics of pumpkinseed (*Lepomis gibbosus*) populations. Canadian Journal of Fisheries and Aquatic Science **48**, 1792–1799.

García-Berthou, E. 2002. Ontogenetic diet shifts and interrupted piscivory in introduced largemouth bass (*Micropterus salmoides*). International Review of Hydrobiology **88**, 353–363.

García-Berthou, E. and R. Moreno-Amich. 2000a. Food of introduced pumpkinseed sunfish: ontogenetic diet shift and seasonal variation. Journal of Fish Biology **57**, 29–40.

García-Berthou, E. and R. Moreno-Amich. 2000b. Introduction of exotic fish into a Mediterranean lake over a 90-year period. Archiv für Hydrobiologie **149**, 271–284.

García-Berthou, E. and R. Moreno-Amich. 2002. Fish ecology and conservation in Lake Banyoles (Spain): the neglected problem of exotic species. Pages 223–231 in I. G. Cowx, editor. Management and Ecology of Lake and Reservoir Fisheries, Blackwell Science, Fishing News Books. Oxford, UK.

García de Jalón, D., M. M. Rustarazo, B. Gallego, and F. Hervella. 1993. Las comunidades piscicolas de los embalses de Madrid. Directrices para su gestion. Ecología **7**, 467–485.

Garvey, J. E., T. P. Herra, and W. C. Leggett. 2002. Protracted reproduction in sunfish: the temporal dimension in fish recruitment revisited. Ecological Applications **12**, 194–205.

Gillespie, G. J. 2000. Eco-morphological and reproductive isolation in populations of pumpkinseed (*Lepomis gibbosus*). MSc. thesis. Trent University, Peterborough, Ontario, Canada.

Gillespie, G. J. and M. G. Fox. 2003. Morphological and life-history differentiation between littoral and pelagic forms of pumpkinseed. Journal of Fish Biology **62**, 1099–1115.

Godinho, F. N. and M. T. Ferreira. 1996. The application of size-structure indices to *Micropterus salmoides* (Lacépède, 1802) and *Lepomis gibbosus* (L., 1758) populations as a management tool for southern Iberian reservoirs. Publicaciones Especiales de Instituciones Espanola Oceanografica **21**, 275–281.

Godinho, F. N. and M. T. Ferreira. 1998a. Spatial variation in diet composition of pumpkinseed sunfish, *Lepomis gibbosus*, and largemouth bass, *Micropterus salmoides*, from a Portuguese stream. Folia Zoologica **47**, 205–213.

Godinho, F. N. and M. T. Ferreira. 1998b. The relative influences of exotic species and environmental factors on an Iberian native fish community. Environmental Biology of Fishes **51**, 41–51.

Godinho, F. N. and M. T. Ferreira. 2000. Composition of endemic fish assemblages in relation to exotic species and river regulation in a temperate stream. Biological Invasions **2**, 231–244.

Godinho, F. N, M. T. Ferreira, and R. V. Cortes. 1997a. Composition and spatial organization of fish assemblages in the lower Guadiana basin, southern Iberia. Ecology of Freshwater Fish **6**, 134–143.

Godinho, F. N., M. T. Ferreira, and R. V. Cortes. 1997b. The environmental basis of diet variation in pumpkinseed sunfish, *Lepomis gibbosus*, and largemouth bass, *Micropterus salmoides*, along an Iberian river basin. Environmental Biology of Fishes **50**, 105–115.

Godinho, F. N., M. T. Ferreira, and M. I. Portugal e Castro. 1998. Fish assemblage composition inrelation to environmental gradients in Portuguese reservoirs. Aquatic Living Resources, **11**, 325–334.

Gross, M. R. 1979. Cuckoldry in sunfishes (*Lepomis*: Centrarchidae). Canadian Journal of Zoology **57**, 1507–1509.

Guti, G., S. Andrikowics, and P. Biró. 1991. Nahrung von Hecht (*Esox lucius*), Hundfisch (*Umbra krameri*), Karausche (*Carassius carassius*), Zwergwels (*Ictalurus nebulosus*) und Sonnenbarsch (*Lepomis gibbosus*) im Ócsa-Feuchtgebiet, Ungarn. Fischökologie, Petersberg, Germany, **4**, 45–66.

Gutiérrez-Estrada, J. C., I. Pulido-Calvo, and J. Prenda. 2000. Gonadosomatic index estimates of an introduced pumpkinseed (*Lepomis gibbosus*) population in a Mediterranean stream, using computational neural networks. Aquatic Sciences **62**, 350–363.

Hall, C. A. S. 1972. Migration and metabolism in a temperate stream ecosystem. Ecology **53**, 585–604.

Hall, S. R. and L. G. Rudstam. 1999. Habitat use and recruitment: a comparison of long-term recruitment patterns among fish species in a shallow eutrophic lake, Oneida Lake, NY. Hydrobiologia **408/409**, 101–113.

Hanson, J. M. and W. C. Leggett. 1985. Experimental and field evidence for inter- and intraspecfic competition in two freshwater fishes. Canadian Journal of Fisheries and Aquatic Sciences **42**, 280–286.

Hanson, J. M. and W. C. Leggett. 1986. Effect of competition between two freshwater fishes on prey consumption and abundance. Canadian Journal of Fisheries and Aquatic Sciences **43**, 1363–1372.

Hart, T. F. and R. G. Werner. 1987. Efects of prey density on growth and survival of white sucker, *Catostomus commersoni*, and pumpkinseed, *Lepomis gibbosus*, larvae. Environmental Biology of Fishes **18**, 41–50.

Hartleb, C. F. and J. F. Haney. 1998. Use of a thermal and light refugium by *Daphnia* and its effects on foraging pumpkinseed. Environmental Biology of Fishes **51**, 339–349.

Heath, D. and D. A. Roff. 1987. Test of genetic differentiation in growth of stunted and nonstunted populations of yellow perch and pumpkinseed. Transactions of the American Fisheries Society **116**, 98–102.

Holčík, J. 1976. Zoznam bezčel'ustnatých a rýb Slovenska. Biológia (Bratislava) **31**, 641–647.

Holčík, J. 1991 Fish introductions in Europe with particular reference to its central and eastern part. Canadian Journal of Fisheries and Aquatic Science **48**, Supplement 1, 13–23.

Holčík, J. 1995. Slnečnica pestrá (*Lepomis gibbosus*). Pages 425–429 in V. Baruš and O. Oliva, editors. Mihulovci *Petromyzontes* a ryby *Osteichthyes*, Fauna ČR a SR, Academia, Praha, Czech Republic.

Houston, A. H. 1980. Components of the hematological response of fishes to environmental temperature change: a review. Pages 241–299 in A. Ali, editor. Environmental Physiology of Fishes. Plenum Press, New York, NY.

Jackson, D. A. and H. H. Harvey. 1989. Biogeographic associations in fish assemblages: local vs. regional processes. Ecology **70**, 1472–1484.

Katina, S. 2003. Application of allometric shape analysis to fish growth with geometrical notes. Tatra Mountains. Mathematical Publications **26**, 323–336.

Keast, A. 1978. Feeding interrelations between age-groups of pumpkinseed (*Lepomis gibbosus*) and comparisons with bluegill (*L. macrochirus*). Journal of the Fisheries Research Board of Canada **35**, 12–27.

Keast, A. and M. G. Fox. 1990. Fish community structure, spatial distribution and feeding ecology in a beaver pond. Environmental Biology of Fishes **27**, 201–214.

Keenleyside, M. H. A. 1978. Reproductive isolation between pumpkinseed (*Lepomis gibbosus*) and longear sunfish (*Lepomis megalotis*) (Centrarchidae) in the Thames river, southwestern Ontario. Journal of the Fisheries Research of Board Canada **35**, 131–135.

Klaar, M., G. H. Copp, and R. Horsfield. 2004. Autumnal habitat use of non-native pumpkinseed Lepomis gibbosus and associations with native fish species in small English streams. Folia Zoologica **53**, 189–202.

Koščo, J. 1997. Ichtyofauna inundačných vôd Laborca. Natura Carpathica **XXXVIII**, 121–132.

Koščo, J., P. Košuth, and I. Ondrej. 2001. Dĺžkový rast slnečnice (*Lepomis gibbosus* Linnaeus, 1758) v inundačných vodách Latorice. Natura Carpathica **XLII**, 191–196.

Krupka, I. 1973. Ďalšie poznatky o veku a raste rýb. Pol'ovníctvo a rybárstvo **25**, 28.

Laughlin, D. R. and E. E. Werner 1980. Resource partitioning in two coexisting sunfish: pumpkinseed (*Lepomis gibbosus*) and northern longear sunfish (*Lepomis megalotis peltastes*). Canadian Journal of Fisheries and Aquatic Sciences **37**, 1411–1420.

Lauder, G. V. 1983. Functional and morphological basis of trophic specialisation in sunfishes (Teleostei, *Centrarchidae*). Journal of Morphology **178**, 1–21.

Mercer, J. L., M. G. Fox, and C. D. Metcalfe. 2000. Changes in benthos and three littoral zone fishes in a shallow, eutrophic Ontario lake following the invasion of the zebra mussel (*Dreissena polymorpha*). Lake and Reservoir Management, Vernon, Madison, **15**, 310–323.

Mittelbach, G. G. 1984. Predation and resource partitioning in two sunfishes (Centrarchidae). Ecology **65**, 499–513.

Mittelbach, G. G. 1986. Predator mediated habitat use: some consequences for species interactions. Environmental Biology of Fishes **16**, 159–169.

Mittelbach, G. G., C. W. Osenberg, and P. C. Wainwright. 1992. Variation in resource abundance affects diet and feeding morphology in the pumpkinseed sunfish (*Lepomis gibbosus*). Oecologia **90**, 8–13.

Mittelbach, G. G., C. W. Osenberg, and P. C. Wainwright. 1999. Variation in feeding morphology between pumpkinseed populations: phenotypic plasticity or evolution? Evolutionary Ecology Research **1**, 111–128.

Morris, J. E. and C. C. Mischke. 2000. A white paper on the status and needs of sunfish aquaculture in the north central region. North Central Regional Aquaculture Center, Ames, IA, 17 pp.

Müller, R. and F. E. J. Fry. 1976. Preferred temperature of fish: a new method. Journal of the Fisheries Research Board of Canada **33**, 1815–1817.

Nash, C. B. 1950. Associations between fish species in tributaries and shore waters of western Lake Erie. Ecology **31**, 561–567.

Neophitou, C. and A. J. Giapis. 1994. A study of the biology of pumpkinseed (*Lepomis gibbosus* (L.)) in Lake Kerkini (Greece). Journal of Applied Ichthyology **10**, 123–133.

O'Hara, J. 1968. The influence of weight and temperature on the metabolic rate of sunfish. Ecology **49**, 159–161.

Osenberg, C. W., E. E. Werner, G. G. Mittelbach, and D. J. Hall. 1988. Growth patterns in bluegill (*Lepomis macrochirus*) and pumpkinseed (*L. gibbosus*) sunfish: environmental variation and the importance of ontogenetic niche shifts. Canadian Journal of Fisheries and Aquatic Sciences **45**, 17–26.

Osenberg, C. W., G. G. Mittelbach, and P. C. Wainwright. 1992. Two-stage life histories in fish: the interaction between juvenile competition and adult performance. Ecology **73**, 255–267.

Parker, R. A. 1958. Some effects of thinning on a population of fishes. Ecology **39**, 304–317.

Pearse, A. S. 1924. Amount of food eaten by four species of fresh-water fishes. Ecology **3**, 254–258.

Pessah, E. and P. M. Powles. 1974. Effect of constant temperature on growth rates of pumpkinseed sunfish (*Lepomis gibbosus*). Journal of the Fisheries Research Board of Canada **31**, 1678–1682.

Poff, L. N. and J. D. Allan. 1995. Functional organization of stream fish assemblages in relation to hydrological variability. Ecology **76**, 606–627.

Povž, M. and S. Šumer. 2005. A brief review of non-native freshwater fishes in Slovenia. Journal of Applied Ichthyology **21**, 316–318.

Rahel, F. J. 1984. Factors structuring fish assemblages along a bog lake successional gradient. Ecology **65**, 1276–1289.

Reid, H. 1930. A study of *Eupomotis gibbosus* (L.) as occurring in the Chamcook Lakes, N. B. Contributions to Canadian Biology and Fisheries **5**, 457–466.

Reshetnikov, Y. S., N. G. Bogutskaya, E. D. Vasil'eva, E. A. Dorofeeva, A. M. Naseka, O. A. Popova, K. A. Savaitova, V. G. Sideleva, and L. I. Sokolov. 1997. An annotated check-list of the freshwater fishes of Russia. Journal of Applied Ichthyology **37**, 687–736.

Robinson, B. W., D. S. Wilson, A. S. Margosian, and P. T. Lotito. 1993. Ecological and morphological differentiation of pumpkinseed sunfish in lakes without bluegill sunfish. Evolutionary Ecology **7**, 451–464.

Robinson, B. W., D. S. Wilson, and G. O. Shea. 1996. Trade-offs of ecological specialization: An intraspecific comparison of pumpkinseed sunfish phenotypes. Ecology **77**, 170–180.

Robinson, B. W., D. S. Wilson, and A. S. Margosian. 2000. A pluralistic analysis of character release in pumpkinseed sunfish (*Lepomis gibbosus*). Ecology **81**, 2799–2812.

Scott, W. B. and E. J. Crossman. 1973. Freshwater fishes of Canada. Bulletin 184. Journal of Fisheries Research Board of Canada, Ottawa, Canada, 730 pp.

Sedlár, J. 1957. Nevítaní hostia v povodí rieky Nitry. Československé rybářství **6**, 91.

Sedlár, J. 1965. Príspevok k súčasnému stavu zarybnenia retenčnej nádrže Virt. Pol'nohospodárstvo **11**, 689–695.

Shao, B. 1997a. Effects of golden shiner (*Notemigonus crysoleucas*) nest association on host pumpkinseeds (*Lepomis gibbosus*). Behavioral Ecology and Sociobiology **41**, 399–406.

Shao, B. 1997b. Nest association of pumpkinseed, *Lepomis gibbosus*, and golden shiner, *Notemigonus crysoleucas*. Environmental Biology of Fishes **50**, 41–48.

Scheuhammer, A. M. and J. E. Graham. 1999. The bioaccumulation of mercury in aquatic organisms from two similar lakes with differing pH. Ecotoxicology **8**, 49–56.

Spătaru, P. 1967. Nutriţia şi unele relaţii trofice la bibanul-soare – *Lepomis gibbosus* (Linnaeus) 1758 - din complexul de bălţi Crapina-Jijila (zona inudabilă a Dunării). Analele Universităţii din Bucureşti, seria Ştiinţele Naturii – Biologie, Bucharest, Romania, **16**, 151–162

Sutter, T. J. and R. M. Newman. 1997. Is predation by sunfish an important source of mortality for the Eurasian watermilfoil biocontrol agent *Euhrychiopsis lecontei*? Journal of Freshwater Ecology **12**, 225–234.

Šumer S, V. Kováč, M. Povž, and M. Slatner. 2005. External morphology of a Slovenian population of pumpkinseed *Lepomis gibbosus* (L.) from a habitat with extreme thermal conditions. Journal of Applied Ichthyology **21**, 306–311.

Tandon, K. K. 1976. Note on the systematics of the pumpkinseed, *Lepomis gibbosus* (Osteichthyes, Perciformes, *Centrarchidae*). Věstník československé společnosti zoologické **XL**, 307–311.

Tandon, K. K. 1977a. Age and growth of pumpkin seed, *Lepomis gibbosus* (Perciformes, *Centrarchidae*) from Hungary. Věstník československé společnosti zoologické **XLI**, 74–79.

Tandon, K. K. 1977b. Morphometric and growth study of *Lepomis gibbosus* (Osteichthyes, *Percidae*) from Italy. Věstník československé společnosti zoologické **XLI**, 211–217.

Thorp, J. H., L. D. Goldsmith, J. A. Polgreen, and L. M. Mayer. 1989. Foraging patterns of nesting and nonnesting sunfish (Centrarchidae: *Lepomis auritus* and *L. gibbosus*). Canadian Journal of Fisheries and Aquatic Sciences **46**, 1342–1346.

Tomeček, J. V. Kováč and S. Katina. 2005. Ontogenetic variability in external morphology of native (Canadian) and non-native (Slovak) populations of pumpkinseed *Lepomis gibbosus* (Linnaeus 1758). Journal of Applied Ichthyology **21**, 1–10.

Turner, A. M. and S. L. Montgomery. 2003. Spatial and temporal scales of predator avoidance: experiments with fish and snails. Ecology **84**, 616–622.

Vander Zanden, M. J. V., G. Cabana, and J. B. Rasmussen. 1997. Comparing trophic position of freshwater fish calculated using stable nitrogen isotope ratios ($\delta^{15}N$) and literature dietary data. Canadian Journal of Fisheries and Aquatic Sciences **54**, 1142–1158.

Vinyard, G. L. 1980. Differential prey vulnerability and predator selectivity: effects of evasive prey on bluegill (*Lepomis macrochirus*) and pumpkinseed (*L. gibbosus*) predation. Canadian Journal of Fisheries and Aquatic Sciences **37**, 2294–2299.

Vila-Gispert, A. and R. Moreno-Amich. 1998. Seasonal abundance and depth distribution of *Blennius fluviatilis* and introduced *Lepomis gibbosus*, in Lake Banyoles (Catalonia, Spain). Hydrobiologia **386**, 95–101.

Vila-Gispert, A. and R. Moreno-Amich. 2000. Fecundity and spawning mode of three introduced fish species in lake Banyoles (Catalunya, Spain) in comparison with other localities. Aquatic Sciences **61**, 154–166.

Villeneuve, F., G. H. Copp, M. G. Fox, and S. Stakenas. 2005. Interpopulation variation in growth and life-history traits of the introduced sunfish, pumpkinseed *Lepomis gibbosus*, in southern England. Journal of Applied Ichthyology **21**, 275–281.

Wainwright, P. C. 1996. Ecological explanation through functional morphology: the feeding biology of sunfishes. Ecology **77**, 1336–1343.

Wainwright, P. C., G. V. Lauder, C. W. Osenberg, and G. G. Mittelbach. 1991a. The functional basis of intraspecific trophic diversification in sunfishes. Pages 515–529 in E. Dudley, editor. The Unity of Evolutionary Biology. Dioscorides Press, Portland, OR.

Wainwright, P. C., C. W. Osenberg, and G. G. Mittelbach. 1991b. Trophic polymorphism in the pumpkinseed sunfish (*Lepomis gibbosus* Linnaeus): effects of environment on ontogeny. Functional Ecology **5**, 40–55.

Welcomme, R. L. 1981. Register of international transfers of inland fish species. FAO Fisheries Technical Paper, Rome, Italy, **213**, x+120 pp.

Werner, E. E. and D. J. Hall. 1976. Niche shifts in sunfishes: experimental evidence and significance. Science **191**, 404–406.

Werner, E. E. and D. J. Hall. 1979. Foraging efficiency and habitat switching in competing sunfishes. Ecology **60**, 256–264.

Witkowski, A. 1979. New locality of sunfish, *Lepomis gibbosus* (L.) (*Osteichthyes: Centrarchidae*) in catchment area of Barycz river. Fragmenta Faunistica **25**, 15–19.

Wolfram-Wais. A., G. Wolfram, B. Auer, E. Mikshi, and A. Hain. 1999. Feeding habits of two introduced fish species (*Lepomis gibbosus, Pseudorasbora parva*) in Neusiedler See (Austria), with special reference to chironomid larvae (Diptera: *Chironomidae*). Hydrobiologia **408/409**, 123–129.

Wren, C. D and H. R. MacCrimmon. 1983. Mercury levels in the sunfish, *Lepomis gibbosus*, relative to pH and other environmental variables of Precambrian Shield lakes. Canadian Journal of Fisheries and Aquatic Sciences **40**, 1737–1744.

Zapata, S. C. and C. Granado-Lorencio. 1993. Age, growth and feeding of the exotic species *Lepomis gibbosus* in a Spanish cooling reservoir. Archiv für Hydrobiologie, Supplement **90**, 561–573.

Patterns and pathways of invasions

The most frequent and strong result . . . was that successful establishment was positively related to propagule pressure. Although this result is intuitively obvious, quantifying it is not of trivial importance with respect to prevention of NIS. For example, accidental introductions of NIS via commerce-related activities might be impossible to halt completely. Reducing the number of individuals released and the frequency of releases, however, reduce the probability of establishment.

Cynthia Kolar and David Lodge (2001)

Chapter seventeen

Marine vs. freshwater invaders: is shipping the key vector for species introductions to Europe?

Stephan Gollasch

INTRODUCTION

Species introductions are of concern as many non-indigenous species (NIS) have been shown to have worldwide significant negative impacts on the recipient ecosystems. Prime introduction vectors are the unintentional transport by ships (i.e. ballast water, tank sediment, and hull fouling), intentional species import for aquaculture purposes (i.e. target species and non-target species, such as disease agents or parasites), accidental releases of species cultured in containment and stocking initiatives (Carlton 1985, 1987, Cohen and Carlton 1995, Eno *et al.* 1997, Reise *et al.* 1999).

Since biological invasions of aquatic species became more and more into focus in Europe in the mid 1990s, regional inventories were prepared for some European coastal waters: North Sea (Gollasch 1996, Reise *et al.* 1999, Nehring 2002), Baltic Sea (Leppäkoski 1994, Gollasch and Mecke 1996, Leppäkoski and Olenin 2000, Olenin *et al.* 2005), British Isles (Eno 1996, Eno *et al.* 1997), Ireland (Minchin and Eno 2002), Azores (Cardigos *et al.* 2006), Mediterranean Sea (Galil and Zenetos, 2002, CIESM 2005). An overall summary of introduced aquatic species in Europe was prepared by Streftaris *et al.* (2005) and Gollasch (2006).

Francesca Gherardi, Biological invaders in inland waters: Profiles, distribution, and threats, 339–345.

This account reviews the relative importance of invasion vectors and also provides a comparison of introduced marine versus freshwater species.

MATERIAL AND METHODS

The earlier mentioned publications were reviewed and summarized. In addition, a comprehensive literature research was undertaken, including publications in scientific journals, meeting reports of international working groups, and Internet databases. Earlier prepared summaries of introduced aquatic species in Europe (Streftaris *et al.* 2005, Gollasch 2006) have been taken as baseline documents and were updated, as new species were introduced since those inventories were completed. Further, these earlier inventories did not separate freshwater from marine-introduced species, but provided an overall account.

Whenever possible, all coastal habitats have been addressed, including inland waters in close proximity to the coast. Noting that shipping is a key introduction vector, looking at marine coastal waters only will result in a biased overview as the busiest European ports, such as Rotterdam, Antwerp, and Hamburg, are located in lower saline waters.

For this comparison of introduced marine versus freshwater species, I used the term introduced species as a synonym of NIS and defined it as any species transported intentionally or accidentally by a human-mediated vector into aquatic habitats outside its native range (see ICES 2005). Freshwater species are those species which complete their entire life cycle in freshwater, whereas those species which do not complete their entire life cycle in freshwater, including brackish water species, are here classified as marine species (see ICES 2005).

RESULTS

Introduced species are known from all European coasts. The highest number of introduced species was found in the Mediterranean Sea and the lowest number in European Arctic waters (Table 1, Fig. 1).

A total of 1069 NIS are known from European coastal and adjacent waters, of which almost 60% are assumed to occur in self-sustaining populations. The dominating type of organisms is composed of zoobenthos invertebrates (575 taxa). Almost 190 fish and 189 phytobenthos species were also observed, while 60 taxa of phytoplankton species are also frequently recorded (Table 2).

Invasion vectors overlap and for several species introduction vectors could only be assumed. Fouling species may be introduced via ship hulls or also in the fouling of species intentionally imported for aquaculture purposes (e.g. on oysters). Further, early life stages of fouling organisms may also be introduced with ballast water releases. For those species the most likely vector was assumed. This approach results in shipping being the most important invasion

Table 1 Freshwater versus marine established non-indigenous species excluding parasites and pathogens in European coastal waters. (Modified from Gollasch 2006)

Region	Salinity: Freshwater species	Salinity: Brackish and marine species	Dominance of brackish and marine species [%]
Mediterranean Sea	10	307	96.8
North Sea	33	99	75.0
Atlantic coast	6	95	94.1
Baltic Sea	41	51	55.4
Black Sea	12	47	79.7
Azores	no data	24	
Irish waters & NW UK	3	33	91.7
Arctic waters	no data	8	
Total	**105**	**664**	**86.3**

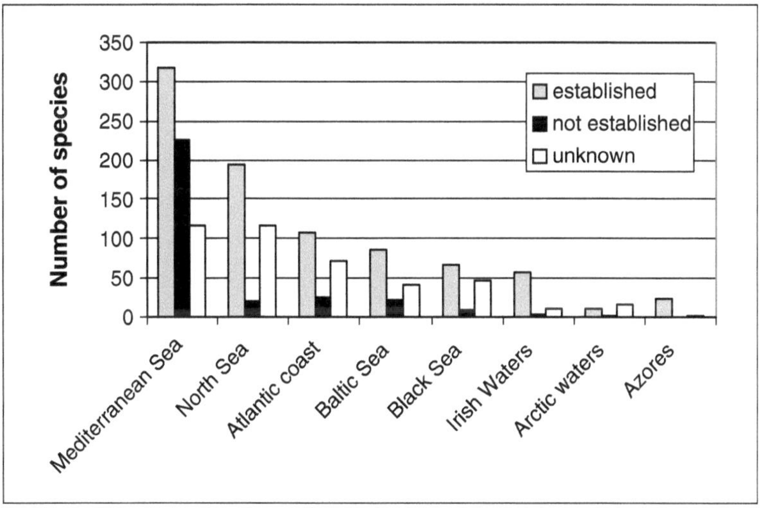

Fig. 1 Number of introduced species per region according to invasion status, i.e. established, not established, or unknown. (Modified from Gollasch 2006)

vector with 215 taxa in ballast water and 131 in hull fouling. Intentional and accidental species introductions for aquaculture purposes account for 134 species and 78 species were intentionally introduced during stocking initiatives. It should be noted that for 179 species the introduction vector could not be

Table 2 Freshwater, marine established non-indigenous species and total number of non-indigenous species in European coastal waters according to life forms. Other = parasites, pathogens, and fungi. (Modified from Gollasch 2006)

	Salinity			
Group	Established freshwater species	Established brackish and marine species	Number of all species	% of all species
Zoobenthos	51	305	575	53.8
Fish	41	50	190	17.8
Phytobenthos	9	168	189	17.7
Phytoplankton	0	101	60	5.6
Zooplankton	2	14	41	3.8
Other	2	26	14	1.3
Total	**105**	**664**	**1,069**	**100.0**

Table 3 Importance of invasion vectors for all species. Lessepsian = species movement through the Suez Canal, range expansion = active and passive species dispersal, aquaculture = species not intended to be placed in open waters, stocking = intentionally released species. Although not being introduced by vectors, those species that have arrived with drift or due to natural range expansion were included here for comparison. (Modified from Gollasch 2006)

Vector	Total	Freshwater	Marine
Aquaculture	134	9	125
Bait	5	0	5
Ballast	215	33	182
Drift	2	0	2
Fouling	131	3	128
Lessepsian	253	0	253
Ornamental	19	12	7
Range expansion	46	0	46
Science	7	0	7
Stocking	78	51	27
Unknown	179	36	143
Total	**1069**	**144**	**925**

identified. Although not being introduced by vectors, those species that have arrived with drift or due to natural range expansion were included here for comparison (Table 3).

Of the established non-indigenous invaders, brackish or marine species are more common (664 species) than freshwater taxa (105 species). In all regions, marine species are dominant. However, the share of freshwater versus marine species is regionally very different. In European seas with a large number of estuaries (e.g. the North Sea) and in the Baltic Sea with its brackish to freshwater environment, the relative number of freshwater species is higher (Table 1).

CONCLUSIONS

The ranking of likely species introduction vectors varies throughout Europe. However, the results show that shipping and aquaculture activities are the dominant invasion vectors. In areas with more shipping activity, more introduced species were found (Gollasch 2006), highlighting the importance of this invasion vector. The large number of invaders in the Mediterranean Sea is almost certainly a result of the removal of the migration barrier between the Red and the Mediterranean seas by the opening of the Suez Canal (Galil 2006). In European Arctic waters, the number of invaders is low. It is assumed that this region is less exposed to the prime invasion vectors. Here, the number of ports engaged in intercontinental trade is limited and also aquaculture facilities are rare.

This summary shows that the majority of aquatic invaders are marine species. However, it should be noted that here inland waters were only considered when they were in close proximity to coastal marine waters, e.g. canals and inner estuaries with shipping engaged in intercontinental traffic. For some regions, inventories on introduced-inland species are unknown. The lack of such data and the approach selected may have resulted in a biased calculation. The dominance of marine species is enormous, i.e. even if future studies on freshwater species in close proximity to the coast reveal additional introduced freshwater species, the dominance of marine species will remain. Regions without larger freshwater habitats in close coastal proximity, e.g. the Mediterranean Sea, are clearly dominated by marine invaders. Even regions with a high number of estuaries and also freshwater ports are dominated by marine species (Table 1). Studies on ballast water arriving in the North Sea revealed that the majority of ships carry ballast water from marine locations (Gollasch 1996). It is therefore more likely that marine species are transported within ballast water. However, it should be noted that freshwater ports are especially at risk of future species invasions when they are engaged in shipping routes to other freshwater ports. This refers, for example, to ports in the North American Great Lakes and also to freshwater ports in the northern and eastern Baltic. The different salinity between those donor and recipient ports makes a natural spread of species impossible, but in ballast water tanks this salinity barrier is absent. As a result, freshwater ballast from a donor port should not be released in recipient freshwater ports. Here a mid-ocean ballast water exchange may be most efficient until other ballast water management or treatment approaches become available.

First records of new invaders are reported with a time lag due sometimes to time consuming publishing procedures. Consequently, the number of first records in this decade will increase in the future. As an example, since the preparation of the last summary of introduced species in Europe (Gollasch 2006), two new invaders were found in the North Sea and its adjacent waters, i.e. *Neogobius melanostomus* (Pallas) in Dutch inland waters (van Beek 2006) and *Rapana venosa* (Valenciennes) in the south-western North Sea (Kerckhof *et al.* 2006). These new records are of concern as both species have the potential to negatively impact on the environment and may also affect resource users. However, it is not clear whether the two species are established in the North Sea region.

ACKNOWLEDGEMENTS

Grateful thanks are expressed to all contributors to this inventory. The list is too numerous to be mentioned here. However, I would especially like to thank Argyro Zenetos and Nikos Streftaris (Hellenic Centre for Marine Research, Anavissos, Greece) for having shared data on another inventory of exotic species in Europe. The author is partner in the EU funded project Delivering Alien Invasive Species Inventories for Europe (DAISIE). This programme will shortly launch a database on alien species in Europe (www.daisie.se).

REFERENCES

Cardigos, F., F. Tempera, S. Avila, J. Goncalves, A. Colaco, and R. S. Santos. 2006. Non-indigenous marine species of the Azores. Helgoland Marine Research **60**, 160–169.

Carlton, J. T. 1985. Transoceanic and interoceanic dispersal of coastal marine organisms: the biology of ballast water. Oceanography and Marine Biology Annual Review **23**, 313–371.

Carlton, J. T. 1987. Patterns of transoceanic marine biological invasions in the Pacific Ocean. Bulletin of Marine Science **41**, 452–465.

CIESM. 2005. Atlas of exotic species in the Mediterranean Sea. www.ciesm.org/atlas visited January 2005.

Cohen, A. N. and J. T. Carlton. 1995. Biological study: Non-indigenous aquatic species in a United States estuary: a case study of the biological invasions of the San Francisco Bay and Delta. US Fisheries and Wildlife and National Sea Grant College Program Report PB96-166525, Springfield, VA, 273 pp.

Eno, N. C. 1996. Non-native species in British waters: effects and controls. Aquatic Conservation: Marine and Freshwater Ecosystems **6**, 215–228.

Eno, N. C., R. A. Clark, and W. G. Sanderson. 1997. Non-native species in British waters: a review and directory. Joint Nature Conservation Committee, Peterborough, UK, 152 pp.

Galil, B. and A. Zenetos 2002. A sea change – exotics in the eastern Mediterranean Sea. Pages 325–336 in E. Leppäkoski, S. Gollasch, and S. Olenin, editors. Invasive aquatic species of Europe: distribution, impacts and management. Kluwer Academic Publishers, Dordrecht, The Netherlands, 583 pp.

Galil, B. 2006. The Suez Canal. Pages 207–300 in S. Gollasch, B. Galil, and A. Cohen, editors. Bridging Divides. Maritime canals as invasion corridors. Springer, Dordrecht, The Netherlands, 315 pp.

Gollasch, S. and R. Mecke. 1996. Eingeschleppte organismen. Pages 146–150 in Warnsignale aus der Ostsee. J. L. Lozan, R. Lampe, W. Matthäus, E. Rachor, H. Rumohr, and H. V. Westernhagen, editors. Parey Buchverlag, Berlin, Germany, 385 pp.

Gollasch, S. 1996. Untersuchungen des Arteintrages durch den internationalen Schiffsverkehr unter besonderer Berücksichtigung nichtheimischer Arten. Diss., Univ. Hamburg; Verlag Dr Kovac, Hamburg, Germany, 314 pp.

Gollasch, S. 2006. Overview on introduced aquatic species in European navigational and adjacent waters. Helgoland Marine Research **60**, 84–89.

ICES. 2005. ICES Code of Practice on the Introductions and Transfers of Marine Organisms 2005, 30 pp.

Kerckhof, F., R. J. Vink, D. C. Nieweg, and J. J. N. Post. 2006. The veined whelk *Rapana venosa* has reached the North Sea. Aquatic Invasions **1**, 35–37.

Leppäkoski, E. 1994. Non-indigenous species in the Baltic Sea. Pages 67–75 in C. F. Boudouresque, F. Briand and C. Nolan, editors. Introduced species in European coastal waters. CEC Ecosystem Research Report **8**, 67–75.

Leppäkoski, E. and S. Olenin. 2000. Non-native species and rates of spread: lessons from the brackish Baltic Sea. Biological Invasions **2**, 151–163.

Minchin, D. and N. C. Eno. 2002. Exotics of coastal and inland waters of Ireland and Britain. Pages 267–275 in E. Leppäkoski, S. Gollasch, and S. Olenin, editors. Invasive aquatic species of Europe: distribution, impacts and management. Kluwer Academic Publishers. Dordrecht, The Netherlands, 583 pp.

Nehring, S., 2002. Biological invasions into German waters: An evaluation of the importance of different human-mediated vectors for nonindigenous macrozoobenthic species. Pages 373–383 in E. Leppäkoski, S. Gollasch, and S. Olenin, editors. Invasive aquatic species of Europe: distribution, impacts and management. Kluwer Academic Publishers. Dordrecht, The Netherlands, 583 pp.

Olenin, S., D. Daunys, and E. Dauniene, E. editors. 2005. Baltic Sea Alien Species Database http://www.ku.lt/nemo/mainnemo.html

Reise, K., S. Gollasch, and W. J. Wolff. 1999. Introduced marine species of the North Sea coasts. Helgoländer Meeresuntersuchungen **52**, 219–234.

Streftaris, N., A. Zenetos, and E. Papathanassiou. 2005. Globalisation in marine ecosystems: the story of non-indigenous marine species across European seas. Oceanography and Marine Biology **43**, 419–453.

van Beek, G. C. W. 2006. The Round goby (*Neogobius melanostomus*) first recorded in the Netherlands. Aquatic Invasions **1**, 42–43.

Chapter eighteen

Modeling biological invasions of inland waters

Hugh J. MacIsaac, Leif-Matthias Herborg, and Jim R. Muirhead

INTRODUCTION

The discipline of invasion biology suffers from a number of problems, not least of which is a lack of a central, organizing paradigm. Ecologists have for decades sought to relate invasion success both to characteristics of invaded communities and to those of the invading species (see references in Davis 2005), the results of which have only served to demonstrate the highly system- or species-specific nature of many invasions. One recent development that holds promise for resolving this problem has been a recognition that successful invasions result from a series of inter-connected stages: introduction effort across biogeographic barriers, suitability of physiochemical conditions in the new environment, and survival in the local biota; each of these stages may pose differing demands on entrained species and may be studied independently (Carlton 1985, Williamson and Fitter 1996, Richardson *et al.* 2000, Kolar and Lodge 2001, 2002, Colautti *et al.* 2006). While many patterns in nature may have alternate explanations, none of which takes logical precedence for testing over any other, the same cannot be said for hypotheses that may account for success of non-indigenous species (NIS). All invasions begin with the introduction of NIS propagules, which may vary in number, quality, sex ratio, age, or number of introduction events (Fig. 1). Propagules are then subject to prevailing conditions of the host lake, which may reduce or eliminate effective propagule pressure depending on the species' ability to tolerate these conditions. Only at this stage does the possibility of positive or negative biological feedback between the NIS and host community occur (Fig. 1). So, for example, an introduced mollusk like *Dreissena polymorpha* Pallas could fail to establish in a new lake owing to stochastic or

Francesca Gherardi, Biological invaders in inland waters: Profiles, distribution, and threats, 347–368.

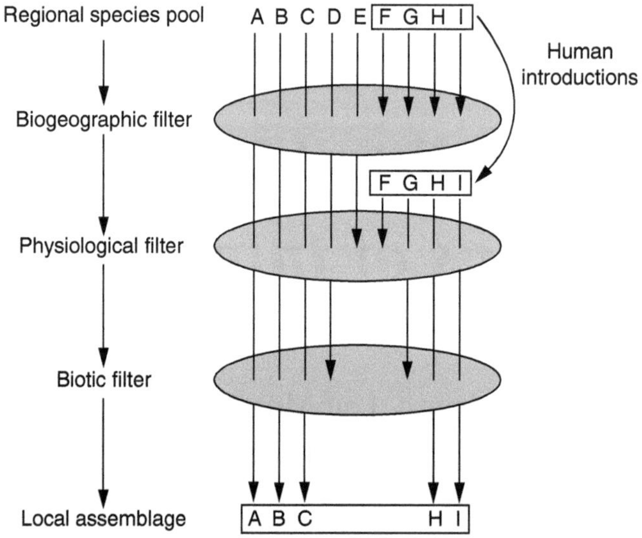

Fig. 1 Transportation of species (F, G, H, I) beyond historic biogeographic barriers to new ecosystems, where sequential filters reduce the number that successfully establish (H, I) as NIS. An introduction effort or 'propagule pressure' filter reduces both the number of species and individuals within particular species that reach the physiological filter. Modeling techniques may help identify physiological and/or constraints that further restrict establishment of NIS in novel ecosystems. (Modified from Rahel 2002)

deterministic processes (e.g. small population size; all larval or post-reproductive propagules), because of insufficient calcium in lake water to allow shell formation, or because predation by molluskivorous waterfowl increased the species' death rate. While this model suggests that introduction effort (i.e. propagule pressure) hypotheses be tested first and biological interaction questions last, we recognize that invasions could fail at any stage (see Colautti *et al.* 2006). Additionally, cases might occur where, under the same biological conditions (e.g. predators), NIS introduction can either succeed or fail depending upon physiological conditions. Indeed, evidence exists supporting the importance of all three stages in failed invasions by various species. If this order of hypothesis testing is accepted, then methodologies must be developed for examining propagule pressure. Unfortunately, formal tests of the importance of propagule pressure have been few in number until quite recently (see Lockwood *et al.* 2005, Colautti *et al.* 2006). For species that pass through the introduction effort filter, other methods are required to determine the suitability of water in newly colonized lakes and ecological conditions therein. In this paper, we begin by reviewing mechanisms of natural dispersal, following which we consider human introductions of species to lakes. As much as possible, we follow the sequence of hypothesis testing outlined above.

NATURAL DISPERSAL

Passive (e.g. wind) and active (e.g. animal-vectored) transport of propagules has long been recognized as important to regional dispersal of species (see reviews Bilton *et al.* 2001, Havel and Shurin 2004, Green and Figuerola 2005). The simplest and most predictable form of dispersal occurs via advection of propagules in currents. Zooplankton and planktonic larval stages of benthic species may be readily dispersed to connected waterbodies, though depending on the life stage involved, mortality during transport may be high. For example, Horvath *et al.* (1996) determined that presence of zebra mussels in upstream lakes was a strong predictor of mussel presence downstream, although abundance of mussel veliger larvae fell exponentially with distance from the source. The same pattern also holds on a broader scale: lakes in the midwestern USA that were connected by streams to invaded lakes were more likely to be invaded than those connected to non-invaded lakes (Bobeldyk *et al.* 2005). A similar pattern was observed in Belarus, where isolated lakes were much less likely to be invaded by zebra mussels than clustered lakes (Kraft *et al.* 2002). Although the generality of these patterns must be tested for species other than zebra mussels, upstream sources of propagules of NIS place downstream ecosystems at high risk of invasion. Western and northern European river systems have become invaded with a succession of NIS from the Black Sea owing to construction of canals that permit passive transfer in currents or active transport in or attached to vessels (Bij de Vaate *et al.* 2002). Likewise, the Volga River has been colonized by a number of species from the Black Sea, placing the Baltic Sea at risk; the Caspian Sea, in turn, has been invaded by NIS from the Black and Azov seas (see Leppäkoski *et al.* 2002). While some of these transfers were undoubtedly made via ships, all of these ecosystems are now linked either directly or indirectly by canals, thus opening the possibility of passive transfers of NIS. Reservoirs are often created by damming rivers, one consequence of which may be greater susceptibility to invasion, in particular for passive dispersing species that cannot persist in strong unidirectional flow (e.g. zooplankton, aquatic weeds, planktonic larvae; Havel *et al.* 2005). These reservoirs may, in turn, serve as 'stepping stones' for secondary spread of NIS to both connected and disconnected aquatic habitats (Havel *et al.* 2005). As an example, the quagga mussel *Dreissena rostriformis bugensis* Andrusov is spreading up the Volga River system after having been introduced in or near Kubyshev Reservior, Russia (Therriault *et al.* 2005).

Wind and rain also may transport NIS locally or between watersheds, particularly those species capable of producing diapausing eggs (Bilton *et al.* 2001). However, the importance of these mechanisms to overall dispersal patterns remains unclear. On the one hand, Louette and De Meester (2005) observed 20 different colonizing species, averaging 4.2 cladoceran species per pond after a 15-month period, in 25 freshly dug and isolated pools in Belgium, and suggested that dispersal of cladocerans was high. Cohen and Shurin (2003) also suggested that zooplankton species were vagile over short distances in their North American study. On the other hand, Jenkins and Underwood (1998) observed only two bdelloid rotifer species in experimental wind socks and four rotifer species from samples of rain, and Havel and Shurin (2004) suggested that zooplankton dispersal might become limiting beyond 20 km.

Local spread of NIS may be affected by strong winds and the availability of resting stages. For example, extensive foam windrows of the NIS waterflea *Bythotrephes longimanus* Leydig have been observed on leeward shores of Lake Huron after strong winds blew across the lake (D. Garton 1986, personal observation). Diapausing eggs contained in this foam could be moved to seed new populations in waters directly proximal to the lake, or, more likely, coat plumage of dabbling waterfowl that could then introduce the species to other visited lakes in the region. Introductions by local movements of waterfowl are likely confined to proximal systems (e.g. Johnson and Carlton 1996, Bilton *et al.* 2001), while long-distance transport would almost certainly be limited to seasonal migrations (e.g. Proctor 1959, 1964, Swanson 1984). Thus even though Green and Figuerola (2005) proposed that transfer of invertebrates by waterfowl may exceed 1,000 km, from a mass-transfer viewpoint, it is far more likely that flocks of migrating waterfowl would effect transfer of species more readily than could single individuals. Ectozoochorous or endozoochorous transfers by seasonally migrating waterfowl could effect both regional and long-distance introductions of NIS in a non-spatially explicit (i.e. not area specific) yet predictable pattern (Bilton *et al.* 2001, Figuerola and Green 2002). For example, Figuerola *et al.* (2003) identified a broad array of invertebrate eggs in digestive tracts of waterfowl arriving to and departing from wetlands in Doñana, Spain. If this information were combined with that on the primary flight paths of the waterfowl species and the distribution of major stopover sites, general models of possible range extensions of zooplankton could be constructed.

Charalambidou *et al.* (2003) suggested that autumnal migrations of waterfowl were more likely to spread *Bythotrephes* via endozoochory from north to south in Europe. However, they also asserted that dispersal probability would drop sharply beyond 60–80 km, based upon waterfowl flight speed and gut retention time for diapausing eggs of the waterflea. The application of molecular markers may allow identification of source–destination relationships (see Bilton *et al.* 2001, Figuerola and Green 2002), although in the case of NIS this is often simplified since there may be very few putative sources from which new populations could be drawn.

In summary, both wind and directed flights by waterfowl are capable of dispersing propagules – especially resting stages of invertebrates – of NIS in a directional manner, although identifying destinations of these propagules can only be done on a relatively crude, regional basis. Development of models to predict the strength of these vectors also has not yet been accomplished. Even though these vectors have likely played important roles in evolutionary changes to species distributions, their respective roles in effecting dispersal of NIS now appear swamped by vectors associated with human activities, in part because the number of propagules potentially transported by an individual waterfowl or by wind is much lower than the bulk transfers associated with ballast water discharges or recreational boats trailered between lakes (e.g. Johnson and Carlton 1996, Hebert and Cristescu 2002). The importance of human vectoring over natural mechanisms like wind or waterfowl has been highlighted for the dispersal patterns of non-indigenous macrophytes in New Zealand (Johnstone *et al.* 1985). Indeed, our understanding of human-mediated dispersal of NIS has been enhanced by studies of inter-lake transfers of invasive macrophytes, which many political jurisdictions recognize as problematic and have taken management actions to prevent.

HUMAN-MEDIATED DISPERSAL

Humans are rapidly changing biological communities of lakes through a combination of habitat change, overexploitation, and species introductions. Biodiversity of lakes appears particularly vulnerable to introduction of NIS (Sala *et al.* 2000, Rahel 2002). Heretics argue that changes to species distributions, including invasions of new habitats, occur naturally. While this is certainly true, the scale by which natural dispersal operates is far lower than that associated with human activities (Hebert and Cristescu 2002, Grigorovich *et al.* 2003), and, in some cases, species are being introduced across biogeographic realms that have, until now, proved resistant to natural dispersal (see Havel and Shurin 2004). For example, the waterflea *Daphnia lumholtzi* Sars was introduced to North America from its native regions of Australasia and northern Africa (Havel and Medley 2006), whereas the signal crayfish *Pacifastacus leniusculus* Dana has been introduced to Europe from North America (Gherardi and Holdich 1999).

Once NIS establish in one location, they may be spread by both natural and human-mediated mechanisms elsewhere in the introduced region. For example, zebra mussels dispersed through the Great Lakes in a pattern consistent with stratified diffusion involving both advective and human-mediated dispersal (Griffiths *et al.* 1991). Buchan and Padilla (1999) developed a diffusion model for zebra mussels that incorporated long-distance movement of trailered boats across Wisconsin and the probability that zebra mussels survive transport and establish in the destination lake. Johnson *et al.* (2001) developed a probability-based model to forecast spread of introduced zebra mussels to inland lakes in Michigan, USA based upon movement of trailered boats between a network of lakes adjacent to, and presumably initially invaded from, the Great Lakes. This effort was made more quantitative by surveying boaters and inspecting their equipment as it was trailered from Lake St. Clair to determine if larval or adult zebra mussels were being transported. These data were combined with estimates of survival of mussels outside water to determine the number of dispersal events from the lake associated with seven different vectors. Application of different modeling tools may allow investigators to identify lakes that will serve as invasion hubs – from which other lakes are colonized – as well as vulnerability of regions or specific lakes to invasion. One of the most promising techniques that links invaded sources with non-invaded destination lakes is gravity modeling. Gravity models can be considered a form of 'propagule pressure' assessment since they seek to quantitatively link sources and destinations.

VECTOR-BASED GRAVITY MODELS

Transportation or gravity models have been applied to many economic and social questions, and have been widely applied over the past decade to assess spatial patterns of aquatic NIS dispersal (Schneider *et al.* 1998, Bossenbroek *et al.* 2001, Drake and Bossenbroek 2004, Drake and Lodge 2004, Leung *et al.* 2004, 2006, MacIsaac *et al.* 2004, Muirhead and MacIsaac 2005). Gravity models of species spread estimate the frequency of dispersal events and are coupled with the explicit nature and spatial arrangement of potential colonization sites in contrast to diffusion models that assume a homogeneous landscape (e.g. Buchan and Padilla 1999). Gravity models provide information on the relative strength of vector movement (i.e. propagule pressure) from an invaded source to non-invaded (or invaded) destination lakes in a manner analogous to Newton's Theory of Gravity for measuring the attractive force between two masses. Perhaps better than for any other type of ecosystem, gravity models hold the promise of accurately forecasting the dispersal of NIS amongst lakes.

Depending on the type of information available and objectives of the study, four classes of gravity models are available: total flow-constrained, production-constrained, attraction-constrained, and production-attraction or doubly-constrained (Haynes and Fotheringham 1984). In the total flow-constrained model, which requires the least knowledge of the system, information is available on only the total number of interactions and we are asked to forecast the interaction pattern among origins and destinations based on their distances. The model is constrained so that the total of the forecast interactions between sources and destinations is equal to the known total number of interactions. In production-constrained models, we have information only on outflows from each of the origins. The inflows into each destination are forecasted based upon the magnitude of outflows from each origin, the pairwise distances between origins and destinations, and an external measure of attractiveness to each destination such as lake area (Bossenbroek *et al.* 2001). During the calculation of predicted inflow, the sole constraint is that the predicted outflow of the balanced model is equal to the measured outflow. Likewise, for attraction-constrained gravity models, we have information on inflows to each of the destinations and are asked to forecast outflows from each of the origins under the constraint that predicted inflows for each destination is equal to measured inflow. The doubly-constrained gravity model, in which information about both outflows and inflows is required, provides the best quality of interaction information and thus is the most spatially-explicit when forecasting dispersal to novel regions. In this model, the forecasted interactions between origins and destinations are constrained in such a manner that the predicted total outflow for each origin and total inflow for each destination are equal to measured outflows and inflows. Schneider *et al.* (1998) and MacIsaac *et al.* (2004) used doubly-constrained gravity models to forecast human-mediated spread of *D. polymorpha* and *B. longimanus*, respectively, in the Great Lakes region.

With the four types of gravity models, predictive ability scales with the amount of effort required to collect the data and the amount of measured information about flows in the system, especially with respect to forecasting the interaction flows between sources and destinations. The total-constrained model requires the least effort and offers the least predictive ability, since we have information only on the total number of interactions within the system (e.g. total number of boaters moving in the area). The summed outflows for each source, summed inflows for each destination, as well as the interaction flows between sources and destinations are unknown and must be solved.

The gain in predictive ability of the production- or attraction-constrained model over the total-flow constrained alternative is large relative to the effort required to collect additional information. For production-constrained models, information of outbound propagule pressure from invaded sources can easily be obtained by surveys of the number of boaters leaving a lake, or the number of owners of boat licenses at regional scales. Bossenbroek *et al.* (2001) modeled the propagule pressure of transporting zebra mussels by the number of boats that could potentially leave a county from five midwestern States and arrive at a particular lake, with lake area as a measure of attractiveness. Leung *et al.* (2004) use a similar production-constrained gravity model approach to model the risk of transporting zebra mussels via boater traffic from Michigan counties to lakes in the State, and again used the number of registered boaters per county as a measure of potential outbound traffic. In their assessment of the utility of using production-constrained gravity models to capture boater traffic within Michigan, Leung *et al.* (2006) used lake area as a proxy for lake attractiveness. The advantage in using these models is that data on the number of boat licenses, which can be used as a proxy for potential outbound traffic, is often available from government databanks. However, these gravity models are limited by the lack of data on the actual inbound vector traffic, which becomes important when forecasting relative risk by contrasting invaded *vs.* currently non-invaded destinations.

In attraction-constrained models, the converse is true. Explicit inbound vector data is required for each destination, and usually will require more intensive survey work, either through contacting officials responsible for monitoring recreational lake use or via mail-based or creel surveys, since the number of non-invaded or invaded destinations is usually greater than the number of invaded sources. This form of model is not normally used as a risk assessment tool for invasions since it only forecasts putative sources for invasion if currently unknown. One advantage of this form of model is its greater confidence in the relative risk of a destination being invaded, but at the cost of lack of information pertaining to the source of the invading propagules. Since survey information for each destination is usually required, the same amount of effort can be used to collect data necessary for the construction of a doubly-constrained model.

In the doubly-constrained model, spatially-explicit knowledge of the vector outflow from invaded sources as well as inflow to currently invaded and non-invaded destinations requires the greatest commitment in terms of data collection. Information on specific trips is necessary to gauge the interaction between sources and destinations, and thus is usually collected in the form of surveys. This form of gravity model allows the best predictive ability since it forecasts interactions between sources and destinations based on distance and external measures of attraction, and is constrained so that summed outflows from sources and summed inflows to destinations are equal to measured outflows and inflows. In other words, the accuracy of the forecast interactions in the doubly-constrained model is likely closest to actual flow than that predicted by the other forms of gravity models. Similar to the production-constrained model, we are able to predict invasions by forecasting inflows to other destinations for each source lake. Likewise, for each destination, we can assess the relative inflow from each of the sources. Schneider *et al.* (1998) used a doubly-constrained gravity model to assess the risk of transporting zebra mussels to inland lakes in Illinois, USA. Based on both creel surveys and contacting officials responsible for boat-access sites, propagule pressure was estimated by the number of boats used per year. In their model, an assumption was made that attractiveness of a particular destination was correlated with boat use. MacIsaac *et al.* (2004) modeled the risk of transporting the spiny waterflea among inland lakes in Ontario using mail and on-site surveys. The doubly-constrained gravity model allows for the greatest confidence in predicting which currently non-invaded lakes are at risk of invasion. Both Schneider *et al.* (1998) and MacIsaac *et al.* (2004) developed prioritized lists of lakes likely to be invaded. Doubly-constrained models may be validated by backcasting the order in which lakes were invaded based on current relative inbound vector traffic if one assumes that human activity patterns have not changed over the time invasions occurred (Fig. 2a). Gravity models may also be validated by comparing forecasts with observed invasions. MacIsaac *et al.* (2004) found that lakes with high inbound traffic were likely to be invaded in subsequent years (Fig. 2b). In addition to providing watershed managers with an assessment of relative inbound traffic to currently non-invaded destinations, doubly-constrained models also can be used to provide an assessment of the relative outbound traffic from invaded sources. This data can be used to identify existing and future "hubs", thereby allowing management efforts to focus on these lakes (Fig. 3) (Muirhead and MacIsaac 2005). By contrast, Drake and Lodge's (2004) assessment of a global network of ship movements amongst ports suggested that

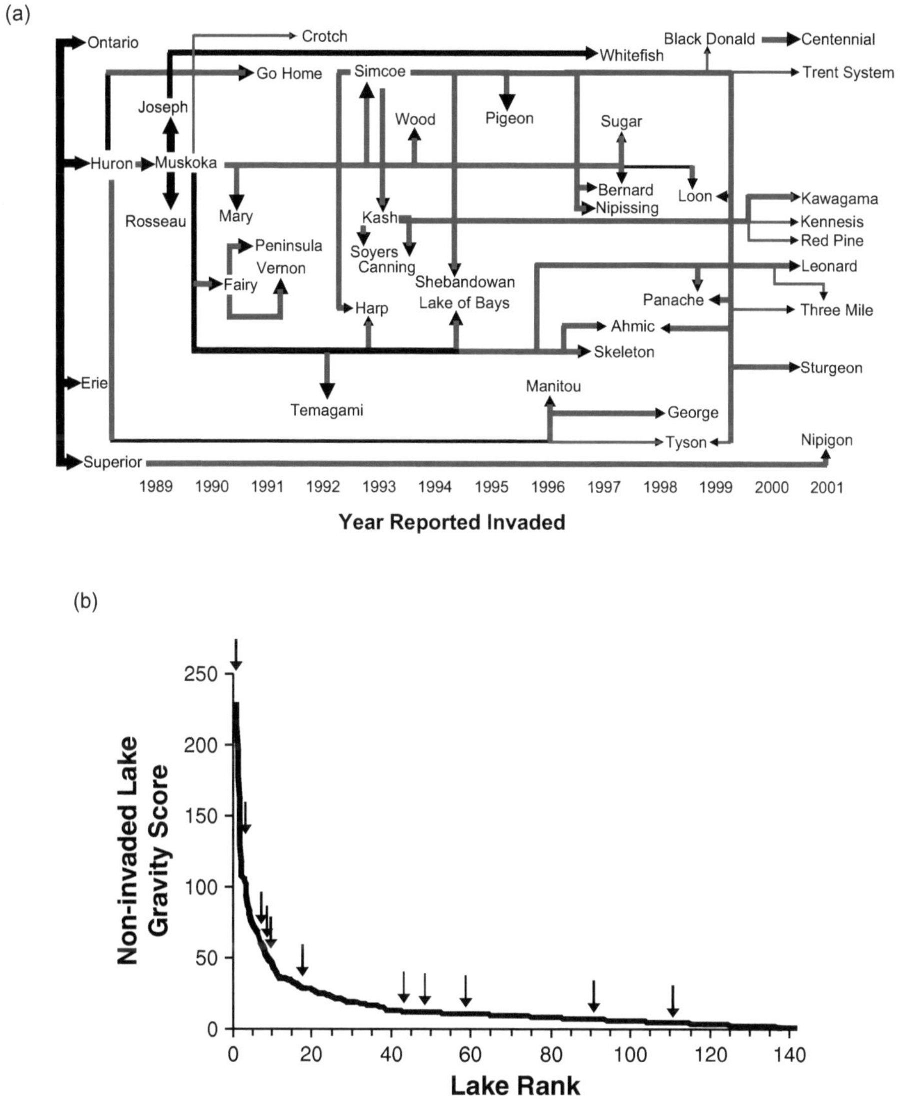

Fig. 2 a) Sequence of introduction of the spiny waterflea *Bythotrephes longimanus* to inland lakes in Ontario from the Great Lakes (Lake Huron). Backcasting sequences are based upon reported order in which lakes were invaded and surveys of boater/angler movements from invaded source lakes to non-invaded destination lakes. Arrow thickness indicates strength of vector from donor lake (left side) to destination lake (right side).
b) Forecast of invasion risk for non-invaded lakes in relation to gravity score (a measure of introduction effort). Lakes that switched from non-invaded to invaded status are indicated with an arrow, and had significantly higher vector inflows than lakes that remained non-invaded. (Modified from MacIsaac *et al.* 2004)

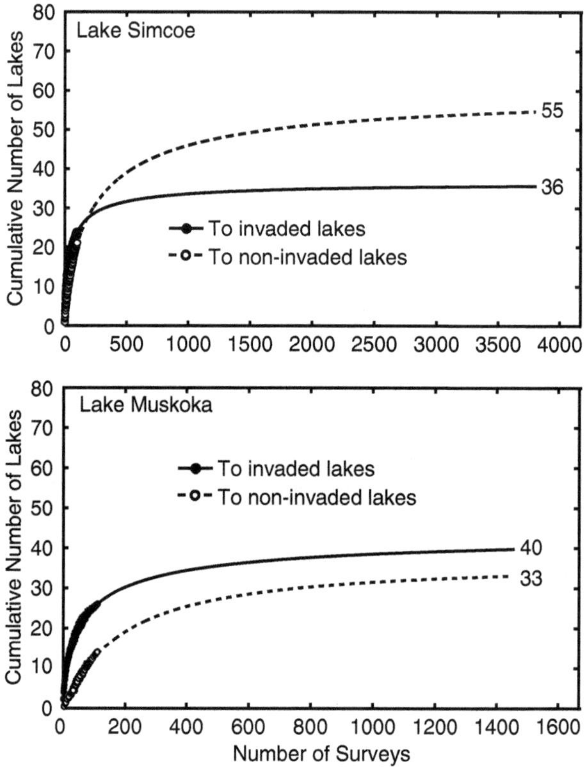

Fig. 3 Cumulative number of lakes visited by boaters that arrived from invaded source lakes (Simcoe, Muskoka) in Ontario. Dotted line represents visitations to non-invaded lakes, solid lines visitations to invaded lakes. Asterisks represent the estimated number of boaters leaving each lake per year. A disproportionate number of boaters leaving Lake Simcoe arrive to lakes that have not yet been invaded (55 *vs.* 36), whereas most outflow from Lake Muskoka is to other lakes that have already been invaded (40 *vs.* 33). Lake Simcoe may develop into a future hub for dispersal of the species, whereas Lake Muskoka, which has already caused many invasions, is less likely to cause future invasions. Results are based upon surveys of boaters leaving the lakes and were modeled using Monte Carlo simulations and nonlinear regression. (Modified from Muirhead and MacIsaac 2005)

a reduced invasion rate could be better obtained through small reductions per ship-invasion risk through ballast control than by knocking out particular "hub" ports. While much remains to be learned about gravity models, these techniques offer insights into how lakes are connected as networks owing to human activities and they ought to be applied much more widely to studies of other aquatic NIS.

Survey methods have also been used to predict spread of other aquatic NIS. Many of these studies involve bait or food fishes that are imported and sold live, following which the purchasers release some or all of their stock. As with their gravity model counterparts, either coarse or fine-scale spatial predictions may be made regarding vulnerability of lakes to NIS introduction depending on the nature of the survey questions. As an example, surveys by Litvak and Mandrak (1993) at bait-fish shops in Toronto, Ontario illustrated that 41% of anglers who purchased bait fishes released unused specimens at destinations up to hundreds of kilometers from their source. The same methodologies may be used to track introduction and potential spread of species in water garden and aquarium trades.

Modeling environmental suitability

Ecologists have used a number of other techniques to model habitat suitability of aquatic NIS. One approach is to empirically determine limnological differences between lakes with and without a particular species. For example, Ramcharan *et al.* (1992) assessed differences in 278 European lakes with and without *D. polymorpha* using discriminant-function analysis (DFA). Their analysis revealed that lakes with *Dreissena* could be separated from those without the species using only two variables: pH and calcium content of lake water. This analysis was extended to demonstrate that lakes with high mussel densities differed from those with low densities based upon nutrient levels (NO_3, PO_4). MacIsaac *et al.* (2000) also used DFA to identify lakes in Europe with and without the waterflea *Bythotrephes*. In this case, lakes with the species tended to be larger, deeper, and have higher Secchi transparency than those without the species. Functions generated in both Ramcharan *et al.*'s (1992) and MacIsaac *et al.*'s (2000) models were then used to predict where these respective species would potentially occur in North America.

Allen and Ramcharan (2001) used logistic regression to differentiate river systems in the USA with and without *Dreissena*. They found that permanent populations tended to occur in systems with moderate ionic strength and some degree of impoundment. Again, based upon this classification, the authors were able to extend their analyses to predict whether *Dreissena* would occur in other, untested river systems.

In a very thorough study, Marchetti *et al.* (2004) used multivariate models to explore determinants of invasion success for all stages of fish invasions – from introduction to integration – in California catchments. They determined that propagule pressure and prior invasion history were significant predictors of fish-establishment success, as were biological features such as parental care and physiological tolerance. During secondary dispersal after establishment, distance from nearest native source and trophic status were important predictors. Maximum size, physiological tolerance, and distance from nearest native source were the best predictors of abundance for established populations. This study clearly highlighted the importance of different factors to success of fish species at different stages of an invasion. In an earlier assessment of assembly rules for aquatic invasions, Moyle and Light (1996) observed that most invasions failed but that all systems are invasible. Among their findings, they noted that piscivores and detritivores/omnivores were most successful in relatively undisturbed systems, and that any species could invade so long as they possessed appropriate environmental tolerances.

Kolar and Lodge (2002) also used multivariate analyses to identify different factors important to different stages of fish introductions to the Great Lakes. Because data were available on failed as well as successful fish introductions, the authors were able to specifically distinguish between characteristics associated with each. Kolar and Lodge (2002) used DFA to determine that successful invaders at the establishment stage grew relatively fast, tolerated greater variation in temperature and salinity, and had more extensive histories of invasiveness than species that failed at this stage. Similar results were obtained when the authors used another methodology, categorical and regression tree analysis (CART). At the spread stage, DFA demonstrated successful species had slower relative growth rates and poor survival in high water temperatures, although they did exhibit a broader temperature range than slow-spreading species. At a final stage of analysis, the authors noted that 'nuisance' species had smaller eggs, wider salinity tolerance, and survive low water temperatures better than nonproblematic fishes.

At this stage we would like to highlight the importance of distinguishing between lakes that are unsuitable for an NIS (e.g. those that have experienced failed invasions) and lakes that possess suitable environmental conditions but which lack sufficient propagule pressure (i.e. they could be viewed as 'not yet colonized'). Both types of lakes have the same characteristic – lack of an established NIS population – yet they differ fundamentally with regard to invasibility. Studies on the invasion in Bellarussian lakes found that even after 200 years, some lakes with suitable environmental conditions are still not invaded by zebra mussels due to limited propagule pressure (Karatayev *et al.* 2003). Similarly, Johnson *et al.* (2006) found that only ~10% of lakes with suitable environmental conditions had been invaded by zebra mussels after 15 years in areas of North America that already support the species. Hence, the distinction between unsuitable lakes and suitable but not-yet-invaded lakes is crucial while studying the invasion process.

Ecological niche modeling

Ecological niche modeling is a methodology that seeks to predict the potential range of NIS based upon matching of environmental characteristics in the native range with those in the introduced range (see Peterson 2003). A wide variety of approaches are available (Elith *et al.* 2006), including two – CLIMEX and Genetic Algorithm for Rule-set Prediction (GARP) – that have been utilized to predict ranges of NIS. CLIMEX seeks to identify the potential distribution and relative abundance of species in relation to climate models, while GARP uses georeferenced environmental and biological information to predict suitable habitat in a species' introduced range. GARP is an iterative machine-learning tool that employs an array of methods including logistic regression, range rules, negated range rules, and atomic rules to identify heterogeneous rule-sets describing a species' environmental niche. Models are constructed using species presence in the native range together with geo-referenced environmental data. GARP models may also be developed using species presence and environmental data from one introduced range to determine where the species could spread in the future. As an example, GARP was used to predict range expansion and boundary limitations for zebra mussels dispersing into the western USA, based upon its distribution patterns in the eastern USA (Drake and Bossenbroek 2004). Resultant models identified environmental, geological, and biological correlates of its occurrence in the east, and forecasts of vulnerable areas in the west. Some areas, including the Columbia and Colorado rivers, were identified as highly vulnerable to establishment, should mussels be introduced, thereby providing managers with spatially-explicit insights into where exclusion programs would be most useful. Similarly, Herborg *et al.* (2007a) developed separate GARP models for Chinese mitten crabs (*Eriocheir sinensis* H. Milne Edwards) in native East Asia and in its introduced range in Europe to develop forecasts of where the species could be expected to establish in North America. GARP may be most useful if combined with measures of introduction effort, which collectively provide managers with information on where species are being introduced as well as the suitability of these habitats.

Below we develop a simple combined model that incorporates a GARP model to identify where Chinese snakehead (*Channa asiatica* Linnaeus) fishes could survive in North America.

An Example of the Chinese snakehead in North America

The Chinese snakehead (*C. asiatica*) is a freshwater fish native to central and southern China (18°–35° N). Its ecology is poorly described, but it is reported to be predaceous and capable of short-distance overland migration. Various species of the genus *Channa* are available on Asian food markets in the USA and Canada, despite bans from import and interstate transport between all American States. Snakeheads are also sold as aquarium fish, but due to their rapid growth and high cost they are not widely used. Despite *C. asiatica's* availability to aquarists in the USA since the early 1900s and occasional reports of the fish in Florida in the 1960s, no established populations have been reported, although other members of this genus have successfully established in the USA (Courtenay and Williams 2004).

We developed a GARP model for *C. asiatica* based on its native distribution (Courtenay and Williams 2004). Polygons representing the native range of the species were converted into 200 occurrence points in a GIS (ArcMap 9.1). The development of ecological niche predictions was consistent with previous applications of GARP (see Herborg *et al.* 2007a). A GARP simulation using all possible combinations of the environmental coverage, allowed determining the effect of each environmental variable on model accuracy using multiple regression analysis. The global climatic and geographic coverages tested included: frost frequency, slope, compound topographic index, precipitation, river discharge, minimum annual air temperature, mean annual air temperature, maximum annual air temperature, and wet day index. Once suitable environmental coverages were determined, 100 models were generated following the best subset method (see for details Anderson *et al.* 2003) and converted into a map of percentage environmental match using ArcMap 9.1. Finally, hierarchical partitioning analysis was applied to test the effect of environmental coverages on predictive accuracy of the final models (Peterson and Cohoon 1999). Once the model was developed and tested, we applied it to identify suitable environments for *C. asiatica* in North America. We refined the analysis by identifying States that are most at risk from *C. asiatica* establishment by using a very crude measure of propagule pressure: the number of individuals of the genus *Channidae* (e.g. snakeheads) reported in the wild by the United States Geological Survey (http://nas.er.usgs.gov/queries/default.asp) (Table 1). We used reports from all species of snakeheads as the same two vectors – aquarium trade and/or

Table 1 Summary of all reports of occurrences of the Asian snakehead genus *Channa* in the contiguous USA, as reported by the United States Geological Survey online database (http://nas.er.usgs.gov/queries/default.asp). Multiple reports from established populations are not included in the number of isolated reports, as they are most likely caused by reproduction rather than human introduction. Occurrence reports are used as a crude measure of potential propagule pressure in a State.

US State	Number of isolated reports	Number of established populations
California	2	
Florida	4	1
Illinois	1	
Massachusetts	3	
Maryland	2	1+1#
Maine	1	
North Carolina	1	
New York	1	
Pennsylvania	2	1
Rhode Island	2	
Tennessee	1	
Virginia	0	1
Wisconsin	1	

Population has been successfully eradicated in 2002

live fish markets – are responsible for introducing individuals of all congeneric species including *C. asiatica*. We recognize that additional sites may have had snakehead releases which either have failed or, if successful, have not been reported.

Hierarchical partitioning of the GARP model revealed that the most important contributors to model accuracy were mean air temperature, ground frost frequency, and precipitation (Table 2). It should be noted that these are not necessarily determinants of distribution, but rather are correlated with predictive accuracy of the model. The GARP model predicted the highest habitat match levels along the west coast of North America (Fig. 4). Specifically, it identified a narrow strip (50–150 km) running from San Francisco Bay to southern Alaska as environmentally suitable. Large areas in the south-eastern USA also were predicted as environmentally suitable in 20 to 30 out of 100 models run (Fig. 4). The level of environmental suitability provides an estimate of the likelihood of survival of the species, as it is based on the number of models out of 100 that predicted a particular area suitable. Nevertheless, there is no defined cutoff point for the level of environmental match below the species cannot survive. For all of these areas deemed suitable, establishment of the species could only occur if an introductory pathway exists. One albeit crude measurement of introduction effort is the number of snakehead introductions reported for each US State (Fig. 5). By combining these data with areas of suitable habitat, locations vulnerable to snakehead establishment can be identified. The combination of propagule supply and potentially suitable habitat places Florida at a high risk of establishment. North Carolina, and to a lesser extent, northern California also have suitable habitat and reports of snakehead occurrences, although they are fewer than in Florida (Table 1). Occurrence data are not available for British Columbia, although the species is sold there and habitat matching suggests that it can survive in that region.

Table 2 Hierarchical partitioning of environmental variables that contributed significantly to predictive models of *Channa asiatica* distribution in North America. The importance of each environmental variable is given as a relative percentage contribution to model accuracy.

Factor	Contribution to Model Accuracy (%)
Ground-frost frequency (number of days)	19
Topographic index (wetness index based on flow accumulation and slope)	1
Precipitation (mm day^{-1})	17
River Discharge (km^3 yr^{-1})	2
Mean temperature (°C)	53
Wet day index (number of days of precipitation)	8

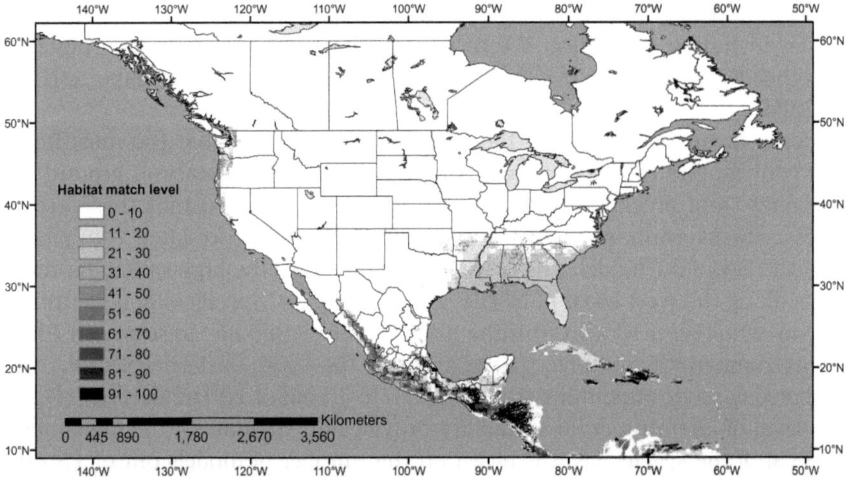

Fig. 4 Predicted occurrence of the Chinese snakehead *Channa asiatica* in North America based on Genetic Algorithm for Rule-set Prediction, an ecological-niche modeling technique, developed using environmental data for its native range in Asia. Areas of greatest habitat suitability include regions of Central America, Jamaica, Puerto Rico, and Hispanola.

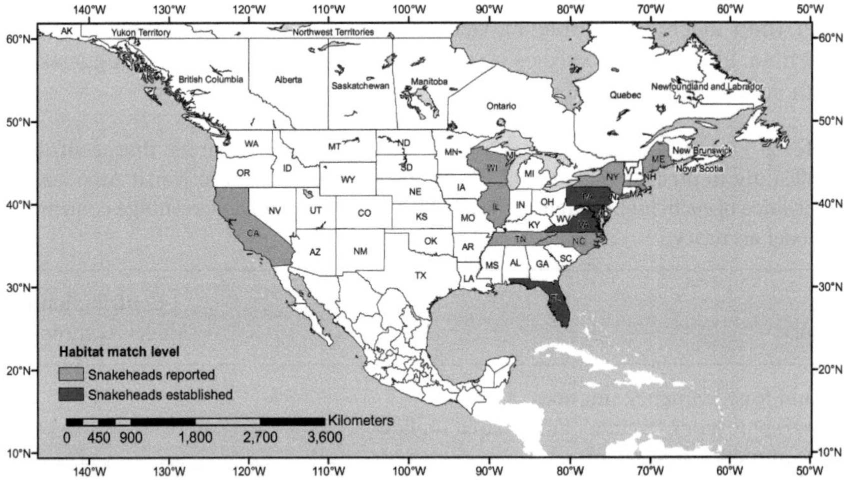

Fig. 5 States in which individual snakeheads (gray stippling) or established populations (dark stippling) were reported, based on the invasive species database of the United States Geological Survey. It is not known whether reports of individual fish-finds reflect established populations.

One of the benefits of environmental niche models is that risk models can be formulated even for those species, like Chinese snakeheads, for which little ecological information exists. While gravity models can provide detailed predictions on vector transport, they do not address the principal question of whether the NIS could survive in the introduced environment. The combination of ecological-niche modeling and basic vector-traffic predictions presented here has the ability to answer and identify locations where a species can survive and a transport vector exists. Clearly a vector-traffic model of higher predictive power (e.g. gravity model) would provide more detailed results, but is outside the scope of this study. However, our prediction identified the highest invasion risk for Florida, followed by North Carolina and California, information that could help policy makers to focus management efforts. One limitation of GARP is that validation tests are rarely conducted for introduced ranges, but two recent studies (Iguchie *et al.* 2004, Herborg *et al.* 2007b) found high predictive accuracy for GARP models in identifying introduced ranges based upon native distribution data. Another issue for prediction of suitable environments of aquatic species is the lack of hydrological datasets (i.e. water temperature, water chemistry, flow speed, etc.) with a wider coverage. The basic output of the GARP also provides no information on the operational procedures by which models are constructed and weightings of environmental criteria are 'black-box' and beyond simple assessment by the investigator. Nevertheless, through additional analysis described in this study (see Herborg *et al.* 2007a for more detail) as well as other recently published methods (Elith *et al.* 2005), relevant environmental parameters and their suitable ranges can be identified, thereby facilitating a deeper understanding of the model.

CONCLUSIONS

Collectively, these studies demonstrate that a broad array of modeling and statistical techniques can be brought to bear on the question of what determines invasion success. It is clear that answers can be far more complex than earlier models of invasion success, as different environmental factors or biological characteristics may be important at different stages of invasion. At the first stage of the invasion sequence, models of natural and human-mediated dispersal have been shown to forecast vulnerability of lakes to invasion. Environmental niche modeling can predict if a potential invaders can pass through the second stage of the invasion process (i.e. physiological filters). Predicting the ability of potential invader to survive the third and final stage – the biological filter – is more complex, and is only possible if biological interactions between the invader and native species are understood, which is typically not the case. Future models that incorporate all three stages of the invasion process will produce spatially-explicit predictions of where an NIS can establish. Lake systems are ideal models for developing and testing invasion theory because they are clearly defined habitats, because natural and human-mediated dispersal can often be quantified, and because they vary in physical–chemical constitution and biological composition. Hence, results of NIS studies conducted on lake ecosystems can be applied to a wider array of habitats in future to predict determinants of invasion success.

ACKNOWLEDGEMENTS

We are grateful for research funding from NSERC's (Canada) Collaborative Research Opportunities and Discovery grants programs, NSF (USA) to David Lodge, an Invasive Species Research Chair to HJM from the Department of Fisheries and Oceans, and an Ontario Graduate Scholarship to JRM. We would like to acknowledge the assistance of Nick Mandrak and Becky Cudmore (DFO Canada) in obtaining the *C. asiatica* distribution data. We are grateful to Dr Francesca Gherardi for hosting an excellent workshop in Florence and for the opportunity to participate in this forum.

REFERENCES

Allen, Y. C. and C. W. Ramcharan. 2001. *Dreissena* distribution in commercial waterways of the US: using failed invasions to identify limiting factors. Canadian Journal of Fisheries and Aquatic Sciences **58**, 898–907.

Anderson, R. P., D. Lew, and A. T. Peterson. 2003. Evaluating predictive models of species' distributions: criteria for selecting optimal models. Ecological Modelling **162**, 211–232.

Bij de Vaate, A., K. Jazdzewski, H. A. M. Ketelaars, and S. Gollasch. 2002. Geographical patterns in range extension of macroinvertebrate Ponto–Caspian species in Europe. Canadian Journal of Fisheries and Aquatic Sciences **59**, 1159–1174.

Bilton, D. T., J. R. Freeland, and B. Okamura. 2001. Dispersal in freshwater invertebrates. Annual Review of Ecology and Systematics **32**, 159–181.

Bobeldyk, A. M., J. M. Bossenbroek, M. A. Evans-White, D. M. Lodge, and G. A. Lamberti. 2005. Secondary spread of zebra mussels (*Dreissena polymorpha*) in coupled lake stream systems. Ecoscience **12**, 339–346.

Bossenbroek, J. M., C. E. Kraft, and J. C. Nekola. 2001. Prediction of long-distance dispersal using gravity models: zebra mussel invasion of inland lakes. Ecological Applications **11**, 1778–1788.

Buchan, L. A. J. and D. K. Padilla. 1999. Estimating the probability of long-distance overland dispersal of invading aquatic species. Ecological Applications **9**, 254–265.

Carlton, J. T. 1985. Transoceanic and interoceanic dispersal of coastal marine organisms: the biology of ballast water. Oceanography and Marine Biology Annual Review **23**, 313–371.

Charalambidou, I., H. A. M. Ketelaars, and L. Santamaria. 2003. Endozoochory by ducks: influence of developmental stage of *Bythotrephes* diapause eggs on dispersal probability. Diversity and Distributions **9**, 367–374.

Cohen, G. M. and J. B. Shurin. 2003. Scale-dependence and mechanisms of dispersal in freshwater zooplankton. Oikos **103**, 603–617.

Colautti, R. I., I. A. Grigorovich, and H. J. MacIsaac. 2006. Propagule pressure: a null model for biological invasions. Biological Invasions **8**, 1023–1037.

Courtenay, W. R. and J. D. Williams. 2004. Snakeheads (Pisces, Channidae) – A biological synopsis and risk assessment. US Department of the Interior, US Geological Survey, Circular 1251, USA, 143 pp.

Davis, M. A. 2005. Invasion biology: 1958–2004: the pursuit of science and conservation. Pages 35–64 in M. W. Cadotte, S. M. McMahon, and T. Fukami, editors. Conceptual ecology and invasions biology: reciprocal Approaches to Nature. Kluwer, London, UK.

Drake, J. M. and J. M. Bossenbroek. 2004. The potential distribution of zebra mussels in the United States. Bioscience **54**, 931–941.

Drake, J. M. and D. M. Lodge. 2004. Global hot spots of biological invasions: evaluating options for ballast-water management. Proceedings of the Royal Society London B **271**, 575–580.

Elith, J., S. Ferrier, F. Huettmann, and J. Leathwick. 2005. The evaluation strip: A new and robust method for plotting predicted responses from species distribution models. Ecological Modelling **186**, 280–289.

Elith, J., C. H. Graham, R. P. Anderson, M. Dudik, S. Ferrier, A. Guisan, R. J. Hijmans, F. Huettmann, J. R. Leathwick, A. Lehmann, J. Li, L. G. Lohmann, B. A. Loiselle, G. Manion, C. Moritz, M. Nakamura, Y. Nakazawa, J. McC. Overton, A. T. Peterson, S. J. Philips, K. Richarson, R. Scachetti-Pereira, R. E. Schapire, J. Soberon, S. Williams, M. S. Wisz, and N. E. Zimmerman. 2006. Novel methods improve prediction of species' distributions from occurrence data. Ecography **29**, 129–151.

Figuerola, J. and A. J. Green. 2002. Dispersal of aquatic organisms by waterbirds: a review of past research and priorities for future studies. Freshwater Biology **47**, 483–494.

Figuerola, J., A. J. Green, and L. Santamaria. 2003. Passive internal transport of aquatic organisms in Doñana, south-west Spain. Global Ecology and Biogeography **12**, 427–436.

Gherardi, F. and D. M. Holdich, editors. 1999. Crayfish in Europe as alien species: How to make the best of a bad situation? A. A. Balkema, Rotterdam, The Netherlands.

Green, A. J. and J. Figuerola. 2005. Recent advances in the study of long-distance dispersal of aquatic invertebrates via birds. Diversity and Distributions **11**, 49–156.

Griffiths, R. W., D. W. Schloesser, J. H. Leach, and W. P. Kovalak. 1991. Distribution and dispersal of the zebra mussel (*Dreissena polymorpha*) in the Great Lakes region. Canadian Journal of Fisheries and Aquatic Sciences **48**, 1381–1388.

Grigorvich, I. A., T. W. Therriault, and H. J. MacIsaac. 2003. History of aquatic invertebrate invasions in the Caspian Sea. Biological Invasions **5**, 103–115.

Havel, J. E., C. E. Lee, and M. J. Vander Zanden. 2005. Do reservoirs facilitate invasions into landscapes. Bioscience **55**, 518–525.

Havel, J. E. and K. A. Medley. 2006. Biological invasions across spatial scales: intercontinental, regional, and local dispersal of cladoceran zooplankton. Biological Invasions **8**, 459–473.

Havel, J. E. and J. B. Shurin. 2004. Mechanisms, effects, and scales of dispersal in freshwater zooplankton. Limnology and Oceanography **49**, 1229–1238.

Haynes, K. E. and A. S. Fotheringham. 1984. Gravity and spatial interaction models. Sage, Beverly Hills, CA.

Hebert, P. D. N. and M. Cristescu. 2002. Genetic perspectives on invasions: the case of the Cladocera. Canadian Journal of Fisheries and Aquatic Sciences **59**, 1229–1234.

Herborg, L. M., C. J. Jerde, D. M. Lodge, G. M. Ruiz, and H. J. MacIsaac. 2007a. Predicting invasion risk using measures of introduction effort and environmental niche models. Ecological Applications **17**, 663–674.

Herborg, L. M., D. Rudnick, Y. Silang, D. M. Lodge, and H. J. MacIsaac. 2007b. Predicting the range of Chinese mitten crabs (*Eriocheir sinensis*) in Europe. Conservation Biology, in revision.

Horvath, T. G., G. A. Lamberti, D. M. Lodge, and W. L. Perry. 1996. Zebra mussels in lake-stream systems: source-sink dynamics? Journal of the North American Benthological Society **15**, 654–575.

Iguchi, K., K. Matsuura, K. M. McNyset, A. T. Peterson, R. Scachetti-Pereira, K. A. Powers, D. A. Vieglais, E .O. Wiley, and T. Yodo. 2004. Predicting invasions of North American basses in Japan using native range data and a genetic algorithm. Transactions of the American Fisheries Society **133**, 845–854.

Jenkins, D. G. and M. O. Underwood. 1998. Zooplankton may not disperse readily in wind, rain, or waterfowl. Hydrobiologia **387/388**, 15–21.

Johnson, L. E., J. M. Bossenbroek, and C. E. Kraft. 2006. Patterns and pathways in the post-establishment spread of non-indigenous aquatic species: the slowing invasion of North American inland lakes by the zebra mussel. Biological Invasions **8**, 475–489.

Johnson, L. E. and J. T. Carlton. 1996. Post-establishment spread in large-scale invasions: dispersal mechanisms of the zebra mussel *Dreissena polymorpha*. Ecology **77**, 1686–1690.

Johnson, L. E., A. Ricciardi, and J. T. Carlton. 2001. Overland dispersal of aquatic invasive species: a risk assessment of transient recreational boating. Ecological Applications **11**, 1789–1799.

Johnstone, L. M., B. T. Coffey, and C. Howard-Williams. 1985. The role of recreational boat traffic in interlake dispersal of macrophytes: a New Zealand case study. Journal of Environmental Management **20**, 263–279.

Karatayev, A. Y., L. E. Burlakova, D. K. Padilla, and L. E. Johnson. 2003. Patterns of spread of the zebra mussel (*Dreissena polymorpha* (Pallas)): the continuing invasion of Belarussian Lakes. Biological Invasions **5**, 213–221.

Kolar, C. S. and D. M. Lodge. 2001. Progress in invasion biology: predicting invaders. Trends in Ecology and Evolution **16**, 199–204.

Kolar, C. S. and D. M. Lodge. 2002. Ecological predictions and risk assessment for alien fishes in North America. Science **298**, 1233–1236.

Kraft, C. E., P. J. Sullivan, A. Y. Karatayev, L. E. Burlakova, J. C. Nekola, L. E. Johnson, and D. K. Padilla. 2002. Landscape patterns of an aquatic invader: assessing dispersal extent from spatial distributions. Ecological Applications **12**, 749–759.

Leung, B., J. M. Bossenbroek, and D. M. Lodge. 2006. Boats, pathways, and aquatic biological invasions: estimating dispersal potential with gravity models. Biological Invasions **8**, 241–254.

Leung, B., J. M. Drake, and D. M. Lodge. 2004. Predicting invasions: propagule pressure and the gravity of Allee effects. Ecology **85**, 1651–1660.

Leppäkoski, E., S. Gollasch, and S. Olenin, editors. 2002. Invasive aquatic species of Europe: distribution, impacts and management. Kluwer Academic Publishers, Dordrecht, The Netherlands.

Litvak, M. K., and N. E. Mandrak. 1993. Ecology of freshwater baitfish use in Canada and the United States. Fisheries **18**, 6–13.

Lockwood, J. L., P. Cassey, and T. Blackburn. 2005. The role of propagule pressure in explaining species invasions. Trends in Ecology and Evolution **20**, 223–228.

Louette, G. and L. De Meester. 2005. High dispersal capacity if cladoceran zooplankton in newly founded communities. Ecology **86**, 353–359.

MacIsaac, H. J., J. Borbely, J. R. Muirhead, and P. Graniero. 2004. Backcasting and forecasting biological invasions of inland lakes. Ecological Applications **14**, 773–783.

MacIsaac, H. J., H. A. M. Ketelaars, I. A. Grigorovich, C. W. Ramcharan, and N. D. Yan . 2000. Modeling *Bythotrephes longimanus* invasions in the Great Lakes basin based on its European distribution. Archiv für Hydrobiologie **149**, 1–21.

Marchetti, M., P. B. Moyle, and R. Levine. 2004. Invasive species profiling: exploring the characteristics of exotic fishes across invasion stages in California. Freshwater Biology **49**, 646–661.

Moyle, P. B. and T. Light. 1996. Biological invasions of freshwater: empirical rules and assembly theory. Biological Conservation **78**, 149–161.

Muirhead, J. R. and H. J. MacIsaac. 2005. Development of lakes as hubs in an invasion network. Journal of Applied Ecology **42**, 80–90.

Peterson, A. T. 2003. Predicting the geography of species' invasions via ecological niche modeling. Quarterly Review of Biology **78**, 419–433.

Peterson, A. T. and K. P. Cohoon. 1999. Sensitivity of distributional prediction algorithms to geographic data completeness. Ecological Modelling **117**, 159–164.

Proctor, V. W. 1959. Dispersal of fresh-water algae by migratory water birds. Science **130**, 623–624.

Proctor, V. W. 1964. Viability of crustacean eggs recovered from ducks. Ecology **45**, 656–658.

Rahel, F. J. 2002. Homogenization of freshwater fish faunas. Annual Review of Ecology and Systematics **33**, 291–315.

Ramcharan, C. W., D. K. Padilla, and S. I. Dodson. 1992. Models to predict potential occurrence and density of the zebra mussel, *Dreissena polymorpha*. Canadian Journal of Fisheries and Aquatic Sciences **49**, 150–158.

Richardson, D. M., P. Pyšek, M. Rejmánek, M. G. Barbour, F. D. Panetta, and C. J. West. 2000. Naturalization and invasion of alien plants: concepts and definitions. Diversity and Distributions **6**, 93–107.

Sala, O. E., F. S. Chapin III, J. Armesto, E. Berlow, J. Bloomfield, R. Dirzo, E. Huber-Sanwald, L. F. Huenneke, R. B. Jackson, A. Kinzig, R. S. Leeman, D. M. Lodge, H. A. Mooney, M. Oesterheld, N. L. Poff, M. T. Sykes, B. H. Walker, M. Walker, and D. H. Wall. 2000. Global biodiversity scenarios for the year 2100. Science **287**, 1770–1774.

Schneider, D. W., C. D. Ellis, and K. S. Cummings. 1998. A transportation model assessment of the risk to native mussel communities from zebra mussel spread. Conservation Biology **12**, 788–800.

Swanson, G. A. 1984. Dissemination of amphipods by waterfowl. Journal of Wildlife Management **48**, 988–991.

Therriault, T. W., M. I. Orlova, M. F. Docker, H. J. MacIsaac, and D. D. Heath. 2005. Invasion dynamics of a freshwater mussel (*Dreissena rostriformis bugensis*) in Eastern Europe as revealed by microsatellite analyses. Heredity **95**, 16–23.
Williamson, M. and A. Fitter. 1996. The characters of successful invaders. Biological Conservation **78**, 163–170.

Chapter nineteen

Biomonotony: definition and assessment for macroinvertebrates in European running waters

Jean-Nicolas Beisel and Simon Devin

INTRODUCTION

Emerging evidence shows that most species are declining as a result of human activity and are being replaced by a much smaller number of expanding species that are successful in human-altered environments (McKinney and Lockwood 1999). The process by which the genetic, taxonomic, or functional similarities of regional biota increase over time is called biotic homogenization (Olden and Rooney 2006). Some researchers view homogenization as a form of community globalization (Lockwood and McKinney 2001, Mooney and Cleland 2001) that causes the "McDonaldization" of the biosphere (Ritzer 1996, Lövei 1997) or the setting of global "McEcosystems" (Enserink 1999). Many studies report on this phenomenon across and within continents, such as the homogenization of the fish fauna across the North America (Rahel 2000, Taylor 2004, Olden *et al.* 2006), and biotic homogenization is now recognized as an important component of the modern biodiversity crisis (Olden *et al.* 2006).

Aquatic ecosystems are not spared this rampant homogenization of the earth's ecosystems, but on the contrary they have some sensitive features that lead to heightened vulnerability (Beisel 2001, Rahel 2002). Several recent

Francesca Gherardi, Biological invaders in inland waters: Profiles, distribution, and threats, 369–379.

reviews based on temporal analyses report range expansions with rapid spreading of aquatic invaders, such as the zebra mussel [*Dreissena polymorpha* (Pallas)] or the Asiatic clam [*Corbicula fluminea* (Muller)] in Europe and in North America (Morton 1997). More worrisome is the introduction rate of non-indigenous species (NIS) which is increasing exponentially across wide ecosystems such as San Francisco Bay (Cohen and Carlton 1998), the North American Great Lakes (Ricciardi and MacIsaac 2000), and the Baltic Sea (Olenin and Leppäkoski 1999).

At the same time, several studies report massive species extinction and profound modifications in freshwater communities (Hughes and Noss 1992, Rahel 2002). When calculating the extinction rate of North American freshwater fauna, Ricciardi and Rasmussen (1999) found a value five times higher than for terrestrial fauna and predicted an extinction rate of 4% per decade, i.e. a depletion of species as high as the alarming rate in tropical forests. Den Hartog *et al.* (1992) have pinpointed the extensive modification of macroinvertebrate communities on the Lower Rhine throughout the 20th century, with the disappearance of 85% of the pollution-sensitive insect species (Plecoptera, Ephemeroptera, Trichoptera), as well as seven of the 14 Lamellibranchia mollusc species originally present. On a global scale, freshwater bivalves are declining (Bogan 1993) and in North America, where this phenomenon has been documented best, more than 70% of the 343 bivalve species occurring are threatened (Neves *et al.* 1997). In Europe, as in the USA, researchers have proclaimed the alarming status of freshwater fish species, the most diverse of all vertebrate groups, which have been threatened or pushed to the local extinction during the last century (Miller *et al.* 1989; Williams and Miller 1990, Den Hartog *et al.* 1992, Rahel 2002, Myers and Worm 2003) and have suffered an overall decline globally (Duncan and Lockwood 2001). In turn, the disappearance of autochthonous species, the poverty of the fauna remaining, and their inability to adapt to new local conditions (Sax and Brown 2000) have favoured the permeability of human-altered freshwater ecosystems to repeated introductions of NIS (decrease in biotic resistance as defined by Elton 1958).

Local extinctions and the spread of NIS are leading to taxonomic changes which need to be described precisely and quantified so that objective comparisons of faunal modifications over time and space can be made. To this end, we propose to define a "biomonotony concept", which allows for an objective assessment of the global modifications in freshwater macroinvertebrate communities and establishes a framework which can be used to predict the consequences an introduced species will have on biodiversity. A practical example, based on the rapid changes over the last few decades in the communities inhabiting a European river, illustrates the relevance of this framework at a local scale for applicative purposes.

DEFINITION OF THE "BIOMONOTONY" CONCEPT

We define biomonotony as a two-dimensional attribute of a modified community, including: (i) the importance of the taxonomic change within the ecosystem due to the loss of indigenous species and the gain of NIS; and (ii) the modification of functional diversity (Fig. 1).

Faunal change consists of biotic homogenization that often replaces unique endemic species with already widespread NIS. Species extinction often affects endemic species and is taxonomically non-random at the local and global scale, with some families reported as containing substantially higher percentages of threatened or declining species than others (McKinney and Lockwood 1999). The clumping of species extinction within particular systematic families accelerates biodiversity loss, meaning that these taxonomic groups will be lost much faster than if extinction had occurred at random (McKinney 1997). Furthermore, the sequence of species loss in communities under pressure from human activity is not random, but it depends on organismic traits which induce, for example, a skewing trophic structure (Duffy 2003). In aquatic ecosystems, several insect families are the most threatened groups (belonging, for example, to Plecoptera, Ephemeroptera or Trichoptera orders, such as Chloroperlidae, Leptophlebiidae, and Brachycentridae, respectively) because their traits (K-selected traits, rare, specialist, poorly-adapted to human activity) do not promote survival in changing ecosystems.

Conversely, introduced species are currently a much smaller fraction of the earth's total biota, with 1–2% of known invasive species and 5–29% of species expanding their ranges locally (McKinney and Lockwood 1999). These reports support in part Williamson's (1996) "tens rule", stating that an average of one in ten successfully introduced species will become invasive. A large heterogeneity

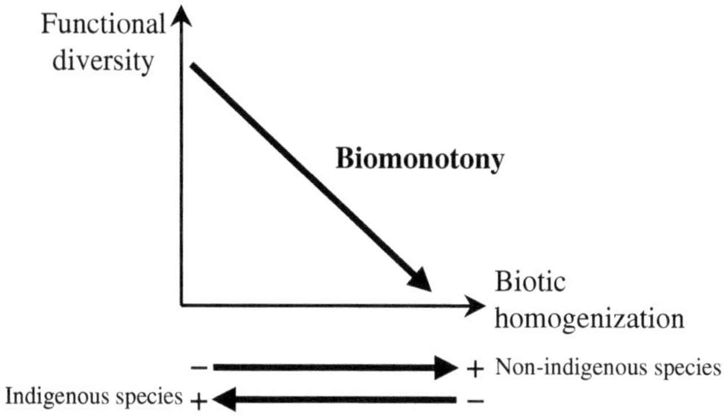

Fig. 1 Hypothetical diagram showing the two biomonotony components: functional diversity and biotic homogenization. The arrow denotes the most extreme (important) biomonotony gradient that can be obtained.

of results was found for the two first steps in the invasion process during which, according to Williamson (1996), one in ten imported species appears settled, and one in ten of settled species becomes established. Nevertheless, these few figures illustrate the replacement importance, whereby a great portion of the fauna is being replaced by a small fraction of the existing species. Long-distance transport via human activity allows these species to spread to isolated or distant areas that they would never have reached by their own means of dispersal (Carlton 2003). In fact, several introduced species have expanded their ranges worldwide and can now be considered cosmopolitan species, such as the Asian tiger mosquito [*Aedes albopictus* (Skuse)] (Eritja *et al.* 2005).

The second aspect of biomonotony concerns modifications in functional diversity. Changes in species diversity have functional consequences because the composition and structure of species in place determine the bioecological traits which in turn influence ecosystem processes (Chapin *et al.* 2000). Species traits may mediate energy or material flux, directly or indirectly, by altering the abiotic conditions which regulate process rates. Successful NIS often possess a homogeneous combination of bio/ecological traits (such as *r*-selected traits, rapid dispersal, eurytopy, and human commensalism) which promotes range extension. Such clustering of invasion-prone traits is apparently a very common result of the non-random way in which closely-related species share common traits. As a result, the biosphere could be increasingly dominated not only by the same few species in terms of systematic units, but also by the same closely-related species (McKinney and Lockwood 1999). Globally, in freshwater ecosystems, the introduction of non-indigenous crustaceans, flatworms, and molluscs has met with more success than for other taxa (Morton 1997, Van der Velde *et al.* 2000, Grigorovitch *et al.* 2002, Devin *et al.* 2005). Such taxonomic clumping of invaders can occur because closely-related species tend to share traits that promote transport, establishment, or massive development in a recipient ecosystem, the three major steps of an invasion process (Vermeij 1996, Kolar and Lodge 2001). This over-representation can have drastic effects on ecosystem functioning by replacing many, specific, complex functional systems with a few, similar, simpler ones. The degree and significance of such alterations depend on the biotic context, as explained by Vaughn and Hakenkamp (2001) for freshwater bivalves.

From a theoretical point of view, if species (local and non-indigenous) perform similar ecological process at similar rates, they are 'functionally redundant' (see Rosenfeld 2002). Thus, extinction of a particular species may make little difference within an ecosystem context, as long as the overall taxon biomass of functionally-redundant species is maintained. Conversely, if a species plays a distinct role (such as a keystone species), the replacement or extinction of that particular species may alter or modify the ecosystem's functioning permanently. For example, a modification of food webs caused by the predatory impact of a single introduced Ponto–Caspian gammarid [*Dikerogammarus villosus* (Sowinski)] may disrupt the dynamics of indigenous species (Dick and Platvoet

2000, Dick *et al.* 2002, Bollache *et al.* 2004) and affect diversity at other trophic levels. Dyer and Letourneau (2003) report that changes in diversity at one trophic level can provoke changes in diversity at other levels, and trophic cascades in aquatic ecosystems are clearly stronger than in terrestrial ones (Shurin *et al.* 2002). In such a case, the introduction, with or without extinction of local fauna, may affect the ecosystem's functioning. Furthermore, on a larger scale and independent of ecosystem functioning which is trophically mediated, a functional consequence of taxonomic changes is that non-indigenous introductions can lead to interactions between species that have never met before (non-additive effects). In particular, the introduction of non-endemic pathogens to indigenous populations may have unpredictable consequences. An example is the unintentional introduction and consequent spread in Europe of the crayfish plague [a virulent oomycetal disease, *Aphanomyces astaci* Schikora, which infested the very susceptible, indigenous crayfish *Astacus astacus* (Linnaeus), Van der Velde *et al.* 2000] following the intentional introduction of North American crayfish [*Orconectes limosus* (Rafinesque), *Pacifastacus leniusculus* (Dana), *Procambarus clarkii* (Girard)] (Chapters 2 and 28). The interdependence between taxonomic change and functional modification is fundamental to our decision to link these two community characteristics in the biomonotony assessment.

THE PRACTICAL ASSESSMENT OF BIOMONOTONY FOR RUNNING-WATERS

Based on the taxonomic changes and on functional diversity, the quantification of biomonotony requires a precise, practical, quantitative method to assess both characteristics. For this study, we focused on biological characteristics at the level of the species' systematic unit, the level best documented as a global diversity descriptor (ranging from genetic diversity within populations to the diversity of ecosystems).

Functional diversity needs to be described objectively. To provide important insights into the functioning of running-water communities, many authors advocate working with groups of organisms that are as similar as possible in terms of biological, physiological, and ecological trait relationships. Following Rosenfeld (2002), we considered the functional effects of a taxon as output variables, or consequences of an organism's traits and resource use. Recently, Usseglio-Polatera *et al.* (2001) developed a functional classification for benthic macroinvertebrates based on both biological and ecological traits. The definition of six groups, nine subgroups, and 43 units according to a biological versus ecological ordination allowed for an innovative assessment of the functional rule of 472 European macroinvertebrate genera. Usseglio-Polatera *et al.* (2001) provided examples illustrating the use of these discrete groups as a valuable tool to track environmental changes over time and space. From a theoretical point of view, species (or genera) belonging to a given group can be considered as 'functionally redundant' for the major processes occurring in that particular ecosystem, if three different factors have been taken into account (Rosenfeld 2002): (i) ecological processes related to morphological and physiological attributes; (ii) demographic characteristics; and (iii) environmental factors influencing where an organism can perform its function. The replacement of a given species belonging to a particular functional group can be counter-balanced, functionally speaking, by the introduction of a NIS belonging to that same group. If each species in a given community is replaced by its functional group, the assessment of community diversity, based on these groups, provides a functional measure of the community. Regardless, this approach does not take into account all the subtleties of ecosystem functioning (e.g. inter-species relationships resulting from co-evolved adaptations such as host–parasite relationships), but it allows for an initial, quantitative assessment of functional diversity in a macroinvertebrate community, and provides baselines to improve this tool. As pointed out by Rosenfeld (2002), even similar species, by definition, will differ to some unknown degree along some functions, but it is still reasonable to expect that some species are more similar than others in terms of their roles in communities and ecosystems.

The second component of biomonotony is more difficult to evaluate, because it requires comparisons of faunal samples with faunal reference lists. For many freshwater ecosystems, it is unrealistic to define a faunal reference list as representing the community as it was several hundred years ago, prior to the human disturbance. However, a relative reference point corresponding to the best-known situation can be obtained in several ways (Nijboer *et al.* 2004), two of them deserving particular attention. The first consists in simply using historical data collected from situations (sites) with the minimum of anthropogenic stress. The second is to model the community as it would be expected to exist in the absence of anthropogenic stress, but this requires a data set to construct, calibrate, and validate the model. An example of this approach is the RIVPACS (River Invertebrate Prediction and Classification Systems) developed in the UK for the assessment of aquatic communities (Wright *et al.* 1984, Wright 1995). In the absence of a modelling approach for many other European eco-regions, we recommend the use of historical data when available. In all cases, either of these approaches will provide a faunal list that can be used to detect the homogenization of taxa considering: (1) the extinction of geographically restricted indigenous species (for reasons such as habitat changes); and (2) the introduction of widespread, highly tolerant species. In most cases, the actual biomonotony will be greater than that observed, the number of non-indigenous introductions and local extinctions in a given region always being underestimated because an absolute faunal reference is unavailable. We proposed, as an initial approach, to evaluate the faunal change using data on the proportion of NIS. The use of a relative measure (number of NIS reported on the total number of species) has the advantage of taking into account both the importance of the introduction of NIS and the extinction of local fauna.

AN ILLUSTRATION OF BIOMONOTONY ASSESSMENT

To illustrate this approach, a faunal analysis of three time periods was performed on the navigable French section of the Moselle River (north-eastern France, Rhine tributary). Three locations were chosen: upstream (near the town of Liverdun), midstream (near the city of Metz), and downstream (between Cattenom and Sierck-les-Bains). For each location (called Upper, Middle, and Lower parts), semi-quantitative data on macroinvertebrate assemblages were collected from different literature sources (grey literature of our institute for 1975; reports from the International Commission for the Protection of the Moselle River; and the authors' unpublished, personal data for 1996 and 2000). Diptera Chironomidae were not considered in this analysis due to the lack of precision in their systematic determination but, ideally, when enough data is available, they should be taken into account in the analyses.

For this example, each species in a community was replaced by their functional group, and functional diversity was then measured as the Shannon-Weaver diversity of functional groups (Usseglio-Polatera *et al.* 2001). Taxonomic change over time was assessed using the proportion of NIS in faunal lists. All taxa considered as non-indigenous (review in Devin *et al.* 2005) were described in various literature sources as having been introduced in the Moselle River during the 20th century, with the exception of the zebra mussel, *D. polymorpha*, a well-known invasive species introduced into the Moselle River in about 1850 (Kinzelbach 1992).

In 1975, the invertebrate communities were the most diversified in terms of functional groups and, logically, least altered by NIS (Fig. 2). Thus, this period can be used as a relative reference, even though the situation was already far from ideal, in other words without anthropogenic disturbance. The whole macroinvertebrate structure prior to 1975 has remained largely ignored, though fragmentary information is available. In 1996, we observed a clear loss of functional diversity and a significant increase in invasive species, representing between 19% and 21% of the collected taxa. The biomonotony increased from upstream to downstream, the Lower part having changed the most, certainly due to a migratory corridor of NIS from the Middle Rhine to its tributary. We can note that the repartition of stations on the scatter plot did not overlap between the three periods. In 2000, the entire section studied is homogeneous in terms of biomonotony, all sections being altered to the same degree. The proportions of NIS in 2000 were above the values noted for 1996, and the functional diversity for the river was slightly below the value of the Upper part for 1996. Functional diversity increased between 1996 and 2000 for the Middle and Lower sections. After the demographic explosion of several new NIS introduced in the beginning of the 1990s (Bachmann *et al.* 1995), densities tend to equilibrate, with, for example, the observed decline of *D. polymorpha* (Bachmann *et al.* 2001), previously over-represented in this ecosystem. This evenness incorporated the relative abundances of taxa,

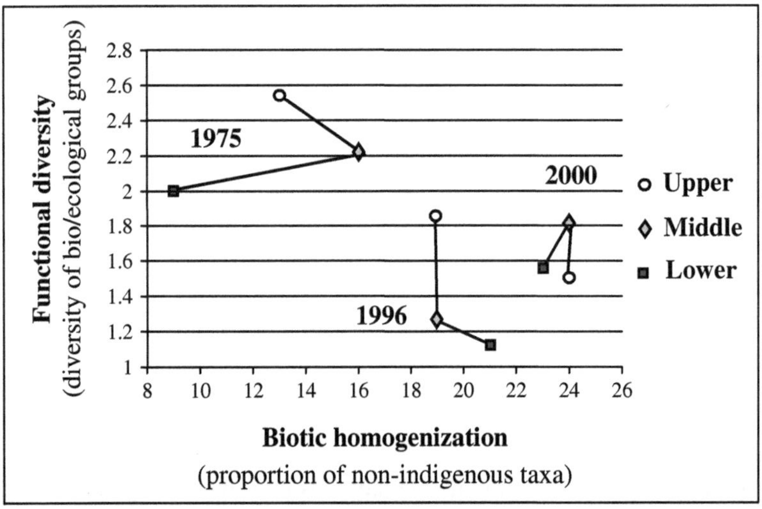

Fig. 2 An example of the practical application of the biomonotony framework: temporal modification of macrobenthic communities between 1975 and 2000 for a navigable sector of a large river subdivided into three sections (Moselle River, France).

and quantified the best representation of different bio/ecological trait combinations, i.e. the highest functional diversity. In other cases, the additions of NIS belonging to original functional groups could increase the functional diversity, especially when they do not lead to the extinction of indigenous species.

This example objectively shows how river integrity can be profoundly modified over a very short period of time. The biomonotony framework can serve as a tool to evaluate these changes and can illustrate the potential effects of rehabilitation measures taken to improve river quality.

CONCLUSIONS

The biomonotony concept is defined as a loss of diversity due to species replacement, species extinction, and the establishment of NIS more-or-less functionally redundant with the indigenous ones. Thus, particular biological or ecological functions could disappear, or be overexpressed, in a biomonotonous ecosystem. This concept provides a framework to objectively assess the degree of ecosystem modification and focuses on functional diversity which is generally not approached fully. These tools will provide an efficient method for the quantitative assessment and study of the ecological consequences of freshwater macroinvertebrate introductions.

ACKNOWLEDGEMENTS

The authors wish to thank Philippe Usseglio-Polatera for his comments on the manuscript and Anna Matheny-Cartier for her revisions. This study is supported by the French Minister of Ecology and Sustainable Development, as part of the 2003–2006 INVABIO – Biological Invasions Program. Financial support from the "Zone Atelier Moselle" by the "Region Lorraine" is gratefully acknowledged.

REFERENCES

Bachmann, V., J. N. Beisel, P. Usseglio-Polatera, and J. C. Moreteau. 2001. Decline of *Dreissena polymorpha* in the River Moselle: biotic and abiotic key factors involved in dynamics of invasive species. Archiv für Hydrobiologie **151**, 263–281.

Bachmann, V., E. Cegielka, P. Wagner, P. Usseglio-Polatera, and J. C. Moreteau. 1995. Establishment of the amphipod *Corophium curvispinum* and the Asiatic clam *Corbicula* sp. in the French part of the River Mosel. Hydroécologie Appliquée **7**, 185–190.

Beisel, J.-N. 2001. The elusive model of a biological invasion process: time to take differences among aquatic and terrestrial ecosystems into account? Ethology, Ecology & Evolution **13**, 193–195.

Bogan, A. E. 1993. Freshwater bivalve extinctions (Mollusca: Unionoida): a search for causes. American Zoologist **33**, 599–609.

Bollache, L., S. Devin, R. Wattier, M. Chovet, J. N. Beisel, J. C. Moreteau, and T. Rigaud. 2004. Rapid range extension of the Ponto–Caspian amphipod *Dikerogammarus villosus* in France: potential consequences. Archiv für Hydrobiologie **160**, 57–66.

Carlton, J. T. 2003. Community assembly and historical biogeography in the North Atlantic Ocean: the potential role of human-mediated dispersal vectors. Hydrobiologia **503**, 1–8.

Chapin, F. S., E. S. Zavaleta, V. T. Eviner, R. L. Naylor, P. M. Vitousek, H. L. Reynolds, D. U. Hooper, S. Lavorel, O. E. Sala, S. E. Hobbie, M. C. Mack, and C. Diaz. 2000. Consequences of changing biodiversity. Nature **405**, 234–242.

Cohen A. N., and Carlton J. T. 1998. Accelerating invasion rate in a highly invaded estuary. Science **279**, 555–558.

Den Hartog, C., F. W. B. Van der Brink, and G. Van der Velde. 1992. Why was the invasion of the river Rhine by *Corophium curvispinum* and *Corbicula* species so successful? Journal of Natural History **26**, 1121–1129.

Devin, S., L. Bollache, P. Y. Noël, and J. N. Beisel. 2005. Patterns of biological invasions in French freshwater systems by non-indigenous macroinvertebrates. Hydrobiologia **551**, 137–146.

Dick, J. T. A. and D. Platvoet. 2000. Invading predatory crustacean *Dikerogammarus villosus* eliminates both native and exotic species. Proceedings of the Royal Society of London B **267**, 977–983.

Dick, J. T. A., D. Platvoet, and D. W. Kelly. 2002. Predatory impact of the freshwater invader, *Dikerogammarus villosus* (Crustacea: Amphipoda). Canadian Journal of Fisheries and Aquatic Sciences **59**, 1078–1084.

Duffy, J. E. 2003. Biodiversity loss, trophic skew and ecosystem functioning. Ecology Letters **6**, 680–687.

Duncan, J. R. and J. L. Lockwood. 2001. Extinction in a field of bullets: a search for causes in the decline of freshwater fishes. Biological Conservation **102**, 97–105.

Dyer, L. A. and D. Letourneau. 2003. Top-down and bottom-up cascades in detrital vs. living food webs. Ecology Letters **6**, 60–68.

Elton, C. S. 1958. The ecology of invasions by animals and plants. University of Chicago Press, Chicago, IL, Edition 2000.

Enserink, M. 1999. Biological invaders sweep in. Science **285**, 1934–1836.

Eritja, R., R. Escosa, J. Lucientes, E. Marquès, D. Roiz, and S. Ruiz. 2005. Worldwide invasion of vector mosquitoes: present European distribution and challenges for Spain. Biological Invasions **7**, 87–97.

Grigorovitch, I. A., H. J. MacIsaac, N. V. Shadrin, and E. L. Mills. 2002. Patterns and mechanisms of aquatic invertebrate introductions in the Ponto–Caspian region. Canadian Journal of Fisheries and Aquatic Sciences **59**, 1189–1208.

Hughes, R. M. and R. F. Noss. 1992. Biological diversity and biological integrity: current concerns for lakes and streams. Fisheries **17**, 11–19.

Kinzelbach, R. 1992. The main features of the phylogeny and dispersal of the zebra mussel *Dreissena polymorpha*. Pages 5–17 in D. Neumann and H. A. Jenner, editors. The Zebra Mussel *Dreissena polymorpha*. Gustav Fischer Verlag, Stuttgart, Germany.

Kolar, C. S. and D. M. Lodge. 2001. Progress in invasion biology: predicting invaders. Trends in Ecology and Evolution **16**, 199–204.

Lockwood, J. L. and M. L. McKinney. 2001. Biotic homogenization. Kluwer Academic Publishers, New York, NY.

Lövei, G. L. 1997. Global change through invasion. Nature **388**, 627–628.

McKinney, M. L. 1997. Extinction vulnerability and selectivity: combining ecological and paleontological views. Annual Review of Ecology and Systematics **28**, 495–516.

McKinney, M. L. and J. L. Lockwood. 1999. Biotic homogenization: a few winners replacing many losers in the next mass extinction. Trends in Ecology and Evolution **14**, 450–453.

Miller, R. R., J. D. Williams, and J. E. Williams. 1989. Extinctions of North American fishes during the past century. Fisheries **14**, 22–38.

Mooney, H. A. and E. E. Cleland. 2001. The evolutionary impact of invasive species. Proceeding of the National Academy Sciences USA **98**, 5446–5451.

Morton, B. 1997. The aquatic nuisance species problem: a global perspective and review. Pages 1–53 in F. D'Itri, editor. Zebra mussels and aquatic nuisance species. Ann Arbor Press, Chelsea, MI.

Myers, R. A. and B. Worm. 2003. Rapid worldwide depletion of predatory fish communities. Nature **423**, 280–283.

Neves, R. J., A. E. Bogan, J. D. Williams, S. A. Ahlstedt, and P. W. Hartfield. 1997. Status of the aquatic mollusks in the southeastern United States: a downward spiral of diversity. Pages 43–86 in G. W. Benz and D. E. Collin, editors. Aquatic fauna in peril: the southeastern perspective. Special Publication 1, Southeast Aquatic Research Institute, Decatur, GA.

Nijboer, R. C., R. K. Johnson, P. F. M. Verdonschot, M. Sommerhäuser, and A. Buffagni. 2004. Establishing reference conditions for European streams. Hydrobiologia **516**, 91–105.

Olden, J. D., N. LeRoy Poff, and M. McKinney. 2006. Forecasting faunal and floral homogenization associated with human population geography in North America. Biological Conservation **127**, 261–271.
Olden, J. D. and T. P. Rooney. 2006. On defining and quantifying biotic homogenization. Global Ecology and Biogeography **15**, 113–120.
Olenin, S. and E. Leppäkoski. 1999. Non-native animals in the Baltic Sea: alteration of benthic habitats in coastal inlets and lagoons. Hydrobiologia **393**, 233–243.
Rahel, F. J. 2000. Homogenization of fish faunas across the United States. Science **288**, 854–856.
Rahel, F. J. 2002. Homogenization of freshwater faunas. Annual Review of Ecology and Systematics **33**, 291–315.
Ricciardi, A. and H. J. MacIsaac. 2000. Recent mass invasion of the North American Great Lakes by Ponto–Caspian species. Trends in Ecology and Evolution **15**, 62–65.
Ricciardi, A. and J. B. Rasmussen. 1999. Extinction rates of North American freshwater fauna. Conservation Biology **13**, 1220–1222.
Ritzer, G. 1996. The McDonaldization of Society – An investigation into the changing character of contemporary social life. Pine Forge Press, Thousand Oaks, CA.
Rosenfeld, J. S. 2002. Functional redundancy in ecology and conservation. Oikos **98**, 156–162.
Sax, D. F. and J. H. Brown. 2000. The paradox of invasion. Global Ecology and Biogeography **9**, 363–371.
Shurin, J. B., E. T. Borer, E. W. Seabloom, K. Anderson, C. A. Blanchette, B. Broitman, S. D. Cooper, and B. S. Halpern. 2002. A cross-ecosystem comparison of the strength of trophic cascades. Ecology Letters **5**, 785–791.
Taylor, E. B. 2004. An analysis of homogenization and differentiation of Canadian freshwater fish faunas with an emphasis on British Columbia. Canadian Journal of Fisheries and Aquatic Sciences **61**, 68–79.
Usseglio-Polatera, P., P. Richoux, M. Bournaud, and H. Tachet. 2001. A functional classification of benthic macroinvertebrates based on biological and ecological traits: application to river condition assessment and stream management. Archiv für Hydrobiologie **139**, 53–83.
Van der Velde, G., S. Rajagopal, B. Kelleher, I. B. Musko, and A. Bij de Vaate. 2000. Ecological impacts of crustacean invaders: general considerations and examples from the river Rhine. Crustacean Issues **12**, 3–33.
Vaughn, C. C. and C. C. Hakenkamp. 2001. The functional role of burrowing bivalves in freshwater ecosystems. Freshwater Biology **46**, 1431–1446.
Vermeij, G. J. 1996. An agenda for invasion biology. Biological Conservation **78**, 3–9.
Williams, J. E. and R. R. Miller. 1990. Conservation status of the North American fish fauna in freshwater. Journal of Fish Biology **37**, 79–85.
Williamson, M. 1996. Biological invasions. Chapman & Hall, London, UK.
Wright, J. F. 1995. Development and use of a system for predicting the macroinvertebrate fauna in flowing waters. Australian Journal of Ecology **20**, 181–197.
Wright, J. F., D. Moss, P. D. Armitage, and M. T. Furse. 1984. A preliminary classification of running-water sites in Great Britain based on macro-invertebrate species and the prediction of community type using environmental data. Freshwater Biology **14**, 221–256.

Chapter twenty

Genetics and invasion biology in fresh waters: a pilot study of Procambarus clarkii *in Europe*

Silvia Barbaresi, Francesca Gherardi, Alessio Mengoni, and Catherine Souty-Grosset

INTRODUCTION

Since Charles Elton, ecologists have struggled to account for the distribution and spread of non-indigenous species (NIS). They now realize that the key factor is "opportunity". The more frequently and persistently a foreign plant or animal is exposed to a new environment, the better its odds of invading. Ecologists call this "propagule pressure" (also termed introduction effort), a composite measure expressing the number of individuals released into a region to which they are non-native. It incorporates estimates of the absolute number of individuals involved in any one release event (propagule size) and the number of discrete release events (propagule number) (Lockwood *et al.* 2005). Propagule pressure has rarely been documented for freshwater organisms except in fish. Duggan *et al.* (2006), for example, using data on aquaria in stores and historical records of fish introduced into Canadian and US waters, showed a clear relationship between their occurrence in shops and the likelihood of their introduction and eventual establishment.

Francesca Gherardi, Biological invaders in inland waters: Profiles, distribution, and threats, 381–400.

The existence of two opposing pathways in Europe has been suggested: virtually all large aquatic ecosystems now support NIS that were unintentionally (e.g. euryhaline invertebrate species) or deliberately (e.g. freshwater fish) introduced by humans (MacIsaac *et al.* 2002). On one hand, geographically remote areas, such as the Black and Caspian seas and the North American Great Lakes (e.g. Ricciardi and MacIsaac 2000, MacIsaac *et al.* 2001), have been found to be biologically linked due to the active dispersal of organisms along artificial canals and their long-distance transfer in ballast water of commercial ships (Cristescu *et al.* 2001, Bij de Vaate *et al.* 2002). On the other hand, human commercial activities, specifically aquaculture, legal or illegal stocking, live food trade, aquarium, and pond trade (Lodge *et al.* 2000), have led to the deliberate introduction of, e.g. several crayfish species from North America to Europe (Gherardi and Holdich 1999).

EVOLUTIONARY GENETICS OF INVASIVE SPECIES

The genetics and evolution of NIS have received far less attention than their ecology but undoubtedly such studies could offer further insights into invasion dynamics (Lee 2002), since genetic and evolutionary processes may be the key features in determining whether invasive species establish and spread (Sakai *et al.* 2001). Of great theoretical and practical importance is the ability to identify the location of origin of NIS and their route of invasion (Wilson *et al.* 1999, Kreiser *et al.* 2000, Cox 2004). Theoretical models of genetic organization and population structure following a founding effect can be described by two different scenarios. The first model predicts subpopulations to show strong genetic structuring and clinal variations, while the second involves extinction and recolonization that enhance gene flow and reduce inter-population differentiation (Alvarez-Buylla and Garay 1994). Migration may be critical, not only as a source of continuing propagule pressure, but also as an important source of genetic variation to the colonizing population if multiple invasions provide the genetic variation necessary for adaptive evolution. Multiple introductions can create invasive populations that are much more genetically diverse than any single source population when the invasive species is highly structured in its native range. Different colonizing populations of the same species are likely to be genetically divergent with different levels of genetic variation and therefore have different capacities to promote invasiveness; characteristics that promote invasiveness might evolve in some populations but not in others. Gene flow between populations could result in the spread of invasive genotypes. Alternatively, gene flow between populations that swamps out locally beneficial alleles could prevent evolution of invasiveness (Kirkpatrick and Barton 1997 in Sakai *et al.* 2001). From this knowledge, we can obtain useful information about the vectors and the number of introductions and, as a consequence, we may attempt to halt or to slow down the invasion process. We can also find effective biological control agents of harmful NIS and understand to which degree the "enemy release" hypothesis (e.g. Keane and Crawley 2002) can explain invasions (Kreiser *et al.* 2000, Patti and Gambi 2001, Allendorf and Lundquist 2003).

Molecular genetic techniques today offer a very powerful set of tools for characterizing populations of NIS and for relating them to the populations of their native and colonized geographical areas (Cox 2004). These molecular markers can provide an indication of the amount of genetic variation lost during a colonization bottleneck and furnish evidence for multiple population sources. They have been successfully used to pinpoint the source areas and the routes of dispersal followed by a number of freshwater non-indigenous crustaceans, including *Cercopagis pengoi* Ostroumov and freshwater Cladocera (Cristescu *et al.* 2001, Hebert and Cristescu 2002). Here we present a pilot study on the red swamp crayfish, *Procambarus clarkii* Girard, in which we used molecular markers with the aim of understanding the dynamics of introductions of such a commercial invasive species that most often follow illegal paths. When historical human records are incomplete, inaccurate, or simply nonexistent, molecular genetic studies often offer a powerful tool for the identification of relationships between introduced populations.

PILOT CASE STUDY: THE RED SWAMP CRAYFISH *PROCAMBARUS CLARKII*

Procambarus clarkii is typically found in marshes, rivers, reservoirs, irrigation systems, and rice fields; the species is a paradigmatic example of a recent freshwater invader (Gherardi 2006). When introduced, it most often becomes established quickly, contributing to the decline of indigenous European species of Astacidae due to its strong competitive pressure and its ability to transmit the oomycete *Aphanomyces astaci* Schikora (Chapter 28).

In Europe, the species was first introduced into Spain in 1973 using specimens from Louisiana (Habsburgo-Lorena 1986). Reasons for its introduction varied and included aquaculture, aquarium and live food trades, and bait (Huner 1977, Huner and Avault 1979). In less than 20 years from this first introduction, new populations of *P. clarkii* have been reported in several countries of Europe, including Portugal, Cyprus, England, France, Germany, Italy, Mallorca, The Netherlands, and Switzerland (Gherardi and Holdich 1999). Populations found in Italy were described for northern and central Italy by Barbaresi and Gherardi (2000). In northern Italy, *P. clarkii* is undergoing a great expansion in both the Po and the Reno drainage basins in Piedmont and in Emilia-Romagna, respectively. In central Italy, the species is widespread in Tuscany, especially in Massaciuccoli Lake after the establishment of a farm in 1990. It is hypothesized that all the *P. clarkii* populations appearing in Tuscany after 1990 probably originated from man-made translocations from this lake. Since its first

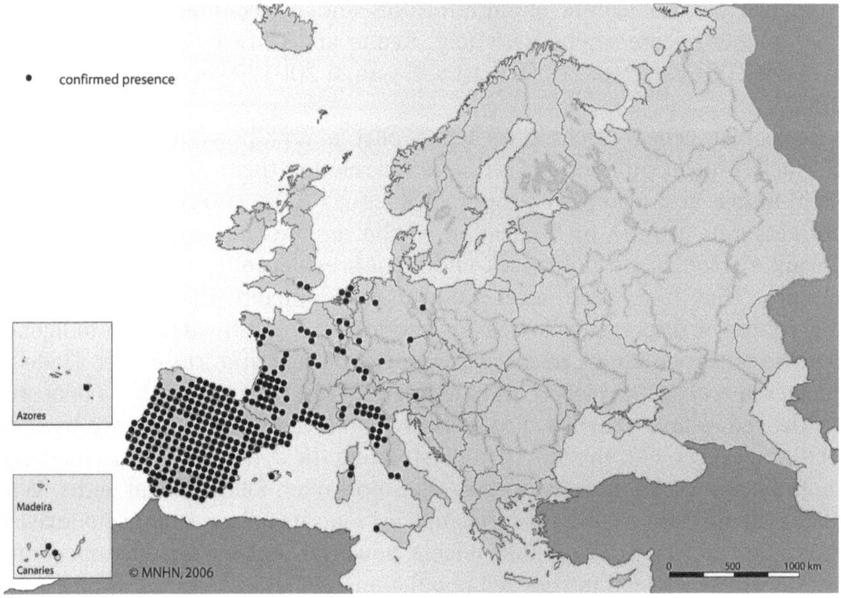

Fig. 1 The map gives details of the current distribution of *Procambarus clarkii* in Europe. Spain, France, and Italy have the majority of populations (Souty-Grosset *et al.* 2006).

introduction, the red swamp crayfish has now also been found in Liguria. As of 2005, *P. clarkii* has invaded 13 countries and is considered a major freshwater pest (Souty-Grosset *et al.* 2006) (Fig. 1).

Different mechanisms play a role in the displacement of the crayfish outside their native ranges: (1) natural, such as active dispersal; (2) accidental, such as escape from holding facilities; or (3) deliberate by the intervention of humans. There are only anecdotal reports about the geographic source of most introduced *P. clarkii* populations. The population genetic approach, on the contrary, is of great help in understanding the contribution of each mechanism to its actual distribution. In order to outline the history of the invasion process throughout Europe, our objectives were to address the role of single versus multiple dispersal events through the comparison of the genetic structure of different European *P. clarkii* populations. The first introduction of *P. clarkii* into Spain from Louisiana is well documented. On the other hand, the events leading to subsequent expansion of the species are only partially known (Laurent 1997, Gherardi and Holdich 1999).

A study was undertaken using a population genetic approach that aimed at analysing the invasion process of this introduced crayfish (Fetzner 1996, Fetzner and Crandall 2001). Because allozyme variability is very low in crayfish (reviewed in Fetzner and Crandall 2001), a preliminary study using random amplification polymorphic DNA (RAPD) markers was initially performed.

It revealed high levels of genetic variability in five European populations, suggesting multiple introductions of individuals coming from different source locations (Barbaresi *et al.* 2003). Subsequently, we analysed the genetic variation at CO1 mtDNA sequences and at five microsatellite loci in *P. clarkii* populations of western Europe to outline their genetic relationships. Our aim was twofold: first, we addressed the role of single versus multiple introduction events and, second, we clarified the effects of the active dispersal of crayfish on the diffusion of this species at a microgeographical scale.

Samples of *P. clarkii* were obtained from two areas of its native range: the first in New Orleans (Louisiana, USA, NOR), the putative source of the European populations, and the second in northern Mexico (Chihuahua, CHI); other populations were sampled from the introduced range in western Europe from May 2000 to January 2003 (Fig. 2 and Table 1).

Localities were selected on the basis of the historical (published) or anecdotal information on their introduction (Gherardi and Holdich 1999, Table 1). In particular, the Spanish site (Veta La Palma, Doñana, Guadalquivir Region, VPA) was the first European locality where the introduction of a 400 kg batch of *P. clarkii* in 1974 was followed by its successful establishment (a 100 kg batch imported a year before in the Spanish province of Badajoz did not prosper). The other sampled populations were collected from Switzerland

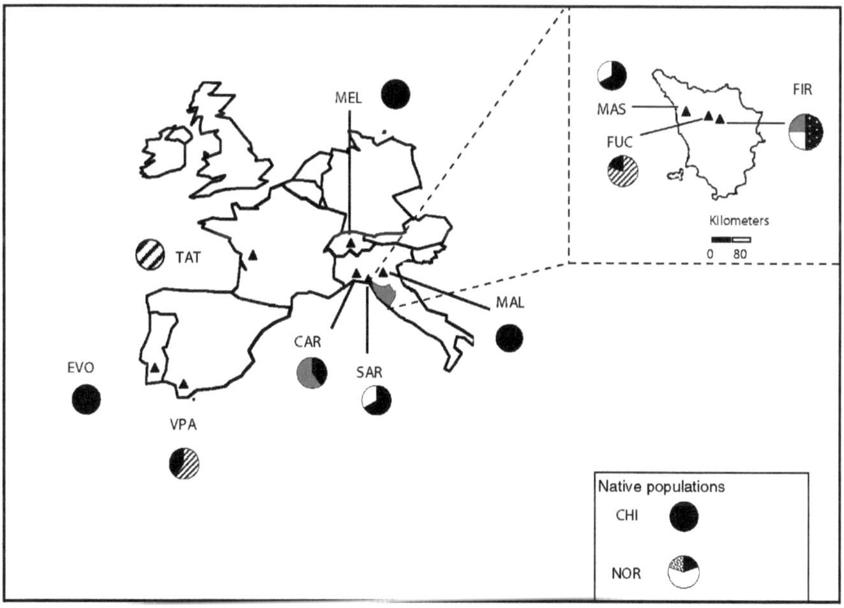

Fig. 2 Geographic location and mtDNA haplotype frequencies of European populations of *Procambarus clarkii* sampled in the present study.

Table 1 *Procambarus clarkii* populations sampled in the present study: sampling location, site code, number of sampled individuals for microsatellite and mtDNA (in parenthesis) analysis, and status of the population (I = introduced, N = native) are indicated. For introduced populations, the table shows the year of introduction and the source population derived from either anecdotic or published (*) data (Habsburgo-Lorena 1986).⁺ indicates populations previously studied using RAPDs (Barbaresi *et al.* 2003).

Collection location	Site code	Sample size	Status	Year	Source
Veta La Palma, Doñana, Sevilla, Spain	VPA	10 (5)	I	1974	Louisiana*
Le Tatre, Givrezac Charente, France	TAT	10 (5)	I	1976	Spain
Évora, Alentejo, Portugal[+]	EVO	10 (5)	I	~1985	Spain
Carmagnola, Torino, Piedmont, Italy	CAR	10 (5)	I	1989	Unknown
Fucecchio, Tuscany, Italy[+]	FUC	10 (5)	I	~1995	Massaciuccoli
Firenze, Tuscany, Italy[+]	FIR	10 (4)	I	~1995	Massaciuccoli
Malalbergo, Bologna, Emilia Romagna, Italy[+]	MAL	10 (5)	I	~1990	Unknown
Mellingen, Aargau, Switzerland	MEL	10 (4)	I	~1990	Unknown
Massaciuccoli, Tuscany, Italy[+]	MAS	10 (3)	I	1992	Spain (Doñana)
Sarzana, La Spezia, Liguria, Italy	SAR	10 (3)	I	1998	Unknown
New Orleans, Louisiana, USA[+]	NOR	10 (5)	N	–	–
Chihuahua, Mexico	CHI	9 (4)	N	–	–

(Tanklager Weiher, Mellingen, MEL), France (Le Tatre, Givrezac Charente, TAT), and southern Portugal (Guadiana River Basin, Évora, Alentejo, EVO). In all these areas, individuals were sampled from the sites where the species was first recorded in these countries. In Italy, we analysed samples from the locality where a wild population of *P. clarkii* was first recorded in 1989 (Delmastro 1992) in Carmagnola (Piedmont, northern Italy, CAR) and from the second known locality of introduction (Massaciuccoli Lake, MAS) in Tuscany.

To explore the hypothesis that some Italian populations could have been founded through either an active dispersal event or human-mediated translocations from Massaciuccoli Lake in Tuscany, additional samplings were performed in Sarzana (SAR, Liguria region) at a distance of about 20 km from the nearest Tuscan population and of 60 km from Massaciuccoli Lake, and in Fucecchio (FUC) and Firenze (FIR) in Tuscany. These latter two populations are 60 km distant from each other but have no direct water connections. Some of the analysed populations (MAS, FUC, FIR, MAL, EVO, and NOR) were previously studied using RAPD markers (Barbaresi *et al.* 2003).

MtDNA analysis

Muscle tissue was removed from a single claw of each specimen and preserved in 96% ethanol until DNA extraction. Total genomic DNA extraction was performed as reported in Barbaresi *et al.* (2003). After precipitation, DNA was stored in the Qiagen-supplied buffer at 4 °C for routine use or at −20 °C for long-term storage. The mitochondrial DNA (mtDNA) sequence analysis included nine individuals from the two localities in the native range and 44 individuals from the 10 localities in the introduced range (Table 1). A 572-base pair (bp) segment of the cytochrome oxidase subunit I gene (COI) was amplified by polymerase chain reaction using the primer Jerry (5′-CAA CAT TTA TTT TGA TTT TTT GG-3′) from Simon *et al.* (1994) and a primer designed by the authors and named ProsSilvia (5′-TCA GGA TAA TCT GAA TAA CG-3′). PCR was carried out in a total volume of 25 μL containing 10X PCR buffer (Gold), 25 mM $MgCl_2$ and 200 μM each dNTP, 10 μM each primer, 2.5 units of Amplitaq Gold *Taq DNA* polymerase and 50–100 ng DNA. An initial 5-min denaturation step at 94 °C was followed by 35 cycles of 30 s at 94 °C for denaturation, 45 s at 48 °C for annealing, 45 s at 72 °C for extension, and by 10 min of final extension at 72 °C. Successful PCR products were purified by the ExoSAP-IT buffer (USB), and then sequenced using the Big Dye Terminator method with an ABI 310 automated sequencer. For most samples the forward and reverse sequences were obtained. Sequences were manually edited with the program CHROMAS version 1.55 (Technelysium, Queensville, Australia). The software ESEE Version 3.2 (based on Cabot and Beckenbach 1989) was used to align sequences by eye. Sequences were trimmed to the same length. There were no gaps in the two regions, making alignment straightforward. The data matrix included the 53 sequences examined in this study.

For each population, we calculated the following genetic diversity indices: the number of haplotypes (*Nhap*), the nucleotide diversity (π/L, the probability that two randomly chosen homologous nucleotides are different), and the gene diversity (*H*, the probability that two randomly chosen haplotypes are different). This latter index was estimated according to Nei (1978). The indices were calculated with Arlequin 2.0 (Schneider *et al.* 2000). Out of the 53 individuals studied, we identified 6 different haplotypes (Table 2). The number of haplotypes found within each population was low, ranging from 1 to 3. One most frequent haplotype was detected in 28 individuals from 10 populations. Few private haplotypes were found, one in the native population from Louisiana (NOR) and three in European populations (FIR: $N = 2$; CAR: $N = 1$). Two haplotypes were shared among few populations (haplotype 1: VPA, FUC, and TAT; haplotype 4: NOR, MAS, and SAR). The amount of mtDNA COI sequence variation

Table 2 *Procambarus clarkii* mtDNA haplotype distribution among the analysed populations. *N* is the number of the analysed individuals; *Nhap* is the number of different haplotypes found in each population. Gene and nucleotide diversities are also reported.

	Haplotypes									
Locality	1	2	3	4	5	6	*N*	N_{hap}	Gene diversity	Nucleotide diversity
FIR			2	1		1	4	3	0.833	0.003
MAS		2		1			3	2	0.667	0.001
MAL		5					5	1	0.000	0.000
FUC	4	1					5	2	0.400	0.001
TAT	5						5	1	0.000	0.000
SAR		2		1			3	2	0.667	0.001
CAR		2	3				5	2	0.600	0.002
VPA	3	2					5	2	0.600	0.002
MEL		4					4	1	0.000	0.000
EVO		5					5	1	0.000	0.000
CHI		4					4	1	0.000	0.000
NOR		1		3	1		5	3	0.700	0.000
Total	12	28	5	6	1	1	53			

within populations is reported in Table 2. Gene diversity (haplotypic diversity, *H*) differed among populations, ranging from 0 to 0.7, in the native NOR population.

To illustrate the relationships between the haplotypes, a minimum spanning network was constructed by the use of Kimura two-parameter genetic distances (Kimura 1980), with Arlequin 2.0 (Schneider *et al.* 2000). All haplotypes were closely related and could be linked in an unambiguous most parsimony network (Fig.3).

As we previously investigated 16S mtDNA and found only one haplotype (Barbaresi 2003), CO1 mtDNA was used to find more variability. However, all haplotypes were closely related, differing by very few mutational steps. This low level of sequence divergence is expected showing that the colonization of *P. clarkii* is recent (35 years); a similar situation has already been described in the Chinese mitten crab *Eriocheir sinensis* Milne-Edwards, which colonized Europe less than 100 years ago (Hänfling and Kollmann 2002)

The number of haplotypes and within-population gene diversity was higher in the source population NOR than in the introduced populations. However, we also found some variation in gene diversity within Europe. All the populations shared at least one haplotype except FIR, which is the population harbouring two haplotypes (4 and 6) deriving from the most frequent haplotype 2. This result sharply contrasts with the anecdotal information stating this population

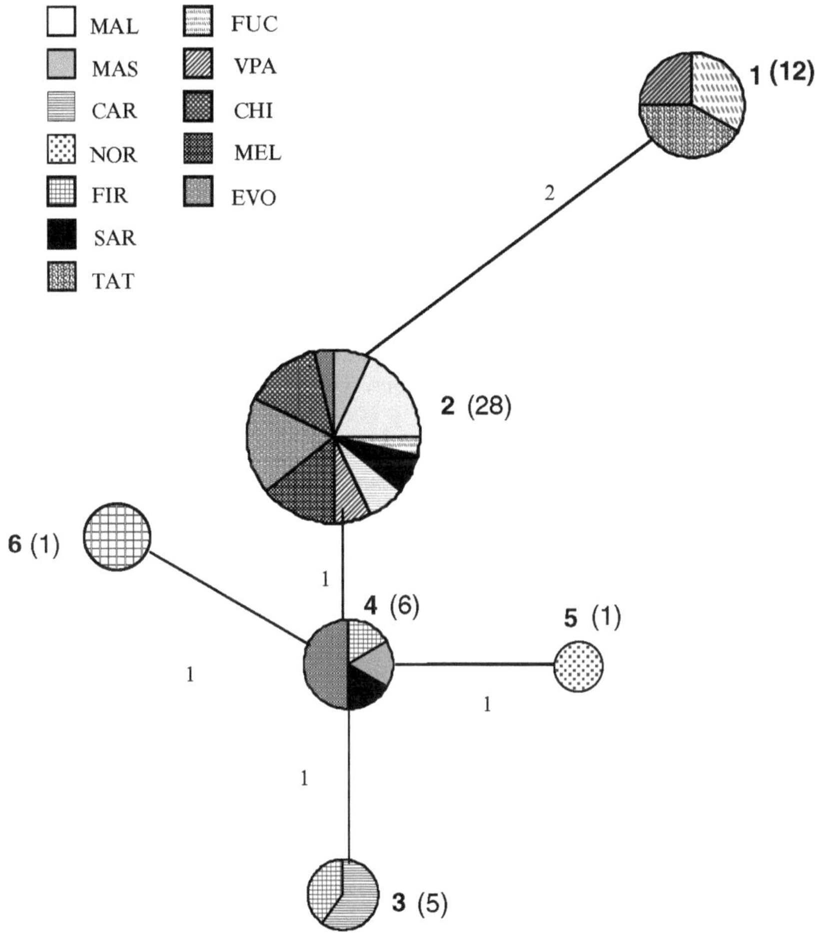

Fig. 3 Haplotype network of the mtDNA sequence data. Numbers along branches denote the number of nucleotide substitutions between haplotypes. The haplotype number (in bold) follows Table 1. The frequency of each haplotype is indicated in parenthesis. Sample names follow Table 1.

is derived from MAS. The detection of unique haplotypes in this population suggests a different origin of source individuals, possibly a source not sampled in this study. These haplotypes could be the result of a different introduction possibly from China following the immigration of a Chinese community to Florence. As the first introduction of *P. clarkii* from Louisiana to Asia was done in 1918 in Japan and in a second step a translocation was made from Japan to China (1948) (Laurent 1997), the introduced individuals in FIR could have evolved differently because of successive translocations and more than 80 years of divergence. This hypothesis should be verified by sampling Japanese and Chinese populations.

Microsatellite analysis

Microsatellite analysis of *P. clarkii* populations was carried out using five polymorphic loci (PclG-07, PclG-15, PclG-26, PclG-28, and PclG-37) described by Belfiore and May 2000. These authors isolated 23 microsatellite loci from the red swamp crayfish but these loci have never been published in any population genetic study. Polymerase chain reactions (PCRs) were carried out in a total volume of 15 μL containing 10X PCR buffer (Finnzyme) with $MgCl_2$ and 200 μM each dNTP, 5 pM each primer, 0.8 units of Dynazyme *Taq DNA* polymerase (Finnzyme) and 50–100 ng DNA. Reactions were performed in a Perkin-Elmer 9600 thermal cycler programmed for an initial melting at 95 °C for 2 min followed by 35 cycles at 95 °C for 30 s, 56 °C for 30 s, and 72 °C for 60 s. A final extension step at 72 °C for 5 min was performed. Amplification products were resolved by electrophoresis on 1.5% agarose gels containing 1 μgmL^{-1} ethidium bromide in TAE buffer. DNA fragments were visualized by image-analysis software for gel documentation (LabWorks Software Version 3.0; UVP, Upland, CA, USA). Sizing of PCR products was carried out using an ABI310 DNA sequencer (PE/ABI, CA, USA) according to the manufacturer's instructions, and the results were analysed using Genescan 3.1 software (PE/ABI, CA, USA). The distribution of allele frequencies, presence of private alleles, single-locus, and average values of observed (*Ho*) and expected heterozygosity (*He*) were evaluated separately for each population using Genepop v. 3.1 (Raymond and Rousset 1995).

Polymorphism of microsatellite loci and genetic diversity

We first tested the variability of the 5 microsatellite loci in the sampled populations. All 5 loci were polymorphic. A total of 99 alleles was observed for the five loci over 12 populations (119 individuals). We detected 16 alleles at locus PclG-07, 18 alleles at locus PclG-15, 26 alleles at locus PclG-26, 11 alleles at locus PclG-28, and 28 alleles at locus PclG-37. The allelic distribution (Fig. 4) varied significantly among loci, with PclG-28 and PclG-07 displaying an approximately normal distribution, and PclG-37, PclG-26, and PclG-15 exhibiting a uniform distribution of alleles. In these latter loci, however, 63%, 58%, and 39% of the alleles, respectively, had a frequency lower than 3%, whereas 18%, 3.8%, and 11%, respectively, had a frequency lower than 1%. Most represented alleles are 241 for locus PclG-26 (18%), 118, 138, 178, and 182 for locus PclG-15 (16%, 11%, 13%, and 12%, respectively), and 107 and 131 for locus PclG-37 (15% and 11%, respectively) (Fig. 4).

The allele number varied from 15 to 45 per population and mean number of alleles within populations (allelic diversity) varied from 3 to 7.8 (Table 3).

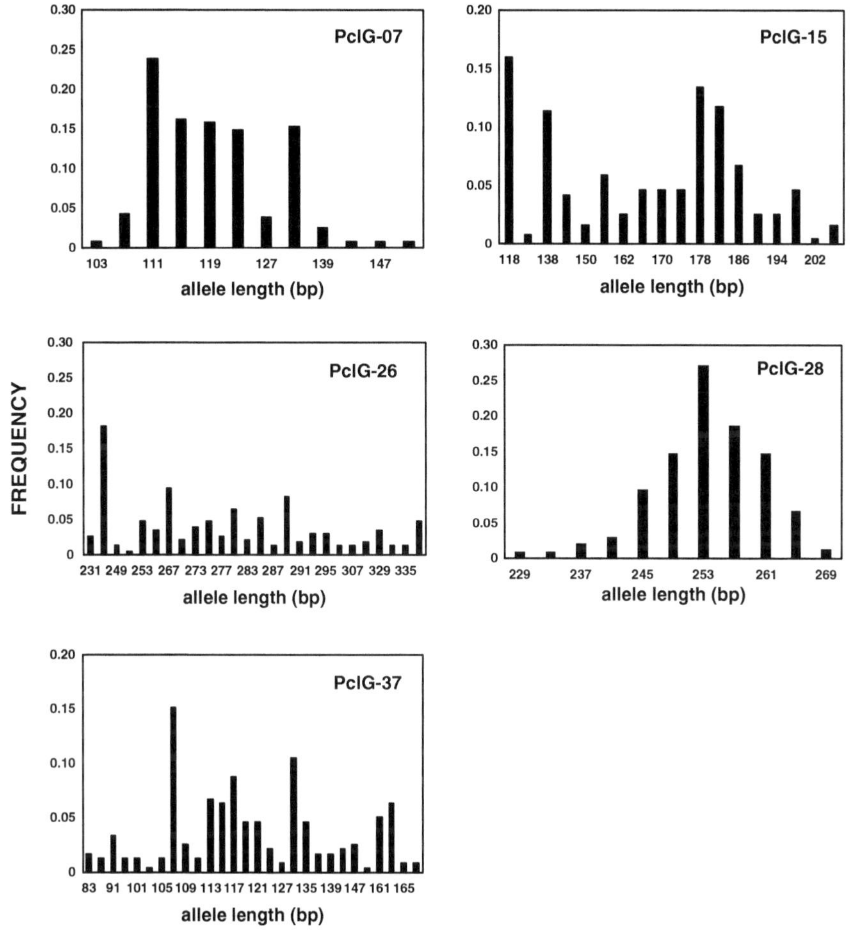

Fig. 4 Distribution of allele frequencies for each microsatellite locus in *Procambarus clarkii*.

Genetic variation within populations has also been expressed as heterozygosity values. Observed heterozygosity (H_o) averaged across loci varied among populations from 0.36 to 1.0 (Table 3). Highly significant deviations from Hardy-Weinberg expectation were found. In particular, four out of 12 sampled populations (SAR, FIR, VPA, and TAT) exhibited a significant heterozygosity excess, while one population (CAR) exhibited a heterozygosity deficiency.

Population differentiation

Only a few alleles were private (6 in locus PclG-37, 4 for PclG-26, 1 for PclG-07, and 3 for PclG-15) and no private allele had a frequency above 5%. With the

Table 3 Microsatellite comparisons of population diversities. For each locus, expected (*He*) and observed (*Ho*) heterozigosity, and number of alleles (*A*) are reported. For each population, mean *He* and *Ho*, deviation from Hardy-Weinberg equilibrium (*P*), mean number of alleles (mean *A*) and number of private alleles detected across all loci are summarized.

		Population											
Locus		SAR	MAS	FIR	FUC	TAT	EVO	MAL	MEL	VPA	CAR	NOR	CHI
PclG-07	H_o	0.800	0.800	0.667	0.900	0.800	1.000	0.889	0.700	0.800	0.000	1.000	0.333
	H_e	0.811	0.810	0.777	0.710	0.736	0.873	0.790	0.763	0.874	0.482	0.852	0.385
	A	8	6	8	4	5	7	5	5	8	3	7	3
PclG-15	H_o	0.700	0.800	0.600	0.600	0.500	0.600	0.700	0.800	0.900	0.800	0.900	0.444
	H_e	0.905	0.889	0.847	0.794	0.778	0.857	0.678	0.742	0.910	0.773	0.921	0.627
	A	8	9	8	9	6	7	7	5	10	4	10	4
PclG-26	H_o	0.900	0.900	0.333	0.600	0.700	0.800	0.556	0.400	0.500	0.400	0.400	0.714
	H_e	0.884	0.847	0.882	0.552	0.773	0.900	0.803	0.689	0.894	0.563	0.363	0.689
	A	10	9	7	4	8	9	5	4	8	3	4	3
PclG-28	H_o	0.900	0.700	1.000	0.700	0.537	0.888	0.800	0.600	0.800	1.000	0.900	0.444
	H_e	0.836	0.726	0.821	0.621	0.700	0.771	0.737	0.773	0.878	0.526	0.831	0.366
	A	6	6	6	6	3	5	4	5	9	2	7	3
PclG-37	H_o	0.500	0.600	0.500	0.900	0.500	0.700	0.600	0.300	0.500	0.100	0.800	0.667
	H_e	0.810	0.864	0.805	0.947	0.757	0.821	0.631	0.742	0.921	0.194	0.931	0.758
	A	6	9	7	12	6	7	6	5	10	3	11	5
Summary													
H_o		0.528	0.528	0.431	0.514	0.444	0.587	0.492	0.431	0.486	1.000	0.556	0.361
H_e		0.590	0.575	0.574	0.503	0.503	0.554	0.506	0.515	0.618	0.319	0.542	0.372
P		<0.02	ns	<0.001	ns	<0.05	ns	ns	0.05	<0.001	<0.0001	ns	ns
Mean A		7.6	7.8	7.2	7	5.6	7	5.4	4.8	9	3	7.8	3.6
Private alleles		0	0	2	2	0	0	3	1	1	2	2	2

exception of SAR, FIR, and MAS, populations showed private alleles ranging from 1 to 3 (Table 3). The inspection of the allelic distribution suggested that the alleles having the highest frequencies generally had wide distributions.

Global χ^2 tests of population differentiation indicated significant heterogeneity in allelic frequencies among the 12 populations ($P < 0.0001$). In an effort to further partition these data, pairwise estimates of genotypic differentiation were calculated. Fifty-nine of 65 pairwise comparisons (90.8%) were significantly different at $P < 0.001$ (χ^2 test) following Bonferroni correction (Rice 1989). Exceptions were the following population pairs: SAR-MAS, SAR-VPA, SAR-EVO, VPA-EVO, VPA-MAS, and NOR-FUC.

Genetic diversity within each population was measured as allelic diversity (mean number of alleles per locus, *A*), observed heterozygosity (*Ho*), and heterozygosity expected from Hardy-Weinberg proportions (*He*). Here we calculated the unbiased estimates of expected heterozygosity proposed by Nei (1978) using the software Genetix V 4.01 (Belkhir *et al.* 1998). Observed heterozygosities within populations were tested for Hardy-Weinberg expectations using Weir and Cockerham's (1984) estimate as implemented in Genepop v. 3.1 (Raymond and Rousset 1995). Since loci exhibited a number of alleles higher than five, we used a 10,000-step 1,000-iteration Markov chain method (10,000 dememorization steps) to calculate estimates of the *P* value.

Genetic evidence for a bottleneck effect in the sampled populations was evaluated using the program BOTTLENECK 1.1 (Cornuet and Luikart 1996), which tests whether *Ho* is significantly different from the expected heterozygosity at mutation drift equilibrium (*Heq*). One tail Wilcoxon sign test (Bottleneck, 10,0000 replications) was used. Significant differences between *Ho* and *Heq* indicate either a recent severe reduction in the effective population size (*Ne*) (if $Ho > Heq$), or a recent expansion in *Ne* or influx of rare alleles from genetically distinct immigrant (if $Ho < Heq$).

Genetic distance between populations was measured by calculating F_{ST} (Weir and Cockerham 1984) between pairs of populations using Arlequin 2.000. The significance of these F_{ST} estimates was tested under the null hypothesis $H_0 =$ 'no difference between populations' by permuting genotypes between populations (10,000 iteractions). The *P* value of the test is the proportion of permutations leading to an F_{ST} value larger or equal to that observed (Schneider *et al.* 2000).

Estimation of fixation index (F_{ST}) of population pairs calculated for all loci (Table 4) showed high levels of genetic differentiation between populations. With the exception of the population pair SAR-MAS, all population pairs were significantly differentiated on the basis of F_{ST}. The F_{ST} values for *P. clarkii* ranged from 0.021 to 0.461. These values indicated low levels of gene flow. However, a low level of genetic differentiation was obtained between MAS and SAR populations.

Both global and pairwise tests of population differentiation were performed using Genepop v. 3.1 (Raymond and Rousset 1995). The significance of the *P* values across the five loci was determined using Fisher's χ^2 method for

Table 4 Fixation index (F_{ST}) for population pairs estimated for all microsatellites loci.* indicates that F_{ST} value is not significant.

	SAR	MAS	FIR	MEL	FUC	TAT	EVO	MAL	NOR	VPA	CAR
SAR	–										
MAS	0.021*	–									
FIR	0.080	0.083	–								
MEL	0.113	0.123	0.148	–							
FUC	0.118	0.144	0.146	0.183	–						
TAT	0.119	0.125	0.101	0.185	0.214	–					
EVO	0.031	0.068	0.084	0.081	0.157	0.121	–				
MAL	0.135	0.123	0.134	0.223	0.195	0.187	0.179	–			
NOR	0.113	0.139	0.122	0.197	0.033	0.199	0.139	0.190	–		
VPA	0.048	0.049	0.076	0.046	0.090	0.137	0.043	0.124	0.110	–	
CAR	0.262	0.281	0.243	0.313	0.343	0.311	0.264	0.354	0.277	0.261	–
CHI	0.231	0.224	0.258	0.289	0.151	0.316	0.290	0.284	0.162	0.188	0.461

Table 5 Microsatellites and analysis of molecular variation (AMOVA; Excoffier *et al.* 1992).* denotes that the variance component is statistically significant ($P < 0.05$).

Source of variation	df	Variation (%)	*F* statistics
Among groups	2	1.72	$F_{CT} = 0.017$
(1) VPA, FUC, TAT, NOR;			
(2) CAR, FIR;			
(3) MEL, MAS, CHI, EVO, SAR, MAL			
Among populations within groups	9	15.81	$F_{SC} = 0.160$*
Among individuals within populations	226	82.47	$F_{ST} = 0.175$*

combining probabilities (Sokal and Rohlf 1981) followed by a sequential Bonferroni correction for multiple tests (Rice 1989).

Differentiation among populations was also assessed by an F_{ST}-based hierarchical AMOVA (Excoffier *et al.* 1992) as implemented in Arlequin 2.000. This analysis (Table 5) allowed us to clarify the partitioning of molecular variance among three levels: variation among groups of populations, among populations within groups, and within populations. Three different parameters were computed (F_{CT}, F_{SC}, and F_{ST}), describing the differentiation at each level.

In order to test correlation between genetic and geographical distances between populations in Italy, a Mantel test (Arlequin 2.000) was performed. One thousand permutations were used to find significance levels. The test was performed for distance estimates based on microsatellite data and no isolation by the distance was found.

CONCLUSIONS

In a previous work, RAPD was used to perform a preliminary screening of the genetic variability in the invasive *P. clarkii* (Barbaresi *et al.* 2003). Results revealed a pronounced level of genetic variation in this species. This finding is in agreement with recent studies conducted in the Australian redclaw crayfish *Cherax quadricarinatus* von Martens (Macaranas *et al.* 1995) and in the native European white-clawed crayfish *Austropotamobius pallipes* Lereboullet, 1858 (Souty-Grosset *et al.* 1999, Gouin *et al.* 2001, 2006).

This study is the first using microsatellite loci characterized by Belfiore and May (2000) in *P. clarkii*. A high genetic variation within sampled populations emerged from the use of microsatellite markers. In addition, microsatellites revealed a high inter-population differentiation with an F_{ST} up to 0.461. For comparison, Gouin *et al.* (2006) found also an F_{ST} of 0.461 in southern French populations of the indigenous white-clawed crayfish *A. pallipes*, indicating an absence of gene flow due to the fragmentation of the populations.

Considering the degree of variability of *P. clarkii* in its native range, contradictory results emerged. In fact, while in the sample from Louisiana a high heterozygosity value was found similar to the major part of the European populations, heterozygosity was low in the Mexican population sampled in the field and naturally dispersed (P. Gutiérrez-Yurrita 2004, personal communication). The level of diversity observed in the Louisianan population could be explained as the result of the intense commercial exploitation of the species in the southern USA (Louisiana accounts for most of the national aquaculture production). In this area, exchanges of crayfish from natural habitats to culture ponds can be common. In addition, stock translocation is reported to be a common practice (Busak 1988).

Heterozygosity found in the European populations was high and could be explained by the fact that the initial genetic structure of a successful invasive population depends on several factors, including the effective population size of the introduction event(s), the genetic diversity of the source population(s), and the number of founding sources. Since this is the first study using microsatellites in this species, no comparison can be made with the results from other authors; in particular, the degree of variability in native populations is unknown. A phylogeographic study from *P. clarkii* from its native range still needs to be undertaken.

The high genetic diversity revealed by microsatellite markers in some introduced populations of *P. clarkii* could be the result of different types of introduction events, i.e. (1) multiple introduction events with individuals from different sources; (2) a single introduction of a large number of individuals from a genetically diverse source population; and (3) a combination of these events. As concerns the population from Doñana, historical data (Table 4) report that this population originated from a stock coming from Louisiana introduced for aquaculture purpose (Gherardi and Holdich 1999). The high genetic variability of the Spanish population could be thus explained by a single introduction event from a very heterogeneous pool coming from that area.

The degree of heterozygosity found in this study would not have been detected in the case of a severe bottleneck. The lack of heterozygote deficiencies is predicted for biological invasions and commonly encountered as, for example, in mussel populations (Holland 2001). An exception is represented by the population from Piedmont, where a bottleneck effect was revealed (according to the BOTTLENECK's test). For this population, the origin from a small number of introduced individuals can be hypothesized.

As shown by pairwise F_{ST}, no differentiation was found between samples from Sarzana (Liguria) and Massaciuccoli (Tuscany). In accordance with the data obtained with mtDNA, this result seems to confirm the hypothesis of a single origin of the population sampled from Liguria, possibly by active dispersal and/or by human translocation of crayfish coming from the nearby Massaciuccoli Lake. From a management point of view, this result points out how dispersal capability could be favoured, at least at a microgeographical scale.

The usefulness of microsatellites as evidence of bottlenecks and gene flow has been outlined by Colautti *et al.* (2005) for the Eurasian spiny waterflea, whose spread was found to depend on long-distance jump dispersal (Suarez *et al.* 2001). In *Dreissena rostriformis bugensis* Andrusov, the use of six microsatellites gave evidence of considerable gene flow (multiple invasions) among populations: its genetic diversity was consistent with the existence of a large metapopulation that has not experienced bottlenecks or founder effects (Therriault *et al.* 2005). *Procambarus clarkii*'s spread may involve both long-distance jump dispersal and natural dispersal at a microgeographical scale. The high level of genetic diversity in introduced population of this species and its corresponding success of establishment support the hypothesis that high genetic variability is an important characteristic of successful invasive populations (e.g. Ehrlich 1986, Holland 2000). However, this result cannot be generalized. For example, Tsutsui *et al.* (2000) used microsatellite markers in the invasive ant *Linepithema humile* and showed that the loss in genetic diversity of the introduced populations of this species was associated with the reduced intraspecific aggression among spatially separate colonies with the formation of interspecifically dominant supercolonies. In that case, genetic bottlenecks have led to widespread ecological success. Our results on *P. clarkii* confirm the model suggested by Barbaresi *et al.* (2003), in which the colonization of Europe by this species derives from subsequent introductions of individuals coming from different source populations. This model is consistent with both the high genetic diversity observed (introduction of different sets of individuals) and the genetic differentiation of populations resulting from the casual bias of introductions.

As shown from a recent work on *Carcinus maenas* (Linnaeus) (Roman 2006), genetic studies can provide significant insights into the colonization pattern of invasive aquatic species even when small samples have been analysed.

However, further work is needed to verify our preliminary results on *P. clarkii* and to identify source populations and pathways of introduction.

ACKNOWLEDGEMENTS

For providing samples we are grateful to: José Juan Chans (Estacion Biologica de Doñana, Spain), Maria Ilhéu (Universidade de Évora, Portugal), Jay Huner (University of Southern Louisiana, USA), David Mazzoni (Università di Bologna, Italy), Roberto Pieraccini (SpeechCycle, New York), Robert Romaire (Louisiana State University, USA), Sebastiano Salvidio (Università di Genova, Italy), and Thomas Stucki (Switzerland). Special thanks are due to Julian Reynolds for improving the English style.

REFERENCES

Allendorf, F. W. and L. L. Lundquist. 2003. Introduction: population biology, evolution, and control of invasive species. Conservation Biology **17**, 24–30.

Alvarez-Buylla, E. R. and A. A. Garay. 1994. Population genetic structure of *Cecropia obtusifolia*, a tropical pionneer tree species. Evolution **48**, 437–453.

Barbaresi, S. 2003. Biological invasions in freshwater: *Procambarus clarkii* as a case study. Ph.D. dissertation, University of Florence, Italy.

Barbaresi, S. and F. Gherardi. 2000. The invasion of the alien crayfish *Procambarus clarkii* in Europe, with particular reference to Italy. Biological Invasions **2**, 259–264.

Barbaresi, S., R. Fani, F. Gherardi, A. Mengoni, and C. Souty-Grosset. 2003. Genetic variability in European populations of an invasive American crayfish: preliminary results. Biological Invasions **5**, 269–274.

Belfiore, N. M. and B. May. 2000. Variable microsatellite loci in red swamp crayfish, *Procambarus clarkii*, and their characterization in other crayfish taxa. Molecular Ecology **9**, 2155–2234.

Belkhir, K, P. Borsa, J. Goudet, L. Chikhi, and F. Bonhomme. 1998. GENETIX, logiciel sous WindowsTM pour la génétique des populations. Laboratoire Génome et Populations, CNRS UPR 9060, Université de Montpellier II, Montpellier, France. http://www.univ-ontp2.fr/~genetix/genetix/genetix.htm

Bij de Vaate, A., K. Jaźdźewski, H. A. M. Ketelaars, S. Gollasch, and G. van der Velde. 2002. Geographical patterns in range extension of Ponto–Caspian macroinvertebrate species in Europe. Canadian Journal of Fisheries and Aquatic Science **59**, 1159–1174.

Busack, C. A. 1988. Electrophoretic variation in the red swamp (*Procambarus clarkii*) and white river crayfish (*P. acutus*) (Decapoda: Cambaridae). Aquaculture **69**, 211–226.

Cabot, E. L. and A. T. Beckenbach. 1989. Simultaneous editing of multiple nucleic acid and protein sequences with ESEE. Comparative Applied Bioscience **5**, 233–234.

Colautti, R. I., M. Mance, M. Viljanen, H. A. M. Ketelaars, H. Bürgi, H. J. MacIsaac, and D. Heath. 2005. Invasion genetics of the Eurasian spini waterflea: evidence for bottleneck and gene flow using microsatellites. Molecular Ecology **14**, 1869–1879.

Cornuet, J. M. and G. Luikart. 1996. Description and power analysis of two tests for detecting recent population bottlenecks from allele frequency data. Genetics **144**, 2001–2014.

Cox, G. W. 2004. Alien species and evolution. Island Press, Washington, Covelo, London.

Cristescu, M. E. A., P. D. N. Hebert, J. D. S. Witt, H. J. MacIsaac, and I. A. Grigorovich. 2001. An invasion history for *Cercopagis pengoi* based on mitochondrial gene sequences. Limnology and Oceanography **46**, 224–229.

Delmastro, G. B. 1992. Sull'acclimatazione del gambero della Lousiana *Procambarus clarkii* (Girard, 1852) nelle acque dolci italiane. Pianura **4**, 5–10.

Duggan, I. C., C. A. M. Rixon, and H. J. MacIsaac. 2006. Popularity and propagule pressure: determinants of introduction and establishment of aquarium fish. Biological Invasions **8**, 377–382.

Ehrlich, P. R. 1986. Which animals will invade? Pages 79–95 in J. A. Drake and H. A. Mooney, editors. Ecology of invasions of North America and Hawaii. Springer, New York, NY.

Excoffier, L., P. E. Smouse, and M. Quattro. 1992. Analysis of molecular variance inferred from metric distances among DNA haplotypes: application to human mitochondrial DNA restriction data. Genetics **131**, 479–491.

Fetzner, J. W. 1996. Biochemical systematics and evolution of the crayfish genus *Orconectes* (Decapoda: Cambaridae). Journal of Crustacean Biology **16**, 111–141.

Fetzner, J. W. and K. A. Crandall. 2001. Genetic variation. Pages 291–326 in D. M. Holdich, editor. Biology of freshwater crayfish. Blackwell Science, Oxford, UK.

Gherardi, F. 2006. Crayfish invading Europe: the case study of *Procambarus clarkii*. Marine and Freshwater Behaviour and Physiology **39**, 175–191.

Gherardi, F. and D. M. Holdich. 1999. Crayfish in Europe as alien species: How to make the best of a bad situation? A. A. Balkema, Rotterdam, The Netherlands.

Gouin, N., F. Grandjean, D. Bouchon, J. D. Reynolds, and C. Souty-Grosset. 2001. Population genetic structure of the endangered freshwater crayfish *Austropotamobius pallipes*, assessed using RAPD markers. Heredity **87**, 80–87.

Gouin, N., F. Grandjean, and C. Souty-Grosset. 2006. Population genetic structure of the endangered crayfish *Austropotamobius pallipes* in France based on microsatellite variation: biogeographical inferences and conservation implications. Freshwater Biology **51**, 1369–1387.

Habsburgo-Lorena, A. S. 1986. The status of the *Procambarus clarkii* population in Spain. Freshwater Crayfish **6**, 131–136.

Hänfling, B. and J. Kollmann. 2002. An evolutionary perspective of biological invasions. Trends in Ecology and Evolution **17**, 545–546.

Hebert, P. D. N. and M. E. A. Cristescu. 2002. Genetic perspectives on invasions: the case of the Cladocera. Canadian Journal of Fisheries and Aquatic Sciences **59**, 1229–1234.

Holland, B. S. 2000. Genetics of marine bioinvasions. Hydrobiologia **420**, 63–71.

Holland, B. S. 2001. Invasion without a bottleneck: microsatellite variation in natural and invasive populations of the brown mussel *Perna perna* (L). Marine Biotechnology **3**, 407–415.

Huner, J. V. 1977. Introductions of the Louisiana red swamp crayfish, *Procambarus clarkii* (Girard): an update. Freshwater Crayfish **3**, 193–202.

Huner, J. V. and J. W. Avault Jr. 1979. Introductions of *Procambarus* spp. Freshwater Crayfish **4**, 191–194.

Keane, R. M. and M. J. Crawley. 2002. Exotic plant invasions and the enemy release hypothesis. Trends in Ecology and Evolution **17**, 164–170.

Kimura, M. 1980. A simple method for estimating evolutionary rates of base substitutions through comparative studies of nucleotide sequences. Journal of Molecular Evolution **16**, 111–120.

Kirkpatrick, M. and N. H. Barton. 1997. Evolution of a species' range. American Naturalist **150**, 1–23.

Kreiser, B. R., J. B. Mitton, and J. D. Woodling. 2000. Single versus multiple sources of introduced populations identified with molecular markers: a case study of a freshwater fish. Biological Invasions **2**, 295–304.

Laurent, P. 1997. Introductions d'écrevisses en France et dans le monde, historique et conséquences. Bulletin Français de la Pêche et de la Pisciculture **344–345**, 345–356.

Lee, C. E. 2002. Evolutionary genetics of invasive species. Trends in Ecology and Evolution **17**, 386–391.

Lockwood, J. L., P. Cassey, and T. Blackburn. 2005. The role of propagule pressure in explaining species invasions. Trends in Ecology and Evolution **20**, 223–228.

Lodge, D. M., C. A. Taylor, D. M. Holdich, and J. Skurdal. 2000. Nonindigenous crayfishes threaten North American freshwater biodiversity: lessons from Europe. Fisheries **25**, 7–20.

Macaranas, J. M., P. B. Mather, P. Hoeben, and M. F. Capra. 1995. Allozyme and RAPD-DNA variation in the redclaw crayfish. Australian Journal of Marine and Freshwater Research **46**, 1217–1228.

MacIsaac, H. J., L. A. Grigorovich, and A. Ricciardi. 2001. Reassment of species invasion concepts: the Great Lakes basin as a model. Biological Invasions **3**, 405–416.

MacIsaac, H. J., T. C. Robbins, and M. A. Lewis. 2002. Modeling ships' ballast water as invasion threats to the Great Lakes. Canadian Journal of Fisheries and Aquatic Sciences **59**, 1245–1256.

Nei, M. 1978. Estimation of average heterozigosity and genetic distance from a small number of individuals. Genetics **89**, 583–590.

Patti, F. P. and M. C. Gambi. 2001. Phylogeography of the invasive polychaete *Sabella spallanzanii* (Sabellidae) based on the nucleotide sequence of internal transcribed spacer 2 (ITS2) of nuclear rDNA. Marine Ecology Progress Series **215**, 169–177.

Raymond, M. and F. Rousset. 1995. GENEPOP (Version 1.2): Population genetics software for exact tests and ecumenicism. Journal of Heredity **86**, 248–249.

Ricciardi, A. and H. J. MacIsaac. 2000. Recent mass invasion of the North American Great Lakes by Ponto–Caspian species. Trends in Ecology and Evolution **15**, 62–65.

Rice, W. R. 1989. Analysing tables of statistical tests. Evolution **43**, 223–225.

Roman, J. 2006. Diluting the founder effect: cryptic invasions expand a marine invader's range. Proceedings of the Royal Society of London B **273**, 2453–2459.

Sakai, A. K., F. W. Allendorf, J. S. Holt, D. M. Lodge, J. Molofsky, K. A. With, S. Baughman, R. J. Cabin, J. C. Cohen, N. C. Ellstrand, D. E. McCauley, P. O'Neil, I. M. Parker, J. N. Thompson, and S. G. Weller. 2001. The population biology of invasive species. Annual Review of Ecology and Systematics **32**, 305–332.

Schneider, S., D. Roessli, and L. Excoffier. 2000. ARLEQUIN: a software for population genetics data analysis. Version 2.000. University of Geneva, Switzerland.

Simon, C., F. Frati, A. Beckenbach, B. Crespi, H. Liu, and P. Flook. 1994. Evolution, weighting and phylogenetic utility of mitochondrial gene sequences and a compilation of conserved polymerase chain reaction primers. Annual Entomological Society of America **87**, 65–701.

Sokal, R. R. and F. J. Rohlf. 1981. Biometry. 2nd edition. W. H. Freeman, New York, NY.

Souty-Grosset, C., F. Grandjean, and N. Gouin. 1999. Molecular genetic contributions to conservation biology of the European native crayfish *Austropotamobius pallipes*. Freshwater Crayfish **12**, 371–386.

Souty-Grosset, C., D. M. Holdich, P. Y. Noël, J. D. Reynolds, and P. Haffner, editors. 2006. Atlas of Crayfish in Europe. Muséum National d'Histoire Naturelle, Collection Patrimoines Naturels, Paris, France.

Suarez, A. V., D. A. Holway, and T. J. Case. 2001. Patterns of spread in biological invasions dominated by long-distance jump dispersal: Insights from Argentine ants. Proceedings of the National Academy of Sciences USA **98**, 1095–1100.

Therriault, T. W., M. I. Orlova, M. F. Docker, J. J. MacIsaac, and D. D. Heath. 2005. Invasion genetics of a freshwater mussel (*Dreissena rostriformis bugensis*) in Eastern Europe: high gene flow and multiple introductions. Heredity **95**, 16–23.

Tsutsui, N. D., A. V. Suarez, D. A. Holway, and T. J. Case. 2000. Reduced genetic variation and the success of invasive species. Proceedings of the National Academy of Sciences USA **97**, 5948–5953.

Weir, B. S. and C. C. Cockerham. 1984. Estimating F-statistics for the analysis of population structure. Evolution **38**, 1358–1370.

Wilson, A. B., K. A. Naish, and E. G. Boulding. 1999. Multiple dispersal strategies of the invasive quagga mussel (*Dreissena bugensis*) as revealed by microsatellite analysis. Canadian Journal of Fisheries and Aquatic Sciences **56**, 2248–2261.

Chapter twenty-one

Do estuaries act as saline bridges to allow invasion of new freshwater systems by non-indigenous fish species?

J. Anne Brown, Dawn M. Scott, and Rod W. Wilson

INTRODUCTION

In Europe, the non-indigenous species (NIS) most frequently introduced are freshwater fish (García-Berthou *et al.* 2005). Once introduced, dispersal is a key element in determining their invasiveness (Rehage and Sih 2004); in order to limit or mitigate impacts, an understanding of likely establishment and rates and routes for dispersion is important (Pihlaja *et al.* 1998, Ricciardi and Rasmussen 1998). Currently, however, we have limited information on the potential for range expansion for most of the non-indigenous freshwater fish species introduced into Europe. Dispersal is likely to be at least partially dependent on the tolerance of these species to variable environmental conditions, but tolerances are often poorly understood, particularly in the new environment. The possibility of using shared estuaries as 'saline bridges' to migrate between contiguous freshwater systems has only very rarely been considered (Brown *et al.* 2001, Bringolf *et al.* 2005), and it has usually been assumed that estuaries will act as barriers (rather than bridges) to the dispersal of freshwater fish. However, the salinity tolerances and physiological responses to salinity are poorly understood in many NIS. This lack of knowledge may lead to their inaccurate classification as strictly freshwater, whereas some species may be

Francesca Gherardi, Biological invaders in inland waters: Profiles, distribution, and threats, 401–414.

at least partially euryhaline and capable of living in a wider range of salinities than previously considered.

Body fluid osmolality in freshwater bony fish is around 300 mOsm kg^{-1}, much higher than that of the surrounding freshwater where osmolalities are around 2–15 mOsm kg^{-1}. This large osmotic gradient leads to a high influx of water across permeable surfaces that must be excreted by the kidneys in order to maintain water balance. Alongside the water influx (mostly across the gills), salts are lost down the concentration gradient, but active uptake of salts by the gills and gut ensures regulation of body fluid salt concentrations (Evans *et al.* 2005). When the external salt concentrations are raised, as in brackish waters, the osmotic gradient between body fluids and the environment is reduced (unless plasma levels of ions are also raised sufficiently). An iso-osmotic point is reached when internal and external osmolalities are equal. At salinities beyond the iso-osmotic point, euryhaline fish continue to maintain relatively stable blood salt levels (and overall osmolality) so that the osmotic gradient is reversed, with higher external concentrations resulting in a state known as hypo-osmoregulation. This is an established feature of euryhaline teleosts living in mesohaline (5–18 psu) and polyhaline environments (18–30 psu) or in ocean seawater (30–35 psu), and leads to water loss by osmosis. Water loss must be balanced by drinking the environmental medium and intestinal water absorption. Hypo-osmoregulation does not, however, occur in strictly freshwater species [such as the common carp *Cyprinus carpio* Linnaeus or channel catfish *Ictalurus punctatus* (Rafinesque)] which show an unabated and ultimately fatal rise in internal body fluid osmolality as external osmolality rises and therefore fail to survive in waters much beyond 400–450 mOsm kg^{-1} (Abo Hegab and Hanke 1982, De Boeck *et al.* 2000, Eckert *et al.* 2001). Hence, physiological studies can give a clear picture of the saline tolerances and acclimation abilities of fish that will influence dispersal via saline waters and have therefore formed an important component of our studies of NIS.

The central hypothesis behind the experimental studies discussed below is that some non-indigenous fish species found in freshwater ecosystems that have previously been considered to be limited to freshwater, may actually have sufficient salinity tolerance to allow them to disperse through estuaries. Using these as "saline bridges" they may reach contiguous freshwater ecosystems, and may disperse between different estuaries through mesohaline/polyhaline coastal waters. In this chapter, we focus upon three invasive fish species that are commonly classified as freshwater fish: pikeperch *Sander lucioperca* (Linnaeus), sunbleak *Leucaspius delineatus* (Heckel), and topmouth gudgeon *Pseudorasbora parva* (Temminck and Schlegel). The results of laboratory investigations on these three species are considered in relation to the likelihood of their dispersal by migration through saline routes. All three of these species are of current concern in the UK and are listed species in legislation dealing with importation and movement of live fish in England and Wales. However, the principles raised apply more generally in appreciating the impact of saline tolerances on the dispersal of non-indigenous fish species.

PIKEPERCH

The pikeperch (also known as zander), *S. lucioperca*, is a piscivorous predator that is now a broadly distributed NIS in the inland waters of mainland Europe, extending from the Ponto–Caspian region, where it originated, to Spain and from Scandinavia to the Mediterranean countries. Pikeperch were first deliberately brought into the UK by the Duke of Bedford who introduced several specimens into two lakes on his Woburn Estates in 1878; subsequent additions to these lakes occurred in the early 1900s (see reviews: Hickley 1986, Hickley and Chare 2004, Copp *et al.* 2005a). Translocations from the lakes into the Great Ouse River system began in 1960 and pikeperch have since dispersed through much of this system. Several specimens were later introduced into Stanborough Lake, adjacent to the River Lee in Hertfordshire. Population growth and dispersal in UK river systems have been slow, but appear to have been helped by anglers. Pikeperch are now found in the Rivers Severn, Lee, and Thames and there have been recent unconfirmed reports of pikeperch in the Welsh Dee (Copp *et al.* 2005a).

In the UK, pikeperch are thought to out-compete the indigenous piscivorous pike *Esox lucius* Linnaeus, and have been blamed for the decline in native populations of cyprinids. However, the range of fish species in the diet of pikeperch appears to depend on their size and the particular locality of the population as well as the available prey populations (Smith *et al.* 1996, Keskinen and Marjomäki 2004, Hickley and Chare 2004). The impacts appear to be of greater significance when conditions are particularly favourable for pikeperch, and indigenous fish are already under pressure. For example, introduction of the pikeperch into Lake Egridir in Turkey had a severely detrimental effect, with rapid total extinction of several indigenous species (Crivelli 1995). Active hunting and predation on salmonid smolts by pikeperch have also been reported (Jepsen *et al.* 2000).

The dispersion of pikeperch through UK river systems has been fairly slow but has now led to the establishment of populations at the tidal limits of the Thames and Severn estuaries and in the tidal reaches of the Ouse (Hickley 1986, Copp *et al.* 2005a). This highlights the possibility of dispersal between river catchments via brackish or salt-water corridors in the Thames or Severn estuaries, or using the more saline reaches of the middle regions of the Wash. Such migrations across saline bridges may already have occurred in other European countries, particularly as pikeperch are known to flourish in lower salinity regions of the Baltic Sea, the Kiel canal, and many European estuaries (Lehtonen *et al.* 1996, Maes *et al.* 1999, Kafemann *et al.* 2000, Ložys 2004). However, these distributions generally suggest that pikeperch populations only occur in low salinities where external osmolality is much less than that of body fluids (as in freshwater). The optimal salinity has been reported to be around 6 psu (Craig 2000), which is in agreement with field observations of pikeperch occupying the Lithuanian coastal region of the Baltic Sea (at salinities of 4.9–6.8 psu) that had a higher condition factor and growth rate than pikeperch in the adjacent freshwater Curonian Lagoon (Ložys 2004).

Laboratory investigations of the osmoregulatory abilities of the pikeperch anecdotally reported fish deaths after transfer from freshwater to salinities of 8 or 12.8 psu (Crăciun *et al.* 1982). However, our studies have revealed a far greater tolerance of saline waters than previously suspected (Brown *et al.* 2001). These studies found that pikeperch could tolerate rapid transfer from freshwater to a salinity of 16 psu, surviving for at least 6 days. An increase in plasma glucose after 24 h in 16 psu water was the result of hormonal stress responses, but was followed within 6 days by restoration of plasma glucose concentrations that were typical of unstressed control fish (Brown *et al.* 2001), indicating that acclimation occurs in the longer term. Pikeperch were also shown to tolerate exposure to a simulated tidal cycles rising to a salinity of 33 psu. Although plasma cortisol and plasma glucose concentrations increased as salinity peaked, showing physiological stress in these fish, there was a rapid recovery after return to freshwater (Brown *et al.* 2001).

The exposure of pikeperch to saline water has been shown to increase blood plasma osmolality (Brown *et al.* 2001). Plasma osmolality was unaffected in pikeperch held in water of 8 psu for 24 h and only slightly increased after 6 days but in pikeperch held in water of 16 psu, a significant elevation in plasma osmolality occurred within 24 h, and further increased after 6 days (Fig. 1). However, these fish in 16 psu were still able to hold plasma osmolality well below that of the external environment, showing that in the short term (up to 6 days at least), pikeperch can hypo-osmoregulate. This is not a typical feature of freshwater fish and studies to further characterize the hypo-osmoregulatory ability of pikeperch are therefore needed. One aspect of this is to examine whether pikeperch can drink the saline environmental medium in order to achieve water balance.

Addition of phenol red to the external environment has allowed us to make the first assessments of whether drinking by pikeperch is a significant component of their volume regulatory mechanisms during hypo-osmoregulation. Collection and analysis of gut fluids showed that drinking rates were minimal in pikeperch held in freshwater (Fig. 2), but increased significantly in saline waters so that the distinct red coloration of imbibed fluid was visible throughout the gut. Drinking increased from 25.9 ± 8.5 μL h^{-1} kg^{-1} in eight pikeperch held in freshwater to 122.9 ± 28.0 μL h^{-1} kg^{-1} in seven pikeperch held in approximately iso-osmotic water (10.5 psu) for 7 days. Similarly, hypo-osmoregulating pikeperch exposed to 12 psu and 16 psu for 5 days showed higher drinking rates than pikeperch held in freshwater (Fig. 2). For the drinking response to play a significant role in body fluid regulation, however,

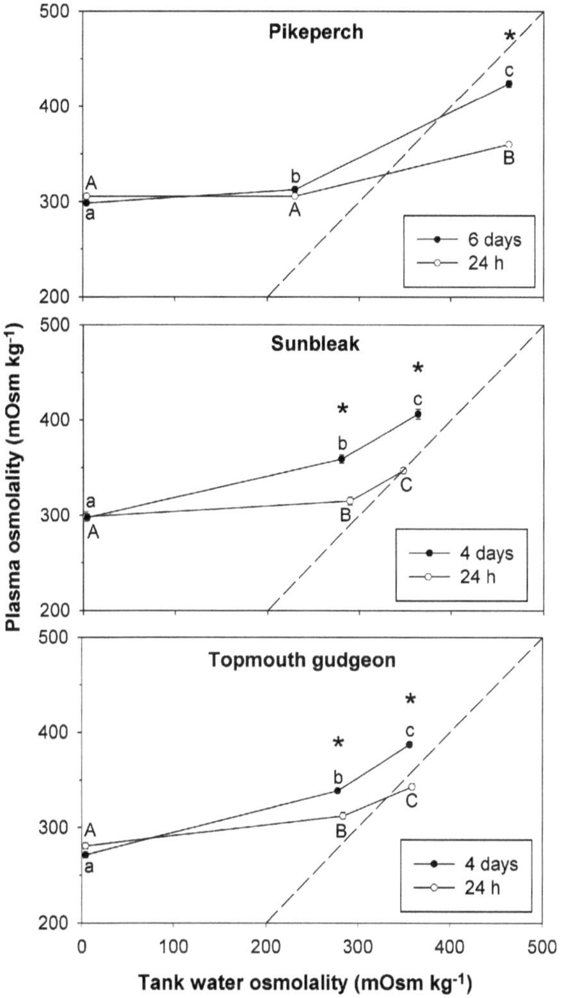

Fig. 1 Plasma osmolality of pikeperch, sunbleak, and topmouth gudgeon held in freshwater and 24 h and 4 days after transfer to higher salinities.
Mean fish masses: pikeperch 266 g; sunbleak 1.73 g; topmouth gudgeon 1.81 g. Salinities employed: pikeperch - 8 psu (230 mOsm kg^{-1}) and 16 psu (462 mOsm kg^{-1}); sunbleak and topmouth gudgeon - 10.8 psu (281 mOsm kg^{-1}) and 13.7 psu (350 mOsm kg^{-1}).
The iso-osmotic line, where plasma osmolality is identical to that of the environment, is shown by a dashed line. Values to the left of this line show plasma that is hyperosmotic to the external medium; values to the right of this line show plasma that is hypo-osmotic. Different letters indicate statistical differences between groups indicated by use of two-way ANOVA and *post-hoc* multiple comparisons (Student-Newman-Keuls Method or Tukey's test). Pikeperch data presented in Brown *et al.* (2001). Data for sunbleak and topmouth gudgeon included in Scott *et al.* (2007). Asterisks (*) indicate significant differences ($P < 0.05$) between 24 h and 4 or 6 day values at each salinity.

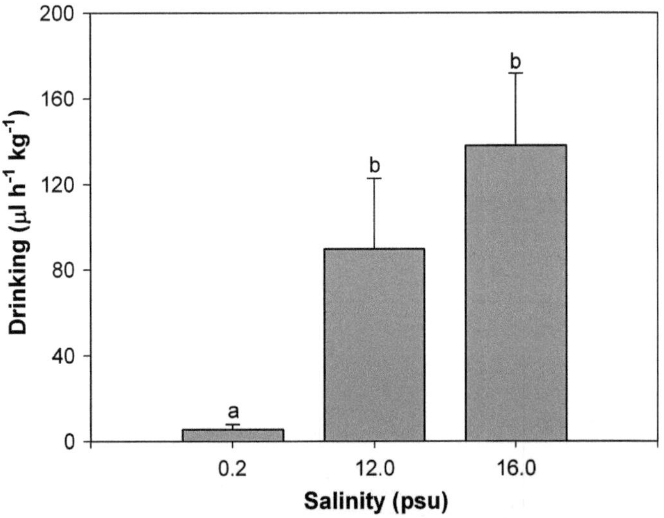

Fig. 2 Drinking rates in pikeperch (mean mass 318 ± 34 g; $n = 8$ per group) measured by addition of phenol red (0.0275 g L^{-1}) to the external environment 16 h before the end of the 5 day salinity exposures. Different letters indicate statistical differences between groups after ANOVA ($P = 0.009$) and linear contrasts: 0.2 psu compared to 12 psu ($P = 0.041$); 0.2 psu compared to 16 psu ($P = 0.003$).

absorption of water by the gut is required. Net water absorption from the gut would result in a higher phenol red concentration in gut fluids than in the environmental medium i.e. a concentration ratio of more than 1.0.

For pikeperch held in freshwater, the phenol red concentration ratio of gut fluids/environmental medium was 0.67 ± 0.24 ($n = 12$), indicating net fluid secretion rather than water absorption. This is not surprising as in freshwater fish, water is already available, in excess. However, in pikeperch held at 12 psu some fish showed intestinal water absorption (concentration ratios above 1.0) after 5 days. At 16 psu a higher level of water absorption was apparent with a mean phenol red concentration ratio of 1.66 ± 0.32 ($n = 8$). Acclimation processes in the gut might be predicted to take several days and this idea is supported by a mean intestinal fluid phenol red concentration ratio of 2.60 ± 0.93 ($n = 7$) with values of up to 6.1 reached in pikeperch kept for 7 days at 10.5 psu.

Experimental studies have shown that pikeperch can tolerate periodic exposure to mixohaline estuarine conditions and mesohaline coastal waters, and are capable of some physiological acclimation that could facilitate their potential migration to new river systems. However, these studies raise the question: how do pikeperch respond behaviourally when faced with changing environmental salinities? The behaviour of pikeperch, monitored using video cameras, has shown that freshwater fish are relatively inactive during daylight hours, but became more active in the dark at night (D. M. Scott, R. W. Wilson and J. A. Brown 2004, unpublished observation). This agrees with recent telemetry studies of pikeperch that showed diel activity patterns with maximal activity at dusk, although there is high individual variability (Poulet *et al.* 2005). The importance of salinity in determining the pikeperch behaviour has been shown in laboratory studies during the normally quiescent daylight hours. Pikeperch were exposed to a 12 h tidal cycle of salinity, rising to a peak of 30 psu in 6 h, then returning to freshwater in the next 6 h. Initially, as salinity rose there was little change in physical activity, but at around 12.5 to 15 psu an increasing proportion of time was spent moving up and down the water column (Fig. 3). Once salinity was reduced, physical activity rapidly returned to the low level typical of pikeperch in daylight hours in freshwater. In the wild, increased vertical mobility would bring pikeperch into surface waters. In estuaries with halocline stratification this could allow selection of the less dense, lower salinity water, and movement towards the mouth of an estuary throughout almost the entire tidal cycle, which may promote migration into new river systems. Potential corroboration of this hypothesis is provided by the repeated reports of pikeperch captures in the tidal Severn and Thames (Hickley 1986, Kirk *et al.* 2002, Copp *et al.* 2005a), and in tidal sections of the lower River Great

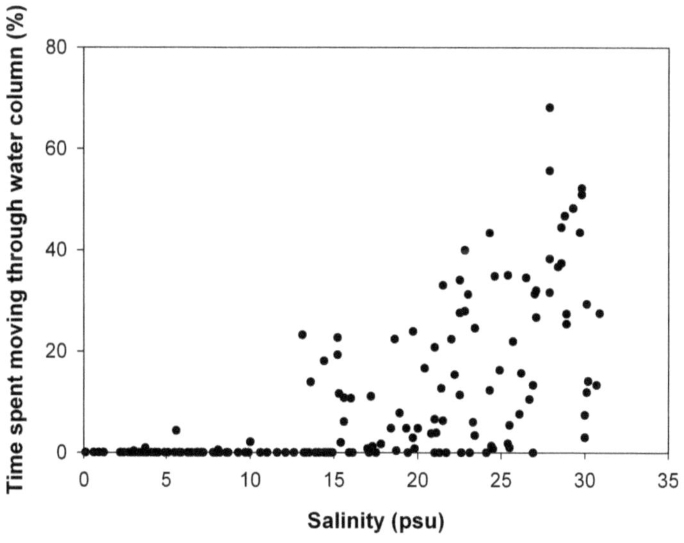

Fig. 3 Time spent moving through the water column in pikeperch (mean mass 325 g; $n = 8$) exposed to a simulated tidal cycle of salinity that reached a peak of 30 psu in 6 h, and returned to freshwater 6 h later. Each point represents an average measurement for an individual fish within a 20-min period. Experiments started at 08.00 am and were conducted entirely in daylight hours.

Ouse (A. C. Pinder 2003, CEH-Dorset, personal communication). Clandestine illegal introductions could account for these occurrences, but in the light of our physiological and behavioural studies, dispersal through saltwater corridors is also a feasible explanation. To assess the behaviour of pikeperch in the wild, acoustic telemetry work (in collaboration with G. H. Copp and S. Stakėnas, Cefas-Lowestoft) is now exploring pikeperch movements in freshwater and tidal reaches of the Rivers Thames and Lee.

SUNBLEAK

The native range of sunbleak *L. delineatus* includes areas of central and eastern Europe, from the Rhine basin in the west to the Volga basin in the east (Maitland 2004), but, while once common, it is now often rare or vulnerable in its native range. This species is amongst the more recent non-indigenous fish to be introduced into the UK and was first introduced in 1986 by an ornamental fish supplier (Farr-Cox *et al.* 1996). Sunbleak rapidly established breeding populations in several locations in southern England, including still waters in Hampshire, waterways in Somerset, and a complex of fishing lakes in Dorset (Gozlan *et al.* 2003a), and more recently has been reported in north-western England (Hickley and Chare 2004).

A detrimental impact of sunbleak on the recruitment of indigenous cyprinids is suggested, possibly because sunbleak share common resources with indigenous cyprinids such as bream, roach, bleak, and rudd (Gozlan *et al.* 2003a). Sunbleak could also aid the spread of non-indigenous copepod parasites to indigenous fish populations (Beyer *et al.* 2005).

A number of unusual characteristics compared to indigenous cyprinids make sunbleak a successful invader. They become sexually mature at one year old and are batch spawners that exhibit male nest guarding (Gozlan *et al.* 2003a, b) hence ensuring a better survival rate for larvae and/or juveniles and rapid establishment of populations. They will also spawn on any flat surface, which could include floating macrophytes and the underside of boats or fishing paraphernalia, increasing their chances of dispersal (Gozlan *et al.* 2003a). Although sunbleak appear to prefer slow-flowing rivers and still waters such as drains and canals, fast-flowing rivers have already been identified as dispersal routes (Pinder and Gozlan 2003), so it is important to consider the salinity tolerance of sunbleak and the influence that salinity may have in limiting or allowing dispersion. This is particularly relevant as sunbleak (like pikeperch) originated in the Ponto–Caspian region that has a geological history of fluctuating water levels and salinities. Consequently, biota with a Ponto–Caspian origin often show some degree of saline tolerance (Dumont 1998, Ricciardi and MacIsaac 2000).

Exposure of sunbleak captured from Stoneham Lakes, Hampshire, UK to 5.4 psu daily increases in salinity have indicated tolerance of 10.8 psu, but at

16.2 psu, reduced feeding was observed and fish rapidly became moribund (D. M. Scott, R. W. Wilson, and J. A. Brown 2004, unpublished observation). In subsequent experiments, sunbleak were exposed to freshwater (0.2 psu), 10.8 psu (iso-osmotic to the body fluids of freshwater-acclimated fish) and 13.7 psu which would be slightly hyper-osmotic to the body fluids of the sunbleak at the beginning of the experiment. Sunbleak tolerated these conditions for at least 4 days with no deaths. Plasma osmolality was significantly increased in sunbleak exposed to elevated salinity regimes, both after 24 h and four days, and with an increase over time (Fig. 1). Mean plasma osmolality matched the external osmolality after 24 h and over the next 3 days rose so as to exceed the external medium. The pattern of changing plasma osmolality in sunbleak is similar to that reported for stenohaline freshwater fish that do not survive in saline water of above ~400–450 mOsm kg^{-1} (~14–16 psu). In these stenohaline species, plasma osmolality is typically ~10% above that of the environment within 24 h of initial exposure and continues to rise thereafter, allowing continued osmotic water influx that aids excretory processes (Abo Hegab and Hanke 1982, Wilkes and McMahon 1986, Van der Linden *et al.* 1999, Eckert *et al.* 2001). However, sunbleak exposed to 13.7 psu for 24 h had lower total water content than that of freshwater control fish (Scott *et al.* 2007). This is presumably because the osmotic gradient for water influx was reduced, coupled with a lack of appropriate compensatory responses, such as reduced renal output or increased drinking of the environmental medium. In sunbleak, the mean plasma osmolality reached 406 mOsm kg^{-1} in fish held for 4 days in water of 13.7 psu (Fig. 1). This is similar to a strictly freshwater fish such as carp, and suggests that tolerance of sunbleak to salinities much higher than this is unlikely. This hypothesis agrees with the reduced feeding observed after exposure to a salinity of 16.3 psu (D. M. Scott, R. W. Wilson, and J. A. Brown 2004, unpublished observation).

Based on these studies we conclude that despite the Ponto–Caspian origin of sunbleak, their limited saline tolerance coupled with adverse effect on feeding will mean that brackish or salt-water corridors are likely to act as barriers to dispersal.

TOPMOUTH GUDGEON

Topmouth gudgeon, *P. parva*, is an Asiatic cyprinid species that was first brought into Europe in 1960, by accidental introduction into Romanian ponds with other fish larvae, and has since spread rapidly across continental Europe (Copp *et al.* 2005b). This may to some extent reflect a high tolerance of environmental variation and phenotypic plasticity (Rosecchi *et al.* 2001), features that are frequently associated with translocation success and rapid invasion (Ricciardi and Rasmussen 1998), as populations of topmouth gudgeon thrive in streams, rivers, and still waters (Pinder and Gozlan 2003).

The high level of dietary overlap of topmouth gudgeon and indigenous species such as roach, and a voracious appetite for fish eggs of other fish species (Xie *et al.* 2000, 2001), indicates a high potential for interspecific interactions (Declerk *et al.* 2002, Pinder and Gozlan 2003, Hickley and Chare 2004). This may explain the rapid expansion of populations of introduced topmouth gudgeon that typically dominate and replace other cyprinids. A part of this invasion may reflect the effects of a recently discovered infectious disease transmitted by topmouth gudgeon (Gozlan *et al.* 2005, Pinder *et al.* 2005) that may infect indigenous fish species.

The first known introduction to the UK was in the mid-1980s at a fish farm in southern England (Gozlan *et al.* 2002). Large geographical distances between current UK populations strongly suggest that humans have played a major role in dispersal of topmouth gudgeon (Pinder and Gozlan 2003, Pinder *et al.* 2005) and that a major dispersal process is inadvertent translocation between fisheries or fish farms, along with other species (Pinder *et al.* 2005). Currently, most topmouth gudgeon populations in Britain are in enclosed lakes and still water bodies distributed throughout England and Wales, but several of these sites are connected to river systems. Topmouth gudgeon have already been identified in the River Itchen and the River Test, in the latter location having escaped from a connected still water site (Pinder *et al.* 2005). Their potential dispersal via saline routes has not yet been considered but will be highly dependent on their tolerance of saline waters.

Topmouth gudgeon, captured at Crampmoor Fish Farm in Hampshire and exposed to freshwater (0.2 psu), an iso-osmotic salinity (10.8 psu), and 13.7 psu, showed 100% survival of all of the experimental conditions for up to 4 days, but feeding on bloodworms was depressed in both 10.8 psu and 13.7 psu. When exposed to iso-osmotic conditions (10.8 psu), topmouth gudgeon showed a similar increase in plasma osmolality as seen in sunbleak (Fig. 1). Exposure to 13.7 psu had a slightly less marked effect than in sunbleak, and after 24 h topmouth gudgeon had a plasma osmolality slightly below that of the environment (hypo-osmotic), with a mean plasma osmolality of 344 mOsm kg^{-1}. However, this was not maintained after 4 days of exposure to 13.7 psu (Fig. 1) and may simply reflect a slight delay in achieving the elevation in plasma osmolality seen in strictly freshwater fish when they are exposed to saline waters (Abo Hegab and Hanke 1982, Van der Linden *et al.* 1999).

Our results suggest that neither topmouth gudgeon nor sunbleak are likely to make significant use of water with salinities greater than 13.7 psu for dispersal in the wild. Certainly, their tolerance of salinities is far less than that of pikeperch. However, even this saline tolerance may permit some dispersal via coastal waters and given that in topmouth gudgeon the plasma osmolality remains below that of the environment for at least 24 h, short migrations could occur in mesohaline coastal regions of the Baltic including the coasts of Lithuania, Latvia, Finland, and Estonia.

SUMMARY AND WIDER CONCLUSIONS

Our physiological and behavioural studies focused on one type of environmental stressor, salinity, that has generally been assumed to limit dispersal of freshwater invasive fishes. Other water quality parameters will also inevitably be involved and ultimately the combined influences of these conditions on dispersal need to be incorporated into models that predict the rate and extent of expansion of invasive species. Of the three invasive species studied (pikeperch, topmouth gudgeon, and sunbleak), pikeperch showed greatest saline tolerance and evidence of osmoregulatory mechanisms, such as stimulation of drinking, to achieve hypo-osmoregulation when fish meet sufficiently saline waters (in excess of approximately 12 psu). Populations of pikeperch have so far only expanded gradually in UK fresh waters, but are now widespread in continental Europe, and saline tolerance is likely to have aided this dispersal. In the UK, pikeperch have now reached the tidal limits of several river systems. Here, we predict that their physiological responses to salinity could facilitate invasion of contiguous freshwater systems. The behavioural responses of pikeperch to increased salinity with increased swimming and vertical movement within the water column at >12.5 psu is likely to take them into the uppermost haloclines with lower salinity and would aid movement downstream.

In contrast to the rather slow dispersal of pikeperch, both sunbleak and topmouth gudgeon have shown rapid expansion once introduced, but our results suggest that both species are likely to be restricted by saline waters, with estuaries presenting a barrier to dispersion. Topmouth gudgeon showed a slightly greater ability than sunbleak to cope with saline water as a slight hypo-osmotic state was initially apparent after exposure to 13.7 psu. However, although both sunbleak and topmouth gudgeon survived this salinity for several days, the inhibition of feeding that we have observed strongly suggests that in the wild, their performance will be severely compromised once salinity reaches ~11 psu.

Although our studies have focused on three NIS of current importance in the UK, the insights gained are relevant in considering dispersal of non-indigenous fish species throughout the world. Indeed, investigation of the salinity tolerance of an American invasive species, the flathead catfish *Pylodictis olivaris* (Rafinesque), has recently identified possible dispersion through estuaries (Bringolf *et al.* 2005). This kind of dispersion has been observed in the non-indigenous gibel (or Prussian) carp, *Carassius gibelio* (Bloch), that was first introduced into ponds and small lakes of Estonia in the late 1940s. The gibel carp was first detected in Estonian brackish waters (Gulf of Riga) in 1985 and has since spread along the entire Estonian Baltic coastline (Vetemaa *et al.* 2005). This provides the potential for gibel carp to disperse to many new freshwater systems. Further studies of NIS are required on a case-by-case basis so as to provide information that can be used in the risk assessments increasingly employed to inform policymakers and regulators.

ACKNOWLEDGEMENTS

This work was partially supported by the Natural Environment Research Council, UK (NE/B502436). We are grateful to Dr. Rudy Gozlan, Adrian Pinder and Kathleen Beyer (Centre for Ecology and Hydrology, Winfrith Technology Centre, Dorset) and Dr. Gordon Copp (Centre for Environment Fisheries and Aquaculture Science, Lowestoft, Suffolk) for helpful discussions regarding the dispersal of sunbleak and topmouth gudgeon in the UK, and to Adrian Pinder for help in capture of sunbleak. We are also grateful to Andy Thomas (Environment Agency) and Mark Stollery (Crampmoor Fish Farm) in enabling us to obtain topmouth gudgeon for these studies. Sue Frankling has provided valuable technical assistance and we have welcomed the participation of several students (Wilaf Moore, Madeleine Brown, Daniel Depledge, Sue Fleming, Greg Langham) in our studies of the physiology and behaviour of non-indigenous fish.

REFERENCES

Abo Hegab, S. and W. Hanke. 1982. Electrolyte changes and volume regulatory processes in the carp (*Cyprinus carpio*) during osmotic stress. Comparative Biochemistry and Physiology **71A**, 157–164.

Beyer, K., D. Kochanowska, M. Longshaw, S. W. Feist, and R. E. Gozlan. 2005. A potential role for invasive sunbleak in the further dissemination of a non-native parasite. Journal of Fish Biology **67**, 1730–1733.

Bringolf, R. J., T. J. Kwak, W. G. Cope, and M. S. Larimore. 2005. Salinity tolerance of flathead catfish: Implications for dispersal of introduced populations. Transactions of the American Fisheries Society **134**, 927–936.

Brown, J. A., W. M. Moore, and E. S. Quabius. 2001. Physiological effects of saline waters on zander. Journal of Fish Biology **59**, 1544–1555.

Copp, G. H., K. J. Wesley, V. Kováč, M. J. Ives, and M. G. Carter 2005a. Introduction and establishment of the pikeperch *Stizostedion lucioperca* (L.) in Stanborough Lake (Hertfordshire) and its dispersal in the Thames. The London Naturalist **82**, 139–154.

Copp, G. H., P. G. Bianco, N. G. Bogutskaya, T. Erős, I. Falka, M. T. Ferreira, M. G. Fox, J. Freyhof, R. E. Gozlan, J. Grabowska, V. Kováč, R. Moreno-Amich, A. M. Naseka, M. Peňáz, M. Povž, M. Przybylski, M. Robillard, I. C. Russell, S. Stakėnas, S. Šumer, A. Vila-Gispert, and C. Wiesner. 2005b. To be, or not to be, a non-native freshwater fish? Journal of Applied Ichthyology **21**, 242–262.

Crăciun, V., M. Crăcium, I. Neacşu, and I. Trandafirescu. 1982. Activity de la pompe de Na^+- K^+ chez le sandre au cours de son acclimatization aux conditions de salinite de la mer noire. Cercetări Marine **15**, 227–233.

Craig, J. F. 2000. Percid Fishes. Systematics, Ecology and Exploitation. Fish and Aquatic Resources Series 3. Blackwell Science, Oxford, UK, 329 pp.

Crivelli, A. J. 1995. Are fish introductions a threat to endemic freshwater fishes in the northern Mediterranean region? Biological Conservation **72**, 311–319.

Declerk, S., G. Louette, T. De Bie, and L. De Meester. 2002. Patterns of diet overlap between populations of non-indigenous and native fishes in shallow ponds. Journal of Fish Biology **61**, 1182–1197.

De Boeck, G., A. Vlaeminck, A. Van der Linden, and R. Blust. 2000. The energy metabolism of common carp (*Cyprinus carpio*) when exposed to salt stress: an increase in energy expenditure or effects of starvation? Physiological and Biochemical Zoology **73**, 102–111.

Dumont, H. J. 1998. The Caspian Lake: History, biota, structure, and function. Limnology and Oceanography **43**, 44–52.

Eckert, S. M., T. Yada, B. S. Shepherd, M. H. Stetson, T. Hirano, and E. G. Grau. 2001. Hormonal control of osmoregulation in the channel catfish *Ictalurus punctatus*. General and Comparative Endocrinology **122**, 270–286.

Evans, D. H., R. M. Piermarini, and K. P. Choe. 2005. The multifunctional fish gill: dominant site of gas exchange, osmoregulation, acid-base regulation, and excretion of nitrogenous waste. Physiological Reviews **85**, 97–177.

Farr-Cox, F., S. Leonard, and A. Wheeler. 1996. The status of the recently introduced fish *Leucaspius delineatus* (Cyprinidae) in Great Britain. Fisheries Management and Ecology **3**, 193–199.

García-Berthou, E., C. Alcaraz, Q. Pou-Rovira, L. Zamora, G. Coenders, and C. Feo. 2005. Introduction pathways and establishment rates of invasive aquatic species in Europe. Canadian Journal of Fisheries and Aquatic Sciences **62**, 453–463.

Gozlan, R. E., A. C. Pinder, and J. Shelley. 2002. Occurrence of the Asiatic cyprinid *Pseudorasbora parva* in England. Journal of Fish Biology **61**, 298–300.

Gozlan, R. E., A. C. Pinder, S. Durand, and J. Bass. 2003a. Could the small size of sunbleak, *Leucaspius delineatus* (Pisces, Cyprinidae) be an ecological advantage in invading British waterbodies? Folia Zoologica **52**, 99–108.

Gozlan, R. E., C. J. Flower, and A. C. Pinder. 2003b. Reproductive success in male sunbleak, a recent invasive fish species in the UK. Journal of Fish Biology **63**, 131–143.

Gozlan, R. E., S. St-Hilaire, S. W. Feist, P. Martin, and M. L. Kent. 2005. Disease threat to European fish. Nature (London) **435**, 1046.

Hickley, P. 1986. Invasion by zander and the management of fish stocks. Philosophical Transactions of the Royal Society London **314**, 571–582.

Hickley, P. and S. Chare. 2004. Fisheries for non-native species in England and Wales: angling or the environment? Fisheries Management and Ecology **11**, 203–212.

Jepsen, N., S. Pedersen, and E. Thorstad. 2000. Behavioural interactions between prey (trout smolts) and predators (pike and pikeperch) in an impounded river. Regulatory Rivers: Research and Management **16**, 189–198.

Kafemann, R., S. Adlerstein, and R. Neukamm. 2000. Variations in otolith strontium and calcium ratios as an indicator of life history strategies of freshwater fish species within a brackish water system. Fisheries Research **46**, 313–325.

Keskinen, T. and T. J. Marjomäki. 2004. Diet and prey size spectrum of pikeperch in lakes in central Finland. Journal of Fish Biology **65**, 1147–1153.

Kirk, R. S., S. Colclough, and S. Sheridan. 2002. Fish diversity in the River Thames. London Naturalist **81**, 75–85.

Lehtonen, H., S. Hansson, and H. Winkler. 1996. Biology and exploitation of pikeperch, *Stizostedion lucioperca* (L.) in the Baltic Sea area. Annales Zoologici Fennici **33**, 525–535.

Ložys, L. 2004. The growth of pikeperch (*Sander lucioperca* L.) and perch (*Perca fluviatilis* L.) under different water temperature and salinity conditions in the Curonian Lagoon and Lithuanian coastal water of the Baltic Sea. Hydrobiologia **514**, 105–113.

Maes, J, J. Pas, A. Taillieu, P. A, Van Damme, and F. Ollevier. 1999. Diel changes in the vertical distribution of juvenile fish in the Zeeschelde Estuary. Journal of Fish Biology **54**, 1329–1333.

Maitland, P. S. 2004. Notes on the distribution and ecology. Pages 145–146 in D. M. Sutcliffe, editor. Keys to the freshwater fish of Britain and Ireland. Freshwater Biological Association Scientific Publication No 62, Ambleside, UK.

Pihlaja, O., E. Niemelae, and J. Erkinaro. 1998. Introduction and dispersion of the bullhead, *Cottus gobio* L., in a sub-Artic salmon river in northern Finland. Fisheries Management and Ecology **5**, 139–146.

Pinder, A. C. and R. E. Gozlan. 2003. Sunbleak and topmouth gudgeon: two new additions to Britain's freshwater fishes. British Wildlife **15**, 77–83.

Pinder, A. C., R. E. Gozlan, and J. R. Britton. 2005. Dispersal of the invasive topmouth gudgeon, *Pseudorasbora parva* in the U.K.: a vector for an emergent infectious disease. Fisheries Management and Ecology **12**, 411–414.

Poulet, N., C. Arzel, S. Messad S. Lek, and C. Argillier. 2005. Diel activity of adult pikeperch *Sander luciperca* (L.) in a drainage canal in the Mediterranean basin during spring. Hydrobiologia **543**, 79–90.

Rehage, J. S. and A. Sih. 2004. Dispersal behaviour, boldness, and the link to invasiveness: a comparison of four Gambusia species. Biological Invasions **6**, 379–391.

Ricciardi, A and J. B. Rasmussen. 1998. Predicting the identity and impact of future biological invaders: a priority for aquatic resource management. Canadian Journal of Fisheries and Aquatic Sciences **55**, 1759–1765.

Ricciardi, A and H. J. MacIsaac. 2000. Recent mass invasion of the North American Great Lakes by Ponto–Caspian species. Trends in Ecology and Evolution **15**, 62–65.

Rosecchi, E, F. Thomas, and A. J. Crivelli. 2001. Can life-history traits predict the fate of introduced species? A case study on two cyprinid fish in southern France. Freshwater Biology **46**, 845–853.

Scott, D. M., R. W. Wilson, and J. A. Brown. 2007. The osmoregulatory ability of the invasive species sunbleak (*Leucaspius delineatus*) and topmouth gudgeon (*Pseudorasbora parva*) at elevated salinities, and their likely dispersal via brackish waters. Journal of Fish Biology **70**, 1604–1614.

Smith, P. A., R. T. Leah, and J. W. Eaton. 1996. Removal of pikeperch (*Stizostedion lucioperca*) from a British canal as a management technique to reduce impact on prey populations. Annales Zoologici Fennici **33**, 537–545.

Van der Linden, A., M. Vanaudenhove, M. Verhoye, G. De Boeck, and R. Blust. 1999. Osmoregulation of the common carp (*Cyprinus carpio*) when exposed to an osmotic challenge assessed in-vivo and non-invasively by diffusion- and T-2-weighted magnetic resonance imaging. Comparative Biochemistry and Physiology A **124**, 343–352.

Vetemaa, M., R. Eschbaum, A. Albert, and T. Saat. 2005. Distribution, sex ratio and growth of *Carassius gibelio* (Bloch) in coastal and inland waters of Estonia (northeastern Baltic Sea). Journal of Applied Ichthyology **21**, 287–291.

Wilkes, P. R. H. and B. R. McMahon. 1986. Responses of a stenohaline fresh-water teleost (*Catostomus commersoni*) to hypersaline exposure. 2. Trans-epithelial flux of sodium, chloride and acidic equivalents. Journal of Experimental Biology **121**, 95–113.

Xie, S., Y. Cui, and T. Z. Zhang. 2000. Seasonal patterns in feeding ecology of three small fishes in the Biandantang Lake, China. Journal of Fish Biology **57**, 867–880.

Xie, S., Y. Cui, and Z. Li. 2001. Dietary-morphological relationships of fishes in Liangzi Lake, China. Journal of Fish Biology **58**, 1714–1729.

Chapter twenty-two

Which factors determine non-indigenous fish dispersal? A study of the red piranha in tropical Brazilian lakes

Anderson O. Latini and Miguel Petrere Jr.

THE FISH INVASION

Fish introductions are common worldwide and are responsible for economic and ecological damage (Welcomme 1988). Non-indigenous fish have been the leading cause of the extinction of fish species in North America (Miller *et al.* 1989) and worldwide (Mooney and Cleland 2001). In Brazil, fish introductions are common; they may cause the local extinction of indigenous fish species and contribute to biodiversity decline and biotic homogenization (Latini and Petrere 2004, Agostinho *et al.* 2005). Nowadays, 134 freshwater fish are endangered in Brazil due to pollution, siltation, impoundments and flood control, fisheries, and introductions of non-indigenous species (NIS) (Agostinho *et al.* 2005).

In the River Doce basin (State of Minas Gerais), there are about 140 natural lakes where some non-indigenous fish have been introduced for sport-fishing purposes. One of these species is the red piranha, *Pygocentrus nattereri* (Kner), a piscivorous characid fish that also feeds on shrimps and other aquatic invertebrates (Sazima and Machado 1990, Uetanabaro *et al.* 1993). The red piranha is

Francesca Gherardi, Biological invaders in inland waters: Profiles, distribution, and threats, 415–422.

naturally distributed in South America in the Amazon River basin, Paraguai–Paraná River basin, north-eastern Brazilian costal rivers, and the Essequibo River basin (Froese and Pauly 2006), and nowadays has a wide distribution (as NIS) on this lake system (Latini *et al.* 2004). It is possible that temporary canals connecting lakes and streams during the rainy season facilitated the spread of this species across lakes.

We used a "range assessment" of the red piranhas in the lakes of this area (Latini *et al.* 2004) and analysed whether abiotic factors are affecting their distribution in this system, testing the following hypotheses:

(i) *The red piranha has a lower probability of colonization on lakes with larger areas.* Larger lakes must have larger species richness, so an inverse relationship between species richness and invasion success is expected (Elton 1958, Levine and D'Antonio 1999, Sakai *et al.* 2001).
(ii) *Lakes with a higher density of aquatic macrophyte mats show decreasing probability of red piranha colonization.* Macrophyte mats provide refuges to a prey, which is less suitable to a predator (Crowder and Cooper 1982, Power *et al.* 1992), so it would be more difficult for the red piranha propagules to persist in lakes with more refuges.
(iii) *There are some limits on the range of physical or abiotic factors in areas where the red piranha propagules may have a successful colonization.* The ecological niche theory (Hutchinson 1957) points to a specific combination of conditions and resources that describe a specific *n*-dimensional limit where species may occur. Thus, the red piranha must have limitations to its ability to colonize some lakes due to these environmental variables.
(iv) *Longer temporary canals between lakes and streams decrease the probability of red piranha colonization of some lakes.* Due to physiological limitations or stochastic factors, increasing canal length would increase the difficulty experienced by the non-indigenous fish accessing new habitats.

METHODS OF STUDY

This study was conducted between coordinates 42°45′ W and 42°25′ W, 19°50′ S and 19°30′ S in the Atlantic Forest biome, one of the most important biodiversity hotspots in Brazil (Myers *et al.* 2000). These lakes are distributed over an area of about 58,000 ha, 36,000 ha of which are in the protected area of Parque Estadual do Rio Doce (PERD), and the remaining 22,000 ha are in *Eucalyptus* plantations of the Companhia Agrícola Florestal (CAF, Arcelor Group). Fishing activities are allowed only for scientific purposes in the PERD area, but in the CAF area fishing is permitted in some lakes for sport. The 57 lakes included in this study are distributed among three stream basins: Belém, Turvo, and Mombaça, which are related to paleo-courses of the River Doce (Fig. 1).

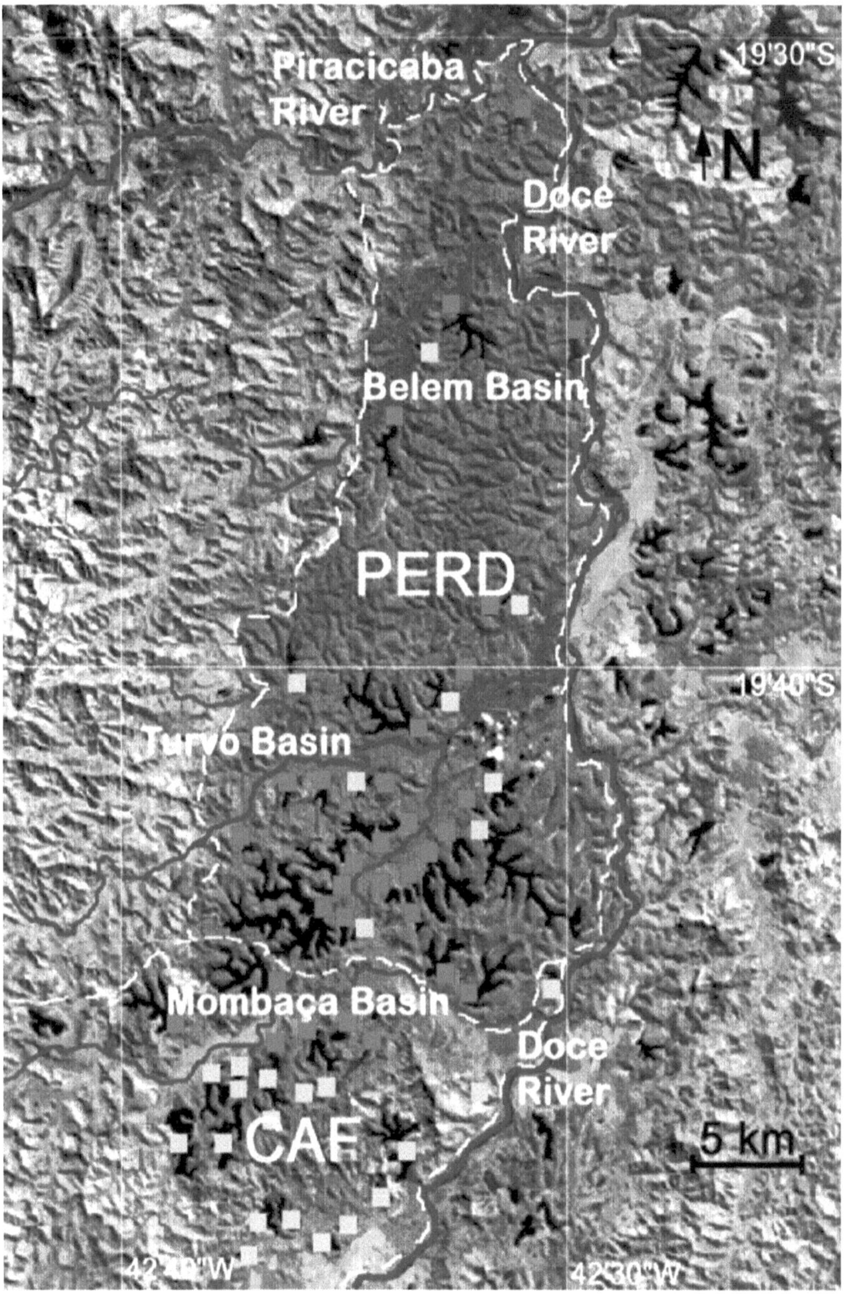

Fig. 1 Study area. The picture shows the lakes (black spots) and the streams (grey lines) in an area of about 35 × 46 km. All 57 study lakes are marked. Gray squares denote where the red piranha is present and white circle where the red piranha is absent. Native Atlantic Forest (PERD) and eucalyptus (CAF) area are illustrated in different shades of gray.

Between February and May 2002, Latini *et al.* (2004) used a rapid assessment programme to obtain the red piranha distribution in this lake system. They used visual methods, line fishing, and gill netting on all these 57 lakes. In that study the red piranha was considered present if it was detected with some of these techniques. This "range assessment" was used in this study to test relations between the red piranha and abiotic factors.

The area of the lakes, the total area with aquatic macrophyte mats, and the temporary canal lengths (between a specific lake and the nearest stream) were measured based on a Landsat 7 (ETM+) satellite image. The total lake perimeter covered by aquatic macrophyte mats was directly estimated on 16 lakes by visual inspection of the shoreline using a boat. Additional measured factors were oxygen content (mg L^{-1}), pH, electrical conductivity (μmho cm^{-1}), and turbidity levels (NTU – nephelometric turbidity units) obtained with specific equipments.

To verify the association between the red piranha distribution (dependent variable) and the sampled variables (independent variables), we fitted the Logistic Regression model and tested this fit using 5% probability of a type I error. This model gives an easy statistical interpretation and comprehensible biological conclusions (Hosmer and Lemeshow 1989).

ARE THERE RELATIONSHIPS BETWEEN NON-INDIGENOUS FISH AND ABIOTIC FACTORS?

The occurrence of the red piranha was not associated with the area of the lakes ($n=57$; $X^2=0.36$; $P=0.55$; $\bar{x}=34.4$, $s=40.6$ to non-invaded lakes and $\bar{x}=77.4$, $s=100.3$ to invaded lakes), or with the density of aquatic macrophytes, i.e. a refuge for indigenous species ($n=16$; $X^2=0.03$; $P=0.87$; $\bar{x}=1.1$, $s=0.63$ to non-invaded lakes and $\bar{x}=0.8$, $s=1.3$ to invaded lakes). Relationships between the species occurrence and oxygen content ($n=17$; $X^2=0.02$; $P=0.88$; $\bar{x}=17.8$, $s=2.26$ to non-invaded lakes and $\bar{x}=17.7$, $s=2.2$ to invaded lakes), pH ($n=15$; $X^2=3.71$; $P=0.06$; $\bar{x}=8.32$, $s=0.99$ to non-invaded lakes and $\bar{x}=7.32$, $s=0.88$ to invaded lakes), electrical conductivity ($n=19$; $X^2=0.30$; $P=0.58$; $\bar{x}=3.25$, $s=1.96$ to non-invaded lakes and $\bar{x}=4.18$, $s=1.45$ to invaded lakes), and water turbidity ($n=18$; $X^2=1.45$; $P=0.22$; $\bar{x}=7.94$, $s=6.89$ to non-invaded lakes and $\bar{x}=8.46$, $s=4.8$ to invaded lakes) were also not significant. So, the occurrence of the red piranha in these lakes seemed not to be affected by the above cited environmental features of the lakes. In contrast, the occurrence of the red piranha was strongly associated with the length of temporary canals ($n=57$; $X^2=17.91$; $P<0.001$; $\bar{x}=3419.3$, $s=3162.1$ to non-invaded lakes and $\bar{x}=1241.1$, $s=1125.4$ to invaded lakes) and the probability of occurrence decreased with an increase in the length of the temporary canals that linked the lakes to the streams ($\beta=-1.879$; Fig. 2).

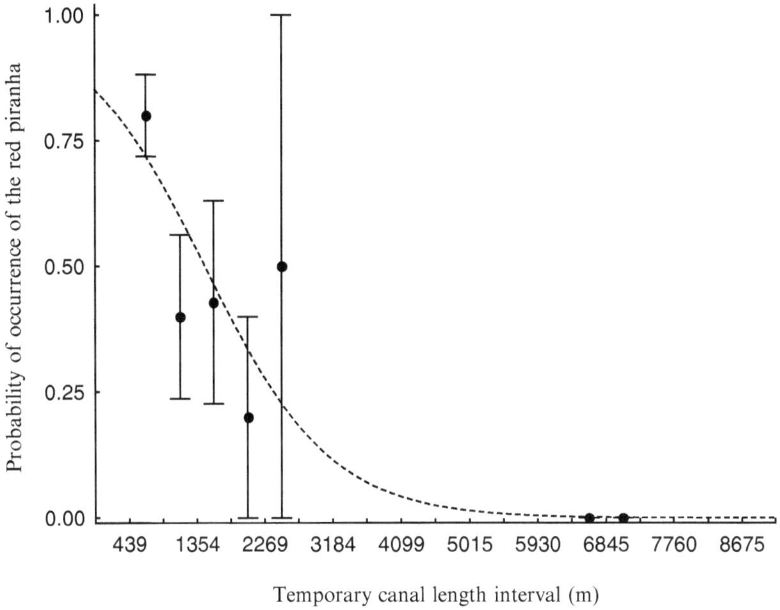

Fig. 2 Probability of occurrence of the red piranha related to the length of temporary canals linking the lakes and the nearest stream. Each point illustrates the probability of occurrence of the red piranha in a lakes' group with a specific temporary canal length interval.

The majority of the invaded lakes are in the PERD area, where the canals are about three times smaller than in CAF area (mean length of 0.92 km in PERD *vs.* 2.82 km in CAF; Student's $t = 3.56$; $df = 24.49$; $P < 0.01$). However, it is possible to identify a specific canal length interval (between 2.27 and 2.73 km) with a higher probability of the occurrence incidence of the red piranha, contrary to the prediction of the model (Fig. 2).

IMPLICATIONS FOR INDIGENOUS FISH CONSERVATION

The species richness of the indigenous community in the lakes is low (Latini and Petrere 2004) which might cause a low resistance to the invasion of NIS (Mack *et al.* 2000). Invasive species success in these lakes, as well as in other systems, may be increased by the life history characteristics of the invader (Lodge 1993, Kolar and Lodge 2001). In fact, the red piranha exhibits territorial behaviour during the reproductive season with strong parental care (Uetanabaro *et al.* 1993) and an efficient predatory habit (Sazima and Machado 1990). These properties might certainly favour its colonization of these lakes (Latini and Petrere 2004).

Several studies showed that environmental variables can facilitate the invasion of NIS (Moyle and Light 1996a, b, Moyle *et al.* 2003, Marchetti *et al.* 2004). In the lakes we studied (either invaded or not by the red piranha), the abiotic factors did not limit the red piranha distribution against expectations from the theory and as reported in several studies on different species (e.g. Jackson *et al.* 2001, Kouamélan *et al.* 2003, Sweka and Hartman 2003). For this reason, it is reasonable to assume that the red piranha success in invading novel habitats is related to its biological potential.

In the river Doce lake system, natural dispersal across the temporary canals is the most important factor determining the red piranha distribution. Invaded lakes are more frequent inside the environmentally protected area than outside due to shorter temporary canals linking the lakes to streams in the protected area. The success of this species decreases with the increasing length of the temporary canals linking lakes to streams. This pattern was also found in other aquatic invasive species, such as *Daphnia lumholtzi* (Sars) in North America (Havel *et al.* 2002) and *Dreissena polymorpha* (Pallas) in both North America and Europe (Kraft *et al.* 2002). Lakes with lower spatial isolation are more susceptible to invasions and consequently deserve special attention and early conservation action.

Lakes with long canals are naturally protected from natural invasion by the non-indigenous red piranha. Probably, human mediated dispersal is the main cause of red piranha introduction and colonization in these lakes. In fact, the vestiges of fishermen (e.g. used hooks, nets, and fishing line) were observed in almost all of these study lakes.

Once established in a new habitat, the control of this invasive species or its eradication is difficult and virtually impossible (Simberloff 2001). If the red piranha is an effective invader of the River Doce lakes and its distribution is determined by the length of canals linking lakes to streams, lakes near streams and not yet invaded will probably be so in the future. On the other hand, more isolated lakes have minor chances of a natural invasion by the red piranha. Therefore, we think that the best option for the conservation of the indigenous fish community would be to concentrate political and conservation efforts on lakes far from streams. Additionally, action to increase awareness by local populations may also be very important to avoid the introduction of the red piranha into these lakes by fishermen.

ACKNOWLEDGEMENTS

This study is part of a Ph.D. thesis presented to the ecology postgraduate programme of the Universidade Estadual de Campinas (UNICAMP) by A.O. Latini, advised by M. Petrere Jr. It was supported by the Fundo Nacional do Meio Ambiente (FNMA 80/2001), Ministério do Meio Ambiente, and by the Conselho Nacional de Desenvolvimento Científico e Tecnológico (CNPq) in association with the Instituto de Pesquisas da Mata Atlântica (IPEMA). The authors thank the Instituto Estadual de Florestas de Minas Gerais (IEF) and the Companhia Agrícola Florestal Santa Bárbara (CAF - Grupo Arcelor) for logistic support, and they are grateful to Dr. Daniela C. Resende who read an early draft of this chapter.

REFERENCES

Agostinho, A. A., S. M. Thomaz, and L. C. Gomes. 2005. Conservation of the Biodiversity of Brazil's Inland Waters. Conservation Biology **19**, 646–652.

Crowder, L. B. and W. E. Cooper. 1982. Habitat structural complexity and the interaction between bluegills and their prey. Ecology **63**, 1802–1813.

Elton, C. S. 1958. The ecology of invasions by animals and plants. Wiley, New York, NY.

Froese, R. and D. Pauly. 2006. FishBase. World Wide Web electronic publication. www.fishbase.org, version (05/2006).

Havel, J. E., J. B. Shurin, and J. R. Jones. 2002. Estimating dispersal from patterns of spread: Spatial and local control of lake invasions. Ecology **83**, 3306–3318.

Hosmer, D. W. and S. Lemeshow. 1989. Applied logistic regression. Wiley, New York, NY.

Hutchinson, G. E. 1957. Concluding remarks. Cold Spring Harbor Symposium of Quantitative Biology **22**, 415–427.

Jackson, D. A., P. R. Peres-Neto, and J. D. Olden. 2001. What controls who is where in freshwater fish communities - the roles of biotic, abiotic and spatial factors. Canadian Journal of Fisheries and Aquatic Sciences **58**, 157–170.

Kolar, C. S. and D. M. Lodge. 2001. Progress in invasion biology: predicting invaders. Trends in Ecology and Evolution **16**, 199–204.

Kouamélan, E. P., G. G. Teugels, V. N'Douba, G. G. Bi, and T. Koné. 2003. Fish diversity and its relationships with environmental variables in a West African basin. Hydrobiologia **505**, 139–146.

Kraft, C. E., P. J. Sullivan, A. Y. Karatayev, L. E. Burlakova, J. C. Nekola, L. E. Johnson, and D. K. Padilla. 2002. Landscape patterns of an aquatic invader: Assessing dispersal extent from spatial distributions. Ecological Applications **12**, 749–759.

Latini, A. O., D. P. Lima-Júnior, H. C. Giacomini, R. O. Latini, D. C. Resende, H. M. V. Espírito-Santo, D. F. Barros, and T. L. Pereira. 2004. Alien fishes in Rio Doce lakes: range, new occurrences and conservation of native community. Lundiana **5**, 135–142.

Latini, A. O., M. Petrere Jr. 2004. Reduction of a native fish fauna by alien species: an example from Brazilian freshwater tropical lakes. Fisheries Management and Ecology **11**, 71–79.

Levine, T. M. and C. M. D'Antonio. 1999. Elton revisited: a review of evidence linking diversity and invasibility. Oikos **87**, 15–26.

Lodge, D. M. 1993. Biological invasions - lessons for ecology. Trends in Ecology and Evolution **8**, 133–137.

Mack, R. N., D. Simberloff, W. M. Lonsdale, H. Evans, M. Cout, and F. A. Bazzaz. 2000. Biotic Invasions: Causes, Epidemiology, Global Consequences and Control. Ecological Applications **10**, 689–710.

Marchetti, M. P., P. B. Moyle, and R. Levine. 2004. Alien fishes in California watersheds: characteristics of successful and failed invaders. Ecological Applications **14**, 587–596.

Miller, R. R., J. D. Williams, and J. E. Williams. 1989. Extinctions of North American fishes during the past century. Fisheries **14**, 22–38.
Mooney, H. A. and E. E. Cleland. 2001. The evolutionary impact of invasive species. Proceedings of the National Academy of Sciences USA **98**, 5446–5451.
Moyle, P. B., P. K. Crain, K. Whitener, and J. F. Mount. 2003. Alien fishes in natural streams: fish distribution, assemblage structure, and conservation in the Cosumnes River, California, USA. Environmental Biology of Fishes **68**, 143–162.
Moyle, P. B. and T. Light. 1996a. Biological invasions of fresh water: empirical rules and assembly theory. Biological Conservation **78**, 149–161.
Moyle, P. B. and T. Light. 1996b. Fish Invasions in California: Do Abiotic Factors Determine Success? Ecology **77**, 1651–1670.
Myers, N., R. A. Mittermeier, C. G. Mittermeier, G. A. B. Fonseca, and J. Kent. 2000. Biodiversity hotspots for conservation priorities. Nature **403**, 853–858.
Power, M. E., J. C. Marks, and M. S. Parker. 1992. Variation in the vulnerability of prey to different predators - community-level consequences. Ecology **73**, 2218–2223.
Sakai, A. K., F. W. Allendorf, J. S. Holt, D. M. Lodge, J. Molofsky, K. A. With, S. Baughman, R. J. Cabin, J. E. Cohen, N. C. Ellstrand, D. E. McCauley, P. O'Neil, I. M. Parker, J. N. Thompson, and S. G. Weller. 2001. The population biology of invasive species. Annual Review of Ecology and Systematics **32**, 302–332.
Sazima, I. and F. A. Machado. 1990. Underwater observations of piranhas in western Brazil. Environmental Biology of Fishes **28**, 17–31.
Simberloff, D. 2001. Inadequate solutions for a global problem? Trends in Ecology and Evolution **16**, 323–324.
Sweka, J. A. and K. J. Hartman. 2003. Reduction of reactive distance and foraging success in smallmouth bass, *Micropterus dolomieu*, exposed to elevated turbidity levels. Environmental Biology of Fishes **67**, 341–347.
Uetanabaro, M., T. Wang, and A. S. Abe. 1993. Breeding behavior of the red-bellied piranha *Pygocentrus nattereri* in nature. Environmental Biology of Fishes **38**, 369–371.
Welcomme, R. L. 1988. International introductions of inland aquatic species. FAO Fisheries Technical Papers **294**, 1–318.

Chapter twenty-three

The relationship between biodiversity and invasibility in central Swedish lakes invaded by Elodea *species*

Daniel Larson and Eva Willén

INTRODUCTION

Since only a fraction of non-indigenous species (NIS) become invasive, it has long been a desirable goal to predict those associated with the largest risks. In addition, some environments have more frequently suffered from biological invasions than others, which has raised the question of whether environment invasibility can be predicted. Many attempts have been made to find general characteristics distinguishing NIS that have become invasive from those that have not, as well as invaded environments from non-invaded environments, but often with ambiguous results (Williamson 1999).

One of the most debated factors suggested to affect the invasibility of an environment is biological diversity. Biodiversity has been proposed to both act as a barrier for invasions (Elton 1958, Tilman 1999) and, conversely, to increase the likelihood that introduced species find suitable environments for establishment (Stohlgren *et al.* 2003). With high biodiversity it is possible that more niches are occupied, which would result in more resistant communities. On the other hand, high biodiversity may be an effect of high niche richness, and, thus, the community would be more likely suitable for a potential invader.

Francesca Gherardi, Biological invaders in inland waters: Profiles, distribution, and threats, 423–433.

To our knowledge, of how species richness affects the invasion process has never previously been assessed for lake ecosystems or aquatic plants. In addition, most studies on biodiversity have correlated the number of NIS within an area with the number of indigenous species, without considering the effect of the match between species and environment. Such an approach may be misleading, since different types of environments both support different number of species and are under different propagule pressure. This chapter will assess how biodiversity affects the establishment of NIS using an approach that includes only two species with similar spreading patterns and environmental demands to avoid the bias correlation studies could experience.

DOES BIODIVERSITY MATTER?

To assess how biodiversity affects the establishment of non-indigenous aquatic plants, invaded and non-invaded lakes were compared. Species data were assembled from a recent investigation of the aquatic plant species composition of 275 lakes situated in a small geographical area in central Sweden, close to Stockholm (Södertörnsekologerna 2001). Within the studied area, only two plant species have been recognized as invaders: the North American elodeids *Elodea canadensis* Michx. and *Elodea nuttallii* (Planch.) H. St. John. These species are not only ecologically similar but also have similar histories of invasion in Europe. In the studied area, *E. canadensis* was first found in the late 19th century, whereas *E. nuttallii* has been noted only during the last 25 years (Larson and Willén 2006). Of the 275 surveyed lakes, 27 were found to have been invaded by one or both *Elodea* species. Here, invaded lakes were compared with non-invaded lakes in terms of species richness measured as number of free-floating, floating-leaved, and submerged plants (except for the non-indigenous *Elodea* species).

The methods used in the studies from which species data were taken differed; sampling effort depended on the surface area of the lake. Hence, the likelihood of detecting rare species within a lake varies with the lake area. For small lakes (<0.01 km^2) only one profile was mapped; for intermediate-sized lakes (0.01–1 km^2) two profiles were mapped; and for large lakes (>1 km^2) three or more profiles were mapped. Further, surveys were made outside the profiles to create a list of all plant species in the lake. Here, species lists both on the whole lake basis and on profile basis ($n = 512$) have been used.

The number of indigenous species was found to be positively related to invasibility. Invaded lakes contained more species than non-invaded lakes (Table 1). Not only the overall number of species, but also the number of elodeids, i.e. species of the same guild as the *Elodea* species, was larger in the invaded lakes. However, when assessing biological diversity it is important to consider the concept of spatial scale, since area is positively correlated with species richness, a relation also showed for aquatic plants (Jones *et al.* 2003,

Table 1 Differences in species-richness, surface area, shore line irregularity, trophic ranking score, maximum depth, and altitude between invaded and non-invaded lakes. Number of lakes (invaded, non-invaded) are as follows: all lakes = 275 (27, 248) and sampled lakes = 38 (23, 15). Numbers of profiles are 79 and 433 for all lakes respectively, and 58 and 53 for the sampled lakes.

	All lakes			Sampled lakes		
	Invaded	Non-invaded		Invaded	Non-invaded	
Species richness, lake level	14	5.5	***	13	7.7	*
Elodeid richness, lake level	8.4	2.2	***	8.1	3.3	**
Species richness, site level	7.9	4.3	***	8.1	4.5	***
Elodeid richness, site level	4.9	1.4	***	4.8	1.8	***
Surface area [km^2]	46	0.26	**	2.2	0.72	NS
Length of shore line [km]	15	2.4	***	14	4.9	NS
Shore line irregularity	3.0	1.6	***	2.8	1.8	*
Maximum depth [m]	15	6.4	**	14	8.2	*
Altitude [m]	18	33	***	18	27	*
Trophic ranking score	8.1	7.3	***	8.1	7.4	*

Student's *t*-test: $^{*}P < 0.05$. $^{**}P < 0.01$. $^{***}P < 0.001$.

Rørslett 1991). Spatial scale has also been reported to affect the outcome when assessing the relationship between biodiversity and invasibility (Herben *et al.* 2004). In addition, since sampling efforts were larger for larger lakes than for smaller lakes and the invaded lakes were larger, the resulting species richness could also be interpreted as an effect of both size and sampling effort. To reduce these possible biases, species richness was also examined at site level. However, at the scale-independent site level, habitats with more aquatic plant species were more frequently invaded, both when comparing all sites with and without *Elodea* spp. and when comparing only invaded and non-invaded sites within the invaded lakes (Table 1). Hence, the results found are interpreted as not just an effect of size. The pattern described above was similar in a subset of the 275 lakes containing 23 invaded and 15 non-invaded lakes, which were compared for water chemistry.

THE ROLE OF HABITAT HETEROGENEITY

The pattern described in the previous section, i.e. habitats with more aquatic plant species were more frequently invaded by the *Elodea* species, needs to be further explored. Our hypothesis was that species richness and invasibility were not affected by each other, but instead that a different common factor affected both processes. One possible explanation was that the invaded lakes had a richer habitat heterogeneity and thereby more suitable niches for occupation, for both indigenous and non-indigenous species. Therefore, in addition to species richness, the lakes were also assigned values for their habitat heterogeneity.

How can habitat heterogeneity be measured?

The larger an area, be it aquatic or terrestrial, the more different types of habitats are likely to be represented, and hence a larger number of species are supported. Moreover, larger areas can also contain several species with similar niches if similar habitats are spatially separated. However, more factors than the size of an area control habitat heterogeneity. For instance, different types of wind and wave exposure create different bottom types in lakes. Hence, lakes with a non-circular shape are likely to contain many gradients of sheltered and exposed areas, and are therefore also more likely to have high habitat heterogeneity. To relate habitat heterogeneity to species richness, and in the long run also to invasibility, both lake area and irregularity of the shore line were calculated (Equation 1; Wetzel 1975).

$$\frac{L}{2 \times \sqrt{A \times \pi}} \tag{1}$$

L = Length of shore line (km)
A = Surface area (km^2)

Differences in heterogeneity

Lake surface area and shore line irregularity are indicators of habitat heterogeneity and both were larger in the invaded lakes. The relationship between habitat heterogeneity and species richness, also found in other studies (e.g. Burnett *et al.* 1998, Nichols *et al.* 1998, Honnay *et al.* 2003), became clear when the number of plant species were plotted against surface area (Fig. 1a) and shore line irregularity (Fig. 1b).

OTHER INVOLVED MECHANISMS

When searching for relationships in nature, it is important to simultaneously consider several mechanisms. Therefore, in addition to our main hypothesis of habitat heterogeneity, consideration was given to other factors affecting the invasion process, i.e. environmental conditions affecting the species, propagule pressure, and anthropogenic disturbances. To get a comprehensive picture, a subset of the 275 lakes was investigated further. This subset of lakes was selected using two criteria followed by random selection. The two criteria

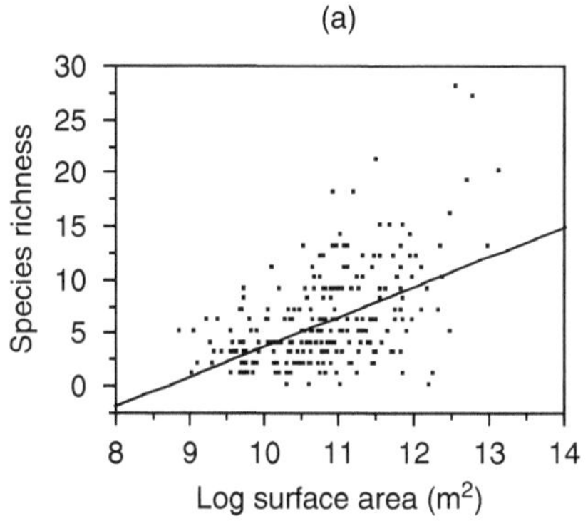

Fig. 1 Relationship between species-richness, surface area and shore line irregularity. a) Linear regression between species-richness and surface area (adj. R^2 = 0.25, $P < 0.001$, $n = 275$).
b) Linear regression between species-richness and shore line irregularity (adj. $R^2 = 0.26$, $P < 0.001$, $n = 275$).

used were surface area (0.05–15 km^2) and distance to the closest invaded lake (< 5 km), which resulted in 23 invaded lakes and 93 non-invaded lakes. The choice of distance was based on the shortage distance by which 95% of the invaded lakes where interconnected. The excluded lakes varied considerably in size or were considered to be too far from a propagule source. Of the 93 non-invaded lakes meeting these criteria, 15 were randomly selected. These non-invaded lakes were compared with the 23 invaded lakes for physical and chemical conditions, propagule pressure, and anthropogenic disturbances.

Habitat match

Water chemistry and light climate were sampled in the 38 lakes and analysed according to international (ISO/EN) standard methods (Wilander *et al.* 2003). Based on the plant species assemblages in each lake, trophic ranking scores (ranging from 1 to 10) were calculated according to Palmer *et al.* (1992). The score assigned to a lake is the mean of the scores for all species present. A low score indicates that a species most often is found in nutrient-poor environments, and a high score indicates the affinity of a species to a nutrient-rich environment. The use of biota as indicators of a nutrient state gives a more comprehensive picture of the available nutrients than that revealed by a few direct water-chemical measurements.

Invaded lakes did not differ from non-invaded lakes in most aspects. When reducing the compared variables to two principal components, separation between the invaded and non-invaded lakes appeared on the second axis for principal component analysis (PCA, Fig. 2a). However, when PCA were performed on chemistry and morphometry variables separately, only morphometry

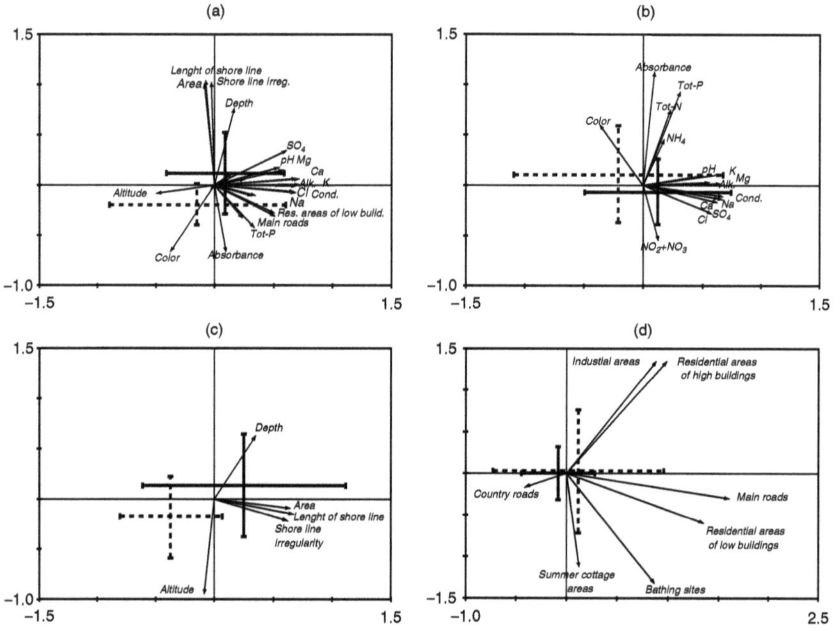

Fig. 2 Principal components analysis performed in CANOCO Version 4.5 (Ter Braak and Smilauer 2002). Mean and SD of invaded lakes (solid crosses, $n = 23$) and non-invaded lakes (dashed crosses, $n = 15$). Differences between invaded lakes and non-invaded lakes tested with *t*-test.

a) Water chemistry, morphometry, and urbanization (variables with a principal length less than 0.5 not shown). One reference lake was excluded prior to analysis due to its extremely high conductivity. Eigenvalues and *P*-values are 0.359 and 0.312 respectively for the first ordination axis, and 0.137 and 0.003 for the second ordination axis.

b) Water chemistry variables. One reference lake was excluded due to its extremely high conductivity. Eigenvalues and *P*-values are 0.531 and 0.238 respectively for the first ordination axis, and 0.149 and 0.259 for the second ordination axis.

c) Morphometry variables. One invaded lake was excluded prior to analysis due to its extremely large area, length of shore line, and shore line irregularity. Eigenvalues and *P*-values are 0.586 and 0.008 respectively for the first ordination axis, and 0.237 and 0.0437 for the second ordination axis.

d) Urbanization variables. Eigenvalues and *P*-values are 0.314 and 0.391 respectively for the first ordination axis, and 0.226 and 0.910 for the second ordination axis.

variables successfully separated invaded lakes from non-invaded lakes (Fig. 2b-c). The results from the PCA are in line with results obtained from testing one variable at the time with the Student's *t*-test. In the *t*-tests, most morphometry variables differed between invaded and non-invaded lakes, e.g. surface area and shore line irregularity (Table 1), whereas the chemistry was similar in the two groups of lakes (Table 2).

Table 2 Means and standard deviations of water chemistry, light climate, and urbanization variables not differentiating invaded from non-invaded lakes (Student's *t*-test, $P > 0.05$). The number of lakes is 38 (23 invaded and 15 non-invaded).

	Invaded lakes		Non-invaded lakes	
	Mean	SD	Mean	SD
pH	8.0	0.63	7.7	0.72
Conductivity [mS cm^{-1}]	0.24	0.11	0.25	0.27
Alkalinity	1.0	0.41	0.85	0.60
Tot-P [μg L^{-1}]	27	16	30	23
Tot-N [mg L^{-1}]	0.63	0.29	0.57	0.19
NO_2+NO_3 [μg L^{-1}]	4.7	1.7	4.3	1.7
NH_4 [μg L^{-1}]	8.6	7.7	9.9	6.6
Ca [mekv L^{-1}]	1.1	0.46	0.93	0.66
Mg [mekv L^{-1}]	0.35	0.091	0.32	0.25
Na [mekv L^{-1}]	0.85	0.52	0.98	0.65
K [mekv L^{-1}]	0.061	0.024	0.060	0.011
SO_4 [mekv L^{-1}]	0.40	0.17	0.36	0.27
Cl [mekv L^{-1}]	0.72	0.49	1.1	1.9
Absorbance. non filtered [420nm $5cm^{-1}$]	0.11	0.055	0.16	0.097
Water colour [mg Pt L^{-1}]	25	14	30	20
Bathing sites (number per km^2 lake area)	1.1	2.0	2.2	5.4
Summer cottage areas (%)	4.7	7.9	4.2	8.1
Residential areas of low-rise buildings (%)	8.3	10	11	21
Residential areas of high-rise buildings (%)	0.50	1.3	0.75	1.5
Industrial areas (%)	0.56	1.6	2.5	8.8
Country roads (km per km^2)	1.4	1.3	1.5	1.5
Main road (km per km^2)	3.8	1.9	4.2	3.0

Although not considered here, other environmental factors might be of importance for the establishment of the *Elodea* species. For instance, adaptation to seasonal nutrient fluctuations has been proposed as one of the key factors explaining the success of invasions by *Elodea* species. (Thiébaut 2005).

Propagule pressure

Propagule pressure was, not surprisingly, larger for the invaded lakes, interpreted by a higher occurrence of *Elodea* species upstream to invaded lakes (9 of 23) than upstream to lakes without *Elodea* species (0 of 15; Pearson Chi-square statistics: $P < 0.01$). An unexpected finding was that not a single non-invaded lake was connected with invaded lakes upstream. This finding could suggest that propagule pressure determines the invasion success of the *Elodea* species.

However, most invaded lakes also lacked invaded upstream lakes. Such a coincidence suggests invasions from other pathways.

The interference of human related actions may also spread *Elodea* species, particularly activities such as boating (Johnstone *et al.* 1985) and fishing, although these were not considered here. However, invaded lakes did not contain more bathing sites, nor did they contain more summer cottage areas (Table 2). Hence, it seems likely that both groups of lakes are equally used for recreational boating, although it is possible that lakes without bathing sites or summer cottage areas are frequently used for recreational boating.

Anthropogenic disturbance

Anthropogenic processes may affect the structure and function of lake ecosystems in numerous ways. It has been proposed to be that invasibility is affected by the presence of humans due to increased environmental disturbance and the transportation of propagules (King and Buckney 2000). Therefore, an estimate of the total disturbance was assigned to each lake by calculating variables such as the percent urban area, length of roads, number of bathing sites in the area surrounding the lakes (500 m from the shore) using Geographical Information System (GIS) software. However, in the studied area, these urbanization variables did not differ for invaded lakes and non-invaded lakes (Table 2, Fig. 2d).

CONTRADICTING THEORIES

Many theories have been developed in the field of invasion biology, some to explain why certain environments suffer from biological invasions more frequently than other environments. Elton (1958) proposed that species-poor communities were more vulnerable to invasions, based on observations from highly invaded oceanic islands. Different kinds of theoretical support for Elton's ideas were soon presented, e.g. by MacArthur (1972) who suggested that the more species present, the more effectively resources would be used, which would result in more resistant communities. Biological diversity has been proposed to act as a barrier for invasions based on the assumption that the more species are present, the more niches are occupied. However, high biological diversity may also indicate high habitat heterogeneity, which in turn increases the likelihood that an introduced species finds a suitable environment for establishment.

Just as different theories predict different outcomes, results from different studies are contradictory; some show cases where the number of biological invasions is negatively correlated with biological diversity, i.e. where the diversity acts as an invasion barrier. This kind of result is often obtained in experimental studies, but many experiments have been criticized for their design (Wardle 2001). Most observational studies, on the other hand, report that the number of invasion is positively correlated with biodiversity (Levine and D'Antonio 1999). In this study, we found no evidence for any direct relationship between invasibility and species richness. However, we suggest that the same underlying factors affect the richness of non-indigenous and indigenous species. Although it is plausible that higher biodiversity leads to more efficient resource use, which in turn could reduce the likelihood of establishment of NIS, the strength of this effect may be small compared to the effect of other factors. Perhaps the weakness of the diversity effect is the reason why a negative relationship only has been detected in controlled experiments whereas field observations show the opposite. Although diversity may exert a too weak effect on the establishment success of NIS (i.e. the invasibility), it may affect their abundance (Levine *et al.* 2004). Thus, diversity should be considered when both the likelihood of establishment and the potential effects of a NIS are assessed.

CONCLUSIONS

Most observational studies on how biodiversity affects the invasion process have compared environments with different degrees of invasion, i.e. correlated biodiversity with invasibility. With such an approach, no consideration is given to the potential influence of different numbers of species in different environment. In this chapter, an approach that includes only two species with similar spreading patterns and environmental demands is used to investigate how lakes invaded and non-invaded by these species differ in species-richness and other aspects, i.e. chemistry, morphometry, and anthropogenic disturbance. The number of lakes sampled for water chemistry and scored for proximity to populated areas ($n = 38$) and the diversity of their water quality would probably have shown significant trends if these types of variables were unambiguous indicators of *Elodea* invasions. The best indicator was instead the diversity of other aquatic plant species, which in turn is an indicator of habitat heterogeneity. We suggest that habitat heterogeneity is the underlying mechanism causing a positive relationship between biodiversity and invasibility.

ACKNOWLEDGEMENTS

We thank Francesca Gherardi for the invitation to contribute to this volume. We also thank Frauke Ecke and two anonymous referees for providing valuable comments on the manuscript. This study is part of the AquAliens project dealing with non indigenous aquatic species in marine and freshwater environments of Sweden. Financial support from the Swedish Environmental Protection Agency is gratefully acknowledged.

REFERENCES

Burnett, M. R., P. V. August, J. H. Brown, and K. T. Killingbeck. 1998. The influence of geomorphological heterogeneity on biodiversity I. A patch-scale perspective. Conservation Biology **12**, 363–370.

Elton, C. S. 1958. The ecology of invasions by animals and plants. Methuen, London, UK.

Herben, T., B. Mandak, K. Bimova, and Z. Munzbergova. 2004. Invasibility and species richness of a community: A neutral model and a survey of published data. Ecology **85**, 3223–3233.

Honnay, O., K. Piessens, W. Van Landuyt, M. Hermy, and H. Gulinck. 2003. Satellite based land use and landscape complexity indices as predictors for regional plant species diversity. Landscape and Urban Planning **63**, 241–250.

Johnstone, I. M., B. T. Coffey, and C. Howard-Williams. 1985. The role of recreational boat traffic in interlake dispersal of macrophytes: A New Zealand case study. Journal of Environmental Management **20**, 263–279.

Jones, J. I., W. Li, and S. C. Maberly. 2003. Area, altitude and aquatic plant diversity. Ecography **26**, 411–420.

King, S. A. and R. T. G. Buckney. 2000. Urbanization and exotic plants in northern Sydney streams. Austral Ecology **25**, 455–461.

Larson, D. and E. Willén. 2006. Non-indigenous and invasive water plants in Sweden. Svensk Botanisk Tidskrift **100**, 5–15. (In Swedish with English abstract)

Levine, J. M., P. B. Adler, and S. G. Yelenik. 2004. A meta-analysis of biotic resistance to exotic plant invasions. Ecology Letters **7**, 975–989.

Levine, J. M. and C. M. D'Antonio. 1999. Elton revisited: a review of evidence linking diversity and invasibility. Oikos **87**, 15–26.

MacArthur, R. 1972. Species packing and competitive equilibrium for many species. Theoretical Population Biology **1**, 1–11.

Nichols, W. F., K. T. Killingbeck and P. V. August. 1998. The influence of geomorphological heterogeneity on biodiversity II. A landscape perspective. Conservation Biology **12**, 371–379.

Palmer, M. A., S. L. Bell, and I. Butterfield. 1992. A botanical classification of standing waters in Britain - Applications for conservation and monitoring. Aquatic Conservation - Marine and Freshwater Ecosystems **2**, 125–143.

Rørslett, B. 1991. Pricipal determinants of aquatic macrophyte richness in northern European lakes. Aquatic Botany **39**, 173–193.

Stohlgren, T. J., D. T. Barnett, and J. T. Kartesz. 2003. The rich get richer: patterns of plant invasions in the United States. Frontiers in Ecology **1**, 11–14.

Södertörnsekologerna. 2001. Vattenväxter i sjöarna på Södertörn och i angränsande områden samt uppbyggnad av en sjödatabas. Södertörnsekologerna, Stockholm, Sweden.

Ter Braak, C. and P. Smilauer. 2002. CANOCO Reference manual and CanoDraw for Windows User's guide: Software for Canonical Community Ordination. Microcomputer Power, Ithaca, NY.

Thiébaut, G. 2005. Does competition for phosphate supply explain the invasion pattern of *Elodea* species? Water Research **39**, 3385–3393.

Tilman, D. 1999. The ecological consequences of changes in biodiversity: A search for general principles. Ecology **80**, 1455–1474.

Wardle, D. A. 2001. Experimental demonstration that plant diversity reduces invasibility – evidence of a biological mechanism or a consequence of sampling effect? Oikos **95**, 161–170.

Wetzel, R. G. 1975. Limnology. Saunders, Philadelphia, PN.

Wilander, A., R. K. Johnson, and W. Goedkoop. 2003. Riksinventering 2000. Swedish University of Agricultural Sciences, Uppsala, Sweden.

Williamson, M. 1999. Invasions. Ecography **22**, 5–12.

Impacts of invaders

I believe that the strongest ethical bases, and possibly the only ethical bases, for concern about introduced species are that they can threaten the existence of native species and communities and that they can cause staggering damage, reflected in economic terms, to human endeavors.

Daniel Simberloff (2003)

Chapter twenty-four

Measuring the impact of freshwater NIS: what are we missing?

Francesca Gherardi

There is no data like more data
(Mercer 1998)

INTRODUCTION

Within the last two decades, the dangers that some non-indigenous species (NIS) pose to indigenous species, ecosystem functioning, economic interests, and public health have been abundantly publicized in both the scientific and the popular literatures. A flood of publications, under the format of both synthetic overviews and detailed accounts of some species, accompanied the heightened interest in biological invasions (e.g. Williamson 1996, Mack *et al.* 2000, Cox 2004). Several underlined the dramatic effects that these species induce to the recipient environment and ascribed them to the potential of NIS to (1) alter and disrupt the biotic structure of ecosystems; (2) affect the wellbeing of other species; (3) push many species toward extinction; (4) reduce the productivity of agriculture and aquaculture; and (5) pose threats to human health and to the health of domesticated or semidomesticated plants and animals (Cox 2004). The media often featured both general problems (e.g. Bright 1998, Devine 1999, Di Justo 2006) and "the invader of the week" (Simberloff 2003a). Many nations (e.g. New Zealand, South Africa, USA, Canada, European Union) and international organizations (e.g. IUCN) began to consider the ecological impact of some NIS as one of the world's most serious conservation issues and several

Francesca Gherardi, Biological invaders in inland waters: Profiles, distribution, and threats, 437–462.

attempted to improve administrative and legal solutions (Simberloff 2003a). Meanwhile, two points have become universally clear: (1) the introduced species already established in natural areas are far more numerous than managers can really control and (2) their number is inevitably destined to increase as a consequence of the exponential growth of transport and commerce (Ewell *et al.* 1999).

Luckily, not "every barrel contains bad apples" (Sagoff 1999) and not necessarily "the worst of species" are "the ones that are alien and numerically successful" (Slobodkin 2001). On the contrary, the enormous benefits of some of the introduced species are universally recognized. Humans, it has been claimed, depend heavily on several non-indigenous organisms for food, shelter, medicine, ecosystem services, aesthetic enjoyment, and cultural identity (Ewel *et al.* 1999). Also among those species that have been inadvertently introduced by man, as many as 80–90%, according to the "tens rule" (Williamson 1996) – or less than 75%, at least for some taxa, according to Jerscke and Strayer (2005) – may actually have minimal detectable effects on the environment.

Indeed, the classification of NIS in function of the good and the bad they provide to both the environment and humans is implicit in the formal definition of "invasive species" first given by the 1992 Rio Convention on Biological Diversity (CBD 2001) – "alien species which threaten ecosystems, habitats, or species" – and then by 1999 President Clinton's Executive Order 13112 – "alien species whose introduction does or is likely to cause economic or environmental harm or harm to human health" (Clinton 1999). Unfortunately, in the "continuum of kaleidoscopic interactions" that characterizes the natural world (Carlton 2002), the attribute "invasive" does not divide conveniently species into taxa that have an impact and those that do not but needs to be assessed on a case-by-case basis.

It is certainly critical to face this recognized difficulty in classifying NIS when the purpose is to stimulate management options and policy actions. Priorities based on assessment of impacts need to be set at all scales, from management of local reserves to national and international policy decisions. So, impact measurements may allow for generating a rank-ordering of major risks to ecosystems and natural communities and for guiding ecosystem restoration efforts. But they also help test hypotheses about how communities function and what factors inhibit or facilitate invasions. Ultimately, the availability of quantitative data will make scientists able to generalize, and even predict, which species will most likely be "invasive" (Parker *et al.* 1999).

MULTILEVEL IMPACTS

Since 1984, the research focused on the impact of freshwater NIS on species, communities, and ecosystems has expanded greatly, especially in North America, New Zealand, Australia, and Western Europe (Fig. 1). Introduced plants, bivalves, and fish were the privileged organisms for impact studies in the 1980s,

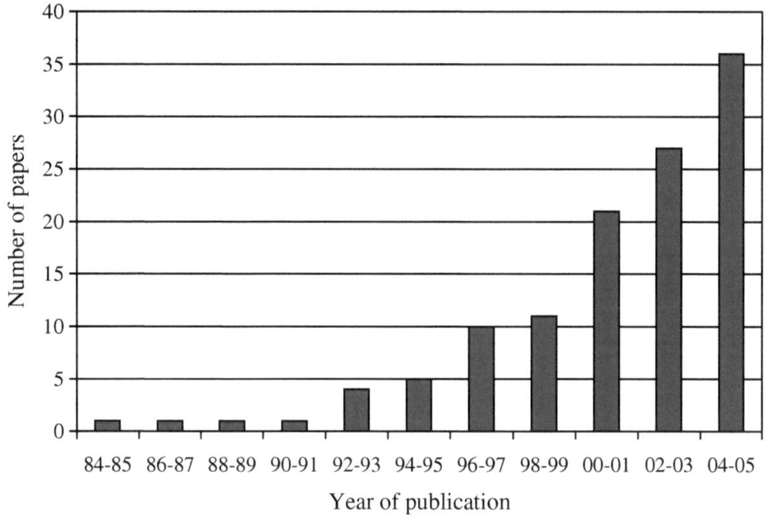

Fig. 1 The rise in the number of publications focused on the impact of freshwater non-indigenous species (NIS). Papers were identified via keywords from Biosis (from 1969 to April 2006) and later selected, reaching a total of 123. Only papers published before 2006 (118) are included here.

but the diversity of the analyzed taxa widened with time, including today also several arthropods (mainly crustaceans) and a few amphibians, e.g. *Bufo marinus* (Linnaeus), reptiles, e.g. *Trachemys scripta elegans* (Wied), and mammals, e.g. *Mustela vison* Schreber (Fig. 2). However, also within the taxa that have been subject to a more extensive research, only a narrow range of genera and families was used as paradigms of freshwater invasions, namely dreissenids (83%), gammarids (38%), and salmonids (50%). Obviously, this concentrated effort on a few organisms is not the reflection of the lack of ecological impacts by other taxa, but the result of some idiosyncratic factors, often interacting (e.g. productivity of laboratories, scientific or economic interest, and easiness of experimentation).

A similar unequal distribution of studies is to be found among the biological levels at which the impact has been measured. Under the classification of Parker *et al.* (1999), five levels of biological complexity may be affected by NIS, namely: (1) individuals (life history, morphology, behavior); (2) population dynamics (abundance, population growth, etc.); (3) genetics (including hybridization); (4) communities (species richness, diversity, trophic structure); and (5) ecosystem processes (nutrient availability, primary productivity, etc.). To these levels a sixth may be added, the "societal" level of impact, in which the focus is on the economic damage that introduced species may inflict to human societies, including social and ethical problems associated with their possible harm to human health.

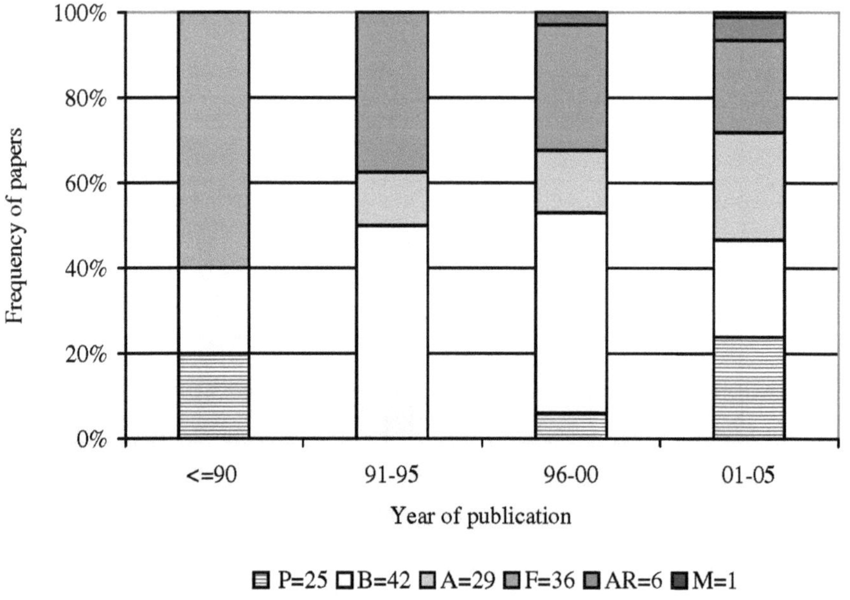

Fig. 2 Distribution per taxon (P = plants, B = bivalves, A = arthropods, F = fish, AR = amphibians and reptiles, and M = mammals) of the articles published before 2006 describing the impact of NIS (see Fig. 1 for the used method of retrieval) classified per periods of 5 years. Before 1990: $n = 5$, 1991–1995: $n = 8$, 1996–2000: $n = 34$, 2001–2005: $n = 92$. Single studies that reported the impact of more than one NIS were tallied more than once.

Population-level effects, eventually leading to a decline in the abundance and diversity of indigenous biota, have been documented much more extensively in the literature than the other biological effects (Fig. 3). Conversely, the most understudied impacts are genetic effects, notwithstanding their potential to provide insights about the still controversial role of NIS to speciation (e.g. Rosenzweig 2001).

There is a nearly unanimous belief that several NIS contribute to the threatening of an ever increasing number of indigenous species. In their analysis of the causes of threat for 1,880 of the nearly 2,500 species now imperiled in the USA, Wilcove *et al.* (1998) clearly showed that introduced species are the second leading factor (after habitat degradation/loss); they endanger, alone or in conjunction with other causes, a total of 49% of species – more than the next three categories (overexploitation, pollution, and disease) combined. This is true also for freshwater indigenous taxa, which were found to suffer from competition with, and/or from predation by introduced species, at percentages that reach 27, 53, 17, and 4 in amphibians, fish, mollusks, and crayfish, respectively. However, NIS are not viewed as the only responsible for biodiversity loss;

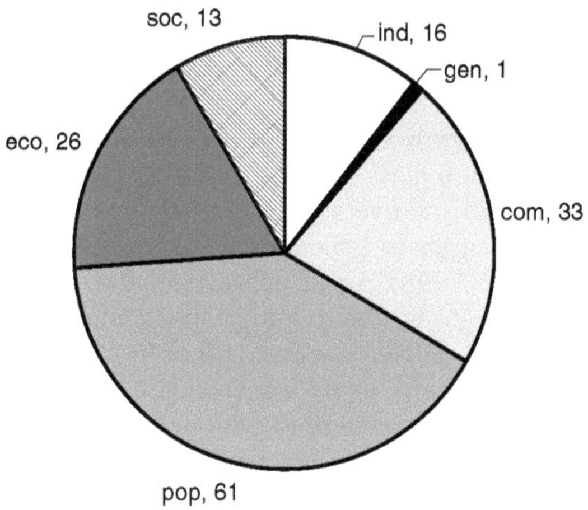

Fig. 3 Number of articles published until April 2006 describing the impact of NIS (see Fig. 1 for the method of retrieval) at six different levels of biological complexity: individual (ind), genetic (gen), population (pop), community (com), ecological (eco), and societal (soc). Single studies that reported the impact of more than one biological level were tallied more than once.

much stress has been recently given to their ancillary role in inducing extinctions or, in the words of Gurevitch and Padilla (2004), to their being "the final nail in the coffin" and "the bouquet at the funeral".

Surprisingly few are those freshwater organisms whose impacts have been analyzed at multiple biological levels. One is the Ponto-Caspian zebra mussel, *Dreissena polymorpha* (Pallas), certainly the most striking example of a species that has been able to change the nature of entire communities. Since the 1980s, after its introduction to the North American Great Lakes, the invasive potential of *D. polymorpha* has attracted much scientific attention, soon becoming the most thoroughly investigated freshwater invader. Today we dispose of an enormous amount of information about its wide-reaching effects on several invaded lakes and rivers in eastern North America (e.g. Nalepa and Schloesser 1993, Strayer *et al.* 1999) and central Europe (Karatayev *et al.* 1997). The invasiveness of this species is widely facilitated by various biological characteristics (i.e. high fecundity, planktonic veliger larvae enabling its fast diffusion, and byssal threads permitting firm attachment to hard substrates) so that, once introduced into a new system, it soon becomes abundant, even exceeding 10 times the biomass of all other indigenous benthic invertebrates. Its rapid population growth makes this species competitively dominant over resident benthic fauna, including several endangered or threatened bivalves. In conjunction with its close relative, the quagga mussel, *Dreissena bugensis* (Andrusov), zebra mussel colonization has highly accelerated the local extinction of unionid species as an effect of fouling (i.e. growing in dense clusters on unionid shells) or competing for seston (Strayer 1999). In Lake St. Clair, all the indigenous freshwater mussels were eliminated by 1997 after the appearance of zebra mussel in the early 1980s (Ricciardi *et al.* 1998, Nalepa *et al.* 2001). By contrast, there are few reports of mass mortalities of resident mussels in the European invaded lakes and rivers (e.g. Lake Balaton in Hungary, Lake Bourget in France, Lake Mikolajskie in Poland, and Lake Hallwill in Switzerland; references in Ricciardi *et al.* 1998). A plausible explanation of this different behavior between continents might be that in Europe indigenous bivalve fauna had been previously exposed to *D. polymorpha* during the Pleistocene era and had already acquired counteradaptations to it (e.g. avoidance behavior through deeper burial in sediment, less ecological sensitivity to fouling) (Ricciardi *et al.* 1998).

Functioning as an "ecosystem engineer" – i.e. a species that "directly or indirectly controls the availability of resources to other organisms by causing physical state changes in biotic or abiotic materials" (Jones *et al.* 1994), *D. polymorpha* can affect all components of the invaded freshwater systems at multiple levels (MacIsaac 1996, Strayer *et al.* 1999, Karatayev *et al.* 2002), as synthesized in Fig. 4. It increases water transparency (by 1.5–2 or more times) and the rate of conversion from organic to inorganic matter; it decreases the amount of seston in the water column (by 1.5–10 times), together with organic matter, biochemical oxygen demand (up to 1.5 times), and the biomass of phytoplankton (1–5–4 times); it affects bacterioplankton by e.g. consuming large bacteria and facilitating growth of small bacteria with excretion and release from protozoan predators; it favors growth and production of macrophytes, periphyton, and benthic algae by increasing water transparency; it changes the abundance and diversity of zooplankton community by preying on microzooplankton and offering refuges to large cladocerans; it alters the abundance and diversity of zoobenthos by offering shelters to e.g. snails and gammarid amphipods, competing for food and space or fouling them (e.g. macrophytes, indigenous mollusks); it enhances abundance of benthivorous and, potentially, planktivorous fish; it transfers the accumulated organic pesticides and polychlorinated biphenyl compounds to their predators, mostly waterfowl, fish, and crayfish; and it diverts production and biomass from pelagic to benthic food webs, shifting ecosystems to an alternative state. A gross estimate of *D. polymorpha*'s monetary cost has been recently provided by Pimentel *et al.* (2005): in the USA alone direct costs, mostly due to its fouling and clogging water intake pipes, water filtration, and electric generating plants, amounted to about US$1 billion per year. If added to the indirect and non-market costs associated with the altered ecosystem processes and reduced indigenous biodiversity, the overall damage inflicted to the human economy by zebra mussel should be enormous. However, notwithstanding this, prevention seems to be currently underfunded. A recent analysis (Leung *et al.* 2002) suggested that it would be beneficial to spend up to US$324,000 per year to obtain a modest reduction in the probability of zebra mussel invasion into a single lake. For

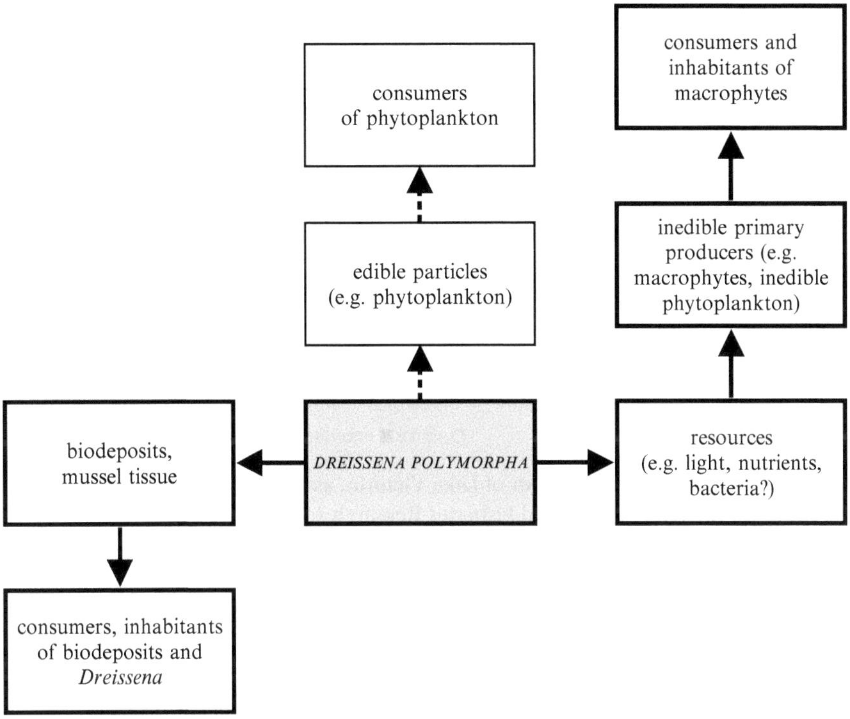

Fig. 4 Ecological changes induced by increasing populations of the zebra mussel (*Dreissena polymorpha*). Bold boxes and lines denote components that tend to increase with the bivalve population, and thin boxes and dotted lines denote components that tend to decrease. (Modified after Strayer *et al.* 1999)

comparison, in 2001 the US Fish and Wildlife Service distributed to all States combined a total of US$825,000 for prevention and control efforts for all aquatic NIS in all lakes.

A second celebrated case of freshwater invader is the Nile perch, *Lates niloticus* (Linnaeus), purposively introduced into Lake Victoria in the 1950s to boost fisheries (Seehausen *et al.* 1997). The dramatic result of this introduction is universally recognized as "the first mass extinction of vertebrates that scientists have never had the opportunity to observe" (Kaufman 1992). Only in the early 1980s was an explosive increase of this species observed leading to a rapid collapse of the species-rich Lake Victoria ecosystem and its replacement by a highly simplified, largely exotic-based community. Nile perch population explosions were accompanied by the disappearance, mostly occurring between 1975 and 1982, of about 200 endemic haplochromine cichlids of the 300+ species that previously were known to occupy a great variety of niches in the lake. By 1983 in Kenya and by 1986 in Tanzania, the indigenous fish community had

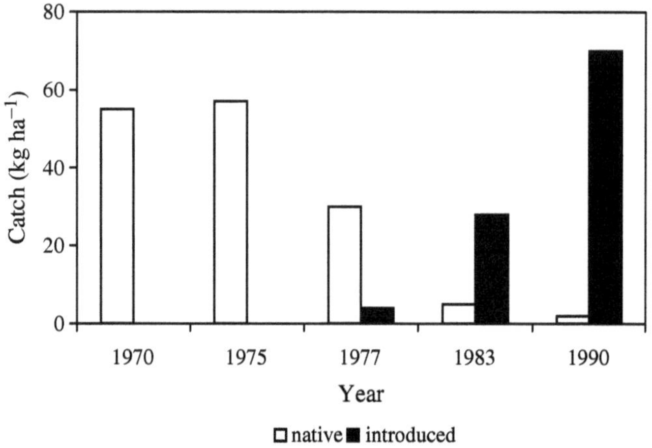

Fig. 5 Demise of the native fish of Lake Victoria, as illustrated by surveys in Kenyan water by the Kenyan Marine and Fisheries Research Institute. Standing stock estimates derived from the following numbers of hauls: 1969–1970 = 19; 1975 = 69; 1982–1990 = 41. (Modified after Kaufman 1992)

been virtually destroyed, while the Nile perch comprised more than 80% of the catch. The remaining 20% consisted of the introduced Nile tilapia, *Oreochromis niloticus* (Linnaeus), an indigenous pelagic minnow, *Rastrineobola argentea* (Pellegrin), and a small remnant of other indigenous fish (Fig. 5).

Coincident with the Nile perch explosion, an abrupt change in the physical environment of the lake was recorded. Today the region between 50 and 25 m in depth is subjected year-round to frequent severe deoxygenation, whereas before 1978 aerobic line penetrated into the lake's deepest waters. So, indigenous fish, seeking reliable refugia in their seasonal moves between shallow and deep water habitats, "may have faced a choice of death by asphyxiation in deeper water or death by Nile perch predation in the more oxygen-rich shallows" (Kaufman 1992). Among the myriad other aspects of the lake's ecology that appear to have changed, productivity and turbidity have both increased, papyrus swamps declined, and snails greatly increased in abundance (Kaufman 1992). The recent blooms of blue-green algae have been imputed to the disappearance of phytoplanktivore haplochromines, while the zooplanktivore haplochromines have been replaced by *R. argentea* (Goldschmidt *et al.* 1993). Finally, the indigenous atyid prawn *Caridina nilotica* (P. Roux), a potential consumer of decaying algae, has replaced the bottom-dwelling detritivorous haplochromines, as the result of several concomitant factors, such as the availability of sheltered habitats for prawn juveniles in the increased submerged vegetation and/or the disappearance of their haplocromine predators (Goldschmidt *et al.* 1993).

The story of the Nile perch also highlights the need for a critical evaluation of data in order to properly understand the role of NIS in species extinctions. The decline of cichlids started long before the introduction of the Nile perch, dating back to the 1920s with the development of railroads, erosion, and shoreline destruction. Then, the urbanization of the 1970s increased eutrophication and decreased lake transparency that affected color vision of many cichlids leading to reduced sexual selection, incorrect mate choice, and breakdown of reproductive isolation; increased nutrients produced anoxic events and favored the invader water hyacinth, *Eichhornia crassipes* (Martius). In its turn, this latter species may have altered nursery areas for juvenile fish (reviewed in Gurevitch and Padilla 2004).

Today, an appraisal of the monetary revenues that local people have obtained from the introduction of the Nile perch may be controversial. Certainly, the importance of biodiversity is often questioned when a commodity of immediate value appears in its place (Kaufman 1992). And there is much evidence that local people have taken real profits from the Nile perch commerce. During 1975–1989, the introduction of the Nile perch into Lake Victoria was followed by (1) production gains amounting to about US$280 million (at 1989 prices); (2) increased number of fishermen and of their dependents by 267% (more than 1.2 million people depend today entirely on fishery); (3) ameliorated food quality for greater numbers of people; (4) intensified exports that reached about 5–10% of the lake's production (Kasulo 2000). However, these estimates do not take into account the changes in the level and distribution of income, and in the ease of entry to fishery. The new fishery that required expensive fishing tools and refrigerating systems has had the effect of concentrating income in the hands of foreign investors and of a small minority of local fishermen (Kasulo 2000).

COMPLEXITY IN THE IMPACTS OF FRESHWATER NIS

Freshwater NIS are well known to exert an immediate impact on the behavior displayed by resident species, which may change their habitat use or activity patterns in response to the new predators or competitors. For instance, in New Zealand, larvae of mayfly *Nesameletus ornatus* Eaton are active on rock surfaces both day and night and drift in the water column when they coexist with indigenous predators (galaxiid fish). Conversely, in rivers invaded by brown trout, they mostly remain beneath rocks during the day and usually only forage on exposed surfaces and drift in the water column at night. This modified behavior was retained when mayflies from brown trout waters were tested in experimental streams without fish or with galaxiids (McIntosh and Townsend 1994). Non-indigenous salmonids also restrict the indigenous fish and amphibians to lesser preferred habitats, where they suffer reduced feeding activity and efficacy (references in Simon and Townsend 2003). Antipredator behaviors may also quickly evolve in the presence of a non-indigenous predator. In Oregon, when red-legged frog (*Rana aurora* Baird and Girard) tadpoles were exposed to chemical cues of the introduced bullfrog (*Rana catesbeiana* Shaw), individuals that had coexisted with bullfrogs showed strong antipredator responses by hiding or reducing their activity, whereas those from populations without bullfrogs did not (Kiesecker and Blaustein 1997).

When no changes at the individual level occur, when they occur but are insufficient, or when organisms become "trapped" by their evolutionarily responses to formerly reliable cues (Schlaepfer *et al.* 2005), the survival rate of resident species may be affected and the abundance of their populations inevitably decreases. Extreme consequences are the endangerment that indigenous species may suffer from the introduction of NIS or their local or global extinctions (Mack *et al.* 2000). The mass extinctions of endemic fish and mussels recorded in Lake Victoria and in North American lakes and rivers, respectively, are not the only environmental catastrophes to which NIS have contributed. Non-indigenous fish are at least partially responsible for the extirpation of 68% of the 40 fish species and subspecies declared extinct over the past 115 years in North America (Miller *et al.* 1989). The combined pressure of urbanization, overexploitation, and introductions of NIS has led to the global extinction of *Pacifastacus nigrescens* (Stimpson) in northern California (Bouchard 1977) and a similar process is ongoing in the same area for Shasta crayfish, *Pacifastacus fortis* (Faxon), today displaced at several locations by habitat loss and competitive interactions with *Pacifastacus leniusculus* (Dana) (Light *et al.* 1995, Chapter 28).

The mechanisms leading to biodiversity loss are many and abundantly explored in inland waters. They range from predation/parasitism, to competition for resources, interference competition, and transmission of parasites. So, the sharp decline in macroinvertebrates that have been recorded in central-European streams after the appearance of the Ponto-Caspian crustacean amphipod *Dikerogammarus villosus* (Sowinsky) in the early 1990s was imputed to the ability of this invasive species to prey upon more macroinvertebrates than the resident species did (Dick and Platvoet 2000, Krisp and Maier 2005). Invasive amphipods, although under the regulatory control of acanthocephalan parasites (MacNeil *et al.* 2003), are also able to kill and consume individuals of close relative species (Dick and Platvoet 2000). Similarly, the decreased abundance of several dominant species of zooplankton [*Daphnia retrocurva* Forbes, *Bosmina longirostris* (O.F. Müller), and *Diacyclops thomasi* (Forbes)] recorded in Lake Ontario (North America) coincided with the increased occurrence of the introduced predator cladoceran *Cercopagis pengoi* (Ostroumov) (Laxson *et al.* 2003).

Food competition with amphibian and reptile NIS – larvae of *R. catesbeiana* and *Osteopilus septentrionalis* (Duméril and Bibron) or adults of *T. scripta elegans* – was hypothesized to cause decreased survivorship, reduced growth rate, and delayed metamorphosis in the larvae of indigenous anurans – *Rana boylii* Baird, *Hyla regilla* Baird and Girard, *Bufo terrestris* (Bonnaterre), and *Hyla cinerea*

(Schneider) – (Kupferberg 1997, Smith 2005) or was found to determine weight loss and high mortality in the European turtle *Emys orbicularis* (Linnaeus) (Cadi and Joly 2004). Similarly, in the US waters, the plankton feeding, non-indigenous alewife, *Alosa pseudoharengus* (Wilson), heavily competes for prey with indigenous fish. Only a few indigenous fish were able to survive to its competition. Among them, the bloater, *Coregonus hoyi* (Milner), was able to coexist with alewife by shifting its diet from small zooplankton to larger benthic prey (Crowder 1984).

NIS may also outcompete indigenous species through direct aggression and/or transmission to them of diseases and parasites. So, the dominance in agonistic interactions of North American crayfish species over the European indigenous species (e.g. Gherardi and Cioni 2004) and the ability to transmit to them the oomycete *Aphanomyces astaci* Schikora – the etiological agent of the crayfish plague (e.g. Alderman and Polglase 1988, Diéguez-Uribeondo and Söderhäll 1993) – have both contributed to the constant contraction of indigenous crayfish biodiversity in Europe (Chapter 28).

The potential for introgressive hybridization is an additional but still elusive threat posed by NIS to closely related indigenous species. Introgression can increase the likelihood of extinction by reducing fitness and the ability of populations to changing conditions. However, the occurrence and consequences of hybridization between indigenous species and NIS have been overlooked in most freshwater species with the exception of fish. Genetic assimilation has led to the extinction of about 38% of North American indigenous fish species (Cox 2004). Cutthroat trout (*Salmo clarkii* Richardson), Apache trout (*Oncorhynchus apache* Miller), and Gila trout (*Oncorhynchus gilae* Miller) have undergone extensive hybridization with the invading rainbow trout, *Oncorhynchus mykiss* (Walbaum) (references in Simon and Townsend 2003). Stocking of conspecifics or escape of farm individuals may result in disruption of local adaptations and reduction of genetic diversity, as shown in brook trout, *Salvelinus fontinalis* (Mitchill) (Hayes *et al.* 1996), and in the Atlantic salmon, *Salmo salar* Linnaeus (Fleming *et al.* 2000). If hybrid individuals show greater fitness or vigor, the pure indigenous species may become extinct by being absorbed into the gene pool of a NIS with high invasive rates. This phenomenon of "genetic assimilation" has been reported in crayfish. For instance, matings between *Orconectes rusticus* (Girard) females and *Orconectes propinquus* (Girard) males yield a fecund and highly competitive progeny, which is replacing the indigenous species in Trout Lake, Wisconsin, USA (Perry *et al.* 2001, 2002, Chapter 28). Indigenous species are threatened by hybridization with a NIS also when the hybrids do not succeed, simply because crossbreeding reduces the number of new offspring added to the species' own population. For instance, females of the European mink, *Mustela lutreola* (Linnaeus), hybridize with males of the introduced North American mink, *Mustela vison* (Schreber); embryos are invariably aborted, but the wastage of eggs exacerbates the decline of the indigenous mink (Rozhnov 1993).

Other subtler evolutionary changes (Cox 2004) may influence several life history characteristics of both NIS and indigenous species in a relatively short-time scale. In fact, once established, NIS are freed from the constraints of gene flow from their parent population and from the biotic pressures of former enemies, are subject to altered selection pressures, and impose strong new evolutionary pressures on indigenous species. As an example, Chinook salmon, *Oncorhynchus tshawytscha* (Walbaum), native to the Pacific coast of North America, was introduced to New Zealand in 1901–1907. From the initial introduction, the species has colonized several river systems along the eastern coast of the South Island, giving rise to isolated populations. These populations, after about 30 generations, now differ genetically among themselves and from their source population in California by several morphological and reproductive features (Quinn *et al.* 2001).

At the community level, the potential by NIS to alter trophic interactions has been abundantly studied in several salmonids. Brown trout (*Salmo trutta* Linnaeus), introduced into New Zealand in 1867 (Townsend 1996), has been responsible for the local extinction of several indigenous galaxiid fish; it also profoundly affects the functioning of stream communities, by reducing the biomass of grazing invertebrates, altering their grazing activity, and ultimately releasing algae from top-down regulation by grazers and therefore indirectly increasing their biomass. Several other NIS are reported to induce trophic cascades, e.g. rainbow trout (*O. mykiss*) (Nyström *et al.* 2001) and crayfish (*P. leniusculus*, Nyström 1999, and *O. rusticus*, Charlebois and Lamberti 1996). Similarly, by preying on crustacean zooplankton, the cladoceran *C. pengoi* seems to cascade down the foodweb in Lake Ontario to increase phytoplankton abundance (Laxson *et al.* 2003).

The impact of NIS on ecosystem processes has rarely been analyzed in freshwater habitats, except for invasive plants. Their induced changes to nutrient cycling were found to range from being inconsistent (e.g. Findlay *et al.* 2002) to causing enormous consequences. For instance, purple loosestrife, *Lythrum salicaria* (Linnaeus) was found to accelerate P turnover (twice that of indigenous cattail *Typha angustifolia* Linnaeus) leading to low porewater nutrient concentrations (Templer *et al.* 1998), whereas common reed *Phragmites australis* (Cav.) Trin. ex Streud. is capable to sequester a large amount of nitrogen and to reduce nutrient availability to other organisms (Templer *et al.* 1998).

THE IMPACT ON HUMAN ACTIVITIES

A limited understanding of the inextricable link between nature and economy may be responsible for the often failed attempts of invasion biologists to arouse public and governmental support for the prevention or control of invasions (Mack *et al.* 2000). Indeed, the direct and indirect economic consequences of the ecological impact of NIS are well acknowledged but as yet poorly quantified. Scientific studies centered on the damage they inflict to human activities are still relatively rare, notwithstanding the increased general interest in the discipline of ecological economics.

Recently, Pimentel *et al.* (2005) attempted to tabulate the damage that NIS inflict to the US economy and the expenses needed for their control. The total cost in 2004 reached US$120 million per year, which appears to be "a formidable loss" even for a productive industrialized society such as the USA (Mack *et al.* 2000). The introduction of about 40 freshwater fish species alone was estimated to cause US$5,400 million in losses each year (Pimentel *et al.* 2005). A similar study has been recently conducted in Canada. Colautti *et al.* (2006) attempted to quantify the economic cost associated with 13 nuisance species (freshwater, marine, and terrestrial), including control costs, reduced yield, reduced land use, trade bans on exported goods, compensation paid to farmers, health care costs, and reduced tourism and tourism-related revenues. Comprehensive data were lacking for virtually all the analyzed species, providing relatively modest costs (an overall of CAN$187 million per year) if compared with two non-indigenous disease outbreaks (SARS and mad cow disease), which posed an additional cost of CAN$2.5 billion in 2003. Zebra mussels, quagga mussels, and the sea lamprey, *Petromyzon marinus* (Linnaeus) made up the bulk of CAN$32.3 million per year in characterized costs to aquaculture and aquatic-related industries. Of these species, dreissenid mussels have affected primarily industries and municipalities with a variety of direct costs. The sea lamprey, which attacks salmonids and other valuable fish, costs CAN$22 million per year to the Great Lakes Fishery Commission – a joint agency administered by the USA and Canadian federal governments – for its control and for research. However, costs incurred through reductions in harvest of commercial and sports fish would likely be much higher without this expenditure.

Other detailed analyses are scarce in the published literature. An example might be the estimated loss of US$30–45 million per year in Lake Tahoe (California) due to the degradation of resources by the introduced Eurasian watermilfoil (*Myriophyllum spicatum* Linnaeus) (Eiswerth *et al.* 2000). A similar analysis was done by Zavaleta (2000): the invasion of riparian tamarisk led to an increased sedimentation in river channels followed by frequent and severe floods. This translates to an overall cost of US$280–450 ha^{-1} that adds to US$7,400 ha^{-1} needed to eradicate the invader and to restore indigenous riparian communities. Conversely, still anecdotal is the monetary damage inflicted to water transportation, recreation activities, and hydraulic systems by other infamous invasive species, e.g. the water hyacinth (Gao and Li 2004) and *Limnoperna fortunei* (Dunker) (Oliveira *et al.* 2006).

Even fewer are the studies that have attempted to assign monetary values to species extinctions and losses in biodiversity, ecosystem services, and aesthetics. One example comes from the analysis by Spencer *et al.* (1991) of the impact of the opossum shrimp, *Mysis relicta* (Loven), on the Flathead Lake in Montana, USA. This species was introduced between 1968 and 1975 by the Montana Department of Fish, Wildlife and Parks to provide a supplemental prey for kokanee salmon, *Oncorhynchus nerka* (Walbaum), the dominant sport fish in the area. However, the shrimp did not become significant components of the kokanee diet, possibly because their vertical migration in deep waters during the daytime precluded fish from exploiting the new prey. On the contrary, they led to marked changes in the community initiating a trophic cascade to higher trophic levels. The density of two cladocerans – *Daphnia longiremis* Sars and *Leptodora kindtii* (Focke) – decreased as an effect of their voracious predation. Consequences were the crash of the population of kokanee, from 26,000–118,000 annual spawners in 1979–1985 to 50 in 1989, and the reduction in the angler catches from 100,000 kokanee through 1985 to fewer than 6000 in 1987 to no reported catches in 1988 and 1989 (Spencer *et al.* 1991). The abundance and diversity of birds and mammals feeding on spawning kokanee, carcasses, and eggs, sharply declined. Among the others, flagship species such as bald eagles and grizzly bears reduced their density. As a result, the number of tourists declined from 46,500 in 1983 to less than 1,000 in 1989 with an obvious economic loss for local activities based on ecotourism (Williamson 1996).

The damage that freshwater NIS may inflict on human health is well exemplified by the 1991 outbreak of cholera in Peru. It caused the death of over 10,000 people after ballast water containing the microbe *Vibrio cholerae* Pacini was released and infected drinking water (Bright 1998). However, conflicting insights emerge from the literature, especially concerning introductions of species into tropical areas. On the one hand, introduced snails may be extremely dangerous, due to their serving as intermediate hosts for some animal or human parasites, as in paragonimiasis. Invasive plants, such as water hyacinth, may also offer protection from predators to some snail species, such as *Biomphalaria sudanica* (Martens), host of *Schistosoma mansoni* Sambon (Plummer 2005). On the other hand, some introduced snails may be also beneficial due to their ability to outcompete indigenous snail vectors of blood and liver flukes (Pointier 1999).

MEASURES OF THE IMPACT

According to Parker *et al.* (1999), much of the discussion about the ecological effects of invasive species has been purely anecdotal in nature. Any conclusions about impacts, or the lack thereof, are often based upon "conjecture, suppositions, and presumptions" (Carlton 2002). The result is that "the case against introducing NIS is often poorly supported even for some of the most infamous invaders" (Parker *et al.* 1999).

This picture seems to be changing in the last few years, at least in the case of freshwater species. More than 100 studies on a total of the 123 here analyzed provide a quantitative estimate of the different categories of the impacts produced by NIS. Several have adopted a correlative approach making useful diachronic and synchronic comparisons between sites. However, in a few of these studies attempts have been made to control for naturally different responses of the invaded community over space and time or for confounding variables in the environment, such as pollution, harvesting, climate change, or other established NIS (Witt *et al.* 2005).

Many drawbacks of quantitative studies derived from the paucity of historical data needed to compare the same area before and after the invasion event. One of the few examples comes from the study of Wei and Chow-Fraser (2006) on the effects of multiple stressors [water level fluctuations, human population growth, and percent cover of nonindigenous *Glyceria maxima* (Hartman) Holmberg 1919 and *P. australis*] on the indigenous *Typha latifolia* Linnaeus marsh community in Lake Ontario (Canada). Studies of historical records, human population census, and field vegetation maps clearly showed that each stressor, including NIS, may explain alone the observed changes in indigenous *T. latifolia* community but that their synergistic interaction induces even greater detrimental effects on it.

The impact of NIS has been more extensively investigated in multiple sites or, most often, in the same site at different stages of invasion. For instance, field and laboratory studies in Northern Ireland showed that the non-indigenous *Gammarus pulex* (Linnaeus) has a greater impact on the composition of the macroinvertebrate community than the indigenous *Gammarus duebeni celticus* Stock and Pinkster due to its more intense predation on ephemeropterans, dipterans, and plecopterans (Kelly *et al.* 2003, Kelly and Dick 2005). These effects seemed to be independent of other environmental factors (Kelly *et al.* 2003) but they also operated at larger scales than those detected within individual rivers (Kelly and Dick 2005). Often comparisons have been made between sites with and without the invader; e.g. some Canadian lakes invaded by the non-indigenous predatory cladoceran *Bythotrephes longimanus* (Leydig) showed a significantly lower richness in crustacean zooplankton species (for 30%) than the non-invaded lakes (Boudreau and Yan 2003). Surprisingly, the ecological role of a species has never been compared between the indigenous and the colonized communities. Certainly, the impacts of potentially invasive species cannot be reliably predicted from their effects in the native ranges. For instance, the virulence that the oomycete *A. astaci* exerts toward the European crayfish could not have been forecasted from its innocuous effects toward North American species (Reynolds 1988). However, information about the biology of an invader in its native range would, on the one hand, provide a useful baseline for unraveling its ecological role and, on the other, greatly improve our understanding of the factors eliciting its invasive behavior in novel areas.

Still today, few studies are adopting stable isotope techniques to quantify the food-web consequences of invasions. Using this method, Vander Zanden *et al.* (1999) found that the indigenous top predator, the lake trout, *Salvelinus namaycush* (Walbaum), had more negative $\delta^{13}C$ values and lower trophic positions (3.3 *vs.* 3.9) in two Canadian lakes invaded by the introduced small-mouth bass, *Micropterus dolomieu* Lacepède, and rock bass, *Ambloplites rupestris* (Rafinesque) than in two non-invaded lakes, as an indication of the NIS-induced shift from a fish-based to a plankton-based diet in lake trout.

Much more numerous are the studies analyzing differences in diet breadth and/or in the position within the food web between invasive and non-invasive related taxa or between invasive taxa. For instance, laboratory experiments revealed that invasive *Gambusia* species consistently fed at higher rates than two non-invasive congeneric species, although displaying similar feeding preferences (Rehage *et al.* 2005). Multiple methods (stable isotope analysis, feeding experiments, mesocosm experiments, and gut content analysis) showed large differences in the potential impact on aquatic food webs of two non-indigenous decapod crustaceans, the Chinese mitten crab, *Eriocheir sinensis* (H. Milne Edwards) and the red swamp crayfish, *Procambarus clarkii* (Girard), in the San Francisco Bay (USA) (Rudnick and Resh 2005). Conversely, using stable isotope techniques, zebra mussel (*D. polymorpha*) and quagga mussel (*D. bugensis*) were found to share suspended detritus as the main food item and to compete with zooplankton and with each other for seston (Garton *et al.* 2005).

Obviously, both the population dynamics of an NIS and the responses by the recipient community (e.g. its species abundance) are expected to vary over time; therefore, any estimate of the impact of an NIS may greatly depend on the temporal scale of the study. Some successful invaders increase steadily to a stable equilibrium density, but others exhibit a more complex behavior, initially reaching very high densities ("boom") but then declining to lower levels ("bust"). And the same NIS may show booms and busts in some areas and steady logistic increase in others, as found for the zebra mussel in eastern Europe (e.g. Karatayev *et al.* 1997). Notwithstanding these well-known idiosyncratic dynamics of NIS populations (Williamson 1996), a minority of studies has been conducted on the impact of freshwater NIS over a long-time scale of analysis (e.g. Yan and Pawson 1997, Johannsson *et al.* 2000, Yan *et al.* 2002, Laxson *et al.* 2003, Barbiero and Tuchman 2004). This is not surprising: long-term monitoring requires a pattern of time and money allocation, which is usually extraneous to the rhythms of both academy and funding agencies, except for few cases (e.g. McCarthy *et al.* 2006). Conversely, long-term data sets would provide the indispensable background to help us assess the biotic resilience of the community, predict the restoration potential of the ecosystem, and finally understand the ecological and evolutionary mechanisms accompanying the integration of NIS in the system. By understanding the long-term feedbacks between invasive species and the invaded communities and ecosystems, "we will be able to evaluate alternative management approaches for well-established invaders, and be better able to identify which new invaders should be targeted for early eradication because of unacceptable acute and chronic impacts"(Strayer et al. 2006, p. 650).

THE NEED FOR PREDICTIVE MODELS

Related to the exiguous number of pre-colonization data sets and the rarity of long-monitoring researches, there is a general lack of theoretical studies aimed at modeling the effects of NIS on resident communities. This is unfortunate, because models might provide significant insights into the impact of NIS by permitting the calculation and comparison of an essentially unlimited range of measures. They may also help design more effective and efficient empirical studies by providing information about which measures are redundant and which measures identify independent effects. Taking advantage of their ability to vary characteristics of either the NIS or the community independently, models would also enhance the accuracy of water quality assessments by decoupling the environmental effects of NIS from those caused by different anthropogenic stressors. In sum, the modeling exercise may put biologists "in a much better position" in order to understand and to predict the impacts of a wide range of invaders (Parker *et al.* 1999).

Following this rationale, Ricciardi (2003) synthesized the data collected in different ecosystems and geographic regions to generate statistical models of the impacts exerted by the zebra mussel (*D. polymorpha*). This exercise permitted the author to determine whether the effects of *D. polymorpha* are consistent in different environments, and therefore whether they are predictable. For instance, by plotting the results of field experiments and surveys from multiple colonized sites, predictable patterns of *D. polymorpha*'s impact on other benthic invertebrates were identified. In the presence of the zebra mussel, the density of other macroinvertebrates was found to increase 2–10 times, whereas they declined in deepwater sites as a possible response to a reduced influx of food particles caused by the filtration activity of near-shore zebra mussel populations (Fig. 6). This information was found to be essential also for the correct interpretation of water quality assessment and of indices of biotic integrity, which are commonly based on the composition of benthic macro-invertebrate communities.

Recently, Vander Zanden *et al.* (2004) developed models aimed at predicting future occurrences and impacts of the non-indigenous smallmouth bass (*M. dolomieu*) into lakes in central Ontario. To identify "vulnerable" lakes, the authors used a conceptual framework for modeling the occurrence and impact of this species. Its colonization was dissected into three steps or filters. The first filter identified which lakes are accessible to colonists, the second filter identified which lakes are capable of supporting a population of NIS, and the third filter identified which lakes would be adversely impacted if an NIS were to become established. Based on this conceptual approach and using a data set composed of 3,046 lakes in central and northern Ontario, the authors developed two separate lake classification models. The first model, based on artificial neural networks, aimed at predicting the location of future bass invasions based on environmental suitability, and the second, based on known food-web

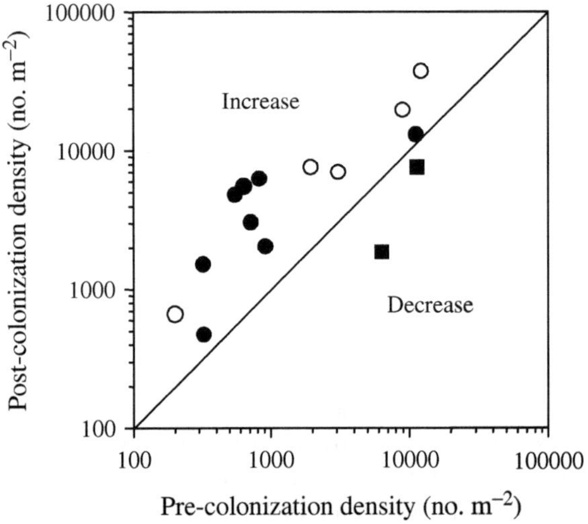

Fig. 6 Changes in the density of macroinvertebrates (excluding *Dreissena*) before and after *Dreissena polymorpha* colonization. Points above the 1:1 line indicate a positive change (increase), points below a negative change (decrease). Data include field surveys of littoral rocky substrata (black circles) and deepwater sediments (black quadrates), and experiments using artificial substrata (white circles). (Modified after Ricciardi 2003)

interactions derived from extensive stable isotope and gut content-based studies, singled out lakes in which bass was expected to have negative impacts on lake trout populations (Vander Zanden *et al.* 1999). By combining the predictions of these two models, along with information about lake remoteness, a subset of lakes in the region was identified, 48 (6%), that were classified as "highly vulnerable", being both likely to be colonized and impacted by smallmouth bass. These lakes, the authors concluded, should have priority for the efforts aimed at minimizing further impacts of bass introductions in Ontario.

Certainly, the inherent difficulty in incorporating in a model the complexity of the community and the manifold biological relationships between NIS and residents has hampered the adoption of the modeling approach also in freshwater systems (Parker *et al.* 1999). Quantification of impacts may be made particularly complex due to a long list of confounding factors that include: the natural variability of the environment in space and time, other anthropogenic disturbances, synergistic effects of other established invaders (the "invasional meltdown", Simberloff and Von Holle 1999, see also Dermott *et al.* 1998), the original composition of the invaded community, the extremely variable time lag between initial introduction and detectable impact (Shigesada and Kawasaki 1997), and the "statistical shrinkage" (i.e. the fit of a regression model to new data is worse than the fit to the original data) lamented by Williamson (1999). All these multiple sources of variability influence the observed responses of a system and may explain why generalizations cannot be made from single experiments at small temporal and spatial scales (Ricciardi 2003). As a consequence, for many, if not most, known NIS, insufficient quantitative data are available to make useful comparisons between systems, times, and geographical areas.

To make the picture worse, when confronted with the increasing number of species that are moved outside their natural range for the first time, most often scientists do not dispose of an invasion history from which to draw predictive information (Ricciardi 2003). An approach might be to predict their impact from the invasion history of functionally similar organisms. For instance, the Asian freshwater mytilid mussel, *L. fortunei*, currently invading South America, displays a number of biological properties that are remarkably similar to those of *D. polymorpha*. Many of these properties are shared with other fouling bivalves, such as *Mytilopsis sallei* (Recluz), *Modiolus striatulus* (Hanley), *Perna viridis* (Linnaeus), and *Xenostrobus secures* (Lamarck). The life history of these species could therefore serve as a template to prioritize other potential pest bivalves without an invasion history (Ricciardi 2003). Taxonomic similarity may be a predictor of impact potential. Indeed, most invasive fouling bivalves belong to the same family, the Mytilidae, and the confamiliar fish, smallmouth bass, *M. dolomieui*, and rock bass, *Ambloplites rupestris* (Rafinesque), alter the food web to the same extent by reducing the diversity and abundance of littoral fish. There are however some remarkable exceptions, also within congeneric species. The rusty crayfish, *O. rusticus*, has greatly expanded its range and displaces indigenous crayfish in North American lakes and streams (Lodge *et al.* 2000), whereas several other congeneric species are declining (Taylor *et al.* 1996).

All the above examples and thoughts may support the pessimistic attitude of several ecologists about the role of scientific research in predicting invaders. Research certainly yields major insights into areas of ecology, evolution, and conservation biology, and the frequent "serendipity" in science ensures that some fraction of these insights will ultimately help management (Simberloff 2003b). But most of these findings will have "little direct relevance to the introduced species problem" (Simberloff 2003b) and precise predictions are often expected to be elusive. Agreeing with Ricciardi (2003), the point here is that lack of precision should not be viewed as a deterrent to developing predictive models where none exist. Even "crude" models, obviously based on reliable data, could be extremely helpful in providing valuable criteria for prioritizing invasion threats (Simon and Townsend 2003).

CONCLUSIONS

Several NIS are today affecting freshwater communities, imperiling indigenous species, altering ecosystem processes, and causing damage to human endeavors. Recognizing these threats certainly represents the strongest and possibly the only "ethical basis" for the concern that scientists, laypeople, and institutions have today about the problem of introduced species (Simberloff 2003a). This general awareness of the detrimental effects of several NIS is expected to translate soon into implemented policies aimed at preventing new undesirable introductions, responding quickly to newly discovered NIS, and controlling the most damaging established NIS. To succeed, however, all management and policy actions should be based on a sound understanding of the impact that target species exert, of their multilevel effects, and of the diverse expression of these effects over space and time. And they should also acknowledge that not all NIS exert a negative impact and the same species may have large effects in some areas and negligible ones in others (Byers *et al.* 2002). The so-often revealed idiosyncratic behavior of NIS can be properly faced – and reliable predictive models of their impact can be developed, only when a large amount of quantitative information is available. And the more idiosyncratic is a species' behavior, i.e. more noise is in the background, the larger is the information required. That is, in invasion biology, as within the entire scientific realm, "there is no data like more data" (Mercer 1998): the limiting factor for predicting the impact of NIS is the chronic scarcity of "numbers".

ACKNOWLEDGEMENTS

This paper was written while the author was acting as a Visiting Scholar at the Columbia University in the City of New York (NY, USA). My warmest thanks are directed to Professor Shahid Nahem for his kind hospitality.

REFERENCES

Alderman, D. J. and J. L. Polglase. 1988. Pathogens, parasites and commensals. Pages 167–212 in D. M Holdich and R. S. Lowery, editors. Freshwater crayfish: biology, management and exploitation. Croom Helm (Chapman & Hall), London, UK.

Barbiero, R. P. and M. L. Tuchman. 2004. Long-term dreissenid impacts on water clarity in Lake Erie. Journal of Great Lakes Research **30**, 557–565.

Bouchard, R. W. 1977. Distribution, systematics status, and ecological notes on five poorly known species of crayfishes in western North America (Decapoda: Astacidae and Cambaridae). Freshwater Crayfish **3**, 409–423.

Boudreau, S. A. and N. D. Yan. 2003. The differing crustacean zooplankton communities of Canadian Shield lakes with and without the nonindigenous zooplanktivore *Bythotrephes longimanus*. Canadian Journal of Fisheries and Aquatic Sciences **60**, 1307–1313.

Bright, C. 1998. Life out of bounds: Bioinvasion in a borderless world. Norton, New York, NY.

Byers, J. E., S. Reichard, J. M. Randall, I. M. Parker, C. S. Smith, W. M. Lonsdale, I. A. E. Atkinson, T. R. Seastedt, M. Williamson, E. Chornesky, and D. Hayes. 2002. Directing research to reduce the impacts of nonindigenous species. Conservation Biology **16**, 630–640.

Cadi, A. and P. Joly. 2004. Impact of the introduction of the red-eared slider (*Trachemys scripta elegans*) on survival rates of the European pond turtle (*Emys orbicularis*). Biodiversity and Conservation **13**, 2511–2518.

Carlton, J. T. 2002. Bioinvasion ecology: assessing invasion impact and scale. Pages 7–17 in E. Leppäkoski, S. Gollasch and S. Olenin, editors. Invasive aquatic species of Europe: Distribution, impacts and management. Kluwer Academic Publishers, Dordrecht, The Netherlands.

CBD. 2001. Invasive alien species: report on existing international procedures, criteria and capacity for assessing risk from invasive alien species, Convention on Biological Diversity. Subsidiary body on scientific, technical and technological advice, Sixth meeting, Montreal, Canada, 12–16 March 2001 UNEP/CBD/SBSTTA/6/INF/5.

Charlebois, P. M. and G. L. Lamberti. 1996. Invading crayfish in a Michigan stream: direct and indirect effects on peryphyton and macroinvertebrates. Journal of the North American Benthological Society **15**, 551–563.

Clinton, W. J. 1999. Executive Order 13112. Invasive species. The White House, Washington, DC.

Colautti, R. I., S. A. Bailey, C. D. A. van Overdijk, K. Amundsen, and H. J. MacIsaac. 2006. Characterised and projected costs of nonindigenous species in Canada. Biological Invasions **8**, 45–59.

Cox, G. W. 2004. Alien species and evolution. Island Press, Washington, Covelo, London.

Crowder, L. B. 1984. Character displacement and habitat shift in a native cisco in southeastern Lake Michigan: evidence for competition? Copeia **1984**, 878–883.

Dermott, R., J. Witt, Y. M. Um, and M. Gonzalez. 1998. Distribution of the Ponto-Caspian amphipod *Echinogammarus ischinus* in the Great Lakes and replacement of native *Gammarus fasciatus*. Journal of Great Lakes Research **24**, 442–452.

Devine, R. S. 1999. Alien invasion: America's battle with non-native animals and plants. National Geographic Society, Washington, DC.

Dick, J. T. A. and D. Platvoet. 2000. Invading predatory crustacean *Dikerogammarus villosus* eliminates both native and exotic species. Proceedings of the Royal Society of London B **267**, 977–983.

Diéguez-Uribeondo, J. and K. Söderhäll. 1993. *Procambarus clarkii* Girard as a vector for the crayfish plague fungus, *Aphanomyces astaci* Schikora. Aquaculture and Fisheries Management **24**, 761–765.

Di Justo, P. 2006. The alien invasion. Wired **14**, 46–47.

Eiswerth, M. E., S. G. Donaldson, and W. S. Johnson. 2000. Potential environmental impacts and economic damages of Eurasian watermilfoil (*Myriophyllum spicatum*) in Western Nevada and Northeastern California. Weed Technology **14**, 511–518.

Ewel, J. J., D. J. O'Dowd, J. Bergelson, C. C. Daehler, C. M. D'Antonio, D. Gomez, D. R. Gordon, R. J. Hobbs, A. Holt, K. R. Hopper, C. E. Hughes, M. Lahart, R. R. B. Leakey, W. G. Lee, L. L. Loope, D. H. Lorence, S. M. Louda, A. E. Lugo, P. B. Mcevoy, D. M. Richardson, and P. M. Vitousek. 1999. Deliberate introductions of species: research needs. BioScience **49**, 619–630.

Findlay, S. E. G., S. Dye, and K. A. Kuehn. 2002. Microbial growth and nitrogen retention in litter of *Phragmites australis* compared to *Typha angustifolia*. Wetlands **22**, 616–625.

Fleming, I. A., K. Hindar, I. B. Mjolnerod, B. Jonsson, T. Balstad, and A. Lamberg. 2000. Lifetime success and interactions of farm salmon invading a native population. Proceedings of the Royal Society of London B **267**, 1517–1523.

Gao, L. and B. Li. 2004. The study of a specious invasive plant, water hyacinth (*Eichhornia crassipes*): Achievements and challenges. Zhiwu Shengtai Xuebao **28**, 735–752.

Garton, D. W., C. D. Payne, and J. P. Montoya. 2005. Flexible diet and trophic position of dreissenid mussels as inferred from stable isotopes of carbon and nitrogen. Canadian Journal of Fisheries and Aquatic Sciences **62**, 1119–1129.

Gherardi, F. and A. Cioni. 2004. Agonism and interference competition in freshwater decapods. Behaviour **141**, 1297–1324.

Goldschmidt, T., F. Witte, and J. Wanink. 1993. Cascading effects of the introduced Nile perch on the detritivorous/phytoplanktvorous species in the sublittoral areas of Lake Victoria. Conservation Biology **7**, 686–700.

Gurevitch, J. and D. K. Padilla. 2004. Are invasive species a major cause of extinctions? Trends in Ecology and Evolution **19**, 470–474.

Hayes, J. P., S. Z. Guffey, F. J. Kriegler, G. F. McCracken and C. R. Rarker. 1996. The genetic diversity of native, stocked and hybrid populations of brook trout in the southern Appalachians. Conservation Biology **10**, 1403–1412.

Jerscke, J. M. and D. L. Strayer. 2005. Invasion success of vertebrates in Europe and North America. Proceedings of the National Academy of Sciences USA **102**, 7198–7202.

Johannsson, O. E., R. Dermott, D. M. Graham, J. A. Dahl, E. S. Millard, D. D. Myles, and J. LeBlanc. 2000. Benthic and pelagic secondary production in Lake Erie after the invasion of *Dreissena* spp. with implications for fish production. Journal of Great Lakes Research **26**, 31–54.

Jones, C. G., J. H. Lawton, and M. Shachak. 1994. Organisms as ecosystem engineers. Oikos **69**, 373–386.

Karatayev, A. Y., L. E. Burlakova, and D. K. Padilla. 1997. The effects of *Dreissena polymorpha* (Pallas) invasion on aquatic communities in eastern Europe. Journal of Shellfish Research **16**, 187–203.

Karatayev, A. Y., L. E. Burlakova, and D. K. Padilla. 2002. Impacts of zebra mussels on aquatic communities and their role as ecosystem engineers. Pages 433–446 in E. Leppäkoski, S. Gollasch and S. Olenin, editors. Invasive aquatic species of Europe: Distribution, impacts and management. Kluwer Academic Publishers, Dordrecht, The Netherlands.

Kasulo, V. 2000. The impact of invasive species in African lakes. Pages 183–207 in C. Perrings, M. Williamson and S. Dalmazzone editors. The Economics of Biological Invasions. Edward Elgar, Cheltenham, UK.

Kaufman, L. 1992. Catastrophic change in species-rich freshwater ecosystems: the lesson of Lake Victoria. BioScience **42**, 846–858.

Kelly, D. W. and J. T. A. Dick. 2005. Effects of environment and an introduced invertebrate species on the structure of benthic macroinvertebrate species at the catchment level. Archiv für Hydrobiologie **164**, 69–88.

Kelly, D. W., J. T. A. Dick., W. I. Montgomery, and C. MacNeil. 2003. Differences in composition of macroinvertebrate communities with invasive and native *Gammarus* spp. (Crustacea: Amphipoda). Freshwater Biology **48**, 306–315.

Kiesecker, J. M. and A. R. Blaustein. 1997. Population differences in responses of red-legged frogs (*Rana aurora*) to introduced bullfrogs. Ecology **78**, 1752–1760.

Krisp, H. and G. Maier. 2005. Consumption of macroinvertebrates by invasive and native gammarids: a comparison. Journal of Limnology **64**, 55–59.

Kupferberg, S. J. 1997. Bullfrog (*Rana catesbeiana*) invasion of a California River: The role of larval competition. Ecology **78**, 1736–1751.

Laxson, C. L., K. N. McPhedran, J. C. Makarewicz, I. V. Telesh, and H. J. MacIsaac. 2003. Effects of the non-indigenous cladoceran *Cercopagis pengoi* on the lower food web of Lake Ontario. Freshwater Biology **48**, 2094–2106.

Leung, B., D. M. Lodge, D. Finoff, J. F. Shogren, M. A. Lewis, and G. Lamberti. 2002. An ounce of prevention or a pound of cure: bioeconomic risk analysis of invasive species. Proceedings of the Royal Society of London B **269**, 2407–2413.

Light, T., D. C. Erman, C. Myrick, and J. Clarke. 1995. Decline of the Shasta crayfish (*Pacifastacus fortis* Faxon) of Northeastern California. Conservation Biology **9**, 1567–1577.

Lodge, D. M., C. A. Taylor, D. M. Holdich, and J. Skurdal. 2000. Nonindigenous crayfishes threaten North American freshwater biodiversity: lessons from Europe. Fisheries **25**, 7–20.

Mack, R. N., D. Simberloff, W. M. Lonsdale, H. Evans, M. Clout, and F. A. Bazzaz. 2000. Biotic invasions: causes, epidemiology, global consequences, and control. Ecological Applications **10**, 689–710.

MacIsaac, H. J. 1996. Potential abiotic and biotic impacts of zebra mussels on the inland waters of North America. American Zoologist **36**, 287–299.

MacNeil, C., N. J. Fielding, J. T. A Dick, M. Briffa, J. Prenter, M. J. Hatcher, and A. M. Dunn 2003. An acanthocephalan parasite mediates intraguild predation between invasive and native freshwater amphipods (Crustacea). Freshwater Biology **48**, 2085–2093.

McCarthy, J. M., C. L. Hein, J. D. Olden, and M. J. Vander Zanden. 2006. Coupling long-term studies with meta-analysis to investigate impacts of non-native crayfish on zoobenthic communities. Freshwater Biology **51**, 224–235.

McIntosh, A. R. and C. R. Townsend. 1994. Interpopulation variation in mayfly anti-predator tactics: differential effects of contrasting predatory fish. Ecology **75**, 2078–2090.

Mercer, R. L. 1998. Language modelling for speech recognition, IEEE Workshop on Speech Recognition, Arden House, Harriman, New York, NY, May 1988.

Miller, R. R., J. D. Williams, and J. E. Williams. 1989. Extinctions of North American fishes during the past century. Fisheries **14**, 22–38.

Nalepa, T. F., D. J. Hartson, D. L. Fanslow, and G. A. Lang 2001. Recent population changes in freshwater mussels (Bivalvia: Unionidae) and zebra mussels (*Dreissena polymorpha*) in Lake St. Clair, USA American Malacological Bulletin **16**, 141–145.

Nalepa, T. F. and D. W. Schloesser. 1993. Zebra mussels: biology, impacts and control. Lewis Publishers, Boca Raton, Florida, FL.

Nyström, P. 1999. Ecological impact of introduced and native crayfish on freshwater communities: European perspectives. Pages 63–84 in F. Gherardi and D. M. Holdich, editors. Crayfish in Europe as alien species: How to make the best of a bad situation? A. A. Balkema, Rotterdam, The Netherlands.

Nyström, P., O. Svensson, B. Lardner, C. Brönmark, and W. Granéli. 2001. The influence of multiple introduced predators on a littoral pond community. Ecology **82**, 1023–1039.

Oliveira de, M. D., A. M. Takeda, L. Fernandes de Barros, D. Savio Barbosa, and E. K. de Resende 2006. Invasion by *Limnoperna fortunei* (Dunker, 1857) (Bivalvia, Mytilidae) of the Pantanal wetland, Brazil. Biological Invasions **8**, 97–104.

Parker, I. M., D. Simberloff, W. M. Lonsdale, K. Goodell, M. Wonham, P. M. Kareiva, M. H. Williamson, B. Von Holle, P. B. Moyle, J. E. Byers, and L. Goldwasser 1999. Impact: toward a framework for understanding the ecological effects of invaders. Biological Invasions **1**, 3–19.

Perry, W. L., J. E. Feder, and D. M. Lodge 2001. Implications of hybridization between introduced and resident *Orconectes* crayfish. Conservation Biology **15**, 1656–1666.

Perry, W. L., D. M. Lodge, and J. E. Feder 2002. Importance of hybridization between indigenous and nonindigenous freshwater species: an overlooked threat to North American biodiversity. Systematic Biology **51**, 255–275.

Pimentel, D., R. Zuniga, and D. Morrison. 2005. Update on the environmental costs associated with alien-invasive species in the United States. Ecological Economics **52**, 273–288.

Pointier, J. P. 1999. Invading freshwater gastropods: Some conflicting aspects for public health. Malacologia **41**, 403–411.

Plummer, M. L. 2005. Impact of invasive water hyacinth (*Eichhornia crassipes*) on snail hosts of schistosomiasis in Lake Victoria, East Africa. EcoHealth **2**, 81–86.

Quinn, T. P., M. T. Kinnison, and M. J. Unwin 2001. Evolution of Chinook salmon (*Oncorhynchus tshawytscha*. populations in New Zealand: pattern, rate, process. Genetica **112–113**, 493–513.

Rehage, J. S., B. K. Barnett, and A. Sih. 2005. Foraging behaviour and invasiveness: do invasive *Gambusia* exhibit higher feeding rates and broader diets than their noninvasive relatives? Ecology of Freshwater Fish **14**, 352–360.

Reynolds, J. D. 1988. Crayfish extinctions and crayfish plague in central Ireland. Biological Conservation **45**, 279–285.

Ricciardi, A. 2003. Predicting the impacts of an introduced species from its invasion history: An empirical approach applied to zebra mussel invasions. Freshwater Biology **48**, 972–981.

Ricciardi, A., R. J. Neves, and J. B. Rasmussen. 1998. Impending extinctions of North American freshwater mussels (Unionoida) following the zebra mussel (*Dreissena polymorpha*) invasion. Journal of Animal Ecology **67**, 613–619.

Rosenzweig, M. 2001. The four questions: what does the introduction of exotic species do to diversity? Evolutionary Ecology Research **3**, 361–367.

Rozhnov, V. V. 1993. Extinction of the European mink: an ecological catastrophe or natural process? Lutreola **1**, 10–16.

Rudnick, D. and V. Resh. 2005. Stable isotopes, mesocosms and gut content analysis demonstrate trophic differences in two invasive decapod crustacean. Freshwater Biology **50**, 1323–1336.

Sagoff, M. 1999. What's wrong with exotic species? Report from the Institute for Philosophy & Public Policy **19**, 16–23.

Schlaepfer, M. A., P. W. Sherman, B. Blossey, and M. C. Runge 2005. Introduced species as evolutionary traps. Ecology Letters **8**, 241–246.

Seehausen, O., F. Witte, E. F. Katunzi, J. Smits, and N. Bouton. 1997. Patterns of the remnant cichlid fauna in southern Lake Victoria. Conservation Biology **11**, 890–904.

Shigesada N and Kawasaki K 1997. Biological invasions: theory and practice. Oxford University Press, Oxford, UK.

Simberloff, D. 2003a. Confronting introduced species: a form of xenophobia? Biological Invasions **5**, 179–192.

Simberloff, D. 2003b. How much information on population biology is needed to manage introduced species? Conservation Biology **17**, 83–92.

Simberloff, D. and B. Von Holle. 1999. Positive interactions of nonindigenous species: invasional meltdown? Biological Invasions **1**, 21–32.

Simon, K. S. and C. R. Townsend. 2003. Impacts of freshwater invaders at different levels of ecological organization, with emphasis on salmonids and ecosystem consequences. Freshwater Biology **48**, 982–994.

Slobodkin, L. B. 2001. The good, the bad and the reified. Evolutionary Ecology Research **3**, 1–13.

Smith, K. G. 2005. Effects of nonindigenous tadpoles on native tadpoles in Florida: evidence of competition. Biological Conservation **123**, 433–441.

Spencer, C. N., B. R. McLelland, and J. A. Stanford. 1991. Shrimp stocking, salmon collapse, and eagle displacement: cascading interactions in the food web of a large aquatic ecosystem. BioScience **41**, 14–21.

Strayer, D. L. 1999. Effects of alien species on freshwater mollusks in North America. Journal of North America Benthological Society **18**, 74–98.

Strayer, D. L., N. F. Caraco, J. J. Cole, S. Findlay, and M. L. Pace. 1999. Transformation of freshwater ecosystems by bivalves: a case study of zebra mussels in the Hudson River. BioScience **49**, 19–27.

Strayer, D. L., V. T. Eviner, J. M. Jeschke, and M. L. Pace. 2006. Understanding the long-term effects of species invasions. Trends in Ecology and Evolution **21**, 645–651.

Taylor, C. A., M. L. Warren Jr, J. F. Fitzpatrick Jr, H. H. Hobbs III, R. F. Jezerinac, W. L. Pflieger, and H. W. Robinson 1996. Conservation status of crayfishes of the United States and Canada. Fisheries **21**, 25–38.

Templer, P., S. Findlay, and C. Wigand 1998. Sediment chemistry associated with native and non-native emergent macrophytes of a Hudson River marsh ecosystem. Wetlands **18**, 70–78.

Townsend, C. R. 1996. Invasion biology and ecological impacts of brown trout *Salmo trutta* in New Zealand. Biological Conservation **78**, 13–22.

Vander Zanden, M. J., J. M. Casselman, and J. B. Rasmussen. 1999. Stable isotope evidence for the food web consequences of species invasions in lakes. Nature **401**, 464–467.

Vander Zanden, M. J., J. D. Olden, J. H. Thorne, and N. E. Mandrak. 2004. Predicting occurrences and impacts of smallmouth bass introductions in north temperate lakes. Ecological Applications **14**, 132–148.

Williamson, M. 1996. Biological invasions Chapman & Hall, London, UK.

Williamson, M. 1999. Invasions. Ecography **22**, 5–12.

Wei, A. and P. Chow-Fraser. 2006. Synergistic impact of water level fluctuation and invasion of *Glyceria* on *Typha* in a freshwater marsh of Lake Ontario. Aquatic Botany **84**, 63–69.

Wilcove, D. S., D. Rothstein, J. Dubow, A. Philips, and E. Losos 1998. Quantifying threats to imperiled species in the United States. BioScience **48**, 607–615.

Witt, A. M., J. M. Dettmer, and C. E. Caceres. 2005. *Cercopagis pengoi* in southwestern Lake Michigan in four years following invasion. Journal of Great Lakes Research **31**, 245–252.

Yan, N. D., R. Girard, and S. Boudreau. 2002. An introduced invertebrate predator (*Bythotrephes*) reduces zooplankton species richness. Ecology Letters **5**, 481–485.

Yan, N. D. and T. W. Pawson. 1997. Changes in the crustacean zooplankton community of Harp Lake, Canada, following invasion by *Bythotrephes cederstroemi*. Freshwater Biology **37**, 409–425.

Zavaleta, E. 2000. The economic value of controlling an invasive shrub. Ambio **29**, 462–467.

Chapter twenty-five

Invasion of the Baltic Sea basin by the Ponto-Caspian amphipod Pontogammarus robustoides *and its ecological impact*

Kęstutis Arbačiauskas and Simona Gumuliauskaitė

INTRODUCTION

Range extensions of amphipods and other aquatic macroinvertebrates in the inland waters of Europe have been mainly facilitated by the interconnection of river basins through artificial canals, intentional introductions, and shipping. Among the most successful and ecologically aggressive aquatic immigrants are Ponto-Caspian amphipods from the family Pontogammaridae. The southern and central water corridors connecting the Black Sea with the North and Baltic seas were used for active westward migration by *Dikerogammarus haemobaphes* (Eichwald) and *Dikerogammarus villosus* (Sowinsky) (see Bij de Vaate *et al.* 2002, Jazdzewski and Konopacka 2002), whereas the expansion across European inland waters outside native ranges by two other pontogammarid species, *Pontogammarus robustoides* (G. O. Sars) and *Obesogammarus crassus* (G. O. Sars) started from the point of their first intentional introduction in the Baltic Sea basin. In Lithuanian fresh waters where the conquest of new areas had begun, *P. robustoides* proved to be the most successful amphipod invader (Arbačiauskas

Francesca Gherardi, Biological invaders in inland waters: Profiles, distribution, and threats, 463–477.

2005). Today, its distribution range is rapidly expanding which may lead to dramatic changes in local communities, whether indigenous or previously altered by earlier invaders. This chapter will describe the spread of *P. robustoides*, along with its life history and ecological impact, and will analyse those factors in the environment that may affect the establishment of its populations.

INVASION HISTORY

During 1960–1961, approximately 1,600 specimens of *P. robustoides*, *O. crassus* and another Ponto-Caspian amphipod *Chaetogammarus warpachowskyi* (G. O. Sars) from the family Gammaridae, were translocated from Dniepr and Simferopol (the Crimea) water reservoirs into the newly constructed Kaunas Water Reservoir (WR) in the middle reach of the Nemunas River. Within a few years, these species reached the Curonian Lagoon by downstream dispersal (Gasiūnas 1963, 1972). After the establishment of pontogammarids and the gammarid *C. warpachowskyi* in Kaunas WR, attempts were made to introduce them into Lithuanian lakes and other reservoirs with the aim of improving fish production. Among others, *P. robustoides* appeared to be the species best adapted to stagnant environments. Currently, this species inhabits three water reservoirs and nine lakes (Fig. 1; Arbačiauskas 2005); for two of those nine lakes no official records of intentional introductions have been found. As natural dispersal into these lakes is not probable, these invasions must have been mediated by deliberate but officially unrecorded translocations. So far, the current distribution of *P. robustoides* across Lithuanian lakes and water reservoirs has definitely resulted from human activity asserted through intentional introductions (Arbačiauskas 2005).

In the 1960s, Ponto-Caspian species, including *P. robustoides*, were also transferred from Kaunas WR into the areas located to the north of Lithuania, in particular to Latvia, Estonia, and the St. Petersburg region (Gasiūnas 1972). Although their survival in the St. Petersburg region was not recorded, *P. robustoides* was detected in the Gulf of Finland (the Neva Bay) in 1999 as a possible result of penetration from the inland waters into which it had been intentionally introduced. Nevertheless, a shipping vector should not be ruled out (Berezina and Panov 2003). Recently, *P. robustoides* is widespread in Latvia; it occurs in the lower reaches and mouths of rivers emptying into the Baltic Sea, in water reservoirs located on the Daugava River, and in the onshore lakes connected to the brackish waters of the sea (Fig. 1; Grudule *et al.* 2007).

The expansion of *P. robustoides* is also ongoing in the southern part of the Baltic Sea basin (Fig. 1). In 1988, a well-established population was found in the lower Oder River and its estuary, Szczecin Lagoon (Gruszka 1999, Wawrzyniak-Wydrowska and Gruszka 2005). In 1996–1998, the species was found in the lower Vistula River and in the Vistula Lagoon (Konopacka 1998, Jazdzewski and Konopacka 2000). Further expanding upstream, the species colonized two

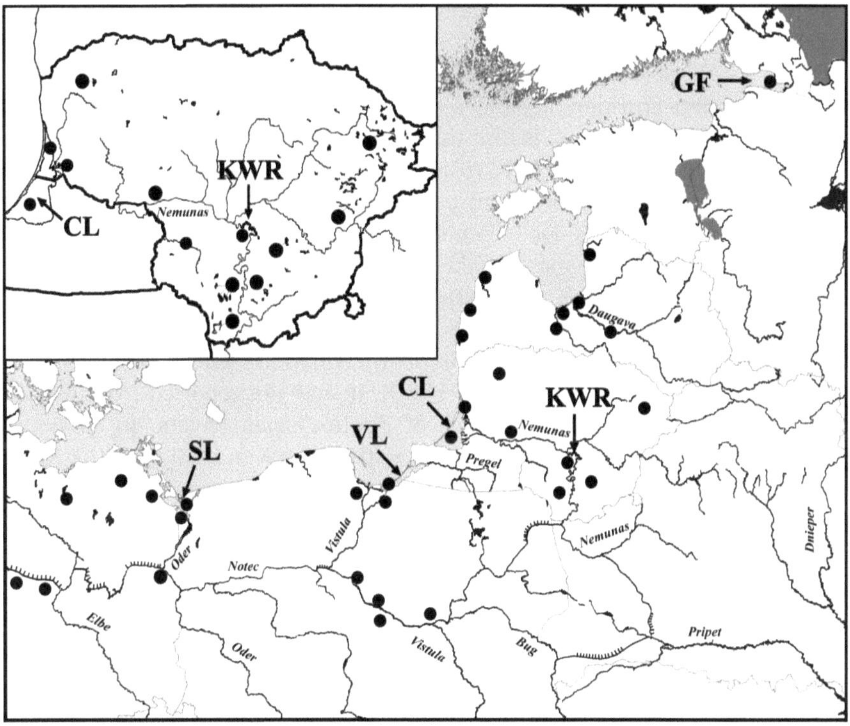

Fig. 1 Distribution of the Ponto-Caspian amphipod *Pontogammarus robustoides* in the Baltic Sea basin. Arrows indicate Kaunas Water Reservoir (KWR), Curonian Lagoon (CL), Vistula Lagoon (VL), Szczecin Lagoon (SL), and the eastern Gulf of Finland (GF). Circles indicate localities.

artificial water reservoirs (Wloclawski and Zegrzynski, WRs) located in the middle reaches of the Vistula River (Jazdzewski *et al.* 2002, Grabowski *et al.* 2006). Finally, for the first time in Poland, *P. robustoides* was recorded in a mesotrophic lake of glacial origin located in the Vistula valley, in central Poland in 2004, but not in the outflow river connecting the lake and the Vistula River, which might have served as a dispersal route (Grabowski and Bacela 2005).

Within north-eastern Germany, *P. robustoides* was first found during 1994 in Peenemundungsgebiet, Mecklenburg-Vorpommern (Rudolph 1997). Since then, the pontogammarid has been reported from the Mittelland Canal and its distribution across the inland waters of Mecklenburg-Vorpommern is increasing (Zettler 1998, 2002, Martens *et al.* 1999).

A few possible methods of invasion of *P. robustoides* into the deltas of the Vistula and Oder rivers have been suggested (Gruszka 1999, Jazdzewski and Konopacka 2000, Bij de Vaate *et al.* 2002, Jazdzewski *et al.* 2004). Firstly, transmission of this pontogammarid from the Curonian Lagoon via ballast waters; secondly, dispersal through the coastal waters of the Baltic Sea; and thirdly, a freshwater route from the Nemunas River basin via the Pregel River system which provides a direct connection between the Curonian and Vistula lagoons. As *P. robustoides* also was detected in the lower Vistula reaches, the central invasion corridor, i.e. the route connecting the Dnieper and Vistula basins via Pripet-Bug canal, is also under consideration.

The actual invasion path of *P. robustoides* into the basins of the Vistula and Oder rivers or whether more than one dispersal vector was operating simultaneously remains unresolved. Penetration through the central corridor by natural spread seems the most unlikely. In its native environment, this species inhabits the lower reaches and deltas of large Ponto-Caspian rivers, lagoons, and some brackish and freshwater Black Sea onshore lakes; its upstream expansion has occurred as a result of introductions into numerous water reservoirs (Dedyu 1980, Jazdzewski 1980). It was not reported from the lotic environments of the higher reaches of Ponto-Caspian rivers. In Lithuania, *P. robustoides* has not managed to colonize the upstream section of the Nemunas River in the 40 years since its introduction into the Kaunas WR, whereas the same upstream section is inhabited by the Ponto-Caspian amphipod *Chelicorophium curvispinum* (G. O. Sars) (Arbačiauskas 2005), which invaded the Baltic Sea basin through the central corridor (Jazdzewski 1980). In the Vistula River, *P. robustoides* dominates only lentic environments of water reservoirs, while the lotic sections are inhabited almost exclusively by another pontogammarid, *D. haemobaphes* (Jazdzewski *et al.* 2002). Hence, the ability of *P. robustoides*, originating from lentic or stagnant water environments, to spread against the flow in lotic waters is probably limited.

The penetration of *P. robustoides* by the freshwater route from the Curonian to the Vistula lagoon seems likely. The same route also is available for the Ponto-Caspian mysid *Paramysis lacustris* (Czerniavskyi), which is abundant in the Curonian Lagoon but, perhaps surprisingly, there is no record of its presence in the Vistula River system, although this mysid proved to have a substantially higher potential for natural dispersal than *P. robustoides* (Arbačiauskas 2005). The salinity of the Baltic coastal waters (7 psu) allows *P. robustoides* to maintain high survival rates and even to reproduce (Berezina and Panov 2003). Hence, expansion through coastal waters might be possible. However, a recent extensive survey of Baltic coastal waters of Poland revealed that this pontogammarid was absent in open-shore waters, although the Vistula and the Oder deltas harbour established populations of *P. robustoides* (Jazdzewski *et al.* 2005). Thus, natural dispersal through Baltic waters seems unlikely.

Within the Baltic Sea, *P. robustoides* can be transferred not only in ballast waters but also in the hull fouling of ships, especially over short distances by slow-speed vessels, such a dispersal method being possibly the most important. Across inland waters, the transmission of non-indigenous amphipods by boats has been documented in Germany (Reinhold and Tittizer 1999). Generally, the expansion of *P. robustoides* in the Baltic Sea basin follows the pattern of jump dispersal, clearly suggesting the involvement of anthropogenic factors.

ENVIRONMENTAL FACTORS THAT MAY LIMIT ESTABLISHMENT

Pontogammarus robustoides is described as a euryhaline freshwater species (Dedyu 1980) or as a brackish water species (Grabowski *et al.* 2005). Although able to endure the salinity of the Baltic Sea, it establishes sustainable populations only in salinities not exceeding 3–4 psu, which are characteristic of Baltic lagoons (Grabowski *et al.* 2006). The increase in freshwater ionic content that large European rivers suffered due to industrial and agricultural pollution has been suggested to be a cause for the explosion of successful aquatic invasions in European inland waters during the last decades of the 20th century (Jazdzewski *et al.* 2002, 2004). A similar hypothesis has been raised for invasions of the North American Great Lakes (MacIsaac *et al.* 2001). Indeed, the increase in salinity of the main European flowing waters may have facilitated the success of crustacean invaders, which are mostly euryhaline species with oligohaline preference (Bij de Vaate *et al.* 2002). This raises the question of what the lower limit of water ionic content might be for the establishment of a sustainable population of *P. robustoides*.

For the Gulf of Finland, it was hypothesized that the low concentration of chloride salts may limit the establishment of populations of *P. robustoides*, which requires at least 17 mg L^{-1} concentration of sodium for its successful reproduction (Berezina and Panov 2003). However, in Lithuanian waters, *P. robustoides* showed the potential to adapt and to establish viable populations under rather a low water ionic content, and even a five-fold lower sodium concentration (Table 1; see also Arbačiauskas 2005).

Many lakes in, at least, the southern part of the Baltic Sea basin satisfy the minimal demands of water ionic content for the establishment of *P. robustoides*. However, in spite of the extensive translocation effort, this species has exhibited long-term survival in relatively few Lithuanian lakes. These are typically large mesotrophic lakes, with the exception of one small eutrophic lake. The latter

Table 1 Water ionic content (mg L^{-1}) of Lithuanian lakes, in which pontogammarids *Pontogammarus robustoides* (Pr) and *Obesogammarus crassus* (Oc), and Ponto-Caspian gammarid *Chaetogammarus warpachowskyi* (Cw) have established sustainable populations. For other lake characteristics see Arbačiauskas (2005).

Lakes	Species	K^+	Na^+	Mg^{2+}	Ca^{2+}	SO_4^{2-}	Cl^-	Total ionic content
Plateliai	Pr	1.6	3.4	5.0	40.1	9.0	6.0	185
Seirijis	Pr, Cw	2.1	4.5	13.4	48.3	15.5	10.0	274
Dusia	Pr, Oc, Cw	3.0	7.7	17.7	44.8	28.3	13.9	296
Daugai	Pr, Oc, Cw	2.8	9.6	13.2	54.0	25.4	15.0	310

is located immediately downstream of a large lake hosting an abundant *P. robustoides* population; thus, the downstream population may be sustained by pontogammarid inflow. The disappearance of the species from three large eutrophic lakes has been recorded. In contrast, the indigenous gammarid *Gammarus lacustris* G. O. Sars, which can tolerate significant oxygen decrease, is widespread across lakes in which the pontogammarid failed to establish (see discussion in Arbačiauskas 2005). In turn, these data have led to the hypothesis that in higher latitudes where stagnant waters are ice-covered for a substantial portion of the year, oxygen content in the water during the winter may be a decisive factor for the long-term survival of *P. robustoides* (Arbačiauskas 2002, 2005). In eutrophic waters, the rise of adverse oxygen conditions is more likely than in waters of a lower nutrient status. Such circumstances under ice-cover, when retreat into a suitable refuge is impossible, may destroy even an established population. During an open-water phase, sufficient oxygen concentrations are always available in the shallowest waters. Therefore, in the Ponto-Caspian region where ice-cover is infrequent, *P. robustoides* also occurs in small eutrophic stagnant water bodies (M. Grabowski 2006, personal communication).

Pontogammarus robustoides is the most successful species across the stagnant fresh waters of Lithuania when compared with other Ponto-Caspian amphipods. Consequently, it shows the widest distribution and the highest population densities (Arbačiauskas 2002, 2005). Although less resistant than *G. lacustris*, *P. robustoides* probably has a greater tolerance of low oxygen in comparison with other Ponto-Caspian amphipod invaders of Lithuanian waters, and this has contributed to its success. According to Dedyu (1980), *P. robustoides* resistance to low oxygen content is greater than of all the other Ponto-Caspian amphipods which have invaded the Baltic Sea basin: *O. crassus*, *Chaetogammarus ischnus* (Stebbing), *C. curvispinum*, *C. warpachowskyi*, *D. haemobaphes*, and *D. villosus* (lethal oxygen concentration: 0.209 *vs.* 0.262, 0.290, 0.300, 0.308, 0.345, and 0.380 mg O_2 L^{-1}, respectively). The importance of oxygen in affecting habitat quality for *P. robustoides* also suggests that a correlation exists between the area of a lake and this species' abundance; in fact, due to wind-induced water motion, the littoral waters of large lakes, especially lakes with considerable fetch and wide littoral zones, contain high concentrations of dissolved oxygen (Arbačiauskas 2005).

Flow velocity is an additional environmental factor that must be discussed with respect to pontogammarid establishment. As already stated, the ability of *P. robustoides* to disperse against the flow in lotic conditions seems to be limited. In its native environment, this species establishes sustainable populations only in lentic or stagnant water environments such as lagoons, deltas, water ways, and reservoirs or lakes, as supported by several reports from the Baltic Sea basin (Jazdzewski *et al.* 2002, Zettler 2002, Berezina and Panov 2003, Grabowski and Bacela 2005, Grudule *et al.* 2007). Across Lithuania, *P. robustoides* only occurs in lotic environments in the Nemunas River downstream from Kaunas WR and in restricted parts of two other rivers, also downstream from abundant pontogammarid populations (see Arbačiauskas 2005). Thus, *P. robustoides* seems to avoid lotic waters.

LIFE HISTORY

The fecundity and body size of *P. robustoides* were estimated to be greatest among amphipods occurring in the fresh waters of Lithuania. For Kaunas WR, the individual weight of egg-bearing females ranged from 24 to 117 mg, and clutch size varied between 34 and 167 eggs (Gasiūnas 1972). This species was found to be one of the most fecund amphipod species occurring in Polish waters, especially in comparison with indigenous species (Bacela and Konopacka 2005). Clutch size varied between 11 and 185 eggs (mean: 65). The smallest observed body length of gravid females was 8.5 mm; the largest, 21 mm. Breeding lasted from the beginning of April until the first week of October, and showed three reproduction peaks, indicating three generations per year. Females of the overwintering generation were the largest in body size and produced the largest clutches, whereas females of the spring and summer generations showed lower values of these traits but they deposited clutches more than once (Bacela and Konopacka 2005).

The three seasonal peaks of pontogammarid progeny observed in the Curonian Lagoon (Jankauskiene 2002) suggest that in Lithuanian waters *P. robustoides* also produces three generations per year. In the Gulf of Finland, its reproduction lasts between May and October, and the number of generations was found to depend upon the seasonal pattern of water temperatures. During "normal" years, this pontogammarid produced three generations, but in 2003, when water temperatures were continuously low, only two generations were observed (N. A. Berezina 2006, personal communication).

So far, the life history characteristics of *P. robustoides* measured in the Baltic Sea basin suggest significant potential for this species to increase in numbers. Indeed, under favourable environmental conditions, its populations are usually more numerous than those of the indigenous amphipod species in similar environments (Gasiūnas 1972, Arbačiauskas 2002).

INTERACTION WITH OTHER AMPHIPODS

Of primary interest is the interaction of *P. robustoides* with the gammarid species indigenous to the Baltic Sea basin. Information from the Lithuanian inland fresh waters indicates competitive exclusion of the indigenous gammarid *G. lacustris* in stagnant water habitats sustaining abundant populations of the pontogammarid (Arbačiauskas 2002, 2005). Four years after the translocation of Ponto-Caspian species into the Kaunas WR, the indigenous species of the *Gammarus pulex* group (probably *Gammarus varsoviensis* Jazdzewski, which inhabits the River Nemunas upstream from the Kaunas WR), was present in different areas of the reservoir, although in low density (Gasiūnas 1972). However, indigenous amphipod species were absent in recently collected samples from the Kaunas WR. Similarly, *Gammarus pulex* (Linnaeus) and *G. lacustris* were reported in the Curonian Lagoon prior to the invasion of the Ponto-Caspian amphipods (Gasiūnas 1959), but not after.

The negative impact of the amphipod invaders (among which primarily the most successful was *P. robustoides*) on other indigenous gammarids, *Gammarus zaddachi* Sexton and *Gammarus duebeni* Liljeborg, has been documented in the Vistula Lagoon, although environmental factors, such as pollution and eutrophication, may also have been involved (Grabowski *et al.* 2006). So far, the negative influence of *P. robustoides* on indigenous *Gammarus* spp. is apparent in habitats which environmentally favour that pontogammarid. The absence of representatives of the genus *Gammarus* when pontogammarids are present has also been reported for the Ponto-Caspian region (Dedyu 1980).

An experimental study of the interaction between *P. robustoides* and *G. lacustris* clearly showed the negative impact that the pontogammarid exerted on the indigenous species (Fig. 2). Highest survival was observed for *G. lacustris* when

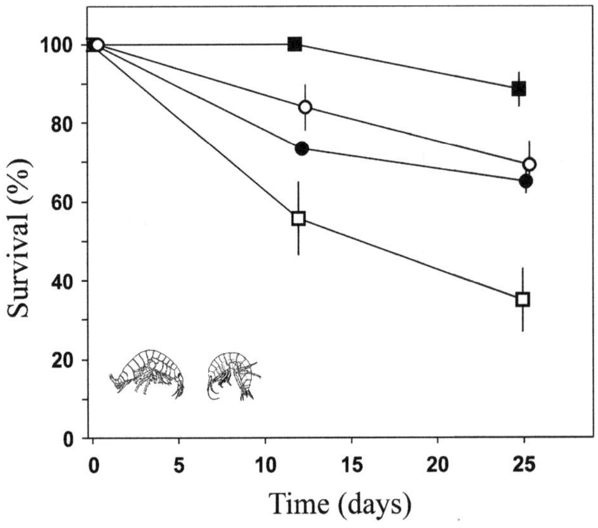

Fig. 2 Survival of *Gammarus lacustris* (squares) and *Pontogammarus robustoides* (circles) under control (closed symbols) and experimental (open symbols) conditions, i.e. when raised separately and together on filamentous algae *Cladophora* sp. as sole food. Error bars indicate SE of 3 (control) or 6 (experiment) replicates. When comparing all treatments, the highest and the lowest survival was found for *G. lacustris* under control and experimental conditions, respectively (Cox-Mantel test, $P \leq 0.009$), and did not differ in *P. robustoides* between treatments (S. Gumuliauskaitė and K. Arbačiauskas 2005, unpublished data).

raised separately, whereas the lowest survival was observed when raised together with the pontogammarid; the survival of *P. robustoides* under control and experimental conditions was similar. These results suggest a predatory impact of *P. robustoides* on *G. lacustris*. In addition, it seems that pontogammarids are aggressive not only towards gammarids, but also towards their congeners.

Thus, asymmetrical intraguild predation (see Polis *et al.* 1989) might be primarily responsible for the displacement of indigenous Baltic Sea basin gammarid species when they encounter the invasive *P. robustoides*. Such an interaction is common between invading and resident gammarid species (Dick 1996, Dick *et al.* 1999, 2002).

The greater aggressiveness of *P. robustoides* in intraguild predation has been shown to affect a dramatic decrease in the numbers of a smaller-sized invasive amphipod, *Gmelinoides fasciatus* (Stebbing), of Baikalian origin, in some habitats of the eastern Gulf of Finland (Berezina and Panov 2003). However, interactions of *P. robustoides* with other amphipod invaders of the Baltic Sea basin are not so clear. The Ponto-Caspian species *O. crassus*, although at low numbers, is capable of long-term survival with *P. robustoides* in the same habitats of lakes or water reservoirs (Arbačiauskas 2005). In lagoons, these pontogammarids share dominance in different areas and usually co-occur (Daunys and Zettler 2006, Grabowski *et al.* 2006). The ecological mechanism of such co-occurrence still remains unknown. In the lentic waters of reservoirs of the Vistula River, *P. robustoides* seems to be superior to another pontogammarid, *D. haemobaphes*, which dominates in the lotic habitats of this river (Jazdzewski *et al.* 2002). In northern Germany, the American invader *Gammarus tigrinus* Sexton and the Ponto-Caspian gammarid *C. ischnus* attained their highest abundance in the mid 1990s, whereas later, *P. robustoides* and, in succession, *D. villosus* have replaced those former amphipod species (Daunys and Zettler 2006).

In contrast, *G. tigrinus* recently appeared to be more successful than *P. robustoides* in the Szczecin and Vistula lagoons (Jazdzewski *et al.* 2005, Wawrzyniak-Wydrowska and Gruszka 2005, Grabowski *et al.* 2006). The displacement of the pontogammarid by the American invader due to asymmetrical intraguild predation seems unlikely, as the former species is highly aggressive and larger in body size, and its predation on *G. tigrinus* has been observed under laboratory conditions (Wawrzyniak-Wydrowska and Gruszka 2005). However, the interactions between amphipod species that invaded the Baltic Sea basin might be environmentally dependent and more complex than supposed, and are of interest for the understanding of the ecological impact of these invaders.

IMPACT ON LITTORAL MACROINVERTEBRATES

Results from an ongoing study of the impact of *P. robustoides* on lake littoral macroinvertebrates, comparing presence or absence of this species in similar lake habitats, suggest a significant pontogammarid effect on community metrics

(Fig. 3). In the habitats where *P. robustoides* is well-established and numerous, it significantly reduces species richness and diversity. However, moderate pontogammarid density in the habitats where the indigenous gammarid *G. lacustris* still survives did not reveal a negative impact on diversity indicators. The

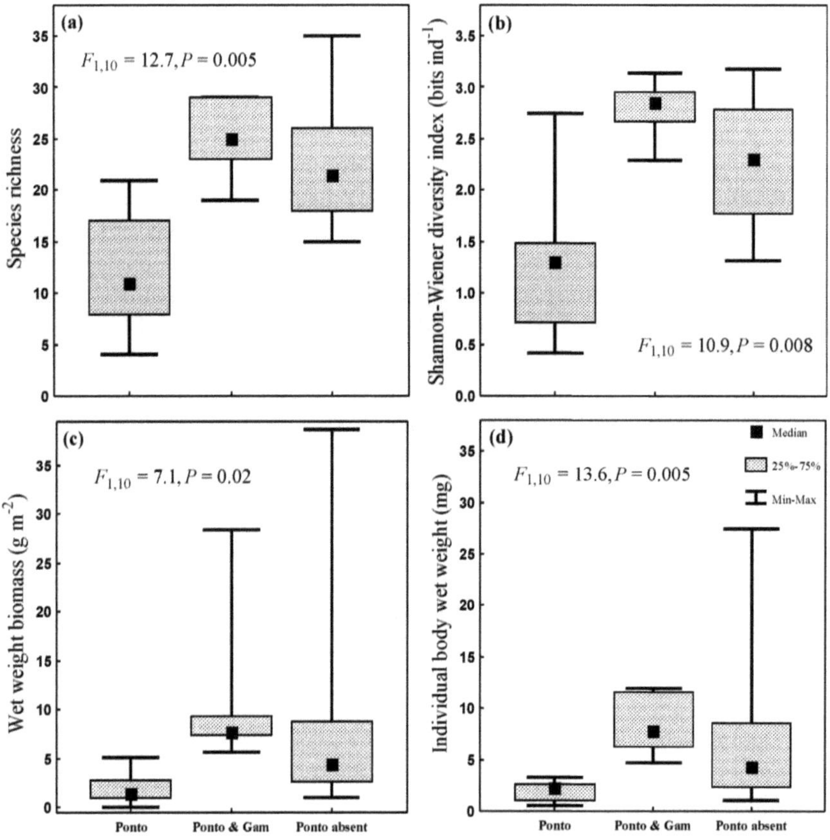

Fig. 3 Variation (median, quartiles, and range) of species richness (a), Shannon-Wiener diversity index (b), wet weight biomass of macroinvertebrates excluding *Pontogammarus robustoides* and chironomids (c), and individual macroinvertebrate body wet weight (pontogammarid excluded) (d) for the three types of lake littoral communities: "Ponto", with well-established and numerous *P. robustoides*; "Ponto & Gam", where *P. robustoides* and *Gammarus lacustris* co-occur; and "Ponto absent", without *P. robustoides*. For a test of well-established *P. robustoides* effect, data for the second and the third community types were merged. Untransformed (species richness and diversity index) or log-transformed (wet weight biomass) data were tested using a mixed model ANOVA with study site as a random factor nested in the *P. robustoides* factor (S. Gumuliauskaitė and K. Arbačiauskas 2004, unpublished data).

individual body size of other macroinvertebrates in habitats dominated by *P. robustoides* was also reduced. The negative impact on benthic biomass was observed only when chironimids (which exhibited high lake-specific biomass variation) were excluded. However, such a negative impact might be expected, as invasive amphipods are capable of negatively affecting both macroinvertebrate diversity and abundance (Kelly *et al.* 2003). The negative impact on chironomids is to be expected because *P. robustoides* is known to prey on them as suggested by gut content analysis of pontogammarids from the Gulf of Finland (Berezina *et al.* 2005).

This ongoing study also suggests that, when abundant, *P. robustoides* negatively affects the freshwater isopod *Asellus aquaticus* (Linnaeus), as this isopod was recorded in the presence of numerous *P. robustoides* at only one of five sites (S. Gumuliauskaitė and K. Arbačiauskas 2004, unpublished data). The detrimental impact that *P. robustoides* has on *A. aquaticus* probably derives from direct predation, as has been suggested elsewhere (Arbačiauskas 2005).

Furthermore, a negative impact of large numbers of non-indigenous amphipods, including *P. robustoides*, on the densities of benthic detritovores was also observed in the stony littoral of Neva Bay, the Gulf of Finland (Berezina and Panov 2003). Notwithstanding recently available information, more research on the effects of *P. robustoides* on macroinvertebrate communities is warranted.

OTHER ECOLOGICAL IMPACTS

Being omnivorous, *P. robustoides* may also affect the composition and abundance of plants. In the littoral zone of Lake Dusia (Lithuania), abundant overgrowths of the filamentous algae *Cladophora* sp. were common prior to the introduction of *P. robustoides*. However, the total extermination of algal overgrowths by pontogammarid grazing was recorded seven years after its introduction (Gasiūnas 1975). When *P. robustoides* attains high densities, the grazing impact exerted on *Cladophora* sp. may cause a dramatic decrease in macroalgal biomass, which has been suggested for the eastern Gulf of Finland (Berezina *et al.* 2005).

The purpose of the first and subsequent deliberate introductions of *P. robustoides* into the Baltic Sea basin was to improve fish production. Indeed, when this species is abundant, its contribution to the diet of various fish species is significant and frequently dominant (Bubinas 1979, K. Arbačiauskas 2002, unpublished data). Nevertheless, the impact that *P. robustoides* and the other introduced Ponto-Caspian peracaridan species exert on fish production in Lithuanian waters has never been precisely quantified, except for an unsubstantiated 20% increase indicated in a few fishery reports.

CONCLUDING REMARKS

The extension in the distribution range of the Ponto-Caspian amphipod *P. robustoides* in the Baltic Sea basin since its deliberate introduction shows a pattern of jump dispersal that indicates the involvement of vectors associated with human activity. In the colonized area, *P. robustoides* has great potential for increasing its abundance under favourable environmental conditions and exhibits the ability to establish sustainable populations where low water ionic content is found. This species prefers lentic or stagnant water environments. Due to its aggressiveness, *P. robustoides* affects indigenous gammarids and in appropriate conditions may be superior to other amphipod invaders. This invasive species is capable of negatively affecting indigenous macroinvertebrates and may dramatically impact other ecosystem components. A future expansion of *P. robustoides* beyond the Baltic Sea basin through vectors associated with human activity seems to be extremely likely. The presence of this pontogammarid in the Mittelland Canal actually means that the species has already invaded the North Sea basin. Successful colonization of *P. robustoides* probably can be predicted across habitats with lentic or stagnant fresh waters.

ACKNOWLEDGEMENTS

A draft of this manuscript profited from comments and suggestions of peer referees and Francesca Gherardi, the editor of this volume. The language was improved by Jonathan Grey. We thank all of them.

REFERENCES

Arbačiauskas, K. 2002. Ponto-Caspian amphipods and mysids in the inland waters of Lithuania: history of introduction, current distribution and relations with native malacostracans. Pages 104–115 in E. Leppäkoski, S. Gollasch and S. Olenin, editors. Invasive aquatic species of Europe. Kluwer Academic Publishers, Dordrecht, The Netherlands.

Arbačiauskas, K. 2005. The distribution and local dispersal of Ponto-Caspian Peracarida in Lithuanian fresh waters with notes on *Pontogammarus robustoides* population establishment, abundance and impact. Oceanological and Hydrobiological Studies **34**, Supplement 1, 93–111.

Bacela, K. and A. Konopacka. 2005. The life history of *Pontogammarus robustoides*, an alien amphipod species in Polish waters. Journal of Crustacean Biology **25**, 190–195.

Berezina, N., S. Golubkov, and J. Gubelit. 2005. Grazing effects of alien amphipods on macroalgae in the littoral zone of the Neva Estuary (Eastern Gulf of Finland, Baltic Sea). Oceanological and Hydrobiological Studies **34**, Supplement 1, 63–82.

Berezina, N. A. and V. E. Panov. 2003. Establishment of new gammarid species in the eastern Gulf of Finland (Baltic Sea) and their effect on littoral communities. Procceedings of the Estonian Academy of Sciences Biology Ecology **52**, 284–304.

Bij de Vaate, A., K. Jazdzewski, H. A. M. Ketelaars, S. Gollasch, and G. Van der Velde. 2002. Geographical patterns in range extension of Ponto-Caspian macroinvertebrate species in Europe. Canadian Journal of Fisheries and Aquatic Sciences **59**, 1159–1174.

Bubinas, A. 1979. Pitanije ryb vodokhranilischa Kaunasskoj GES akklimatizirovannymi v nem mizidami Kaspijskogo reliktovogo kompleksa *Mesomysis kowalewskyi* Czern. (*Paramysis lacustris*) [Feeding of fish in the Kaunas Hydro-Electric Plant Water Reservoir by acclimatized in it Mysidae of *Mesomysis kowalewskyi* Czern. (*Paramysis lacustris*) of the Caspian complex]. Lietuvos TSR Mokslų Akademijos Darbai Serija B **88** (4), 89–96.

Daunys, D. and M. L. Zettler. 2006. Invasion of the North American amphipod (*Gammarus tigrinus* Sexton, 1939) into the Curonian Lagoon, south-eastern Baltic Sea. Acta Zoologica Lituanica **16**, 20–26.

Dedyu, I. I. 1980. Amfipody presnykh i solonovatykh vod jugo-zapada SSSR [Amphipods of fresh and salt waters of the South-West part of the USSR], Shtiintsa Publishers, Kishivev, Moldova.

Dick, J. T. A. 1996. Post-invasion amphipod communities of Lough Neagh, N. Ireland: influences of habitat selection and mutual predation. Journal of Animal Ecology **65**, 756–767.

Dick, J. T. A., W. I. Montgomery, and R. W. Elwood. 1999. Intraguild predation may explain an amphipod replacement; evidence from laboratory populations. Journal of Zoology London **249**, 463–468.

Dick, J. T. A., D. Platvoet, and D.W. Kelly. 2002. Predatory impact of the freshwater invader, *Dikerogammarus villosus* (Crustacea: Amphipoda). Canadian Journal of Fisheries and Aquatic Sciences **59**, 1078–1084.

Gasiūnas, I. 1959. Kormovoj zoomakrobentos zaliva Kurshju Mares [Fodder macrozoobenthos of the Curonian Lagoon]. Pages 191–291 in K. Jankevičius, editor. Kurshju Mares [The Curonian Lagoon], Vilnius, Lithuania.

Gasiūnas, I. 1963. Akklimatizacija kormovykh rakoobraznykh (Kaspijskogo reliktovogo tipa) v vodokhranilische Kaunasskoj GES i vozmozhnosti ikh pereselenija v drugie vodojemy Litvy [The acclimatization of fodder crustaceans (of the Caspian relict type) into the reservoir of Kaunas Hydroelectric power station and the possibility of their transference into other water bodies]. Lietuvos TSR Mokslų Akademijos Darbai Serija C **30**(1), 79–85.

Gasiūnas, I. 1972. Obogoschenie kormovoj bazy ryb vodoemov Litvy akklimatizirovannymi rakoobraznymi Kaspijskogo kompleksa [Enrichment of fodder basis of water bodies of Lithuania by acclimatized crustaceans-like organisms from the Caspian Sea complex]. Pages 57–68 in J. Maniukas and J. Virbickas, editors. Voprosy razvedenija ryb i rakoobraznykh v vodoemakh Litvy [On the breeding of fish and crustacean-like organisms in the water bodies of Lithuania], Vilnius, Lithuania.

Gasiūnas, I. 1975. Peracarida oz. Dusya (bas. Baltijskogo morja) [Peracarida from Lake Dusya (Baltic Sea basin)]. Gidrobiologicheskij Zhurnal **11**(1), 46–50.

Grabowski, M. and K. Bacela. 2005. First finding of the Ponto-Caspian gammarid species *Pontogammarus robustoides* and *Dikerogammarus haemobaphes* (Crustacea: Amphipoda) in the post-glacial lake of the Vistula valley. Lauterbornia **55**, 107–111.

Grabowski, M., A. Konopacka, K. Jazdzewski, and E. Janowska. 2006. Invasions of alien gammarid species and retreat of natives in the Vistula Lagoon (Baltic Sea, Poland). Helgoland Marine Research **60**, 90–97.

Grudule, N., E. Parele, and K. Arbačiauskas. 2007. Distribution of Ponto-Caspian amphipod *Pontogammarus robustoides* in Latvian waters. Acta Zoologica Lituanica **17**, 28–32.

Gruszka, P. 1999. The River Odra estuary as a gateway for alien species immigration to the Baltic Sea basin. Acta Hydrochimica et Hydrobiologica **27**, 374–382.

Jankauskienė, R. 2002. Ponto-Kaspijos aukštesniųjų vėžiagyvių ir žuvų lervų gausumo ir biomasės sezoninė dinamika ir paros vertikalios migracijos Kuršių marių litoralėje [The density-biomass seasonal dynamics and diurnal vertical migrations of Ponto-Caspian higher crustaceans and fish larvae in the littoral zone of the Curonian Lagoon]. Ekologija Vilnius **4**, 21–28.

Jazdzewski, K. 1980. Range extension of some gammaridean species in European inland waters caused by human activity. Crustaceana Supplement **6**, 84–107.

Jazdzewski, K. and A. Konopacka. 2000. Immigration history and present distribution of alien crustaceans in Polish waters. Crustacean Issues **12**, 55–64.

Jazdzewski, K. and A. Konopacka. 2002. Invasive Ponto-Caspian species in waters of the Vistula and Oder basins and of the Southern Baltic Sea. Pages 384–398 in E. Leppäkoski, S. Gollasch and S. Olenin, editors. Invasive aquatic species of Europe. Kluwer Academic Publishers, Dordrecht, The Netherlands.

Jazdzewski, K., A. Konopacka, and M. Grabowski. 2002. Four Ponto-Caspian and one American gammarid species (Crustacea, Amphipoda) recently invading Polish waters. Contributions to Zoology **71**, 115–122.

Jazdzewski, K., A. Konopacka, and M. Grabowski. 2004. Recent drastic changes in the gammarid fauna (Crustacea, Amphipoda) of the Vistula River deltaic system in Poland caused by alien invaders. Diversity and Distributions **10**, 81–87.

Jazdzewski, K., A. Konopacka, and M. Grabowski. 2005. Native and alien malacostracan Crustacea along the Polish Baltic Sea coast in the twentieth century. Oceanological and Hydrobiological Studies **34**, Supplement 1, 175–193.

Kelly, D. W., J. T. A. Dick, W. I. Montgomery, and C. MacNeil. 2003. Differencies in the composition of macroinvertebrate communities with invasive and native *Gammarus* spp. (Crustacea: Amphipods). Freshwater Biology **48**, 306–315.

Konopacka, A. 1998. Nowy dla Polski gatunek kielza, *Dikerogammarus haemobaphes* (Eichwald, 1841) i dwa inne rzadkie gatunki skorupiakow obunogich w Wisle. Przeglad Zoologiczny **3–4**, 211–218.

MacIsaac, H. J., I. A. Grigorovich, and A. Ricciardi. 2001. Reassessment of species invasions concepts: Great Lake basin as a model. Biological Invasions **3**, 405–416.

Martens, A., T. O. Eggers, and K. Grabow. 1999. Erste Funde von *Pontogammarus robustoides* (Sars) im Mittellandkanal (Crustacea: Amphipoda). Lauterbornia **35**, 39–42.

Polis, G. A., C. A. Myers, and R. D. Holt. 1989. The ecology and evolution of intraguild predation: potential competitors that eat each other. Annual Review of Ecology and Systematics **20**, 297–230.

Reinhold, M. and T. Tittizer. 1999. Verschleppung von Makrozoen durch Kuhlwasserfilter eines Schiffes. Wasser und Boden **51**, 61–66.

Rudolph, K. 1997. Zum Vorkommen des Amphipoden *Pontogammarus robustoides* Sars, 1894 im Peenemundungsgebiet. Natur und Museum **127**, 306–312.

Wawrzyniak-Wydrowska, B. and P. Gruszka. 2005. Population dynamics of alien gammarid species in the River Odra estuary. Hydrobiologia **539**, 13–25.

Zettler, M. L. 1998. Zur Verbreitung der Malacostraca (Crustacea) in den Binnen- and Kustengewasser von Mecklenburg-Vorpommern. Lauterbornia **32**, 49–65.

Zettler, M. L. 2002. Crustaceologische Neuigkeiten aus Mecklenburg-Vorpommern. Archiv der Freunde der Naturgeschichte in Mecklenburg **41**, 15–36.

Chapter twenty-six

Changes in the aquatic systems of north-eastern Europe after invasion by Gmelinoides fasciatus

Nadezhda A. Berezina

INTRODUCTION

The main pathways which have caused an increase in the dispersal rate of amphipods in Europe are associated with human activity (Jażdżewski 1980). Destruction of natural geographic barriers has resulted in range expansions of many amphipod species in different directions. For example, since the 20th century, eight amphipod species of Ponto-Caspian, Baikalian, and Atlantic origin have expanded widely within Russia due to both several human-mediated vectors (shipping, intentional and accidental introductions, or natural migration via the constructed waterways of Europe) and the elevated plasticity of some non-indigenous species (Alimov and Bogutskaya 2004).

The successful establishment of non-indigenous amphipod species at new locations and often the consequent local extinction of indigenous species are the main causes of structural alterations of natural habitats. According to Elton (1958), each introduction of a new species, whether followed or not by an explosion of its density, exerts some influence on the indigenous communities and ecosystem stability. Even a relatively small number of specimens may be sufficient to cause severe damage in the recipient ecosystem (Mack *et al.* 2000). Especially in the cases of addition of functionally dominant species (keystone

Francesca Gherardi, Biological invaders in inland waters: Profiles, distribution, and threats, 479–493.

species, ecosystem engineers, or species with many trophic relations), the impact on the recipient ecosystem is strong, inducing rapid changes in the structure and functioning of communities (e.g. Olenin and Leppäkoski 1999, Alimov and Bogutskaya 2004, Orlova *et al.* 2006).

The strong predatory impact of non-indigenous amphipods on indigenous invertebrates and the related shifts in communities were found to occur in some of the aquatic ecosystems in Europe (Dick *et al.* 2002, Van der Velde *et al.* 2002, Kelly and Dick 2005, Kelly *et al.* 2006). Although the possible effects of invasive amphipods on invertebrate species and communities were described, there is a lack of detailed analyses of related changes in the recipient systems that require long-term observations.

In the 1960s and 1970s, the amphipod *Gmelinoides fasciatus* (Stebbing) has been abundantly introduced from Siberia into lakes and reservoirs in the former USSR with the purpose of enhancing fish production. These introductions resulted in a rapid expansion of this species' geographic range (Alimov and Bogutskaya 2004). While *G. fasciatus* before the 1960s was limited to the basins of Siberian rivers (Angara, Barguzin, Irtysh, Lena, Pyasina, Tunguska, Selenga, and Yenisey), it has spread into many waterbodies of Russia and adjacent countries ranging between latitudes 48°–74° N and longitudes 25°–110° E (Panov and Berezina 2002). *Gmelinoides fasciatus* is today recorded in many aquatic systems of European Russia (Fig. 1), where it has become a keystone species.

This chapter reviews the pathways taken by *G. fasciatus* from the Lake Baikal basin to the Baltic Sea basin (Lake Peipsi-Pihkva, Lake Ladoga, Lake Onega, and the Neva Estuary), describes the biological traits that make this amphipod a successful invader, and attempts to identify changes in the invertebrate communities of the diverse recipient ecosystems of north-eastern Europe after this species became established. Also, it assesses the risk of further invasions by *G. fasciatus*.

A HISTORY OF THE EXPANSION OF *G. FASCIATUS*

Study of the malacostracan fauna in large lakes of north-eastern Europe (the Baltic Sea basin) had begun at the close of the 19th century (Yarzhinskyi 1870). Before the 1970s, four amphipod species, the glacial relict *Gammaracanthus lacustris* Sars, *Monoporeia affinis* (Lindstrom), *Pallasea quadrispinosa* Sars, and the freshwater *Gammarus lacustris* Sars, inhabited Lake Ladoga and Lake Onega (Gerd 1946, Segerstråle 1954, Stalmakova 1968). *Pallasea quadrispinosa* and *G. lacustris* were common amphipods in Lake Peipsi-Pihkva (Timm and Timm 1993). First data on zoobenthos of Neva Bay (easternmost part of the Baltic Sea) connected with these large lakes were obtained at the beginning of the 20th century (Deryugin 1923, 1925), when *M. affinis*, *P. quadrispinosa*, and *Gammarus pulex* (Linnaeus) were reported as the most abundant crustacean

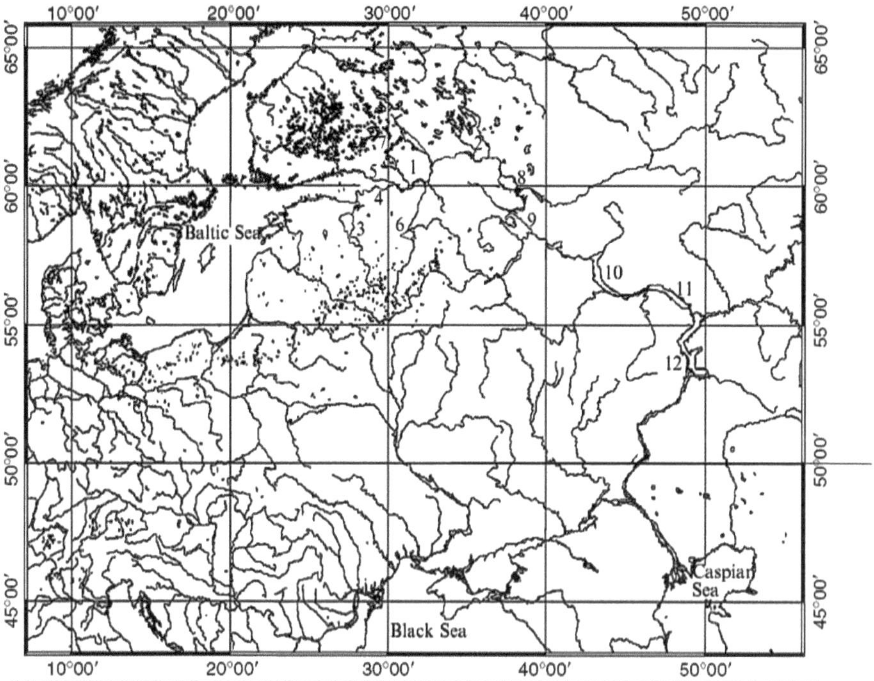

Fig. 1 Records of *Gmelinoides fasciatus* in the aquatic ecosystems of European Russia: Lake Ladoga (1), Lake Onega (2), Lake Peipsi-Pihkva (3), Neva Bay (4), eastern Gulf of Finland (5), Lake Ilmen (6), Lake Otradnoe (7), Lake Beloe (8), Rybinsky Reservoir (9), Gor'kovsky Reservoir (10), Cheboksarsky Reservoir (11), and Kuibyshevsky Reservoir (12).

species. According to the results of investigations of 1982–1984, two amphipods, *G. lacustris* and *P. quadrispinosa*, and the isopod *Asellus aquaticus* (Linnaeus) were common species in the littoral zone of the northern Neva Bay (Winberg and Gutelmakher 1987). While the density of *G. lacustris* did not exceed 700 m^{-2} in this period, the density of *A. aquaticus* was locally very high (10,000 m^{-2}). Considerable changes had been recorded by the beginning of the 21st century. The relict amphipods and *G. lacustris* had disappeared from Neva Bay; *G. pulex* became rare and some new crustacean species had appeared (Berezina and Panov 2003a, 2004b).

The newcomer *G. fasciatus* originated from the Lake Baikal basin. It was shown to be a successful invader, colonizing the majority of aquatic ecosystems in the Baltic Sea basin. *Gmelinoides fasciatus* was first found in Lake Peipsi-Pihkva in 1972 (Timm and Timm 1993). By 1988 it had invaded Lake Ladoga (Panov 1996). In the Neva Estuary this amphipod was discovered in 1996 (Berezina and Panov 2003a) and later (in 2001) it was recorded in Lake Onega (Berezina and Panov 2003b). The deliberate introduction of *G. fasciatus* and its subsequent natural expansion were the main causes of its spread into these systems.

In Lake Peipsi-Pihkva, *G. fasciatus* was introduced accidentally at the beginning of the 1970s during several attempts to enrich the indigenous population of *G. lacustris* by addition of specimens from a Siberian population (Timm and Timm 1993, Alimov and Bogutskaya 2004). These introductions were "contaminated", because the released material (several million amphipod specimens) contained a mixture of *G. fasciatus* (1–2% in density). The accidentally introduced *G. fasciatus* survived and was first recorded in Lake Peipsi-Pihkva in 1972 (Timm and Timm 1993). By 1990, it was established in the whole littoral zone of this lake (Timm *et al.* 1996, Panov *et al.* 2000).

An intentional introduction was the main cause of *G. fasciatus* expansion to Lake Ladoga. Between 1971 and 1976 this species (from the Posolskiy Bay of Lake Baikal) was introduced in six small lakes of the Karelian Isthmus (Arkhiptseva *et al.* 1977), linked to the lake-river system of the Vuoksa River connecting via the River Burnaya with Lake Ladoga. Earlier, in the late 1960s, several attempts to introduce another gammarid *G. lacustris* from Siberian lakes to some of these lakes were made (Karpevich and Lukonina 1972), but failed. *Gmelinoides fasciatus* was the more tolerant species and it became successfully established in some of these recipient lakes (Nilova 1976, Mitskevich 1981); it migrated via the lake-river system to other lakes of north-western Russia. In the late 1980s, *G. fasciatus* was already found at several sites in Lake Ladoga (Panov 1996, Kurashov *et al.* 1996, Slepukhina *et al.* 2000). By 2000 it had colonized the whole littoral zone of this lake, becoming a keystone species in the majority of sites (Berezina and Panov 2004a). The *G. fasciatus* population from Lake Ladoga became the source for secondary introductions of this species to other aquatic systems. In a short period, the Baikalian amphipod had expanded its range over several hundred kilometres to the west (Neva Estuary) and east (Lake Onega).

By the 1990s, *G. fasciatus* had reached the Neva Bay (the freshwater part of the Gulf of Finland). In 1999, it was recorded in the oligohaline Neva Estuary, which is the very first record of the Baikalian amphipod in brackish waters. Today it had become a common species in many habitats in the eastern Gulf of Finland dispersing in the area with a salinity of 0.05–2.0‰ (Berezina and Panov 2003a, Berezina *et al.* 2005, Haahti and Kangas 2006).

In 2001, *G. fasciatus* was found along the western shore of Lake Onega, where it has established dense populations (Berezina and Panov 2003b). This species was not recorded in Lake Onega during intensive survey before the mid-1990s (Polyakova 1999). It is likely that introduction of *G. fasciatus* occurred in the late 1990s. The migration route of this species to Lake Onega is most likely via the Svir' River (where it was also found), connecting the lake with Lake Ladoga. The second possible source of this invasion may be part of the population of Lake Beloe (where it was first recorded in 1994), migrating upstream through the Volga-Baltic waterway (the so-called northern invasion corridor, according to Bij de Vaate *et al.* 2002).

THE BIOLOGY OF *G. FASCIATUS*

Gmelinoides fasciatus belongs to the opportunistic ecological category or *r*-strategist: it has a short time for reproduction of one generation, fast growth and maturation, high resistance to environmental factors (including eutrophication and some pollutants), high genetic variability, and a wide food spectrum (Alimov and Bogutskaya 2004). These traits permit it to adapt and increase in density during a short period in the recipient ecosystems, becoming an abundant species and expanding to new areas.

It is a middle-sized amphipod. The sizes of newborn specimens range from 1.2 to 1.4 mm, reaching a maximum of 16 mm in body length (Bekman 1962, Panov and Berezina 2002). The sizes of ovigerous females range from 3.4 to 10 mm at different temperatures, with the smallest sizes found in Neva Bay during a hot summer period of 27–31 °C (Alimov and Bogutskaya 2004). Clutch sizes vary from 3 to 45 eggs depending on female body length. Embryogenesis lasts 180–220 degree-days, i.e. 10–12 days at 18 °C. Developmental time from hatched juveniles of *G. fasciatus* to maturation correlates to temperature and takes about 1,000–1,200 degree-days or 50–60 days at 18–20 °C.

Being omnivorous (Berezina 2005), *G. fasciatus* has many trophic links. While juveniles and specimens with body lengths up to 7 mm are preferably detritivorous (70–90% in relative abundance), the larger specimens are omnivores with about 35% of animal food such as infusorians, oligochaetes, nematodes, and larvae of chironomids, trichopterans, and ephemeropterans (Berezina 2007).

The environmental barriers for its successful establishment include hypoxia in water and sediments (oxygen $< 1\text{–}2\ \text{mgL}^{-1}$), pH < 6, a high content of humic acids in water (brown water), and soft water with calcium content less than 5–7 mgL^{-1} (Bekman 1962, Nilova 1976, Berezina 2001, 2003, Berezina *et al.* 2005). For example, *G. fasciatus* was either absent or rare (0–500 individuals m^{-2}) in some locations of Lake Ladoga, with "brown water" (i.e. water enriched by humic compounds) or near streams with discharged acidic water from wetlands and bogs. A water salinity of around 5‰ is the upper limit for survival of *G. fasciatus* adults, while salinity below 2‰ is needed for successful embryogenesis and the survival of neonates (Berezina *et al.* 2001).

Gmelinoides fasciatus was among the first invertebrates to recolonize the previously lifeless pulp-mill discharges in Lake Ladoga and was a very abundant component in some waterbodies under hard eutrophication. However, it did not colonize (or decreased considerably in its density) the habitats influenced by the waste water discharge of oil and aluminum plants. Also, strong chemical pollution of habitats by heavy metals or the release of warm water by power stations may cause a significant decrease in the density of *G. fasciatus* (Skalskaya 1998). Low densities of *G. fasciatus* were reported from the Novosibirsky Reservoir (the Ob' River basin) during abrupt fluctuations in water level (Vizer 2006).

CHANGES IN THE RECIPIENT COMMUNITIES

The deliberate introductions of *G. fasciatus* in many lakes and reservoirs of Russia and adjacent areas have had initially unexpected, long-term, and undesirable consequences. This amphipod, which was introduced with the aim of food enhancement for fish, occurred as the secondary food item in diets of marketable fish (pike-perch, dace, whitefish, and bream). However, being the main or a regular item in the diet of perch, roach, ruff, ide, and eel-pot, it facilitated population growth of the weed fish (Mitskevich 1981, Vizer 2006). At the same time, it became a keystone species in the majority of the recipient systems, affecting indigenous species and invertebrate communities.

Lake Ladoga

Lake Ladoga, the largest lake in Europe (18,135 km^2, mean depth 46.9 m), has a mesotrophic status in most areas but it is subject to intensive eutrophication in numerous shallow bays. From the 1960s to the 1980s, the indigenous *G. lacustris* was a common species in the lake, inhabiting macrophyte beds and exposed stones (Kuzmenko 1964, Panov 1996). In spite of being widespread, *G. lacustris* was not very abundant at the majority of sites, varying from 20 to 300 individuals m^{-2}, but it reached high densities (1,500–4,500 m^{-2}) at several sites. The indigenous isopod *A. aquaticus* occurred at densities of 100–4,500 m^{-2} (Berezina and Panov 2004a). In comparison, in 1988–90 the density and biomass of the newcomer *G. fasciatus* established at several sites ranged from 2,500 to 7,500 m^{-2}, exceeding 53,800 m^{-2} at one site (Panov 1996).

Mid-summer surveys in the littoral zone of Lake Ladoga in 2000 and 2005 showed that *G. fasciatus* was the dominant species in terms of density and biomass. Today its density is similar to the late 1980s level, varying from 2,000 to 10,000 m^{-2} in macrophyte beds and in exposed sand-stony littoral (Berezina and Panov 2004a), while the very high density (above 30,000 m^{-2}) was recorded in more eutrophic parts of the lake. The contribution of *G. fasciatus* to the total biomass of zoobenthos is very high, averaging 86% in the majority of habitats.

The range of *G. lacustris* has decreased considerably during last decades. In 2000, the species was recorded at only three locations (at the south-eastern and northern shores of Lake Ladoga and near Valaam Island) with densities of 16–254 m^{-2}. In 2005, *G. lacustris* was not recorded during sampling of whole coast of the lake except for one location at the northern coast, where *G. lacustris* is not endangered by the invader due to the low density of *G. fasciatus* at this site.

Lake Peipsi-Pihkva

Lake Peipsi-Pihkva is a large but shallow (3,558 km^2, up to 15.3 m deep) eutrophic lake. Before the invasion by *G. fasciatus*, the average density of the

Table 1 Maximum density and biomass of *Gmelinoides fasciatus* in the littoral zone of the studied waterbodies.

Waterbodies	Observation period	Maximum density, individuals m^{-2}	Maximum biomass, $g\ m^{-2}$	Author
Lake Peipsi-Pihkva	1980	20,000	80	Timm and Timm 1993
	1992	24,500	71	Timm and Timm 1993
	1993	29,500	137	Melnik 2000
	1996	17,000	102	Panov *et al.* 2000
	2000	83,000	225	Berezina and Panov 2004a
Lake Ladoga	1990	53,800	158.6	Panov 1996
	1996	54,500	135.8	Berezina and Panov 2004a
	2000	37,400	66	Berezina and Panov 2004a
	2005	34,600	48.9	Berezina *et al.*, unpubl. data
Lake Onega	2001	8,300	34	Berezina and Panov 2003b
Neva Estuary	1999	13,100	–	Berezina and Panov 2003a
	2001	17,500	40.9	Berezina and Panov 2003a
	2002	6,000	10.7	N. A. Berezina, unpubl. data
	2004	6,700	13.2	N. A. Berezina, unpubl. data
	2005	2,400	7.8	N. A. Berezina, unpubl. data

"small" animals of macrozoobenthos (excluding large molluscs) was 2,600 m^{-2}; and its biomass reached 12 g m^{-2} (Timm *et al.* 1996). Already in 1980, the density and biomass of zoobenthos had increased essentially on account of established *G. fasciatus*. Its densities had reached high values (up to 20,000 m^{-2}) at several sites in the lake (Table 1). In 1992, a gradual decrease in the abundance of the Ephemeroptera, Hydrachnellae, Oligochaeta, Pisidiidae, and the mollusc *Bithynia tentaculata* (Linnaeus) occurred in the shallow-water zone, mainly due to progressing eutrophication, while the abundance of *G. fasciatus* further increased (Table 1).

According to the data recorded since 1993 (Melnik 2000), the density of *G. fasciatus* reached its peak in the exposed sandy littoral of the lake (29,500 m^{-2}). Also, it was abundant in *Dreissena polymorpha* (Pallas) beds and among accumulations of the molluscs *Unio tumidus* Philipsson, *Anodonta anatina* (Linnaeus), and *Viviparus viviparus* (Linnaeus), ranging from 2,000 to 11,000 individuals m^{-2}. In 1996 the densities of *G. fasciatus* were in the range of 7,600–17,000 m^{-2} (Panov *et al.* 2000). In 2000 mid-summer densities of *G. fasciatus* varied from 10,000 to 20,000 m^{-2}, reaching locally a very high value (Table 1). The relative abundance of the Baikalian invader reached a maximum of 98% of the total zoobenthic biomass (Berezina and Panov 2004a).

In Lake Peipsi-Pihkva, *G. fasciatus* inhabits the *G. lacustris* habitats, replacing the former dominant indigenous species. While a rare finding of *G. lacustris* was recorded in the late 1990s, it was not found during an intensive survey in 2000 (Berezina and Panov 2004a).

Lake Onega

The mid-summer density and biomass of *G. fasciatus* in the oligotrophic Lake Onega (9,943 km^2, mean depth 26.8 m) varied between 2,200–5,200 m^{-2} and 4–27 g m^{-2} at different sites (Berezina and Panov 2003b). The maximum abundance of *G. fasciatus* was recorded in macrophyte beds (*Potamogeton gramineus* Linnaeus) in the south-western part of the lake (Table 1). The share of the species in bottom communities was considerable, constituting 50–83% of the total density and 46–79% of the total biomass. High densities and biomasses were recorded also for chironomids, oligochaetes, molluscs, stone flies, caddis flies, mayflies [*Baetis rhodani* (Pictet) and *Heptagenia sulfurea* (Müller)], and for the isopod *A. aquaticus*. At some locations, indigenous *G. lacustris* was found (50 individuals m^{-2}).

Neva Estuary

The Neva Estuary, the largest estuary (3,600 km^2) of the Baltic Sea, is one of the most eutrophic parts of the sea, impacted by a number of human activities. After the invasion by *G. fasciatus* in 1990s, densities and biomasses of benthos had increased two- to sevenfold. During the last several years, these values have been recorded as 17,000–20,000 m^{-2} and 30–40 g m^{-2} (Berezina 2005). At present, non-indigenous amphipods have played significant roles in this ecosystem, contributing 36–90% to the total biomass of benthos. Differences in the maximal densities of *G. fasciatus* in different years (Table 1) relate to changing salinity, the negative influence of macroalgal blooms, and predation by another invasive amphipod (Berezina and Panov 2003a, Berezina *et al.* 2005). The highest density of *G. fasciatus* (17,500 m^{-2}) was recorded in 2001 at the northern coast of Neva Bay. The indigenous amphipod *G. lacustris* disappeared from the fauna of Neva Bay after invasion by *G. fasciatus* (Berezina and Panov 2004b).

The phenomenon of the replacement of one amphipod species by another is common in this group of invertebrates (see Alimov and Bogutskaya 2004 and literature therein). *Gmelinoides fasciatus* typically replaces indigenous *G. lacustris* in Lake Peipsi-Pihkva, Lake Onega, Lake Ladoga, and the Neva estuary as the result of predation (N. A. Berezina 2006, unpublished data). It seems likely that it has also affected other indigenous crustaceans (*A. aquaticus* and *P. quadrispinosa*). In many cases, eutrophication facilitated the more rapid disappearance of more sensitive indigenous species and their replacement by the more tolerant invader *G. fasciatus*.

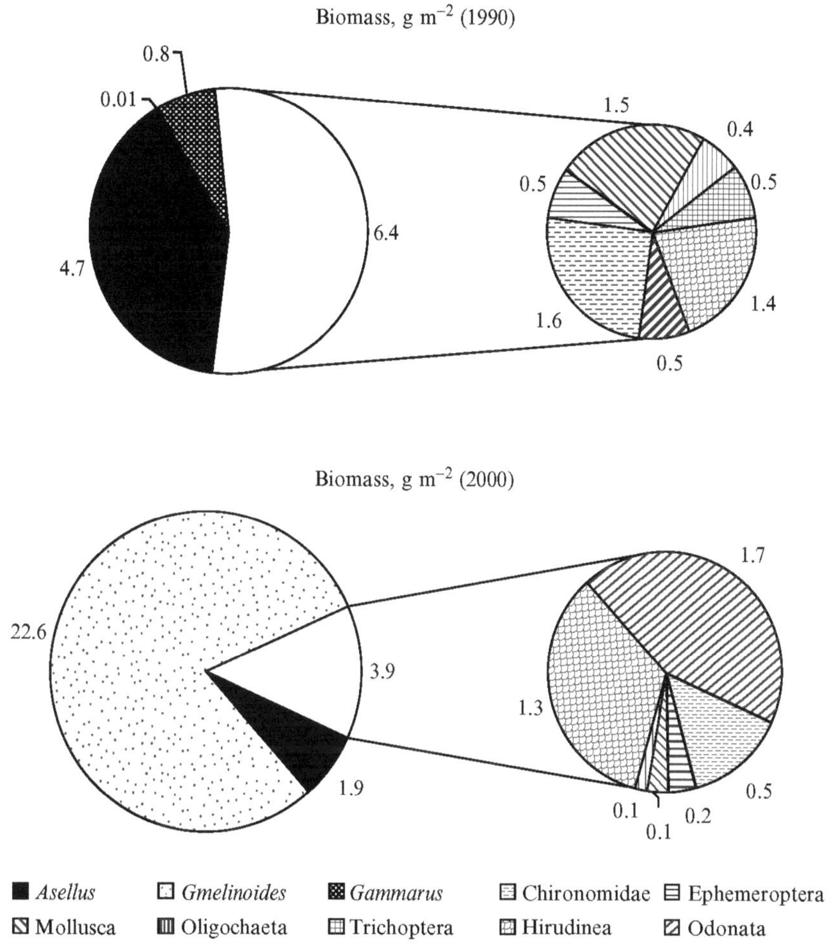

Fig. 2 Changes in the taxonomic composition and biomasses of macroinvertebrates during a decade (1990–2000) after *Gmelinoides fasciatus'* invasion at an undisturbed benthic community (Lake Ladoga, cape Bykovets, site 1).

Gmelinoides fasciatus may prey on macroinvertebrates (Berezina *et al.* 2005) that can result in community alterations. For example, observations of benthic communities in the coastal zone of Lake Ladoga after the invasion by *G. fasciatus* (in 1990 and 2000) showed that the taxonomic composition of the benthic community and the density of invertebrate groups have changed abruptly. Two examples of changing benthic communities from undisturbed Site 1 and disturbed eutrophic Site 2 can be considered as possible scenarios of community alteration caused by the invasive amphipod *G. fasciatus* (Figs. 2 and 3). As shown in Fig. 2, in 1990 (at the beginning of the invasion by *G. fasciatus*), site 1 in Lake Ladoga was characterized by a high taxonomical diversity with a prevalence of the isopod *A. aquaticus* (4.7 g m^{-2}) and the presence of the indigenous amphipod *G. lacustris* (0.8 g m^{-2}). In 2000, the total biomass of benthos was 2.2 times higher than in 1990 on account of *G. fasciatus* population growth that increased its biomass from 0.02 to 22.6 g m^{-2}. The predatory insect larvae (Odonata) and leeches (Erpobdellidae) remained at around 3 g m^{-2}, whereas the biomass of isopods and of other groups (trichopterans, chironomids, ephemeropterans, and oligochaetes), the favourite food items of *G. fasciatus* (Berezina *et al.* 2005), decreased at least two-fold. Similarly, the invasion of another amphipod, *G. pulex*, dramatically altered macroinvertebrate and fish communities through strong predatory and competitive interactions (Kelly *et al.* 2003, Kelly and Dick 2005). In particular, in the invaded sites of Irish rivers fewer taxa of macroinvertebrates, including a lower abundance of ephemeropterans, dipterans, and plecopterans, were found.

Another situation is shown by an example of community change at site 2, which was influenced by pulp-mill discharges until the 1980s that facilitated the decrease in the abundance of the benthic community and the disappearance of some crustaceans and insects before colonization by *G. fasciatus* in 1988 (Fig. 3). The more tolerant *G. fasciatus* rapidly established at this site, reaching a high biomass (42.2 g m^{-2}) as early as 1990. Other invertebrates formed a biomass of 8.5 g m^{-2}; among them, chironomids (3.4 g m^{-2}) and ephemeropterans (2.5 g m^{-2}) were the more abundant, whereas the biomass of *A. aquaticus* was insignificant (0.1 g m^{-2}). In the next decade, *G. fasciatus* increased its biomass slightly from 42.2 to 47.5 g m^{-2}, without however any modification in the overall biomass of the other benthic species (around 8 g m^{-2}). However, changes in the structure of the benthic community were observed: isopods and oligochaetes disappeared, ephemeropterans and chironomids decreased in biomass due to predation by *G. fasciatus*, while the biomass of predaceous trichopterans increased.

RISK OF FURTHER INVASION BY *G. FASCIATUS*

Nowadays, *G. fasciatus* range expansion in Europe is fast. For example, this species continues to expand its range to the west in the Baltic Sea and in 2005 it was recorded outside the Neva Estuary near Estonia and Finland, from which it may penetrate the Baltic Sea and other regions probably via ship ballast. This amphipod may invade the coastal waters of the Baltic Sea with salinities below 5 ppt, such as the Bothnian Bay (2–4 ppt), Curonian Lagoon (0.5–7.0 ppt), and Vistula Lagoon (1–7 ppt) (Berezina *et al.* 2001). However, its establishment will be confined to those habitats where salinity can decrease to 1–2 ppt at least temporarily.

The aquatic system of Lake Ladoga – Neva River and its estuary – Gulf of Finland is a hub for water transport to Europe, North America, and the inland waters of Russia. Recently, in the Gulf of Finland some new terminals in the

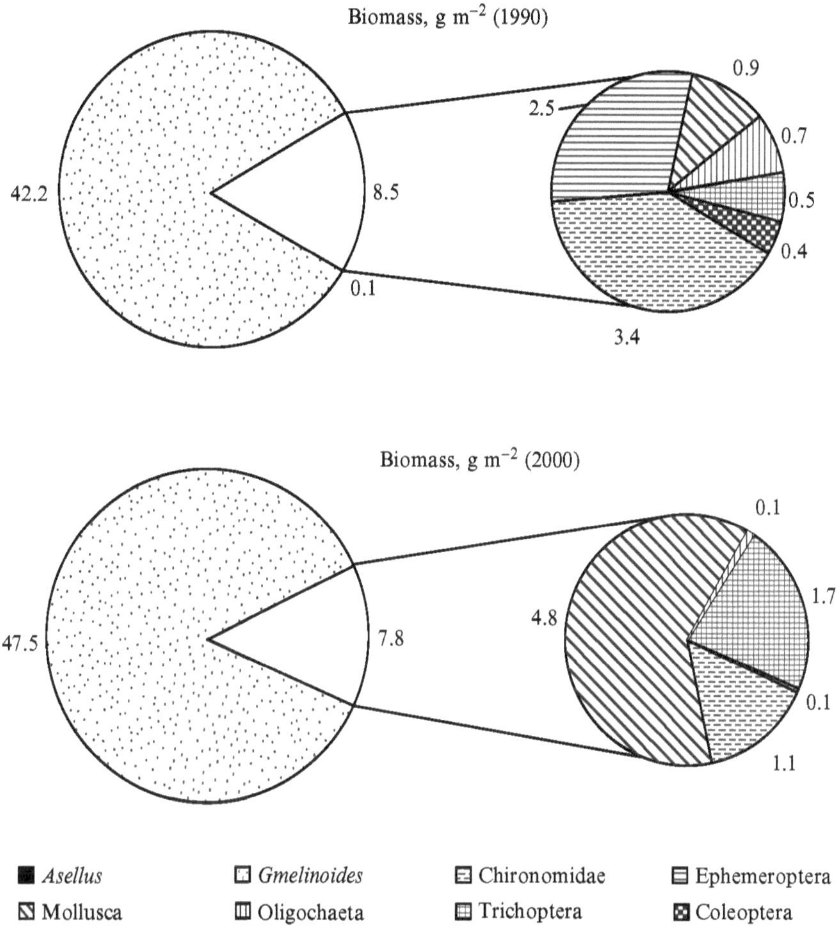

Fig. 3 Changes in the taxonomic composition and biomasses of macroinvertebrates during a decade (1990–2000) after *Gmelinoides fasciatus'* establishment at a disturbed benthic community (Lake Ladoga, Shchuchij Bay, site 2).

St. Petersburg area have started to function and new terminals are planned for construction that are likely to increase traffic and associated species invasions.

The risk of further invasions should be assessed on the basis of possible vector pathways, the invasibility of systems, and the ecological requirements of the invader, including information on its tolerance to environmental factors, life cycle traits, reproduction rate, food habits, energy requirements, and strength of its interactions with other species. Such a risk assessment of invasions was provided for the Finnish Lake District connecting to the Russian part of the eastern Gulf of Finland and Lake Ladoga area via canal-river systems (Pienimäki and Leppäkoski 2004). Low calcium and acidic conditions may limit the establishment of *G. fasciatus* in these lakes. Nevertheless, its spread from the Lake Ladoga basin to other areas is very likely.

CONCLUSIONS

Non-indigenous amphipod species cause dramatic alterations in aquatic communities because of their key role. The consequences of their invasions need a thorough assessment at genetic, species, community, and biotope levels to face the loss of biodiversity and disruption of the system stability that they may cause.

ACKNOWLEDGEMENTS

I warmly thank the editor for the invitation to contribute to this book and three reviewers for valuable comments. This study was supported by Russian Academy of Sciences Program ("Biodiversity"), President's Program "Scientific Schools of Russia" (grants # 5577.2006.4), the St. Petersburg's Scientific Centre of Russian Academy of Sciences and European Commission 6th Framework Programme Integrated Project ALARM.

REFERENCES

Alimov, A. F. and N. G. Bogutskaya, editors. 2004. Biological invasions in aquatic and terrestrial ecosystems. KMK Scientific Press, Moscow – St. Petersburg, Russia. (In Russian)

Arkhiptseva, N. T., I. V. Baranov, G. M. Zabelina, V. V. Pokrovsky, S. A. Serebrova, I. I. Tereshenkov, and G. A. Tsybaleva 1977. Lakes of the North Vuoksa basin. Proceedings of State Research Institute on Lake and River Fisheries **124**, 83–134. (In Russian)

Bekman, M. Y. 1962. Ecology and production of *Micruropus possolskyi* Sow. and *Gmelinoides fasciatus* Stebb. Proceedings of Limnological Institute of Siberian Department of Academy of Sciences of USSR **2**, 141–155. (In Russian)

Berezina, N. A. 2001. Influence of ambient pH on freshwater invertebrates under experimental conditions. Russian Journal of Ecology **32**, 343–351.

Berezina, N. A. 2003. Tolerance of freshwater invertebrates to changing salinity of water. Russian Journal of Ecology **4**, 261–266.

Berezina, N. A. 2005. Seasonal dynamics and fecundity of *Gmelinoides fasciatus* (Stebbing 1899) (Amphipoda: Crustacea) in macrophyte zone of northern Neva Bay (Gulf of Finland, Baltic Sea). Zoologichesky Zhurnal **84**, 411–419. (In Russian).

Berezina, N. A. 2007. Food spectra and consumption rates of four amphipod species from the North-West of Russia. Fundamental and Applied Limnology/Archiv für Hydrobiologie **168**, in press.

Berezina, N. A., S. M. Golubkov, and J. I. Gubelit. 2005. Grazing effects of alien amphipods on macroalgae in the littoral zone of the Neva Estuary (eastern Gulf of Finland, Baltic Sea). Oceanological and Hydrobiological studies **34**, 63–82.

Berezina, N. A., V. V. Khlebovich, V. E. Panov, and N. V. Zaporozhets. 2001. Salinity resistance in the amphipod *Gmelinoides fasciatus* (Stebb.) introduced into the Gulf of Finland basin (Baltic Sea). Doklady Biological Sciences **379**, 366–368.

Berezina, N. A. and V. E. Panov. 2003a. Establishment of new gammarid species in the eastern Gulf of Finland (Baltic Sea) and their effects on littoral communities. Proceedings of Estonian Academy of Sciences, Biology, Ecology **52**, 284–304.

Berezina, N. A. and V. E. Panov. 2003b. Establishment of the Baikalian amphipod *Gmelinoides fasciatus* (Amphipoda, Crustacea) in Lake Onega. Entomological Review **83**, S171–S174.

Berezina, N. A. and V. E. Panov. 2004a. Amphipod populations in littoral zones of the Neva estuary and large lakes of the Baltic Sea basin. Pages 179–189 in A. F. Alimov and M. B. Ivanova, editors. Hydrobiological regime of different aquatic ecosystems. Nauchnyi mir, Moscow, Russia. (In Russian)

Berezina, N. A. and V. E. Panov. 2004b. Distribution, population structure and salinity resistance of the invasive amphipod *Gmelinoides fasciatus* (Stebbing) in the Neva Estuary (Gulf of Finland, Baltic Sea). Hydrobiologia **514,** 199–206.

Bij de Vaate, A., K. Jażdżewski, H. A. M. Ketelaars, S. Gollasch, and G. van der Velde. 2002. Geographical patterns in range extension of Ponto-Caspian macroinvertebrate species in Europe. Canadian Journal of Fisheries and Aquatic Sciences **59**, 1159–1174.

Deryugin, K. M. 1923. Hydrology and benthos. Study of Neva River and its basin 2. Hydrological and hydrobiological investigations in the Neva Bay 1. Proceedings of Hydrological Institute **63**, 3–38. (In Russian)

Deryugin, K. M. 1925. Hydrology and benthos of the eastern part of the Gulf of Finland. Study of Neva River and its basin 2. Hydrological and hydrobiological investigations in the Neva Bay 4. Proceedings of the State Hydrological Institute **103,** 3–48. (In Russian)

Dick, J. T. A., D. Platvoet, and D. W. Kelly. 2002. Predatory impact of the freshwater invader *Dikerogammarus villosus* (Crustacea: Amphipoda). Canadian Journal of Fisheries and Aquatic Sciences **59**, 1078–1084.

Elton, C. S. 1958. The ecology of invasions by animals and plants. Methuen, London.

Gerd, S. V. 1946. Review of hydrobiological investigations of Karelian lakes. Proceedings of Karelian-Finnish Department of VNIIORKh, 26–140. (In Russian)

Haahti, H. and P. Kangas, editors. 2006. State of the Gulf of Finland in 2004. MERI. Report Series of the Finnish Institute of Marine Research **55,** 3–25.

Jażdżewski, K. 1980. Range extensions of some gammaridean species in European inland waters caused by human activity. Crustaceana Supplement **6**, 84–106.

Karpevich, A. F. and N. K. Lukonina. 1972. Transfer of fish and aquatic invertebrates in 1968. Voprosy Ikhtiologii, **12** 364–380. (In Russian)

Kelly, D. W., R. J. Bailey, C. MacNeil, J. T. A. Dick, and R. A. McDonald. 2006. Invasion by the amphipod *Gammarus pulex* alters community, composition of native freshwater macroinvertebrates. Diversity and Distributions **12,** 525–534.

Kelly, D. W., J. T. A. Dick, W. I. Mortgomery, and C. MacNeil. 2003. Differences in composition of macroinvertebrate communities with invasive and native *Gammarus* spp. (Crustacea: Amphipoda). Freshwater Biology **48**, 306–315.

Kelly, D. W. and J. T. A. Dick. 2005. Effects of environment and an introduced invertebrate species on the structure of benthic macroinvertebrate species at the catchment level. Archiv für Hydrobiologie **164**, 69–88.

Kurashov, E. A., I. V. Telesh, V. E. Panov, N. V. Usenko, and M. A. Rychkova. 1996. Invertebrate communities associated with macrophytes in Lake Ladoga: effects of environmental factors. Hydrobiologia **322**, 49–55.

Kuzmenko, K. N. 1964. To biology of the gammarid *Gammarus lacustris* (Sars) in Lake Ladoga. Pages 57–66 in S. V. Kalesnik, editor. Elements of Lake Ladoga regime. Nauka, Moscow-Leningrad, Russia. (In Russian)

Olenin, S. and E. Leppäkoski. 1999. Non-native animals in the Baltic Sea: alteration of benthic habitats in coastal inlets and lagoons. Hydrobiologia **393**, 233–243.

Orlova, M. I., I. V. Telesh, N. A. Berezina, A. E. Antsulevich, A. A. Maximov, and L. F. Litvinchuk. 2006. Effects of nonindigenous species on diversity and community functioning in the eastern Gulf of Finland (Baltic Sea). Helgoland Marine Research **2**, 98–105.

Mack, R. N., D. Simberloff, W. M. Lonsdale, H. Ewans, M. Clout, and F. A. Bazzaz. 2000. Biotic invasions: causes, epidemiology, global consequences, and control. Ecological Applications **10**, 689–710.

Melnik, M. M. 2000. Current status of macrozoobenthos in Lake Peipsi-Pihkva. PhD Thesis. St. Petersburg, Russia. (In Russian)

Mitskevich, O. I. 1981. Feeding on acclimatized gammarids by fish in Lake Otradnoe. Proceedings of State Research Institute on Lake and River Fisheries **173**, 71–76. (In Russian)

Nilova, O. I. 1976. Some features of ecology and biology of *Gmelinoides fasciatus* Stebb., acclimatized in Lake Otradnoe in Leningrad Region. Proceedings of State Research Institute on Lake and River Fisheries **110**, 10–15. (In Russian)

Panov, V. E. 1996. Establishment of the Baikalian endemic amphipod *Gmelinoides fasciatus* in Lake Ladoga. Hydrobiologia **322**, 187–192.

Panov, V. E. and N. A. Berezina. 2002. Invasion history, biology and impacts of the Baikalian amphipod *Gmelinoides fasciatus*. Pages 96–103 in E. Leppäkoski, S. Gollasch and S. Olenin, editors. Invasive aquatic species of Europe. Distribution, impacts and management. Kluwer Academic Publishers, Dordrecht, The Netherlands.

Panov, V. E., T. Timm, and H. Timm. 2000. Current status of an introduced Baikalian amphipod *Gmelinoides fasciatus* Stebbing, in the littoral communities of Lake Peipsi. Proceedings of the Estonian Academy of Sciences, Biology, Ecology **49**, 71–80.

Pienimäki, M. and E. Leppäkoski. 2004. Invasion pressure on the Finnish Lake District, invasion corridors and barriers. Biological Invasions **6**, 331–346.

Polyakova, T. N. 1999. Bottom biocenoses under conditions of anthropogenic eutrophication. Pages 211–227 in N. N. Filatov, editor. Lake Onega. Ecological Problems. Rossiiskaya Akademya Nauk, Petrozavodsk, Russia. (In Russian)

Segerstråle, S. 1954. The freshwater amphipods *Gammarus pulex* and *Gammarus lacustris* in Denmark and Fennoscandia – a contribution to the late and postglacial immigration history of the aquatic fauna in Northern Europe. Societas Scientiarum Fennica. Commentationes Biologicae **15**, 1–91.

Skalskaya, I. A. 1998. Establishment of the Baikalian amphipod in the Upper Volga basin. Pages 244–245 in Proceedings of International conference of environmental problems of large rivers basins. Toljatti, Russia. (In Russian)

Slepukhina, T. D., M. A. Barbashova, and G. F. Raspletina. 2000. Long-term successions and fluctuations of macrozoobenthos in different locations of Lake Ladoga. Pages 249–255 in N. N. Filatov, editor. Lake Ladoga. Monitoring, investigations of current state and management problems of Lake Ladoga and other large lakes. Rossiiskaya Akademiya Nauk, Petrozavodsk, Russia. (In Russian)

Stalmakova, G. A. 1968. Zoobenthos of Lake Ladoga. Pages 4–70 in Biological resources of Lake Ladoga. Zoology. Nauka, Leningrad, Russia. (In Russian)

Timm, V. and T. Timm. 1993. The recent appearance of Baikalian crustacean, *Gmelinoides fasciatus* (Stebbing, 1899) (Amphipoda, Gammaridae) in Lake Peipsi. Proceedings of the Estonian Academy of Sciences. Biology. Ecology **42**, 144–153.

Timm, T., K. Kangur, H. Timm, and V. Timm. 1996. Macrozoobenthos of Lake Peipsi-Pihkva, long-term biomass changes. Hydrobiologia **338**, 139–154.

van der Velde, G., I. Nagelkerken, S. Rajagopal, and A. bij de Vaate. 2002. Invasions by alien species in inland freshwater bodies in western Europe, the Rhine Delta. Pages 360–372 in E. Leppäkoski, S. Gollasch and S. Olenin, editors. Invasive aquatic species of Europe. Distribution, impacts and management. Kluwer Academic Publishers, Dordrecht, The Netherlands.

Vizer, A. M. 2006. Acclimatization of Baikalian gammarids and Far-Eastern mysids in Novosibirsky Reservoir. Ph D Thesis, Tomsk, Russia, 21 pp. (In Russian)

Winberg, G. G. and B. L. Gutelmakher, editors. 1987. Neva Bay, Hydrobiological investigations. Nauka, Leningrad, Russia. (In Russian)

Yarzhinskyi, F. F. 1870. On fauna of crustaceans of Lake Onega. Proceedings of Sankt-Petersburg Society of Scientists **1**, 61–63. (In Russian)

Chapter twenty-seven

The predatory impact of Dikerogammarus villosus *on fish*

Sandra Casellato, Alessandra Visentin, and Giovanni La Piana

INTRODUCTION

Dikerogammarus villosus (Sowinsky), an amphipod gammarid crustacean, is an extremely aggressive species to the point that it has been nicknamed "killer shrimp" for its behaviour towards indigenous invertebrate species (Bij de Vaate and Klink 1995, Nesemann *et al.* 1995, Tittizer 1996a, 1996b, Grabow *et al.* 1998, Devin *et al.* 2001, 2004, Müller *et al.* 2001, Jażdżewski and Konopacka 2002, Mürle *et al.* 2003, MacNeil and Platvoet 2005). From its original Ponto-Caspian area, it had invaded central and western Europe (Fig. 1) through the southern corridor connecting the Danube with the Rhine and the central corridor connecting the Dnieper with the Vistula, Oder, and Elbe basins (Bij de Vaate *et al.* 2002).

What is especially worrying is that this small crustacean, only a few centimetres long when fully grown, shortly after its arrival in new habitats eliminates all other gammarids, competing with them not only for food resources and space, but also attacking and devouring them. It often kills or simply bites off much more prey than it eats (Dick *et al.* 2002) and may severely damage the natural food webs of rivers and lakes. Numerous recent studies show that its aggressive behaviour is impressive: in all the central European hydro-systems in which it had settled, it has mostly replaced indigenous species (Dick

Francesca Gherardi, Biological invaders in inland waters: Profiles, distribution, and threats, 495–506.

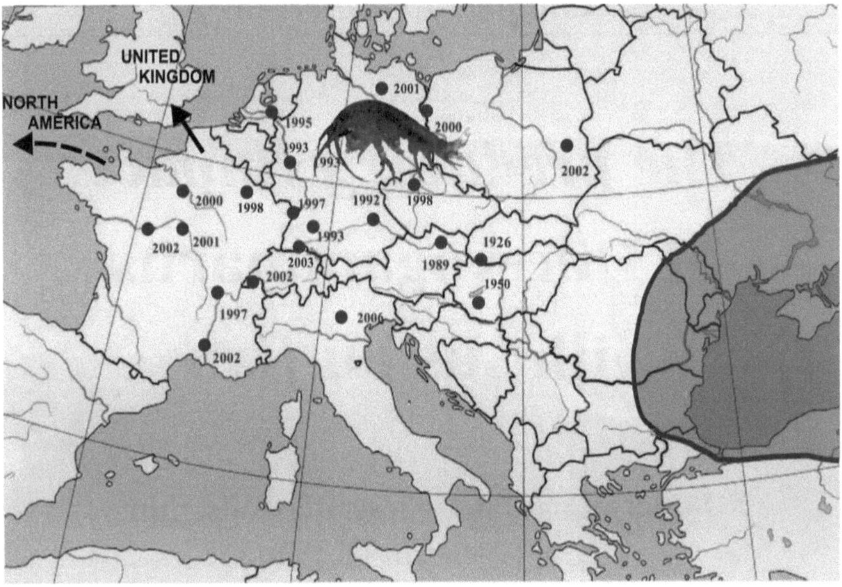

Fig. 1 Distribution range of *Dikerogammarus villosus* in Europe; the origin area is shown in grey with the year of the first record. Arrows indicate predicted further invasions. (From Bollache *et al.* 2004, modified).

and Platvoet 2000, Whitfield 2000, Van der Velde *et al.* 2002, Kinzler and Maier 2003, Bollache *et al.* 2004, MacNeil and Platvoet 2005). The decreased numbers of invertebrates recorded in many lakes and rivers are suspected to be due to the increasing numbers of *D. villosus* (Haybach *et al.* 2003). It is also regarded as a risk for the vulnerable early life-stages of some vertebrates, because it has been observed attacking small fish and fish larvae. The size of these prey, which are sometimes much larger than *D. villosus*, does not seem to pose an obstacle to its voracity (Müller *et al.* 2002, Devin *et al.* 2003, Schmidt and Josens 2004). Sometimes, *D. villosus* appears simply to attack and injure prey, indicating that its impact on prey is not limited by predator satiation (Dick *et al.* 2002).

Successful invasion by *D. villosus*

The species has a surprising capacity for adaptation, explaining the ease with which it settles in an enormous variety of both freshwater and brackish environments in Europe. Discussing the biological attributes associated with successful Ponto-Caspian invaders into Europe, Bij de Vaate *et al.* (2002) reported that the invasive potential of this species is related to its relatively short life-span and generation time, non-specific food preferences, and the capacity to tolerate wide ranges of water salinity. It can also tolerate wide ranges of temperature and water oxygenation, and adapts to several types of substrate (Devin *et al.* 2003, Wijnhoven *et al.* 2003). Finally, it can reproduce all year round (Kley and Maier 2003, Devin *et al.* 2004) and is capable to form good ecological segregation between adults and juveniles, thus lowering intraspecific competition between medium-sized and larger specimens (Devin *et al.* 2003, Casellato *et al.* 2006).

Not only a shredder and detritus feeder

Dikerogammarus villosus is widely distributed in the Ponto-Caspian area and is very common in the lower reaches and delta of the Danube (Kinzelbach 1995, Neseman *et al.* 1995). It was accidentally introduced into Lake Balaton in 1950, through the canal link with the Danube, where it replaced the indigenous species *Gammarus roselii* Gervais (Musko 1994). Its presence was reported by Musko (1989, 1990), together with *Chelicorophium curvispinum* (G. O. Sars) and *Dikerogammarus haemobaphes* (Eichwald), another Ponto-Caspian species. In this environment, it is not the most abundant species and no aggressive behaviour was observed, although it had been suspected of being more predatory than other gammarids (Marguiller *et al.* 1998), which have traditionally been viewed as principally herbivorous shredders and detritivores (Cummings and Klug 1979). In fact, this gammarid had already been mentioned in less recent literature as omnivorous and not simply as a shredder (Dedju 1980). Other gammarid species, such as *Gammarus pulex* (Linnaeus) and *Gammarus duebeni celticus* (Liljeborg), have been found to act as effective predators and not only as shredders or detritus feeders (Dick *et al.* 1993, 1995).

The associated species

A strong association has been often observed in many aquatic environments between *D. villosus* and the mollusc bivalve *Dreissena polymorpha* (Pallas) (zebra mussel), another Ponto-Caspian species widely distributed throughout the world (Minchin *et al.* 2002). In the River Moselle (Devin *et al.* 2004), in Lake Garda (Casellato *et al.* 2006), and in Lake Balaton (Musko and Bako 2005), they belong to the same assemblage. These two species have co-evolved over a long period of time; moreover, the presence of *D. polymorpha* seems to favour the amphipod by increasing the habitat complexity. This is consistent with the "invasional meltdown" theory (Simberloff and Von Holle 1999) according to which positive interactions between species with a common past increase the probability of the successful establishment of exotic species of the same origin. So, the previous establishment of *D. polymorpha* in European rivers and lakes may also have favoured the settlement of *D. villosus*. The zebra mussel has long been present in many European rivers and lakes (Giusti and Oppi 1972, Bedulli and Franchini 1978, Binelli *et al.* 1997, Minchin *et al.* 2002, Musko and Bako' 2005) and in the Great Lakes of North America (Stewart and Haynes 1994,

Ricciardi *et al.* 1997). In Lake Eire, in particular, it became firmly established and, after its arrival, an increase in the local amphipod fauna was observed with a reduction in copepods and rotifers. The reasons for these structural modifications in lacustrine communities were attributed to changes in the habitat caused by the presence of zebra mussel and its faeces deposition, which became another food resource for gammarids and for other macroinvertebrates, potential prey for *D. villosus* (Ricciardi *et al.* 1997, Gonzáles and Downing 1999). The arrival of *D. villosus* may have found the way already paved in that environment, as well as in the lakes Geneva, Constance, and Garda, where *D. polymorpha* had long been established.

To verify the hypothesized capability of *D. villosus* to prey, not only on other aquatic microinvertebrates but also on fish eggs, laboratory experiments were carried out in January 2005, when some fish spawn along the shorelines of the lakes. *Coregonus lavaretus* Linnaeus, an endemic species of Lake Garda, is a salmonid which lays many eggs in the gravelly bottoms. These bottoms are also preferred by *D. villosus* that has been observed in abundance near the areas where *C. lavaretus* lays its eggs.

THE FOOD PREFERENCE OF *D. VILLOSUS*

Experimental design

Specimens of *D. villosus* used for the experiments were collected by the "kick sampling" method (Barbour *et al.* 1999) (kick net of 1 mm mesh frame, aperture 250 mm) in sites of Lake Garda where a population of the invader can now be found (Casellato *et al.* 2006). Only healthy, visibly non-gravid and non-parasitized adults (approximately 20–25 mm body length) were selected. Fertilized whitefish eggs, all more or less at the same stage of embryonic development, were furnished by the Incubatoio Ittiogenico of Bardolino (Verona, Italy).

Laboratory experiments were performed in a thermostatic room at 15 °C. In experiment 1, small glass tanks (diameter: 14 cm) were filled with filtered aerated water taken from the collection sites, previously treated with a light antibacterial and antimycotic. To provide shelter for gammarids and to reproduce the natural environment, small stones and weeds were collected along the shores of the lake, washed, sterilized, and placed on the bottom of the tanks. Before the experiment, the eggs of *C. lavaretus* were kept in a Zug bottle, to avoid their agglutination and to preserve them in optimal conditions until their use for laboratory experiments.

In the first set of experiments, five replicates were performed for each of the following compositions: (a) one *D. villosus* offered with 10 fish eggs, (b) one *D. villosus* offered with 20 eggs, (c) two *D. villosus* offered with 10 eggs, and (d) two *D. villosus* offered with 20 eggs. Each test lasted four days; the number of opened eggs was checked twice a day and carefully observed under a binocular microscope to ascertain predation.

In the second set of experiments, five mesocosms were set up in small aquaria (30 × 30 × 10 cm), to verify whether *D. villosus* was able to choose among a few available prey, including *C. lavaretus* eggs. Three individuals of *D. villosus* with 25 adult specimens of each of the following species, *Asellus aquaticus* Linnaeus, *Echinogammarus stammerii* (S. Karaman), and 25 larvae of *Chironomus* sp. were kept together in each of the five replicates. Macroinvertebrates and eggs were introduced into the mesocosms and allowed to settle for 2 h before *D. villosus* specimens were added. The mesocosms were then monitored for a week. A control blank (i.e. a mesocosm without *D. villosus*) was also set up, to verify prey survival.

In experiment set l, data were analysed by a two-way ANOVA for the type-by-type comparisons [factor 1 and factor 2 being the composition of microcosm (a, b, c) and time (repeated measure: days 1, 2, 3, and 4), respectively] and by a one-way ANOVA for the day-by-day comparisons, the factor being the composition of microcosm. Data of experiment set 2 were analyzed by a one-way ANOVA, the factor being the type of prey (*Chironomus*, *A. aquaticus*, *E. stammeri*, fish eggs). A post-hoc analysis by Tukey's honest significance difference (Hsu 1996) was applied to evaluate differences among groups of the *n*-prey.

Whitefish eggs are consumed

In experiment set 1, significant differences were found among the four experiments ($F_{3,16} = 8.03$, $P = 0.0017$) but not among days ($F_{3,48} = 2.3$, $P = 0.089$). There are significant differences among the four types in the first day ($F_{3,16} = 3.6885$, $P < 0.05$), especially in the second day ($F_{3,16} = 11.193$, $P < 0.001$), whereas no significant differences about the eaten prey can be observed in the third ($F_{3,16} = 0.9237$, $P = 0.452$) and in the fourth day ($F_{3,16} = 0.8444$, $P = 0.489$) (Fig. 2). The role of available number of eggs does not seem to influence the behaviour of a single specimen of *D. villosus* ($F_{1,8} = 1.42$, $P = 0.267$). Instead, when two specimens are present, there is a very significant change in their attitude when the number of available eggs increases ($F_{1,8} = 9.87$, $P = 0.0138$). If the amount of available eggs is small, there is no significant difference in considering one or two *D. villosus* ($F_{1,8} = 0.36$, $P = 0.5651$). Instead, when more prey are available, there is a significant change in the behaviour of *D. villosus*; in fact, there is a clear increase ($F_{1,8} = 7.99$, $P = 0.0223$) in the number of the eggs eaten.

To ascertain the true extent of predation by the gammarids, the broken eggs were removed from the aquaria and observed under a binocular microscope in order to separate the eggs hatched by fries and those eaten by the gammarids. In the first case of hatching fry, the margin of eggs is smooth and the rim is curled on the inside, while the eggs shredded by the jaws and mandibles of *D. villosus* show the rim with serrated edges with some pieces missing (Fig. 3).

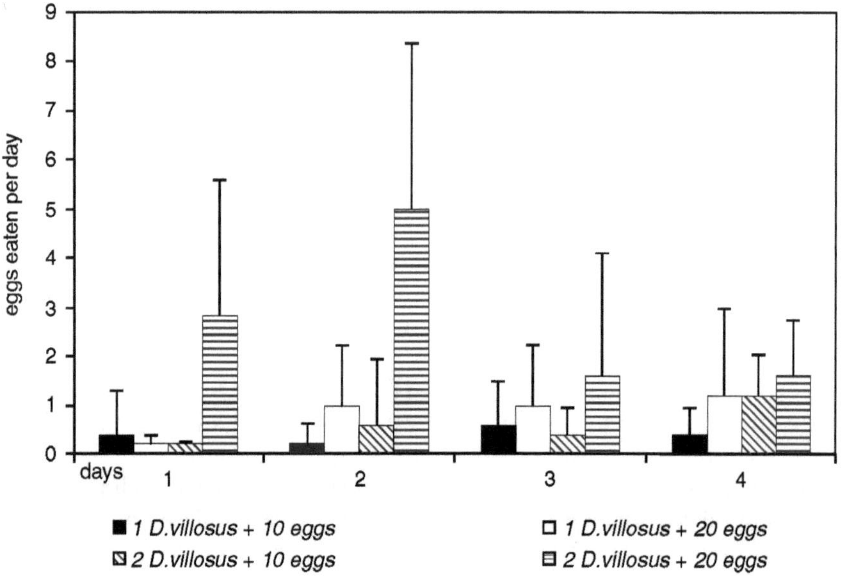

Fig. 2 Mean number of the whitefish eggs eaten in a day by one or two individuals of *Dikerogammarus villosus* when offered with 10 or 20 eggs in a laboratory experiment.

Whitefish eggs are preferred

At the end of experiment set 2, the specimens of *D. villosus* in each replicate had consumed most of the eggs and of chironomid larvae, but only a small part of the other prey (Fig. 4). Analyses of data showed strong evidence against the null hypothesis that the available kind of prey are equally eaten by *D. villosus* ($F_{3,16} = 28.192$, $P = 0$). Post-analysis test, applied to determine which prey is preferred by *D. villosus* (Table 1), showed that consumption of *Chironomus* larvae (55.2% eaten) and fish eggs (63.2% eaten) is higher than that of *A. aquaticus* (3.2% eaten) and *E. stammeri* (5.6% eaten).

In both experiments, all gammarids survived throughout the experiments in all replicates, indicating that the animals were not negatively affected by the laboratory conditions. The complete survival of each type of prey was also observed in the controls.

WHAT IS THE FUTURE FOR *D. VILLOSUS* AND WHITEFISH?

Our experiments show that whitefish eggs are most highly appreciated by *D. villosus*, together with chironomid larvae, at least in a laboratory mesocosm. This may explain its observed crowding in the spawning period of fish along the shores of Lake Garda, which are also the preferred habitats for the gammarid.

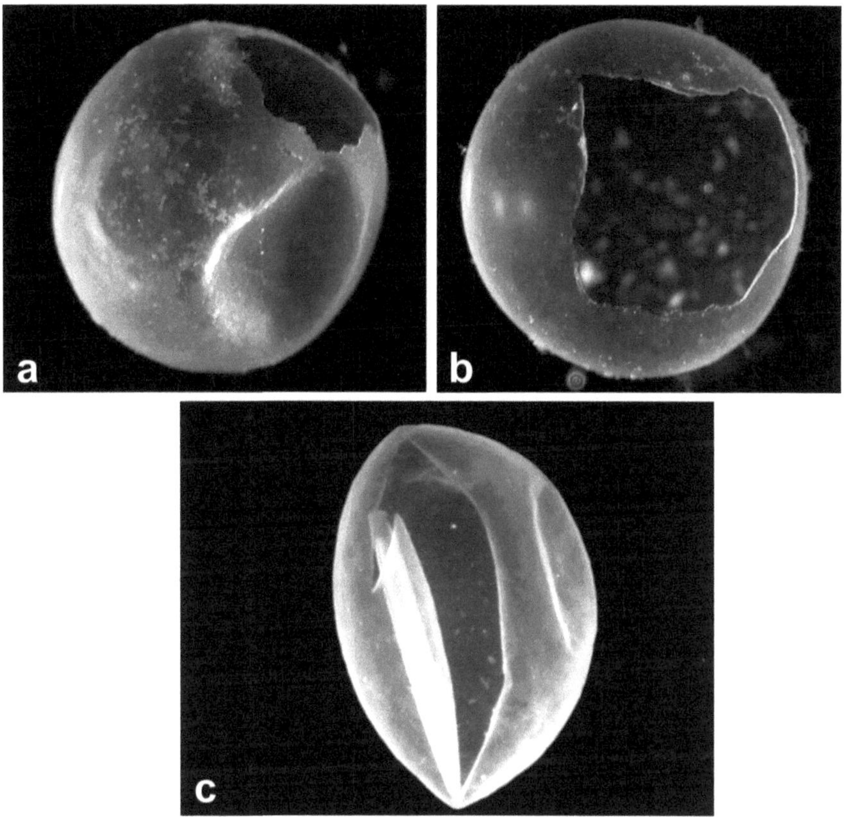

Fig. 3 Eggs opened (a) and eaten (b) by *Dikerogammarus villosus*; egg from which larvae had hatched (c).

Alarm about the "irresistible rise" of the "killer shrimp" will increase if our concern is not only directed to other invertebrates that share the spatial niche with *D. villosus*, but also to fish. Our laboratory experiments demonstrate that this species is able to break egg shells, using its mandibles and gnatopods. Fish production in Lake Garda, and possibly of other European lakes invaded by this species (Bollache 2004), could be seriously threatened if *D. villosus* populations continue to increase.

Experts on aquatic ecology predict that *D. villosus* will soon invade Great Britain and the North American Great Lakes, where the zebra mussel had already settled, a circumstance which seems to favour the arrival of compatriots (Ricciardi *et al.* 1997, Ricciardi and Rasmussen 1998, Ricciardi 2001).

What can stop the invasion?

It is difficult to imagine how the advance of *D. villosus* can be halted. There were at least three main vectors for its spread westwards: (i) its easy mobility along

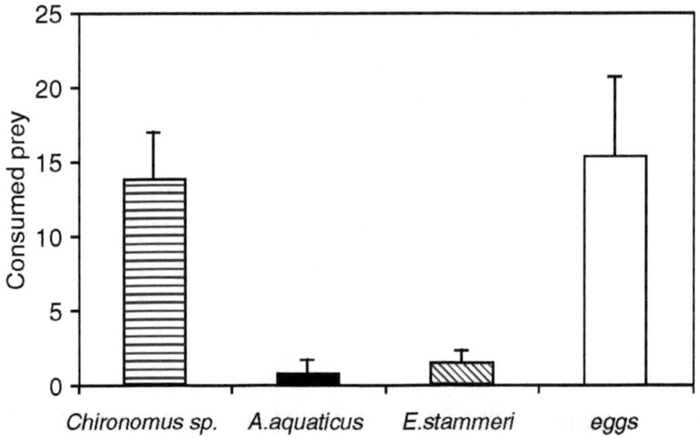

Fig. 4 Mean numbers ($n = 5$) of the prey consumed by *Dikerogammarus villosus* in a week.

Table 1 Multiple comparisons among types of prey; 1: *Chironomus* larvae, 2: *Asellus aquaticus* 3: *Echinogammarus stammeri*, 4: fish eggs. The first column indicates which groups are being compared, the 2nd, 3rd, and 4th columns represent respectively the observed mean differences and their confidence interval; the last column reports the *P*-value of each comparison controlling for multiplicity.

Type of comparison	Difference	Lower	Upper	Adjusted *P*
2 *vs.* 1	−13	−19.062	−6.938	0.000
3 *vs.* 1	−12.4	−18.462	−6.338	0.000
4 *vs.* 1	2	−4.062	8.062	0.782
3 *vs.* 2	0.6	−5.462	6.662	0.992
4 *vs.* 2	15	8.938	21.062	0.000
4 *vs.* 3	14.4	8.338	20.462	0.000

newly constructed canals connecting rivers and lake basins in Europe; (ii) passive transport in ship ballast; (iii) aquaculture management. There is no remedy for the first vector. For the second, *D. villosus* is known to dislike high levels of water salinity and proposals to discharge ballast water into salt water before ships reach their destinations may prevent the shrimp from arriving alive. On ships, ballast water and residues in tanks could be filtered or exposed to heat, chemicals, ultraviolet light, ozone, or other treatments to kill unwanted organisms (Taylor *et al.* 2002). However, although easy mobility and ballast water have been the most important carriers to central and western Europe for *D. villosus*, they were not the vectors to Italy. Lake Garda, the largest lake in Italy, is not linked in any way with the European fluvial-lacustrine network, but

D. villosus most probably reached it in consignments of fish larvae coming from fish-farms in the Caspian area; less probably, with boats transported to Lake Garda from other European lakes. The species is rapidly spreading to all lake basins and is in expansion because juveniles have the highest densities in the entire population over the year (Casellato *et al.* 2006). At the present time, it is impossible to arrest its invasion. Clearly, this species will shortly also spread throughout the Po basin, as a few individuals have already been found in the outlet river of Lake Garda, the Mincio (Casellato *et al.* 2006). We hypothesize that, in a short time, this non-indigenous species will be found in other aquatic environments in northern Italy. Studies on its biology indicate that just a few individuals are sufficient to establish new populations rapidly in a new aquatic system – an enormous risk, not only for other aquatic invertebrates but also for fish.

ACKNOWLEDGEMENTS

We thank Dr. Dario Basso for help and suggestions for the statistical analyses of experimental data and Dr. Ivano Confortini for having kindly supplied the *C. lavaretus* eggs from the Incubatoio Ittiogenico of Lake Garda.

REFERENCES

Barbour, M. T., J. Gerritsen, B. D. Snyder, and J. B. Stribling. 1999. Rapid bioassessment, protocols for use in streams and wadeable rivers: periphyton, macroinvertebrate and fish. 2nd edition. EPA 841-B-99-002. EPA, Washington, DC.

Bedulli, D. and D. Franchini. 1978. *Dreissena polymorpha* (Pallas, 1771): primi rinvenimenti sul fiume Po. Quaderni della Stazione Idrobiologia di Milano **6**, 85–92.

Bij de Vaate, A., K. Jażdżewski, H. A. M. Ketelaars, S. Gollasch, and G. Van Der Velde. 2002. Geographical patterns in range extension of Ponto-Caspian macroinvertebrate species in Europe. Canadian Journal of Fisheries and Aquatic Sciences **59**, 1159–1174.

Bij de Vaate, A. and A. G. Klink. 1995. *Dikerogammarus villosus* (Sowinsky, 1894) (Crustacea: Gammaridae) a new immigrant in the Dutch part of the Lower Rhine. Lautebornia **20**, 51–54.

Binelli, A., A. Provini, and S. Galassi. 1997. Throphic modification in Lake Como (N Italy) caused by zebra mussel (*Dreissena polymorpha*). Water, Air and Soil Pollution **99**, 633–640.

Bollache, L. 2004. *Dikerogammarus villosus* (Crustacea: Amphipoda): another invasive species in Lake Geneva. Revue Suisse de Zoologie **111**, 309–313.

Bollache, L., S. Devin, R. Wattler, M. Chovet, J. N. Beisel, J. C. Moreteau, and T. Rigaud. 2004. Rapid range extension of the Ponto-Caspian amphipod *Dikerogammarus villosus* in France: potential consequences. Archiv für Hydrobiologie **160**, 57–66.

Casellato, S., G. La Piana, L. Latella, and S. Ruffo. 2006. *Dikerogammarus villosus* (Sowinsky, 1984) (Crustacea, Amphipoda, Gammaridae) for the first time in Italy. Italian Journal of Zoology **73**, 97–104.

Cummins, K. W. and M. J. Klug. 1979. Feeding ecology of stream invertebrates. Annual Review of Ecology and Systematics **10**, 147–172.

Dedju, I. 1980. Amphipods of fresh and salt water of the south-west part of USSR Ya I. Tarobogatov, editor. Shtiintsa Publisher, Kishinev, Moldavia, 220 pp. (In Russian)

Devin, S., J. N. Beisel, V. Bachmann, and J. C. Moreteau. 2001. *Dikerogammarus villosus* (Amphipoda: Gammaridae): another invasive species newly established in the Moselle river and French hydrosystems. Annales de Limnologie **37**, 21–27.

Devin, S., C. Piscart, J. N. Beisel, and J. C. Moreteau. 2003. Ecological traits of the amphipod invader *Dikerogammarus villosus* on a mesohabitat scale. Archiv für Hydrobiologie **158**, 43–56.

Devin, S., C. Piscart, J. N. Beisel, and J. C. Moreteau. 2004. Life history Traits of the Invader *Dikerogammarus villosus* (Crustacea: Amphipoda) in the Moselle River, France. International Review of Hydrobiology **89**, 21–34.

Dick, J. T. A., R. W. Elwood, and I. Montgomery. 1995. The behavioural basis of a species replacement: differential aggression and predation between the introduced *Gammarus pulex* and the native *G. duebeni celticus* (Amphipoda). Behavioral Ecology and Sociobiology **37**, 393–398.

Dick, J. T. A., I. Montgomery, and R. W. Elwood. 1993. Replacement of the indigenous amphipod *Gammarus duebeni celticus* by the introduced *G. pulex*: differential cannibalism and mutual predation. Journal of Animal Ecology **62**, 79–88.

Dick, J. T. A. and D. Platvoet. 2000. Invading predatory crustacean *Dikerogammarus villosus* eliminates both native and exotic species. Proceedings of the Royal Society of London B **267**, 977–983.

Dick, J. T. A., D. Platvoet, and D. W. Kelly. 2002. Predatory impact of the freshwater invader *Dikerogammarus villosus* (Crustacea: Amphipoda). Canadian Journal of Fisheries and Aquatic Sciences **59**, 1078–1084.

Giusti, F. and E. Oppi. 1972. *Dreissena polymorpha* (Pallas) nuovamente in Italia (Bivalvia,Dreissenidae) Memorie del Museo Civico di Storia Naturale di Verona **20**, 45–49.

Gonzales, M. and A. Downing. 1999. Mechanism underlying amphipod responses to zebra mussel (*Dreissena polymorpha*) invasion and implications for fish-amphipod interactions. Canadian Journal of Fisheries and Aquatic Sciences **56**, 679–685.

Grabow, K., T. O. Eggers, and A. Martens. 1998. *Dikerogammarus villosus* Sowinsky (Crustacea, Amphipoda) in navigable canals and rivers of Northern Germany. Lautebornia **33**, 103–107.

Haybach, A., M. Schleuter, and T. Tittizer. 2003. Current distribution of mayflies (Insecta): Ephemeroptera in German Federal Waterways. Pages 313–315 in E. Gaino, editor. Research Update on Ephemeroptera & Plecoptera, University of Perugia, Italy.

Hsu, J. C. 1996. Multiple Comparisons. Theory and Methods, Chapman & Hall. Kinzelbach, R. 1995. Neozoans in European waters exemplifying the word process of invasion and species mixing. Experientia **51**, 526–538.

Jażdżewski, K. and A. Konopacka. 2002. Invasive Ponto-Caspian species in waters of the Vistula and Oder basins and Southern Baltic Sea. Pages 384–398 in E. Leppäkoski, S. Gollash and S. Olenin, editors. Invasive aquatic species of Europe. Distribution, impacts and managements. Kluwer Academic Publisher, Dordrecht, The Netherlands.

Kinzelbach, R. 1995. Neozoans in European waters-Exemplifying the worldwide process of invasion and species mixing. Experientia **51**, 526–538.

Kinzler, W. and G. Maier. 2003. Assymetry in mutual predation: possible reason for the replacement of native gammarids by invasives. Archiv für Hydrobiologie **157**, 473–481.

Kley, A. and G. Maier. 2003. Life history characteristics of the invasive freshwater gammarids *Dikerogammarus villosus* and *Echinogammarus ischnus* in the river Main and the Main-Donau canal. Archiv für Hydrobiologie **156**, 457–469.

MacNeil, C. and D. Platvoet. 2005. The predatory impact of the freshwater invader *Dikerogammarus villosus* on native *Gammarus pulex* (Crustacea: Amphipoda); influences of differential microdistribution and food resources. Journal of Zoology, London **267**, 31–38.

Marguiller, S., F. Dehairs, G. Van der Velde, B. Kelleher, and S. Rajagopal. 1998. Initial results on the trophic relationships based on *Corophium curvispinum* in the Rhine traced by stable isotopes. Pages 171–177 in P. H. Nienhuis, R. S. E. W. Leuven, and A. M. J. Ragas, editors. New concepts for sustainable management of river basins. Backuys Publisher, Leiden, The Netherlands.

Minchin, D., L. Frances, and M. Sullivan. 2002. Zebra Mussel: impact and spread. Pages 135–146 in E. Leppäkoski, S. Gollash, and S. Olenin, editors. Invasive aquatic species of Europe. Distribution, impacts and managements, Kluwer Academic Publisher, Dordrecht, The Netherlands.

Müller, J. C., S. Schramm, and A. Seitz. 2002. Genetic and morphological differentiation of *Dikerogammarus* invaders and their invasion history in Central Europe. Freshwater Biology **47**, 2039–2045.

Müller, O., M. I. Zettler, and P. Gruszka. 2001. Distribution and status of *Dikerogammarus villosus* (Sowinsky) (Crustacea, Amphipoda) in the middle and in the upper part of the River Oder and in the adjacent waterways. Lauterbornia **41**, 105–112. (In German)

Mürle, U., A. Becker, and P. Rey. 2003. Ein neur Flohkrebs im Bodensee. http//www.bodensee-ufer.de.

Musko, I. B. 1989. Amphipoda (Crustacea) in the littoral zone of Lake Balaton (Hungary) Qualitative and quantitative studies. Internationale Revue der gesamten Hydrobiologie **74**, 195–205.

Musko, I. B. 1990. Qualitative and quantitative relationship of Amphipoda (Crustacea) living on macrophytes in lake Balaton (Hungary). Hydrobiologia **191**, 269–3274.

Musko, I. B. 1994. Occurence of Amphipoda in Hungary since 1853, Crustaceana **66**, 144–152.

Musko, I. B. and B. Bako. 2005. The density and biomass of *Dreissena polymorpha* living on submersed macrophytes in Lake Balaton (Hungary). Archiv für Hydrobiologie **162**, 229–251.

Neseman, H., M. Pockl, and K. J. Wittmann. 1995. Distribution of epigean Malacostraca in the middle and upper Danube (Hungary, Austria, Germany). Miscellanea Zoologica Hungarica **10**, 49–68.

Ricciardi, A. 2001. Facilitative interactions among aquatic invaders: is an "Invasional meltdown" occurring in the Great Lakes? Canadian Journal of Fisheries and Aquatic Sciences **58**, 2513–2525.

Ricciardi, A. and J. B. Rasmussen. 1998. Predicting the identity and impact of future biological invaders: a priority for aquatic resource management. Canadian Journal of Fisheries and Aquatic Sciences **55**, 1759–1765.

Ricciardi, A., F. G. Whoriskey, and J. B. Rasmussen. 1997. The role of the zebra mussel (*Dreissena polymorpha*) in structuring macroinvertebrate communities on hard substrata. Canadian Journal of Fisheries and Aquatic Sciences **54**, 2596–2608.

Schmidt, O. and G. Josens. 2004. Preliminary study of the scars borne by Gammaridae (Amphipoda, Crustacea) Belgian Journal of Zoology **134**, 75–78.

Simberloff, D. and B. Von Holle. 1999. Positive interaction of non-indigenous species: Invasional Meltdown? Biological Invasions **1**, 21–32.

Stewart, T. W. and J. M. Haynes. 1994. Benthic macroinvertebrate community of Southwestern Lake Ontario following invasion of *Dreissena*. Journal of Great Lakes Research **20**, 479–493.

Taylor, A., G. Rigby, S. Gollash, M. Voigt, G. Hallegraeff, T. Mccollin, and A. Jelmert. 2002. Preventive treatment and control techniques for ballast water. Pages 484–507 in E. Leppäkoski, S. Gollash, and S. Olenin, editors. Invasive aquatic species of Europe. Distribution, impacts and managements, Kluwer Academic Publishers, Dordrecht, The Netherlands.

Tittizer. T. 1996a. Danube canal now a short cut for fauna. Danube Watch **2**, 7–8.

Tittizer, T. 1996b. Vorkommen und Ausbreitung acquatischer Neozen (Makrozoobenthos) in den Bundeswasserstraisen. Pages 49–86 in H. Gebhardt, R. Kinzelbach, and S. Schmidt, editors. Gebietsfremde Tierarten. Fischer. Ecomed, Landsberg, Bavaria, Germany.

Van der Velde, G., I. Nagelkerken, S. Rajagopal, and A. Bij de Vaate. 2002. Invasion by alien species in inland freshwater bodies in Western Europe: The Rhine Delta. Pages 360–372 in E. Leppäkoski, S, Gollash, and S. Olenin, editors. Invasive aquatic species of Europe. Distribution, impacts and managements, Kluwer Academic Publishers, Dordrecht, The Netherlands.

Whitfield, J. 2000. Ecology: Shrimp-eat-shrimp. Nature **405**, 895.

Wijnhoven, S., M. G. van Riel, and G. Van der Velde. 2003. Exotic and indigenous freshwater gammarid species: physiological tolerance to water temperature in relation to ionic content of the water. Aquatic Ecology **37**, 151–158.

Chapter twenty-eight

Understanding the impact of invasive crayfish

Francesca Gherardi

INTRODUCTION

In a rather provocative article, Parker *et al.* (1999) claimed that, up to then, little scientific attention had been placed on developing either theoretical or operational generalizations about the impact of invasive species. Specifically, the authors lamented the lack of a general framework in which to discuss "what impact is, or how we decide that the non-indigenous species exceeds that of another, or how we decide that the impact of a particular species is greater in one place than in another" (Parker *et al.* 1999, p. 4). Today, this scenario seems to have changed for several freshwater non-indigenous species (NIS), e.g. the zebra mussel *Dreissena polymorpha* (Pallas) (e.g. Karatayev *et al.* 2002, Ricciardi 2003), but it has remained practically unaltered for other widely diffused bioinvaders that have, however, attracted much scientific attention in these latest years, such as freshwater crayfish.

Crayfish are the largest and amongst the longest lived invertebrate organisms in temperate freshwater environments, and often exist at high densities. Most species are keystone consumers (Nyström *et al.* 1996), feeding on benthic invertebrates, detritus, macrophytes, and algae in lotic and lentic waters (e.g. Whitledge and Rabeni 1997). They also constitute the main prey of several species, including otter (Slater and Rayner 1993), fish (e.g. Blake 1995), and birds (Rodríguez *et al.* 2005). Because of their capability to integrate into the food web at many levels and to persist on the substantial energy reserves of the detrital pool, crayfish are good candidates for invading aquatic systems (Moyle and Light 1996).

Francesca Gherardi, Biological invaders in inland waters: Profiles, distribution, and threats, 507–542.

Crayfish native biodiversity is large but unequally distributed throughout the continents. Over 600 species have been described worldwide (Crandall 2002), the majority occurring in North and Central America (75% of the world's total). About 100 indigenous species live in the southern hemisphere, five in Europe, four in Asia, and none in continental Africa and in Antarctica (Hobbs 1988, Fitzpatrick 1995, Taylor *et al.* 1996, Taylor 2002). We know today that much of their present distribution is due to human intervention, even for the most celebrated "indigenous" species. For instance, as reported by Linnaeus (1746, p. 358) and confirmed by Pontoppidan (1775, p. 175), the noble crayfish *Astacus astacus* (Linnaeus) seems to have been imported into Sweden from Russia by John III (King of Sweden since 1568) and later into Finland (Westman 1973). Similarly, the white-clawed crayfish *Austropotamobius pallipes* (Lereboullet) might have been introduced into Ireland by monastic orders from France in the 12th century, as suggested by Reynolds (1997). Genetic evidence for this hypothesis has recently been obtained by Gouin *et al.* (2003).

In the last few decades, as a result of the exponential growth in the volume and complexity of the international trade, the world distribution of crayfish has dramatically changed (e.g. Pérez *et al.* 1997) mostly due to transcontinental or interstate translocations of a still relatively small proportion of crayfish species (hereafter referred to as NICS, i.e. non-indigenous crayfish species) – 20 according to Hobbs *et al.* (1989), but increasing all the time. Human-mediated translocations have been often accidental (in ballast, via canals, by escapes from holding facilities) but the introduction of some crayfish species was deliberate (for aquaculture and stocking, for food, as aquarium pets, as live bait, and for snail and weed control; see Lodge *et al.* 2000, Chapter 2). In Africa, the introduction of North American and Australian species from the 1970s onwards (Arrignon *et al.* 1990) was aimed to broaden the range of commercial fisheries in lakes and dams (Mikkola 1996) and to control freshwater snails that carry human schistosomiasis (Hofkin *et al.* 1991, Chapter 4). Indeed, most crayfish introductions were motivated by our desire to eat them (Holdich 1999a) that in its turn generates economic interests. Some species are in fact highly valued as food, and in countries like Scandinavia and Louisiana feasting on them has become a cultural icon. As a consequence, for instance in Europe, wild stocks are managed, leading to annual catches of more than 4,000 tons, and some species are cultivated with a total production of about 150 tons (Ackefors 1999).

Hence, crayfish introductions have certainly provided economic benefits to several countries, because of (1) restoring the productivity of indigenous stocks (e.g. in Sweden), (2) compensating for their lack (e.g. in Spain), or (3) developing extensive or semiintensive cultivation systems (e.g. in the People's Republic of China) (Ackefors 1999). However, once introduced for stocking and aquaculture and kept in outdoor ponds, crayfish of several species almost inevitably escape (Hobbs *et al.* 1989) and a proportion of them is able to establish self-sustaining populations in the colonized habitats. The invasion process may continue: while some populations remain localized around the point of introduction, others spread widely, becoming invasive (Kolar and Lodge 2001). Compared to other aquatic species, the spread of this taxon is favored by the ability of some species to stay for relatively long periods out of water and to travel long distances, sometimes overland [e.g. *Procambarus clarkii* (Girard), Gherardi and Barbaresi 2000]. Contradicting the predictions made by the tens rule (Williamson 1996), the majority of the naturalized populations has the potential to become noxious. For instance, of the eight NICS imported into Britain since 1970 five have been found in the wild and four of these have become established, three being now officially classified as pests (Holdich 1999b). Today, the list of species that are causing concern in the introduced areas includes *Pacifastacus leniusculus* (Dana) in California (USA), Europe, and Japan, *Orconectes limosus* (Rafinesque) in Europe, *Orconectes rusticus* (Girard) in North America, *P. clarkii* in Africa, California, Europe, and Japan, *Astacus leptodactylus* Eschscholtz in some European countries, and *Cherax destructor* Clark in Africa and Australia (Holdich 1999a). Other species, like *Cherax quadricarinatus* (von Martens) in Ecuador, are expected to lead problems in the near future (Romero 1997).

Concerns relate to the modifications that NICS may induce in the structure of freshwater food webs (e.g. Lodge *et al.* 1998) and the economic damage they may inflict through, for instance, reduced recruitment of commercially valuable fish stocks (Nyström 1999). In the short term, they may decrease the biomass and species richness of macroinvertebrates, macrophytes, and periphyton (e.g. Lodge and Lorman 1987) and, in the long term, they may contribute to the decline of several invertebrate taxa, including indigenous crayfish species, amphibians, and fish (e.g. Guan and Wiles 1997, Chapter 29). Once added to a system, NICS have the potential to impose "considerable environmental stress" and, in most instances, they may induce "irreparable shifts in species diversity" (Hobbs *et al.* 1989, p. 309).

In this chapter, the literature focusing on the multiform "environmental stress" that NICS have been found to pose to the colonized areas around the world will be reviewed, as well as the "shifts in species diversity" that they have caused. In addition, information about their effects upon human economy and health will be discussed. By identifying several gaps in the existing literature, I will suggest the directions that research should take to extend and strengthen the current knowledge of the impact of NICS and to help prioritize interventions in freshwater systems.

THE LITERATURE ON THE IMPACT OF NICS

Since the 1980s, the effects exerted by NICS on the environment, indigenous species, national economies, and human health have been abundantly publicized (e.g. Holdich 1988, 1999a, Hobbs *et al.* 1989, Gherardi and Holdich

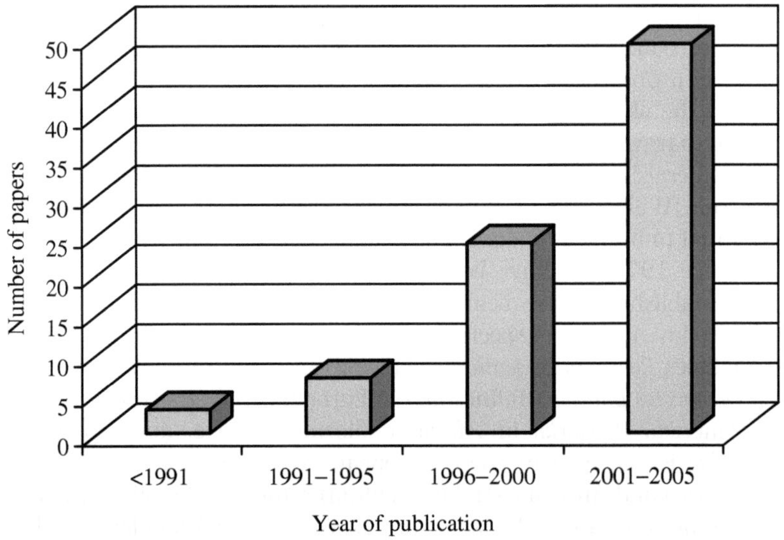

Fig. 1 The rise in the number of publications that described the impact of non-indigenous crayfish species (NICS). Papers (reaching a total of 103) were identified via keywords from Biosis (1967–February 2006), ASFA (1960–March 2006), and Zoological Record (1978–March 2006). Only papers reporting quantitative data published before 2006 (83) are included here.

1999, Lodge *et al.* 2000). The International Association of Astacology largely contributed in heightening awareness about the negative impact of NICS by adopting a resolution in 1987 (VII Symposium Lausanne, Switzerland), in which governments were recommended to "find the means to stop the importation of living crayfish into their countries for any purpose ... except for governmentally approved research, restockings or introductions". This resolution was subsequently renewed (in: 1997, Florence, Italy; 1998, Amsterdam, Holland and Augsburg, Germany, and 2002, Querétaro, Mexico).

The consequent surge of interest in NICS encouraged more proactive research and led, since 1987, to a sharp increase, mostly in Europe, in the number of publications, that included 18 reviews (e.g. Holdich 1999a, Lodge *et al.* 2000, Nyström 2002, Westman 2002, Geiger *et al.* 2005), focused on describing the impact of some NICS (Fig. 1), particularly *P. clarkii* and *P. leniusculus* (Fig. 2). The studies that have attempted to quantify impacts have often done so using an experimental approach (reviewed in Chapter 30). However, short-term researches largely prevailed: only 12% of the 85 quantitative studies here analyzed monitored the effects of NICS on a system for more than 1 year. Some of these studies adopted classical *in situ* enclosure/exclosure experiments to analyze the effects of NICS on the abundance of several components of the

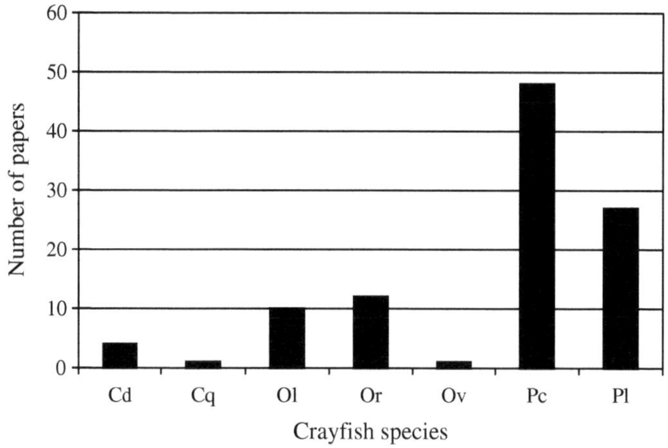

Fig. 2 The number of publications describing the impact of NICS (103, see Fig. 1) distributed among crayfish species. Cd = *Cherax destructor*; Cq = *C. quadricarinatus*; Ol = *Orconectes limosus*; Or = *O. rusticus*; Ov = *O. virilis*; Pc = *Procambarus clarkii*; Pl = *Pacifastacus leniusculus*.

resident community. Lodge *et al.* (1994) pioneered this experimental approach by revealing the control of *O. rusticus* on the trophic interactions existing in the littoral zone of some North American lakes.

Several other studies applied a correlative approach by comparing NICS and indigenous crayfish species for growth rate, densities, vulnerability to predators, predatory ability, and behavioral dominance over food and shelter. On the one hand, the impact exerted on macrophytes and benthic invertebrate abundance has been elegantly compared between invasive and indigenous species through a combination of field and laboratory experiments (e.g. Olsen *et al.* 1991). On the other, to test the ability of several NICS to outcompete indigenous crayfish, experiments were mostly run in the laboratory (e.g. Söderbäck 1991, Holdich *et al.* 1995, Vorburger and Ribi 1999, Gherardi and Cioni 2004).

Ideally, much knowledge of the effects of NICS on the community should derive from the comparison (1) of one site before and after their invasion and/or (2) of different sites, with and without an invader present, at the same time. Unfortunately, only a small minority of studies (e.g. Rodríguez *et al.* 2005, McCarthy *et al.* 2006) adopted this comparative approach, in large part because the interest in an invasive species most often arises after it has spread extensively and only when it has already had an impact. An obvious drawback of these studies is that results may be confounded by temporal or spatial trends in the environment such as pollution, harvesting, or climate change; they should therefore be accompanied by a thorough documentation of other contemporary modifications in the habitat.

Finally, there is a surprising lack in the literature on NICS of any quantitative estimate of the relative impact of a species between its native and invaded habitats. Indeed, this comparison would be critical to assess the varying success of an invasive species. Evolutionary history in fact plays a major role in this process and invasive species may display greater impacts in systems without co-evolved predators, parasites, and prey (see e.g. the "enemy release" hypothesis, Torchin *et al.* 2003). This is well illustrated by the impact of the North American *P. leniusculus* in the UK (Holdich *et al.* 2004).

THE FIVE LEVELS OF ECOLOGICAL IMPACTS

According to Parker *et al.* (1999), the impact of an invader can be measured at five levels of biological complexity: (1) effects on individuals (life history, morphology, behavior); (2) population dynamic effects (abundance, population growth, etc.); (3) genetic effects (including hybridization); (4) community effects (species richness, diversity, trophic structure); and (5) effects on ecosystem processes (nutrient availability, primary productivity, etc). An overview of the literature on NICS clearly shows that some of these effects are documented much more than others (Fig. 3). The most understudied impacts were genetic changes that, on the contrary, should require much more attention by invasion biologists. They may in fact give rise to new invasive forms, induce decline of the

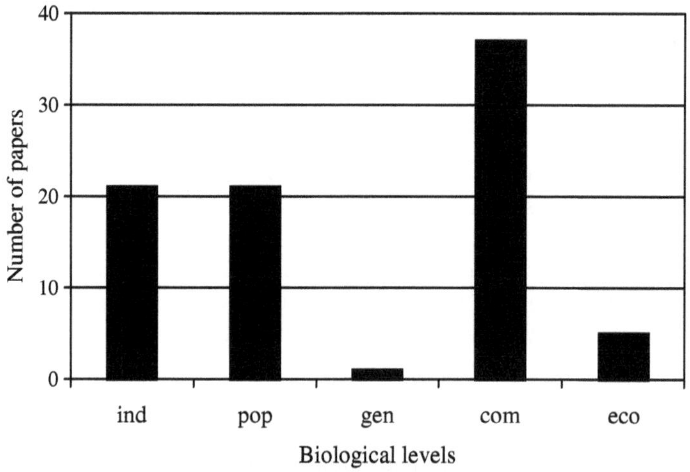

Fig. 3 The number of studies (85) reporting quantitative data on the impact of NICS at five different biological levels: individual (ind), population (pop), genetic (gen), community (com), and ecosystem function (eco). Single studies that reported impacts at more than one level were classified on the basis of the most prominent one.

indigenous species through e.g. introgressive hybridization, and eventually lead to long-term evolutionary shifts (Cox 2004).

The best examples of studies, classified at the biological level, which have quantified the impact of NICS, are reviewed below.

Effects on individuals and populations

The development of predator-prey or competitive interactions is expected to induce changes in the life history, morphology, and behavior of the interacting species. These changes are obviously adaptive as they should reduce the risks for the "weaker" species to be preyed upon or to be outcompeted by the "stronger" species (e.g. Lima 1998). When, on the contrary, no adaptation appears or the changes undergone do not suffice to develop a balance between the prey and the new predator or between competitors, the abundance, population dynamics, or distribution of the "weaker" species – usually the indigenous species – may be strongly affected. The most extreme effect at the population level is extinction, at the local or even at the global level.

Several studies have focused on an NICS acting as a predator upon a naïve species, either invertebrate or vertebrate. Most often they showed lethal or sublethal effects. For instance, *P. leniusculus* easily consumes *Lymnaea* snails and induces tail-cuts to *Rana temporaria* (Linnaeus) tadpoles (Nyström *et al.* 2001); *P. clarkii* (1) is able to prey upon the embryos and free swimming larvae of up to 13 amphibian species (Cruz and Rebelo 2005); (2) takes a significantly shorter time than the indigenous *A. pallipes* to capture *Triturus vulgaris* (Linnaeus) larvae and *Rana esculenta* Linnaeus tadpoles (Renai and Gherardi 2004); or (3) is not deterred by toxins contained in Californian newt [*Taricha torosa* (Rathke)] eggs (Gamradt and Kats 1996).

Obviously, lethal or sublethal effects exerted by NICS are mostly due to the relatively short coevolutionary history between them and the naïve prey that does not allow the latter to develop efficient antipredator behaviors, morphological structures, or chemical repellents. The absence of these mechanisms has been evoked to explain, for instance, the sharp decline of the California newt recorded in three streams in the Santa Monica Mountains of southern California after about 10 years from the introduction of *P. clarkii* (Gamradt and Kats 1996).

However, historical coexistence in the area of origin seems not to be the only prerequisite for the exhibition of efficient antipredator behaviors: notwithstanding their common history in the area of origin, naturalized mosquitofish [*Gambusia holbrooki* (Girard)], once exposed to visual and chemical stimuli of *P. clarkii* in a confined environment in Portugal, did not decrease their activity and did not avoid spaces with high predation risks (Leite *et al.* 2005). Mosquitofish were therefore subject to a strong predation pressure by the NICS as confirmed from the frequent occurrence of their remains in crayfish gut content (Gutiérrez-Yurrita *et al.* 1998, Correia 2003).

Often, prior experience with a predator species is not required by the prey to exhibit avoidance behaviors. These are on the contrary stimulated by chemical cues released by injured conspecifics (alarm odors). For instance, when exposed to alarm odors, the snail *Physa gyrina* (Say) significantly increases its use of spatial refugia by moving to the waterline, floating at the surface, or crawling out of the water (McCarthy and Dickey 2002). These behaviors were independent of the predator's identity and therefore were exhibited in the presence of both the known *Orconectes juvenilis* (Hagen) and the unknown *P. clarkii*.

In only one case was it found that the prey, although naïve, had quickly acquired the ability to recognize an NICS as a predator. An ecological study, conducted in Pennsylvania (USA), revealed that the snail *P. gyrina* moved under cover in the presence of pumpkinseed sunfish but moved to the water surface in the presence of *O. rusticus*; the combined effects of fish and crayfish were intermediate to their individual effects (Turner *et al.* 2000). On the one hand, these results can explain why periphyton standing crop in covered habitats was the lowest in the fish treatment when covered habitat use by snails was the highest, and the highest in the control and in the crayfish treatment when covered habitat use by snails was the lowest. On the other hand, they showed that the prey has learned some stimuli emitted by the NICS (possibly its odor) and behaves accordingly to avoid it. This is not the typical response shown by the snail to a generalized predator. On the contrary, the snail seems to be able to distinguish the specific identity of *O. rusticus* with respect to fish predators. Other behavioral responses to predatory NICS by potential prey, leading, for instance, to changes in their microhabitat selection, diurnal activity or foraging patterns, have never been studied.

As regards to competitive interactions, few studies have analyzed the effects induced on non-crayfish species by resource competition with NICS. In the laboratory setting, Carpenter (2005) conducted density manipulation experiments to analyze the competition between *O. virilis*, established in the Colorado River basin, and two endemic fish species, the Gila chub, *Gila intermedia* (Girard), and the flannelmouth sucker, *Catostomus latipinnis* (Baird and Girard). The results showed that growth of flannelmouth suckers, but not of Gila chub, was affected by crayfish more than by intraspecific competition, leading to the conclusion that each case of competitive interaction should be taken as a separate case. At a population level, the ability of crayfish to outcompete some fish species, for instance by expelling individuals from their shelters and therefore by making them more vulnerable to piscivorous fish, may have detrimental effects. In six riffles of the River Great Ouse (England), the numbers of *P. leniusculus* and of the two dominant benthic fish, bullhead [*Cottus gobio* (Linnaeus)] and stone loach [*Noemacheilus barbatulus* (Linnaeus)], were inversely correlated (Guan and Wiles 1997; other examples in Chapter 29).

The most commonly documented impact of NICS on competitive interactions is the agonistic dominance they show over indigenous crayfish species (Chapter 31), with the only known exception of *Astacopsis franklinii* (Gray), endemic to

Tasmania, which dominated equally sized specimens of the non-indigenous *C. destructor* (Elvey *et al.* 1996). When exposed in the laboratory to an NICS, indigenous crayfish changed their posture and behavior, quickly assuming the role of subordinates [e.g. *Procambarus acutus acutus* (Girard) *vs. P. clarkii*, Gherardi and Daniels 2004]. When, however, stable dominance hierarchies between NICS and indigenous crayfish species cannot be formed, as observed in the dyad *P. clarkii* and *A. pallipes* (Gherardi and Cioni 2004), the repetition of escalated fights may lead to injuries usually suffered by the weaker indigenous crayfish followed by its likely death. Indeed, NICS, e.g. *P. leniusculus*, may prey upon the indigenous *A. pallipes* and *A. leptodactylus*, and predation may have contributed to the gradual decline of *A. pallipes*, at least in England (Holdich and Domaniewski 1995).

Dominance hierarchies usually translate into a differential capability to compete for resources, such as shelters. It is known that shelters act as a "principal resource bottleneck" in crayfish populations (Hobbs 1991) serving to minimize predation risks and in some species to attract mates (Gherardi 2002). They are usually limited so that an inferior ability to compete for them may lead to a higher susceptibility to predation in subordinate species, being therefore the cause of their increased mortality (Gherardi and Cioni 2004). When tested in the laboratory, dominant NICS were most often found to competitively exclude subordinate indigenous crayfish from the offered shelter [in: *O. rusticus vs.* other *Orconectes* species, Hill and Lodge 1999; *P. leniusculus vs. A. astacus*, Söderbäck 1991; and *P. leniusculus vs. Cambaroides japonicus* (De Haan), Usio *et al.* 2001, Nakata and Goshima 2003] also when NICS showed a low preference for the shelter in a noncompetitive context [*P. clarkii vs. A. pallipes*, Gherardi and Cioni 2004, and *vs. P. acutus acutus*, Gherardi and Daniels 2004; but not *P. leniusculus vs. Austropotamobius torrentium* (Schrank), Vorburger and Ribi 1999]. Although conducted in an artificial setting, the results of these experimental studies are both informative and predictive, being a reflection of the documented range expansion of NICS. In fact, those crayfish species that were found to replace other crayfish species are also dominant in interspecific contests staged in the laboratory (e.g. Capelli 1982, Söderbäck 1991, Hill and Lodge 1994).

Predation and competition, both acting at the individual level and coupled with the potential for reproductive interference (i.e. females or males of a species may choose mates of the other species; Söderbäck 1994), enhance the effects of habitat loss, overexploitation, and pollution in inducing a dramatic decline of crayfish biodiversity. Of the 67 threatened species in North America, 5% were subject to interference by NICS, 62% to habitat degradation or habitat loss, and 33% to pollution (Wilcove *et al.* 1998). NICS have already contributed to the global extinction of other crayfish species: *Pacifastacus nigrescens* (Stimpson), once common in the creeks of the San Francisco Bay area in northern California, is now extinct due to the combined pressures of urbanization, overexploitation, and introductions of NICS (Bouchard 1977). In the same area, a similar process is ongoing. Shasta crayfish [*Pacifastacus fortis* (Faxon)], designated endangered in 1988, is now limited to small, isolated populations, having been displaced at several locations in its native watershed by habitat loss and interactions with *P. leniusculus* (Light *et al.* 1995). Subject to similar threats from NICS, the European indigenous species *A. astacus*, *A. pallipes*, and *A. torrentium* have been designated as vulnerable by Groombridge (1993) and as protected in Appendix III of the Bern Convention. The European Habitats Directive (Directive 92/43/EEC) considers them as species whose exploitation and harvesting should be subject to control. *Austropotamobius pallipes* and *A. torrentium* are also listed in Appendix V, as species that require the setting up of special areas of conservation for their protection (Souty-Grosset *et al.* 2006).

Population effects on indigenous species may also be caused by indirect mechanisms, for instance through the transmission of pathogens and diseases. There is an extensive literature showing that North American crayfish species carry a subclinical infection of the oomycete *Aphanomyces astaci* Schikora, the aetiological agent of the crayfish plague (e.g. Alderman and Polglase 1988, Diéguez-Uribeondo and Söderhäll 1993). This disease does not require its host in order to spread, as the spores can become attached to damp surfaces and be transported in this manner. So, crayfish plague has spread and is still spreading via the hundreds of thousands of crayfish trappers and their gear. A large number of European crayfish populations has been decimated by the plague since 1860, leading to reduced production of *A. astacus* and *A. leptodactylus* by up to 90% particularly in Scandinavia, Germany, Spain, and Turkey. For example, in Sweden 90 tons were exported in 1908 (from a total catch of 200 tons), but export dropped to 30 tons by 1910 (Holdich 1999a). In Finland exports declined from 16 million *A. astacus* in 1890 to less than 2 million in 1910 (Westman 1991). When the plague spread to Turkey in the 1980s, the annual catch of *A. leptodactylus* plunged from 7,000 to 2,000 tons (Köksal 1988). It declined even further in the early 1990s, reaching 200 tons in 1991, which virtually eliminated exports from Turkey to western Europe. However, since then there has been a steady increase in stocks (Harlioğlu 2004). The impact of the crayfish plague is still high in Europe; to mitigate this, it has even been suggested that plague-resistant strains of native species should be created through genetic engineering (Westman 2002), which, in their turn, might pose additional threats to the original biodiversity.

The large emphasis that researchers have given to *A. astaci* has created, as Edgerton *et al.* (2004) lamented, a kind of "inertia" in astacology, "which has curtailed researchers, state fish-disease diagnosticians, and resource managers from fully assessing and considering the existence of other serious pathogens of crayfish and the ensuing consequences for the native species" (p. 1473). Indeed, notwithstanding that the range of crayfish diseases is repeatedly assumed to be much wider than previously thought (e.g. Vogt 1999), especially in North America, little attention has been given to commensals or parasites other than *A. astaci*. It seems unlikely that these pathogens are species-specific; so, introduced crayfish might bring a host of organisms that may profoundly affect indigenous species. To make the picture worse, some commensals or parasites of crayfish may affect other animals, humans included (helminth parasites of vertebrates; Hobbs *et al.* 1989; bacterial fish diseases, enteric redmouth, Dunlin *et al.* 1976; infectious pancreatic necrosis, Halder and Ahne 1988). As a consequence, there is an urgent need for researchers to address the issue of crayfish diseases from a broader perspective than has been done up to now.

Genetic effects

Invaders may exert indirect genetic impacts on indigenous species, resulting in altered patterns of natural selection or gene flow within indigenous populations. They may change selection regimes or, in a subtler way, disrupt gene flow due, for instance, to their fragmenting populations of indigenous species and lead them to risky bottlenecks. Hybridization between an invader and an indigenous species is a direct effect that may have three possible consequences: (1) the creation of a new invasive genotype; (2) the production of sterile hybrids with the resulting waste of gametes and resource competition with indigenous species; and (3) the production of a hybrid swarm and widespread introgression, leading to a virtual extinction of indigenous taxa through "genetic pollution" (Parker *et al.* 1999).

There has been little mention in the literature of the occurrence of hybridization between indigenous and invading crayfish, e.g. *A. astacus* and *A. leptodactylus* (Cukerzis 1968), although in that case both belonged to the same family and were of European origin. Unique genetic documentation has been provided by Perry *et al.* (2001, 2002) for hybridization between indigenous and invading *Orconectes* species. Using diagnostic nuclear and mitochondrial DNA markers along with morphological data, these authors compared crayfish from allopatric and sympatric populations of the invasive *O. rusticus* and of the indigenous *Orconectes propinquus* (Girard) and *O. virilis* (Hagen) in Wisconsin (USA). Hybridization occurred between *O. rusticus* and *O. propinquus* in sympatric sites, whereas *O. virilis* hybridized with neither of these species. A detailed study of the dynamics of hybridization conducted in Trout Lake showed that over 6% of the crayfish were F_1 hybrids, 4% were F_2 individuals (hybrid × hybrid origin), and 13% were backcrosses (product of hybrid × parental matings). The majority of F_1 hybrids (95%) were the result of *O. rusticus* females mating with *O. propinquus* males; only 1% of the total crayfish population was the product of F_1 hybrids backcrossing to *O. propinquus*, whereas 13% represent backcrosses to *O. rusticus*. The F_1 hybrids, therefore, appeared to mate disproportionately with pure *O. rusticus* that led to much greater genetic introgression of nuclear DNA from *O. propinquus* to *O. rusticus* than in the reverse direction. A consequence of this is the gradual elimination of *O. propinquus* genes from the population.

These results add to the wide morphological evidence of putative hybrids among crayfish species (Perry *et al.* 2002). The implication of Perry *et al.*'s (2002) study is clear that hybridization and introgression pose a substantial threat to the conservation of crayfish biodiversity and that further research is required to test the potential for hybridization among resident and invasive species. These studies have the potential to predict species at risk of losing their genetic identity.

Effects on communities

While NICS clearly cause direct impacts on prey or on competitors and alter their abundance and distribution, they may also indirectly change interactions within communities that sometimes result in trophic cascades.

A large variety of articles has shown that, because of their omnivorous nature, introduced crayfish can profoundly modify the trophic structure of freshwater communities at several levels, often acting as keystone species (*sensu* Paine 1966). They also display a wide plasticity in their feeding behavior, switching from detritivore/scavenger to herbivore/carnivore habits in response to food availability (Hobbs *et al.* 1989). For instance, the large niche breadth and trophic diversity shown in the gut contents of a naturalized *P. clarkii* population in Portugal denoted that a wide variety of aquatic macroinvertebrates is consumed by this species and that the specific composition of the prey changes according to its seasonal availability (Correia 2002).

Several experimental studies, most often conducted in the littoral zone of lentic systems (an exception in Stenroth and Nyström 2003), have detailed the direct and indirect trophic effects that NICS have on the food web structure. On the contrary, little is known about the influence directly or indirectly exerted by NICS on e.g. the pelagic compartments of the invaded systems. Figure 4 is an attempt to summarize the trophic links that NICS may produce in littoral zones of ponds and lakes (see also Nyström 1999, 2002). All these effects are most often density dependent (e.g. Charlebois and Lamberti 1996) even if relatively low densities of the potentially invasive *O. virilis* (Chambers *et al.* 1990), and of the invasive *O. rusticus* (e.g. Lodge *et al.* 1994) and *P. clarkii* (Gherardi and Acquistapace 2007) can dramatically reduce biomass and species richness of submersed macrophytes and the abundance of invertebrate prey. The impact on food web structure may also be influenced by the digestion rate of the crayfish species (e.g. Bernardo and Ilhéu 1994), the species composition of the community (Nyström 2002), and the presence of a second introduced predator (Nyström *et al.* 2001), all conditions that make any generalizations provisional.

Young and adult NICS are represented in Fig. 4 as different components of the food web, as supported by the literature (Nyström 1999). In fact, although adult crayfish, at least in the laboratory, can efficiently prey upon juvenile or small fish, amphibian eggs and tadpoles, and macroinvertebrates (e.g. Gherardi *et al.*

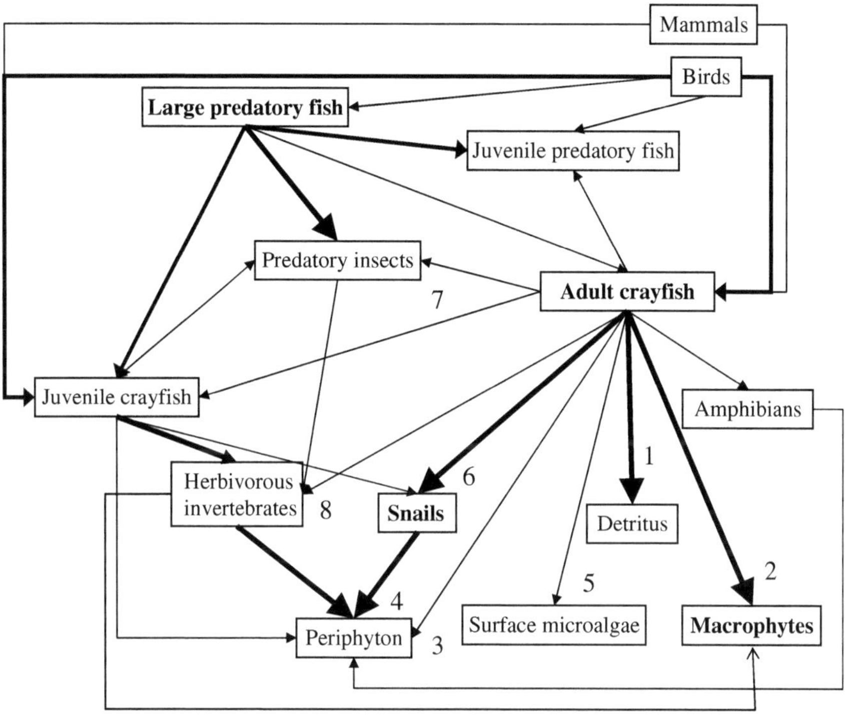

Fig. 4 Food web links (those numbered are discussed in the text) in a littoral zone of a lake or pond. Arrow thickness denotes the hypothesized strength of the interactions. The most important interactors in the food web are in bold.

2001, Renai and Gherardi 2004) including juvenile crayfish, the guts of many species contain a large proportion of organic detritus and plants, especially macrophytes, whereas juveniles feed predominantly upon macroinvertebrates (Momot *et al.* 1978, Momot 1995). For instance, in a Swedish stream, 46% of the adult *P. leniusculus* were found to have consumed invertebrates as opposed to 87% of the juveniles (Nyström 1999). In a laboratory study, adult *P. clarkii* even exhibited feeding preference for plant food over animal food when the energetic cost of active predation was high (Ilhéu and Bernardo 1993a, 1995). When kept together, juveniles may be easily cannibalized by adults. However, the cannibalistic behavior of crayfish has not been confirmed in field studies (Momot 1995), although it is expected to be most severe upon molting individuals in the wild (Reynolds and O'Keeffe 2005). The presence of crayfish remains in crayfish stomachs in enclosures was assumed to derive from the crayfish consumption of their exuviae (Stenroth and Nyström 2003). Additionally, both large predatory invertebrates and fish can reduce the abundance of juvenile crayfish (Nyström 1999). Subsequently, young crayfish have less influence on the food web structure than adult crayfish due to the strong negative effects exerted on them by vertebrate and invertebrate predators.

Typically, adult NICS affect the biomass of organic detritus (link 1, Fig. 4) and of macrophytes (link 2, Fig. 4). In their turn, changes in detritus and macrophyte biomass have multiple non-trophic effects on the community, because of their role of either protective cover or substrate and breeding sites for a multitude of organisms. Detritus is considered to be a highly nutritive food for crayfish for its "microbial conditioning" or "chemical-defense leaching" (Newman 1991) and for their easier handling by crayfish (Cronin *et al.* 2002). It may be preferred to green plants in laboratory experiments (Ilhéu and Bernardo 1995) and has been often found in gut contents of *P. clarkii* (Ilhéu and Bernardo 1993b, Gutiérrez-Yurrita *et al.* 1998) and *P. leniusculus* (Stenroth and Nyström 2003).

The intense grazing on aquatic macrophytes by *O. rusticus*, *P. leniusculus*, and *P. clarkii* (e.g. Lodge *et al.* 1994, Nyström *et al.* 1996, Gherardi and Acquistapace 2007), coupled with their non-consumptive plant clipping and uprooting (Lodge *et al.* 1994, Nyström and Strand 1996, Smart *et al.* 2002, Gherardi and Acquistapace 2007) and their preference for seedlings rather than for older plants (Nyström and Strand 1996), may induce a significant decline in macrophyte abundance. In enclosure/exclosure experiments, *O. rusticus* abundance of 19 g m^{-2} reduced total macrophyte biomass of 64%, while abundance >140 g m^{-2} eliminated all macrophytes in 12 weeks (Lodge and Lorman 1987). More than 80% of macrophyte biomass was lost about 20 years after *P. clarkii*'s introduction in the Doñana National Park (Spain) (Gutiérrez-Yurrita *et al.* 1998). The destruction of much more plant tissue than the crayfish can eat (Lodge 1991) may induce a positive effect to the system, because fragmentation could produce nutritious coarse particulate organic matter (CPOM) (Vannote *et al.* 1980). However, macrophyte destruction in nutrient-rich conditions, particularly in eutrophic shallow lakes, is generally followed by a switch from a clear to a turbid state dominated by surface microalgae, like *Microcystis*, growth (Rodríguez *et al.* 2003). In its turn, this may lead to a decrease in primary production of macrophytes and periphyton due to the reduced light penetration.

By feeding selectively (Lodge and Lorman 1987, Nyström *et al.* 1996), crayfish may reduce the biodiversity of macrophytes and even control aquatic weeds (Warner and Green 1995). Cronin *et al.* (2002) studied feeding preferences of *P. clarkii* among 14 species of freshwater macrophytes (including macroscopic algae) and measured this species' response to manipulation of the combined plant traits of morphology, toughness, and surface features, and their response to chemistry. The preference of crayfish was related to plant phenolics, protein, nitrogen, fiber, lignin, cellulose, ash, and carbon. Like most generalist herbivores or omnivores (reviewed in Cirujano *et al.* 2004, Anastácio *et al.* 2005a), *P. clarkii* seems to base its feeding decisions on multiple plant traits, such as morphology, structure, chemical defenses, and nutritive value. Often, it feeds upon plants whose finely branched or filamentous morphologies make them easier to handle and to consume, and not because of their nutritional value. Similarly, *O. rusticus* consumes single-stemmed species more than rosulate or highly branched forms (Lodge and Lorman 1987). NICS usually avoid plants, which are chemically defended by multiple compounds, notwithstanding their high concentrations of protein, nitrogen, and dry mass (Bolser *et al.* 1998). The preference for some macrophyte species (e.g. *Chara hispida* Linnaeus *vs. Ceratophyllum submersum* Linnaeus) can even influence *P. clarkii*'s distribution and abundance, as found in a Spanish wetland (Cirujano *et al.* 2004).

However, no single macrophyte trait appears to be a good predictor of its palatability and factors other than plant tissue quality and morphology, such as cover or protection from predators afforded by the plant (e.g. Damman 1987), the consumer's state or hunger (e.g. Cronin and Hay 1996), and the consumer's prior feeding experiences (e.g. Dorn *et al.* 2001), may also be important in determining feeding decisions. Differences in the impact of crayfish on macrophyte growth were also found to be related to crayfish sex and activity, and to the abundance of alternative foods (Chambers *et al.* 1990).

NICS may affect periphytic algae (link 3, Fig. 4) in a number of ways that may result in positive (+) or negative (−) effects by: (1) consuming and dislodging periphyton during feeding, movement, or burrowing (−), (2) reducing the abundance of algivorous invertebrates (or vertebrates), which can indirectly increase algal abundance (+) (Luttenton *et al.* 1998); (3) fertilizing periphyton with their faeces (+) (Charlebois and Lamberti 1996); and (4) consuming or destroying macrophytes on which some algae grow (−) (Lodge *et al.* 1994). Therefore, periphyton responses to the crayfish impact are expected to vary in function of the effect that prevails in the system. Due to the morphology of their feeding apparatus, *O. rusticus* and *P. leniusculus* are not as efficient grazers on microalgae as snails are (Lodge *et al.* 1994, Luttenton *et al.* 1998, Nyström *et al.* 1999, Nyström *et al.* 2001) and their consumption is not so intense as to outweigh the positive indirect effect of the crayfish-induced reduction in snail densities (the crayfish–snail–periphyton link is strong). Other potential periphyton grazers, e.g. amphibian tadpoles (Nyström 1999) and herbivorous insects, may be preyed on by crayfish but probably their consumption is not so extensive to produce an indirect positive effect on microalgal abundance (the crayfish–amphibian–periphyton and the crayfish–insect grazers–periphyton links are weak; Lodge *et al.* 1994, Nyström *et al.* 1999). As a result, because snails are both the prey group primarily affected by crayfish and the functionally most important grazer group among the many other grazing taxa, crayfish may indirectly generate an increased abundance of microalgae by relaxing them from the grazing pressure of snails (link 4, Fig. 4), thus inducing – although being omnivorous – a trophic cascade in the food webs of the littoral zones of lakes or ponds. However, in the habitats characterized by abundant filamentous algae, the intense grazing from crayfish seems not to be sufficient to compensate for the reduced grazing from snails (Nyström *et al.* 1996).

A correlation between the density of *P. clarkii* and the abundance and composition of surface (but not pelagic) microalgae (link 5, Fig. 4) has been recently found in an *in situ* experiment conducted in a Mediterranean wetland (Gherardi and Lazzara 2006). Six areas (10 × 7 m) were delimited along a channel in the "Padule di Fucecchio" (Italy) to host crayfish populations at either low (1 m^{-2}) or high (14 m^{-2}) density. The analysis of chlorophylls *a*, *b*, and *c*, and phaeopigments showed that biomass of surface microalgae, mostly composed of Cyanobacteria, was strongly affected by the presence of dense populations of *P. clarkii*. The potential of their direct consumption by *P. clarkii* was confirmed by behavioral studies that showed crayfish on macrophytes feeding on the floating film.

In both lentic and lotic systems, crayfish can have direct and indirect negative effects on the biomass and species richness of macroinvertebrates (links 6, 7, and 8, Fig. 4) as the result of several mechanisms (Charlebois and Lamberti 1996), i.e.: (1) consumption; (2) increased drift through prey escape and incidental dislodgment by their foraging; and (3) possible inhibition of invertebrate colonization. Each of these mechanisms could have different consequences (e.g. direct mortality *vs.* displacement to downstream areas) for the local macroinvertebrate assemblage. When crayfish become abundant in lentic systems, species composition of invertebrates may change towards less vulnerable prey species.

Gastropoda are the taxon most affected by NICS (link 6, Fig. 4) and are sometimes eliminated. In Trout Lake, Wisconsin, snails declined from >10,000 to <5 individuals m^{-2} after 19 years of colonization by *O. rusticus* (Wilson *et al.* 2004). Also bivalves may be affected by NICS; for instance, predation by *O. rusticus* may have a significant impact on the colonization rate of zebra mussels (Perry *et al.* 1997, 2000).

NICS are selective in their choice of snails, thin-shelled snails being preferred to thick-shelled species because they are easier to handle (Nyström and Pérez 1998). Hence, in systems invaded by NICS, snail species composition may be altered from the thin-shelled *Lymnaea* spp. to the thick-shelled *Bithynia* spp. (Nyström *et al.* 2001). Large snails are also less profitable, because handling time of shells increases exponentially: adult *P. leniusculus* took less than a minute to feed upon small *Lymnaea stagnalis* (Linnaeus) but spent more than 27 minutes handling and consuming large snails. However, in complex habitats such as macrophyte beds crayfish may be less indiscriminate in their selection (Nyström and Pérez 1998). Also crayfish naïveté to a mollusk, although never definitively proved, may influence consumption. In a laboratory experiment Correia *et al.* (2005) showed that *P. clarkii* is able to capture the exotic snail *Physa acuta* (Draparnaud), but never preyed upon the Asian clam *Corbicula fluminea* (Muller). Possibly, prior experience coupled with the availability of alternative prey may be decisive as to whether the predator will proceed with or abort an attack.

The direct impact of NICS on non-snail macroinvertebrates (links 7 and 8, Fig. 4) largely depends on the life style and behavior of any single species.

In lentic waters, crayfish predation is weak on species that: (1) move quickly enough to escape tactile-feeding crayfish (e.g. isopods, amphipods, some Diptera, Heteroptera, and Coleoptera); (2) circumvent crayfish recognition ability by living in cases (e.g. Trichoptera); or (3) avoid contact by living in the sediment (e.g. some Diptera) (Lodge *et al.* 1994, Nyström *et al.* 1996). In streams, crayfish may have less predictable effects on invertebrate communities than in lentic waters (Stenroth and Nyström 2003) even if, also in these systems, slow-moving species are expected to decline (i.e. leeches, dragonflies, caddisflies, isopods, and mollusks) whereas more mobile prey or prey living in sediments seem to be less affected (i.e. chironomids and stoneflies).

The indirect impact of NICS on zoobenthic communities is particularly complex. Through consumption and destruction of macrophytes, crayfish can alter littoral habitats, leading to declines in macrophyte-associated taxa (Nyström *et al.* 1996). Crayfish may also influence detrital substrates through bioturbidation and feeding: the reduction in detritus has potential consequences for zoobenthic communities, particularly collector-gatherers (e.g. some Ephemeroptera, Trichoptera, and Diptera). Finally, crayfish predation upon, or competition with other zoobenthic predators such as Odonata larvae could reduce their occurrence, subsequently leading to an increase in the abundance of their prey (McCarthy *et al.* 2006).

Recently, the effects of *O. rusticus* – and of other NICS – on benthic invertebrate densities were examined by McCarthy *et al.* (2006) by applying a fixed-effect model meta-analysis on pre-existing data. A comparison was made among cage experiments conducted in 14 studies that excluded crayfish as the control and maintained a given density of crayfish as the treatment. The negative effect of crayfish on the total invertebrate abundance appeared to be driven by the significant decline of a few taxonomic orders, specifically Gastropoda and Diptera (Fig. 5). However, the abundance of Amphipoda, Coleoptera, Ephemeroptera, Oligochaeta, and Trichoptera was also reduced in crayfish treatments, although results for these taxa were not significant. Comparisons across orders showed that the greatest negative impact was exerted on Gastropoda (followed by Diptera, Amphipoda, and Ephemeroptera).

Finally, NICS can be prey items for fish, birds, and mammals, constituting a new resource for higher trophic levels in several areas of their introduction. In the Lower Guadalquivir Basin (Spain), before the introduction of *P. clarkii*, eels mostly preyed upon fish species (mosquitofish and carp). After crayfish introduction, only 17% of their stomachs (*vs.* 50% before crayfish introduction) contained other fish species, whereas the dominant prey item was *P. clarkii* reaching 67% of occurrence (Montes *et al.* 1993). As they readily feed upon *P. clarkii*, eels were proposed as selective biological control organisms (Mueller and Frutiger 2001). Still in southern Spain, *P. clarkii* is also an important part of the diet of at least six bird species, in particular white storks, night herons, and little egrets, whose diet is composed of up to 80% of crayfish (Rodríguez *et al.* 2005). Though no quantitative study has been yet made,

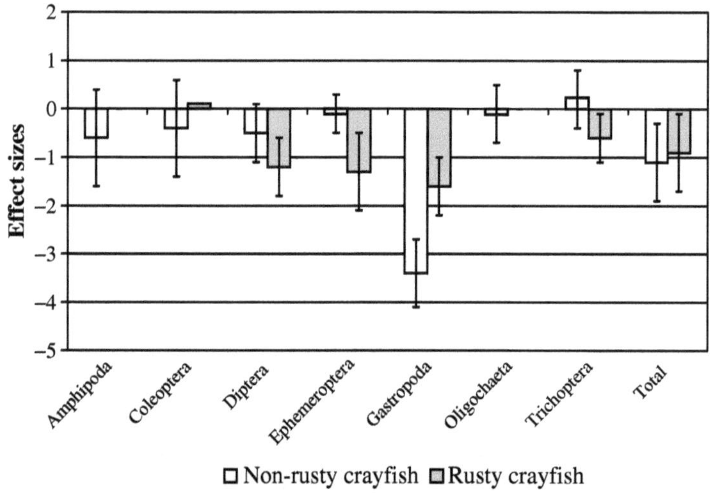

Fig. 5 Results from a fixed-effect model meta-analysis applied by McCarthy *et al.* (2006) on data obtained from 14 enclosure/exclosure experiments from the published literature. The effect size is the difference between the effects of crayfish on zoobenthic abundance in normal versus elevated densities, measured in units of standard deviation. Effect sizes (±95% Confidence Interval) are shown for total invertebrate and individual order densities. Negative values indicate a negative effect of crayfish on invertebrate density.

the appearance of *P. clarkii* has been considered responsible for the increase in a number of avian species, like some Ardaeidae, together with cormorants, in some European areas, such as in Massaciuccoli Lake (Tuscany) (Barbaresi and Gherardi 2000). In Doñana National Park it has also become the most common prey category of the otter, *Lutra lutra* (Linnaeus) (Delibes and Adrian 1987).

Notwithstanding the large amount of available data, much information derives from short-term studies and therefore the long-lasting strength of each link can only be hypothesized. A few long-term studies have attempted to analyze – all the other factors being equal – changes in species richness after the introduction of NICS. The study of Rodríguez *et al.* (2005) reviewed the existing information on the communities before the appearance of *P. clarkii* in Chozas Lake (Spain) and compared it with the data collected after its invasion. For macroinvertebrates, data from proximal non-invaded wetlands were also used in the analysis. The results (Table 1) were clear in showing the decline in submerged vegetation and the decrease of macroinvertebrate populations, both leading to a dramatic depletion of food resources, shelters, and breeding sites for fish, amphibians, and birds.

In a northern temperate lake in Wisconsin, USA (Sparkling Lake), McCarthy *et al.* (2006) conducted a 24-year time series analysis of *O. rusticus* and zoobenthos abundances. As a confirmation of a companion study (see above),

Table 1 Changes in the vegetation cover and faunal abundance as occurred in Chozas Lake (Spain) after the introduction of an NICS (*Procambarus clarkii*). (Modified after Rodríguez *et al.* 2005)

	Before	After	Loss (%)
Vegetation cover (%)	95	<3	99
Macroinvertebrates (genera, number)	31	9	71
Waterfowl (species, number)	50	26	52
Amphibians (species, number)	6	1	83

this crayfish species was found to exert a significant, although variable, impact on zoobenthic communities, its abundance being negatively correlated with the abundance of total zoobenthos and of Diptera, Ephemeroptera, Odonata, and Trichoptera. However, the authors proved that the invertebrate taxa had much resilience in the long term. Nearly all invertebrate taxa were negatively affected by the crayfish within a 1-year lag, but no taxa exhibited sustained declines over the course of the *O. rusticus* invasion. Several factors may explain this pattern. Large-bodied invertebrate taxa, such as snails and crayfish, have slower generation times and are not as productive as smaller taxa, and are thus less able to respond numerically following population reductions of crayfish. In addition, aquatic insect larvae have a winged adult stage and, as a consequence, great dispersal ability – thus facilitating quick recovery. Whatever the explanation might be, McCarthy *et al.*'s (2006) results are illustrative in underlining the importance of a long-term monitoring of the invaded community to capture the complex picture of the impact of NICS.

Effects on ecosystems

As compared with the individual, population, and community levels, documented changes to ecosystem processes have been rarely reported in the literature on NICS. This probably reflects a lack of measurements of ecosystem processes rather than a lack of impact on these processes. The abundantly documented community effects are on the contrary expected to determine significant changes in energy flux and nutrient cycling in the invaded systems.

Essentially, the introduction of NICS may alter the pathways of the energy flux in two ways, i.e. through augmenting connectance by feeding at several trophic levels and through increasing the availability of autochthonous carbon as a food source for higher trophic levels (Stenroth and Nyström 2003). This was clearly proved by Geiger *et al.* (2005) in temporary freshwater marshes in Spain. Before the introduction of *P. clarkii* (Fig. 6a), macrophytes and the associated periphyton were the dominant primary producers. Only a small portion of the energy was transmitted from them to herbivores, whereas most of

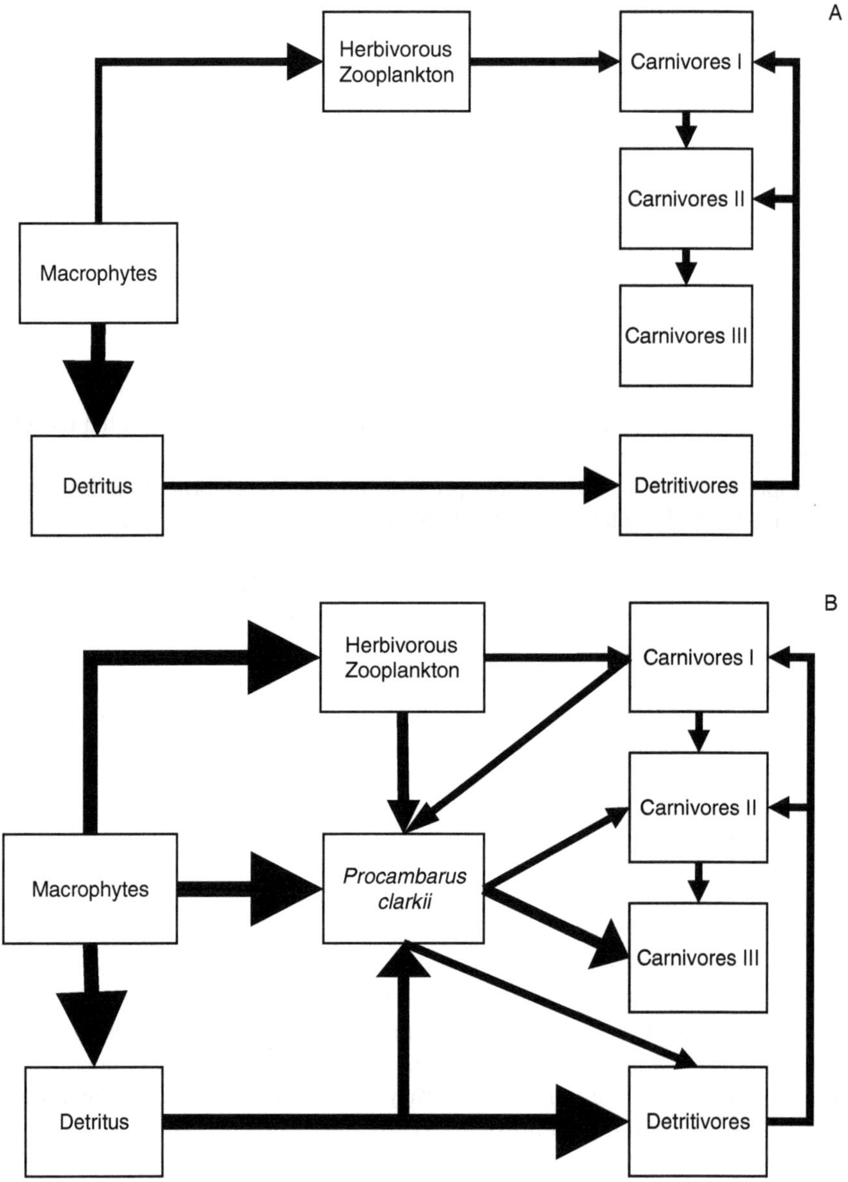

Fig. 6 Hypothesized energy flows in a marshland system before (A) and after (B) the introduction of an NICS (*Procambarus clarkii*). (Modified after Geiger *et al.* 2005)

it was lost to the detritus pool, which accumulated large amounts of organic matter. Detritivores, mainly macroinvertebrates (oligochaetes, chironomids) and meiofauna (nematodes, ostracods), used only a small fraction of the

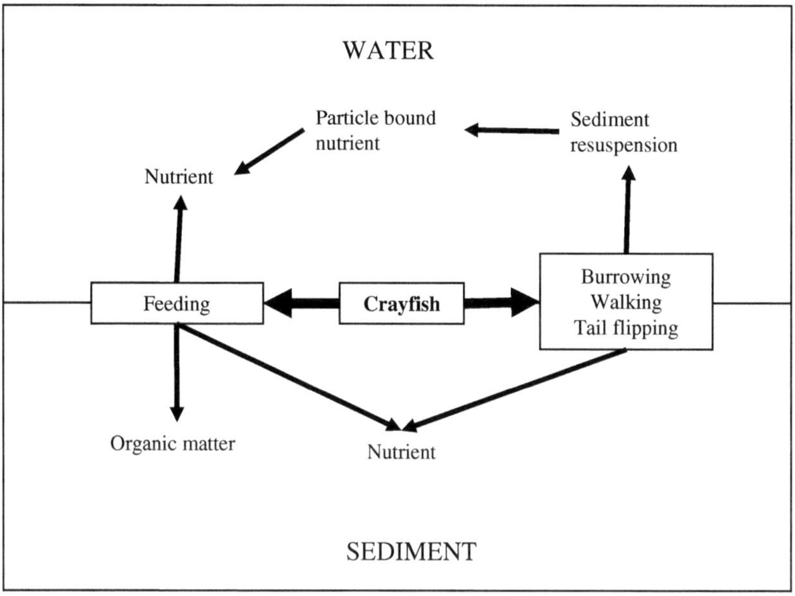

Fig. 7 Effects of the benthic activity of an NICS (*Procambarus clarkii*) on physical and chemical characteristics of water and sediments in a floodplain wetland in Spain. (Modified after Angeler *et al.* 2001)

deposited material. This system was characterized by a high diversity of herbivores and consisted of a minimum of four levels of consumers. Due to the large number of trophic levels and losses of energy to the detritus pool, the energy transferred to top predators such as birds and mammals was comparatively low. After the introduction of crayfish (Fig. 6B), much of the detritus was consumed by *P. clarkii* and the energy gained was directly transferred to the top predator level (fish, birds, and mammals). This resulted in a decreased importance of macrophytes, herbivores, and primary carnivores but offered a larger availability of energy for vertebrate predators.

The role that NICS may play through their benthic activity on physical and chemical characteristics of water and sediments was investigated by Angeler *et al.* (2001) in a floodplain wetland in Spain. *Procambarus clarkii* was hypothesized to affect the ecosystem processes by: (1) recycling sediment bound nutrients and (2) resuspending sediments associated with crayfish foraging, burrowing, and locomotory activity (walking, tail flipping) (Fig. 7). Compared to the control, the enclosures with crayfish showed a significant increase in both dissolved inorganic nutrients (soluble reactive phosphorus and ammonia) and total suspended solids as a result of crayfish bioturbation. At the same time, crayfish reduced the content of organic matter in the sediment and slightly increased total phosphorus and nitrogen content in sediments as the effect of its benthic activity.

Crayfish-mediated bioturbation has the general effect of impoverishing water quality by increasing total suspended solids and hence turbidity and by reducing light penetration and plant productivity (Anastácio and Marques 1997, Rodríguez *et al.* 2003). Crayfish may also alter sediment characteristics as a result of oxygen supply due to their activity. Bioturbation is often associated with crayfish constructing burrows, as commonly observed in *P. clarkii*, *P. leniusculus*, and *C. destructor* (e.g. Sommer and Goldman 1983, Gutiérrez-Yurrita and Montes 1999). However, crayfish, like *O. limosus*, can also move bottom sediments due to the friction between the extremities of its pereopods and the bottom during walking, the force exerted above the bottom by the uropods and the telson of escaping crayfish, and the pressure of the uropods and claws into the substrate to slow its movement while walking in fast currents (Maude and Williams 1983). In experimental flumes, Statzner *et al.* (2000, 2003) showed that *O. limosus* at a fixed biomass (174 g m^{-2}) significantly affected sand and gravel erosion. Its effect of bioturbator varied in function of the presence of refugia and aggression: sediment erosion averaged 2.8 and 1.4 kg dry weight m^{-2} d^{-1} when crayfish hierarchies were established and refugia were available. Once the refugia were removed, these rates increased to 4.0 and 3.2 kg dry weight m^{-2} d^{-1}. Additionally, bioturbation by crayfish was found to change bedform roughness, physical particle consolidation, proportion of sand in gravel interstices, sand cover by gravel, and the cover of filamentous algae. Such changes, in turn, may affect the abundance and structure of the entire benthic community, e.g. by modifying their substrate or by reducing algae and biofilm available for grazers. Also, sand reduction among gravel might alter the egg survival of gravel-breeding fish, like salmonids.

WHY DO NICS CAUSE MORE PROBLEMS THAN INDIGENOUS CRAYFISH?

In areas without any indigenous ecological equivalent, the changes caused by the introduction of NICS may be complex and usually affect all levels of ecological organization. Impacts range from subtle behavioral modifications of resident species to altered energy and nutrient fluxes in the ecosystem. Impacts at the community level can be strong when the introduced crayfish experience little predation or competition from native predators and have prey that lack efficient defense adaptations to them (Nyström *et al.* 2001). The modes of resource acquisition by NICS and their capacity to develop new trophic relationships, coupled with their action as bioturbator, may lead to dramatic direct and indirect effects on the ecosystem.

When NICS replace an indigenous ecological equivalent, their resource-acquisition mode should not be novel to the colonized community and therefore the resulting impact is expected to be weak. But the overall effect of NICS can be strong if, once introduced, they are capable of building high densities and/or of reaching large size. Indeed, several introduced crayfish often reach much higher densities than indigenous crayfish, e.g. *P. leniusculus vs. A. pallipes* (Guan and Wiles 1996). Densities of *O. limosus* ranged between 0.4 and 77 m^{-2} in a reservoir in Poland, of *O. rusticus* between 0.18 and 21 m^{-2} in North American streams, and of *P. leniusculus* between 3.7 and 21.7 m^{-2} in an English river (reviewed in Nyström 1999). On the contrary, densities of the indigenous species have been estimated to reach a maximum of 14.3 m^{-2} for *A. astacus* (in a Swedish lake), 3 m^{-2} for *A. pallipes* (in France), 3.6 m^{-2} for *C. japonicus* (in Japan), 0.83 m^{-2} for *P. fortis* (in USA), and 3.3 m^{-2} for *Paranephrops planifrons* White (in New Zealand) (references in Nyström 1999). The drastic decline of biota recorded by Wilson *et al.* (2004) in Trout Lake, Wisconsin (USA), when *O. rusticus* catches reached a threshold of nine crayfish $trap^{-1}$, suggests that its high abundance is the primary reason for its large impact, even in lakes with pre-existing crayfish populations.

Several biological traits contribute to the achievement by crayfish of high densities/large size. Relatively to indigenous crayfish, some NICS are characterized by higher fecundity, faster growth rates, and better physiological tolerances to changing environmental conditions (Lindqvist and Huner 1999, Chapter 12). They also might be better at coping with changes induced by human activities that cause pollution and habitat destruction. For instance, *P. clarkii* is a good colonizer of disturbed aquatic habitats and can survive in anoxic conditions in burrows (Gherardi 2006). Also, higher survival rate, hence leading to higher densities and/or larger sizes, is expected when a species is introduced without a full complement of specific parasites, pathogens, and enemies. And large sizes, in their turn, make crayfish both resistant to gape-size limited predators (such as many fish) and agonistically superior in resource fights. As a consequence, because of their large numbers, coupled with their wide trophic plasticity, NICS exert a greater direct (through consumption) or indirect (through competition) effect on the other biota, particularly on crayfish species, benthic fish, mollusks, and macrophytes (Nyström *et al.* 1996). This is also true for those species that apparently have not caused much environmental degradation, such as *A. leptodactylus* in England whose high numbers are producing considerable problems for anglers (Holdich 1999b). Obviously, large size usually translates into an overall higher energy and nutrient demand, but NICS may also be more efficient energy converters and may display higher metabolic rate when compared with similarly sized crayfish species.

As a consequence, a combination of larger size and greater weight-specific consumption of macrophytes and snails may explain the greater ecological impact suffered by those North American lakes where *O. rusticus* had replaced its congeners (Olsen *et al.* 1991) (Table 2), whereas the more efficient grazing by *P. leniusculus* than *A. astacus* seems to be the cause of the dramatic decrease

Table 2 Summary of test results after Olsen *et al.* (1991) for mechanisms governing the greater impact of *Orconectes rusticus* (Or) relative to *O. propinquus* (Op) and *O. virilis* (Ov) on a benthic community structure.

Mechanisms	Replace Op	Replace Ov
1. Individual size	Or > Op	Or < Ov
2. Population density	Or < Op	Or < Ov
3. Population biomass	Or < Op	Or < Ov
4. Weight-specific consumption, C, and sublethal damage, D, of snails	C: Or > Op D: Or = Op	C: Or > Ov D: Or > Ov
5. Weight-specific selection for single-stemmed macrophytes	Or = Op	Or = Ov
6. Weight-specific consumption, C, and destruction, D, of macrophytes	C: Or > Op D: Or = Op	C: Or = Ov D: Or = Ov

of macrophytes biomass (including the extinction of several rare *Chara* species), as recorded by Nyström and Strand (1996) in Sweden.

EFFECTS ON HUMAN ECONOMY AND HEALTH

From an anthropocentric perspective, a useful approach to measure impacts is to assess the damage caused by NICS by calculating the economic cost they induce to human societies, also in terms of the social and ethical problems associated with their potential harm to human health. A review of the existing literature in the matter shows that much of the discussion has been purely anecdotal. A negative consequence is the often encountered difficulty in justifying to decision makers the need of containing the spread of NICS and of mitigating the risks they pose.

The introduction of NICS has been often assumed to have contributed in a positive way to human economy by: (1) restoring some traditions proper to the cultural heritage of a country, e.g. crayfishing in Sweden and Finland (Kirjavainen and Sipponen 2004); (2) producing some economic benefits for many families in poorly developed areas, e.g. in Andalusia, Spain (Geiger *et al.* 2005); (3) leading to a diversification of agriculture to include astaciculture, e.g. by crayfish farmers in Britain and in Spain (Holdich 1999a); and (4) increasing trade between countries inside Europe as well as between European and extra-European countries (Ackefors 1999).

There are, however, several examples showing that often the introduction of commercially valuable crayfish has also led to negative results in the marketplace. Despite the original aim of crayfish farmers in Britain to produce crayfish for export to the Scandinavian market where they fetch a high price (Holdich 1993), most of the exports are now being made with crayfish harvested from natural waters and not from farming (Holdich 1999b). Similarly, in Africa, very few of the several projects that led to crayfish importations since the 1960s can be regarded as successful (de Moor 2002). For instance, in Lake Naivasha, Kenya, only about 40 metric tons of *P. clarkii* are now caught annually exclusively for local consumption (mainly tourism), after their first harvests in 1975 of several hundred tons per year (Smart *et al.* 2002). Crayfish were reported to spoil valuable fish (tilapia and largemouth bass) caught in gillnets (up to 30% of the catch) and to damage fish nets (de Moor 2002). They are also responsible for the decline of the rooted vegetation and therefore of the increase of phytoplankton, decrease of water transparency, and reduction of fish, including commercial species, in the littoral zone (de Moor 2002).

In several countries, introduced crayfish have today much lower commercial values than indigenous species, as exemplified by the Scandinavian market where consumers are willing to pay substantially higher prices for the indigenous *A. astacus* than for the naturalized *P. leniusculus* (Holdich 1999a). Crayfishing, in its turn, may produce extensive environmental impacts and associated costs, as the result of the continuous roaming of fishermen, causing a physical alteration of the habitat and the capture of non-target organisms (e.g. turtles) in the crayfish traps (Geiger *et al.* 2005).

There are several examples of damage to other human activities. *Procambarus clarkii* is a recognized pest in rice cultures in various parts of the world. Although in Louisiana (USA) double cropping crayfish and rice are practiced with success (Chien and Avault 1980), the species produces economic costs in rice fields in Portugal (Correia 1993). As an example, economic losses per ha of Euro 43.40 of the 2004 rice production (a 6.3% decrease in profits) was reported in the "Baixo Mondego" area exclusively due to crayfish (Anastácio *et al.* 2005a). Damage to rice production primarily consists of crayfish consuming seedlings, but negative effects derive from the increased turbidity and decreased dissolved oxygen content due to the crayfish bioturbation (Anastácio *et al.* 2005a, b, c).

Burrowing by several NICS (e.g. *P. clarkii, P. leniusculus*, and *C. destructor*) can be a problem in areas other than agricultural, e.g. lawns, golf courses, levees, dams, dykes, and in rivers and lakes (e.g. Anastácio and Marques 1997). A few authors have lamented the damage caused by *C. destructor* burrowing to dam walls and irrigation canals (de Moor 2002). More often cited is the effect of *P. clarkii* to "honeycomb" banks, with their consequent structural damage (Huner 1977) that seriously affects areas with extensive canal irrigation systems and water control structures (Adão and Marques 1993). Although not recorded as a burrowing species in its native North American habitat (Holdich 1999a), *P. leniusculus* causes considerable damage to river banks by burrowing in the UK (Sibley 2000).

Indeed, if a monetary value were to be assigned to species extinctions and losses in biodiversity, ecosystem services, and aesthetics, the total economic damage of introduced crayfish might be enormous. Neither do we have records of the financial costs of control and remediation, except an estimate of US$4.5 million for the restitution of *P. fortis* in California (USA) (US Fish and Wildlife Service) and the £100,000 spent trying, unsuccessfully, to eradicate *P. leniusculus* from rivers in Scotland (Collins 2006).

Up to today, little attention has been paid to the potential harm that NICS pose to human health. NICS, such as *P. clarkii*, often live in areas contaminated by sewage and toxic industrial residues and may have high heavy metal concentrations in their tissues (Geiger *et al.* 2005); they were found to bioaccumulate metals such as nickel, lead, and zinc in their tissues and organs at a significantly higher rate than the indigenous species (Gherardi *et al.* 2002). Their potential to transfer contaminants to their consumers, including man, is obviously high. Measurements of accumulation of heavy metals in waterfowl and other wetland birds living and feeding in a toxic spill area of Spain showed that heavy metals like arsenic have entered the food chain and were detected in some bird species, such as white storks, spoonbills and grey herons, all predators of crayfish (Geiger *et al.* 2005).

The finding that *P. clarkii* may consume Cyanobacteria is of increasing concern for human health (Gherardi and Lazzara 2006). Several Cyanobacteria release a wide range of toxins and BMAA (β-*N*-methylamino-L-alanine) that may produce lethal animal and human intoxications (e.g. Carmichael 1988, Cox *et al.* 2005). Among the few organisms so far investigated, *P. leniusculus* and *P. clarkii* were found to accumulate such toxins in their tissues (Lirås *et al.* 1998, Vasconcelos *et al.* 2001), being therefore able to transfer them to more sensitive organisms, man included.

The other side of the coin is the ability of *P. clarkii* to control, through predation and competition, populations of the pulmonate snails *Biomphalaria* and *Bulinus* known to host *Schistosoma mansoni* and *S. haematobium*, the agents of human schistosomiasis (Chapter 4). Schistosomiasis is one of the most widespread diseases in Africa: in Kenya alone, it is known to affect 3.5 million individuals with 12 million more at risk of infection. As suggested by Mkoji *et al.* (1999), due to the quick spread of this crayfish in African waterbodies, the epidemiology of schistosomiasis is expected to be significantly altered with time although the possibilities remain that African snails will soon evolve measures to avoid crayfish predation before their extinction or that the parasite will change its host.

WHAT COMES NEXT?

Despite the considerable attention that NICS have received since 1987, a review of the many papers published in the last few years has revealed the absence in the literature of a general framework about their impact. Nonetheless, a global view is required to help us coordinate and improve both control and research efforts for the existing NICS and hopefully predict the impact of future introductions.

First, this review lamented the general lack of simultaneous measurements of the impact of NICS at multiple ecological levels. This is unfortunate, because information about responses at several levels of organization is required to fully understand the range of the impact observed (Simon and Townsend 2003). Also, little effort has been made to study the impact of NICS at the genetic level. We cannot even guess therefore what their role might be in determining the future evolutionary pathway that the invaded system will follow after the introduction of NICS. Neither do we have quantitative estimates about direct or indirect monetary (and social) costs induced by NICS to the human economy and health. And only a few paradigmatic invasive species, namely *P. leniusculus*, *P. clarkii*, and *O. rusticus*, have been extensively studied, whereas information is scanty or even absent about other species, like *C. destructor*.

Short-term laboratory and field studies have revealed a myriad of direct and indirect effects of crayfish on some compartments of the invaded ecosystem (most often the littoral zone). Their results were certainly able to provide mechanistic insights into the interactions between crayfish and the other organisms in the community and to quantify the immediate impact of NICS. However, these sometimes complex interactions have been only seldom investigated at larger levels. Few comparative field studies (e.g. Charlebois and Lamberti 1996) and even fewer long-term studies (Wilson *et al.* 2004, McCarthy *et al.* 2006) have been conducted to examine the dynamic relationships among organisms, including the prolonged effects of crayfish invaders on native communities over multiple generations.

Indeed, coupling the results of small- and large-scale studies may help capture a more realistic picture of the impact of NICS. As held by McCarthy *et al.* (2006), each method is not without its own biases. Along with characterizing only short-term ecological responses, small-level studies may suffer from experimental artifacts. In contrast, long-term studies certainly provide important insights into the long-term threats of an NICS and the potential adaptability of the native taxa to it, but their results may also be confounded by environmental factors that cannot be controlled (Parker *et al.* 1999). Because the disadvantage of one approach is the advantage of another, coupling multiple levels of analyses certainly provides a method in which interactions at one level can be successfully translated to another.

Finally, we need much more intensive work on the whole-system impact of crayfish invasion; specifically, we need to learn more about the natural variability among systems and the extent to which the impact of an NICS depends on the community or ecosystem where it is measured. So, on what level can we generalize across systems? And how often does the impact of a crayfish invader depend on the presence of other NIS? Not before having answered to these and to the other questions above will we be able to generalize, and even predict, which species is more likely to have the greatest impact in aquatic systems (Parker *et al.* 1999).

ACKNOWLEDGEMENTS

This paper was written while the author was acting as a Visiting Scholar at the Columbia University in the City of New York (NY, USA) and was partly supported by MIUR (ex-60%). My warmest thanks are directed to Professor Shahid Nahem for his kind hospitality.

REFERENCES

Ackefors, H. 1999. The positive effects of established crayfish. Pages 49–61 in F. Gherardi and D. M. Holdich, editors. Crayfish in Europe as alien species: How to make the best of a bad situation? A. A. Balkema, Rotterdam, The Netherlands.

Adão, H. and J. C. Marques. 1993. Population biology of the red swamp crayfish *Procambarus clarkii* (Girard, 1852) in southern Portugal. Crustaceana **65**, 336–345.

Alderman, D. J. and J. L. Polglase. 1988. Pathogens, parasites and commensals. Pages 167–212 in D. M. Holdich and R. S. Lowery, editors. Freshwater crayfish: biology, management and exploitation. Croom Helm (Chapman & Hall), London, UK.

Anastácio, P. M., A. M. Correia, and J. P. Menino. 2005a. Processes and patterns of plant destruction by crayfish: effects of crayfish size and developmental stages of rice. Archiv für Hydrobiologie **162**, 37–51.

Anastácio, P. M., A. M. Correia, J. P. Menino, and L. Martins da Silva. 2005b. Are rice seedlings affected by changes in water quality caused by crayfish? Annales de Limnologie – International Journal of Limnology **41**, 1–6.

Anastácio, P. M. and J. C. Marques. 1997. Crayfish, *Procambarus clarkii*, effects on initial stages of rice growth in the lower Mondego River valley (Portugal). Freshwater Crayfish **11**, 608–617.

Anastácio, P. M., V. S. Parente, and A. M. Correia. 2005c. Crayfish effects on seeds and seedlings: identification and quantification of damage. Freshwater Biology **50**, 697–704.

Angeler, D. G., S. Sánchez-Carrillo, G. García, and M. Alvarez-Cobelas. 2001. The influence of *Procambarus clarkii* (Cambaridae, Decapoda) on water quality and sediment characteristics in a Spanish floodplain wetland. Hydrobiologia **464**, 9–98.

Arrignon, J. C. V., J. V. Huner, and P. J. Laurent. 1990. L'écrevisse rouge des marais. Maisonneuve et Larose, Paris, France, 87 pp.

Barbaresi, S., F. Gherardi. 2000. The invasion of the alien crayfish *Procambarus clarkii* in Europe, with particular reference to Italy. Biological Invasions **2**, 259–264.

Bernardo, J. M. and M. Ilhéu. 1994. Red swamp crayfish (*Procambarus clarkii*): contribution to material cycling. Verhandlungen Internationale Vereinigung Limnologie **25**, 2447–2449.

Blake, M. A. 1995. The vulnerability of juvenile signal crayfish to perch and eel predation. Freshwater Biology **33**, 233–244.

Bolser, R. C., M. E. Hay, N. Lindquist, W. Fenical, and D. Wilson. 1998. Chemical defenses of freshwater macrophytes against crayfish herbivory. Journal of Chemical Ecology **24**, 1639–1657.

Bouchard, R. W. 1977. Distribution, systematics status, and ecological notes on five poorly known species of crayfishes in western North America (Decapoda: Astacidae and Cambaridae). Freshwater Crayfish **3**, 409–423.

Capelli, G.M. 1982. Displacement of northern Wisconsin crayfish by *Orconectes rusticus* (Girard). Limnology and Oceanography **27**, 741–745.

Carmichael, W. W. 1988. Toxins of freshwater algae. Pages 121–147 in A. T. Tu, editor. Handbook of natural toxins, Vol. 3, Marine toxins and venoms. Marcel Dekker, New York, NY.

Carpenter, J. 2005. Competition for food between an introduced crayfish and two fishes endemic to the Colorado River basin. Environmental Biology of Fishes **72**, 335–342.

Chambers, P. A., J. M. Hanson, J. M. Burke, and E.E. Prepas. 1990. The impact of the crayfish *Orconectes virilis* on aquatic macrophytes. Freshwater Biology **24**, 81–91.

Charlebois, P. M. and G. L. Lamberti. 1996. Invading crayfish in a Michigan stream: direct and indirect effects on peryphyton and macroinvertebrates. Journal of the North American Benthological Society **15**, 551–563.

Chien, Y. H. and J. W. Avault Jr. 1980. Production of crayfish in rice fields. Progress in Fish Cultivation **42**, 67–71.

Cirujano, S., J. A. Camargo, and C. Gómez-Cordovés. 2004. Feeding preferences of the red swamp crayfish *Procambarus clarkii* (Girard) on living macrophytes in a Spanish wetland. Journal of Freshwater Ecology **19**, 219–226.

Collins, V. 2006. £100,000 battle against alien crayfish ends in defeat. Crayfish News - International Association of Astacology Newsletter **28**, 10–11.

Correia, A. M. 1993. Situation de l'acclimatation de l'écrevisse rouge des marais *Procambarus clarkii* au Portugal. L'Astaciculteur de France **35**, 2–9.

Correia, A. M. 2002. Niche breadth and trophic diversity: feeding behaviour of the red swamp crayfish (*Procambarus clarkii*) towards environmental availability of aquatic macroinvertebrates in a rice field (Portugal). Acta Oecologica **23**, 421–429.

Correia, A. M. 2003. Food choice by the introduced crayfish *Procambarus clarkii*. Annales Zoologici Fennici **40**, 517–528.

Correia, A. M., N. Bandeira, and P. M. Anastácio. 2005. Predator-prey interactions of *Procambarus clarkii* with aquatic macroinvertebrates in single and multiple prey systems. Acta Oecologica **28**, 337–343.

Cox, G. W. 2004. Alien species and evolution. Island Press, Washington, DC.

Cox, P. A., S. A. Banack, S. J. Murch, U. Rasmussen, G. Tien, R. R. Bidigare, J. S. Metcalf, L. F. Morrison, G. A. Codd, and B. Bergman. 2005. Diverse taxa of Cyanobacteria produce β-N-methylamino-L-alanine, a neurotoxic amino acid. Proceedings of the National Academy of Sciences USA. **102**, 5074–5078.

Crandall, K. A. 2002. Crayfish as model organisms. Freshwater Crayfish **13**, 3–10.

Cronin, G. and M. E. Hay. 1996. Seaweed-herbivore interactions depend on recent history of both the plant and animal. Ecology **77**, 1531–1543.

Cronin, G., D. M. Lodge, M. E. Hay, M. Miller, A. M. Hill, T. Horvath, R. C. Bolser, N. Lindquist, and M. Wahl. 2002. Crayfish feeding preferences for fresh water macrophytes: The influence of plant structure and chemistry. Journal of Crustacean Biology **22**, 708–718.

Cruz, M. J. and R. Rebelo. 2005. Vulnerability of Southwest Iberian amphibians to an introduced crayfish, *Procambarus clarkii*. Amphibia-Reptilia **26**, 293–303.

Cukerzis, J. 1968. Interspecific relations between *Astacus astacus* L. and *A. leptodactylus* Esch. Ekologia Polska Ser. A **16**, 629–636.

Damman, H. 1987. Leaf quality and enemy avoidance by the larvae of a pyralid moth. Ecology **68**, 88–97.

Delibes, M. and I. Adrián. 1987. Effects of crayfish introduction on otter *Lutra lutra* in the Doñana National Park, SW Spain. Biological Conservation **42**, 153–159.

Diéguez-Uribeondo, J. and K. Söderhäll. 1993. *Procambarus clarkii* Girard as a vector for the crayfish plague fungus, *Aphanomyces astaci* Schikora. Aquaculture and Fisheries Management **24**, 761–765.

Dorn, N. J., G. Cronin, and D. M. Lodge. 2001. Feeding preferences and performance of an aquatic lepidopteran on macrophytes: plant hosts as food and habitat. Oecologia **128**, 406–415.

Dunlin, M., T. Huddleston, R. E. Larson, and G. W. Klontz. 1976. Enteric redmouth disease. University of Idaho Forest, Wildlife and Range Experiment Statistic Bulletin **8**, 1–15.

Edgerton, B. F., P. Henttonen, J. Jussila, A. Mannonen, P. Paasonen, T. Taugbøl, L. Edsman, and C. Souty-Grosset. 2004. Understanding the causes of disease in European freshwater crayfish. Conservation Biology **18**, 1466–1474.

Elvey, W., A. M. M. Richardson, and L. Bermuta. 1996. Interactions between the introduced yabby, *Cherax destructor*, and the endemic crayfish, *Astacopsis franklinii*, in Tasmanian streams. Freshwater Crayfish **11**, 349–363.

Fitzpatrick Jr, J. F. 1995. The Eurasian Far-eastern crawfishes: a preliminary review. Freshwater Crayfish **8**, 1–11.

Gamradt, S. C. and L. B. Kats. 1996. Effect of introduced crayfish and mosquito fish on California newts (*Taricha torosa*). Conservation Biology **10**, 1155–1162.

Geiger, W., P. Alcorlo, A. Baltanás, and C. Montes. 2005. Impact of an introduced Crustacean on the trophic webs of Mediterranean wetlands. Biological Invasions **7**, 49–73.

Gherardi, F. 2002. Behaviour. Pages 258–290 in D. M. Holdich, editor. Biology of freshwater crayfish. Blackwell, Oxford, UK.

Gherardi, F. 2006. The red swamp crayfish (*Procambarus clarkii*) in Europe: a lesson for biodiversity conservation. Marine and Freshwater Behaviour and Physiology **39**, 175–191.

Gherardi, F. and P. Acquistapace. 2007. Invasive crayfish in Europe: the impact of *Procambarus clarkii* on the littoral community of a Mediterranean lake. Freshwater Biology, **52**, 1249–1259.

Gherardi, F. and S. Barbaresi. 2000. Invasive crayfish: activity patterns of *Procambarus clarkii* in the rice fields of the Lower Guadalquivir (Spain). Archiv für Hydrobiologie **150**, 153–168.

Gherardi, F., S. Barbaresi, O. Vaselli, and A. Bencini. 2002. A comparison of trace metal accumulation in indigenous and alien freshwater macro-decapods. Marine and Freshwater Behaviour and Physiology **35**, 179–188.

Gherardi, F. and A. Cioni. 2004. Agonism and interference competition in freshwater decapods. Behaviour **141**, 1297–1324.

Gherardi, F. and W. H. Daniels. 2004. Agonism and shelter competition between invasive and indigenous crayfish species. Canadian Journal of Zoology **82**, 1923–1932.

Gherardi, F. and D. M. Holdich, editors. 1999. Crayfish in Europe as alien species: How to make the best of a bad situation? A. A. Balkema, Rotterdam, The Netherlands.

Gherardi, F. and L. Lazzara. 2006. Effects of the density of an invasive crayfish (*Procambarus clarkii*) on pelagic and surface microalgae in a Mediterranean wetland. Archiv für Hydrobiologie **165**, 401–414.

Gherardi, F., B. Renai, and C. Corti. 2001. Crayfish predation on tadpoles: a comparison between a native (*Austropotamobius pallipes*) and an alien species (*Procambarus clarkii*. Bulletin Français de la Pêche et de la Pisciculture **361**, 659–668.

Gouin, N., F. Grandjean, S. Pain, C. Souty-Grosset, and J. Reynolds. 2003. Origin and colonization history of the white-clawed crayfish, *Austropotamobius pallipes*, in Ireland. Heredity **9**, 70–77.
Groombridge, B. 1993. 1994 IUCN Red list of threatened animals. World Conservation Monitoring Centre, Cambridge, UK.
Guan, R.-Z. and P.R. Wiles. 1996. Growth, density and biomass of crayfish, *Pacifastacus leniusculus*, in a British lowland river. Aquatic Living Resource **9**, 265–272.
Guan, R.-Z. and P.R. Wiles. 1997. Ecological impact of introduced crayfish on benthic fishes in a British lowland river. Conservation Biology **11**, 641–647.
Gutiérrez-Yurrita, P. J. and C. Montes. 1999. Bioenergetics and phenology of reproduction of the introduced red swamp crayfish, *Procambarus clarkii*, in Doñana National Park, Spain, and implications for species management. Freshwater Biology **42**, 561–574.
Gutiérrez-Yurrita P. J., G. Sancho, M. Á. Bravo, Á. Baltanás, and C. Montes. 1998. Diet of the red swamp crayfish *Procambarus clarkii* in natural ecosystems of the Doñana National Park temporary fresh-water marsh (Spain). Journal of Crustacean Biology **18**, 120–127.
Halder, M. and W. Ahne. 1988. Freshwater crayfish *Astacus astacus* – a vector for infectious pancreatic necrosis virus (IPVN). Diseases of Aquatic Organisms **4**, 205–209.
Harlioğlu, M. M. 2004. The present situation of freshwater crayfish, *Astacus leptodactylus* (Eschscholtz, 1823) in Turkey. Aquaculture **230**, 181–187.
Hill, A. M. and D. M. Lodge. 1994. Diel changes in resource demand: competition and predation in species replacement among crayfishes. Ecology **75**, 2118–2126.
Hill, A. M. and D. M. Lodge. 1999. Replacement of resident crayfishes by an exotic crayfish: the roles of competition and predation. Ecological Applications **9**, 678–690.
Hobbs Jr, H. H. 1988. Crayfish distribution, adaptive radiation and evolution. Pages 52–82 in D. M. Holdich and R. S. Lowery, editors. Freshwater crayfish: biology, management and exploitation. Croom Helm, London, UK.
Hobbs, H. H. III. 1991. Adaptations and convergence in North American crayfishes. Freshwater Crayfish **2**, 541–549.
Hobbs, H. H. III, J. P. Jass, and J. V. Huner. 1989. A review of global crayfish introductions with particular emphasis on two North American species (Decapoda, Cambaridae). Crustaceana **56**, 299–316.
Hofkin, B. V., G. M. Mkoji, D. K. Koech, and E. S. Loker. 1991. Control of schistosome-transmitting snails in Kenya by the North American crayfish *Procambarus clarkii*. Journal of Tropical Medicine and Hygiene **45**, 339–344.
Holdich, D. M. 1988. The dangers of introducing alien animals with particular reference to crayfish. Freshwater Crayfish **7**, XV–XXX.
Holdich, D. M. 1993. A global review of astaciculture – freshwater crayfish farming. Aquatic Living Resources **6**, 307–317.
Holdich, D. M. 1999a. The negative effects of established crayfish introductions. Pages 31–61 in F. Gherardi and D. M. Holdich, editors. Crayfish in Europe as alien species: How to make the best of a bad situation? A. A. Balkema, Rotterdam, The Netherlands.
Holdich, D. M. 1999b. The introduction of alien crayfish species into Britain for commercial exploitation – An own goal? Pages 85–97 in J. C. Von Vaupel Klein and F. R. Schram, editors. The biodiversity crisis and Crustacea. A. A. Balkema, Rotterdam, The Netherlands.

Holdich, D. M. and J. C. J. Domaniewski. 1995. Studies on a mixed population of the crayfish *Austropotamobius pallipes* and *Pacifastacus leniusculus* in England. Freshwater Crayfish **10**, 37–45.

Holdich, D. M., J. P. Reader, W. D. Rogers, and M. Harlioglu. 1995. Interactions between three species of crayfish (*Austropotamobius pallipes, Astacus leptodactylus* and *Pacifastacus leniusculus*). Freshwater Crayfish **10**, 46–56.

Holdich, D. M., P. J. Sibley, and S. Peay. 2004. The white-clawed crayfish – a decade on. British Wildlife **15**, 153–164.

Huner, J. V. 1977. Introductions of *Procambarus clarkii* (Girard) – an update. Freshwater Crayfish **3**, 93–202.

Ilhéu, M. and J. M. Bernardo. 1993a. Experimental evaluation of food preference of redswamp crayfish, *Procambarus clarkii*: vegetal versus animal. Freshwater Crayfish **9**, 359–364.

Ilhéu, M. and J. M. Bernardo. 1993b. Aspects of trophic ecology of red swamp crayfish (*Procambarus clarkii* Girard) in Alentejo, South of Portugal. Pages 417–423 in Proceedings VI Spanish Meeting on Limnology, Granada, Spain.

Ilhéu, M. and J. M. Bernardo. 1995. Trophic ecology of red swamp crayfish *Procambarus clarkii* (Girard) – preference and digestability of plant foods. Freshwater Crayfish **10**, 132–139.

Karatayev, A. Y., L. E. Burlakova, and D. K. Padilla. 2002. Impacts of zebra mussels on aquatic communities and their role as ecosystem engineers. Pages 433–446 in E. Leppäkoski, S. Gollasch, and S. Olenin, editors. Invasive aquatic species of Europe. Distribution, impacts and management. Kluwer, Dordrecht, The Netherlands.

Kirjavainen, J. and M. Sipponen. 2004. Environmental benefit of different crayfish management strategies in Finland. Fisheries Management and Ecology **11**, 213–218.

Kolar, C. S. and D. M. Lodge. 2001. Progress in invasion biology: predicting invaders. Trends in Ecology and Evolution **16**, 199–204.

Köksal, G. 1988. *Astacus leptodactylus* in Europe. Pages 365–400 in D. M. Holdich and R. S. Lowery, editors. Freshwater crayfish: biology, management and exploitation. Croom Helm, London, UK.

Leite, E. P., P. M. Anastácio, M. Ferreira, L. Vicente, and A. M. Correia. 2005. Do Eastern mosquitofish exhibit anti-predator behavior towards red swamp crayfish? Zoological Studies **44**, 513–518.

Light, T., D. C. Erman, C. Myrick, and J. Clarke. 1995. Decline of the Shasta crayfish (*Pacifastacus fortis* Faxon) of Northeastern California. Conservation Biology **9**, 1567–1577.

Lima, S. L. 1998. Nonlethal effects in the ecology of predator-prey interactions. What are the ecological effects of anti-predator decision-making? BioScience **48**, 25–33.

Lindqvist, O. V. and J. V. Huner. 1999. Life history characteristics of crayfish: what makes some of them good colonizers? Pages 23–30 in F. Gherardi and D. M. Holdich, editors. Crayfish in Europe as alien species: How to make the best of a bad situation? A. A. Balkema, Rotterdam, The Netherlands.

Linnaeus, C. 1746. Fauna Suecica. Stockholm, Sweden, 411 pp.

Lirås, V., M. Lindberg, P. Nyström, H. Annadotter, L. A. Lawton, and B. Graf. 1998. Can ingested cyanobacteria be harmful to the signal crayfish (*Pacifastacus leniusculus*)? Freshwater Biology **39**, 233–242.

Lodge, D. M. 1991. Herbivory on freshwater macrophytes. Aquatic Botany **41**, 195–224.

Lodge, D. M. and J. G. Lorman. 1987. Reductions in submersed macrophyte biomass and species richness by the crayfish *Orconectes rusticus*. Canadian Journal of Fisheries and Aquatic Sciences **44**, 591–597.

Lodge, D. M., M. W. Kershner, J. E. Aloi, and A. P. Covich. 1994. Effects of an omnivorous crayfish (*Orconectes rusticus*) on a freshwater littoral food web. Ecology **75**, 1265–1281.

Lodge, D. M., R. A. Stein, K. M. Brown, A. P. Covich, C. Brönmark, J. E. Garvey, and S. P. Klosiewski 1998. Predicting impact of freshwater exotic species on native biodiversity: challenges in spatial scaling. Australian Journal of Ecology **23**, 53–67.

Lodge, D. M., C. A. Taylor, D. M. Holdich, and J. Skurdal 2000. Non-indigenous crayfishes threaten North American freshwater biodiversity: lessons from Europe. Fisheries **25**, 7–20.

Luttenton, M. R., M. J. Horgan, and D. M. Lodge. 1998. Effects of three *Orconectes* crayfish on epilithic microalgae: a laboratory experiment. Crustaceana **71**, 845–855.

Maude, S. H. and D. D. Williams. 1983. Behavior of crayfish in water currents: Hydrodynamics of eight species with reference to their distribution patterns in southern Ontario. Canadian Journal of Fisheries and Aquatic Sciences **40**, 68–77.

McCarthy, J. M., C. L. Hein, J. D. Olden, and M. J. Vander Zanden. 2006. Coupling long-term studies with meta-analysis to investigate impacts of non-native crayfish on zoobenthic communities. Freshwater Biology **51**, 224–235.

McCarthy, T. M., and B. F. Dickey. 2002. Chemically mediated effects of injured prey on behavior of both prey and predators. Behaviour **139**, 585–602.

Mikkola, H. 1996. Alien freshwater crustacean and indigenous mollusk species with aquaculture potential in eastern and southern Africa. South African Journal of Aquatic Sciences **22**, 90–99.

Mkoji, G. M., B. V. Hofkin, A. M. Kuris, A. Stewart-Oaten, B. N. Mungai, J. H. Kihara, F. Mungai, J. Yundu, J. Mbui, J. R. Rashid, C. H. Kariuki, J. H. Ouma, D. K. Koech, and E. S. Loker. 1999. Impact of the crayfish *Procambarus clarkii* on *Schistosoma haematobium* transmission in Kenya. American Journal of Tropical Medicine and Hygiene **61**, 751–759.

Momot, W. T. 1995. Redefining the role of crayfish in aquatic ecosystems. Reviews in Fisheries Science **3**, 33–63.

Momot, W. T., H. Gowing, and P. D. Jones. 1978. The dynamics of crayfish and their role in ecosystems. American Midland Naturalist **99**, 10–35.

Montes, C., M. A. Bravo-Utrera, A. Baltanás, C. Duarte, and P. J. Gutiérrez-Yurrita. 1993. Bases ecológicas para la gestión del Cangrejo Rojo de las Marismas en el Parque Nacional de Doñana. ICONA, Ministerio de Agricultura y Pesca, Madrid, Spain.

Moor, I., de 2002. Potential impacts of alien freshwater crayfish in South Africa. African Journal of Aquatic Sciences **27**, 125–139.

Moyle, P. B. and T. Light. 1996. Biological invasions of fresh water: empirical rules and assembly theory. Biological Conservation **78**, 149–161.

Mueller, R. and A. Frutiger. 2001. Effects of intensive trapping and fish predation on an (unwanted) population of *Procambarus clarkii*. Abstracts of Annual Meeting of the North American Benthological Society, LaCrosse, WI, 3–8, June 2001.

Nakata, K. and S. Goshima. 2003. Competition for shelter of preferred sizes between the native crayfish species *Cambaroides japonicus* and he alien crayfish species *Pacifastacus leniusculus* in Japan in relation to prior residence, sex difference, and body size. Journal of Crustacean Biology **23**, 897–907.

Newman, R. M. 1991. Herbivory and detritivory on freshwater macrophytes by invertebrates: a review. Journal of the North American Benthological Society **10**, 89–114.

Nyström, P. 1999. Ecological impact of introduced and native crayfish on freshwater communities: European perspectives. Pages 63–84 in F. Gherardi and D. M. Holdich, editors. Crayfish in Europe as alien species: How to make the best of a bad situation? A. A. Balkema, Rotterdam, The Netherlands.

Nyström, P. 2002. Ecology. Pages 192–235 in D. M. Holdich, editor. Biology of freshwater crayfish. Blackwell, Oxford, UK.

Nyström, P., C. Brönmark, and W. Granéli. 1996. Patterns in benthic food webs: a role for omnivorous crayfish? Freshwater Biology **36**, 631–646.

Nyström, P. and J. R. Pérez. 1998. Crayfish predation on the common pond snail (*Lymnaea stagnalis*): the effect of habitat complexity and snail size on foraging efficiency. Hydrobiologia **368**, 201–208.

Nyström, P. and J. A. Strand. 1996. Grazing by a native and an exotic crayfish on acquatic macrophytes. Freshwater Biology **36**, 673–682.

Nyström, P., O. Svensson, B. Lardner, C. Brönmark, and W. Granéli. 2001. The influence of multiple introduced predators on a littoral pond community. Ecology **82**, 1023–1039.

Olsen, T. M., D. M. Lodge, G. M. Capelli, and R. J. Houlihan. 1991. Mechanisms of impact of an introduced crayfish (*Orconectes rusticus*) on littoral congeners, snails, and macrophytes. Canadian Journal of Fisheries and Aquatic Sciences **48**, 1853–1861.

Paine, R. T. 1966. Food web complexity and species diversity. American Naturalist **100**, 65–75.

Parker, I. M., D. Simberloff, W. M. Lonsdale, K. Goodell, M. Wonham, P. M. Kareiva, M. H. Williamson, B. Von Holle, P. B. Moyle, J. E. Byers, and L. Goldwasser 1999. Impact: toward a framework for understanding the ecological effects of invaders. Biological Invasions **1**, 3–19.

Pérez, J. R., J. M. Carral, J. D. Celada, M. Sáez-Royuela, C. Munoz, and A. Sierra. 1997. Current status of astaciculture production and commercial situation of crayfish in Europe. Aquaculture Europe **22**, 6–13.

Perry, W. L., J. E. Feder, and D. M. Lodge. 2001. Implications of hybridization between introduced and resident *Orconectes* crayfish. Conservation Biology **15**, 1656–1666.

Perry, W. L., D. M. Lodge, and J. E. Feder 2002. Importance of hybridization between indigenous and nonindigenous freshwater species: an overlooked threat to North American biodiversity. Systematic Biology **51**, 255–275.

Perry, W. L., D. M. Lodge, and G. A. Lamberti. 1997. Impact of crayfish predation on exotic zebra mussels and native invertebrates in a lake-outlet stream. Canadian Journal of Fisheries and Aquatic Sciences **54**, 120–125.

Perry, W. L., D. M. Lodge, and G. A. Lamberti. 2000. Crayfish (*Orconectes rusticus*) impacts on zebra mussel (*Dreissena polymorpha*) recruitment, other macroinvertebrates and algal biomass in a lake-outlet stream. American Midland Naturalist **144**, 308–316.

Pontoppidan, E. 1775. The natural history of Norway. Linde, London, UK.

Renai, B. and F. Gherardi. 2004. Predatory efficiency of crayfish: comparison between indigenous and non-indigenous species. Biological Invasions **6**, 89–99.

Reynolds, J. D. 1997. The present status of freshwater crayfish in Ireland. Bulletin Française de la Pêche et de la Pisciculture **347**, 693–700.

Reynolds, J. D. and C. O'Keeffe. 2005. Dietary patterns in steam- and lake-dwelling populations of *Austropotamobius pallipes*. Bulletin Française de la Pêche et de la Pisciculture **376–377**, 715–730.

Ricciardi, A. 2003. Predicting the impacts of an introduced species from its invasion history: an empirical approach applied to zebra mussel invasions. Freshwater Biology **48**, 972–981.

Rodríguez, C. L., E. Bécares, and M. Fernández-Aláez. 2003. Shift from clear to turbid phase in Lake Chozas (NW Spain) due to the introduction of American red swamp crayfish (*Procambarus clarkii*). Hydrobiologia **506–509**, 421–426.

Rodríguez, C. L., E. Bécares, M. Fernández-Aláez, and C. Fernández-Aláez. 2005. Loss of diversity and degradation of wetlands as a result of introducing exotic crayfish. Biological Invasions **7**, 75–85.

Romero, X. M. 1997. Redclaw crayfish aquaculture in Ecuador: the new boom. NAGA, The ICLARM Quarterly **20**, 18–21.

Sibley, P. 2000. Signal crayfish management in the River Wreake catchment. Pages 95–108 in D. Rogers and J. Brickland, editors. Crayfish Conference Leeds. Environment Agency, Leeds, UK.

Simon, K. S. and C. R. Townsend. 2003. Impacts of freshwater invaders at different levels of ecological organization, with emphasis on salmonids and ecosystem consequences. Freshwater Biology **48**, 982–994.

Slater, F. M. and G. Rayner. 1993. *Austropotamobius pallipes* in otter diet in the mid-Wye catchment of central Wales. Freshwater Crayfish **9**, 365–367.

Smart, A. C., D. M. Harper, F. Malaisse, S. Schmitz, S. Coley, and A.-C. Gourder de Beauregard. 2002. Feeding of the exotic Louisiana red swamp crayfish, *Procambarus clarkii* (Crustacea, Decapoda), in an African tropical lake: Lake Naivasha, Kenya. Hydrobiologia **488**, 129–142.

Söderbäck, B. 1991. Interspecific dominance and aggressive interaction in the freshwater crayfishes *Astacus astacus* (L.) and *Pacifastacus leniusculus* (Dana). Canadian Journal of Zoology **69**, 1321–1325.

Söderbäck, B. 1994. Reproductive interference between two co-occurring crayfish species, *Astacus astacus* L. and *Pacifastacus leniusculus* Dana. Nordic Journal of Freshwater Research **69**, 137–143.

Sommer, T. R. and C. R. Goldman. 1983. The crayfish *Procambarus clarkii* from California ricefields: ecology, problems and potential harvest. Freshwater Crayfish **5**, 418–428.

Souty-Grosset, C., D. M. Holdich, P. Y. Noël, J. D. Reynolds, and P. Haffner, editors. 2006. Atlas of Crayfish in Europe. Muséum national d'Histoire naturelle, Paris, France (Patrimoines naturels, 64), 188 pp.

Statzner, B., E. Fièvet, J.-Y. Champagne, R. Morel, and E. Herouin. 2000. Crayfish as geomorphic agents and ecosystem engineers: biological behavior affects sand and gravel erosion in experimental streams. Limnology and Oceanography **45**, 1030–1040.

Statzner, B., O. Peltret, and S. Tomanova. 2003. Crayfish as geomorphic agents and ecosystem engineers: effect of a biomass gradient on baseflow and flood-induced transport of gravel and sand in experimental streams. Freshwater Biology **48**, 147–163.

Stenroth, P. and P. Nyström. 2003. Exotic crayfish in a brown water stream: effects on juvenile trout, invertebrates and algae. Freshwater Biology **48**, 466–475.

Taylor, C. 2002. Taxonomy and conservation of native crayfish stocks. Pages 236–257 in D. M. Holdich, editor. Biology of freshwater crayfish. Blackwell Science, Oxford, UK.

Taylor, C. A., M. L. Warren Jr, J. F. Fitzpatrick Jr, H. H. Hobbs III, R. F. Jezerinac, W. L. Pflieger, and H. W. Robinson. 1996. Conservation status of crayfishes of the United States and Canada. Fisheries **21**, 25–38.

Torchin, M. E., K. D. Lafferty, A. P. Dobson, V. J. McKenzie, and A. M. Kuris. 2003. Introduced species and their missing parasites. Nature **421**, 628–630.

Turner, A. M., R. J. Bernot, and C. M. Boes. 2000. Chemical cues modify species interactions: the ecological consequences of predator avoidance by freshwater snail. Oikos **88**, 145–158.

Usio, N., M. Konishi, and S. Nakano. 2001. Species displacement between an introduced and a 'vulnerable' crayfish: the role of aggressive interactions and shelter competition. Biological Invasions **3**, 179–185.

Vannote, R. L., G. W. Minshal, K. W. Cummins, J. R. Sedell, and C. E. Cushing 1980. The river continuum concept. Canadian Journal of Fisheries and Aquatic Sciences **37**, 130–137.

Vasconcelos, V., S. Oliveira, and F. O. Teles. 2001. Impact of a toxic and a non-toxic strain of *Microcystis aeruginosa* on the crayfish *Procambarus clarkii*. Toxicon **39**, 1461–1470.

Vogt, G. 1999. Diseases of European freshwater crayfish, with particular emphasis on interspecific transmission of pathogens. Pages 87–103 in F. Gherardi and D. M. Holdich, editors. Crayfish in Europe as alien species. How to make the best of a bad situation? A. A. Balkema, Rotterdam, The Netherlands.

Vorburger, C. and G. Ribi. 1999. Aggression and competition for shelter between a native and an introduced crayfish in Europe. Freshwater Biology **42**, 111–119.

Warner, G. F. and E. I. Green. 1995. Choice and consumption of aquatic weeds by signal crayfish (*Pacifastacus leniusculus*). Freshwater Crayfish **8**, 360–363.

Westman, K. 1973. The population of the crayfish *Astacus astacus* in Finland and the introduction of the American crayfish *Pacifastacus leniusculus* Dana. Freshwater Crayfish **1**, 41–55.

Westman, K. 1991. The crayfish fishery in Finland – its past, present and future. Finnish Fisheries Research **12**, 187–216.

Westman, K. 2002. Alien crayfish in Europe: negative and positive impacts and interactions with native crayfish. Pages 76–100 in E. Leppäkoski, S. Gollasch and S. Olenin, editors. Invasive aquatic species of Europe. Distribution, impacts and management. Kluwer, Dordrecht, The Netherlands.

Whiteledge, G. W. and C. F. Rabeni. 1997. Energy sources and ecological role of crayfishes in an Ozark stream: insights from stable isotopes and gut analysis. Canadian Journal of Fisheries and Aquatic Sciences **54**, 2555–2563.

Wilcove, D. S., D. Rothstein, J. Dubow, A. Philips, and E. Losos 1998. Quantifying threats to imperiled species in the United States. BioScience **48**, 607–615.

Williamson, M. 1996. Biological invasions Chapman & Hall, London, UK.

Wilson, K. A., J. J. Magnuson, D. M. Lodge, A. M. Hill, T. K. Kratz, W. L. Perry, and T. V. Willis. 2004. A long-term rusty crayfish (*Orconectes rusticus*) invasion: dispersal patterns and community change in a north temperate lake. Canadian Journal of Fisheries and Aquatic Sciences **61**, 2255–2266.

Chapter twenty-nine

Predation of invasive crayfish on aquatic vertebrates: the effect of Procambarus clarkii *on fish assemblages in Mediterranean temporary streams*

Maria Ilhéu, João Manuel Bernardo, and Sílvia Fernandes

INTRODUCTION

Non-indigenous species (NIS) have various effects on indigenous communities. In some cases, species invasion can result in substantial loss of biodiversity through competitive interactions and predation (e.g. Diamond and Case 1986, Ashton and Mitchell 1989, Brown 1989, Vitousek *et al.* 1996). In fresh water, in particular, the extinction of indigenous species is becoming common as

Francesca Gherardi, Biological invaders in inland waters: Profiles, distribution, and threats, 543–558.

aggressive species invade degraded ecosystems. The loss of biodiversity has often been caused by the spread of species assisted by man. Endemic fish and amphibians are among the taxa most vulnerable to intentional introductions throughout the world. As a result of introductions of NIS, several species have disappeared or have been subject to major reductions in their numbers to the extent that they are now threatened with extinction (Lloyd 1990, Arthington 1991, Minckley and Douglas 1991, Townsend and Crowl 1991, Witte *et al.* 1992).

Typically, successful invaders are characterized by a tolerance to wide environmental conditions, omnivory, rapid growth, dispersal, breeding in ephemeral habitats, and other traits associated with opportunism. The high predation efficiency and the lack of predators frequently make them the originators of important changes to the original biota.

Crayfish are among the most notorious invasive aquatic species in freshwater ecosystems and have been reported not only to displace indigenous crayfish but also to impact a number of other aquatic organisms (Chapter 28). Crayfish feed omnivorously on detritus, algae, plants, invertebrates (including other crayfish), and vertebrates (e.g. Ilhéu and Bernardo 1993a, 1995, Momot 1995, Smart *et al.* 2002, Chapter 30). Low-protein resources such as plants, detritus, and algae are important energy sources for maintenance (e.g. Ilhéu and Bernardo 1995, Rudnick and Resh 2005) but they might be expected to feed predominantly on macroinvertebrates when these are available (Ilhéu and Bernardo 1993b, Correia 2003, Nyström 2005). A large number of studies have shown that, because of their omnivorous character, introduced crayfish can profoundly modify the trophic structure of freshwater communities at several levels, often acting as keystone species (e.g. Lodge *et al.* 1994, Nyström *et al.* 1996, 2001, Covich *et al.* 1999, Dorn and Wojdak 2004, Usio and Townsend 2004).

Crayfish display a wide plasticity in their feeding behaviour, switching from detritivore/herbivore to scavenger/carnivore habits in response to food availability (Ilhéu and Bernardo 1993a, 1993b, 1995, Nyström 2002, Alcorlo *et al.* 2004). This is in accordance with previous studies that provided evidence of the crayfish predation effects of crayfish particularly on benthic invertebrates (Ilhéu *et al.* 2002, Smart *et al.* 2002, Dorn and Wojdak 2004, Wilson *et al.* 2004, Correia *et al.* 2005, Geiger *et al.* 2005, McCarthy *et al.* 2006, Roth *et al.* 2006), including bivalves (e.g. Perry *et al.* 1997, 2000, Reynolds and Donohoe 2001), amphibian species (e.g. Renai and Gherardi 2004, Cruz and Rebelo 2005), and fish (e.g. Savino and Miller 1991, Guan and Wiles 1997). The aim of this chapter is to review the predation effects of invasive crayfish on indigenous vertebrates, both amphibians and fish, and present a case study on the impact of the red swamp crayfish, *Procambarus clarkii* (Girard), on fish assemblages in dry-season pools of temporary streams in southern Portugal.

CRAYFISH PREDATION ON VERTEBRATES

Crayfish predation on amphibians

There are numerous reports of crayfish impact on amphibians (reviewed in Nyström 1999), but few studies focus on the effect of crayfish predation on amphibian populations.

Recently, both experimental studies and correlative field surveys implicated non-indigenous predators, such as crayfish, as major contributors to amphibian population decline and, in some instances, to local extinction (Kats and Ferrer 2003). The introduction of the American signal crayfish, *Pacifastacus leniusculus* (Dana), in Sweden produced a negative impact on the embryos and larvae of the indigenous amphibians (Axelsson *et al.* 1997, Nyström and Abjörnsson 2000, Nyström *et al.* 2002). The effects of this species on the breeding habitat selection by anurans were also reported (Nyström *et al.* 2001).

Laboratory studies showed *P. clakii* preying upon tadpoles of several indigenous species of *Rana* from Italy (Gherardi *et al.* 2001, Renai and Gherardi 2004), along with embryos and tadpoles of *Bufo bufo* (Linnaeus) and larvae of *Triturus vulgaris* (Linnaeus) (Renai and Gherardi 2004). The introduction of this species into Californian mountain streams was a cause of the decline of the newt *Taricha torosa* (Rathke) as a result of predation by the crayfish on embryos and larvae (Gamradt and Kats 1996).

Many amphibians from the south of Portugal reproduce mainly in temporary ponds where typical aquatic predators, such as fish, are absent. However, these habitats are now occupied by the red swamp crayfish, and a negative correlation between the distribution of some amphibian species and the presence of the crayfish was found (Beja and Alcazar 2003). After the introduction of this crayfish into a lagoon in NW Spain, five amphibian species previously reproducing in the lagoon eventually disappeared (Rodríguez *et al.* 2005). In the Paul de Boquilobo, a wetland in central Portugal, apparently for no other reason than the impact of the red swamp crayfish, there was a collapse of the amphibian community, with drastic reductions of the populations of *Pleurodeles waltl* (Michahelles), *Triturus marmoratus* (Latreille), and *Rana perezi* (Seoane), and the probable extinction of previously abundant species, such as *Hyla arborea* (Linnaeus) and *Pelodytes punctatus* (Daudin) (Cruz 2006).

There are numerous reports of crayfish consuming amphibian eggs. Cruz and Rebelo (2005) found a very high consumption rate of amphibian egg masses by the red swamp crayfish in mesocosm experiments, even when alternative vegetable food items were available. Cruz *et al.* (2006a) assessed the effects of the presence of *P. clarkii* in the breeding site distribution of 13 amphibian species in SW Portugal. Amphibian species richness was lower in places where crayfish presence was a negative predictor of the breeding probability of all urodeles [*P. waltl, Salamandra salamandra (*Linnaeus), *Triturus boscai* (Lataste), and *T. marmoratus*] and of two anurans [*Pelobates cultripes* (Cuvier) and *B. bufo*]. In the temporary ponds of Doñana Natural Park, red swamp crayfish had a strong effect on the reproductive success of *Bufo calamita* (Laurenti), the survival of its embryos being strikingly reduced when they were directly exposed to crayfish (Cruz *et al.* 2006b).

Dorn and Wojdak (2004) studied the effects of the virile crayfish, *Orconectes virilis* (Hagen), on the development and structure of littoral communities in newly established freshwater ponds in south-western Michigan (USA) and found no bullfrog (*Rana catesbiana* Shaw) tadpoles in any of the crayfish ponds although tadpoles were abundant in the control ponds.

In response to indigenous predators, amphibians frequently develop adaptive mechanisms, such as toxic or unpalatable eggs and larvae (e.g. Eklöv and Werner 2000). Behavioural or morphological features may also be effective to escape predators (e.g. Rödel 1999, Lardner 2000, Nyström and Abjörnsson 2000). When no co-evolutionary process takes place, prey species may lack suitable responses to crayfish predation. However, some crayfish species, such as *P. clarkii*, are apparently resistant to certain amphibian toxic compounds and are able to predate eggs with highly protective gelatine layer (Gamradt and Kats 1996, Renai and Gherardi 2004).

In general, invasive crayfish are efficient predators and a significant threat to amphibians. Furthermore, in several instances, habitat degradation caused by crayfish (e.g. Ilhéu *et al.* 2002, Geiger *et al.* 2005, Rodríguez *et al.* 2005) may severely affect the amphibian populations as well.

Crayfish predation on fish

Crayfish have been implicated in the decline of fish populations mainly indirectly due to competition for food and shelter (Guan and Wiles 1997, Miller *et al.* 1992, Gherardi 2002, Nyström 2002, Light 2005) and through the destruction of macrophyte beds, which are important habitats for juvenile fish (Rubin and Svensson 1993, Scheidegger and Bain 1995, Shoup *et al.* 2003).

However, several studies also suggest that crayfish readily prey on all the life stages of fish, but the effects and mechanisms of predation are less well-studied than the crayfish-fish competitive interactions (reviewed in Dorn and Mittlelbach 1999). Even though the number of studies involving the role of crayfish as an omnivorous predator is increasing, most knowledge is based on experimental work and the magnitude of these effects is very much dependent on the particular environmental context.

In northern Wisconsin lakes (USA), the decline of game fish populations has been attributed to the invasions of the crayfish *Orconectes rusticus* (Girard) (Hobbs *et al.* 1989). Egg predation by crayfish has been suggested to cause declines in bass *Micropterus dolomieu* Lacépède, pumpkinseed sunfish *Lepomis gibbosus* (Linnaeus), walleye *Stizostedion vitreum* (Mitchill), and lake trout *Salvelinus namaycush* (Walbaum). Savino and Miller (1991) report the predation of *O. rusticus* on lake trout eggs in experimental conditions. Egg consumption

(2–5 eggs day^{-1} per crayfish) was dependent upon temperature, substrate, and crayfish species. The authors concluded that, in conditions of high crayfish density and/or low egg density in cobble habitats, crayfish predation may affect trout recruitment. Similar effect may occur with bass and pumpkinseed sunfish, as the crayfish potential to consume eggs of warm-water species is greater (Horns and Magnunson 1981). Further, most bass and pumpkinseed sunfish concentrate their nests in shallow littoral zones, which make eggs more vulnerable to crayfish predation. In experimental ponds, Dorn and Mittlebach (2004) and Dorn and Wojdak (2004) observed virile crayfish preying heavily on fish eggs and found that crayfish presence significantly affected the successful reproduction of bluegill sunfish, *Lepomis macrochirus* Rafinesque, reducing the recruitment of young-of-the-year. Fitzsimons *et al.* (2002) estimated a crayfish (*O. virilis*) consumption of 82% of the potential egg numbers at lake trout spawning reefs based on both the published literature (Miller *et al.* 1992) and experimental work, concluding that high crayfish density and low egg abundance (<100 eggs m^{-2}) may also limit natural recruitment of trout in Lake Ontario. Corkum and Cronin (2004) also found that consumption of rainbow trout eggs by crayfish depended on several factors, including crayfish density, food levels, and habitat complexity. Other studies report that *O. virilis*, in its native range, is a significant egg predator of the fathead minnow, *Pimephales promelas* Rafinesque. Matity *et al.* (1994) found that breeding male fathead minnows had significantly more scars from crayfish pinches than non-breeding males or females, which may be due to their involvement in nest defence. The fathead minnows exposed to chemical cues from these crayfish presented variations in hatching time, egg morphology, and fry size, as shown by Kusch and Chivers (2004), who also suggested that the recognition of crayfish as an egg predator is innate in this species. However, Rubin and Svensson (1993) studied predation by the indigenous crayfish *Astacus astacus* (Linnaeus) on trout eggs and fry and found no evidence of this species' ability to consume trout eggs laid in reeds.

Evidence for crayfish predation on juvenile or adult fish is scarce and few data are available to judge their importance. Light (2005) noted that sculpin *Cottus beldingi* (Linnaeus) and signal crayfish may be considered as intraguild predators (*sensu* Polis and Holt 1992) based on observations of each species consuming the young-of-the-year of each other. Guan and Wiles (1997) found evidence of predation by signal crayfish on sculpin as well as on bullhead *Cottus gobio* (Linnaeus) and stone loach *Noemacheilus barbatulus* (Linnaeus). Nevertheless, because such predation was rare, it seems unlikely that it has significant population consequences.

In fact, only in high crayfish densities is the impact more perceptible. Rogowski and Stockwell (2006) assessed the potential impacts of non-indigenous crayfish on the threatened white sands pupfish, *Cyprinodon tularosa* (Miller and Echelle), and found that fish biomass was significantly lower at high densities of *O. virilis*. On several occasions virile crayfish were seen consuming adult pupfish in a high density crayfish experiment.

Other studies are not so conclusive on predation rates. Xinya (1995) studied the effect of *P. clarkii* on the survival of fry and fingerlings of cultivated fishes in China [carp, *Cyprinus carpio* (Linnaeus); grass carp, *Ctenopharyngodon idella* (Valenciennes); silver carp, *Hypophthalmichthys molitrix* (Valenciennes); and tilapia, *Oreochromis niloticus* (Linnaeus)] and found no significant negative impacts; fry grew well with or without crayfish. Stenroth and Nyström (2003) studied the effect of signal crayfish on brown trout using enclosures in a Swedish stream and found no impact on growth or survival of juvenile trout. Virile crayfish were found to reduce the activity rate and movement in and out of shelter of the Little Colorado spinedace, *Lepidomeda vittata* Cope, a threatened indigenous minnow of the south-western USA, but the predation rate was not significant (Bryan *et al.* 2002).

Several studies reported the presence of fish in the crayfish diet (e.g. Lorman and Magnuson 1978, Ilhéu and Bernardo 1993a, Gutiérrez-Yurrita *et al.* 1998, Correia 2003, Pérez-Bote 2005). In experimental laboratory studies, Ilhéu and Bernardo (1993b) demonstrated that crayfish predation success is very much dependent on the ability of the prey to escape. The choice of the feeding strategies may be interpreted in terms of cost-benefit analysis. In spite of a preference for animal food items, such as fish, the high costs involved in active predation may explain the low consumption of high mobility animals. This conclusion leads us to hypothesise that crayfish feeding on fish occurs mainly when prey is vulnerable and thus low costs of predation are involved. This situation only happens in strongly confined conditions, which is the case of the temporary aquatic systems during the dry period.

PREDATION EFFECTS OF INVASE CRAYFISH ON FISH OF DRY-SEASON STREAM POOLS IN SOUTHERN PORTUGAL

Lowland streams in Mediterranean-climate regions are shaped by predictable seasonal events of flooding and drying over an annual cycle. During the summer and early autumn, streams show a marked pattern of zero flow. When the flow ceases, and because of the high temperatures and evaporation, many rivers show long dry reaches and the surface waters are reduced to isolated pools (Bernardo and Alves 1999). Throughout the dry months, the wet area and volume of the summer pools decrease, and environmental conditions become more critical to the biota (Ilhéu 2004). Many receding pools will eventually dry and the remaining ones are important refuges for the aquatic organisms. During the following run-off period, streams start to flow again and longitudinal connectivity is re-established (Bernardo and Alves 1999, Ilhéu 2004).

During the drying phase, the combination of nutrients, organic detritus, shallow waters, favourable temperature, and dense populations of algae and water plants form the basis of a highly productive food web capable of supporting large populations of aquatic consumers. Throughout the summer period, the aquatic biota, namely fish and invertebrates including the red swamp crayfish, become progressively more concentrated in the receding pools.

The red swamp crayfish is an example of a successful invader in the temporary streams of the Mediterranean region. This species is very well adapted to the natural flow variation of this type of streams and frequently develops high density populations (Ilhéu 1994, Ilhéu and Bernardo 1996). Moreover, these waterbodies show no indigenous crayfish species and thus *P. clarkii* occupies a vacant niche.

The potential impacts of NIS on temporary aquatic systems are of special concern because such systems can have relatively high levels of endemism (e.g. Williams *et al.* 1985, Doadrio 2001, Cabral 2005). Many fish extinctions have been associated with the introduction of NIS (Miller *et al.* 1989, Richter *et al.* 1997, Minckley *et al.* 2002). The potential impact of NIS is apparently higher when endemic fish evolved in an environment with few aquatic predators and competitors (Minckley and Douglas 1991). Moreover, regions with Mediterranean climate are especially susceptible to invasions, as suggested by Macdonald *et al.* (1988).

Assessing the impact of NIS in the rivers of southern Portugal is of extreme importance because of the high conservation value of the indigenous fish fauna, which show a high proportion of endemism most of which are threatened (Cabral 2005). This fauna is mostly composed of cyprinids and no indigenous piscivorous species exists. Fish assemblages are dominated by indigenous species, although NIS, such as pumpkinseed sunfish *L. gibbosus* and mosquitofish *Gambusia holbrooki* Girard, also occur very frequently (Ilhéu 2004).

Crayfish predation on fish was studied in 17 isolated pools in the south of Portugal (Degebe stream, 2nd and 3rd order) during the summer of 2002. Crayfish and fish were collected with electrofishing. In the shallow pools crayfish were also captured by hand.

Pool volumes were highly variable, ranging from 0.06 to 700 m^3, with mean water depth from 0.03 to 0.7 m. Very shallow pools showed a high percentage of aquatic vegetation, mainly filamentous algae. Many pools presented critical conditions for the aquatic fauna because of the extreme temperature and low dissolved oxygen (DO). Maximum water temperature was 28.6 ± 2.9 °C. Minimum oxygen concentrations were lower than 1 mgL^{-1} in 41% of pools and very large daily ranges were observed in 53% of the pools, mostly shallow ones.

Fish density in the pools ranged from 1.9 to 80 fish m^{-2}. Shallow pools (<0.15 m mean water depth) had high fish densities, ranging from 20.1 to 80.2 m^{-2}, with a mean value of 50.7 $m^{-2} \pm 19.5$ (SD). Fish density was inversely correlated to the pool depth (after Spearman correlation: $r = -0.82$, $P < 0.001$) which may be interpreted as an effect of fish concentration as pools shrink.

The most abundant species in the pools were mosquitofish (33.6%) and the Iberian roach *Squalius alburnoides* (Steindachner) (30.4%), with the former being more abundant in the very shallow pools (<0.15 m water depth), reaching up to 60% of total fish specimens. Lower frequencies were observed for pumpkinseed (10.1%), arched-mouth nase *Chondrostoma lemmingii* (Steindachner) (5.1%), Iberian dace *Squalius pyrenaicus* (Günther) (3.2%), stoneloach *Cobitis paludica* (De Buen) (2.6%), barbels *Barbus* spp. (1.1%), Guadiana nase, *Chondrostoma willkommii* Steindachner (0.3%), and largemouth bass *Micropterus salmoides*, Lacépède (0.03%). Specimens with 0–30 mm total length were dominant in shallow pools (<40 cm depth) where no fish larger than 50 mm occurred. In the deeper pools, very small fish were less representative and larger fish (50–80 mm) were observed.

Crayfish density in the pools ranged from 0 to 39.3 m^{-2} and, as for fish, was inversely correlated to the water depth (after Spearman correlation: $r = -0.46$, $P < 0.05$). Apparently, crayfish concentrate in the pools where fish are more abundant and vulnerable because of the spatial confinement. In fact, choice tests showed that fish is the preferred food type if the costs involved in the capture are relatively low (Ilhéu and Bernardo 1993b).

A total of 409 crayfish stomach contents were analysed. For fish specimens, the identification was based on skeleton remains, skin, and scales. The number of eaten organisms was estimated by the skeleton parts. The food types in the crayfish diet for each pool were characterized through a Modified Main Food Index: MFI = (frequency of occurrence × proportion of each food item to the total volume of the gut content)$^{1/2}$ (Bernardo 1990).

Crayfish consumed a broad diversity of food items, the most common being vegetal detritus, invertebrates, and fish. Vegetal detritus was dominant in 76% of the pools. Overall, invertebrates were the second most important item in the crayfish diet. Fish was the third most consumed item, present in 24% of crayfish stomach contents; this type of food was dominant in 24% of the pools.

Consumption of animal material was negatively correlated with water depth (after Spearman correlation: $r = -0.63$, $P < 0.01$; Fig. 1) and pool area (after Spearman correlation: $r = 0.69$, $P < 0.01$). Fish consumption by crayfish increased significantly with fish density in pools (after Spearman correlation: $r = 0.68$, $P < 0.05$) (Fig. 2). This fact clearly supports the hypothesis that crayfish concentrate in the shallow pools because of the high availability and vulnerability of the fish. The most abundant prey, mosquitofish, was also the more consumed, representing 76% of total fish, thus corroborating the opportunistic character of *P. clarkii*. Pumpkinseed sunfish represented only 9% of fish prey and some indigenous species, such as Iberian roach and barbel juveniles, were also observed in the stomach contents but in very low frequencies and volumes.

As the larger proportions of fish in the crayfish diet were observed in the shallower (< 0.15 m water depth) and densely populated pools (both by crayfish and fish), the predation pressure of crayfish on fish was estimated on the basis of the conditions of those pools. For the evaluation of the predation pressure, values from this and previous studies on crayfish ecology were used.

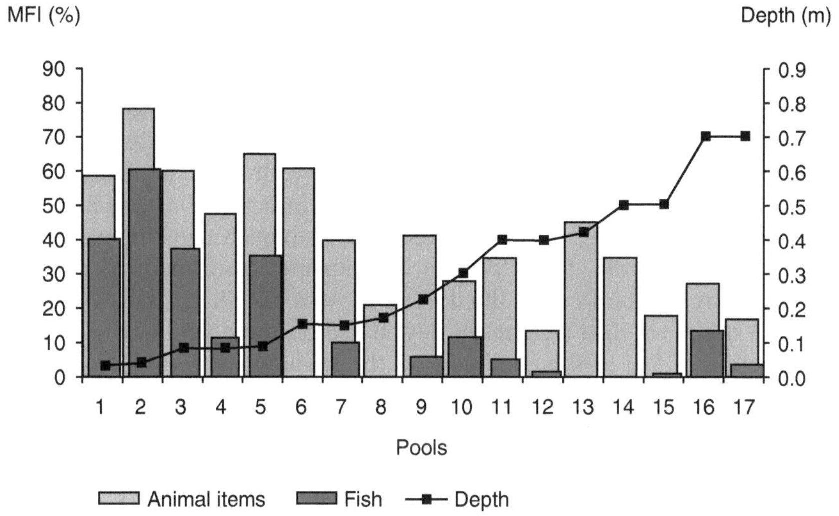

Fig. 1 Main Food Index (MFI) of total animal items and fish consumed by crayfish in each pool and pool water depth. (After Bernardo 1990, modified)

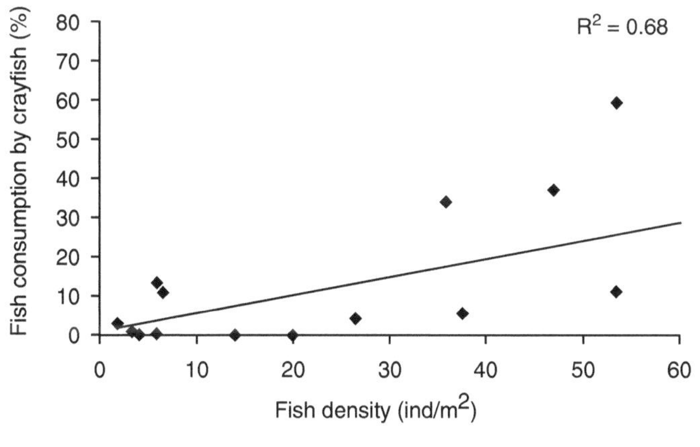

Fig. 2 Relation between fish consumption by crayfish and fish density in the studied pools.

To estimate crayfish predation pressure on fish, crayfish food consumption rate and the proportion of fish in the stomach contents were used. Total crayfish biomass per area was assessed using the crayfish density and crayfish mean weight. The consumption rate of fish by crayfish (CONS, g of fish m^{-2}) was calculated as:

$$CONS = DENS \times CW \times DCR \times PF$$

in which DENS is the crayfish density (N m^{-2}), CW is the mean crayfish weight (g), DCR is the daily consumption rate (g food/g crayfish per day), and PF is the proportion of fish in the crayfish stomach contents. Weights are wet weights.

A crayfish density of 10 m^{-2} was adopted; this density was observed in some pools in this study and is a mean value for these stream types (Ilhéu 1994). Crayfish size was homogenous, the average weight being 30.0 g (±5.3 SD).

The mean proportion of fish observed in the stomach contents was 69.7%. The consumption rate, based on the experimental assessment in *ad libitum* conditions by Bernardo and Ilhéu (1994), was 0.088 g/g crayfish day^{-1}, i.e. 8.8% of the crayfish weight per day. Based on these values, the estimated consumption of fish in the conditions of the shallow summer pools is 18.4 g fish m^{-2} day^{-1}.

In order to have a relative measure of the magnitude of this consumption rate, this value was compared to the fish density. The estimated mean biomass of fish in the shallow and densely populated pools was 126.3 g m^{-2}. Based on this value and on the consumption rate of fish, a prediction of the approximate time for crayfish to consume all fish present in pools at the time of sampling would be 6.9 days.

Results clearly demonstrate that the crayfish performs a significant role in the removal of fish but only in the last phase of the vanishing summer pools. As pools recede, confinement and concentration of the aquatic fauna causes the intensification of biological interactions, in particular predation. High vulnerability of fish makes them the ideal prey only during the low-water conditions associated with the temporary character of these streams. As the surface water disappears, in the extreme confinement of the very shallow pools, fish are totally predated.

CONCLUDING REMARKS

Crayfish have been described as an opportunistic species feeding on various types of animal and vegetal food items (e.g. Ilhéu and Bernardo 1993a, 1995, Gherardi *et al.* 2001, Nyström 2002, Correia 2003). As the crayfish grow, diet shifts from a more zoophagous to an almost exclusively phytophagous-detritophagous one (Ilhéu and Bernardo 1993a, Pérez-Bote 2005). Usually, fully grown adults feed almost entirely on vegetal matter, either fresh or detritic (Ilhéu and Bernardo 1995).

As previous experimental laboratory studies predicted (Ilhéu and Bernardo 1993b), when prey face confined conditions, as in shallow pools, crayfish profit from their vulnerability and in such an advantageous cost–benefit ratio crayfish behave as opportunistic predators. In the large pools, fish display strong escape ability, and predation efficiency by crayfish is low. Moreover, large persistent pools tend to show higher habitat complexity, which is also reflected in the crayfish predation success.

Both invasive and indigenous crayfish generally are well adapted for dispersal and can move from habitats with low prey availability to other locations where prey densities are higher and foraging success is greater. They often occupy "ephemeral home ranges" and move from one location to another (Ilhéu *et al.* 2003). The short-term movement is well studied in crayfish (e.g. Gherardi *et al.* 1998, Gherardi 2002, Bubb *et al.* 2006).

In general, the cost-benefit analysis – the basic principle of theories of optimal foraging strategies (Schoener 1971) – is consistent with the opportunistic character of the red swamp crayfish. To consume a certain quantity and quality of food (i.e. the benefit) the forager has to spend a cost involved in food searching, pursuit, handling or catching, and eating. The balance of costs and associated benefits will determine the choice among the available options. Thus, depending on the availability and ease of capture of the food categories, crayfish may be a phyto/detritophagous or a predator.

When highly populated receding pools reach very low water depths, crayfish face the ideal conditions of prey availability/vulnerability to perform an easy and rewarding predation. This pattern is likely to be found in other temporary, ephemeral, or intermittent aquatic systems of temperate or tropical regions where the dramatic circumstances of the vanishing waterbodies provide a feast for the predators.

ACKNOWLEDGEMENTS

The authors thank Ana Manuel Costa for assistance with the literature research and two anonymous referees for their valuable comments on the manuscript.

REFERENCES

Alcorlo, P., W. Geiger, and M. Otero. 2004. Feeding preferences and food selection of the red swamp crayfish, *Procambarus clarkii*, in habitats differing in food item diversity. Crustaceana **77**, 435–453.

Arthington, A. H. 1991. Ecological and genetic impacts of introduced and translocated freshwater fishes in Australia. Canadian Journal of Fisheries and Aquatic Science **48**, Supplement 1, 33–43.

Ashton, P. J. and D. S. Mitchell. 1989. Aquatic plants: patterns and modes of invasion, attributes of invading species and assessment of control programmes. Pages 111–154 in J. A. Drake, H. A. Mooney, F. Di Castri, R. H. Groves, F. J. Kruger, M. Rejmànek and M. Williamson, editors. Biological invasions: a global perspective. Wiley, Chichester, UK.

Axelsson, E., P. Nystrom, and C. Brönmark. 1997. Crayfish predation on amphibian eggs and larvae. Amphibia-Reptilia **18**, 217–228.

Beja, P. and R. Alcazar. 2003. Conservation of Mediterranean temporary ponds under agricultural intensification: an evaluation using amphibians. Biological Conservation **114**, 317–326.

Bernardo, J. M. 1990. Dinâmica de uma lagoa costeira eutrófica (Lagoa de Santo André). PhD Thesis, Universidade de Lisboa, Portugal.

Bernardo, J. M. and M. H. Alves. 1999. New perspectives for ecological flow determination in semiarid regions: a preliminary approach. Regulated Rivers: Research and Management **15**, 221–229.

Bernardo, J. M. and M. Ilhéu. 1994. Red swamp crayfish (*Procambarus clarkii*) contribution to material cycling. Verhandlungen der Internationale Vereinigung Limnologie **25**, 2447–2449.

Brown, J. H. 1989. Patterns, modes, and extents of invasions by vertebrates. Pages 85–110 in J. A. Drake, H. A. Mooney, F. Di Castri, R. H. Groves, F. J. Kruger, M. Rejmánek, and M. Williamson, editors. Biological invasions: a global perspective. Wiley, Chichester, UK.

Bryan, S. D., A. T. Robinson, and M. G. Sweetser. 2002. Behavioral responses of a small native fish to multiple introduced predators. Environmental Biology of Fishes **63**, 49–56.

Bubb, D. H., T. J. Thom, and M. C. Lucas. 2006. Movement, dispersal and refuge use of co-occurring introduced and native crayfish. Freshwater Biology **51**, 1359–1368.

Cabral, M. J., editor. 2005. Livro vermelho dos vertebrados de Portugal. Instituto de Conservação da Natureza, Lisboa, Portugal.

Corkum, L. D. and D. J. Cronin. 2004. Habitat complexity reduces aggression and enhances consumption in crayfish. Journal of Ethology **22**, 23–27.

Correia, A. M. 2003. Food choice by the introduced crayfish *Procambarus clarkii*. Annales Zoologici Fennici **40**, 517–528.

Correia, A. M., N. Bandeira, and P.M. Anastácio. 2005. Predator–prey interactions of *Procambarus clarkii* with aquatic macroinvertebrates in single and multiple prey systems. Acta Oecologica **28**, 337–343.

Covich, A. P., M. A. Palmer, and T. A. Crowl. 1999. The role of benthic invertebrate species in freshwater ecosystems. BioScience **49**, 119–127.

Cruz, M. J. 2006. Impacts of an exotic species, the red swamp crayfish (*Procambarus clarkii*) on Amphibian communities of Southwest Iberian Peninsula. Ph.D. thesis, Universidade de Lisboa, Portugal. (In Portuguese)

Cruz, M. J., S. Pascoal, M. Tejedo, and R. Rebelo. 2006a. Predation by an exotic crayfish, *Procambarus clarkii*, on natterjack toad, *Bufo calamita*, embryos: its role on the exclusion of this amphibian from its breeding ponds. Copeia **2006**, 274–280.

Cruz, M. J. and R. Rebelo. 2005. Vulnerability of Southwest Iberian amphibians to an introduced crayfish, *Procambarus clarkii*. Amphibia-Reptilia **26**, 293–304.

Cruz, M. J., R. Rebelo, and E. G. Crespo. 2006b. Effects of an introduced crayfish *Procambarus clarkii*, on the distribution of South-western Iberian amphibians in their breeding habitats. Ecography **29**, 329–338.

Diamond, J. and T. J. Case. 1986. Community ecology. Harper & Row, New York.

Doadrio, I., editor. 2001. Atlas y libro rojo de los peces de España. Dirección General de Conservación de la Naturaleza, CSIC, Madrid, Spain.

Dorn, N. J. and G. G. Mittlebach. 1999. More than predator and prey: a review of interactions between fish and crayfish. Vie et Milieu **49**, 229–237.

Dorn, N. J. and G. G. Mittlebach. 2004. Effects of native crayfish (*Orconectes virilis*) on the reproductive success and nesting behaviour of sunfish (*Lepomis* spp.) Canadian Journal of Fisheries and Aquatic Sciences **61**, 2135–2143.

Dorn, N. J. and J. M. Wojdak. 2004. The role of omnivorous crayfish in littoral communities. Oecologia **140**,150–159.

Eklöv, P. and E. E. Werner. 2000. Multiple predator effects on size-dependent behavior and mortality of two species of anuran larvae. Oikos **88**, 250–258.

Fitzsimons, J. D., D. L. Perkins, and C. C. Kruger. 2002. Sculpins and crayfish in lake trout spawning areas in Lake Ontario: estimates of abundance and egg predation on lake trout eggs. Journal of Great Lakes Research **28**, 421–436.

Gamradt, S. C. and L. B. Kats. 1996. Effect of introduced crayfish and mosquitofish on California newts. Conservation Biology **10**, 1155–1162.

Geiger, W., P. Alcorlo, A. Baltanás, and C. Montes. 2005. Impact of an introduced Crustacean on the trophic webs of Mediterranean wetlands. Biological Invasions **7**, 49–73.

Gherardi, F. 2002. Behavior. Pages 258–290 in D.M. Holdich, editor. Biology of freshwater crayfish. Blackwell Scientific, Oxford, UK.

Gherardi, F., S. Barbaresi, and F. Villanelli. 1998. Movement patterns of the white-clawed crayfish *Austropotamobius pallipes*, in a Tuscan stream. Journal of Freshwater Ecology **13**, 413–424.

Gherardi, F., B. Renai, and C. Corti. 2001. Crayfish predation on tadpoles: a comparison between a native (*Austropotamobius pallipes*) and an alien species (*Procambarus clarkii*). Bulletin Français de la Pêche et de la Pisciculture **361**, 659–668.

Guan, R. Z. and P. R. Wiles. 1997. Ecological impact of introduced crayfish on benthic fishes in a British lowland river. Conservation Biology **11**, 641–647.

Gutiérrez-Yurrita, P. J., G. Sancho, M. Á. Bravo, Á. Baltanás, and C. Montes. 1998. Diet of the red swamp crayfish *Procambarus clarkii* in natural ecosystems of the Doñana National Park temporary freshwater marsh (Spain). Journal of Crustacean Biology **18**, 120–127.

Hobbs, H. H. III, J. P. Pass, and J. V. Huner. 1989. A review of global crayfish introductions with particular emphasis on two North America species (Decapoda: Cambaridae). Crustaceana **56**, 299–316.

Horns, W. H. and J. J. Magnunson. 1981. Crayfish predation on lake trout eggs in Trout Lake, Wisconsin. Rapport et Procès-Verbaux des Réunions du Conseil International pour l'Exploration de la Mer **178**, 299–303.

Ilhéu, M. 1994. Ecologia e impacto trófico do Lagostim Vermelho dos Pântanos (*Procambarus clarkii*). Master Sc. Thesis, Faculdade de Ciências e Tecnologia da Universidade de Coimbra, Portugal.

Ilhéu, M. 2004. Padrões de ocupação de habitat da ictiofauna de curso de tipo Mediterrânico. PhD Thesis, Universidade de Évora, Portugal.

Ilhéu, M., P. Acquistapace, C. Benvenuto, and F. Gherardi. 2003. Shelter use of the red-swamp crayfish (*Procambarus clarkii*) in dry-season stream pools. Archiv für Hydrobiologie **157**, 535–546.

Ilhéu, M. and J. M. Bernardo. 1993a. Aspects of trophic ecology of red swamp crayfish (*Procambarus clarkii* Girard) in Alentejo, South of Portugal. Pages 417–423 in Actas del VI Congreso Español de Limnologia, Granada, Spain.

Ilhéu, M. and J. M. Bernardo. 1993b. Experimental evaluation of food preference of red swamp crayfish, *Procambarus clarkii*: vegetal versus animal. Freshwater Crayfish **9**, 359–364.

Ilhéu, M. and J. M. Bernardo. 1995. Trophic ecology of red swamp crayfish *Procambarus clarkii* (Girard) – preferences and digestibility of plant foods. Freshwater Crayfish **10**, 132–139.

Ilhéu, M. and J. M. Bernardo. 1996. Life-history and population biology of red swamp crayfish (*Procambarus clarkii*) in a reservoir at the South of Portugal. Freshwater Crayfish **11**, 54–59.

Ilhéu, M., P. Guilherme, and J. M. Bernardo. 2002. Impact of red swamp (*Procambarus clarkii)* on aquatic invertebrates and macrophytes assemblages: a case study in the South of Portugal. Verhandlungen der Internationale Vereinigung Limnologie **28**, 144–147.

Kats, L. B. and R. P. Ferrer. 2003. Alien predators and amphibian declines: review of two decades of science and the transition to conservation. Diversity and Distributions **9**, 99.

Kush, R. C. and D. P. Chivers. 2004. The effects of crayfish predation on phenotypic and life history variation in fathead minnows. Canadian Journal of Zoology **82**, 917–921.

Lardner, B. 2000. Moorphological and life history responses to predators in larvae of seven anurans. Oikos **88**, 169–180.

Light, T. 2005. Behaviour effects of invaders: alien crayfish and native sculpin in a California stream. Biological Invasions **7**, 353–367.

Lloyd, L. 1990. Ecological interactions of *Gambusia holbrooki* with Australian native fishes. In D. A. Pollard, editor. Introduced and translocated fishes and their ecological effects. Bureau of Rural Resources Proceedings **8**, 94–97.

Lodge, D. M., M. W. Kershner, J. E. Aloi, and A. P.Covich. 1994. Effects of an omnivorous crayfish (*Orconectes rusticus*) on a freshwater littoral food web. Ecology **75**, 1265–1281.

Lorman, J. and J. J. Magnuson. 1978. The role of crayfish in aquatic systems. Fisheries **3**, 8–10.

Macdonald, I. A. W., D. M. Graber, S. DeBenedetti, R. H. Groves, and E. R. Fuentes 1988. Introduced species in nature in Mediterranean-type climatic regions of the world. Biological Conservation **44**, 37–66.

Matity, J. G., D. P. Chivers, and R. J. F. Smith. 1994. Population and sex differences in antipredator responses of breeding fathead minnows (*Pimephales promelas*) to chemical stimuli from garter snakes (*Thamnophis radix* and *T. sirtalis*). Journal of Chemical Ecology **20**, 2111–2121.

McCarthy, J. M., C. I. Hein, J. D. Olden, and M. J. V. Zanden. 2006. Coupling long-term studies with meta-analysis to investigate impacts of non-native crayfish on zoobenthic communities. Freshwater Biology **51**, 224–235.

Miller, R. R., J. D. Williams, and J. E. Williams. 1989. Extinction of North American fishes during the past century. Fisheries **14**, 22–38.

Miller, J. E., J. F. Savino, and R. K. Neely. 1992. Competition for food between crayfish (*Orconectes virilis*) and the slimy sculpin (*Cottus cognatus*). Journal of Freshwater Ecology **7**,127–137.

Minckley, W. L. and M. E. Douglas. 1991. Discovery and extinction of Western fishes: a blink of the eye in geologic time. Pages 7–17 in W. L. Minckley and J. E. Deacon, editors. Battle Against Extinction: Native Fish Management in the American Southwest. University of Arizona Press, Tucson, AR.

Minckley, W. L., R. R. Miller, and S. M. Norris. 2002. Three new pupfish species, Cyprinodon (Teleostei, Cyprinodontidae), from Chihuahua, Mexico, and Arizona, USA. Copeia **2002**, 687–705.

Momot, W. T. 1995. Redefining the role of crayfish in aquatic ecosystems. Reviews in Fisheries Science **3**, 33–63.
Nyström, P. 1999. Ecological impact of introduced and native crayfish on freshwater communities. Pages 63–85 in F. Gherardi and D. M. Holdich, editors. Crayfish in Europe as alien species: How to make the best of a bad situation? A. A. Balkema, Rotterdam, The Netherlands.
Nyström, P. 2002. Ecology. Pages 192–235 in D. H. Holdich, editor. Biology of freshwater crayfish. Blackwell Scientific, Oxford, UK.
Nyström, P. 2005. Non-lethal predator effects on the performance of a native and an exotic crayfish species. Freshwater Biology **50**, 1938–1949.
Nyström, P. and K. Abjörnsson. 2000. Effects of fish chemical cues on the interactions between tadpoles and crayfish. Oikos **88**, 181–190.
Nyström, P., L. Birkedal, C. Dahlberg, and C. Brönmark. 2002. The declining spadefoot toad *Pelobates fuscus*: calling site choice and conservation. Ecography **25**, 488–498.
Nyström, P., C. Brönmark, and W. Granéli. 1996. Patterns in benthic food webs: a role for omnivorous crayfish? Freshwater Biology **36**, 631–646.
Nyström, P., O. Svensson, B. Lardner, C. Brönmark, and W. Granéli. 2001. The influence of multiple introduced predators on a littoral pond community. Ecology **82**, 1023–1039.
Pérez-Bote, J. L. 2005. Feeding ecology of the exotic red swamp crayfish, *Procambarus clarkii* (Girard, 1852) in the Guadiana River (SW Iberian Peninsula). Crustaceana **77**, 1375–1387.
Perry, W. L., D. M. Lodge, and G. A. Lamberti. 1997. Impact of crayfish predation on exotic zebra mussels and native invertebrates in a lake-outlet stream. Canadian Journal of Fisheries and Aquatic Sciences **54**, 120–125.
Perry, W. L., D. M. Lodge, and G. A. Lamberti. 2000. Crayfish (*Orconectes rusticus*) impacts on zebra mussel (*Dreissena polymorpha*) recruitment, other macroinvertebrates and algal biomass in a lake-outlet stream. American Midland Naturalist **144**, 308–316.
Polis, G. A. and R. D. Holt. 1992. Intraguild predation: the dynamics of complex trophic interactions. Trends in Ecology and Evolution **7**, 151–154.
Renai, B. and F. Gherardi. 2004. Predatory efficiency of crayfish: comparison between indigenous and non-indigenous species. Biological Invasions **6**, 89–99.
Reynolds, J. D. and R. Donohoe. 2001. Crayfish predation experiments on the introduced zebra mussel, *Dreissena polymorpha*, in Ireland, and their potential for biocontrol. Bulletin Français Pêche et Pisciculture **361**, 669–681.
Richter, B. D., D. P. Braun, M. A. Mendelson, and L. L. Master. 1997. Threats to imperilled freshwater fauna. Conservation Biology **11**, 1081–1093.
Rödel, M. O. 1999. Predation on tadpoles by hatchtlings of the freshwaterturtle *Pelomedusa subrufa*. Amphibia-Reptilia **20**, 173–183.
Rodríguez, C. F., E. Bécares, M. Fernandez-Aláez, and C. Fernandez-Aláez C. 2005. Loss of diversity and degradation of wetlands as a result of introducing exotic crayfish. Biological Invasions **7**, 75–85.
Rogowski, D. L. and A. C. Stockwell. 2006. Assessment of potential impacts of exotic species on populations of a threatened species, White Sands pupfish, Cyprinodon Tularosa. Biological Invasions **8**, 79–87.
Roth, B. M., C. L. Hein, and M. J. V. Zanden. 2006. Using bioenergetics and stable isotopes to assess the trophic role of rusty crayfish (*Orconectes rusticus*) in lake littoral zones. Canadian Journal of Fisheries and Aquatic Sciences **63**, 335–344.

Rubin, J. F. and M. Svensson. 1993. Predation by the noble crayfish, *Astacus astacus* (L.), on emerging fry of sea trout, *Salmo trutta* (L.). Nordic Journal of Freshwater Research **68**, 100–104.

Rudnick, D. and V. Resh. 2005. Stable isotopes, mesocosms and gut content analysis demonstrate trophic differences in two invasive decapod Crustacea. Freshwater Biology **50**, 1323–1336.

Savino, J. F. and J. E. Miller. 1991. Crayfish (*Orconectes virilis*) feeding on young lake trout (*Salvelinus namaycush*): effect of rock size. Journal of Freshwater Ecology **6**, 161–170.

Scheidegger, K. J. and M. B. Bain. 1995. Larval fish distribution and microhabitat use in free-flowing and regulated rivers. Copeia **1995**, 125–135.

Schoener, T. W. 1971. Theory of feeding strategies. Annual Review of Ecology and Systematics **11**, 369–404.

Shoup, D. E., R. E. Carlson, and R. T. Heath. 2003. Effects of risk and foraging return on the diel use of vegetated habitat by two size-classes of bluegills. Transactions of the American Fisheries Society **132**, 590–597.

Smart, A. C., D. M. Harper, F. Malaisse, S. Schmitz, S. Coley, and A. C. Gourder de Beauregard. 2002. Feeding of the exotic Louisiana red swamp crayfish, *Procambarus clarkii* (Crustacea, Decapoda), in an African tropical lake: Lake Naivasha, Kenya. Hydrobiologia **488**, 129–142.

Stenroth, P. and P. Nyström. 2003. Exotic crayfish in a brown water stream: effects on juvenile trout, invertebrates and algae. Freshwater Biology **48**, 466–475.

Townsend, C. R. and T. A. Crowl. 1991. Fragmented population structure in a native New Zealand fish: an effect of introduced brown trout? Oikos **61**, 347–354.

Usio, N. and C. R. Townsend. 2004. Role of crayfish: consequences of predation and bioturbation for stream invertebrates. Ecology **85**, 807–822.

Vitousek, P. M., C. M. D'Antonio, L. L. Loope, and R. Westbrooks. 1996. Biological invasions as global environmental change. American Scientist **84**, 468–478.

Williams, J. E., D. D. Bowman, J. E. Brooks, A. A. Echelle, R. J. Edwards, D. A. Hedrickson, and J. J. Landye. 1985. Endangered aquatic ecosystems in North American deserts with a list of vanishing fishes of the region. Journal of the Arizona-Nevada Academy of Science **20**, 1–62.

Wilson, K. A., J. J. Magnuson, D. M. Lodge, A. M. Hill, T. K. Kratz, W. L. Perry, and T. V. Willis. 2004. A long-term rusty crayfish (*Orconectes rusticus*) invasion: dispersal patterns and community change in a north temperate lake. Canadian Journal of Fisheries and Aquatic Sciences **61**, 2255–2266.

Witte, F., T. Goldschmidt, P.C. Goudswaard, W. Lightvoet, M. J. P. Van Oijen, and J. H. Wanink. 1992. Species extinction and concomitant ecological changes in Lake Victoria. Netherlands Journal of Zoology **42**, 214–232.

Xinya, S. 1995. Effects of the crayfish *Procambarus clarkii* on the survival of the fry and fingerlings of fishes cultivated in China. Freshwater Crayfish **8**, 528–532.

Chapter thirty

Assessing the trophic ecology of crayfish: a case study of the invasive Procambarus clarkii

María Crehuet, Paloma Alcorlo, Miguel Angel Bravo-Utrera, Angel Baltanás, and Carlos Montes

INTRODUCTION

Freshwater crayfish, among the largest invertebrate members of biological communities in non-marine ecosystems throughout the world, have a functional role that is tightly linked to their trophic ecology both as consumers of a large variety of food sources and as prey for a number of vertebrate predators ranging from fish to mammals (Hogger 1988). Although the low diversity of crayfish species in Europe, as compared to that in North America, might suggest a low significance of this group to the functioning of the ecosystems, the recent introduction of several non-indigenous species has changed this view dramatically (Chapter 28).

Severe environmental impacts at both the population and the ecosystem level followed those introductions. Impacts produced on other (indigenous) crayfish should be distinguished from those impacts affecting non-crayfish species. The former ones are frequently related to the spread of diseases (mainly the crayfish plague, *Aphanomyces astaci* Schikora) and to competitive interactions, whereas

Francesca Gherardi, Biological invaders in inland waters: Profiles, distribution, and threats, 559–576.

the damage of introduced crayfish to non-crayfish populations and to ecosystem functioning is related to trophic interactions (Chapter 28). Non-indigenous crayfish are known to alter significantly the biomass and species richness of their trophic resources, mainly submerged macrophytes and macroinvertebrates (Lodge *et al.* 1994, Momot 1995, Charlebois and Lamberti 1996, Nyström *et al.* 1996, Parkyn *et al.* 1997). Such changes, however, do not always result from direct consumption by crayfish but frequently are the by-product of non-consumptive activities. For instance, the abundances of aquatic plants and invertebrate prey are reduced simply when crayfish feed actively on them. However, population declines are also caused because crayfish, while hunting or digging their burrows, alter the structure of the surrounding habitat and reduce food resources and shelter for many other invertebrate species (Carpenter and Lodge 1986, Hart 1992, Nyström *et al.* 1996).

There is a general agreement that detailed information on the trophic requirements of introduced crayfish species is fundamental in order to identify and to manage their likely impact on native ecosystems. Many studies devoted to fulfil that aim have been traditionally based on the analysis of crayfish gut contents (e.g. Alcorlo *et al.* 2004). Such an approach, although useful, also proved to overestimate detritus and vegetation consumption by crayfish, as compared to food of animal origin, thus leading to a misconception of the role of crayfish in the aquatic food webs (Whitledge and Rabeni 1997, Correia 2003, Alcorlo *et al.* 2004). Other methodological approaches have been successfully applied to crayfish research, improving our understanding of crayfish trophic ecology (e.g. enclosure experiments in the field, prey choice tests in the laboratory, and stable isotopes analysis).

The aim of this paper is to review these methodological approaches, with special attention to their pros and cons, as well as to support the use of mixed research strategies that combine the several approaches available. That aim is addressed in the following sections. First, examples from the literature will illustrate the several methods currently applied and their drawbacks. Second, we will refer to a case study, in which two of such methods (stable isotopes analysis plus food preference tests) have been combined to provide information on the trophic ecology of the juvenile red swamp crayfish (*Procambarus clarkii* Girard).

ASSESSMENT OF THE TROPHIC ECOLOGY OF CRAYFISH

Laboratory *vs.* field (enclosure) experiments

Feeding activity and food preferences of crayfish, the causal links to their impact on aquatic food webs, have been repeatedly assessed using field observations (Feminella and Resh 1989, Nyström *et al.* 1999) and laboratory experiments (Covich *et al.* 1981, Ilhéu and Bernardo 1993a, b, 1995, Lochmann *et al.* 1995,

Cronin 1998, Cronin *et al.* 2002, Correia *et al.* 2005). Laboratory experiments are frequently performed using glass aquaria or individual containers that allow for both individual based observations under controlled conditions and the use of measuring/recording devices difficult to operate in outdoor conditions. Because of those advantages, laboratory experiments have been extensively used for the study of crayfish growth (Brown *et al.* 1992, McClain *et al.* 1992a,b, Oliveira and Fabião 1998, Bondar *et al.* 2005); consumption rates (Rundquist and Goldman 1981, Ilhéu and Bernardo 1995, Correia *et al.* 2005); and food digestibility (Brown *et al.* 1990, Reigh *et al.* 1990). Brown *et al.* (1990), for instance, used faecal collection chambers to estimate digestibility of various macrophytes by *Orconectes virilis* (Hagen), concluding that digestibility coefficients were significantly different between macrophytes but not between male and female crayfish. Food preferences have been addressed in the laboratory as well. Covich (1977), Ilhéu and Bernardo (1993b, 1995), and Cronin *et al.* (2002) used choice tests in which alternative food items were offered in similar quantities to crayfish and their preferences were evaluated according to the amount of resources left after some time. Wiernicki (1984) also used a laboratory experimental approach to assess assimilation efficiency of juvenile and adult *P. clarkii* feeding on plant detritus to conclude that assimilation efficiency was greater when crayfish were fed plant material previously colonized by micro-organisms.

Despite their advantages, laboratory experiments are not free from operational difficulties, among which to obtain and keep in good conditions an adequate stock (diverse and abundant) of crayfish prey items is not trivial. In addition, this approach suffers from lack of realism and a restricted scope (Diamond 1986). Field studies, the counterpart to laboratory experiments which focus mainly at the population level, provide realistic data but have serious drawbacks like site replications and the lack of regulation of independent variables (temperature, rainfall, light, etc.) (Diamond 1986). Those disadvantages, nevertheless, can be significantly reduced with the use of cage or enclosure experiments (Lodge and Lorman 1987, Feminella and Resh 1989, Lodge *et al.* 1994, Angeler *et al.* 2001, Stenroth and Nyström 2003, Bondar *et al.* 2005).

The aim of enclosures is to isolate the biological communities that occur on a given surface of aquatic benthic habitat while keeping the fluxes above it (water, nutrients, oxygen, plankton, etc.) as natural as possible. Accordingly, a variety of enclosure designs (boxes, cages, tubes), mesh sizes (1–15 mm), cage areas (0.2–6 m^2), and crayfish density (0–20 m^2) have been used in both lotic (Charlebois and Lamberti 1996, Parkyn *et al.* 1997, Perry *et al.* 1997, Stelzer and Lamberti 1999, Perry *et al.* 2000) and lentic habitats (Lodge and Lorman 1987, Feminella and Resh 1989, Lodge *et al.* 1994, Angeler *et al.* 2001, Rodríguez *et al.* 2003, Stenroth and Nyström 2003, Bondar *et al.* 2005). Because of the skills of crayfish at burrowing, enclosures fixed in lentic habitats with muddy substrate are recommended to have their walls buried 15–40 cm down into the sediment to avoid crayfish movements from or into the experimental area (Lodge and Lorman 1987). With hard substrate (boulders, gravel, and cobbles), as used to happen in lotic habitats, enclosures are frequently designed as tubes or boxes embedded in the stream bed and with top, upstream, and downstream walls with silicon mesh to allow water to pass through, and the bottom covered with sediment (Charlebois and Lamberti 1996, Perry *et al.* 1997, Bondar *et al.* 2005).

Within enclosures, community composition and prey abundances can be precisely assessed before and after interacting with crayfish in order to test its effects (as changes in abundance, biomass, and species composition) on prey populations. That was the approach used by Feminella and Resh (1989) to document the impact of *P. clarkii* grazing on pondweed (*Potamogeton pectinatus* Linnaeus) in a Californian freshwater marsh; and by Lodge and Lorman (1987) and Charlebois and Lamberti (1996) to show the influence of the non-indigenous *Orconectes rusticus* (Girard) on benthic food webs in three northern Wisconsin lakes and a Michigan stream, respectively. Angeler *et al.* (2001) also used enclosures to estimate the effect of a non-indigenous crayfish (*P. clarkii*) on the water quality of a wetland in central Spain (Tablas de Daimiel National Park). Their results show that crayfish feeding activity worsens water quality by mobilizing to the water column nutrients otherwise stored in the sediments and that crayfish alter sediment features directly by recycling organic matter in the sediment (Angeler *et al.* 2001). Similarly, Rodríguez *et al.* (2003) found that while fish activity had no significant effect on plant growth in a shallow lake in north-western Spain, *P. clarkii* feeding activities were responsible for the reduction of the plant biomass, forcing the system from a mesotrophic macrophyte-dominated clear water state into a turbid and eutrophic one. Other crayfish studies using enclosure experiments to assess the effect of indigenous/non-indigenous crayfish on benthic communities include Lodge *et al.* (1994), Parkyn *et al.* (1997), Perry *et al.* (1997), Stewart *et al.* (1998), Evans-White *et al.* (2001), and Ságová-Marecková (2002). A remarkable work in this context is the attempt of McCarthy *et al.* (2006) to provide a novel insight on a broad scale by combining a meta-analysis of enclosure experimental studies and a long-term observational study.

Useful as they are to identify overall crayfish impacts on biological communities, enclosure experiments meet difficulties when trying to distinguish the specific mechanisms behind such impacts. For instance, although macrophyte destruction has been repeatedly linked to the presence of crayfish, it remains disputed how much is due to direct consumption (i.e. crayfish herbivory) and how much to indirect non-consumptive activities (e.g. crayfish predation on epiphytic snails or mechanical destruction of macrophytes) (Lodge and Lorman 1987). In laboratory experiments, such a problem can be partially solved by keeping track of the macrophyte biomass destroyed but not ingested (for instance, stems left as floating pieces in the aquaria), but in field studies such an approach is almost impossible to follow. To address that aim, two alternative, but complementary, methods stand out in studies of crayfish trophic ecology: stomach contents analysis and the analysis of stable isotopes.

Stomach contents analysis

This is a straightforward method – and the most widely applied – to identify which trophic resources are exploited as food by crayfish (e.g. Ilhéu and Bernardo 1993a, b, Whitledge and Rabeni 1997, Gutiérrez-Yurrita *et al.* 1998, Verhoef *et al.* 1998, Parkyn *et al.* 2001, Correia 2002, 2003, Hollows *et al.* 2002, Stenroth and Nyström 2003, Alcorlo *et al.* 2004, Bondar *et al.* 2005, Rudnick and Resh 2005).

Due to its role as an invasive species in many areas, the diet of *P. clarkii* has been studied intensively, a main issue being whether the species is an opportunistic or a selective feeder. Ilhéu and Bernardo (1993a, b) found that the gut contents of *P. clarkii* living in a series of Portuguese aquatic habitats (from streams to a reservoir) were dominated by vegetal biomass as compared to macroinvertebrates or detritus. Similar results were obtained by Gutiérrez-Yurrita *et al.* (1998) while studying *P. clarkii* (502 stomachs analysed) in the lower Guadalquivir Basin (southern Spain). Moreover, the analysis of 80 additional stomachs from the same geographic area confirmed previous observations and showed that food item diversity in crayfish stomachs is correlated with prey availability (Alcorlo *et al.* 2004). Correia (2002, 2003) also found that crayfish resource use is adjusted to the availability of aquatic prey, but in the Portuguese rice field populations that she studied there was a reverse dominance of food items in crayfish guts; those of animal origin, not plants or detritus, were the most abundant. Gut contents analysis has also been used to assess crayfish diet in enclosure experiments by Stenroth and Nyström (2003) and Bondar *et al.* (2005) working with *Pacifastacus leniusculus* (Dana) in stream habitats in Sweden and British Columbia, respectively.

Data provided by stomach contents analysis are extremely useful but tedious to obtain and the method requires some taxonomic skills for the identification of the partially digested prey pieces. This approach provides a snapshot of the feeding habits of the studied organism and can hardly reflect trophic relationships in the long term (Alcorlo *et al.* 2004). In addition, information obtained through this method is frequently biased because diverse food items are preserved at different rates inside the crayfish gut passage. Whereas soft prey are digested easily and fast, some others (or their hard parts) can remain longer in crayfish stomachs, thus increasing their possibilities of being found and recorded (Correia 2003). Plant tissues, for instance, are more difficult to assimilate than food resources of animal origin, and, because they can remain inside the crayfish gut for a long period, crayfish preference for vegetation is frequently overestimated (Brown *et al.* 1986, 1990). Indeed, some probable prey of crayfish are unlikely to be found in crayfish stomachs because they are digested so quickly that their identification becomes difficult, if not impossible (Hanson *et al.* 1990, Momot 1995).

Finally, when consumption data are intended for the quantitative assessment of crayfish energy sources, the additional estimation of assimilation efficiencies and net production efficiencies is required (Whitledge and Rabeni 1997).

Stable isotopes analysis

Although stable isotopes analysis has been used for the study of food webs for nearly 30 years, its application to the study of crayfish trophic ecology is relatively new (France 1996a,b), focusing on the elucidation of major energy pathways passing through crayfish in wetland and stream ecosystems (Whitledge and Rabeni 1997). The technique is based on simple grounds: the isotopic signature of a consumer (i.e. the ratio of stable isotopes of a given element in the tissues of an organism, e.g. crayfish) reflects the isotopic signature of all the trophic resources it had assimilated (Anderson *et al.* 1987). Carbon and nitrogen isotopes ($\delta^{13}C$ y $\delta^{15}N$) are the stable isotopes most frequently used in food web analysis. The technique does not normally provide information about specific taxa in the diet of the consumer, but it can discriminate between alternative food sources (plant *vs.* animal; inland *vs.* seashore, etc.). Moreover, isotopic signals integrate information on resource use over longer time periods (Whitledge and Rabeni 1997) [for a more detailed account of stable isotopes analysis applied to food web studies see, for instance, DeNiro and Epstein (1978), Rau (1980), Fry and Sherr (1984), Minagawa and Wada (1984), Kling *et al.* (1992), Michener and Schell (1994), Hobson and Wassenaar (1999)].

Stable isotopes analyses have been applied to several crayfish species with contrasting results. *Orconectes punctimanus* (Creaser) and *Orconectes luteus* (Creaser), two indigenous species in an Ozark stream (Missouri), mainly behave as predators of macroinvertebrates (Whitledge and Rabeni 1997), whereas *Orconectes neglectus* (Faxon) and *Orconectes nais* (Faxon), occurring in a stream in Kansas, function more as algal and detrital processors than as predators (Evans-White *et al.* 2001). Similarly, isotopic signals found in *P. leniusculus* in its homeland of British Columbia and in the non-indigenous crayfish *P. clarkii* in California suggest that those species feed mainly on detrital biofilm and on terrestrially derived detritus, respectively (Bondar *et al.* 2005, Rudnick and Resh 2005).

The use of stable isotopes analysis allows us to differentiate between the resources ingested (i.e. those found in the gut) and those really assimilated by the crayfish, a very important difference from an energetic point of view. This point is clearly illustrated in two studies of the trophic ecology of the New Zealand crayfish *Paranephrops planifrons* (White) showing that most of the energy used for crayfish growth comes from invertebrates despite finding that the stomach contents were frequently dominated by detritus (Parkyn *et al.* 2001, Hollows *et al.* 2002).

Compared to stomach contents analysis, stable isotopes analysis is less time-consuming, especially when working with juveniles that are small and difficult to handle. On the other hand, measuring isotope ratios is more complex than stomach contents analysis. Some complexities refer to tissue selection (exoskeleton, hepatopancreas, gills, or muscle) and sample treatment (acidification or lipid extraction) which have been recently evaluated by Stenroth *et al.* (2006), whereas others are instrumental (the method requires a mass spectrometer, a more expensive and less available piece of equipment than a binocular microscope for examining gut contents).

Mixed strategies

As usually happens in other research areas, competing techniques for the study of the trophic ecology of crayfish provide better results when used in combination. Most of the studies quoted in the previous sections do indeed apply that complementary approach. Feminella and Resh (1989), for instance, combined *in situ* enclosure experiments with gut contents analysis in their effort to determine whether or not *P. clarkii* reduce pondweed (*P. pectinatus*) abundance at Coyote Hills Marsh (California, USA). Enclosure experiments, with crayfish densities ranging from 0 to 3 m^{-2}, showed the kind of strong negative relationship that can result from either an herbivore—plant interaction or mechanical destruction of pondweed due to crayfish activity. Gut contents analysis confirmed that the decline of *P. pectinatus* was mainly due to direct trophic interactions.

Ilhéu and Bernardo (1993b) studied the diet of the red swamp crayfish (*P. clarkii*) by checking the gut contents of 164 animals collected with traps in several locations in Alentejo (southern Portugal). Although their results suggested a preferential consumption of vegetal material rather than other types of food, laboratory experiments revealed a crayfish preference for benthic invertebrates. Parkyn *et al.* (2001) and Hollows *et al.* (2002) combined gut contents analysis and stable isotopes analysis to stress discrepancies in the results derived from each method and to show shifts in the crayfish diet with changing environmental conditions (from native forest streams to non-indigenous pasture settings).

A study that combines all the approaches (field enclosure experiments, gut contents analysis, laboratory experiments, and stable isotopes analysis) was made by Bondar *et al.* (2005), who studied the trophic ecology of *P. leniusculus* in a Canadian stream within its native distribution range. Enclosures were used to control for food type, food availability, and crayfish density, to prevent predation pressures on the crayfish and to run replicates in a standard way. The diet of the crayfish was first identified through gut contents analysis and it was later adjusted using stable isotopes analysis to account for the assimilation of each food type, thereby eliminating overestimation of those diet items having a long latency in the gut. Laboratory experiments were finally used to assess growth granted by each food type. It was concluded that, although growth would be far more rapid if they feed on invertebrates, juveniles and adults do not differ in their diet. Besides, allochthonous detritus is their main food source (Bondar *et al.* 2005). The same species, however, was studied by Stenroth and Nyström (2003) to address its impact on the communities of a stream in southern Sweden where the crayfish has been introduced. Using enclosures (see above) as well as gut contents analysis, they concluded that *P. leniusculus* is responsible for the decline of dominant predatory invertebrate populations which seemed to be affected not because of their trophic status (i.e. there was no direct competition with the crayfish) but because of their reduced mobility (i.e. crayfish can easily capture them) (Stenroth and Nyström 2003).

A CONTRIBUTION TO THE KNOWLEDGE OF THE TROPHIC ECOLOGY OF AN INVASIVE CRAYFISH IN DOÑANA MARSHLANDS

In this section, we present a case study that aims to contribute to the existing knowledge of the trophic ecology of the red swamp crayfish (*P. clarkii*) in freshwater marshlands in southern Spain. Given that previous efforts focused on crayfish population abundance and distribution and on gut contents analysis (Montes *et al.* 1993, Gutiérrez-Yurrita 1998, Alcorlo *et al.* 2004), here we address food preference experiments and stable isotopes analysis.

Attempts to introduce six species of non-indigenous crayfish (*Astacus astacus* Linneus, *Astacus leptodactylus* Eschscholtz, *Orconectes limosus* Rafinesque, *Cherax destructor* Clark, *P. leniusculus*, and *P. clarkii*) were made in the Iberian Peninsula over the last half-century. Only the last three species thrived to establish wild populations, but *P. clarkii* is undoubtedly the most successful of them all. This species was first introduced in the Iberian Peninsula in 1973 using a stock from Louisiana (USA) (Habsburgo-Lorena 1978), as a challenge to the indigenous populations and aimed at commercial use. The success of that attempt was so great that it immediately prompted additional introductions in rice fields placed in the Lower Guadalquivir basin (south-western Spain). Since then, the distribution range of the species continued to grow due either to natural dispersal or to deliberate dissemination by fishermen (Gaudé 1983).

The introduction of *P. clarkii* in the area was initially supported by scientific arguments based on the 'empty niche' concept: large invertebrates like crayfish were absent from the whole area and it was assumed that no competition effects would appear and that the system would not suffer any harmful change (Molina and Cadenas 1983). On the contrary, the red swamp crayfish has produced severe changes in the structure and functioning of ecosystems in the area, and now plays a central role in the trophic webs of Doñana National Park (Gutiérrez-Yurrita 1997, Gutiérrez-Yurrita *et al.* 1998, Geiger *et al.* 2005).

Field and laboratory studies typify *P. clarkii* as an omnivore species that feeds on large quantities of invertebrates, plants, and detritus (Feminella and Resh 1989, Huner and Barr 1991, Ilhéu and Bernardo 1993a, b, 1995, Gutiérrez-Yurrita *et al.* 1998, Correia 2002, 2003, Alcorlo *et al.* 2004, Rudnick and Resh 2005). Within that context, this study focuses on the trophic activity of juvenile crayfish in the food webs of Doñana National Park (southern Spain). Choice test experiments for detecting food preferences combined with stable isotopes analysis ($\delta^{13}C$ y $\delta^{15}N$) have been carried out.

Food preference experiments

Red swamp crayfish have been repeatedly blamed as responsible for the local extinction of several snail species in the marshlands of Doñana. In order to test crayfish preference for mollusc prey, a choice experiment was set up. This time, juvenile crayfish were used as experimental subjects instead of adults. Although omnivory in *P. clarkii* is widely assumed, it seems that shifts in diet might occur during its ontogenetic development (Correia 2003) with juveniles showing preference for food of animal origin and adults behaving mainly as herbivores (Ilhéu and Bernardo 1993a, Montes *et al.* 1993). Accordingly, if the functional role of *P. clarkii* changes with ontogeny, the impact of crayfish on the ecosystem strongly depends on its population structure.

All experimental crayfish were raised in the laboratory from eggs laid by 80 adult crayfish collected in July 2004 from Lucio Martinazo, a site in the freshwater marsh of Doñana National Park. Juvenile crayfish fed exclusively on chironomid larvae for nearly four months in order to assure homogeneous, single origin isotopic signatures [note, nevertheless, that this protocol can alter posterior crayfish trophic response due to training (Martin and Bateson 1991)]. Once crayfish reached a size close to 3 cm (mean total length $= 2.98 \pm 0.35$ cm), 84 juvenile crayfish were selected, weighted, and starved for 24 hours before proceeding with the experiments. Two food items were offered, both of animal origin, as alternatives: gastropods (*Physa acuta* Draparnaud) and chironomid larvae (Diptera, Chironomidae).

All the experiments were run on single crayfish to avoid competition for food and no individual was used more than once. Three treatments, with 12 replicates each, were carried out as follows: (1) only chironomid larvae were offered; (2) only snails were offered; and (3) both chironomids and snails were offered. Prey items were weighed before each experiment and offered to crayfish in groups of 12 individuals. After one hour, the number of prey consumed and their corresponding biomass were recorded. Twenty-four hours later, experimental crayfish were frozen to proceed with stable isotopes analysis. That lag was presumed to allow the non-assimilated prey biomass to be excreted out of crayfish gut.

The treatment involving the simultaneous offer of chironomids and snails implies that experimental crayfish have access to a lager amount of food ($12 + 12$ prey) as compared to treatments with a single item offered. Accordingly, changes in resource use, if existing, might also be explained because of a change in the amount (not in the quality) of food offered. In order to control for such a possibility, an additional pair of treatments was added to the

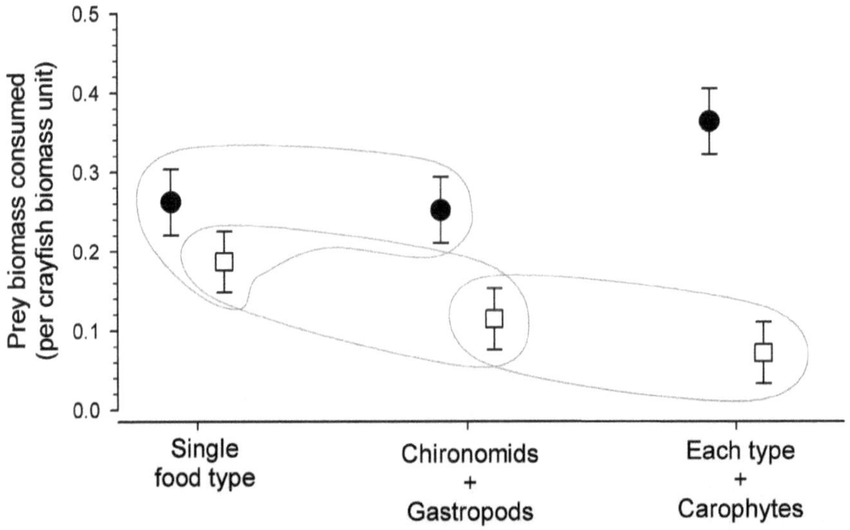

Fig. 1 Results of the ANOVA performed on consumption rates of juvenile *Procambarus clarkii* on chironomids (full circles) and gastropods (open squares) in experimental treatments. Mean values (± S.E.) of the response variable, transformed to account for normality, are plotted. Non-significantly different treatments (Tukey test) are enclosed in a grey line.

experiments: chironomids and snails were each presented to juvenile crayfish together with stems of carophytes (*Chara* spp.) offered in excess (i.e. carophytes, which are one dominant constituent of crayfish diet in the area, were provided in large enough quantities so that crayfish never run out of them).

Consumption rate was computed as prey biomass consumed per crayfish biomass unit (g). This consumption rate was transformed to account for normality and analysed using a one-way ANOVA followed by a Tukey test for *a posteriori* comparisons (Zar 1999). The ratio of the number of prey consumed to the number of prey available was estimated for each individual crayfish.

Mean consumption rates for chironomids and gastropods were similar when offered as single resources but significantly different when offered in combination ($F_{(5,\ 66)} = 29.286$, $P < 0.00001$; Fig. 1). Whereas chironomids are always heavily consumed, preference for snails decreases when other food resources (chironomids and carophytes) are involved. Such an overwhelming crayfish preference for chironomids may be due to training while cultured in the laboratory, but the smaller handling times involved in capturing and eating the chironomids, as compared to the hard-shelled gastropods, might also explain this pattern. In other words, snails may not be negatively selected by juvenile *P. clarkii* but simply consumed at different rates. Indeed, snail biomass was consumed in amounts comparable to those of chironomids when no other

Table 1 Average consumption rates of the different food resources offered to juvenile *Procambarus clarkii* in the experiments. Rates refer to the prey biomass consumed relative to the crayfish biomass (both in grams). Consumption rates of alternative food sources in brackets.

Resource		Consumption rate	
Chironomids	(alone)	0.0723	
	(+ gastropods)	0.0670	(0.0183)
	(+ charophytes)	0.1278	(0.7539)
Gastropods	(alone)	0.0388	
	(+ chironomids)	0.0183	(0.0670)
	(+ charophytes)	0.0065	(0.0339)

food item was present, and crayfish never waited until chironomids (and carophytes) were exhausted to feed on gastropods when they were all present in the experiment (Table 1). When discussing the likely preference of *P. clarkii* for gastropods, Gutiérrez-Yurrita (1997) suggested that benefits other than metabolic energy must be at play in this interaction (e.g. calcium carbonate required for crayfish growth). Of similar concern was the argument made by Nyström *et al.* (1999) and Stenroth and Nyström (2003) that crayfish are prone to predate on slow moving invertebrates, mostly on gastropods, because that food source is expected to provide more energy in the long run. That hypothesis is difficult to test with our experiments because of their short-term nature and the high mobility of juvenile crayfish as compared to the adults. However it points out the need to include different time scales and crayfish ontogenetic stages (or some other estimate of population structure) into future studies addressing the impact of crayfish food preferences in wild habitats.

Stable isotopes analysis

Crayfish were frozen immediately at the end of the experiments to prevent further digestion of food items. Hence, all crayfish carried the same isotopic baseline signature (due to the chironomid-based diet held for 4 months) plus an additional signal added by the food items consumed and assimilated during the 24 h experiments. Prey items were also analysed in order to characterize their isotopic signature. Therefore, crayfish stable isotopes analysis was intended as a test for the sensibility of the method to detect diet shifts over short-term periods.

All samples were analysed using a Micromass CF-Isochrom mass spectrometer for measuring isotope ratios of carbon and nitrogen according to standard methods and run by the SIdI (Servicio Interdepartamental de Investigación) at the UAM (Universidad Autónoma de Madrid).

Isotopic fractionation coefficient, i.e. the change on isotopic signature between the diet and the consumer tissues because of the digestion and

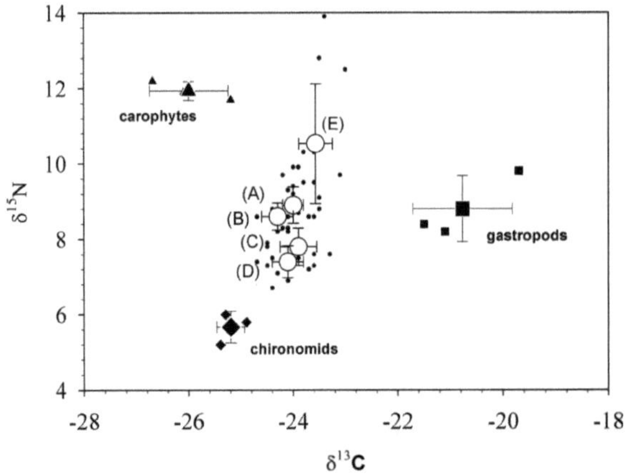

Fig. 2 Scatter plot of isotopic signals measured in juvenile *Procambarus clarkii* [open circles (average values); dots (individual data)] and in several food sources [large closed symbols (average values); small closed symbols (individual data)]. Codes for crayfish isotopic signals refer to the different treatments (food provided): (A) chironomids only; (B) gastropods only; (C) chironomids plus gastropods; (D) chironomids plus carophytes; and (E) gastropods plus carophytes. Variability expressed as standard deviation.

assimilation process ($\Delta_{tissue\text{-}diet} = D_{tissue} - D_{diet}$; where D is $\delta^{13}C$ or $\delta^{15}N$), was estimated from data as 1.2‰ ($\delta^{13}C$) and 3.2‰ ($\delta^{15}N$) for a chironomid-based diet. These figures are in agreement with standard values for a food supply of animal origin 0.5–1‰ for $\delta^{13}C$ (De Niro and Epstein 1978, Fry and Sherr 1984) and 3–4‰ for $\delta^{15}N$ (Minagawa and Wada 1984)], but differ from those estimated for *P. clarkii* feeding on water-worms (Oligochaeta) (2.0‰ for $\delta^{13}C$ and 1.0‰ for $\delta^{15}N$) in streams flowing into San Francisco bay (Rudnick and Resh 2005). The difference suggests that coefficients of fractionation are context-dependent and must be estimated independently for each ecosystem studied.

Dual isotope plots (Fig. 2) show isotopic signals measured in both food items and crayfish in each treatment. Data from the experiment including gastropods plus charophytes were excluded because resource consumption was negligible, leading to very large within-treatment variability. There were significant differences in nitrogen isotope signals between treatments (*t*-test comparisons between each treatment and the control), but almost no changes were found in carbon, except for the experiment with gastropods only. Such differences mainly reflect, in addition to food preferences, differential assimilation rates over a 24 h period. As expected from fractionation between trophic levels, the $\delta^{15}N$ signal increased in crayfish feeding exclusively on chironomids as compared to the signature found in its food supply (Fig. 2). Although similar effects are also expected with other food items (i.e. the consumer should show a signal enriched in the heavier isotope due to fractionation), a decrease is observed instead. This result is most likely due to the short study period (24 h) which might affect the chances for the differential excretion of the lighter nitrogen isotope. Still, the isotopic signal of juvenile crayfish reflects systematically and coherently the small differences in diet observed during the short-term experiments (Fig. 2).

Although much work is still to be done on crayfish trophic ecology using isotopic signals, this approach seems to be precise and sensitive enough to detect both dominant (ruling ecosystem functioning) as well as sporadic (frequently of interest from a conservation perspective) food resources.

CONCLUDING REMARKS

As expected, all methods applied to the study of the trophic ecology of freshwater crayfish have advantages and limitations. Hence, the most efficient way to address that topic must rely in a wise combination of them all. Laboratory experiments, with their capability to control environmental variables and to measure crayfish responses precisely, even at the individual level, allow for the study of growth, consumption rates, food digestibility, and prey preferences. However, such experiments lack realism and require adequate facilities in order to keep alive both the crayfish and their prey. Field experiments, frequently performed as enclosure/exclosure designs, are the reasonable alternative to laboratory experiments. They provide a higher degree of realism but make it difficult to track individual crayfish responses or to make close-up observations of the feeding process.

Grown either in the laboratory or in the wild, the analysis of gut contents is the most direct approach to the study of crayfish diet. This procedure, however, is not free from problems. Because prey are digested by crayfish at different rates, observational data from gut analysis are necessarily biased towards heavily sclerotized prey (or their parts) and food items hard to assimilate (e.g. plant material). In addition, quantitative estimates of prey abundances inside crayfish guts are difficult to make because of the variety of prey remains (e.g. claws, legs, cephalic capsules, and carapaces). Stable isotopes analysis recently joined the collection of methods for the study of the trophic ecology of crayfish. That technique is more expensive than gut contents analysis or direct observation in the laboratory, and cannot provide information on crayfish diet down to the specific level. However, it is very accurate in identifying main energy sources and their contribution to crayfish growth and survival in the long term.

Finally, a case study has been introduced including a selection experiment and a stable isotopes analysis. The case focuses on the trophic ecology of *P. clarkii*, an invasive species in European wetlands. Several studies already addressed food preferences in adult crayfish, so that juveniles have been used here instead. Although crayfish showed a clear preference for chironomids, aquatic snails were also consumed. Because snail consumption diminishes with the presence of alternative prey, it is argued that crayfish, although able to feed efficiently on gastropods, prefer to use snails as a secondary food source.

Indeed, it is suggested that, in the presence of other prey, gastropods might be exploited not for metabolic energy but mainly as sources of minerals (e.g. calcium carbonate) needed for the crayfish growth.

Stable isotopes analysis allowed for the estimation of fractionation coefficients for chironomids. Those fractionation coefficients seem to be context-dependent, as demonstrated by comparisons with coefficients estimated for the same crayfish species in different habitats and using different food sources. Thus, site-specific reference studies are strongly recommended. In addition, the technique proved to be sensitive enough to detect changes in crayfish diet that occur over short periods of time (24 h), opening new possibilities for the application of this approach in short-term studies.

ACKNOWLEDGEMENTS

We thank the editor for her kind invitation to contribute to this volume and three anonymous referees for their helpful comments. This study was supported by Grant No. 095A/2002 (Ministerio de Medio Ambiente).

REFERENCES

Alcorlo, P., W. Geiger, and M. Otero. 2004. Feeding preferences and food selection of the red swamp crayfish, *Procambarus clarkii*, in habitats differing in food item diversity. Crustaceana **77**, 435–453.

Anderson, R. K., P. L. Parker, and A. Lawrence. 1987. A $^{13}C/^{12}C$ tracer study of the utilization of presented feed by commercially important shrimp *Penaeus vannamei* in a pond growout system. Journal of the World Aquaculture Society **18**, 148–155.

Angeler, D. G., S. Sánchez-Carrillo, G. García, and M. Álvarez-Cobelas. 2001. The influence of *Procambarus clarkii* (Cambaridae, Decapoda) on water quality and sediment characteristics in a Spanish floodplain wetland. Hydrobiología **464**, 89–98.

Bondar, C. A., K. Bottriell, K. Zeron, and J. S. Richardson. 2005. Does trophic position of the omnivorous signal crayfish (*Pacifastacus leniusculus*) in a stream food web vary with life history stage or density? Canadian Journal of Fisheries and Aquatic Sciences **62**, 2632–2639.

Brown, P. B., P. Tazik, M. L. Hooe, and W. G. Blythe. 1990. Consumption and apparent dry matter digestibility of aquatic macrophytes by male and female crayfish (*Orconectes virilis*). Aquaculture **89**, 55–64.

Brown, P. B., J. E. Wetzel, and A. Spacie. 1992. Evaluation of naturally-occuring organisms as food for juvenile crayfish *Procambarus clarkii*. Journal of the World Aquaculture Society **23**, 211–216.

Brown, P. B., C. D. William, E. H. Robinson, D. M. Akiyama, and A. L. Lawrence. 1986. Evaluation of methods for determining in vivo digestion coefficients for adult red swamp crayfish *Procambarus clarkii*. Journal of the World Aquaculture Society **17**, 19–24.

Carpenter, S. R. and D. M. Lodge. 1986. Effects on submerged macrophytes on ecosystem processes. Aquatic Botany **26**, 341–370.

Charlebois, P. M. and G. A. Lamberti. 1996. Invading crayfish in a Michigan stream: direct and indirect effects on periphyton and macroinvertebrates. Journal of the North American Benthological Society **15**, 551–563.

Correia, A. M. 2002. Niche breadth and trophic diversity: feeding behaviour of the red swamp crayfish (*Procambarus clarkii*) towards environmental availability of aquatic macroinvertebrates in a rice field (Portugal). Acta Oecologica **23**, 421–429.

Correia, A. M. 2003. Food choice by the introduced crayfish *Procambarus clarkii*. Annales Zoologici Fennici **40**, 517–528.

Correia, A. M., N. Bandeira, and P. M. Anastácio. 2005. Predator-prey interactions of *rocambarus clarkii* with aquatic macroinvertebrates in single and multiple prey systems. Acta Oecologica **28**, 337–343.

Covich, A. P. 1977. How do crayfish respond to plants and mollusca as alternate food sources? Freshwater Crayfish **3**, 165–179.

Covich, A. P., L. L. Dye, and J.S. Mattice. 1981. Crayfish predation on *Corbicula* under laboratory conditions. The American Midland Naturalist **105**, 181–188.

Cronin, G. 1998. Influence of macrophyte structure, nutritive value, and chemistry on the feeding choices of a generalist crayfish. Pages 307–317 in E. Jeppesen, M. A. Sondegaard, M. O. Sondegaard and K. Christoffersen, editors. The structure role of submersed macrophytes in lakes. Springer-Verlag, New York, NY.

Cronin, G., D. M. Lodge, M. E. Hay, M. Miller, A. M. Hill, T. Horvath, R. C. Bolser, N. Lindquist, and M. Wahl. 2002. Crayfish feeding preferences for freshwater macrophytes: the influence of plant structure and chemistry. Journal of Crustacean Biology **22**, 708–718.

DeNiro, M. J. and S. Epstein. 1978. Influence of diet on the distribution of carbon isotopes in animals. Geochimica et Cosmochimica Acta **45**, 341–351.

Diamond, J. 1986. Overview: laboratory experiments, field experiments, and natural experiments. Pages 3–22 in J. Diamond and T.J. Case, editors. Community Ecology. Harper & Row, New York, NY.

Evans-White, M., W. K. Dodds, L. J. Gray, and K. M. Fritz. 2001. A comparison of the trophic ecology of the crayfishes (*Orconectes nais* (Faxon) and *Orconectes neglectus* (Faxon)) and the central stoneroller minnow (*Campostoma anomalum* (Rafinesque)): omnivory in a tallgrass prairie stream. Hydrobiologia **462**, 131–144.

Feminella, J. W. and V. H. Resh. 1989. Submersed macrophytes and grazing crayfish: An experimental study of herbivory in a California freshwater marsh. Holartic Ecology **12**, 1–8.

France, R. 1996a. Ontogenetic shift in crayfish $\delta^{13}C$ as a measure of land-water ecotonal coupling. Oecologia **107**, 239–242.

France, R. 1996b. Absence or masking of metabolic fractionation of ^{13}C in a freshwater benthic food web. Freshwater Biology **36**, 1–6.

Fry, B. and E. B. Sherr. 1984. $\delta^{13}C$ measurements as indicators of carbon flow in marine and freshwater ecosystems. Contributions in Marine Science **27**, 13–47.

Gaudé, A. P. III. 1983. The Louisiana red swamp crawfish (*Procambarus clarkii*) in Coto Doñana National Park (Sevilla, Huelva; Spain). Thesis. University of Southern Louisiana, LA, 98 pp.

Geiger, W., P. Alcorlo, Á. Baltanás, and C. Montes. 2005. Impact of an introduced Crustacean on the trophic webs of Mediterranean wetlands. Biological Invasions 7, 49–73.

Gutiérrez-Yurrita, P. J. 1997. El papel ecológico del cangrejo rojo (*Procambarus clarkii*) en los ecosistemas acuáticos del Parque Nacional de Doñana. Una perspectiva ecofisiológica y bioenergética. Tesis Doctoral (Inédita). Departamento de Ecología, Universidad Autónoma de Madrid, Spain, 348 pp.

Gutiérrez-Yurrita, P. J., G. Sancho, M. Á. Bravo, Á. Baltanás, and C. Montes. 1998. Diet of the red swamp crayfish *Procambarus clarkii* in natural ecosystems of the Doñana National Park temporary fresh-water marsh (Spain). Journal of Crustacean Biology **18**, 120–127.

Habsburgo-Lorena, A. S. 1978. Present situation of exotic species of crayfish introduced into Spanish continental waters. Freshwater Crayfish **4**, 175–184.

Hanson J. M., P. A. Chambers, and E. E. Prepas. 1990. Selective foraging by the crayfish *Orconectes virilis* and its impact on the macroinvertebrates. Freshwater Biology **24**, 69–80.

Hart, D. D. 1992. Community organization in streams: the importance of species interactions, physical factors, and chance. Oecologia **91**, 220–228.

Hobson, K. A. and L. I. Wassenaar. 1999. Stable isotope ecology: an introduction. Oecologia **120**, 312–313.

Hogger, J. B. 1988. Ecology, population and behaviour. Pages 114–144 in D. M. Holdich and R. S. Lowery, editors. Freshwater crayfish: Biology, Management and Exploitation. Croom Helm, London, UK.

Hollows, J. W., C. R. Townsend, and K. J. Collier. 2002. Diet of crayfish *Paranephrops zealandicus* in bush and pasture streams: insights from stable isotopes and stomach analysis. New Zealand Journal of Marine and Freshwater Research **36**, 129–142.

Huner, J. V. and J. E. Barr. 1991. Red swamp crayfish: biology and exploitation. Sea Grant College Program, Centre of Wetland Resources, Louisiana State University, Baton Rouge, LA, 128 pp.

Ilhéu, M. and J. M. Bernardo. 1993a. Aspects of trophic ecology of red swamp crayfish (*Procambarus clarkii* Girard) in Alentejo, South of Portugal. Actas VI Congreso Español de Limnología, 417–423.

Ilhéu, M. and J. M. Bernardo. 1993b. Experimental evaluation of food preferences of red swamp crayfish, *Procambarus clarkii*: vegetal versus animal. Freshwater Crayfish **9**, 359–364.

Ilhéu, M. and J. M. Bernardo. 1995. Trophic ecology of red swamp crayfish *Procambarus clarkii* (Girard) – preferences and digestibility of plant foods. Freshwater Crayfish **10**, 132–139.

Kling, G. W., B. Fry, and W. J. O'Brien. 1992. Stable isotopes and planktonic trophic structure in Artic lakes. Ecology **73**, 561–566.

Lochmann, R., W. R. McClain, and D. M. Gatlin III. 1995. Response of the red crayfish, *Procambarus clarkii*, to different quantities and sources of dietary carbohydrates and lipids. Freshwater Crayfish **8**, 364–375.

Lodge, D. M. and J. G. Lorman. 1987. Reductions in submersed macrophyte biomass and species richness by the crayfish *Orconectes rusticus*. Canadian Journal of Fisheries and Aquatic Sciences **44**, 591–597.

Lodge, D. M., M. W. Kershner, and J. E. Aloi. 1994. Effects of an omnivorous crayfish (*Orconectes rusticus*) on a freshwater littoral food web. Ecology **75**, 1265–1281.

Martin, P. and P. Bateson. 1991. La medición del comportamiento. Alianza, Madrid, Spain, 237 pp.

McCarthy, J. M., C. L. Hein, J. D. Olsen, and M. J. Vander Zanden. 2006. Coupling long-term studies with meta-analysis to investigate impacts of non-native crayfish on zoobenthic communities. Freshwater Biology **51**, 224–235.

McClain, W. R., W. H. Neill, and D. M. III Gatlin. 1992a. Nutrient profiles of green and decomposed rice-forages and their utilization by juvenile crayfish (*Procambarus clarkii*). Aquaculture **101**, 251–265.

McClain, W. R., W. H. Neill, and D. M. III Gatlin. 1992b. Partitioning the contributions of forage-based production system components to weight gain of juvenile crayfish (*Procambarus clarkii*). Aquaculture **101**, 267–281.

Michener, R. H. and D. M. Schell. 1994. Stable isotopes ratios as tracers in marine aquatic food webs. Pages 138–157 in K. Lajtha and R. H. Michener, editors. Stable isotopes in ecology and environmental science. Blackwell Scientific Publications, Oxford, UK, 316 pp.

Minagawa, M. and E. Wada. 1984. Stepwise enrichment of $\delta^{15}N$ along food chains: further evidence and the relation between $\delta^{15}N$ and animal age. Geochimica et Cosmochimica Acta **48**, 1135–1140.

Molina, F. and R. Cadenas. 1983. Impacto de la pesca del cangrejo rojo americano (*Procambarus clarkii*) en los ecosistemas marismeños del Parque Nacional de Doñana durante la campaña de 1983. ICONA. Informe Inédito, Madrid, Spain, 35 pp.

Momot, W. T. 1995. Redefining the role of crayfish in aquatic ecosytems. Reviews in Fisheries Sciences **3**, 33–63.

Montes, C., M. Á. Bravo-Utrera, Á. Baltanás, C. Duarte, and J. P. Gutiérrez-Yurrita. 1993. Bases ecológicas para la gestión integral del cangrejo de rojo de las marismas del Parque Nacional de Doñana. ICONA, Madrid, Spain, 269 pp.

Nyström, P., C. Brönmark, and W. Granéli. 1996. Patterns in benthic food webs: a role for omnivorous crayfish? Freshwater Biology **36**, 631–646.

Nyström, P., C. Brönmark, and W. Granéli. 1999. Influence of an exotic and native crayfish species on a littoral benthic community. Oikos **85**, 545–553.

Oliveira, J. and A. Fabião. 1998. Growth responses of juveniles red swamp crayfish, *Procambarus clarkii* Girard, to several diets under controlled conditions. Aquaculture Research **29**, 123–129.

Parkyn, S. M., K. J. Collier, and B. J. Hicks. 2001. New Zealand stream crayfish: functional omnivores but trophic predators? Freshwater Biology **46**, 641–652.

Parkyn, S. M., C. F. Rabeni, and K. J. Collier. 1997. Effects of crayfish (*Paranephrops planifrons: Parastacidae*) on in- stream processes and benthic faunas: a density manipulation experiment. New Zealand Journal of Marine and Freshwater Research **31**, 685–692.

Perry, W. L., D. M. Lodge, and G. A. Lamberti. 1997. Impact of crayfish predation on exotic zebra mussels and native invertebrates in a lake-outlet stream. Canadian Journal of Fisheries and Aquatic Science **54**, 120–125.

Perry, W. L., D. M. Lodge, and G. A. Lamberti. 2000. Crayfish (*Orconectes rusticus*) impacts on zebra mussel (*Dreissena polymorpha*) recruitment, other macroinvertebrates and algal biomass in a lake-outlet stream. American Midland Naturalist **144**, 308–316.

Rau, G. H. 1980. Carbon-13/Carbon-12 variation in subalpine lake aquatic insects: food source implications. Canadian Journal of Fisheries and Aquatic Sciences **37**, 742–746.

Reigh, R. C., S. L. Braden, and R. J. Craig. 1990. Apparent digestibility coefficients for common feedstuffs in formulated diets for red swamp crayfish, *Procambarus clarkii*. Aquaculture **84**, 321–334.

Rodríguez, C. F., E. Becarés, and M. Fernández-Aláez. 2003. Shift from clear to turbid phase in lake Chozas (NW Spain) due to the introduction of American red swamp crayfish (*Procambarus clarkii*). Hydrobiologia **506–509**, 421–426.

Rudnick, D. and V. Resh. 2005. Stable isotopes, mesocosms and gut content analysis demonstrate trophic differences in two invasive decapod Crustacea. Freshwater Biology **50**, 1323–1336.

Rundquist, J. C. and C. R. Goldman. 1981. Methodological considerations for quantitative determination of consumption for the crayfish *Pacifastacus leniusculus*. Freshwater Crayfish **5**, 27–55.

Ságová-Marecková, M. 2002. Interactions between crayfish, benthic invertebrates, macrophyte roots and sediment in a littoral zone. Archiv für Hydrobiologie **155**, 645–665.

Stelzer, R. S. and G. A. Lamberti. 1999. Independent and interactive effects of crayfish and darters on a stream benthic community. Journal of the North American Benthological Society **18**, 524–532.

Stenroth, P. and P. Nyström. 2003. Exotic crayfish in a brown water stream: effects on juvenile trout, invertebrates and algae. Freshwater Biology **48**, 466–475.

Stenroth, P., N. Holmqvist, P. Nyström, O. Berglund, P. Larsson, and W. Granéli. 2006. Stable isotopes as an indicator of diet in omnivorous crayfish (*Pacifastacus leniusculus*): the influence of tissue, sample treatment, and season. Canadian Journal of Fisheries and Aquatic Sciences **63**, 821–831.

Stewart, T. W., J. G. Miner, and R. L. Lowe. 1998. An experimental analysis of crayfish (*Orconectes rusticus*) effects on a *Dreissena*-dominated benthic macroinvertebrate community in western Lake Erie. Canadian Journal of Fisheries and Aquatic Sciences **55**, 1043–1050.

Verhoef, G. D., P. L. Jones, and C. M. Austin. 1998. A comparison of natural and artificial diets for juveniles of the Australian freshwater crayfish *Cherax destructor*. Journal of the World Aquaculture Society **29**, 243–248.

Whitledge, G. W. and C. F. Rabeni. 1997. Energy sources and ecological role of crayfishes in an Ozark stream: insights from stable isotopes and gut analysis. Canadian Journal of Fisheries and Aquatic Sciences **54**, 2555–2563.

Wiernicki, C. 1984. Assimilation efficiency by *Procambarus clarkii* fed Elodea (*Egera densa*) and its products of decomposition. Aquaculture **36**, 203–215.

Zar, J. H. 1999. Biostatistical Analysis, 4th Edition. Prentice-Hall, Englewood Cliffs, NJ.

Chapter thirty-one

Introduction and spread of crayfish (Parastacidae) in Western Australia and their potential to displace indigenous species

Jessica Lynas, Andrew Storey, and Brenton Knott

INTRODUCTION

On a global scale, crayfish have been translocated extensively beyond their natural range, both within and between continents, due largely to demand for their commercial culture (Holdich 1987, Horwitz 1990, Elvey *et al.* 1996, Gherardi and Holdich 1999). Deleterious impacts on the receiving environment typically include predation on and competition with indigenous species (Holdich 1987, Horwitz 1990, Elvey *et al.* 1996, Gherardi and Holdich 1999, Chapter 28), alteration to food webs resulting in changes to nutrient and energy flow (e.g. Holdich 1987, Nyström *et al.* 1999, Chapter 28), and the introduction of diseases (e.g. Horwitz 1990, Gherardi and Holdich 1999, Vogt 1999, Chapter 28). The potential for a non-indigenous crayfish species (NICS) to replace an indigenous species has been recognized by, for example, Capelli and Munjal (1982), Butler and Stein (1985), Momot and Leering (1986), Söderbäck (1991), and Vorburger and Ribi (1999), with competitive exclusion being cited as the mechanism for such species replacements (Bovbjerg 1952, 1970, Aiken

Francesca Gherardi, Biological invaders in inland waters: Profiles, distribution, and threats, 577–596.

1965, Capelli 1982). Under competitive exclusion two, geographically sympatric, non-interbreeding populations sharing ecological attributes cannot coexist indefinitely; eventually, the population with the superior competitive ability will displace the other (Cole 1960, Hardin 1960). In the case of typically aggressive crayfish, this may occur through interference competition when one species is able to inhibit another's access to a common limiting resource through territoriality or aggression (Jaeger 1974).

In contrast to the evidence from Europe, currently there are no reports in Australia of displacement neither of indigenous crayfish by an NICS nor of introductions of crayfish non-indigenous to the continent. However, there has been widespread translocation of crayfish within Western Australia (WA), and the yabby, introduced from Victoria, currently is expanding its distribution within the State. This is of particular concern given the conservation significance of the WA crayfish fauna (Whiting *et al.* 2000). Due to the ancient separation of freshwater systems of this region from the rest of Australia, the crayfish fauna has remained effectively isolated, resulting in an endemic biota (Figgis 1993, Myers *et al.* 2000) that now is threatened by displacement and/or replacement by non-indigenous species.

This paper summarizes the current situation of NICS in WA and specifically discusses the potential for the deleterious impact of yabbies to indigenous crayfish species and to the ecology of local natural freshwater ecosystems generally.

LEGISLATION

Australia has a long history of introductions of animals and plants that have had a deleterious effect on indigenous fauna. This has lead, in part, to the development of legislation to prevent unauthorized introductions of non-indigenous species (NIS); however, movement of species within the continent is not so well regulated.

Within Australia there are two levels of jurisdiction governing the importation and exportation of fauna. At the Commonwealth level, the Wildlife Protection (Regulation of Exports and Imports) Act (1982) regulates imports of plants and goods that may have an "adverse effect on, or on the habits of, native Australian animals" (Part 1: 3E). If an NICS were imported into Australia and became established, it then could be included in the List of Key Threatening Processes under section 183 of the Environment Protection and Biodiversity Conservation (EPBC) Act (1999). NICS currently do not appear in the list since none have been imported from overseas to date (see www.deh.gov.au/cgi-bin/sprat/public/publicgetkeythreats.pl).

Regulations for the importation and exportation of crayfish within Australia are State-specific. In general terms, the exportation of crayfish from each State or territory is not regulated (with the exception of Tasmania); movement within the State is not regulated (except in the Northern Territory), but importations are regulated (Horwitz 1990). It is acknowledged, however, that it is very difficult to prevent unregulated transport of crayfish, particularly given the interest in establishing populations for aquaculture and recreational fisheries.

CRAYFISH IN WA

The crayfish fauna of WA is represented by two genera (Family Parastacidae), *Cherax* Clark and *Engaewa* Riek. In discussing *Cherax* spp. in the State, it is easier to use the commonly expressed vernacular terms to avoid confusion over which scientific names formally are correct. The indigenous peoples of south-western Australia recognized three forms of crayfish that now are classified within the genus *Cherax*: marron, gilgies, and koonacs. Translocations have occurred both within WA and into WA from other States (Table 1). Whilst this paper focuses principally on the likely impacts of increases in geographical ranges of the marron, redclaw, and the yabby, the potential for extinction of one or more of the five currently recognised species of *Engaewa* (Horwitz and Adams 2000) caused by the yabby also should be anticipated. *Engaewa* spp. have very restricted coastal distributions in the extreme south-western corner of the State.

Marron

Nicholl and Horwitz (2000) have recognized the marron as a flagship species for river conservation within WA. Since these large crayfish are much sought after for human consumption, considerable effort has been expended in their aquaculture, locally, interstate, and overseas (Morrissy *et al.* 1990). It comes as a surprise to many that the current distribution of marron in WA represents a geographical range significantly increased post-European colonization of the State. Morrissy (1978), based on historical accounts and the State Fisheries Department records, concluded that the pre-European distribution of marron was south of Mandurah in coastal lakes, creeks, and rivers from the Harvey to the Kent rivers (Fig. 1). The restricted distribution of marron resulted from their habitat requirement of permanent pools and their limited powers of dispersal (Shipway 1951, Morrissy 1978, Morrissy and Fellows 1990).

Whilst it is not known if the historical expansion of marron into rivers and permanent wetlands north of Mandurah have had deleterious impacts on the indigenous crayfish and other stream fauna of this new range, the movement of marron by humans within the south-west corner of the State may not have been without casualty. Recent data indicate that the widespread smooth marron morph is having a markedly negative impact on the 'hairy' marron morph! *Cherax tenuimanus* (Smith), the 'hairy' marron, was described on specimens from Margaret River but marron from other rivers beyond this catchment recently have been separated into a second species, *Cherax cainii* Austin, the 'smooth' marron (Austin and Ryan 2002). The more ubiquitous *C. cainii*, which

Table 1 Species of *Cherax* spp. in WA, together with their indigenous and non-indigenous distributions.

Species name	Common name	Indigenous distribution	Non-indigenous distribution
Cherax cainii	Smooth marron	The high rainfall region between just west of Albany and just south of Perth (Riek 1967, Morrissy 1978).	Hutt River in the north to Esperance in the east (Lawrence and Morrissy 2000).
Cherax destructor	Yabby	Over 2 million km^2, from South Australia and southern Northern Territory in the west, to south-west Victoria and the extreme south-east of SA in the east (Riek 1967).	WA – Hutt River in the north to Esperance in the south-east (Morrissy and Cassells 1992, Horwitz and Knott 1995). Isolated populations inland at Cue and Leonora. Tasmania–Midlands
Cherax preissii	Koonac	Widespread throughout much of the south-west and Wheatbelt, from north of Perth to east of Albany, mainly inland	
Cherax quadricarinatus	Redclaw	Far northern Queensland and northern and eastern parts of the Northern Territory (Riek 1969, Curtis and Jones 1995). Known only from limited area, in exorheic drainage systems from Daly River, NT, to Normanby River, NE QLD.	WA – Kimberley Drainage Division in Western Australia. QLD – populations in lake systems in northern and south-eastern Queensland.
Cherax quinquecarinatus	Gilgie	North of Perth to west of Albany, mainly coastal	
Cherax tenuimanus	Hairy marron	Restricted to the Margaret River	

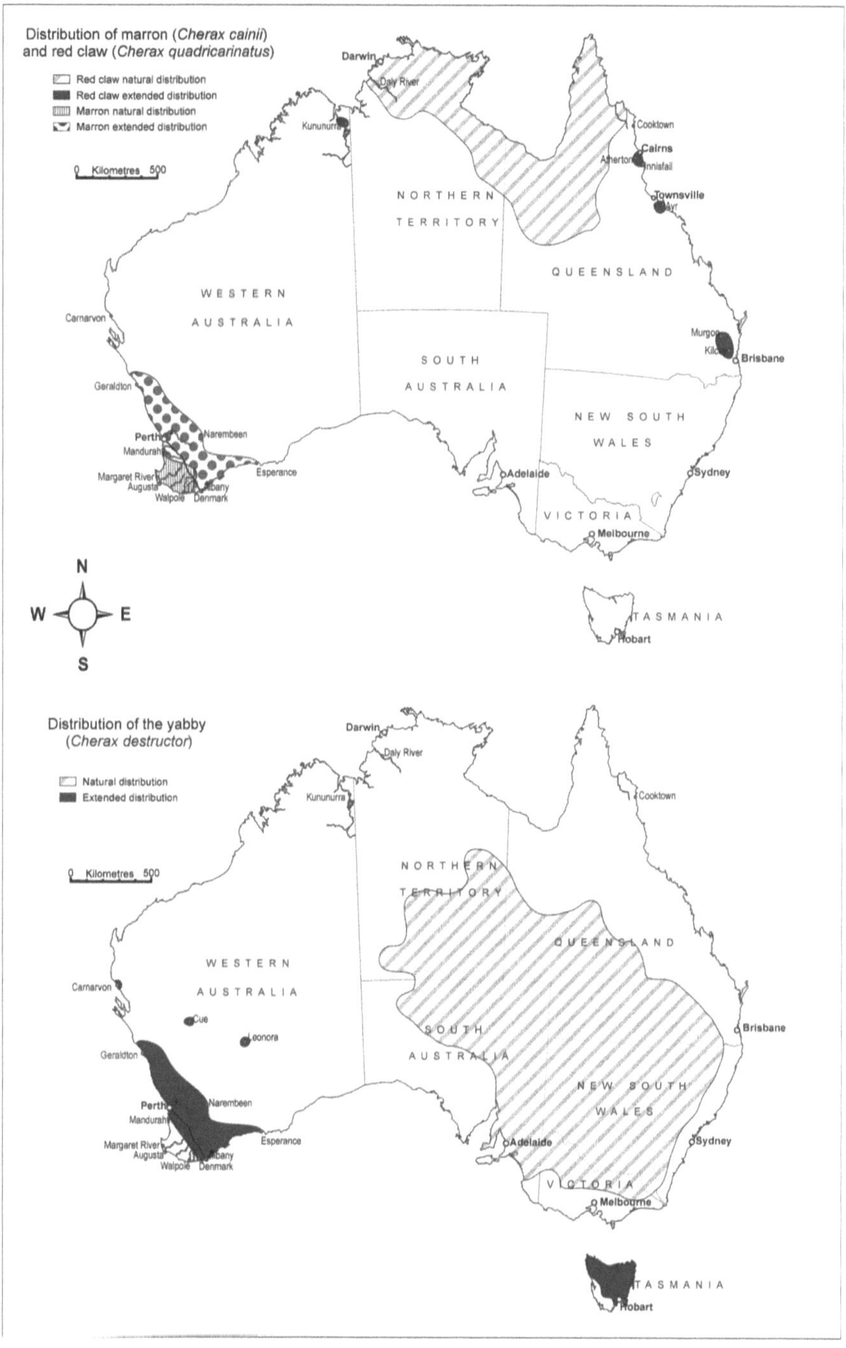

Fig. 1 Map showing the current known distributions of yabbies and redclaw in Western Australia. Distributions are not necessarily continuous and represent best current information. There are likely to be other occurrences which have not yet been discovered.

is the subject of most marron research, is the species widely used in aquaculture, including within the Margaret River catchment itself where smooth marron escapees have bred with hairy marron. Consequently, the distribution of the hairy marron phenotype has been restricted predominantly to forested headwater reaches of the Margaret River, and is now formally gazetted under the WA Wildlife Conservation Act 1950, following IUCN Red List Categories and Criteria version 3.1, as critically endangered (Bunn 2004). The potential for species replacement through competitive exclusion or reproductive interference are yet to be evaluated. There is, however, some evidence that *C. cainii* has a greater growth rate and earlier spawning (Bunn 2004). A management plan to enhance recovery of the 'hairy' morph within the Margaret River system will rely heavily on local community involvement to be effective (Bunn 2004).

The biological impacts of the northern spread of marron into areas with drier, warmer climate and different geological conditions from those wetter, cooler conditions characteristic of their 'pre-European' range have not been evaluated, although there is clear evidence of phenotypic plasticity. Individual biomass, corrected for body length, shows statistically significant reduction on a south to north axis (M. Bennet-Chambers 2007, unpublished data). Given this significant gradient, it is appropriate to raise the question of whether aquaculture of marron in the northern areas where evaporation substantially exceeds rainfall constitutes the wisest use of a limited groundwater resource. It is unlikely that the ecological costs of growing marron depending on groundwater have ever been included in calculating the full costs of production.

Redclaw, *Cherax quadricarinatus*

The redclaw *Cherax quadricarinatus* von Martens is endemic to ephemeral catchments of the Gulf of Carpentaria of northern Australia (Riek 1969, Curtis and Jones 1995, Jones *et al.* 2000). Because of its attraction for aquaculture, the species has been translocated widely within northern Australia and overseas (Horwitz 1990, Curtis and Jones 1995), and not always legally. Redclaw is classed as a restricted fish species for importation into WA (Anonymous 1997). Even so, the potential for aquaculture in the Ord Valley in the East Kimberley of the 'Walkamin' strain was assessed under quarantine (Doupé *et al.* 2004). A limited number of aquaculture licenses subsequently were issued, and shortly after wild populations of redclaw were found established in Lake Kununurra, a Ramsar wetland formed through impoundment of the Ord River (Fig. 1) within the Kimberley Drainage Division of Western Australia (Morgan *et al.* 2004). Since redclaw used in Ord River aquaculture are a genetically different strain to those now found in Lake Kununurra, the source of their introduction is unknown. Doupé *et al.* (2004), however, suggest it is the result of illegal translocations by recreational fishermen. Redclaw has subsequently spread downstream from Lake Kununurra into the lower Ord River, where local recreational anglers have reported redclaw in the stomach of barramundi [*Lates calcarifer* (Bloch)] and catfish (*Arius* spp.). It is quite likely that the species will soon reach Parry Lagoon on the Ord River floodplain, another Ramsar wetland.

The implications for the ecology of these systems are unknown; however, three indigenous species of *Macrobrachium* prawns [*Macrobrachium australiense* Holthuis, *M. bullatum* Fincham, and *M. rosenbergii* (de Man)] and three species of atyiid shrimp (*Caridina* cf *longirostris*, *C. ?nilotica*, and *C. serratirostris* de Man) may all come under competitive pressure. Given their rapid growth rate and tolerance of a wide range of environmental conditions (Jones and Ruscoe 2001), redclaw are likely to thrive in fresh waters of northern WA and rapidly expand their range into the many aquatic systems in this region. Recently, the State Department of Fisheries provided funding to elucidate both the genetic origins and extent of redclaw throughout the Ord River and the reproductive biology, parasitology, and trophic interactions of this with other decapod species.

The yabby, *Cherax destructor*

Since being introduced in 1932 into WA, from a farm dam in western Victoria, the yabby, *Cherax destructor* Clark, has spread into natural river systems within the south-west of WA where it now co-occurs with indigenous crayfish species. Although morphologically distinguishable from the white yabby, *Cherax albidus* Clark, also from eastern Australia, allozyme evidence provides little support for genetic separation between the two species and it has been suggested that the two species should be synonymized (Austin 1986, Campbell *et al.* 1994). *Cherax destructor* is the senior synonym by virtue of page priority; hence, zoologists in WA use the species epithet *destructor*, for example, as used here, but for essentially commercial reasons, fisheries personnel use the epithet *albidus* (e.g. Morrissy and assells 1992). Austin (1985) reported little allozyme diversity from yabbies in the State, but recent studies now indicate considerable variation in WA yabby populations. This perhaps reflects the expansion of yabby aquaculture in the 1990s with farmers introducing multiple strains from eastern Australia.

Although introduced initially to a farm dam at Narembeen, 280 km east of Perth, into a landscape that would not have facilitated easily the natural spread of the crayfish, its hardiness and ability to grow even in stagnant farm dams, together with the human interest in crayfish as a food item, meant that crayfish were spread quickly and widely. Many farmers actually thought they were culturing the indigenous koonac [*Cherax preissii* (Erichson)]. By 1985, most known yabby sites were still east of the Albany Highway (i.e. east of the typical range of indigenous WA crayfish; Austin 1985). Yabby populations were first reported in natural waterbodies west of the 'yabby exclusion zone' by Lynas *et al.* (2004). They have since then shown a continuing strong spread northwards, to the north-east and south-east, with their current distribution being from the Hutt River in the north to Esperance in the south-east (Morrissy and Cassells 1992, Horwitz and Knott 1995). They have also colonized cave streams *via* temporary, short streams on coastal sand-plain (Jasinska *et al.* 1993), occur in coastal plain rivers to the west of the Darling Scarp, and occur in the arid northern Goldfields Region near Leonora (see Fig. 1). The spread of yabbies into natural habitats has generated potential for interactions with other WA indigenous fauna. For example, the spread of yabbies into the Swan-Avon catchment has led to their potential interaction with the critically endangered western swamp tortoise, *Pseudemydura umbrina* Siebenrock, near the Ellen Brook Nature Reserve proclaimed for preservation of the tortoise (Bradsell *et al.* 2002). Yabbies showed strongly aggressive and predatory behaviour toward tortoise hatchlings in a laboratory study using hatchlings of a non-endangered species of tortoise (Bradsell *et al.* 2002).

It seems reasonable to suggest, given the extent of their range expansion to date and apparent capacity to colonize a wide diversity of waterbodies, that yabbies will continue to progress into larger river systems of the Swan Coastal Plain, making their move further into the south-west relatively simple, as has occurred with other invasive species, such as the mosquitofish *Gambusia holbrooki* Girard.

The extensive spread of yabbies appears facilitated by their biology. They are an *r*-selected species with a short life cycle, multiple spawning events, a high spawning frequency, fast growth rates and high fecundity (Lawrence and Jones 2002, Beatty *et al.* 2005). Under suitable conditions, yabbies are able to breed year-round (Morrissy *et al.* 1984). Such life-history traits would allow the successful colonization of disturbed habitats and areas which have undergone anthropogenic modification, such as many of the rivers of south-western Australia. This effectively enables yabbies to become the most abundant crayfish species in many of the freshwater systems throughout its translocated range in south-western Australia (Beatty *et al.* 2005). In addition, yabbies are burrowing crayfish adapted to long-term population survival in the fluctuating environments of relatively impermanent and often highly eutrophic still waters (Morrissy *et al.* 1984). They are also more tolerant than indigenous species of extremes in temperature (Morrissy 1990), hypoxia (Morrissy *et al.* 1984, Holdich and Lowery 1988), and salinity (Department of Fisheries website http://www.wa.gov.au/westfish/aqua/broc/aqwa/marron/).

Human activity has also aided their spread through misguided information and recreational carelessness. Typically, the extent of the problem has gone largely unnoticed due to the common misnaming of this crayfish as the koonac. Yabbies are commonly used as bait for redfin perch and trout fishing in the large government irrigation dams to the south of Perth, with unused live bait often being discarded directly into these waters (Morrissy and Cassells 1992).

Impacts to marron and gilgies

Despite their range expansion and suitability as colonisers, little research has been undertaken, until recently, to ascertain the possible ecological impacts of the invasive yabby in WA. Consequently, studies were undertaken to examine

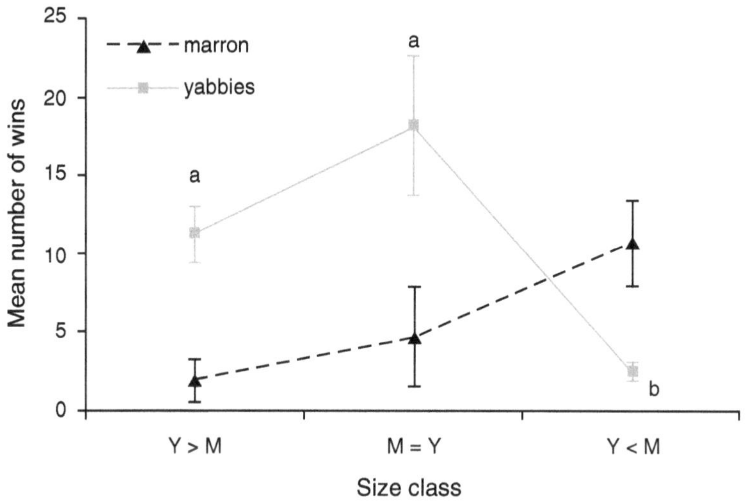

Fig. 2 The interaction between species and size in relation to the mean number of tension contacts 'won' (±SE) by marron and yabbies (n = 28). Letters denote equal means determined using the least significant range test.

the potential for competitive exclusion of two indigenous species, the smooth marron, *C. cainii* and the gilgie, *Cherax quinquecarinatus* (Gray).

In aggressive behaviour trials designed to predict the likely outcomes of competitive interactions in natural systems, aggressive dominance was found to be strongly influenced by size (Lynas 2002). When yabbies were larger or equal in body mass to marron they 'won' a significantly higher number of tension contacts (Fig. 2). Alternatively, when yabbies were of smaller body mass than marron, the marron 'won' a significantly higher number of contacts (Fig. 2; Lynas 2002). Given the small size of gilgies, aggressive experiments with yabbies utilised only individuals of similar body mass. In such interactions, yabbies were found to be equal in aggression to gilgies (Lynas 2002). The importance of size in establishing aggressive dominance is well reported in studies of crustaceans generally (e.g. Hartnoll 1974, Stein 1976), as well as crayfish (Bovbjerg 1956, Lowe 1956, Horwitz 1980, Momot and Leering 1986, Vorburger and Ribi 1999).

Laboratory-based competition experiments were also conducted with sediment as the limiting resource. Sediment was chosen because of its documented significance in influencing crayfish distributions (Bovbjerg 1952, 1970, Suter and Richardson 1977, Grow 1981, Capelli and Magnuson 1983), and its necessity for protection from predators and cannibalism. In these trials, both indigenous species demonstrated a preference for clay substrate, perhaps because they were able to bury themselves in this sediment (Fig. 3; Lynas *et al.* 2006). In the presence of yabbies, however, the number of both gilgies and

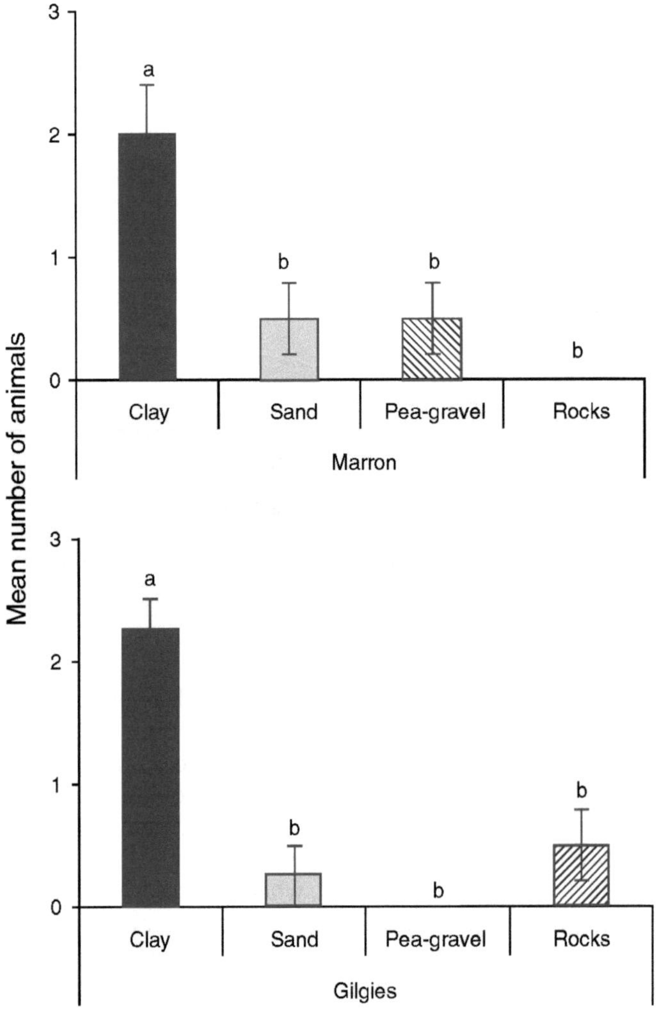

Fig. 3 Mean number (±SE) of marron (above) or gilgies (below) on each sediment when alone, i.e. not in the presence of yabbies. Letters denote equal means determined using the least significant range test.

marron found on the clay was significantly reduced (Fig. 4; Lynas *et al.* 2006). Agonistic behaviour in the form of tension contacts (fight, strike, threat, and avoidance) was common during the trials, with interspecific contests being most commonly recorded on the clay substrate in both marron and gilgie trials (Lynas *et al.* 2006). Therefore, it seems that yabbies used agonistic behaviour to control access to the limiting resource and effectively excluded marron and gilgies from clay substrates in the laboratory (Lynas *et al.* 2006).

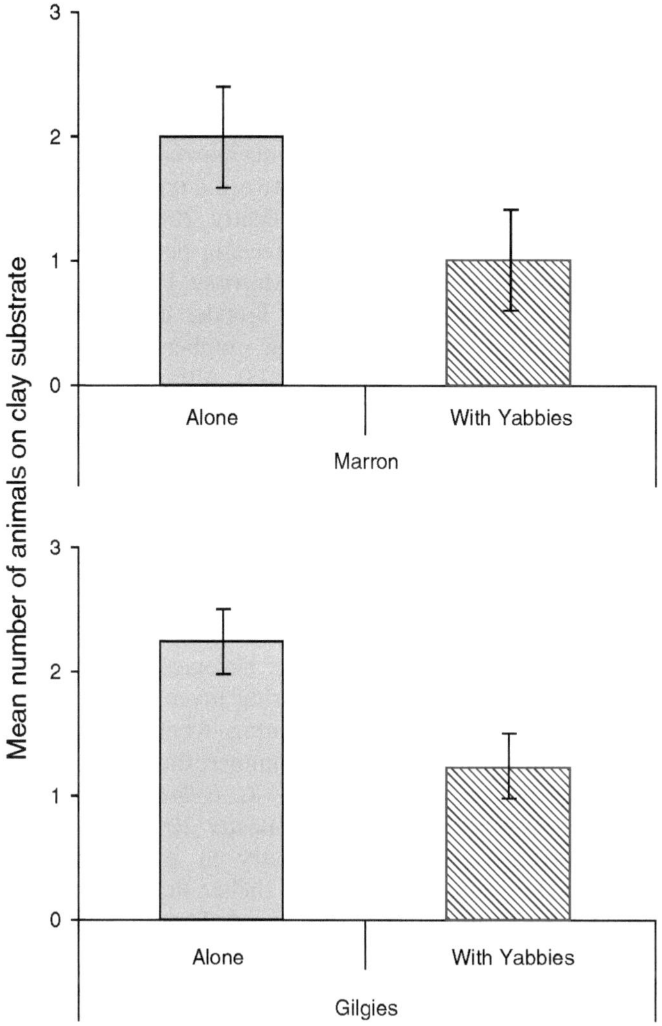

Fig. 4 Mean number (±SE) of marron (above) or gilgies (below) on the "preferred" clay substrate when alone and when in trial with yabbies.

Results from both aggressive behaviour and sediment competition trials indicate that, in habitats of co-occurrence where there is substantial overlap in resource use, the potential for exclusion of marron and/or gilgies by the invasive yabby is high (Lynas 2002, Lynas *et al.* 2006). The experiments demonstrated that aggressive behaviour could be used to procure a limiting resource. Superior competitive ability manifested itself through interference, with the displacement of subordinate species from preferred substrates

(Lynas *et al.* 2006). Yabbies were found to be capable of evicting both marron and gilgies from suitable substrates, thereby indicating the exclusion of these species from the use of a limiting resource under laboratory situations (Lynas *et al.* 2006).

It is suggested that in natural environments marron are likely at a disadvantage. Marron have a life-history strategy between a typical *r*- (summer brooder) and K-selected (winter brooder) species (Beatty 2005). Traits analogous to a winter brooder include synchrony of breeding period, long life cycle, and reliance on permanent aquatic systems (Morrissy 1975, 1983, Beatty *et al.* 2003, Beatty 2005). Traits of a summer brooder include a relatively short brooding period during summer, high egg number per brood, and a rapid growth rate (Beatty *et al.* 2003, Beatty 2005). Although capable of attaining a larger size than yabbies, 2 kg *vs.* 90–100 g (Lawrence and Jones 2002) and having a similar growth rate (Beatty *et al.* 2005), marron mature later and at a much larger size than yabbies (Beatty *et al.* 2003). In a study of marron established in the Hutt River (outside its natural range), Beatty (2005) reported release of juveniles earlier than that previously noted by Morrissy (1975) for the more southerly Warren River populations. Plasticity in biological parameters was considered due to environmental conditions, namely temperature and photoperiod (Beatty *et al.* 2003, Beatty 2005).

There is considerable asynchrony in the biological cycles of these species, with yabbies reaching maturity and releasing juveniles earlier than marron. Therefore, yabbies would have the size advantage when members of both species come into contact in natural systems. In summer, the diet of both marron and yabbies in the Hutt River is dominated by *G. holbrooki*, but yabbies show a dietary shift towards herbivory in winter (Beatty 2007). Consequently, when yabbies and marron co-occur in sympatry in natural habitats, yabbies may dominate food resources and suitable shelter sites over marron juveniles. Marron would have little likelihood of successfully establishing stable populations where yabbies already exist. Predicting the outcome of which species would 'win' when yabbies invade a river system with a stable marron population already in occupation is much more difficult and probably depends upon the initial conditions. Larger marron may dominate access to limiting resources over smaller yabbies. Nevertheless, yabbies would persist due to their high fecundity and ability to withstand environmental fluctuations resulting in a more unpredictable future. The survival of marron populations would be subject to increased uncertainty since their juveniles would be unable to compete successfully with yabbies. Furthermore, given the importance of size in determining the outcome of aggressive interactions, we predict that in natural environments where yabbies attain a much larger size than gilgies ($\leq$30 g in biomass), yabbies generally would have a size advantage, thereby controlling access to limiting resources such as food and suitable shelter sites when the two species occur in sympatry. Beatty (2007) further suggested the dietary switch reported in yabbies from the Hutt River leads to the potential for competition with the smaller gilgies in unproductive freshwater systems common to the south-west.

As well as competition from NIS, indigenous WA crayfish are at risk from infection from the microsporidian *Thelohania parastaci* Moodie known to be carried by the yabby (Horwitz 1990, Moodie *et al.* 2003). WA crayfish species had not been exposed previously to the disease and therefore are likely to be susceptible. *Thelohania* was found in WA farm dam populations of yabbies in the 1990s (Jones and Lawrence 2002), and has since been reported in yabby populations in the Hutt River (Beatty 2005). This microsporidian could be transmitted to indigenous species by sympatric yabbies. Infection of crayfish by this parasite leads to the destruction of striated and cardiac muscle tissue, resulting in reduced locomotor activity (Henneguy and Thélohan 1892, Cossins and Bowler 1974, Quilter 1976). Survival time of infected individuals has been reported to range from a few months (in the New Zealand *Paranephrops zealandicus* White in Quilter 1976) to two years (in the astacid *Austropotamobius pallipes* Lereboullet in Brown and Bowler 1977), although whether death is always inevitable for infected individuals has yet to be ascertained (Moodie *et al.* 2003). *Thelohania*, therefore, may increase the risk of predation of infected crayfish and reduce their ability to compete with healthy individuals.

Impacts to koonacs

Koonacs, the third main indigenous crayfish species in south-western Australia, have suffered a reduction in distribution post European settlement due to fragmentation and loss of swamps feeding into the headwater streams. The potential impact of the yabby on the endemic koonac, *C. preissii*, is currently unknown; although, Horwitz (1980) noted the importance of size in aggressive interactions between gilgies and koonacs. Juvenile gilgies were found to be dominant over juvenile koonacs but no aggressive differences were detected in the adults of these species (Horwitz 1980). Since yabbies are strong burrowers, they likely would be able to invade swamps inhabited by koonacs. The indigenous species breed in spring within capped burrows and unplug them at the end of the dry season to forage within the waterbody. Given the importance of size in aggressive and competitive interactions, juvenile koonacs are likely to be out-competed by larger yabbies when they emerge from burrows.

Impacts to *Engaewa*

Finally, the importance for conservation of the five currently recognised species of *Engaewa* (Horwitz and Adams 2000) cannot be ignored. The small, strongly burrowing forms of *Engaewa*, with typical burrowing characteristics *sensu* Holdich (2002), have a very restricted coastal distribution in permanently moist acid peat swamps from Dunsborough to Albany; such swamp habitats are more continuous along the south coast from Walpole to Augusta, but become much more fragmented between Augusta and Dunsborough (Burnham 2005). Initially thought to be closely related to the genus *Engaeus* from south-eastern Australia, the study on a region of the 16S mitochondrial gene by Crandall *et al.* (1999) concluded that the genus represents a major, distinctive clade within the Parastacidae. Such relict forms, with ancestry dating from Gondwanic origins, are particularly susceptible to increasing pressures from human activities, including habitat fragmentation and loss (Burnham 2005). Whilst *Engaewa* spp. and indigenous *Cherax* species currently coexist in various levels of sympatry throughout the geographic range of *Engaewa*, it will be important to evaluate the impact on *Engaewa* spp. of the further spread of the yabby into the south-western corner of the State, should it occur. Potential impacts are likely from changes to the structure and function of the ecosystem, including habitat alteration, changes to food web dynamics, and the introduction of disease.

THE FUTURE UNDER A DRYING CLIMATE AND GLOBAL WARMING

Current climatic trends in the south-west of WA likely afford a further advantage to the invasive yabby, particularly over marron. With an increasing drying climate and reduced rainfall across the south-west of WA, groundwater levels are decreasing (Allan and Haylock 1993, Anonymous 2002). There are likely to be numerous ecological consequences. Wetlands and streams driven by groundwater inputs may become ephemeral or permanently dry. Those crayfish able to burrow to the water table may survive dry periods (Bovbjerg 1952, 1970, Taylor 1983). The yabby is a strong burrower (Morrissy *et al.* 1984) and has been recorded alive from burrows beneath lake beds that have been dry for eight years (Holdich and Lowery 1988). Marron, however, inhabit permanent freshwater systems and are not strong burrowers but show a preference for sheltering under logs or stones in the bed of streams (Shipway 1951). Indeed, Riek (1969) suggested the poor development of chelae muscles in marron restricted their burrowing ability. Further, marron do not burrow to escape drought (Maguire *et al.* 1999, Lawrence and Jones 2002). However, in a laboratory trial designed to determine the response of crayfish to lowering groundwater, a single marron did construct an angular pit (j-shaped burrow) approximately 30 cm deep when water levels were artificially lowered (Lynas 2002). This refutes the suggestion that marron are physiologically unable to construct burrows. The experiment further highlights the plasticity of marron and suggests that further investigation on the drought-response mechanisms of WA crayfish and effects of lowering groundwater are required. Gilgies and koonacs are both strong burrowers and able to survive in temporary environments. Therefore, marron would likely be more severely impacted by lowering groundwater levels, with yabbies having a considerable advantage.

With an increase in water temperatures associated with global warming, yabbies would again be at an advantage. The maximum growth of yabbies is at 28 °C (Holdich and Lowery 1988), but for smooth marron, maximum growth is at 24 °C. No research has been undertaken on temperature responses of koonacs and gilgies, but an increasing temperature would likely reduce fitness given they are both indigenous to the colder waters of south-western WA.

CONCLUSIONS

Within the vast area of WA, crayfish occupied a comparatively small area within the reliably high rainfall zone of the south-western corner of the State, where they were widely disjunct from the crayfish of eastern Australia. Because of the interest for their consumption by humans, marron (*C. cainii*), redclaw (*C. quadricarinatus*), and yabbies (*C. destructor*) have been translocated, probably in many cases by recreational fishermen, but also by aquaculturalists, and without regulatory sanction. It is not known if the early spread of marron has affected the ecology of invaded aquatic systems. Similarly, the effects of the recent introduction of redclaw to the Ord River system are unknown. The spread of smooth marron into the Margaret River, however, is threatening the persistence of the endemic hairy marron, *C. tenuimanus*. The impact of the invasive yabby is likely to be much more detrimental to the indigenous crayfish of WA (marron, gilgies, and koonacs), particularly given the drying climate in the south-west, with burrowing crayfish more likely to survive periods of drought. Yabbies, too, have greater tolerance of increased salinity than indigenous crayfish, with dryland salinity being a major landuse and ecological problem in the southern half of WA. In all cases of interactions with indigenous crayfish, the aggressive yabby is likely to be competitively superior, particularly in the juvenile stages because of the earlier release of young into streams.

The spread of the yabby seems inexorable, given commercial interest, the current widespread distribution, the extent of unregulated movement (for example populations inland at Leonora and Cue), and developing climate change. It is difficult to foresee any effective control impeding the continuing spread of *C. destructor* in WA and consequent impacts on indigenous species.

ACKNOWLEDGEMENTS

We wish to acknowledge, with gratitude, assistance from Dr. Craig Lawrence (Department of Fisheries, Government of Western Australia), Dr. Kyle Armstrong, and Dr. Jane Prince (School of Animal Biology, The University of Western Australia). Funding was provided by the Department of Zoology, The University of Western Australia.

REFERENCES

Aiken, D. E. 1965. Distribution and ecology of three species of crayfish from New Hampshire. American Midland Naturalist **73**, 240–244.

Allan, R. J. and Haylock M. R. 1993. Circulation features associated with the winter rainfall decrease in southwestern Australia. Journal of Climate **7**, 1356–1367.

Anonymous. 1997. Aquaculture of non-endemic species in Western Australia, redclaw crayfish. Fisheries Management Paper No. 100. Department of Fisheries of Western Australia.

Anonymous. 2002. Climate variability and change in south west Western Australia. Indian Ocean Climate Initiative, Perth, Western Australia, September 2002, 36 pp.

Austin, C. M. 1985. Introduction of the yabbie, *Cherax destructor* (Decapoda: Parastacidae) into southwestern Australia. Western Australian Naturalist **16**, 78–82.

Austin, C. M. 1986. Electrophoretic and morphological systematic studies of the genus *Cherax* (Decapoda: Parastacidae) in Australia. Ph.D. thesis, Department of Zoology, The University of Western Australia, Western Australia.

Austin, C. M. and S. G. Ryan. 2002. Allozyme evidence for a new species of freshwater crayfish of the genus *Cherax* Erichson (Decapoda: Parastacidae) from the south-west of Western Australia. Invertebrate Systematics **16**, 357–367.

Beatty, S. 2005. Translocations of freshwater crayfish: Contributions from life histories, trophic relations and diseases of three species in Western Australia. Ph.D. thesis, Murdoch University, Western Australia.

Beatty, S. J. 2007. The diet and trophic positions of translocated, sympatric populations of *Cherax destructor* and *Cherax cainii* in the Hutt River, Western Australia: Evidence of resource overlap. Marine and Freshwater Research, in press.

Beatty, S. J., D. L. Morgan, and H. S. Gill. 2003. Reproductive biology of the large freshwater crayfish *Cherax cainii* in south-western Australia. Marine and Freshwater Research **54**, 597–608.

Beatty, S., D. Morgan, and H. Gill. 2005. Role of life history strategy in the colonisation of Western Australian aquatic systems by the introduced crayfish *Cherax destructor* Clark, 1936. Hydrobiologia **549**, 219–237.

Bovbjerg, R. V. 1952. Comparative ecology and physiology of the crayfish *Orconectes propinquus* and *Cambarus fodiens*. Physiological Zoology **25**, 34–56.

Bovbjerg, R. V. 1956. Some factors affecting aggressive behaviour in crayfish. Physiological Zoology **26**, 173–178.

Bovbjerg, R. V. 1970. Ecological isolation and competitive exclusion in two crayfish (*Orconectes virilis* and *Orconectes immunis*). Ecology **51**, 225–236.

Bradsell, P., J. Prince, G. Kuchling, and B. Knott. 2002. Aggressive interactions between freshwater turtle, *Chelodina oblonga*, hatchlings and freshwater crayfish, *Cherax* spp.: Implications for the conservation of the critically endangered western swamp turtle, *Pseudemydura umbrina*. Wildlife Research **29**, 295–301.

Brown, D. J. and K. Bowler. 1977. A population study of the British freshwater crayfish *Austropotamobius pallipes* (Lereboullet). Freshwater Crayfish **3**, 33–49.

Bunn, J. J. S. 2004. Investigation of the replacement of Margaret River hairy marron *Cherax tenuimanus* (Smith) by smooth marron *C. cainii* Austin. MSc thesis, Centre for Ecosystem Management, Edith Cowan University, Perth, Western Australia.

Burnham, Q. F. 2005. The systematics of the reducta complex of the burrowing freshwater crayfish *Engaewa* Riek. Honours thesis, Centre for Ecosystem Management, Edith Cowan University, Perth, Western Australia.

Butler, M. J. and R. A. Stein. 1985. An analysis of the mechanisms governing species replacements in crayfish. Oecologia **66**, 168–177.

Campbell, N. J. H., M. C. Geddes, and M. Adams. 1994. Genetic variation in yabbies, *Cherax destructor* and *C. albidus* (Crustacea: Decapoda: Parastacidae), indicates the presence of a single, highly sub-structured species. Australian Journal of Zoology **42**, 745–760.

Capelli, G. M. 1982. Displacement of Northern Wisconsin crayfish by *Orconectes rusticus* (Girard). Limnology and Oceanography **27**, 741–745.

Capelli, G. M. and J. J. Magnuson. 1983. Morphoedaphic and biogeographic analysis of crayfish distribution in northern Wisconsin. Journal of Crustacean Biology **3**, 548–564.

Capelli, G. M. and B. L. Munjal. 1982. Aggressive interactions and resource competition in relation to species displacement among crayfish of the genus *Orconectes*. Journal of Crustacean Biology **2**, 486–492.

Cole, L. C. 1960. Competitive exclusion. Science **132**, 348–349.

Cossins, A. R. and K. Bowler. 1974. An histological and ultrastructural study of *Thelohania contejeani* Henneguy 1892 (Nosematidae), microsporidian parasite of the crayfish *Austropotamobius pallipes* Lereboullet. Parasitology **68**, 81–91.

Crandall, K. A., J. W. J. Fetzner Jr, S. H. Lawler, M. Kinnersley, and C. M. Austin. 1999. Phylogenetic relationships among the Australian and New Zealand genera of freshwater crayfishes (Decapoda: Parastacidae). Australian Journal of Zoology **47**, 199–214.

Curtis, M. C. and C. M. Jones. 1995. Overview of redclaw crayfish *Cherax quadricarinatus* farming practices in northern Australia. Freshwater Crayfish **10**, 447–455.

Doupé, R. G., D. L. Morgan, H. S. Gill, and A. J. Rowland. 2004. Introduction of redclaw crayfish *Cherax quadricarinatus* (von Martens) to Lake Kununurra, Ord River, Western Australia: Prospects for a 'yabby' in the Kimberley. Journal of the Royal Society of Western Australia **87**, 187–191.

Elvey, W., S. J. Chilcott, and A. C. Sanger. 1996. The distribution and potential ecological impact of the introduced Yabby, *Cherax destructor* Clark 1936, in Tasmania. Inland Fisheries Commission. Report prepared for the Australian Nature Conservation Agency Feral Pests Program. Project Number FPP 71.

Figgis, P. 1993. Southwest Western Australia. Pages 298–317 in P. Figgis and G. Mosley, editors. Australia's Wilderness Heritage. Volume 1. World Heritage Areas. Landsdowne Press, Sydney, Australia.

Gherardi, F. and D. M. Holdich, editors. 1999. Editor's summary. Page 7 in F. Gherardi and D. M. Holdich, editors. Crayfish in Europe as alien species: How to make the best of a bad situation? A. A. Balkema, Rotterdam, The Netherlands.

Grow, L. 1981. Burrowing behaviour in the crayfish *Cambarus diogenes diogenes* Girard. Animal Behaviour **29**, 351–356.

Hansard, 2006. Parliamentary Records of the State Government of Western Australia, 6th March 2006.

Hardin, G. 1960. The competitive exclusion principle. Science **131**, 1292–1297.

Hartnoll, R. G. 1974. Variation in growth patterns between some secondary sexual characteristics in crabs (Decapoda: Brachyura). Crustaceana **27**, 131–136.

Henneguy, F. and P. Thélohan. 1892. Myxosporidies parasites des muscles chez quelques crustacés decapodes. Annales de Micrographie **4**, 617–641.

Holdich, D. M. 1987. The dangers of introducing alien animals with particular reference to crayfish. Freshwater Crayfish **7**, 15–30.

Holdich, D. M., editor. 2002. Biology of freshwater crayfish. Blackwell Science, Oxford.

Holdich, D. M. and R. S. Lowery, editors. 1988. Freshwater crayfish: biology, management and exploitation. Croom Helm London, UK.

Horwitz, P. 1980. Aspects of behavioural ecology of the two species of freshwater crayfish, *Cherax quinquecarinatus* and *Cherax plebejus* (Decapoda: Parastacidae). Honours thesis, Zoology Department, the University of Western Australia, Western Australia.

Horwitz, P. 1990. The translocation of freshwater crayfish in Australia: Potential impact, the need for control and global relevance. Biological Conservation **54**, 291–305.

Horwitz, P. and M. Adams. 2000. The systematics, biogeography and conservation status of species in the freshwater crayfish genus *Engaewa* Riek (Decapoda: Parastacidae) from south-western Australia. Invertebrate Taxonomy **14**, 655–680.

Horwitz, P. and B. Knott. 1995. The distribution and spread of the yabby *Cherax destructor* complex in Australia: Speculations, hypotheses and the need for research. Freshwater Crayfish **10**, 81–91.

Jaeger, R. G. 1974. Competitive exclusion: Comments on survival and extinction of species. Bioscience **24**, 33–39.

Jasinska, E. J., B. Knott, and N. Poulter. 1993. Spread of the introduced yabby *Cherax* spp. (Crustacea: Decapoda: Parastacidae), beyond the natural range of freshwater crayfishes in Western Australia. Journal of the Royal Society of Western Australia **76**, 67–69.

Jones, C. M. and I. M. Ruscoe. 2001. Assessment of five shelter types in the production of redclaw crayfish *Cherax quadricarinatus* (Decapoda: Parastacidae) under earthen pond conditions. Journal of the World Aquaculture Society **32**, 41–52.

Jones, C. M., C. P. McPhee, and I. M. Ruscoe. 2000. A review of genetic improvement in growth rate in redclaw crayfish *Cherax quadricarinatus* (von Martens) (Decapoda: Parastacidae). Aquaculture Research **31**, 61–67.

Jones, J. B. and C. S. Lawrence. 2002. Diseases of yabbies (*Cherax albidus*) in Western Australia. Aquaculture **194**, 221–232.

Lawrence, C. S. and N. M. Morrissy. 2000. Genetic improvement of marron *Cherax tenuimanus* (Smith) and yabbies *Cherax* spp. in Western Australia. Aquaculture Research **31**, 69–82.

Lawrence, C. and C. Jones. 2002. *Cherax*. Pages 635–669 in D. M. Holdich, editor. Biology of freshwater crayfish. Blackwell Science, Oxford, UK.

Lowe, M. E. 1956. Dominance-subordinance relationships in the crawfish *Cambarellus shufeldtii*. Tulane Studies in Zoology **4**, 135–170.

Lynas, J. 2002. Is the introduced yabby, *Cherax destructor*, an ecological threat to species of local freshwater crayfish in Perth, Western Australia? Honours thesis, Department of Zoology, University of Western Australia, Perth, Western Australia.

Lynas, J., P. Lindhjem, A. W. Storey, and B. Knott. 2004. Is the yabby, *Cherax destructor* (Parastacidae) in Western Australia an ecological threat? Freshwater Crayfish **14**, 37–44.

Lynas, J., A. W. Storey, K. Armstrong, J. Prince, and B. Knott. 2006. Invasion by the exotic crayfish, *Cherax destructor* Clark (Parastacidae) into habitats of local crayfish near Perth, Western Australia. Freshwater Crayfish **15**, 176–188.

Maguire, G., G. Cassells, and C. Lawrence. 1999. Aquaculture WA: Farming marron. Fisheries Western Australia Information Sheet.

Momot, W. T. and G. M. Leering. 1986. Aggressive interaction between *Pacifastacus leniusculus* and *Orconectes virilis* under laboratory conditions. Freshwater Crayfish **6**, 87–93.

Moodie, E. G., L. F. Le Jambre, and M. E. Katz. 2003. *Thelohania parastaci* sp. nov. (Microspora: Thelohaniidae), a parasite of the Australian freshwater crayfish, *Cherax destructor* (Decapoda: Parastacidae). Parasitology Research **91**, 151–165.

Morgan, D. L., A. J. Rowland, H. S. Gill, and R. G. Doupé. 2004. The implications of introducing a large piscivore (*Lates calcarifer*) into a regulated northern Australian river (Lake Kununurra, Western Australia). Lakes and Reservoirs: Research and Management **9**, 181–193.

Morrissy, N. M. 1975. Spawning variation and its relationship to growth rate and density in the marron, *Cherax tenuimanus* (Smith). Fisheries Research Bulletin of Western Australia **16**, 1–32.

Morrissy, N. M. 1978. The past and present distribution of marron, *Cherax tenuimanus* (Smith) in Western Australia, New Guinea and New Zealand. Freshwater Crayfish **5**, 534–544.

Morrissy, N. M. 1983. Crayfish research and industry activities in Australia, New Guinea and New Zealand. Freshwater Crayfish **5**, 534–544.

Morrissy, N. M. 1990. Optimum and favourable temperatures for growth of *Cherax tenuimanus* (Smith) (Decapoda: Parastacidae). Australian Journal of Marine and Freshwater Research **41**, 735–746.

Morrissy, N. M., N. Caputi, and R. R. House. 1984. Tolerance of marron (*Cherax tenuimanus*) to hypoxia in relation to aquaculture. Aquaculture **41**, 61–74.

Morrissy, N. M. and G. Cassells. 1992. Spread of the introduced yabbie, *Cherax albidus* Clark 1936, in Western Australia. Fisheries Research Report. Fisheries Department, Western Australia **92**, 1–27.

Morrissy, N. M., L. H. Evans, and J. V. Huner. 1990. Australian freshwater crayfish: Aquaculture species. World Aquaculture **21**, 113–122.

Morrissy, N. M. and C. J. Fellows. 1990. The recreational marron fishery in Western Australia, summarized research statistics, 1971–1987. Fisheries Research Report, Fisheries Department Western Australia **87**, 1–27.

Myers, N., R. A. Mittermeier, C. G. Mittermeier, G. A. B. da Foneca, and J. Kents. 2000. Biodiversity hotspots for conservation priorities. Nature **403**, 853–858.

Nickoll, R and P. Horwitz. 2000. Evaluating flagship species for ecosystem restoration: a case study involving freshwater crayfish and a river in south-western Australia. Pages 557–564 in J. Craig, N. Mitchell and O. Saunders, editors. Nature Conservation 5: Nature Conservation in Production Environments: Managing the Matrix. Surrey, Beatty and Sons, Chipping North, NSW, Australia.

Nyström, P., C. Brönmark, and W. Granéli. 1999. Influence of an exotic and a naïve crayfish species on a littoral benthic community. Oikos **85**, 545–553.

Quilter, C. G. 1976. Microsporidian parasite *Thelohania contejeani* Henneguy from New Zealand freshwater crayfish. Journal of Marine and Freshwater Research **10**, 225–231.

Riek, E. F. 1967. The freshwater crayfish of Western Australia (Decapoda: Parastacidae). Australian Journal of Zoology **15**, 103–121.

Riek, E. F. 1969. The Australian freshwater crayfish (Crustacea: Decapoda: Parastacidae) with descriptions of new species. Australian Journal of Zoology **17**, 855–918.

Shipway, B. 1951. The natural history of the marron and other freshwater crayfishes of south-western Australia. Part 1. Western Australian Naturalist **3**, 7–12.

Söderbäck, B. 1991. Interspecific dominance relationship and aggressive interactions in the freshwater crayfishes *Astacus astacus* and *Pacifastacus leniusculus* (Dana). Canadian Journal of Zoology **69**, 1321–1325.

Stein, R. A. 1976. Sexual dimorphism in crayfish chelae: Functional significance linked to the reproductive activities. Canadian Journal of Zoology **54**, 220–227.

Suter, P. J. and A. M. M. Richardson. 1977. The biology of two species of *Engaeus* (Decapoda: Parastacidae) in Tasmania III. Habitat, food, associated fauna and distribution. Australian Journal of Marine and Freshwater Research **28**, 95–103.

Taylor, R. C. 1983. Drought-induced changes in crayfish populations along a stream continuum. American Midland Naturalist **110**, 286–298.

Vogt, G. 1999. Diseases of European freshwater crayfish, with particular emphasis on interspecific transmission of pathogens. Pages 87–103 in F. Gherardi and D. M. Holdich, editors. Crayfish in Europe as alien species: How to make the best of a bad situation? A. A. Balkerma, Rotterdam, The Netherlands.

Vorburger, C. and G. Ribi. 1999. Aggression and competition for shelter between a native and an introduced crayfish in Europe. Freshwater Biology **42**, 111–119.

Whiting, A. S., S. H. Lawler, P. Horwitz, and K. A. Crandall. 2000. Biogeographic regionalisation of Australia: Assigning conservation priorities based on endemic freshwater crayfish phylogenetics. Animal Conservation **3**, 155–163.

Chapter thirty-two

The zebra mussel Dreissena polymorpha: *reproduction and competition with the sponge* Ephydatia fluviatilis

Tisza Lancioni and Elda Gaino

INTRODUCTION

The rapid expansion of *Dreissena polymorpha* (Pallas) in freshwater environments has resulted in serious problems from both an economic and an ecological standpoint. In this regard, investigations have been focused on the life-cycle and possible competitors able to counteract the spreading of this mollusc for management and control programmes.

Histological gonadal observations produced a wide literature on the reproduction of zebra mussels (Walz 1973, 1978, Borcherding 1986, 1991, 1995, Bielefeld 1991, Denson and Wang 1993, Garton and Haag 1993, Wang and Denson 1995, Gist *et al.* 1997, Mantecca *et al.* 2000, Bacchetta *et al.* 2001, Vailati *et al.* 2001, Juhel *et al.* 2003). This species has one (Haag and Garton 1992, Gist *et al.* 1997, Juhel *et al.* 2003) or more than one reproductive event per year (Walz 1978, Borcherding 1991, Wang and Denson 1995, Jantz and Neumann 1998, Bacchetta *et al.* 2001). Among the external factors affecting gonadal activity, temperature has a strong influence (Ram *et al.* 1993,

Francesca Gherardi, Biological invaders in inland waters: Profiles, distribution, and threats, 597–611.

Borcherding 1995, Fong *et al.* 1995, Wacker and Von Elert 2003), 12–13 °C being the value that triggers gamete development and spawning (Borcherding 1991, Neumann *et al.* 1993, Domaglia 1997, Bacchetta *et al.* 2001).

Competition among species has been extensively studied in marine environments (Jackson 1977, Russ 1982, Hirata 1987, Pansini and Pronzato 1990), whereas less attention has been paid to freshwater ecosystems (Lancaster *et al.* 1988, Dudley *et al.* 1990, Lauer and Spacie 2000, 2004). In fresh waters, sponges are well known as competitors (Molloy *et al.* 1997), because they colonize hard substrata usually occupied by other sessile invertebrates, including the non-indigenous zebra mussel *D. polymorpha* (Ricciardi *et al.* 1995, Early and Glenoek 1999, Lauer and Spacie 2000, 2004).

In particular, Ricciardi *et al.* (1995) suggested that sponges [*Ephydatia mülleri* (Lieberkühn), *Spongilla lacustris* (Linnaeus), and *Eunapius fragilis* (Leidy)] involved in an overgrowth situation may control mussel abundance locally. Investigation an sponge–zebra mussel interactions in south-western Lake Michigan revealed mussel cellular energy deprivation, death, and digestion most probably due to anoxia, caused by the growing sponge *E. fragilis*, which interrupts the mussel oxygenated water supply (Early and Glonek 1999).

In southern Lake Michigan, Lauer and Spacie (2000) studied the effects of freshwater sponge overgrowth on the fitness of adult zebra mussels using three criteria: biochemical impact, wet weight, and lethal effect. The authors found a significant reduction in glycogen content, soft tissue reduction, and reduced survivability of sponge-covered zebra mussels when compared with the non-sponge-covered population. In addition, Lauer and Spacie (2004), on the basis of experimental tests on zebra mussel growth in response to *E. fragilis* and *E. mülleri* removal and *vice versa*, suggested that sponges could eventually displace zebra mussels from hard vertical substrata, whereas zebra mussels did not overgrow sponge colonies, even though they were able to recolonize hard substrata if they were yielded by sponges.

Dreissena polymorpha was first found in Lake Trasimeno (Umbria, central Italy) in summer 1999 (Spilinga *et al.* 2000); this report showed sporadic abundances along the northern and eastern shores as well as in the main islands of the lake. The coexistence of the zebra mussel and the sponge *Ephydatia fluviatilis* (Linnaeus) was based on a survey carried out from June 2003 to May 2004 (Lancioni and Gaino 2005). Photographic monitoring showed that the sponge has a suppressive influence and can be considered a natural enemy of *D. polymorpha*, as both species are sessile and filter-feeders.

The main objectives of this study are: to determine the distribution of *D. polymorpha* on various substrata along the shore of Lake Trasimeno; to study the reproduction of the mollusc in this ecosystem; and to describe the spatial interaction between *D. polymorpha* and *E. fluviatilis*. Observations were carried out for 1 year from June 2003 to May 2004 at 12 sampling sites located along the shore of the lake, 10 with a hard substratum and 2 with soft sediments (Fig. 1).

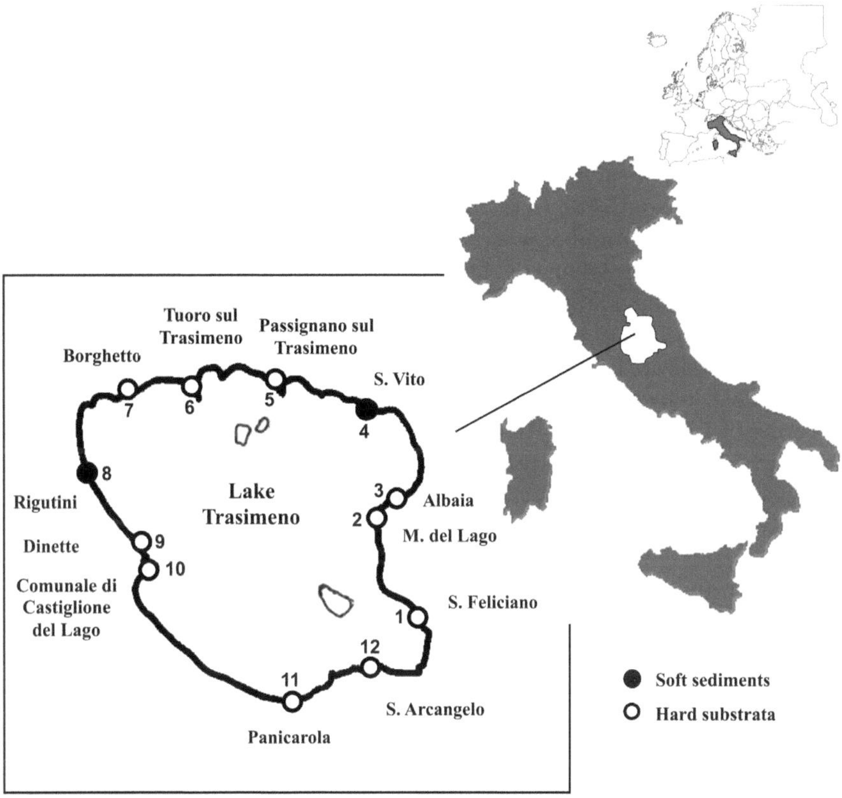

Fig. 1 Map showing sampling sites (1–12) along the shores of Lake Trasimeno (central Italy).

At each sampling site, a 5 m transect along the shore was inspected monthly and the presence/absence of *D. polymorpha* and *E. fluviatilis* was recorded together with the type of substratum. The selectivity of the two investigated species with respect to the type of substratum was evaluated by Pearson's Chi-squared test of independence by comparing the aggregated data of presence on the various substrata (observed frequency) with the overall frequency of substrata encountered during sampling (on the basis of which the expected frequency for a random colonization was calculated). The similarity of substrata selection between *D. polymorpha* and *E. fluviatilis* was also statistically tested by a Chi-squared test.

DREISSENA POLYMORPHA IN LAKE TRASIMENO: DISTRIBUTION AND REPRODUCTION

In Lake Trasimeno, the shells of *D. polymorpha* are highly polymorphic in their colour, pattern, and shape. Shell length varies from less than 1 mm to a maximum of 38 mm. According to Sprung (1992), sexually mature individuals are usually ≥ 8 mm in length. Therefore, we investigated the reproductive activity of this species in all adult individuals, with an average shell length of 15.2 mm. Our data are consistent with previous observations conducted in North America (Miller and Payne 1996, Tucker *et al.* 1996), in the Euphrates basin (Bobat *et al.* 2004), in Ireland (Juhel *et al.* 2003), and in northern Italian lakes (Galassi and Provini 2000).

While in the southern areas of the lake *D. polymorpha* was only sporadic (Panicarola), huge numbers were found: (a) at the western stations (Dinette and Comunale di Castiglione del Lago sampling sites); (b) at the eastern stations (San Feliciano and Monte del Lago); and (c) at Passignano sul Trasimeno, among the northern ones.

Zebra mussels are known to gather and form three-dimensional druses (Burks *et al.* 2002), which consist of a surprisingly high number of individuals (in the order of 10^5–10^6) (Griffiths *et al.* 1991, Kovalak *et al.* 1993, Stanczykowska and Lewandowski 1993, Schloesser *et al.* 1994, Ricciardi *et al.* 1997). The substratum strongly affects colonization by *D. polymorpha* (Lyakhnovich *et al.* 1994, Karatayev and Burlakova 1995, Karatayev *et al.* 1998, Bobat *et al.* 2002). In Lake Trasimeno, *D. polymorpha* was found to be attached to various hard substrata, varying from concrete landing stages (reaching the highest density of 200,000 m^{-2}) to pebbles; from rocks (density varied from 114 m^{-2} to 140,000 m^{-2}) to artificial substrata (tires, tiles, plastic buoys, pumps, and boat keels), and to the submerged portion of reed stalks of *Phragmites australis* (Cav.) Trin. ex Steud. (maximum density of 2,036 m^{-2}). In contrast, soft substrata are unsuited to its colonization: no individuals were observed on silt (San Vito) and only gathered in clumps on sandy substrata (Rigutini beach). Clumps consist of mixed mussel populations, where the individuals adhere to one another or attach to sand-grain sediments and filamentous algae by means of their byssal threads. The same phenomenon has been observed in other lakes (Karatayev 1983, Burlakova 1998).

The results of the Chi-squared statistics performed on frequency data (Table 1) show that the distribution of *D. polymorpha* on different substrata is significantly different ($P = 0.002$) to that expected on the basis of the substratum availability. As expected, *D. polymorpha* positively selects hard substrata, with a preference for concrete, pebbles, and rocks, whereas sandy and silty bottoms are avoided. Also, the stalks of *P. australis* do not seem to represent a preferred substratum for this species.

Because a knowledge of the life-cycle of an invasive species is needed when management and control programmes are to be defined (Denson and Wang 1994, Ram *et al.* 1996a), the reproductive biology of *D. polymorpha* in Lake Trasimeno was investigated. We analysed histological sections of the gonads, in line with previous investigations conducted elsewhere (Walz 1973, 1978, Borcherding 1986, 1991, 1995, Bielefeld 1991, Denson and Wang 1993, Garton and Haag 1993, Wang and Denson 1995, Gist *et al.* 1997, Mantecca *et al.* 2000, Bacchetta *et al.* 2001, Vailati *et al.* 2001, Juhel *et al.* 2003).

Table 1 Overall observed frequencies of *Dreissena polymorpha* and *Ephydatia fluviatilis* on different substrata and frequencies of encountered substrata along the transects investigated in Lake Trasimeno.

	Observed frequency		
Substratum	*Dreissena polymorpha*	*Ephydatia fluviatilis*	Substratum
Concrete	31	14	48
Rocks and pebbles of various composition	53	18	98
Reed stalks of *Phragmites australis*	2	0	12
Tire	1	0	1
Pump filter	6	4	6
Iron gate	0	0	12
Sand	2	0	12
Silt	0	0	12
Boat keel	2	2	2
D. polymorpha shells *	–	(6)	(24)
Total	97	44	227

* Not included for Chi-squared statistics

The population was composed of 44% of sexually mature males and 38% of sexually mature females. Although all the examined mussels were adults, 18% of samplings had inactive gonads, a feature that did not allow us to attribute them a sex. Sexual dimorphism was not seen except that for the gametes. The *sex ratio* of females/males was 1:1.16. No hermaphrodites were observed.

The method of Gist *et al.* (1997) was used to define the phases of gonadal maturation through the shape of the gonadal cells: stage 0, inactive; stage 1, developing; stage 2, pre-spawn; stage 3, post-spawn.

At the beginning of the reproductive activity, gonadal acini were present within the connective tissue and we could not determine the sex. As gonadal maturation proceeded, gonadal acini (tubules and follicles in the male and female line, respectively) proliferated and increased their size. Gamete differentiation occurred synchronously at both sexes and in all sampling sites, thus confirming observations for other populations (Borcherding 1991, Haag and Garton 1992, Denson and Wang 1993, Garton and Haag 1993, Mantecca *et al.* 2000, Bacchetta *et al.* 2001). Although many authors have described more than one reproductive event in a year (Walz 1978, Borcherding 1991, Wang and Denson 1995, Jantz and Neumann 1998, Bacchetta *et al.* 2001), in Lake Trasimeno *D. polymorpha* showed an annual reproductive cycle, as already observed in other areas (Haag and Garton 1992, Gist *et al.* 1997, Juhel *et al.* 2003).

In males, gametogenesis (Fig. 2A) ends with the differentiation of sperm with the flagella oriented towards the centre of the lumen (Fig. 2B). Spawning causes the tubules to empty. In females, the undifferentiated cells transform into primary oogonia, which develop into oocytes by means of meiotic division. The beginning of this process is shown by the presence of bigger and markedly darker cells. Developing oocytes usually remain attached by a stalk to the internal surface of the follicle (Fig. 2C). Mature oocytes (averaging 40–80 μm in diameter) detach from the follicular surface (Fig. 2D) and project into the lumen of the acinus. Mature oocytes have a germinal vesicle and a nucleus with two nucleoli. After gamete release, ovaries show tissue degeneration and resting mature oocytes can often be observed in the lumen (Fig. 2E). Male and female spawning causes the gonad to empty (Fig. 2F).

Gonadal activity depends on both internal and external factors (Ram *et al.* 1996b). Among the latter, temperature exerts a strong influence (Ram *et al.* 1993, Borcherding 1995, Fong *et al.* 1995, Wacker and Von Elert 2003), 12–13 °C being the value that triggers gamete development and spawning (Borcherding 1991, Neumann *et al.* 1993, Domaglia 1997, Bacchetta *et al.* 2001). This value is confirmed in our study, which shows that gonadal maturation occurs in April, when water temperature increases from 4.3 °C to 13.4 °C (Fig. 3). We found a large number of dead zebra mussels in July and August when the water temperature reached 27.8 °C and 28.4 °C, respectively.

An exhaustive review on the life-history of the zebra mussel is reported in Ackerman *et al.* (1994), who outlined the key events from fertilization to adulthood. This representation provides a reasonable model of the life-history of *D. polymorpha*. In Lake Trasimeno, spawning takes place from April to June. In this last month, the monitoring of the hard substrata colonized by *D. polymorpha* shows an astonishing number of very small individuals (less than 1 mm in length) interspersed among larger individuals or adhered to them by the byssal threads. Their small size and abundance are coherent by their belonging to the new generation derived from the larval metamorphosis.

DREISSENA POLYMORPHA/EPHYDATIA FLUVIATILIS INTERACTION FOR SPACE

In order to prevent *D. polymorpha* spreading in freshwater ecosystems, several projects have focused on chemical compounds and physical processes reducing the survival of this species (reviews in: Shtegman 1986, McMahon 1990, Jenner and Janssen-Mommen 1993). However, molluscicides have harmful consequences for the environment. Consequently, many efforts have been carried out on finding natural enemies, such as predators, parasites, and competitors (Molloy *et al.* 1997, Molloy 1998). The latter include organisms able to interfere with *D. polymorpha*'s biological functions, such as adhesion to the substratum and filtration. In this regard, sponges act as valid enemies, since

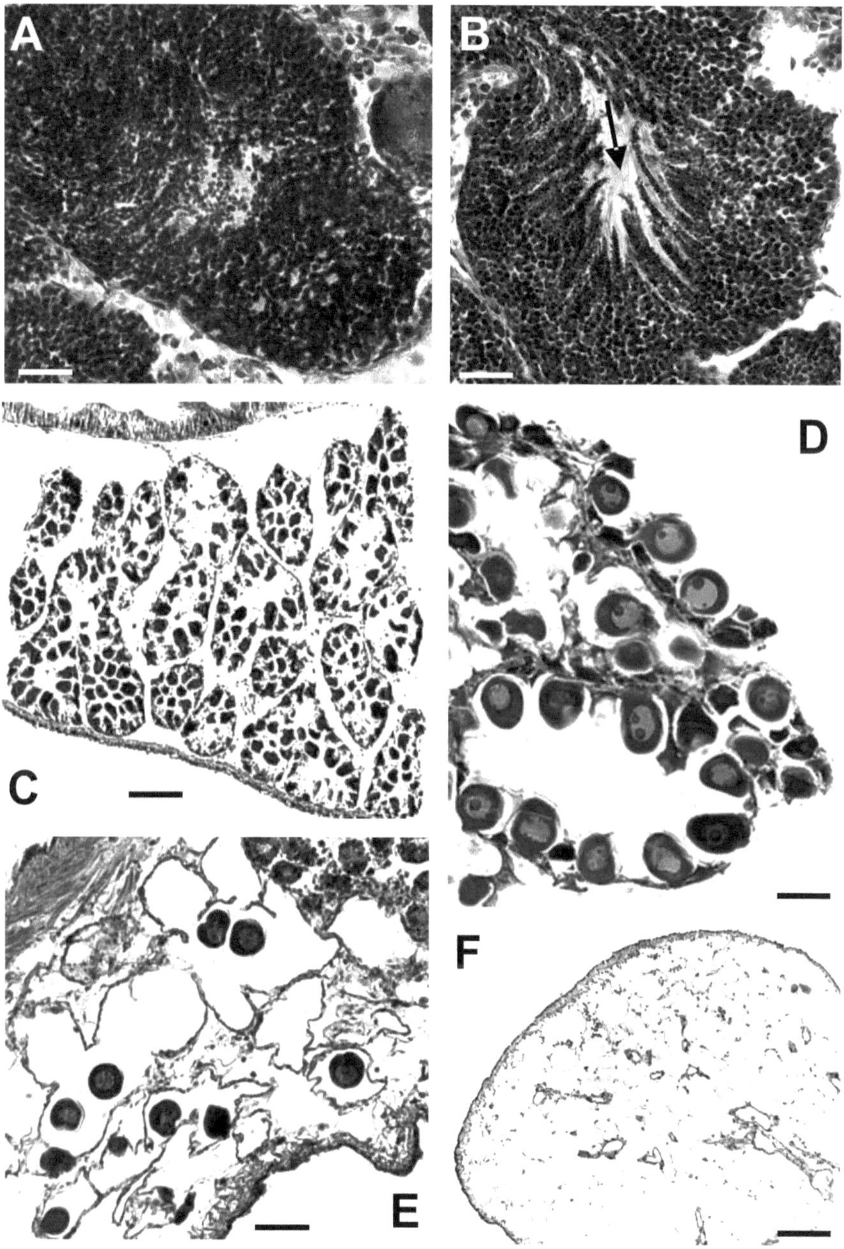

Fig. 2 Mature (male, A–B and female, C–E) and immature (F) gonads. **A**, initial stage of sperm maturation (bar = 40 μm); **B**, spermatozoa showing flagella (arrow) oriented towards the lumen of the tubule (bar = 68 μm); **C**, early stage of egg differentiation (bar = 30 μm); **D**, mature eggs showing their gradual detachment from the ovary (bar = 43 μm); **E**, the post-spawning stage showing resting eggs (bar = 83 μm); **F**, general view of the gonadal inactive stage (bar = 110 μm).

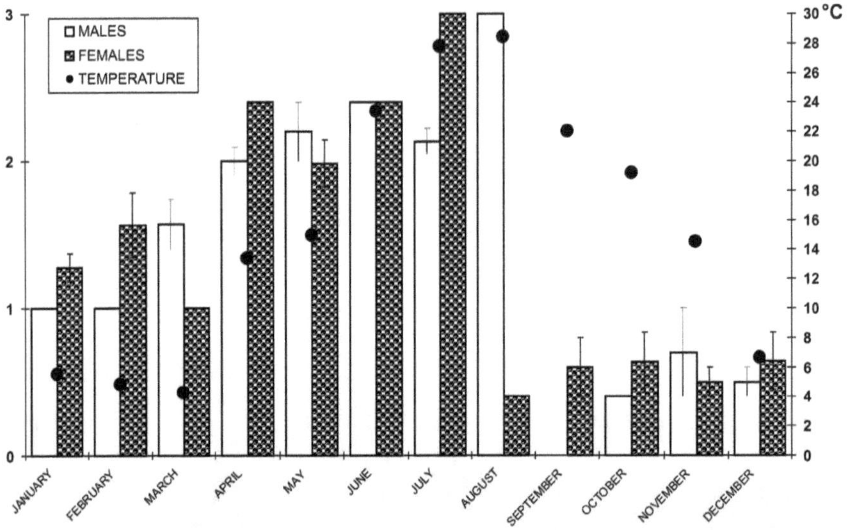

Fig. 3 Gonadal stages (males and females) of *Dreissena polymorpha* over one year of monitoring. Histograms show the mean value (± standard error of the mean) computed for all sampling sites in each month. The values on the main *y*-axis represent the stage of gonadal maturation (0 = inactive; 1 = developing; 2 = pre-spawn; 3 = post-spawn). Black disks denote temperature values (expressed in °C, related to secondary *y*-axis).

they are able to encapsulate and kill zebra mussels (Ricciardi *et al.* 1995, Early and Glonek 1999, Lauer and Spacie 2000, 2004, Ricciardi and Reiswig 2000), as already discovered in the last century (Arndt 1937, Sebestyen 1937, Zhadin 1946).

Investigations on *E. fluviatilis* living in Lake Piediluco (Terni, central Italy), showed that *P. australis* reed belts constitute an elective substratum on which sponges grow (Moretti and Corallini Sorcetti 1980, Gaino *et al.* 2003) and host other organisms (Gaino *et al.* 2004). A lowering of water levels occurred in Lake Trasimeno in the first half of the 20th century and this led to the regression of the reed belt (Di Giovanni 1961, 1968), a process that forced the sponge population to compete with zebra mussels for the colonization of the same substrata.

In a preliminary study on the competition between *D. polymorpha* and *E. fluviatilis* in Lake Trasimeno, it was found that in some areas, where these organisms coexisted, sponges interfered with the molluscs for filter-feeding activity and substrata colonization (Lancioni and Gaino 2005). It has been calculated that a single zebra mussel can filter at least 1 L of water in 24 h (Lei 1993). Zebra mussels select suspended material with a preference for edible algae and micro-organisms (Berg *et al.* 1996, Baker *et al.* 1998, Dionisio Pires and Van Donk 2002, Baker and Levinton 2003, Dionisio Pires *et al.* 2004).

In sponges, according to Frost (1991), a specimen of *S. lacustris* as big as a finger can filter up to 125 L/day. Even though we have no data on the filtration rate in *E. fluviatilis*, it seems likely that a volume of the sponge equal to that of *S. lacustris* can reach similar effective clearance rate values. We experimentally tested that a specimen of *E. fluviatilis* of such a size corresponds to 20 zebra mussels in wet weight.

Another advantage of *E. fluviatilis* over mussels during filtration is that the sponge can also feed by ingesting particulate matter by means of the cells delimiting its external surface (Willenz and Van de Vyver 1982). On this account, this sponge can bypass the dimensional constraint imposed by choanocytes, the specialized cells of the sponge aquiferous system, which capture suspended matter less than 1–2 μm in diameter (Francis and Poirrier 1986).

The results of the Chi-squared statistics performed on frequency data (Table 1) show that *E. fluviatilis* is not evenly distributed on the different substrata ($P = 0.001$) and that, similarly to zebra mussels, the sponge positively selects hard bottoms (concrete landing stages, and pebbles and rocks composed of various materials). On these substrata, the growth and size of sponges vary according to the season, reaching the maximum in warmer months (300 × 240 mm in length; 60 mm in thickness), while during autumn and winter sponges survive as thin encrusting specimens.

The comparison between the mollusc and sponge distribution (Table 1) shows that, although quite different in presence, they have statistically equivalent preferences ($P = 0.66$) with respect to the type of substratum. In addition, *D. polymorpha* was found in 93% cases in the same portions of the transects as *E. fluviatilis* (Panicarola, Passignano sul Trasimeno, Monte del Lago and Comunale di Castiglione del Lago sampling sites). In two of these sampling sites we followed the time progression of the *E. fluviatilis*/*D. polymorpha* interaction and found that the sponge outcompetes the mussels. We observed that, whereas *E. fluviatilis* specimens are clearly separated from the assemblages composed of medium-sized bivalves (about 15 mm in length) (Fig. 4A), the sponge invades the surface occupied by larger zebra mussels (Fig. 4B) and extends over the valves gradually enveloping the shell (Fig. 4C), thus leading to the death of the mollusc (Fig. 4D).

Sponges produce resistant bodies (gemmules) elaborated in different sampling sites and in different months. Their differentiation is triggered by both desiccation and drastic changes in water temperature (warming or cooling).

CONCLUSIONS

The patchy distribution of *E. fluviatilis* in Lake Trasimeno restricts active competition with *D. polymorpha* in those areas markedly colonized by the sponge population. In addition, whereas temperature values and desiccation represent a constraint that hinder *D. polymorpha*, sponges can face such stressful environmental conditions by the production of gemmules, which are able to survive and give rise to new specimens. The flexibility in feeding and reproduction

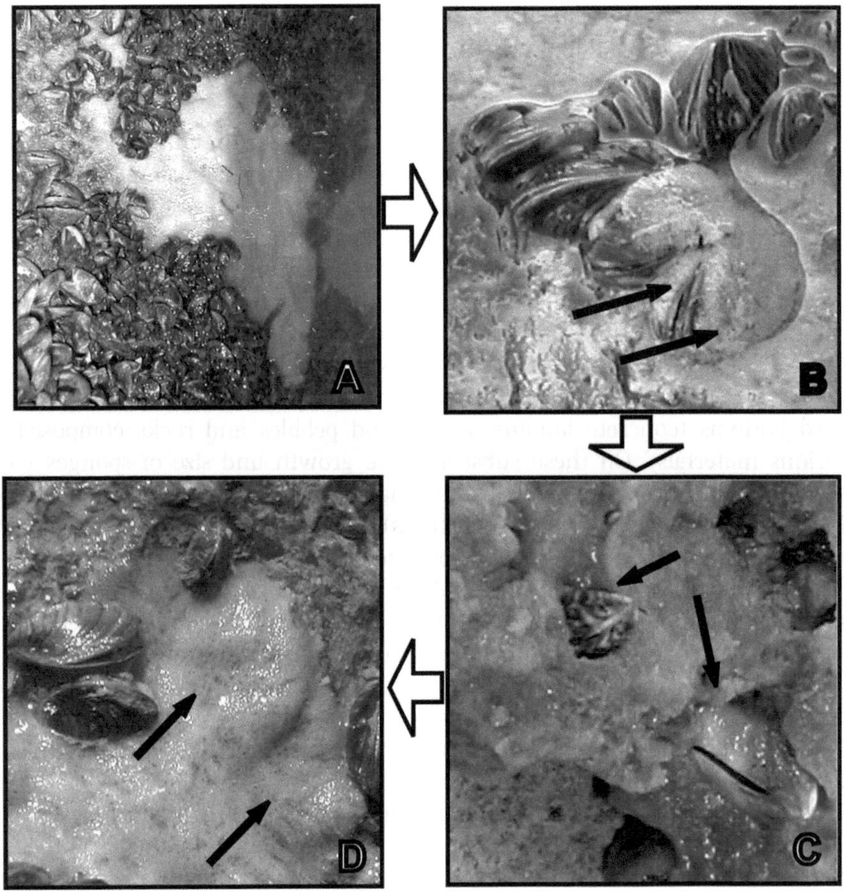

Fig. 4 *Dreissena polymorpha*/*Ephydatia fluviatilis* interaction: large arrows denote the pathway of sponge overgrowth. **A**, spatial separation between the zebra mussel clumps and the sponges; **B**, encrusting sponges form laminar extensions that invade the surface occupied by zebra mussels (arrows); **C**, sponge growth leads to a gradual covering of the shell (arrows); **D**, encapsulation of the shell (arrows), causing the death of *D. polymorpha*.

of the indigenous *E. fluviatilis* can be seen as a winning strategy in the competition with the non-indigenous *D. polymorpha*. Therefore, also in Lake Trasimeno, sponges represent a natural enemy of this mollusc.

ACKNOWLEDGEMENTS

We are very grateful to Dr. Alessandro Ludovisi for his helpful collaboration in the statistical analysis of the data. The study was partially supported by the National Integrated Fund for Research (FISR-MICENA).

REFERENCES

Ackerman, J. D., B. Sim, S. J. Nichols and R. Claudi. 1994. A review of the early history of zebra mussels (*Dreissena polymorpha*): comparisons with marine bivalves. Canadian Journal of Zoology **72**, 1169–1179.

Arndt, W. 1937. *Ochridaspongia rotonda* n.g. n.sp., ein neur Susswassershwamm aus dem Ohridasee. Archiv für Hydrobiolgie **31**, 636–677.

Bacchetta, R., P. Mantecca, and G. Vailati. 2001. Reproductive behavior of the freshwater mussel *Dreissena polymorpha* in Italy: a comparison between two populations. Archiv für Hydrobiolgie **151**, 247–262.

Baker, S. M. and J. S. Levinton. 2003. Selective feeding by three native North American freshwater mussels implies food competition with zebra mussels. Hydrobiologia **505**, 97–105.

Baker, S. M., J. S. Levinton, J. P. Kurdziel, and S. E. Shumway. 1998. Selective feeding and biodeposition by zebra mussels and their relation to changes in phytoplankton composition and seston load. Journal of Shellfish Research **17**, 1207–1213.

Berg, D. J., S. W. Fisher, and P. F. Landrum. 1996. Clearance and processing of algal particles by zebra mussels (*Dreissena polymorpha*). The International Association for Great Lakes Research **22**, 779–788.

Bielefeld, U. 1991. Histological observations of gonads and digestive gland in starving *Dreissena polymorpha* (Bivalvia). Malacologia **33**, 31–42.

Bobat, A., M. O. Hengirmen, and W. Zapletal. 2002. Problems of the Zebra mussel at dams and hydro projects on the Euphrates river. Pages 475–485 in Proceedings of the International Conference on Hydro-2002: Development, Management, Performance. Kirifl, Turkey.

Bobat, A., M. O. Hengirmen, and W. Zapletal. 2004. Zebra mussel and fouling problems in the Euphrates basin. Turkish Journal of Zoology **28**, 161–177.

Borcherding, J. 1986. Gonadenreifung und Fortpflanzugszyklen bei *Dreissena polymorpha* (Pall.). Diplomarbeit an der Universität Köln. Malacologia **36**, 15–27.

Borcherding, J. 1991. The annual reproductive cycle of the freshwater mussel *Dreissena polymorpha* Pallas in lakes. Oecologia **87**, 208–218.

Borcherding, J. 1995. Laboratory experiments on the influence of food availability, temperature and photoperiod on gonad development in the freshwater mussel *Dreissena polymorpha*. Malacologia **36**, 15–27.

Burks, R. L., N. C. Tuchman, C. A. Call, and J. E. Marsden. 2002. Colonial aggregates: effects of spatial position on zebra mussel responses to vertical gradients in interstitial water quality. Journal of North American Benthological Society **21**, 64–75.

Burlakova, L. E. 1998. Ecology of *Dreissena polymorpha* (Pallas) and its role in the structure and function of aquatic ecosystems. Candidate Dissertation, Zoology Institute of the Academy of Science, Republic Belarus, Minsk, Belarus, 168 pp.

Denson, D. R. and S. Y. Wang. 1993. Histological observations on the reproductive cycle of the zebra mussel, *Dreissena polymorpha*. American Zoologist **33**, 49A.

Denson, D. R. and S. Y. Wang. 1994. Reproductive cycle of zebra mussel. U.S. Army Corps of Engineers, Technical Note ZMR-4-07, pp. 1–2.

Di Giovanni, M. V. 1961. Il problema delle idrofite e delle alghe nei confronti della pesca nel Lago Trasimeno. Pages 1–18 in Atti delle VII Giornate Veterinarie sui prodotti della Pesca. Pucci, Ancona, Italy.

Di Giovanni, M. V. 1968. Il controllo idraulico delle piante acquatiche nei confronti della pescosità del Lago Trasimeno. Pages 1–12 in Atti delle VIII Giornate Veterinarie sui prodotti della Pesca. Pucci, Ancona, Italy.

Dionisio Pires, L. M., R. R. Jonker, E. Van Donk, and H. J. Laanbroek. 2004. Selective grazing by adults and larvae of the zebra mussel (*Dreissena polymorpha*): application of flow cytometry to natural seston. Freshwater Biology **49**, 116–126.

Dionisio Pires, L. M. and E. Van Donk. 2002. Comparing grazing by *Dreissena polymorpha* on phytoplankton in the presence of toxic and non-toxic cyanobacteria. Freshwater Biology **47**, 1885–1865.

Domaglia, J. 1997. Annual reproductive cycle of the females of *Dreissena polymorpha* (Pall.) from Lake Inulec. Polskie Archiwum Hydrobiologii **44**, 467–475.

Dudley, T. L., C. M. D'Antonio, and S. D. Cooper. 1990. Mechanisms and consequences of interspecific competition between two stream insects. Journal of Animal Ecology **59**, 849–866.

Early, T. A. and T. Glonek. 1999. Zebra mussel destruction by a Lake Michigan sponge: populations, *in vivo* super ^{31}P nuclear magnetic resonance, and phospholipid profiling. Environmental Science & Technology **33**, 1957–1962.

Fong, P. P., K. Kyozuka, J. Duncan, S. Rynkowski, D. Mekasha, and J. L. Ram. 1995. The effect of salinity and temperature on spawning and fertilization in the zebra mussel *Dreissena polymorpha* (Pallas) from North America. Biological Bulletin **189**, 320–329.

Francis, J. C. and M A. Poirrier. 1986. Patrticle uptake in two fresh-water sponge species, *Ephydatia fluviatilis* and *Spongilla alba* (Spongillidae). Transactions of the American Microscopical Society **105**, 11–20.

Frost, T. M. 1991. Porifera. Pages 95–124 in J. H. Thorp and A. P. Covich, editors. Ecology and classification of North American freshwater invertebrates. Academic Press, St Louis, MO.

Gaino, E., T. Lancioni, G. La Porta, and B. Todini. 2004. The consortium of the sponge *Ephydatia fluviatilis* (L.) living on the common reed *Phragmites australis* in Lake Piediluco (central Italy). Hydrobiologia **520**, 165–178.

Gaino, E., M. Rebora, C. Corallini, and T. Lancioni. 2003. The life-cycle of the sponge *Ephydatia fluviatilis* (L.) living on the reed *Phragmites australis* in an artificially regulated lake. Hydrobiologia **495**, 127–142.

Galassi, S. and A. Provini. 2000. Studio della *Dreissena polymorpha* nel lago di Como. Pages 112–121 in R. De Bernardi, M. Gatto, R. Mosello, and L. Segre, editors. Atti della "Giornata di studio" I laghi come risorsa per lo sviluppo. Milano, Italy.

Garton, D. W. and W. R. Haag. 1993. Seasonal reproductive cycles and settlement patterns of *Dreissena polymorpha* in Western Lake Erie. Pages 111–128 in T. F. Nalepa and D. Schloesser, editors. Zebra mussels: Biology, impacts, and control. Lewis Publishers, Ann Arbor, MI.

Gist, D. H., M. C. Miller, and W. A. Brence. 1997. Annual reproductive cycle of the zebra mussel in the Ohio River: a comparison with Lake Erie. Archiv für Hydrobiologie **138**, 335–379.

Griffiths, R. G., D. W. Schloesser, J. H. Leach, and W. P. Kovalak. 1991. Distribution and dispersal of the zebra mussel (*Dreissena polymorpha*) in the Great Lakes region. Canadian Journal of Fisheries and Aquatic Sciences **48**, 1381–1388.

Haag, W. R. and D. W. Garton. 1992. Synchronous spawning in a recently established population of the zebra mussel, *Dreissena polymorpha*, in western Lake Erie, U.S.A. Hydrobiologia **234**, 103–110.

Hirata, T. 1987. Succession of sessile organisms on experimental plates immersed in Nabeta Bay, Izu Peninsula, Japan. II. Succession of invertebrates. Marine Ecology Progress Series **38**, 25–35.

Jackson, J. B. C. 1977. Competition on marine hard substrata: the adaptive significance of solitary and colonial strategies. American Naturalist **111**, 743–767.

Jantz, B. and D. Neumann. 1998. Growth and reproductive cycle of the zebra mussel in the River Rhine as studied in a river bypass. Oecologia **114**, 213–225.

Jenner, H. A. and J. P. M. Janssen-Mommen. 1993. Monitoring and control of *Dreissena polymorpha* and other macrofouling bivalves in the Netherlands. Pages 537–554 in T. F. Nalepa and D. Schloesser, editors. Zebra mussels: Biology, impacts and controls. Lewis Publishers, Ann Arbor, MI.

Juhel, G., S. C. Culloty, R. M. O'Riordan, J. O'Connor, L. De Faoite, and R. McNamara. 2003. A histological study of the gametogenic cycle of the freshwater mussel *Dreissena polymorpha* (Pallas, 1771) in Lough Derg, Ireland. Journal of Molluscan Studies **69**, 365–374.

Karatayev, A. Yu. 1983. Ecology of *Dreissena polymorpha* Pallas and its effects on macrozoobenthos of the thermal power plant's cooling reservoir. Candidate Dissertation, Zoology Institute of Academy of Science Belarussian SSR, Minsk, Belarus, 153 pp.

Karatayev, A. Yu and L. E. Burlakova. 1995. The role of *Dreissena* in lake ecosystems. Russian Journal of Ecology **26**, 207–211.

Karatayev, A. Yu, L. E. Burlakova, and D. K. Padilla. 1998. Physical factors that limit the distribution and abundance *of Dreissena polymorpha* (Pall.). Journal of Shellfish Research **17**, 1219–1235.

Kovalak, W. P., G. D. Longton, and R. D. Smithee. 1993. Infestation of power plant water systems by the zebra mussel (*Dreissena polymorpha*). Pages 359–380 in T. F. Nalepa and D. Schloesser, editors. Zebra mussels: Biology, impacts and control. Lewis Publishers, Ann Arbor, MI.

Lancaster, J., A. G. Hilldrew, and C. R. Townsend. 1988. Competition for space by predators in streams: field experiments on a net-spinning caddisfly. Freshwater Biology **20**, 185–193.

Lancioni, T. and E. Gaino. 2005. Competition between the freshwater sponge *Ephydatia fluviatilis* (L.) and the zebra mussel *Dreissena polymorpha* (Pallas) in Lake Trasimeno (central Italy). Italian Journal of Zoology **72**, 27–32.

Lauer, T. E. and A. Spacie. 2000. The effects of sponge (Porifera) biofouling on zebra mussel (*Dreissena polymorpha*) fitness: reduction of glycogen, tissue loss, and mortality. Journal of Freshwater Ecology **15**, 83–92.

Lauer, T. E. and A. Spacie. 2004. Space as a limiting resource in freshwater systems: competition between zebra mussels (*Dreissena polymorpha*) and freshwater sponges (Porifera). Hydrobiologia **517**, 137–145.

Lei, J. 1993. Estimation of filtration rate of zebra mussels. U.S. Army Corps of Engineers Technical Note, ZMR-4-06, pp. 1–3.

Lyakhnovich, V. P., A. Yu Karatayev, N. I. Andreev, S. I. Andreeva, S. A. Afanasiev, A. K. Dyga, V. P. Zakutskiy, V. I. Zolotareva, A. A. Lvova, M. Y. Nekrasova, V. F. Osadchikh, Y. V. Pligin, A. A. Protasov, and G. M. Tishchikov. 1994. Living conditions. Pages 109–119 in J. I. Starobogatov, editor. Freshwater zebra mussel *Dreissena polymorpha* (Pall.) (Bivalvia. Dreissenidae): systematics, ecology, practical meaning. Nauka Press, Moscow, Russia.

Mantecca, P., G. Vailati, and R. Bacchetta. 2000. Studio del ciclo riproduttivo di *Dreissena polymorpha* (Mollusca: Bivalvia) nel Lago di Como e nel Fiume Adda mediante l'analisi istologica delle gonadi. Rivista di Idrobiologia **39**, 185–200.

McMahon, R. F. 1990. The zebra mussel: U. S. utility implications, EPRI GS-6995, Electric Power Research Institute, Palo Alto, CA, 70 pp.

Miller, A. C. and B. S. Payne. 1996. Density and size demography of newly established populations of *Dreissena polymorpha* in the U. S. inland waterway system. Pages 99–116 in F. D'itri, editor. Zebra mussels and other aquatic nuisance species. Ann Arbor Press, Ann Arbor, MI.

Molloy, D. P. 1998. The potential for using biological control technologies in the managment of *Dreissena* Spp. Journal of Shellfish Research **17**, 177–183.

Molloy, D. P., A. Yu Karatayev, L. E. Burlakova, D. P. Kurandina, and F. Laurelle. 1997. Natural enemies of zebra mussels: predators, parasites, and ecological competitors. Reviews in Fisheries Science **5**, 27–97.

Moretti, G. P. and C. Corallini Sorcetti. 1980. Al lago Trasimeno è arrivata *Ceraclea fulva* Ramb. (Trichoptera - Leptoceridae). Rivista di Idrobiologia **19**, 1–10.

Neumann, D., J. Borcherding, and B. Jantz. 1993. Growth and seasonal reproduction of *Dreissena polymorpha* in the Rhine river and adjacent waters. Pages 95–109 in D. Neumann and H. A. Jenner, editors. The zebra mussel *Dreissena polymorpha*. Gustav Fischer Verlag, Jena, Germany.

Pansini, M. and R. Pronzato. 1990. Observations on the dynamics of a Mediterranean sponge community. Pages 403–415 in K. Rützler, editor. New perspectives in sponge biology. Smithsonian Institution Press, Washington, DC.

Ram, J. L., G. W. Crawford, J. U. Walker, J. J. Mojares, N. Patel, P. P. Fong, and K. Kyozuka. 1993. Spawning in the zebra mussel (*Dreissena polymorpha*): activation by internal or external application of serotonin. Journal of Experimental Zoology **265**, 587–598.

Ram, J. L., P. P. Fong, and D. W. Garton. 1996a. Physiological aspects of zebra mussel reproduction: maturation, spawning, and fertilization. American Zoologist **36**, 326–338.

Ram, J. L., P. P. Fong, and D. W. Garton. 1996b. Serotonergic mechanisms mediating spawning and oocyte maturation in the zebra mussel, *Dreissena polymorpha*. Invertebrate Reproduction & Development **30**, 29–37.

Ricciardi, A. and H. M. Reiswig. 2000. Comment on "Zebra mussel destruction by a Lake Michigan sponge: populations, in vivo ^{31}P nuclear magnetic resonance and phospholipid profiling". Environmental Science & Technology **34**, 1379–1380.

Ricciardi A, F. L. Snyder, D. O. Kelch, and H. M. Reiswig. 1995. Lethal and sublethal effects of sponge overgrowth on introduced dreissenid mussels in the Great Lakes – St. Lawrence River system. Canadian Journal of Fisheries and Aquatic Sciences **52**, 2695–2703.

Ricciardi, A., F. G. Whoriskey, and J. B. Rasmussen. 1997. The role of the zebra mussel (*Dreissena polymorpha*) in structuring macroinvertebrate communities on hard substrata. Canadian Journal of Fisheries and Aquatic Sciences **54**, 2596–2608.

Russ, G. R. 1982. Overgrowth in a marine epifaunal community: competitive hierarchies and competitive networks. Oecologia **53**, 12–19.

Schloesser, D. W., A. Bij De Vaate, and A. Zimmerman. 1994. A bibliography of *Dreissena polymorpha* in European and Russian waters: 1964–1993. Journal of Shellfish Research **13**, 243–267.

Sebestyen, O. 1937. Colonization of two new fauna-elements of Pontus-origin (*Dreissena polymorpha* Pall. and *Corophium curvispinum* G. O. Sars forma *devium* Wundsch) in Lake Balaton. Verhandlungen International Vereinigung Limnologie **8**, 169–182.

Shtegman, B. K., editor. 1986. Biology and control of *Dreissena*. Israel Program for Scientific Translations, IPST Cat. No. 1774, Jerusalem, Israel, 145 pp.

Spilinga, C., U. Chiappafreddo, and Q. Pirisinu. 2000. *Dreissena polymorpha* (Pallas) al Lago Trasimeno. Rivista di Idrobiologia **39**, 145–152.

Sprung, M. 1992. Observations on shell growth and mortality of *Dreissena polymorpha* in lakes. Pages 19–28 in D. Neumann and H. A. Jenner, editors. The zebra mussel *Dreissena polymorpha*. Limnologie Aktuell vol 4, E. Schweizerbart'sche Verlagsbuchhandlung. Science Publishers, Stuttgart, Germany.

Stanczykowska, A. and K. Lewandowski. 1993. Thirty years of studies *of Dreissena polymorpha* ecology in Mazurian Lakes of Northeastern Poland. Pages 3–33 in T. F. Nalepa and D. W. Schloesser, editors. Zebra mussels: biology, impacts and control. Lewis Publishers, Ann Arbor, MI.

Tucker, J. K., F. A. Cronin, D. W. Soergel, and C. H. Theiling. 1996. Predation on zebra mussel (*Dreissena polymorpha*) by common carp (*Cyprinus carpio*). Journal of Freshwater Ecology **11**, 363–372.

Vailati, G., R. Bacchetta, and P. Mantecca. 2001. Marcatori istologici della contaminazione da DDT in *Dreissena polymorpha*: il caso del Lago Maggiore. Rivista di Idrobiologia **40**, 289–303.

Wacker, A. and E. Von Elert. 2003. Food quality controls reproduction of zebra mussel (*Dreissena polymorpha*). Oecologia **135**, 332–338.

Walz, N. 1973. Untersuchungen zur Biologie von *Dreissena polymorpha* Pallas in Bodensee. Archiv für Hydrobiologie Supplement **55**, 106–119.

Walz, N. 1978. The energy balance of the freshwater mussel *Dreissena polymorpha* in laboratory experiments and in Lake Constance. II. Reproduction. Archiv für Hydrobiologie Supplement **55**, 106–119.

Wang, Y. and D. R. Denson. 1995. A histological study of the reproductive pattern of zebra mussels. U. S. Army Corps of Engineers, Technical Report EL-95–20, 29 pp.

Willenz, Ph. and G. Van de Vyver. 1982. Endocytosis of latex beads by the exopinacoderm in the freshwater sponge *Ephydatia fluviatilis*: An *in vitro* and *in situ* study in SEM and TEM. Journal of Ultrastructural Research **79**, 294–306.

Zhadin, V. I. 1946. The travelling shellfish *Dreissena*. Priroda **5**, 29–37.

Chapter thirty-three

Seasonal effects on the antioxidant response and metal accumulation of Dreissena polymorpha

Antonia C. Elia, Ambrosius J. M. Dörr, Marino Prearo, and Maria C. Abete

INTRODUCTION

Mussels are employed as sentinel organisms in biomonitoring programmes for the assessment of environmental quality. They are sessile, euryhaline, filtering organisms which can bioaccumulate by direct exposure or through the food chain when chronically exposed to different pollutants, even if these are present in low concentrations (Sheehan *et al.* 1995). The evaluation of contaminants in aquatic organisms can estimate and quantify the bioavailable fraction which might have the potential to induce an effect. However, the determination of body concentrations alone does not provide useful indications about the effects. Therefore, biomonitoring programmes require the quantification of some biological responses in order to assess the health state of contaminant-exposed organisms.

Several classes of contaminants, including heavy metals, can cause oxidative stress by enhancing the reactive oxygen species (ROS) formation in living organisms (Winston and Di Giulio 1991). These compounds, which can also be produced by mollusc haemocytes during an inflammatory process in

Francesca Gherardi, Biological invaders in inland waters: Profiles, distribution, and threats, 613–623.

response to foreign cells (Tiscar and Mosca 2004), are potentially very toxic as they are a source of hydroxyl radicals and need to be inactivated in order to limit the cellular damage.

The evaluation of oxidative stress is generally performed by following the changes of enzymatic activities (superoxide dismutase, catalase, glutathione peroxidases, and glutathione reductase) and of the rates of the molecules (reduced glutathione) involved in antioxidant defence. These variations have already been employed as biomarkers of contaminant-mediated oxidative stress in aquatic organisms (Winston and Di Giulio 1991). However, the antioxidant defence mechanism of aquatic organisms is subjected to seasonal control. Thus, it may fluctuate throughout the year, depending on the availability of nutrients, reproductive cycle, temperature, and other factors. The metabolic status of aquatic invertebrates can influence its defence capacity against changing environmental parameters and/or xenobiotics. Oxidative stress is thus a seasonal event in bivalve molluscs (Sheehan and Power 1999).

The non-indigenous zebra mussel (*Dreissena polymorpha* Pallas) was recorded for the first time in Lake Trasimeno (Italy) in 2000 (Spilinga *et al.* 2000). It is an invasive freshwater mussel native to Eurasia and is widely distributed in aquatic environments. It is a sedentary species with a relatively long lifespan; its high filtration rate facilitates the bioaccumulation of contaminants (Kraak *et al.* 1991, Hendriks *et al.* 1998). Recent studies have shown that zebra mussels can tolerate temperatures of up to 30 °C, indicating that many warm-water systems may not be excluded from its invasion (Iwanyzki and McCauley 1992).

This freshwater species fulfils the requirements of a good biomonitor for persistent organic pollutants (POPs) (Binelli *et al.* 2001a, b, 2004, 2005) and for trace metal availability and accumulation (Gundacker 1999, de Lafontaine *et al.* 2000, Camusso *et al.* 2001). Several biochemical parameters have been evaluated in zebra mussels as biomarkers of inorganic contamination. For example, de Lafontaine *et al.* (2000) indicated a significant variation in the levels of MT, EROD, DNA strand breaks, LPO, and VG, most probably as a result of variable levels of metal contamination. Giambérini and Cajaraville (2005) suggested that the digestive lysosomal response of zebra mussel, experimentally exposed to cadmium, could have the potential to be employed as a biomarker in freshwater biomonitoring.

Until now, few studies have been performed on the glutathione system and its associated enzymes in *D. polymorpha* (Pflugmacher *et al.* 1998). Therefore, in the present research, total glutathione, glutathione S-transferase, glutathione peroxidase, glutathione reductase, glyoxalase I and glyoxalase II, and catalase were evaluated on specimens collected seasonally from Lake Trasimeno (Italy). The purpose of this study was to investigate the influence of the seasons on the detoxificant response of *D. polymorpha* and on heavy metal accumulation (lead, chromium, and cadmium) in the same specimens. These surveillances might be useful to establish the correct sampling time for this freshwater species, when field studies are performed.

HEAVY METALS

Sample preparation and metal determination

Specimens of *D. polymorpha* of both sexes of about 1.5–2 cm shell length were collected from the southern part of Lake Trasimeno, between southern Arcangelo and Isola Polvese, in autumn 2000 (late September and early November), winter (January and early March), spring (late May and middle of June), and summer (July) 2001. The mean lake water temperature was 17.9 ± 1.4 °C (autumn), 7.5 ± 2.1 °C (winter), 22.8 ± 1.5 °C (spring), and 24.8 ± 0.4 °C (summer). The samples were placed in thermally insulated boxes, previously filled with lake water and transferred immediately live to the laboratory. They were deprived of the foot and the byssus gland, rinsed with ice-cold saline, blotted dry, divided in pools of 50 specimens each, and immediately stored at −80 °C, until used.

Each pooled sample of *D. polymorpha* was analysed for the determination of cadmium, chromium, and lead. The sample tissue (2 g) was mineralized in a microwave oven with a nitric acid and hydrogen peroxide mixture. The sample was filtered through filter paper into a 10 mL volumetric flask followed by rinsing of both the flask and the filter paper with ultrapure water and made to volume with ultrapure water. The determination of cadmium, chromium, and lead was carried out with an atomic absorption spectrophotometer with electrothermal atomization (GFAAS) Perkin–Elmer Analyst 600 (Norwalk, CT, USA). The Zeeman effect was used to counteract aspecific absorption. Single element hollow–cathode lamps were used for lead ($\lambda = 283.3$ nm), cadmium ($\lambda = 228.8$ nm), and chromium ($\lambda = 357.9$ nm). The calibration curves were made using standard solutions (Standard Merck – Darmstadt Germany) of each metal. The quantification limit was 0.01 ppm (μg/g) for cadmium, 0.05 ppm for chromium, and 0.04 ppm for lead (Baldini *et al.* 1996).

Kruskal Wallis ANOVA was applied to the data and pairwise comparisons were tested using a Mann–Whitney *U* test to discriminate differences for each sampling season. The chemical analysis was conducted on 2 (autumn), 6 (winter), and 3 (spring and summer) pools per season. Significance was tested at the 5% level ($P < 0.05$).

Heavy metal accumulation in *D. polymorpha*

It is already known that total metal concentration in the water column is a poor predictor for the metal that is really accumulated by organisms and which might induce toxic effects. Until now, the study of metal accumulation in aquatic species from Lake Trasimeno has been conducted in only one invertebrate species, *Procambarus clarkii* (Girard) (Elia *et al.* 2006). No data are available for heavy metal accumulation in zebra mussels from Lake Trasimeno. It was reported by Gundacker (1999) that *D. polymorpha* collected from urban

Table 1 Cadmiun, chromium, and lead content in *Dreissena polymorpha* sampled seasonally from Lake Trasimeno.

Metals	Autumn	Winter	Spring	Summer
Cd (μg/g)	0.21 ± 0.01^{ab}	0.17 ± 0.09^{b}	0.71 ± 0.08^{a}	0.45 ± 0.23^{a}
Cr (μg/g)	0.11 ± 0.01	0.19 ± 0.06	0.32 ± 0.27	0.15 ± 0.01
Pb (μg/g)	0.06 ± 0.01	0.11 ± 0.02	0.10 ± 0.04	0.10 ± 0.02

Mean values $\pm$ SD. Statistical comparison: different letters (a, b) denote significant differences between sampling periods (in the same row).

waters of Vienna (Austria) can be considered a poor accumulative indicator of near-background heavy metal contamination. The author stated also that those specimens showed a tissue-specific metal accumulation: cadmium was at the highest concentration in soft body parts, while lead, copper, and zinc contents were at the highest in the byssal threads. The heavy metal concentrations of *D. polymorpha* sampled seasonally from Lake Trasimeno were evaluated only in the soft tissues (Table 1).

The results showed that metal levels in these specimens were generally low and almost constant for Pb and Cr, as indicated by its lack of statistically significant differences, even if higher Cr content was achieved in samples collected in late spring. Additionally, specimens collected during the warmer seasons, mainly in spring, exhibited the highest Cd level. The results of chemical analyses of water, performed in October 2000, February and June 2001, indicated a very low Cd, Cr, and Pb level which was generally below the detection limits (Monarca *et al.* 2004).

The changing of seasons can affect the tissue metal concentration in *D. polymorpha*. In fact, the accumulation of As, Cd, Ni, Se, Cr, and Hg in specimens of zebra mussel sampled from the St. Lawrence River was at a minimum in June, when gonad maturation was complete and spawning was in progress, and increased in the post-spawning period, from July through August, when the fresh weight and the body lipid content of the organism decreased (Knaw *et al.* 2003). Dilution effects caused by rapid development of gonadal tissues in mussels can result in an apparent reduction in element concentrations.

Data about the reproductive stage of *D. polymorpha* collected from Italian biotopes are reported by some authors, which indicated for this species a gonadal developmental phase in winter, followed by maturation and deposition phases during spring and summer. In autumn, zebra mussels have inactive gonads (Binelli *et al.* 2004). This species from Lake Trasimeno, sampled during the years 2000–2001, had almost the same life cycle. In fact, when the sampling was carried out in July, we observed the presence of juvenile specimens of about 1 mm shell length, suggesting that spawning occurred most probably in late spring (unpublished data). From our present results it might be assumed that the higher Cd level in specimens sampled in spring, when water temperature was about 22 °C, could be explained by the increased concentration of this metal during the postspawning period of the species.

ANTIOXIDANT RESPONSES

Sample preparation and biochemical assays

For sampling see section on heavy metals. The pools of zebra mussel tissue (1 g) were homogenized and the cytosolic fractions obtained were employed for the study of levels of total glutathione content (GSH + 2GSSG) and enzymatic activities using spectrophotometric methods (Elia *et al.* 2001). The total thiol content (GSH + 2GSSG) was determined by the GR recycling assay at 412 nm on the deproteinized supernatant and it is reported as μmol per g of tissue (wet weight).

Catalase activity (CAT; EC 1.11.1.6) was measured following the decrease in absorbance at 240 nm due to H_2O_2 consumption ($\varepsilon = -0.04$ $mM^{-1}cm^{-1}$). Glutathione peroxidase activity (GPx Se-dependent enzyme, EC 1.11.1.9) toward H_2O_2 as the substrate was determined and the oxidation of NADPH was followed at 340 nm ($\varepsilon = -6.22$ $mM^{-1}cm^{-1}$). Glutathione reductase activity (GR; EC. 1.8.1.7) was assayed following the decrease in absorbance at 340 nm ($\varepsilon = -6.22$ $mM^{-1}cm^{-1}$) due to oxidation of NADPH. Glutathione S-transferase activity (GST; EC 2.5.1.18), with the substrate 1-chloro-2, 4-dinitrobenzene (CDNB), was measured. at 340 nm ($\varepsilon = 9.6$ mM^{-1} cm^{-1}). Glyoxalase I activity (GI; EC 4.4.1.5) was determined at 240 nm ($\varepsilon = 3.37$ $mM^{-1}cm^{-1}$) using 1.0 mM GSH/methylglyoxal hemithioacetal as the substrate. Glyoxalase II (GII; EC 3.1.2.6) was determined at 412 nm ($\varepsilon = 13.6$ $mM^{-1}cm^{-1}$) by monitoring GSH formation in the presence of 5,5′-dithio-bis-2-nitrobenzoic acid (DTNB) and S-D-lactoylglutathione (LSG). Protein concentration of supernatant fractions was determined according to Lowry *et al.* (1951) employing bovine serum albumin as a protein standard.

Kruskal Wallis ANOVA and a Mann–Whitney *U* test were conducted on about 8–10 pools per season and run in triplicate to discriminate differences for each sampling season.

Seasonal changes

Dreissena polymorpha collected from Lake Trasimeno showed significant differences in biochemical parameters according to the season (Table 2). All specimens exhibited higher levels of biochemical parameters in autumn and winter than in spring and summer, except for Se-GPx and GST. Biochemical and chemical parameters in the zebra mussel tissue were examined using nonparametric Spearman's rank correlations. Statistically significant correlations were

Table 2 Antioxidant enzymes (CAT, Se-GPx, GR, GST, GI, and GII) and total glutathione (GSH+2GSSG) in specimens of *Dreissena polymorpha* sampled seasonally from Lake Trasimeno.

	Autumn	Winter	Spring	Summer
[1]CAT	54.20 ± 19.40^a	67.01 ± 18.66^a	36.36 ± 6.19^b	29.60 ± 4.47^c
[2]Se-GPx	33.29 ± 11.88^b	25.78 ± 6.16^b	78.21 ± 5.53^a	33.55 ± 13.16^b
[2]GR	8.74 ± 3.75	5.16 ± 0.89	4.27 ± 1.63	4.89 ± 1.82
[2]GST	189.97 ± 33.61^a	121.62 ± 56.51^a	68.43 ± 19.21^b	154.38 ± 45.30^a
[2]GI	219.64 ± 58.95^a	397.66 ± 103.98^a	98.86 ± 18.19^b	80.30 ± 23.79^c
[2]GII	44.48 ± 10.13^a	49.86 ± 12.62^a	27.61 ± 5.32^b	22.26 ± 6.05^b
[3]GSH+2GSSG	98.73 ± 10.96^a	65.34 ± 7.28^a	37.36 ± 13.32^b	31.76 ± 7.04^b

Mean values ± SD. Enzymatic activities are reported in [1]μmol/min/mg prot and [2]nmol/min/mg prot; [3]glutathione content is reported in nmol/g of wet weight. Statistical comparison: different letters (a, b, c) indicate significant differences between sampling periods (in the same row).

Table 3 Spearman correlation for biological data and metal content in *Dreissena polymorpha* from Lake Trasimeno.

	Cd	Cr	Pb
CAT	−0.75*	−0.01	0.10
Se-GPx	0.73*	0.63*	0.09
GR	−0.34	−0.51	−0.30
GST	−0.05	−0.54	0.33
GI	−0.71*	0.03	0.19
GII	−0.73*	−0.11	−0.04
CAT	−0.75*	−0.01	0.10

Significant correlations are reported with the asterisk (*) at $P < 0.05$.

found for some of these biochemical parameters that were positive for Se-GPx with Cd and Cr and negative for GI, GII and CAT with Cd (Table 3).

Catalase is an inducible enzyme which allows the breakdown of hydrogen peroxide to water and oxygen. In the current research, the enzymatic activity of CAT in zebra mussels was at the lowest in summer and it was slightly different from spring and markedly dissimilar from the other seasons (about 50%). The enzyme was not positively correlated with heavy metals. Probably, the significant reduction in CAT activity is correlated with the biological cycle of this species. Thus, the seasonal variations of CAT activity recorded in *D. polymorpha* from Lake Trasimeno indicated a greater prooxidant challenge during the warmer seasons. Other authors observed the lowest enzymatic response of

CAT, SOD, and GPx in the digestive gland of the bivalve *Saccostrea cucullata* (Born) exposed to polyaromatic hydrocarbons (PAH) in winter. According to those authors, the decrease of enzymatic activity could be explained by a shortage of food in the winter and by the bivalve reproductive cycle which starts in the autumn (Niyogi *et al.* 2001). Glutathione peroxidase catalyses the reduction of hydrogen peroxides. The enzymatic activity is coupled with the oxidation of GSH and generates alcohols. A marked induction of Se-GPx enzymatic activity was observed in *D. polymorpha* collected in spring from Lake Trasimeno which was about 2.4 times higher than that of specimens from other seasons. Generally, an increase in antioxidant enzyme activities could be related to reproductive events in invertebrate organisms or to the higher metabolic activity of the organisms during the warmer season (Vidal *et al.* 2002).

The biochemical parameters herein investigated are the first data evaluated in this freshwater species and therefore we cannot assess if the highest Se-GPx activity results from stress conditions owing to Cr and mainly Cd exposure or to the biological cycle, so further studies are required to investigate this aspect. Glutathione reductase is involved in the regeneration of reduced GSH from the oxidized glutathione (GSSG) and the balance between GSSG and GSH is necessary for the maintenance of cellular homeostasis. Even if the seasonal progress of this enzyme in zebra mussels from Lake Trasimeno did not show significant variations, the lower GR activity observed in samples collected during the warmer seasons may result in a depletion of total glutathione content in this freshwater species. In fact, the modification of the rate of glutathione as well as the balance between the rates of reduced and oxidized glutathione (GSH/GSSG) can be correlated with the variation in GR activity. The reduction in the rate of GSH has been related to the enhancement of lipid peroxidation in aquatic organisms which is generally correlated with the decrease of antioxidant enzymatic activities (Doyotte *et al.* 1997, Cossu *et al.* 2000).

Glutathione S-transferase is involved in conjugation processes of the reactive electrophilic centres of different substrates with the thiol group of GSH. In the current research, GST activity in specimens of *D. polymorpha* was at the lowest in spring and then constant through the other seasons. The enzyme was not positively correlated with metals and thus this lowest activity could most likely be due to the different regulation of GST expression during the biological cycle. Another, potentially invasive, freshwater species such as *Corbicula fluminea* (Muller) displayed no clear seasonal trend for GST activity (Vidal *et al.* 2002), while *Perna perna* (Linnaeus) showed seasonal variability of GST, SOD, and CAT in accordance with temperature changes and reproductive cycle (Wilhelm Filho *et al.* 2001). In other freshwater invertebrates, such as bryozoans, a reduction of antioxidant defence was reported during the spring coinciding with the end of their biological cycle (Elia *et al.* 2001, 2007). These results suggest that the defence capacity is related to the metabolic status of the organisms and thus might reflect a susceptibility to environmental parameters and/or contaminants.

Glyoxalase I catalyses the formation of the thioester S-D-lactoylglutathione from methylglyoxal and GSH, while glyoxalase II catalyses the hydrolysis of the thioester to regenerate GSH and liberate D-lactate. Glyoxalases enzymes were also subjected to seasonal variation in other freshwater invertebrates (Elia *et al.* 2001, 2007) and also in another non-indigenous species of Lake Trasimeno such as *P. clarkii* (Elia *et al.* 2006). Both enzymes, when evaluated on the soft tissues of *D. polymorpha*, exhibited their lowest activity in spring and summer at about 70% (GI) and 50% (GII). These enzymatic activities have been found to be markedly influenced by the seasonal changes and in this respect, the lowest glyoxalases activities in spring and summer could represent a reduction in this detoxification pathway. As a result, the depleted enzymatic system is not able to counterbalance the increased toxicity of 2-ketoaldehydes, which might be formed during the lipid peroxidation process.

The total glutathione level in the zebra mussel of Lake Trasimeno was lower than those observed in other freshwater invertebrates (Cossu *et al.* 1997, Elia *et al.* 2001, 2007). Sheehan and Power (1999) reported a low thiol level in mussels collected in winter, which could be produced by fluctuations of pollutants and by variations of temperature, food availability, and reproductive cycle. In this study, thiol content was at the lowest level in specimens of zebra mussel collected during the warmer seasons (about 70%). Generally, the decreased thiol level in organisms might reflect a weakened defence ability which may facilitate the onset of oxidative damage and a prooxidant status. The water temperatures in Lake Trasimeno (20–25 °C during the warmer seasons) might have induced a lowering of total glutathione and of some enzymatic activities. A recent laboratory study has been performed on specimens of *D. polymorpha*, collected from Lake Trasimeno in the years 2000–2001, exposed to different temperatures (4, 18, 28, and 37 °C) for 15 h (Buschini *et al.* 2003). According to those authors, the haemocytes showed increasing DNA damage at the highest temperature, as a result of general stress in this species. In the same experiment, the same authors also found that this species showed a higher resistance to sodium hypochlorite (NaClO) at low dosage at 18 °C, probably due to more efficient detoxifying and repair mechanisms at that temperature.

CONCLUSIONS

Specimens of *D. polymorpha* collected seasonally from Lake Trasimeno displayed low metal concentrations in their soft tissues. From the correlation analysis it emerged that only glutathione peroxidase was positively related with Cd content as measured in these specimens. Therefore, the seasonal variability of the antioxidant parameters in the zebra mussel seems to be affected by its biological status. These marked differences indicated that comparisons among different biotopes or sites are significant only if the sampling of the zebra mussel is carried out at the same period of its biological cycle. The baseline condition of the antioxidant response of *D. polymorpha* might be useful to assess the contaminants impact on this non-indigenous species when employed in biomonitoring studies.

REFERENCES

Baldini, M., F. Fabietti, S. Gianmarioli, R. Onori, L. Orefice, and A. Stacchini. 1996. Metodi di analisi utilizzati per il controllo chimico degli alimenti. Rapporti ISTISAN 96/34, Italy.

Binelli, A., R. Bacchetta, P. Mantecca, F. Ricciardi, A. Provini, and G. Vailati. 2004. DDT in zebra mussels from Lake Maggiore (N Italy): level of contamination and endocrine disruptions. Aquatic Toxicology **69**, 175–188.

Binelli, A., R. Bacchetta, G. Vailati, S. Galassi, and A. Provini. 2001b. DDT contamination in Lake Maggiore (N Italy) and effects on zebra mussels spawning. Chemosphere **45**, 409–415.

Binelli, A., S. Galassi and A. Provini. 2001a. Factors affecting the use of *Dreissena polymorpha* as a bioindicator: the PCB pollution in lake Como (N Italy). Water, Air and Soil Pollution **125**, 19–32.

Binelli, A., F. Ricciardi, C. Riva and A. Provini. 2005. Screening of POP pollution by AChE and EROD activities in Zebra mussels from the Italian Great Lakes. Chemosphere **61**, 1074–1082.

Buschini, A., P. Carboni, A. Martino, P. Poli, and C. Rossi. 2003. Effects of temperature on baseline and genotoxicant-induced DNA damage in haemocytes of *Dreissena polymorpha*. Mutation Research **537**, 81–92.

Camusso, M., R. Balestrini, and A. Binelli. 2001. Use of zebra mussel (*Dreissena polymorpha)* to assess trace metal contamination in the largest Italian subalpine lakes. Chemosphere **44**, 263–270.

Cossu, C., A. Doyotte, M. Babut, A. Exinger, and P. Vasseur. 2000. Antioxidant biomarkers in freshwater bivalves, *Unio tumidus*, in response to different contamination profiles of aquatic sediments. Ecotoxicology and Environmental Safety **45**, 106–121.

Cossu, C., A. Doyotte, M. C. Jacquin, M. Babut, A. Exinger, and P. Vasseur. 1997. Glutathione reductase, selenium dependent glutathione peroxidase, glutathione levels, and lipid peroxidation in freshwater bivalves, *Unio tumidus*, as biomarkers of aquatic contamination in field studies. Ecotoxicology and Environmental Safety **38**, 122–131.

de Lafontaine, Y., F. Gagné, C. Blaise, G. Costan, P. Gagnon, and H. M. Chan. 2000. Biomarkers in zebra mussels (*Dreissena polymorpha*) for the assessment and monitoring of water quality of the St Lawrence River (Canada). Aquatic Toxicology **50**, 51–71.

Doyotte, A., C. Cossu, M. C. Jacquin, M. Babut, and P. Vasseur. 1997. Antioxidant enzymes, glutathione and lipid peroxidation as relevant biomarkers of experimental or field exposure in the gills and the digestive gland of the freshwater bivalve *Unio tumidus*. Aquatic Toxicology **39**, 93–110.

Elia, A. C., A. J. M. Dörr, C. Mastrangelo, M. Prearo, and M. C. Abete. 2006. Glutathione and antioxidant enzymes in the hepatopancreas of crayfish *Procambarus clarkii* (Girard, 1852) of Lake Trasimeno (Italy). Bulletin Francais de la Pêche et de la Pisciculture **380–381**, 1351–1361.

Elia, A. C., R. Galarini, A. J. M. Dörr, and M. I. Taticchi. 2007. Heavy metal contamination and antioxidant response of a freshwater bryozoan (*Lophopus crystallinus* Pall., Phylactolaemata). Ecotoxicology and Environmental Safety **66**, 188–194.

Elia, A. C., A. Ludovisi, and M. I. Taticchi. 2001. Study of seasonal variations of glutathione and detoxificant enzymes in *Lophopus crystallinus* Pallas (Briozoa) from Piediluco lake (Umbria, Italy). Italian Journal of Zoology **68**, 291–297.

Giambérini, L. and M. P. Cajaraville. 2005. Lysosomal responses in the digestive gland of the freshwater mussel, *Dreissena-polymorpha*, experimentally exposed to cadmium. Environmental Research **98**, 210–214.

Gundacker, C. 1999. Tissue-specific heavy metal (Cd, Pb, Cu, Zn) deposition in a natural population of the zebra mussel *Dreissena-polymorpha* Pallas. Chemosphere **38**, 3339–3356.

Hendriks, A. J., H. Pieters, and J. de Boer. 1998. Accumulation of metals, polycyclic (halogenated) aromatic hydrocarbons and biocides in zebra mussel and eel from the Rhine and Meuse River. Environmental Toxicology and Chemistry **17**, 1885–1898.

Iwanyzki, S. and R. W. McCauley. 1992. Upper lethal temperatures of adult zebra mussels (*Dreissena polymorpha*). Pages 667–675 in T. F. Nalepa and D. W. Schloesser, editors. Zebra mussels: biology, impacts and control. Lewis Publishers, Chelsae, MI.

Knaw, K. H. M., H. M. Chan, and Y. de Lafontaine. 2003. Metal contamination in zebra mussel (*Dreissena-polymorpha*) along the St. Lawrence River. Environmental Monitoring and Assessment **88**, 193–219.

Kraak, M. H. S., M. C. T. Scholten, W. H. M. Peeters, and W. C. Dekock. 1991. Biomonitoring of heavy metals in the western-european rivers Rhine and Meuse using the fresh-water mussel *Dreissena-polymorpha*. Environmetal Pollution **74**, 101–114.

Lowry, O. H., N. J. Rosebrough, A. L. Farr, and R. J. Randall. 1951. Protein measurement with the Folin phenol reagent. Journal of Biological Chemistry **193**, 265–275.

Monarca, S., C. Zani, S. D. Richardson, A. Thruston Jr, M. Moretti, D. Feretti, and M. Villarini. 2004. A new approach to evaluating the toxicity and genotoxicity of disinfected drinking water. Water Research **38**, 3809–3819.

Niyogi, S., S. Biswas, S. Sarker, and A. G. Datta. 2001. Antioxidant enzymes in brackish-water oyster, *Saccostrea cucullata* as potential biomarkers of polyaromatic hydrocarbon pollution in Hooghly Estuary (India): seasonality and its consequences. The Science of the Total Environment **281**, 237–246.

Pflugmacher, S., C. Wiegand, A. Oberemm, K. A. Beattie, E. Krause, G. A. Codd, and C. E. W. Steinberg. 1998. Identification of an enzymatically formed glutathione conjugate of the cyanobacterial hepatotoxin microcystin-LR: the first step of detoxication. Biochimica et Biophysica Acta **1425**, 527–533.

Sheehan, D., J. Mcintosh, A. Power, and P. J. Fitzpatrick. 1995. Environmental biochemistry. Biochemical Society Transactions **23**, 419–422.

Sheehan, D. and A. Power. 1999. Effects of seasonality on xenobiotic and antioxidant defence mechanisms of bivalve molluscs. Comparative Biochemistry and Physiology C **123**, 193–199.

Spilinga, C., U. Chiappafreddo, and Q. Pirisinu. 2000. *Dreissena polymorpha* (Pallas) al Lago Trasimeno. Rivista di Idrobiologia **39**, 145–152.

Tiscar P. G. and F. Mosca. 2004. Defense mechanisms in farmed marine molluscs. Veterinary Research Communications **28**, 57–32.

Vidal, M. L., A. Basseres, and J. F. Narbonne. 2002. Seasonal variation of pollution biomarkers in two population of *Corbicula fluminea* (Müller). Comparative Biochemistry and Physiology C **131**, 133–151.
Wilhelm Filho, D., T. Tribess, C. Gáspari, F. D. Claudio, M. A. Torres, and A. R. M. Magalhães. 2001. Seasonal changes in antioxidant defenses of the digestive gland of the brown mussel (*Perna perna*). Aquaculture **203**, 149–158.
Winston, G. W. and R. T. Di Giulio. 1991. Prooxidant and antioxidant mechanisms in aquatic organisms. Aquatic Toxicology **19**, 137–161.

Managing invasions

Each invasion has a certain degree of specificity. Yet, generalities are emerging yielding encouraging insights into how invasions operate and how they may be best addressed by conservation managers and policymakers. Although complicated by economic, social, and political concerns, IAS policy decisions must also be based on clear, scientific reasoning.

Jeb Byers *et al.* (2002)

Chapter thirty-four

Towards a European strategy to halt biological invasions in inland waters

Piero Genovesi

INTRODUCTION

A major portion of biodiversity in the world depends on inland waters; about 8,500 of the 20,000 species of fish live in this medium, and many species of reptiles, amphibians, birds, semiaquatic mammals, along with invertebrates and plants, depend on freshwater ecosystems. Of the 117 threatened species in Europe, 28 live in freshwater ecosystems (IUCN 2006). The biological integrity of inland waters is also crucial for people: it provides water and food to a large majority of world populations (rice alone is the main part of the diet of over 50% of the globe's inhabitants).

Inland waters also include particularly fragile ecosystems; modification of the physical characters of basins, loss and fragmentation of habitats, overexploitation of plant and animal species, chemical pollution, and biological invasions threaten most basins around the world, posing risks not only to the biological diversity, but also to the wellbeing of a vast majority of humans.

Of the threats listed above, biological invasions are one of the most rapidly growing causes of biodiversity loss. If non-indigenous species (NIS) are threatening all regions of the world, the effects are particularly high in the wetlands of Europe, where the percentage of the recorded introductions is the highest in the world (Europe 25.1%, Asia 16.4%, Africa 14.7%, Oceania 14.7%, South and Central America 14.1%, Middle East 8.4%, North America 6.3%; DIAS 1997). The particular susceptibility of European inland waters to invasions is a direct

Francesca Gherardi, Biological invaders in inland waters: Profiles, distribution, and threats, 627–637.

consequence of the rapid growth of transport, trade, and tourism. The most relevant cause of the introduction of aquatic NIS is the movement of vessels (both through ballast waters and hull fouling). The major pathways of species translocations (i.e. for aquaculture, horticulture, aquarium trade, fishing, sport angling) are a direct or indirect cause of the globalization of economies.

The high number of NIS in European fresh waters is a major threat to several indigenous species; for example, the critically endangered white-headed duck [*Oxyura leucocephala* (Scopoli)] is threatened by hybridization with the introduced ruddy duck [*Oxyura jamaicensis* (Gmelin)] (Hughes *et al.* 1999); the endemic European mink [*Mustela lutreola* (Linnaeus)] is being out-competed by the introduced American mink (*Mustela vison* Schreber) (Sidorovich *et al.* 1999); the endemic white-clawed crayfish [*Austropotamobius pallipes* (Lereboullet)] is affected by several non-indigenous crayfish introduced into the European basins; the introduced muskrat [*Ondatra zibethicus* (Linnaeus)] is responsible for dramatically reducing the diversity of invertebrate fauna (Nummi *et al.* 2006); the endangered European pond turtle [*Emys orbicularis* (Linnaeus)] is threatened by the introduced red-eared terrapin [(*Trachemys scripta elegans* (Weid)] through competition (Cadi and Joly 2003); and the entire fish fauna of several regions is being profoundly altered by introduced species (e.g. Bianco and Ketmaier 2001).

Apart from the impact on biodiversity, freshwater NIS may also cause high losses to the European economy; *Gyrodactylus salaris* Malmberg and *Anguillicola crassus* (Kuwahara, Niimi and Hagakihave) led to dramatic decreases in the incomes of the fisheries sector in several Nordic States (Weidema 2000), and the muskrats and coypu [*Myocastor coypus* (Kerr)] severely damage river banks through digging and increase the risk and severity of floods in many central and southern European States; in 1995–2000, the management of the coypu in Italy caused losses of over €11 million; it has been estimated that in the future this species will cost the country over €12 million per year (Panzacchi *et al.* 2007).

There is therefore an urgent need to develop measures to prevent further invasions of freshwater systems and to mitigate the impacts of NIS already present in these fragile ecosystems. This is also required in the respect of the commitment of European Countries to halt the loss of biodiversity in the region by 2010, as agreed under the Kyiv Biodiversity Resolution.

OBSTACLES TO A EUROPEAN STRATEGY ON BIOINVASIONS IN FRESH WATERS

Responding to the ambitious commitments of both preventing new invasions in European fresh waters and mitigating the impacts of invasive NIS is a complex task and much effort will be required from all the European States and institutions.

There are in fact many and different obstacles to the development of actions on biological invasions, including technical constraints and political issues.

These add to the limited support of the public and of decision-makers because of a widespread lack of awareness of the threats posed by invasive NIS. In the following sections of this chapter, I will describe the main constraints that need to be taken into account when developing a regional policy on the issue, basing my argument on some case studies.

Lack of transboundary cooperation

Many basins are shared between two or more countries so that actions carried out at the national level may be undermined by a lack of coordination between neighboring countries. An example of the need for a supranational approach is the case of the attempts to recover the population of the white-clawed crayfish in Portugal; by the middle 1970s, only three small residual populations of *A. pallipes* survived in the north-eastern portion of the country. Between 1977 and 1986, a program aimed at recovering these populations, through farming and restocking, was started, with the support of the European Commission through a LIFE program. In the same period, several intentional introductions of non-indigenous crayfish species [*Pacifastacus leniusculus* (Dana) and *Procambarus clarkii* (Girard)] were made by the bordering Spanish authorities, with the aim of improving the crayfish harvest industry. These introductions, carried out in basins shared with Portugal, caused severe outbreaks of the crayfish plague that contributed to the failure of the recovery efforts in Portugal (Gutiérrez-Yurrita *et al.* 1999, Cuellar and Cuellar 2000, Holdich 2003).

Limited ability to promptly detect incursions

When prevention of unwanted introductions fails and a new species enters a country, it is critical to rapidly detect the introduction and to promptly implement responses. When the response is prompt, even marine invasions can be halted; for example, an infestation of the seaweed *Caulerpa taxifolia* (Vahl) C. Ag. – a non-indigenous alga that has invaded a large portion of the Mediterranean – has been successfully treated in California because adequate responses started 17 days after detection (Anderson 2005).

Unfortunately, most European countries seem to lack an effective alarm system and responses to new invasions are often delayed not only because of the difficulty of detecting them, but often because of a lack of communication between the competent authorities and whoever first identified the new species, whether private or public entities.

Ineffective responses to new incursions

Early detection is wasted when it is not followed by a prompt action. Unfortunately, even when identification of a new invasion is rapid and competent authorities have been promptly informed, responses may be delayed for different reasons. The example of the already mentioned seaweed *C. taxifolia* is paradigmatic. The species was detected in 1984 in the waters of Monaco by an expert on algae who immediately reported the case to the local authorities; at that time the invaded area was about 1 m^2 and the invasion could have been halted in two days by scuba divers. Unfortunately, no action was started and by 1989 the invaded area had increased to 3 ha, reaching 31 ha in 1991. Even then eradication was not started, mostly because a debate arose within the academic world on the origin of the alga, the severity of the potential impacts, and the expected long-term effects of this invasion. Only in 1995, a recommendation was approved by the Bern convention calling States to control proliferation of the alga (recommendation 45, 1995). The decision arrived far too late, when the alga had already expanded, invading many areas of the Mediterranean, and its eradication was no longer feasible (Meinesz 1999).

Patterns of spread of invasive NIS

Freshwater ecosystems often facilitate the spread of NIS, because larvae, seeds, and propagules can be passively transported by the water current, NIS may be passively transported by boats even for long distances (e.g. Buchan and Padilla 1999) and many aquatic or semiaquatic species are able to swim using water as a corridor of expansion. Invasion speed and the distance spread can thus be much greater in aquatic than in terrestrial ecosystems, making a response in this habitat more difficult.

Inappropriate technical tools for controlling most freshwater invaders

Control in freshwater systems is often more difficult than in terrestrial environments. The only cases of successful eradications in inland waters concern a few semiaquatic mammals (muskrat, coypu, Canadian beaver, American mink; see Genovesi 2005 for a review), one waterbird species (the ongoing attempt to eradicate the ruddy duck from the Palearctic), and a few fish (Copp *et al.* 2005); no eradication of amphibians, reptiles, plants, or invertebrates has been completed so far in Europe. This is the case of non-indigenous crayfish, despite the many attempts and efforts devoted to controlling their impacts. Many control programs so far attempted have required the dispersion of chemical toxicants into the water, with risks of causing undesired impacts on nontarget species. For example, in an attempt to remove the parasite *G. salaris* from their rivers, Sweden and Norway have dispersed in the last 10 years high quantities of a chemical toxicant (rotenone) without succeeding in eradicating the pathogen.

Legal and organizational aspects

The efficient control of NIS has been in some cases limited by inadequate legal or organizational frameworks; for example, several European States have legislation that automatically protects NIS when they become established in a country (e.g. the Italian L. 157/92 protects all naturalized species without distinguishing between indigenous and non-indigenous ones). Furthermore, several laws ban the use of removal methods that may be essential for controlling NIS, as in the case of the use of herbicides in shellfish designated areas that is a major obstacle to the eradication of the marine plant *Spartina anglica* C. E. Hubbard from Ireland (Hammond and Cooper 2002). An additional critical aspect is the need to coordinate all the relevant sectors of both the governmental (e.g. fishing, transport, trade, management) and the nongovernmental bodies (e.g. angling associations, pet trade associations, horticulture organizations).

Misperception of the costs/benefits of control

The technical, legal, and organizational difficulties in eradicating freshwater NIS are not the only reasons that may explain the limited action to address inland water invasions in Europe. There is also a general (and erroneous) perception that eradication is generally very costly and that in most cases fails. For example, the successful eradication of the coypu from East Anglia in the 1980s at a cost of about €5 million in 11 years was considered a very costly investment; but, if we compare this investment with the cost of permanent control of the coypu in Italy (over €3.7 million in 2000, estimated to exceed €12 million/year in the future), eradication – whenever feasible – seems to be a worthwhile alternative.

Trade regulations

The difficulties associated with the control or eradication of freshwater invasive species and the ease and speed of spread of the invaders make the development of efficient prevention measures particularly urgent. It is clear that to mitigate the impacts of aquatic invasive species it is essential to prevent the introduction into countries that have not yet been invaded through trade regulations. This is often in conflict with the free market policy of Europe, as also stated in the European Community Treaty that prohibits quantitative restrictions on imports and exports by Articles 28 and 29. A major problem with this legal conflict is that there is no clear understanding of which States can and cannot do. In fact, the European Court of Justice has examined two cases of import regulations introduced by European States to prevent the impact caused by invasive species, and the decisions have been somehow contradictory. One case concerned a ban to the import of non-indigenous crayfish adopted by Germany to prevent the risk of introducing the crayfish plague into its territory; in 1994, the Court considered the ban unjustified and ruled to remove the trade restriction (case C-131/93). An opposite example (not concerning freshwater species) is the case of the restriction on importing non-indigenous bees into the island of Læsø to protect an indigenous breed of bees (case: C-67/97). In this case the Court found that the Danish decision was justified by the need to protect the health and life of animals; however, also in this case, in early 2005 Denmark decided to remove the restriction because of the necessity of avoiding conflicts with the free market policy of Europe.

These cases show the general uncertainty in the European legal framework on what can be and what cannot be done. So, several European States have introduced bans on importing non-indigenous crayfish (e.g. Norway, Ireland), others have banned their keeping (e.g. Great Britain), and most renounced the regulation of trade in any form. It appears that, given the unclear European legal framework, States do not even implement the legislative measures that they have the power to take and are reluctant to impose import bans even when these are clearly justified.

Several European institutions have attempted to develop trade regulations for preventing unwanted introductions. The European Plant Protection Organisation (EPPO) and the EC Regulation no. 338/97 have introduced lists of plant and animal species whose introduction into Europe should be strictly regulated. Unfortunately, the mechanisms for including new species on both these list are rather slow and complex (nine species so far included in the EPPO quarantine list; four in the EC reg. no. 338) and the system is far from being adequate, given the rapidly growing increase of new introductions.

Inadequate circulation of information

Another cause of delay and inefficacy in the responses to new invasions is the lack of information on the impact of the newly established species at a national level and of expertise on management options. For these reasons, information-sharing between States is very important; data on the impacts caused by the newly recorded species and information on the possible management tools are often available elsewhere in the world. Mechanisms for making this information rapidly available to local decision makers are thus essential to reduce the time for a response and to increase the ability to act.

In Europe, databases on invasive NIS have been developed for specific sectors (e.g. the EPPO/EC plant health system) or areas (e.g. NOBANIS for Nordic Countries; www.nobanis.org), but no comprehensive regional information system on invasive NIS is currently available. To bridge this gap, the European Commission has supported, under the Sixth Framework Programme for Research, Technological Development, and Demonstration Activities, the production of a pan-European database on invasive NIS aimed also at providing a registry of experts for the different taxonomic groups and areas (DAISIE, www.europe-aliens.org).

To ensure the flow of information regarding species not yet known in Europe, regional databases need to be integrated with international instruments, such as the Global Invasive Species Database (www.issg.org/database/) developed by IUCN or the Global Invasive Species Information Network (http://www.gisinetwork.org), an international web portal on biological invasions.

TOWARD A EUROPEAN STRATEGY ON AQUATIC NIS

In order to effectively minimize the adverse impact on European biodiversity, economy, and human health and well-being inflicted by invasive NIS, it seems urgent and critical to develop and implement coordinated measures and cooperative efforts. This requires that European States revise their policies – through the development of national strategies or action plans – and that the European institutions (including the European Commission and the Bern Convention) develop a regional policy on the issue, with the common aim of increasing awareness, strengthening national and regional capacity and cooperation, preventing new invasions, supporting rapid response, and mitigating the adverse impacts caused by those invasive NIS that are already established in Europe.

The key elements of a European policy on invasive species have been defined by the "European Strategy on Invasive Alien Species" (Genovesi and Shine 2004); this document has been formally approved in 2003 by the European States that also committed to implement the Strategy by 2008 in at least half of the European countries through national strategies and action plans. The document has also been welcomed by the conference of the parties of the Convention on Biological Diversity and its implementation by European States and institutions has been recommended by the World Conservation Conference in 2004. The European Strategy applies to all environments and taxonomic groups and covers well the specific problem of the European inland waters. Below I will provide a synthesis of the main elements of the Strategy – with particular reference to freshwater ecosystems:

1. Building awareness and support.

The limited understanding by the European public and decision-makers of the magnitude of the threats posed by biological invasions limits the commitment of the relevant agencies and other stakeholders in efforts to prevent and mitigate the impacts posed by NIS. The first aim of a pan-European strategy is hence to raise awareness of the Europe's public, decision-makers, scientists, and other stakeholders of the risks posed by invasive NIS, and the benefits of prevention and mitigation for native biodiversity. This requires vigorous information and education programs, and the involvement of key stakeholders (e.g. travel departments, shipping authorities, aquaculture and fishing associations).

2. Collecting, managing, and sharing information.
Considering the difficulty of responding to freshwater invasions once they have occurred, it seems critical to create national inventories of NIS, develop tools for sharing information at the national and regional levels, and support research in order to strengthen the capacity to predict the consequences of introductions. Also, lists of experts and information on the techniques to respond to invasions should be implemented and effectively circulated.

3. Strengthening national policy, legal, and institutional frameworks.
Any action to prevent unwanted introductions into inland waters necessarily requires the involvement of many different sectors, including trade, water resource management, infrastructure development, horticulture, aquaculture, tourism, and recreation (e.g. angling, pet trade). A clear partitioning of roles and responsibilities among the different agencies and departments is therefore crucial, and States should start a review of their institutional and legal frameworks. The creation of an authority, with the role of coordinating and leading the efforts of all the responsible agencies, can be very effective. All the relevant bodies should identify focal points to liaise with other departments on this specific matter. National action plans or strategies should be developed, addressing all aspects of prevention and mitigation of invasions.

4. Regional cooperation and responsibility.
The above cited case of the introduction of non-indigenous crayfish species in Spain – affecting the efforts to recover endemic white-clawed crayfish in Portugal – shows the need for a coordinated approach to biological invasions at a regional level. States should recognize the risk that activities within their jurisdiction or control may pose to other States as a potential source of invasions and take appropriate individual and cooperative actions to minimize that risk. Mechanisms for the exchange of information, notification of relevant invasions, and consultation should be developed, also by strengthening the role of the regional authorities (e.g. Bern Convention, EPPO).

5. Prevention.
If in all environments prevention of unwanted introduction is far more cost-effective and environmentally desirable than eradication and control, for aquatic ecosystems prevention is in most cases the only means to address invasions. It is therefore important to prioritize available resources on prevention, beginning at the place of origin or export of species, focusing on the most relevant pathways (i.e. ballast water management), and facilitating the application of agreed standards and practices (i.e. on aquaculture, fisheries, etc). A system of lists should be developed at the European level, based on a blacklist approach. The involvement of the most relevant societal sectors (e.g. trade, tourism, transport, pet trade, aquaculture) is also essential.

6. Early detection and rapid response.
Rapid identification of new invasions is a critical element of prevention, and the development of effective surveillance procedures is thus essential, for example, by monitoring the most vulnerable areas or the areas adjacent to containment facilities (fish farms, aquaculture facilities, aquaria, etc.). Control and eradication in freshwater ecosystems are often very difficult, but many examples show that rapid response can permit effective eradication of even invertebrate aquatic species. The time between documenting an introduction and implementing a response should thus be reduced by streamlining the authorization processes, ensuring that all competent authorities have adequate powers, and by having in place contingency plans for specific groups of organisms (e.g. algae, invertebrates, vertebrates).

7. Management of impacts.
When prevention fails, eradication should be considered, when this is a feasible option. Where eradication is no longer feasible, containment and control should be evaluated, especially when there is a risk of spread of the invading species to neighboring countries.

8. Restoration of native biodiversity.
Any strategy needs to go further than the defensive three-stage hierarchical approach based on prevention-eradication-control and should support restoration measures for species, natural habitats, and ecosystems that have been affected by biological invasions. Increased resilience of native biodiversity can in turn provide greater protection against re-invasion or new incursions. In this regard, the use of indigenous plants and animal species in landscaping, revegetation, fish farming, aquaculture, etc. should be promoted.

CONCLUSIONS

Biological invasions in European freshwater ecosystems are a major threat to the biological diversity, economy, and human health and wellbeing. Aquatic NIS are difficult to detect, tend to disperse rapidly, and are extremely difficult to eradicate and control. So, to address this threat it is essential to focus on prevention, by improving the capacity of European States and institutions to exchange information, address pathways of introduction, regulate importation and introduction of species, and mitigate the impacts once prevention has failed. To reach these aims each European State should develop a holistic and comprehensive national action plan/strategy, consistent and harmonized with the European Strategy; and European States and institutions are expected to work jointly toward a consistent regional policy on invasive NIS.

REFERENCES

Anderson, L. W. J. 2005. California's reaction to *Caulerpa taxifolia*: a model for invasive species rapid response. Biological Invasions **7**, 1003–1016.

Bianco, P. G. and V. Ketmaier. 2001. Anthropogenic changes in the freshwater fish fauna of Italy, with reference to the central region and *Barbus graellsii*, a newly established alien species of Iberian origin. Journal of Fish Biology **59**, 190–208.

Buchan, L. A. J. and D. K. Padilla. 1999. Estimating the probability of long-distance overland dispersal of invading aquatic species. Ecological Applications **9**, 254–265.

Cadi, A. and P. Joly. 2003. Competition for basking places between the endangered European pond turtle (*Emys orbicularis*) and the introduced red-eared slider (*Trachemys scripta elegans*). Canadian Journal of Zoology **81**, 1392–1398.

Copp, G. H, K. J. Wesley, and L. Vilizzi. 2005. Pathways of ornamental and aquarium fish introductions into urban ponds of Epping Forest (London, England): the human vector. Journal of Applied Ichthyology **21**, 1–12.

Cuellar, L. and M. C. Cuellar. 2000. Evolution de l'aphanomycose et repeuplements avec l'écrevisse signal (*Pacifastacus leniusculus*) en Espagne. L'Astaciculteur de France **65**, 2–9.

DIAS. 1997. Database on Introductions of Aquatic Species. Internet addresses. World Aquaculture Information Center, Food and Agriculture Organization of the United Nations, Rome, Italy. http://www.fao.org/waicent/faoinfo/fishery/statist/fisoft/dias/index.htm.

Genovesi, P. 2005. Eradications of invasive alien species in Europe: a review. Biological Invasions **7**, 127–133.

Genovesi, P. and C. Shine. 2004. European strategy on invasive alien species. Nature and environment, n. 137. Council of Europe Publishing, Strasbourg, France, 67 pp.

Gutiérrez-Yurrita, P. J., J. M. Martínez, M. A. Bravo, C. Montes, M. Ilhéu, and J. Bernardo. 1999. The status of crayfish populations in Spain and Portugal. Pages 161–192 in F. Gherardi and D. M. Holdich, editors. Crayfish in Europe as alien species: How to make the best of a bad situation? A. A. Balkema, Rotterdam, The Netherlands.

Hammond, M. E. R. and A. Cooper. 2002. *Spartina anglica* eradication and inter-tidal recovery in Northern Ireland estuaries. Pages 124–131 in D. Veitch and M. Clout, editors. Turning the tide: the eradication of invasive species. IUCN SSC Invasive Species Specialist Group. IUCN, Gland, Switzerland, viii + 414 pp.

Holdich, D. M. 2003. Crayfish in Europe – an overview of their taxonomy, legislation, distribution, and crayfish plague outbreaks. Pages 15–34 in D. M. Holdich and P. J. Sibley, editors. Management & conservation of crayfish. Proceedings of a conference held in Nottingham on 7th November, 2002. Environment Agency, Bristol, UK.

Hughes, B., J. Criado, S. Dalany, U. Gallo-Orsi, A. Green, M. Grussu, C. Perennou, and J. A. Torres. 1999. The status of the ruddy duck (*Oxyura jamaicensis*) in the western Palearctic: towards an action plan for eradication. Report by the Wildlfowl & Wetlands Trust to the Council of Europe, Council of Europe Publishing, Strasbourg, France.

IUCN. 2006. IUCN Red List of Threatened Species. <www.iucnredlist.org>. Downloaded on 01 June 2006.

Meinesz, A. 1999. Killer algae. University of Chicago Press, Chicago, IL.

Nummi, P., V. M. Väänänen, and J. Malinen. 2006. Alien grazing: indirect effects of muskrats on invertebrates. Biological Invasions **8**, 993–999.

Panzacchi, M., S. Bertolino, R. Cocchi, and P. Genovesi. 2007. Population control of coypu in Italy vs. eradication in U.K.: a cost/benefit analysis. Wildlife Biology **13**, in press.

Sidorovich, V., H. Kruuk, and D. W. Macdonald. 1999. Body size, and interactions between European and American mink (*Mustela lutreola* and *M. vison*) in Eastern Europe. Journal of Zoology **248**, 521–527.

Weidema, I., editor. 2000. Introduced species in the Nordic Countries. Nord 2000:13. Nordic Council of Ministers, Copenhagen, Denmark.

Chapter thirty-five

A risk assessment of biological invasions in the inland waterways of Europe: the Northern Invasion Corridor case study

Vadim E. Panov, Yury Yu. Dgebuadze, Tamara A. Shiganova, Andrew A. Filippov, and Dan Minchin

INTRODUCTION

Inland waterways have provided opportunities for the spread of aquatic non-indigenous species (NIS) for many centuries (reviewed in Ketelaars 2004, Galil and Minchin 2006, Galil *et al.* 2007). Canals connecting different river basins have allowed for range extensions of many species, either by active movement and/or by ship transport. Over the past century, the potential for species to expand their range has been enhanced due to increasing trade and the construction of canals. The waterways occur at low altitudes and presently the main European corridor routes consist of an interlinked network of 30 main canals with more than 100 branches, and more than 350 ports exist in low-altitude Europe.

The European Agreement on Main Inland Waterways of International Importance (AGN) was signed under the framework of the United Nations

Francesca Gherardi, Biological invaders in inland waters: Profiles, distribution, and threats, 639–656.

Economic Commission for Europe (UNECE) in Geneva in 1996. The AGN sets down standards for a uniform infrastructure and operational procedures for the European inland waterway network (EWN). With the adoption of the AGN by the Russian Federation in 2002, the international network of European waterways defined in the Agreement now consists of approximately 28,000 km of main navigable rivers and canals, extending from the Atlantic Ocean to the Ural mountains and connecting 37 countries in Europe and beyond. Currently, the network includes Austria, Bulgaria, Croatia, the Czech Republic, Hungary, Finland, France, Germany, Greece, Italy, Lithuania, Luxembourg, The Netherlands, the Republic of Moldova, Romania, the Russian Federation, Slovakia, and Switzerland. The EWN will be further developed in the future, as more than 80 missing links and bottlenecks have recently been identified (Anonymous 2005).

Such development will enable further opportunities for NIS to extend their ranges to other river basins previously separated over geological time. The first opportunity for NIS to spread began in the late 1700s with the construction of canals connecting previously isolated river basins and linkages to European seas: Mediterranean, Black, and Azov seas to the south, and the Baltic and White seas to the north. Presently, there are four main inland water trading routes (invasion corridors) that enable in particular the range expansion of several Ponto-Caspian species through Europe (Jażdżewski 1980, Panov *et al.* 1999, Bij de Vaate 2002, Nehring 2002, Slynko *et al.* 2002, Van der Velde *et al.* 2002, Ketelaars 2004, Pienimäki and Leppäkoski 2004, Galil *et al.* 2007).

These invasion corridors include the Northern Invasion Corridor which links the southern seas (Black and Azov seas) with the Caspian Sea via the Azov Sea–Caspian waterway (the main European inland waterway number E90, including the Volga–Don Canal opened in 1952), and with the Baltic and White seas via the Volga–Baltic waterway (the main European inland waterway number E50 with the Volga–Baltic Canal first opened in 1810 and reopened after major reconstruction in 1964) and via the White Sea–Baltic Sea waterway (the main European inland waterway number E60 with the White Sea–Baltic Sea Canal opened in 1932). This largest inland European invasion corridor consists of approximately 6,500 km of waterways, representing the so-called United Deepwater System of Russia with 21 inland ports of international importance, and linking four main watersheds in European Russia (Black, Caspian, Baltic, and White seas basins) (Fig. 1). The Volga River represents the longest section in the Northern Invasion Corridor. This river is 3,530 km long and includes 12 large and more than 300 medium and small reservoirs (Slynko *et al.* 2002).

In 2003, the Russian Government adopted the national Concept of Development of the Inland Water Transport, which is likely to increase trade along the Northern Invasion Corridor. According to this concept, the United Deepwater System of Russia should be fully open for international shipping by 2010 and be integrated into the European inland waterway network. The main focus is to

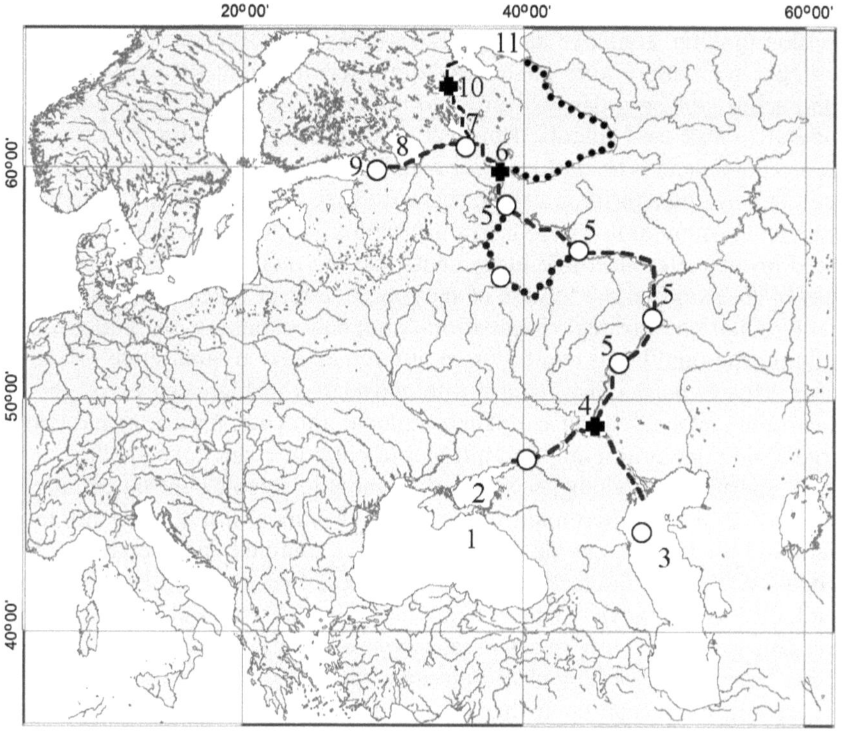

Fig. 1 The European Northern Invasion Corridor. Numbers on the map indicate: 1 – Black Sea, 2 – Azov Sea, 3 – Caspian Sea, 4 – Volga–Don Canal, 5 – Volga River reservoirs, 6 – Volga–Baltic Canal, 7 – Onega Lake, 8 – Ladoga Lake, 9 – Gulf of Finland, 10 – White Sea-Baltic Canal, and 11 – White Sea. Dashed lines indicate the main navigable waterways of the Northern Invasion Corridor, dotted line – the secondary waterways, opened cycles – the monitoring stations.

provide connections between the Volga, Don, and Danube rivers to link more than 15 European countries. These future developments may highly facilitate the transfer of NIS across European inland waters and coastal ecosystems, which require appropriate risk assessment-based management options to address risks posed by human-mediated introductions of these species.

Ecological risk assessment is a process that evaluates the likelihood that adverse ecological effects may occur or are occurring as a result of exposure to one or more stressors. The process is used to systematically evaluate and organize data, information, assumptions, and uncertainties in order to help understand and predict the relationships between stressors and ecological effects in a way that is useful for environmental decision making (Anonymous 1998). Generally speaking, the risk assessment is a part of the process of managing risks, and there are many different risk assessment approaches in different decision-making contexts and levels ranging from specific case studies to strategic regulation and policy making (Gerrard and Petts 1998). These approaches can be separated into two major distinct types: quantitative and qualitative risk assessments. However, because quantification of risks is not always possible, it is better to convey conclusions (and associated uncertainties) qualitatively than to ignore them, because they are not easily understood or estimated (Anonymous 1998). In our opinion, quantitative risk assessments, based on objective scientific judgements, can be more applicable for the local level of decision-making in case of site-specific and/or species-specific management, while the strategic regulation- and policy-making on both national and international levels can be based in large extent on qualitative risk assessment. This is particularly true if one considers the high degree of scientific uncertainty when dealing with such a global and complex ecological issue as large-scale intercontinental and intracontinental introductions of NIS.

The specific methodologies of risk assessment of shipping-mediated introductions of NIS include two main types: the environmental matching risk assessment and the species-specific risk assessment (Pienimäki and Leppäkoski 2004, Leppäkoski and Gollasch 2006). Based on these two principal approaches, we conducted a qualitative risk assessment of NIS introductions along the Northern Invasion Corridor, with the general purpose to develop a conceptual model of risk assessment of biological invasions for the European inland navigable waterways, which can be further used as a tool for management purposes.

CONCEPTUAL MODEL OF QUALITATIVE RISK ASSESSMENT FOR INLAND WATERWAYS

The qualitative risk assessment was based on the analyses of data from the national database on aquatic NIS in European Russia (Panov *et al.* 2007a), the AquaInvader information system (Panov *et al.* 2006), and other relevant sources (Slynko *et al.* 2002, Panov *et al.* 2007b). The national database includes both published information and primary field data on aquatic NIS distributions from the national monitoring network established along the whole Northern Invasion Corridor (Fig. 1).

There are five main components to the risk assessment of NIS for the navigable inland waterways we have made:

1. Identification of the principle recipient and donor areas of NIS (risk areas) and invasion routes.
2. Identification of the main vectors of NIS introductions.
3. Assessment of inoculation rates (propagule pressure).
4. Assessment of the vulnerability of potential recipient areas to invasions from past patterns and likely environmental suitability.

5. Assessment of the invasiveness of NIS both in the recipient risk area and in the potential donor areas based on known dispersal abilities, establishment success, and ecosystem impacts.

Qualitative estimations of inoculation rates and ecosystem vulnerability to invasions and species invasiveness were ranked *low*, *medium*, and *high*, and these estimations were further used for an assessment of the integrated ecosystem risk level for the main risk areas within the Northern Invasion Corridor. The analysis of these five main components was conducted for initial predictive risk assessment for selected recipient areas (risk recipient areas).

Identifying recipient and donor areas of NIS and invasion routes

There are four principle high risk areas along the Northern Invasion Corridor, which act as recipient and also as donor areas of aquatic NIS: Azov Sea (41 established NIS), Caspian Sea (46 established NIS), Volga River reservoirs (85 established NIS), and Gulf of Finland of the Baltic Sea (23 established NIS). The White Sea can be considered as a *low*-risk area, as only 10 NIS (both marine and freshwater) have become established in its basin, and currently there is no evidence of its possible role as donor area (Fig. 2). In this water system, Lake Ladoga and Lake Onega (the largest European lakes), have the lowest number of established NIS (only three), and is considered as *low*-risk area (see discussion below).

The principle donor areas of NIS for the four main recipient ecosystems include: Black Sea (25 species), Caspian Sea (four species), and Asia (three species) for the Azov Sea; Black and Azov seas (44 species) for the Caspian Sea; Lower Volga (42 species), Black and Azov seas (16 species), and Baltic Sea basin (18 species) for the Volga River reservoirs; Ponto–Caspian basin (10 species) and western Baltic (11 species) for the Gulf of Finland (Fig. 3). The principle donor areas of NIS are from southern regions. This may reflect climate change with the concomitant advantage of an available route for southern species to spread northwards.

We identified 10 patterns of NIS dispersal along the Northern Invasion Corridor, considered here to be invasion routes (Fig. 2). Along routes 1–7 we found 122 northward invasions but only 24 southward invasions along routes 8–10. Each route links the basins, seas, reservoirs, and canals as follows: route 1, the Black and Azov seas basins with the Caspian Sea via the Volga–Don Canal, resulting in 46 invasions into the Caspian Sea (Grigorovich *et al.* 2003, Panov *et al.* 2007a); route 2, the Black and Azov seas to the Volga River reservoirs, resulting in 16 invasions; route 3, the Black and Azov seas basin across two geographic barriers via Volga River reservoirs to the Baltic Sea basin (Gulf of Finland), resulting in five invasions (e.g. two species of predatory onychopod cladocerans, *Cercopagis pengoi* and *Cornigerius maeoticus*; Panov *et al.* 2007b); route 4, the Caspian Sea basin, from the Caspian and Lower

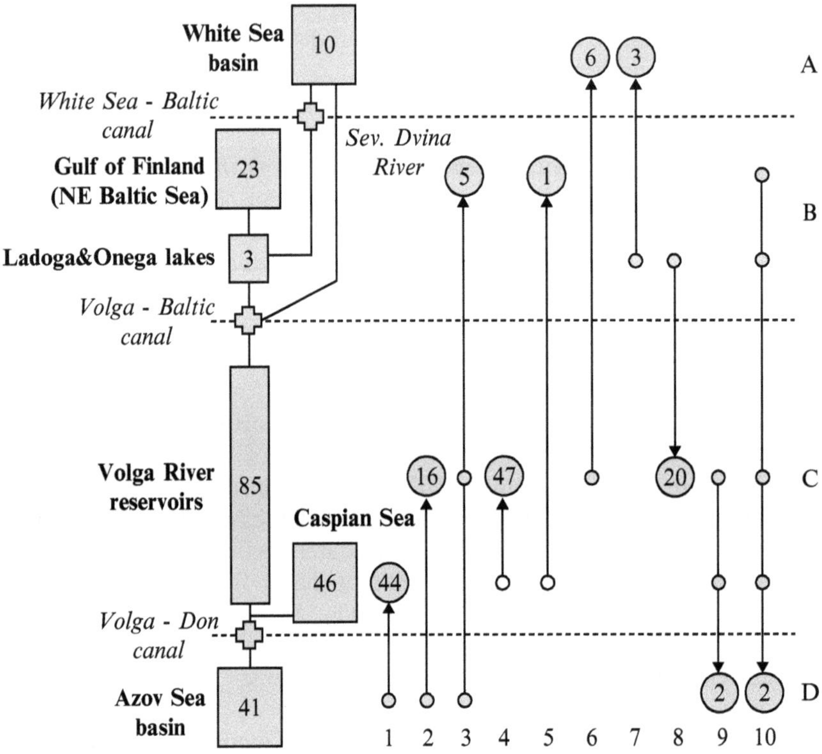

Fig. 2 Main recipient areas of aquatic NIS and specific invasion routes of their introductions within the Northern Invasion Corridor. Numbers in boxes and circles indicate the number of established NIS in risk areas (see Fig. 1) and by the invasion route, respectively. Dashed lines indicate the geographic barriers between previously isolated basins of: A – White Sea, B – Baltic Sea, C – Caspian Sea, and D – Black and Azov seas.

Volga, to the Middle and Upper Volga reservoirs, resulting in the spread of 47 species attributed, in part, to recent climate changes (Slynko *et al.* 2002); route 5, the Caspian Sea directly to the Baltic Sea, resulting in the invasion of *Evadne anonyx* (Rodionova and Panov 2006); route 6, the Caspian basin to the White Sea basin via canals linking the Upper Volga with the Severnaya Dvina River basin, the northernmost part of the Northern Invasion Corridor, resulting in the invasion of *Dreissena polymorpha* (Panov *et al.* 2007a); and route 7, the Baltic Sea basin to the White Sea basin, resulting in the invasion of two fish species and of the Chinese mitten crab, *Eriocheir sinensis* (Panov *et al.* 2007a).

Routes 8–10 are north-south movements from the Baltic basin via the Volga–Baltic Canal to the Volga River reservoirs (20+ species), from the Caspian basin via the Volga–Don Canal to the Azov Sea basin (two species),

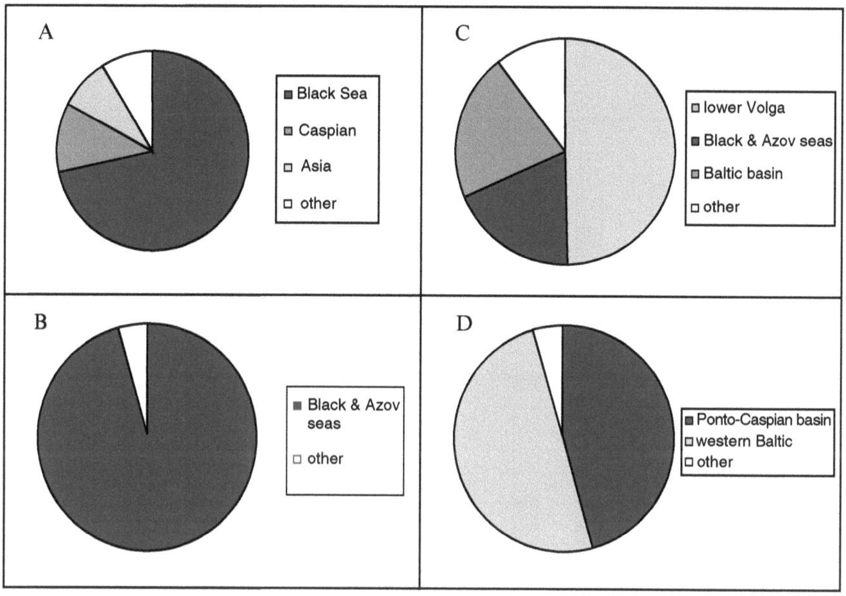

Fig. 3 Donor areas of aquatic NIS in the main risk areas of the Northern Invasion Corridor (A – Azov Sea, B – Caspian Sea, C – Volga River reservoirs, and D – eastern Gulf of Finland).

and long-distance transfer from the Baltic basin to the Azov Sea basin across two main geographical barriers via the Volga–Baltic Canal, Volga River, and Volga–Don Canal of the invasive crustaceans *Bythotrephes longimanus* and *E. sinensis* (Panov *et al.* 2007a).

Identifying vectors of introductions of NIS

In all main recipient areas within the Northern Invasion Corridor, shipping-related activities are the most important vectors of introductions of NIS. This includes direct transfers of NIS with ballast water, sediments, and hull fouling, and with migrations of NIS via navigable canals. The relative importance of shipping-related vectors has increased since the 1950s in all risk areas (Fig. 4). Long-term changes in the relative importance of different vectors indicate that shipping-mediated accidental introductions were the primary pathway of introductions into the semi-enclosed Azov Sea and the Gulf of Finland prior to 1950, whereas in the geographically isolated Caspian Sea basin and in the Volga River reservoirs other vectors, such as intentional species introductions, were prominent agents of introductions (Fig. 4). However, after the opening of the Baltic–Volga and Volga–Don canals in the middle of the 20th century, shipping also started to play the most important role for introductions of NIS

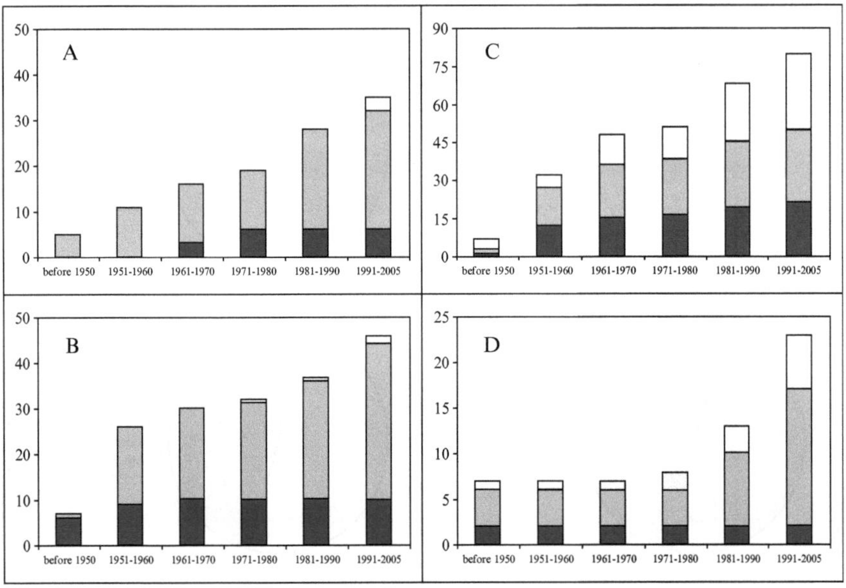

Fig. 4 Dynamics of the introductions of NIS in the four main risk areas (A – Azov Sea, B – Caspian Sea, C – Volga River reservoirs, and D – eastern Gulf of Finland) along the Northern Invasion Corridor for different vectors (black bars – intentional introductions, grey bars – shipping-mediated accidental introductions, and open bars – other vectors). Note the different scales for panels C and D.

for the Caspian Sea and the Volga River reservoirs. There is a clear trend in increasing importance of shipping-related vectors over time in all main risk areas, with greatest increase in rates of shipping-mediated introductions in the Gulf of Finland during the last 15 years (Fig. 4). The latter phenomenon can mainly be attributed to the effects of climate change that facilitates the establishment of warm-water species in the gulf (Panov *et al.* 2007b). Climate changes may also have facilitated the range expansion of some NIS during the last decades by other vectors, including natural migrations of NIS from adjacent southern areas, specifically for the Gulf of Finland and Volga River (Slynko *et al.* 2002, Panov *et al.* 2007b) (Fig. 4).

Assessing inoculation rates

The inoculation rates by propagules of aquatic NIS (including their resting stages) of the main risk areas within the Northern Invasion Corridor can be assessed only indirectly from shipping statistics that include information of volumes of discharged ballast water. However, such detailed information is often lacking, and only general statistics on the number of ships entering the

Caspian Sea and the Gulf of Finland and the volume of transported cargo are available. In 2003 and 2004, approximately 400 ships entered the Caspian Sea (mainly via the E90 water route) in each year transporting ~1 million tons of cargo. In the same two years, approximately 7,000 ships with 18 million tons of cargo entered the Gulf of Finland via the E50 water route each year. However, we do not have information on the likely volumes of ballast transported and the voyage durations; such information is important in relation to survival and inoculation potential. The use of such statistics for the estimation of inoculation rates of NIS without consideration of ballast water history and the duration of ship voyage allows for only preliminary qualitative estimates and may be associated with significant uncertainty.

Indirect estimations of inoculation rates using data on the long-term dynamics of invasion rates (Fig. 4) are even more questionable and uncertain. For instance, significant increase in the number of shipping-mediated introductions of NIS for the Caspian Sea during the last 15 years (Fig. 4B) can be attributed to the increased ship traffic in this period and to the related increases in inoculation rates. However, for the Gulf of Finland, even more profound increases in the number of shipping-mediated introductions of NIS (mostly Ponto-Caspian crustaceans) have been observed during the last 15 years (Fig. 4D). These introductions were most likely mediated by climate changes, as shipping intensity in the gulf via the E50 water route did not increase compared to earlier time periods (Panov *et al.* 2007b). In contrast to the inland ports, cargo turnover in the marine ports in the Gulf of Finland increased several-fold over the last 10–15 years, and currently exceeds 100 million tons per year (Panov *et al.* 2003). This has resulted in extremely high volumes of released ballast water and, consequently, in high inoculation rates by propagules of NIS. Taking into account available data on shipping, estimated inoculation rates within the Northern Invasion Corridor are relatively *low* for the Volga River reservoirs, Ladoga and Onega lakes, and White Sea, are *medium* for the Azov and Caspian seas, and must be considered *high* for the Gulf of Finland (Table 1).

Assessing the vulnerability to invasions of potential recipient areas

Ecosystem vulnerability to invasions may depend on abiotic and biotic resistance of the specific ecosystem to the establishment of NIS. Abiotic resistance is related to the environmental match of potential donor and recipient ecosystems. Considering main donor areas of NIS outside and within the Northern Invasion Corridor (see section on identification of recipient and donor areas of NIS and invasion routes above), abiotic resistance can be roughly estimated as *low* for the Azov and Caspian seas, *medium* for the Volga River reservoirs and Gulf of Finland, and *high* for the lakes Ladoga and Onega and for the White Sea basin.

Table 1 Ranking of three factors of risk for the invasions of aquatic NIS (inoculation rates, area invasibility, and species invasiveness) and the resulting integrated risk level for six risk areas along the Northern Invasion Corridor.

Risk area	Inoculation rates	Area invasibility	Species invasiveness	Integrated risk level
Azov Sea	medium	medium	medium	medium
Caspian Sea	medium	high	high	high
Volga River Reservoirs	low	high	medium	medium
Ladoga and Onega lakes	low	low	high	medium
Gulf of Finland	high	medium	high	high
White Sea	low	low	high	medium

Limnological conditions in the glacial lakes Ladoga and Onega, such as the low conductivity of the water, make these lakes unsuitable for the progressive expansion of NIS from the Caspian Sea basin to the Gulf of Finland and White Sea region via the Volga–Baltic Canal and White Sea–Baltic Canal. Dispersal of NIS from the Caspian Sea basin to the White Sea basin is possible only via waterways connected to the Severnaya Dvina River (invasion route 6; see Fig. 2). Any non-indigenous transmissions through lakes Ladoga and Onega are likely to have taken place with human-aided processes. Presently, only two invasive crustaceans are known from these lakes, the Baikalian zebra amphipod, *Gmelinoides fasciatus* (Stebbing), and the Chinese mitten crab, *E. sinensis* (Panov 1996, 2006). Thus, the lakes Ladoga and Onega can be considered as natural barriers to the dispersal of NIS that expand their ranges along waterways, including newly built canals. Successful transmission through these lakes may only occur for certain life history stages that attach to the hull of ships and that are able to tolerate short periods of exposure to unfavourable water conditions, or via transport in the ballast water of ships.

The biotic resistance of an ecosystem is related to the strength of interspecies relationships for any new invasive species, including food supply, competition, predator-prey, and parasite–host relationships. At the present stage, estimations of biotic resistance of aquatic ecosystems towards biological invasions are largely lacking (but see Ricciardi and MacIsaac 2000, Colautti *et al.* 2004, DeRivera *et al.* 2005, Fenieva *et al.* 2006), and the development of approaches to such estimations requires further study and was not considered in our qualitative risk assessment.

Considering rough qualitative estimates of abiotic resistance (i.e. environmental matching in terms of salinity and temperature regimes), vulnerability of the studied ecosystems to biological invasions can be estimated as *low* for the lakes Ladoga and Onega and the White Sea basin, *medium* for the Azov Sea basin, and *high* for the Caspian Sea, Volga River reservoirs, and the Gulf of Finland (Table 1).

Assessing the invasiveness of NIS

Because risk areas within the Northern Invasion Corridor are serving both as donors and recipients of NIS, it is important to assess the invasiveness of NIS in these areas according to their potential to spread, establish in new environments, and affect potential recipient ecosystems (species-specific risk assessment). We used available data on the main life history traits (salinity and temperature tolerance, fecundity and patterns of reproduction, ability to produce resting stages, etc.), invasion history, and known ecological impacts (Panov *et al.* 2006, 2007a,b) for an assessment of species invasiveness in different taxonomic groups of NIS in the main risk areas.

In all main risk areas, apart from the Caspian Sea where plants dominate as NIS, crustaceans were the largest group of NIS with the highest proportion of *medium* and *high* level of invasiveness (Fig. 5). Estimated proportions of established *medium* and *high*-risk NIS for these areas were highest for the Gulf of Finland (86%), and were somewhat lower for the Azov Sea basin, Volga River reservoirs, and Caspian Sea (58%, 48%, and 39%, respectively). However, the known level of negative impacts of NIS on the ecosystem biodiversity and functions is certainly the highest for the Caspian Sea, currently experiencing severe consequences of the *Mnemiopsis leidyi* A. Agassiz invasion (Shiganova

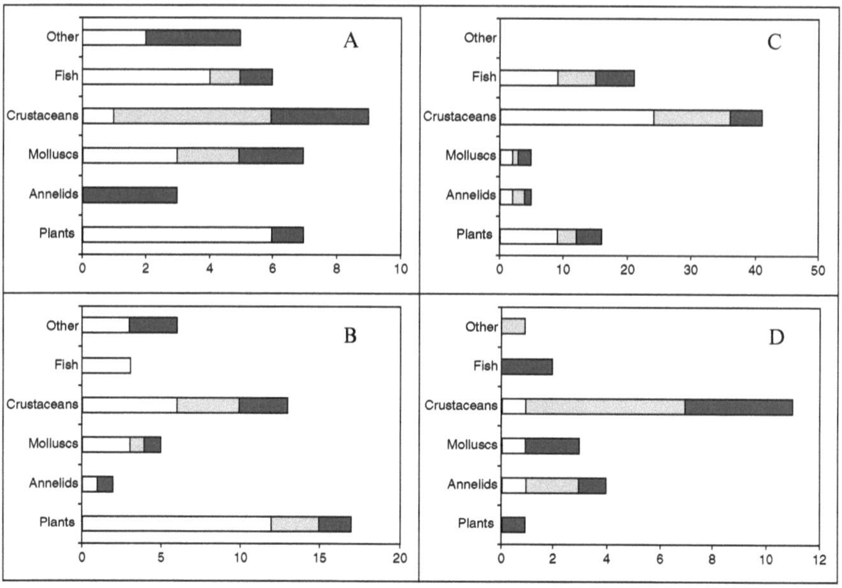

Fig. 5 Taxonomic composition of NIS in main risk areas along the Northern Invasion Corridor (A– Azov Sea, B – Caspian Sea, C – Volga River reservoirs, and D – eastern Gulf of Finland). Black bars indicate proportion of high-risk species.

et al. 2004). Compared to the Caspian Sea, the level of known negative impacts of invasive species in other main risk areas can be estimated as *medium*.

RESULTS OF THE QUALITATIVE RISK ASSESSMENT

Qualitative estimations of inoculation rates, ecosystem vulnerability to invasions, and species invasiveness (see sections above) were used for an overall assessment of the integrated risk level for each risk area within the Northern Invasion Corridor (also ranked as *low*, *medium*, and *high*). The estimated integrated ecosystem risk level was considered *high* for the Caspian Sea and Gulf of Finland, and *medium* for the rest of the risk areas (Table 1).

From the combination of environmental matching and species-specific risk assessments, we estimated likely levels of establishment for 34 key high-risk target NIS in risk areas within the Northern Invasion Corridor as a predictive risk assessment (Table 2). We deduced that the highest number of these will appear within the Gulf of Finland (17 species), followed by the Caspian Sea with seven potential new high-risk invaders. This assessment generally corresponded with the independently estimated *high* integrated ecosystem risk level for these two risk areas (Table 1).

The qualitative approach adopted here follows the predictive risk assessment first used in the Nordic Council-supported project on risk assessment of NIS in Nordic coastal waters (Gollasch and Leppäkoski 1999) for the eastern Gulf of Finland area, with a prediction of invasion of two invasive Ponto-Caspian onychopod species, *C. maeoticus* and *Podonevadne trigona* (Sars) into the eastern gulf (Panov *et al.* 1999). Subsequently, *C. maeoticus* was found in the Gulf of Finland in 2003 (Rodionova *et al.* 2005). However, the first new onychopod invader (i.e. after publication of the initial risk assessment in 1999) was another onychopod species, *Evadne anonyx*. This species was first recorded in the zooplankton of the gulf in 2000 (Rodionova and Panov 2006). *Evadne anonyx* was not considered as a high-risk species, because it had no previous invasion history and was not considered to be able to develop sustainable populations at salinities below 9 ppt (Panov *et al.* 2007b). In the eastern Gulf of Finland, *E. anonyx* successfully established in areas with water salinities as low as 1–3 ppt (Rodionova and Panov 2006). This unexpected invasion of *E. anonyx* into the eastern Baltic Sea may indicate that the most common Ponto-Caspian onychopods, not listed among our 34 high-risk species in the Table 2 [i.e. *Podonevadne camptonyx* (Sars), *Podonevadne angusta* (Sars), *Polyphemus exiguus* Sars, *Evadne prolongata* Behning], may pose some risks of long-distance intra-continental transfer if appropriate vectors of introduction are available (shipping along the Volga–Baltic inland waterway). In general, the Ponto-Caspian onychopods pose the highest risks for the Gulf of Finland, as all the non-indigenous cladocerans established in the eastern Baltic Sea belong to this group, and the rapid and successful establishment of three onychopod species

Table 2 Aquatic NIS with a high potential to become established with self-reproducing populations in risk areas along the Northern Invasion Corridor.

Taxa	Azov Sea	Caspian Sea	Volga River Reservoirs	Ladoga and Onega Lakes	Gulf of Finland	White Sea
Plants						
Pseudosolenia calcar-avis (Schultze) Sundström	1924	1934	**high**	low	**high**	low
Pseudo-nitszchia seriata (Hasle) Hasle	indigenous	1990	**high**	low	**high**	low
Skeletonema subsalsum (Cleve-Euler) Bethge	indigenous	indigenous	1957	low	**high**	low
Thalassiosira incerta Makarova	indigenous	indigenous	1967	low	**high**	low
Actinocyclus normanii (W. Gregory) Hustedt	indigenous	indigenous	1986	low	**high**	low
Chroomonas acuta Utermöhl	indigenous	indigenous	1988	low	**high**	low
Cnidaria						
Blackfordia virginica Mayer	1930s	1956	low	–	**high**	low
Bougainvillia megas Kinne	1950s	1961	low	–	**high**	low
Ctenophora						
Mnemiopsis leidyi A. Agassiz	1988	1999	–	–	medium	low
Annelida						
Ficopomatus enigmaticus (Fauvel)	1960	1961	low	–	**high**	medium
Hypania invalida (Grube)	indigenous	**high**	1960	low	**high**	medium
Marenzelleria neglecta (Sikorski and Bick)	**high**	**high**	–	–	1996	medium
Mollusca						
Rapana venosa (Valenciennes)	1956	low	–	–	low	low
Mytilaster lineatus (Gmelin)	indigenous	1919	–	–	medium	low
Teredo navalis Linnaeus	1953	medium	–	–	low	medium
Dreissena polymorpha (Pallas)	indigenous	indigenous	1953	low	1986	1970s
Dreissena bugensis (Andrusov)	**high**	**high**	1992	low	**high**	**high**
Mytilopsis leucophaeata (Conrad)	**high**	**high**	–	–	2004	low

Table 2 Continued.

Taxa	Azov Sea	Caspian Sea	Volga River Reservoirs	Ladoga and Onega Lakes	Gulf of Finland	White Sea
Crustacea						
Acartia tonsa Dana	2000	1981	medium	low	1934	**high**
Bythotrephes longimanus Leydig	1960s	low	1957	indigenous	indigenous	indigenous
Cercopagis pengoi (Ostroumov)	indigenous	indigenous	1960	low	1992	low
Podonevadne trigona (Sars)	indigenous	indigenous	1966	low	**high**	low
Cornigerius maeoticus (Pengo)	indigenous	indigenous	1970	low	2003	low
Evadne anonyx Sars	indigenous	indigenous	low	–	2000	low
Rhithropanopeus harrisii (Gould)	1960	1958	medium	low	**high**	medium
Eriocheir sinensis H. Milne-Edwards*	medium	**high**	–	–	low	medium
Callinectes sapidus M. J. Rathbun	1967	medium	–	–	low	low
Fish						
Pseudorasbora parva (Linnaeus)	1970	**high**	high	low	medium	low
Clupeonella cultriventris (Nordmann)	1970	indigenous	1964	low	**high**	**high**
Neogobius melanostomus (Pallas)	indigenous	indigenous	1968	low	**high****	medium
Neogobius fluviatilis (Pallas)	indigenous	indigenous	1960	low	**high**	medium
Neogobius iljini Vasiljeva et Vasiljev	indigenous	indigenous	1970	low	**high**	medium
Proterorhinus marmoratus (Pallas)	indigenous	indigenous	1981	low	**high**	medium
Perccottus glenii Dybowski	**high**	**high**	1981	**high**	1950	**high**
Total number of HRS	**4**	**7**	**3**	**1**	**17**	**4**

Numbers in cells indicate the year of first record; empty cells indicate that establishment is not considered possible
* adults of *E. sinensis* were recorded in all risk areas
** single specimens of *N. melanostomus* were recorded in the western Gulf of Finland in 2005 (Ojaveer 2006)

has occurred in this region during the last 15 years, facilitated, most likely, by climate changes in the region (Panov *et al.* 2007b).

It is important to note the primary importance of the Northern Invasion Corridor for most recent invasions of NIS from southern regions to the Gulf of Finland (high-risk area) and the Baltic Sea. This importance exists despite the fact that the corridor currently contributes less than 20% to the volume of cargo transported to the Gulf of Finland (approximately 20 million tons compared to around 100 million tons per year from other transport corridors) and even much less in terms of ballast water (most ballast water released in the Gulf of Finland ports originates from areas other than the Ponto-Caspian). The role of other invasion corridors in biological invasions of the gulf is minor. During the last 15 years all other invasions corridors contributed to only one successful establishment of NIS in the gulf. In 2003, the Atlantic species Conrad's false mussel, *Mytilopsis leucophaeata*, was discovered in an area affected by cooling water discharges from a nuclear power plant (Laine *et al.* 2006).

The Gulf of Finland can also be considered as a high-risk donor area of NIS for aquatic systems outside the Northern Invasion Corridor area, specifically for the adjacent inland water ecosystems (Pienimäki and Leppäkoski 2004), and the North American Great Lakes, which are connected with eastern Baltic by an intercontinental invasion corridor (see Panov *et al.* 1999, 2003, 2007b). Ponto-Caspian invasive onychopods producing large numbers of resting eggs may pose a very high risk of introduction with ballast waters, even when regular ballast water management procedures are carried out (e.g. ballast water exchange during oceanic ship voyages), which might be ineffective for the resting eggs accumulating in the sediments of ballast tanks (MacIsaac *et al.* 1999, Bailey *et al.* 2005).

Managing the dispersal of species by shipping is a priority throughout the Northern Invasion Corridor. Taking into account the most important invasion routes within this system (Fig. 2), the risk-reducing management options for ballast water and other shipping-mediated vectors should be implemented at the main entrances to the Northern Invasion Corridor, in the ports of the lower Don River (entrance to the European inland waterway E90 from the Azov Sea) and in the ports of eastern Gulf of Finland, specifically the Port of St. Petersburg (entrance to the main European inland waterway E50 from the Baltic Sea). These management options should include treatment of ballast water and sediments, and hull fouling.

CONCLUSIONS

The Northern Invasion Corridor is playing an important role in the introductions of NIS in eastern Europe; its significance may increase over time with further construction and/or improvement of navigable watercourses and with their integration into the European network of inland waterways. Past patterns of transmission of NIS are likely to be repeated in the future, resulting in a further expansion of Ponto-Caspian species and in increased invasions within the Caspian Sea basin itself. Specifically, we expect a range extension of several invasive species already established in the ecosystems along the corridor.

The qualitative approach to risk assessment of aquatic invasions, tested for the Northern Invasion Corridor in the present study, can be considered as a useful tool for management purposes and is also applicable to other main European invasion corridors. The Northern Invasion Corridor case study indicated that relevant management options should be first implemented at the entrances to the inland waterways. Examples of "abiotic resistance" to biological invasions, such as the soft-water lakes Ladoga and Onega, with effective natural barrier to the dispersal of Ponto-Caspian species, demonstrate the potential effectiveness of such barriers along other European inland waterways for preventing the dispersal of actively migrating NIS. The application of more accurate quantitative methodology of risk assessment of biological invasions will be possible only after the development and implementation of comprehensive information systems on shipping statistics and after the collection of detailed information on the biological traits of invasive NIS established in the recognized donor areas (including potentially invasive indigenous species). To obtain such information further studies are required.

ACKNOWLEDGEMENTS

We are grateful to three anonymous reviewers for their valuable comments and to Dr. Wolfgang Jansen (North/South Consultants Inc., Canada) for careful reading and editing of a previous draft of the manuscript. This study has been supported by the European Commission Sixth Framework Programme Integrated Project ALARM (Assessing LArge scale environmental Risks for biodiversity with tested Methods, contract GOCE-CT-2003-506675), Strategic Targeted Research Project DAISIE (Delivering Alien Invasive Species Inventories for Europe, contract SSPI-CT-2003-511202), the Ministry of Education and Science of the Russian Federation (contract 02.435.11.4003), and Program "Biodiversity" of Russian Academy of Sciences (Area 5).

REFERENCES

Anonymous. 1998. Guidelines for ecological risk assessment. U.S. Environmental Protection Agency, Risk Assessment Forum, Washington, DC, U.S.A., EPA/630/R095/002F.

Anonymous. 2005. Inventory of most important bottlenecks and missing links in the E waterway network. Resolution 49. United Nations, New York and Geneva.

Bailey, S.A., I. C. Duggan, P. T. Jenkins, and H. J. MacIsaac. 2005. Invertebrate resting stages in residual ballast sediment of transoceanic ships. Canadian Journal of Fisheries and Aquatic Sciences **62**, 1090–1103.

Bij de Vaate, A., K. Jazdzewski, H. A. M. Ketelaars, S. Gollasch, and G. Van der Velde. 2002. Geographical patterns in range extension of Ponto-Caspian macroinvertebrate species in Europe. Canadian Journal of Fisheries and Aquatic Sciences **59**, 1159–1174.

Colautti, R. I., A. Ricciardi, I. A. Grigorovich, and H. J. MacIsaac. 2004. Is invasion success explained by the enemy release hypothesis? Ecology Letters **7**, 721–733.

DeRivera, C. E., G. M. Ruiz, A. H. Hines, and P. Jivoff. 2005. Biotic resistance to invasion: native predator limits abundance and distribution of an introduced crab. Ecology **86**, 3364–3376.

European Agreement on Main Inland Waterways of International Importance (AGN). 1996. Geneva, Switzerland, 19 January 1996. Internet: http://www.unece.org/trans/conventn/agn.pdf

Fenieva, I. Yu., S. V. Budaev, and Yu. Yu. Dgebuadze. 2006. Simulation of Cladoceran survival strategy under conditions of food depletion. Russian Journal of Ecology **37**, 28–34.

Galil, B. S., S. Nehring, and V. E. Panov. 2007. Waterways as invasion highways – Impact of climate change and globalization. In W. Nentwig, editor. Pages 59–74 in Biological invasions. Ecological Studies n 193, Springer, Berlin, Germany.

Galil, B. and D. Minchin. 2006. Snakes and ladders: navigable inland waterways as invasive corridors. Pages 71–75 in J. Davenport and J. L. Davenport, editors. The ecology of transportation: Managing mobility for the environment. Environmental Pollution Series n 10. Springer, Dordrecht, The Netherlands.

Gerrard, S. and J. Petts. 1998. Isolation or integration? The relationship between risk assessment and risk management. Pages 1–19 in R. E. Hester and R. M. Harrison, editors. Risk assessment and risk management. Royal Society of Chemistry, Cambridge.

Grigorovich, I. A., T. W. Therriault, and H. J. MacIsaac. 2003. History of aquatic invertebrate invasions in the Caspian Sea. Biological Invasions **5**, 103–115.

Jażdżewski, K. 1980. Range extensions of some gammaridean species in European inland waters caused by human activity. Crustaceana Supplement **6**, 84–106.

Ketelaars, H. A. M. 2004. Range extensions of Ponto-Caspian aquatic invertebrates in continental Europe. Pages 209–236 in H. J. Dumont, T. A. Shiganova and U. Niermann, editors. Aquatic invasions in the Black, Caspian, and Mediterranean seas. Kluwer Academic Publishers, Dordrecht, The Netherlands.

Laine, A. O., J. Mattila, and A. Lehikoinen. 2006. First record of the brackish water dreissenid bivalve *Mytilopsis leucophaeata* in the northern Baltic Sea. Aquatic Invasions **1**, 38–41.

Leppäkoski, E. and S. Gollasch. 2006. Risk assessment of ballast water mediated species introductions – a Baltic Sea approach. Report prepared for HELCOM, Helsinki, Finland. 111 pp.

MacIsaac, H. J., I. A. Grigorovich, J. A. Hoyle, N. D. Yan, and V.E. Panov. 1999. Invasion of Lake Ontario by the Ponto-Caspian cladoceran predator *Cercopagis pengoi*. Canadian Journal of Fisheries and Aquatic Sciences **56**, 1–5.

Nehring, S. 2002. Biological invasions into German waters: an evaluation of the importance of different human-mediated vectors for nonindigenous macrozoobenthic species. Pages 373–383 in E. Leppäkoski, S. Gollasch and S. Olenin, editors. Invasive aquatic species of Europe. Distribution, impacts and management. Kluwer Academic Publishers, Dordrecht, The Netherlands.

Ojaveer, H. 2006. The round goby *Neogobius melanostomus* is colonising the NE Baltic Sea. Aquatic Invasions **1**, 44–45.

Panov, V. E. 1996. Establishment of the Baikalian endemic amphipod *Gmelinoides fasciatus* Stebb. in Lake Ladoga. Hydrobiologia **322**, 187–192.

Panov, V. E. 2006. First record of the Chinese mitten crab, *Eriocheir sinensis* H. Milne Edwards, 1853 (Crustacea, Decapoda, Varunidae) from Lake Ladoga, Russia. Aquatic Invasions **1**, 28–31.

Panov, V. E., D. E. Bychenkov, N. A. Berezina, and A. A. Maximov. 2003. Alien species introductions in the eastern Gulf of Finland: current state and possible management options. Proceedings of the Estonian Academy of Sciences, Biology and Ecology **52**, 254–267.

Panov, V. E., Yu.Yu. Dgebuadze, T. A. Shiganova, and A. A. Filippov. 2007a. Checklist for aquatic alien species in inland waterways of European Russia. Aquatic Invasions 2, in press.

Panov, V. E., P. I. Krylov, and I. V. Telesh. 1999. The St. Petersburg harbour profile. Pages 225–244 in S. Gollasch and E. Leppäkoski, editors. Initial risk assessment of alien species in Nordic coastal waters. Nord 1999:8. Nordic Council of Ministers, Copenhagen, Denmark.

Panov, V. E., N. V. Rodionova, P. V. Bolshagin, and E. A. Bychek. 2007b. Invasion biology of Ponto-Caspian onychopod cladocerans (Crustacea: Cladocera: Onychopoda). Hydrobiologia, in press.

Panov, V. E., V. S. Shestakov, and M. B. Dianov. 2006. Database of Aquatic Invasive Species of Europe (AquaInvader). Internet: http://www.zin.ru/rbic/projects/aquainvader/

Pienimäki, M. and E. Leppäkoski. 2004. Invasion pressure on the Finnish Lake District: invasion corridors and barriers. Biological Invasions **6**, 331–346.

Ricciardi, A. and H. J. MacIsaac. 2000. Recent mass invasion of the North American Great Lakes by Ponto-Caspian species. Trends in Ecology and Evolution **15**, 62–65.

Rodionova, N. V., P. I. Krylov, and V. E. Panov. 2005. Invasion of the Ponto–Caspian predatory cladoceran *Cornigerius maeoticus maeoticus* (Pengo, 1879) into the Baltic Sea. Oceanology **45**, 66–68.

Rodionova, N. V. and V. E. Panov. 2006. Establishment of the Ponto-Caspian predatory cladoceran *Evadne anonyx* in the eastern Gulf of Finland, Baltic Sea. Aquatic Invasions **1**, 7–12.

Shiganova, T. A., H. J. Dumont, A. F. Sokolsky, A. M. Kamakin, D. Tinenkova, and E. K. Kurasheva. 2004. Population dynamics of *Mnemiopsis leidyi* in the Caspian Sea, and effects on the Caspian ecosystem. Pages 71–111 in H. J. Dumont, T. A. Shiganova and U. Niermann, editors. Aquatic invasions in the Black, Caspian, and Mediterranean seas. Kluwer Academic Publishers, Dordrecht, The Netherlands.

Slynko, Yu. V., L. G. Korneva, I. K. Rivier, K. H. Shcherbina, V. G. Papchenkov, M. I. Orlova, and T. W. Therriault. 2002. Caspian-Volga-Baltic invasion corridor. Pages 399–411 E. in Leppäkoski, S. Gollasch and S. Olenin, editors. Invasive aquatic species of Europe. Distribution, impacts and management. Kluwer Academic Publishers, Dordrecht, The Netherlands.

Van der Velde, G., I. Nagelkerken, S. Rajagopal and A. Bij de Vaate. 2002. Invasions by alien species in inland freshwater bodies in Western Europe: the Rhine delta. Pages 360–372 in E. Leppäkoski, S. Gollasch and S. Olenin, editors. Invasive aquatic species of Europe: Distribution, impacts and management. Kluwer Academic Publishers, Dordrecht, The Netherlands.

Chapter thirty-six

Logistics of shipboard and dockside testing of ballast water treatment systems in the United States

David A. Wright

INTRODUCTION

It is now widely accepted that ballast water, used to maintain ships' trim and stability, is a major vector for the unintentional introduction of non-indigenous organisms into coastal waters (Carlton and Geller 1993, Carlton *et al.* 1995). Ballast capacities range from several cubic meters (m^3 or tonnes) in the case of fishing boats to hundreds of thousands of tonnes in very large bulk carriers, where ballasting rates can be as high as 15,000–20,000 tonnes h^{-1}. Ballast water discharges were recognized as an international concern as early as 1973, when the United Nations requested the World Health Organization to investigate the spread of epidemic disease spread by ballast water. The U.N. International Maritime Organization (IMO) first adopted a voluntary ballast water exchange standard in 1991 and in 1997; after several subsequent revisions and enhancements, the IMO Marine Environmental Protection Committee (MEPC) adopted the IMO guidelines for management of ships' ballast water. This culminated in the February 2004 UN. *International Convention for the Control and Management of Ships Ballast Water and Sediments*, whose stated goal was to control adverse ecological, economic, and human health effects caused

Francesca Gherardi, Biological invaders in inland waters: Profiles, distribution, and threats, 657–668.

by the global dispersal of aquatic nuisance species resulting from ballasting/de-ballasting procedures worldwide. The Convention is pending ratification by 30 States, representing 35% of the world merchant shipping tonnage. Criteria for 'successful' management or treatment have been published as Regulation D-1, relating to Open Ocean Ballast Water Exchange, and Regulation D-2, described as the Ballast Water Performance Standard pertinent to the efficacy of ballast water treatment.

In short, Regulation D-1 reads: The criterion for successful exchange is defined as a better than 95% volumetric replacement of water either through an empty-refill procedure or a pass-through procedure involving 3× the volume of the tank (or less if the 95% exchange is satisfactorily met).

In short, Regulation D-2 reads: The criterion for successful treatment has been defined as the discharge of less than 10 viable organisms/m^3 greater than or equal to 50 μm in minimum dimension and less than 10 viable organisms/mL less than 50 μm in minimum dimension and greater than or equal to 10 μm in minimum dimension. Specific bacteria are included as 'indicator microbes', namely strains O1 and O139 of *Vibrio cholerae* (standard: <1 cfu/100 mL or <1 cfu/gm wet weight zooplankton); *Escherichia coli* (standard: <250 cfu/100 mL); and intestinal enterococci (<100 cfu/100 mL).

Under regulation D-5 of the Convention, these standards are subject to review, taking into account safety considerations; environmental acceptability, i.e. not causing more or greater environmental impacts than it solves; practicability, i.e. compatibility with ship design and operations; cost effectiveness; and biological effectiveness in terms of removing, or otherwise rendering inactive harmful aquatic organisms and pathogens in ballast water. The Convention states that such a review process "should include a determination of whether appropriate technologies are available to achieve the standard, an assessment of the above mentioned criteria, and an assessment of the socio-economic effect(s) specifically in relation to the developmental needs of developing countries, particularly small island developing States".

In 1990, the US Congress enacted the Non-Indigenous Aquatic Nuisance Prevention and Control Act which required ships to take measures to reduce the risk of invasive species transfers to the Great Lakes and Hudson River. This was reauthorized and expanded nationally as the US National Invasive Species Act (NISA) of 1996 that became effective in July 1999. The National Aquatic Invasive Species Act (NAISA) was introduced in the US Congress in 2003. If passed, it will strengthen the original NISA and make voluntary aspects of that legislation mandatory. Problems associated with ballast water exchange are well known and are perhaps best illustrated by the near loss of the car transporter *Cougar Ace* off Alaska in July 2006. Pending legislation makes provision for ballast water treatment to standards that are at least as stringent as those associated with the 2004 IMO convention.

Current research initiatives related to ballast water treatment have, therefore, focused on the development of the most cost-effective and environmentally sound technologies for achieving these standards. An important, related goal has also been the refinement of appropriate biological end points that effectively define such standards. It is, for example, a difficult task to assess the viability of resting spores or dormant stages of organisms, including several members of the phytoplankton community, such as dinoflagellates, of which many species are toxic to marine life and therefore cause severe economic damage and can be even of an immediate threat to human health. Beyond the immediate aim of establishing 'dose–response' parameters for treatment technologies, an important secondary goal has been to establish a realistic set of measurements that can be universally applied to compliance monitoring once ballast water treatment technologies are in common usage. The UN 2004 Ballast Water Convention makes provision for the inspection of ships by port State control officers, who can verify whether the vessels are carrying valid certification and have up-to-date ballast water log books. Inspectors will have the authority to collect ballast water samples and may subject these to detailed inspection to determine whether appropriate management or treatment has been carried out. Universal standardization of this inspection process will require reliable end point determinations that are reasonably standardized and not too onerous or costly to apply.

This paper describes progress to date in two ongoing initiatives in the USA designed to achieve these goals. The Baltimore Harbor Ballast Water Treatment Technology Demonstration Program is one of several dockside testing facilities that are operational or in the planning stages in the USA. Similar initiatives are being pursued internationally in Norway, Singapore, and Germany, and at the Royal Netherlands Institute for Sea Research. The Baltimore facility is the culmination of more than 15 years continuous funding by the US National Oceanic and Atmospheric Administration (NOAA) into invasive species research and mitigation technologies, and currently represents a partnership between NOAA, Maryland Port Administration, the US Maritime Administration, and the University of Maryland Center for Environmental Science. The facility is not affiliated with the testing center located at Key West under the auspices of the US Environmental Protection Agency's Environmental Technology Verification (ETV) program, but represents a potential estuarine alternative to that facility and a freshwater testing facility proposed for the Great Lakes. The Shipboard Technology Evaluation (STEP) Program, under the auspices of the US Coast Guard, addresses a need expressed by technology developers and vessel owners for incentives that will encourage the development of prototype treatment systems and shipboard testing. However, vessel owners have been reluctant to invest the resources to install and operate experimental treatment systems that might not meet discharge standards mandated by future regulations. To address this concern, vessels accepted into this program may be granted an equivalency to future ballast water discharge standard regulations, for up to the life of the vessel or the system, while they operate satisfactorily.

DOCKSIDE TESTING – BALTIMORE HARBOR BALLAST WATER TREATMENT TECHNOLOGY DEMONSTRATION PROGRAM

The Baltimore Ballast Water Treatment (BWT) Test Facility was constructed at the Maryland Port Administration Clinton St. Terminal in 2004–2005. The site is located on the north shore of the Patapsco River estuary less than a mile from Baltimore's Inner Harbor. The system is fully operational and is currently being used to test a range of Ballast Water Treatment (BWT) technologies at shipboard, ship-ready scale (Figs. 1 and 2). The system is located in a locked, secure indoor environment, a disused, covered dock that is shielded from the direct sunlight and not subject to large fluctuations in temperature. Experimental biocide discharge permits at the site have been negotiated with Maryland Department of the Environment on a case-by-case basis.

The facility has several advantages as a test facility for the estuarine environment. It is centrally located in the Chesapeake Bay at a working dock immediately adjacent to active loading/unloading facilities involving ballasting/de-ballasting operations. It is located in an area of high productivity that easily meets the July 2005

Fig. 1 Baltimore Ballast Water Treatment Technology Testing Facility (Maryland Port Administration, Clinton St. Terminal). Tank layout and biocide dosing equipment.

Fig. 2 Baltimore Ballast Water Treatment Technology Testing Facility (Maryland Port Administration, Clinton St. Terminal). From top left, clockwise. Self-priming pump (1500–2000 gpm ≡ 250–350 $m^3\ h^{-1}$); Sea strainer; UV irradiation system (Aquionics Inc.); 20 cm discharge + 10 cm backwash discharge; 3 (treatment) × 3 (replicate) sampling tanks; Primary filter (Arkal Inc.) with backwash system.

modifications to the International Maritime Organization's standards for minimum testing requirements at dockside sites with respect to numbers of organisms, i.e. "test organisms greater than or equal to 10 μm and less than 50 μm in minimum dimension should be present in a total density of preferably 10^4, but not less than 10^3 per mL and should consist of at least five species from at least three different phyla/divisions" (MEPC\53\24\Add.1 ANNEX 3, p. 20). TSS, POC, and DOC conditions are also met at this site (MEPC\53\24\Add.1 ANNEX 3, p. 19). Significantly, such conditions are not universally met in the US coastal waters. At the Clinton St. site, plankton densities in the 10–50 μm range occasionally approached 10^5 per mL during the late summer months with a broad range of phyla represented.

At the July 2005 meeting of the International Maritime Organization (IMO) Marine Environmental Protection Committee, the need was established for BWT testing to be performed at two (or more) salinities separated by at least 10 psu. The naturally occurring salinities at the site vary seasonally between <5–>15 psu, although some salinity manipulation is possible through access to a freshwater supply at the test site. While testing over a range of salinities remains an option at the Baltimore site, its primary application is perceived as a mid-salinity alternative to pending or established US testing sites at Key West, Florida (fully saline), and Duluth, Minnesota (freshwater). Its ambient temperature is also intermediate between those two sites, although, with temperatures closer to the Gulf Coast during the summer, it is also seen as a warm water alternative to a potential cold water test site such as Puget Sound.

SHIPBOARD TECHNOLOGY EVALUATION (STEP) PROGRAM

This program provides shipping companies and vendors of treatment technologies with an opportunity to install ship-ready equipment, subject to a tolerably successful record of pre-installation trials, either at full-scale or with appropriate scaling parameters. In return for essentially providing test platforms for best available technologies and the acquisition of important information on biological end points and the feasibility and durability of BWT technologies under 'real-world' conditions, shippers and suppliers who successfully complete the STEP program qualify for an 'equivalency' to any future standards that may apply to BWT. Details of application requirements for the US Coast Guard STEP program may be found at http://www.uscg.mil/hq/g-m/mso/step.htm. The basic components of a STEP application are: (1) a letter of commitment from the shipping company, the suppliers of the BWT system, and the principal scientists involved in the shipboard trials; (2) appropriate environmental compliance documentation relating to any shipboard hazards associated with the BWT technology and environmental consequences of discharged, treated water; (3) documentation of preliminary experiments demonstrating efficacy of the applicant's treatment system – these can include bench-scale or mesocosm studies essentially demonstrating that the technology will be effective an acceptable level, e.g. 98% removal of (zoo)plankton >50 μm in the smallest dimension; (4) study plan demonstrating an appropriate degree of endpoint replication during each trial and the conduct of at least one trial in each distinct water body where the vessel operates; (5) flow chart indicating the movement of correspondence, decision points, and STEP time line.

Necessary components of the STEP process itself include:

- Comparison of treatment system performance with ballast water exchange (although this is being reconsidered).
- Biological experiments documenting viability/mortality of entrained organisms.
- Submission of quarterly and annual reports on the system's operation and performance. Presentation of the physical design and engineering of the treatment system, and plans to maintain reliable operation and monitor the system's performance for a three year period (years 2–4) following initial trial(s). In year 5, a final shipboard test is to be conducted to provide detailed long-term performance data.

It should be noted that STEP procedures and requirements are similar, but not necessarily identical to specifications listed by IMO. For a summary of the latter, the reader is directed to MEPC\53\24\Add.1 ANNEX 3 (pp. 15–17).

SUMMARY OF PROGRESS TO DATE IN DOCKSIDE AND SHIPBOARD TESTING

Dockside testing

Dockside tests performed at the Baltimore test site since 2004 represent a continuation of earlier trials carried out aboard the US reserve fleet vessel, the USS *Cape May* in 2001. Most of the information from earlier tests has been reported elsewhere (Wright *et al.* 2005a, 2007b) and concentrated on the performance of UV and biocide treatments applied singly. The focus of more recent trials has been the efficacy of combination treatments, including the effect of primary filtration in tandem with secondary treatments such as UV irradiation and biocides. The test system was also used to investigate subsidiary questions associated with individual treatments. These included the relative efficacy of the nominal 55 μm *vs.* the nominal 100 μm configurations of the filter, the effect of backwashing on filter performance, and the relative efficacy of UV irradiation applied "in series" *vs.* "in parallel". The configuration of the system offers the option of directing water through all four UV systems in parallel or directing a stream of water first through one pair, then through the second pair of UV units. A summary of system configurations/combinations tested in 2004/5 is shown below:

- 100% *vs.* 50% UV irradiation (half-strength UV obtained by turning off one unit on either side of the system)
- 100% and 50% UV irradiation in series *vs.* in parallel

- 55 μm filter configuration + 50% or 100% UV irradiation
- 55 μm *vs.* 100 μm filter configuration of filter in combination with UV
- 100% and 50% UV irradiation in parallel in combination with 0.5 mg L^{-1} and 1 mg L^{-1} Seakleen
- 55 μm filter configuration filter + 100% *vs.* 50% UV + 0.5 mg L^{-1} and 1 mg L^{-1} Seakleen

Results from these studies were presented in detail at the March 2006 World Maritime Technology Conference (Wright 2006) and are briefly summarized here. Using the disk filter configured for a 55 μm cut-off in series with the UV system employed at maximum output (30 kW, delivering ~200 mW s^{-1} cm^{-2}), a 95% removal of plankton >50 μm was achieved, although in another test the same performance (95% removal) was seen with the UV system run at 50% output (~100 mW s^{-1} cm^{-2}). Only limited comparison can be made between the results of these trials and data obtained by Waite *et al.* (2003) in a filter-UV trial conducted at similar flow rates. Waite *et al.* (2003) reported a 90% removal of zooplankton in the >50 μm range using a 50 μm screen filter, compared with an 81% removal of this size fraction obtained from the Baltimore study (Wright 2006). However, zooplankton viabilities after filter + UV treatment were not recorded by Waite *et al.* (2003). Based on observations of bacterial growth in their studies, Waite *et al.* (2003) recommend a UV dose substantially higher than the 60 mW s^{-1} cm^{-2} achieved by the system they tested. Such a conclusion is supported by a series of scaling studies (Wright *et al.* 2006, 2007b) that indicated a minimum dose of 200 mW s^{-1} cm^{-2} was needed for the broad spectrum kill that would be required for successful ballast water treatment.

Shipboard trials

As a precursor to formal STEP trials, a test cruise was performed aboard the Princess Cruise Lines ship *Coral Princess* by the same scientific team responsible for the Baltimore dockside trials. In a report by Royal Haskoning Environmental Management published by the Northeast Midwest Institute, it was postulated that the cruise ship industry should play a leading role in the development of ballast water treatment technology and the suggestion was made that "the cruise ship industry may act as a testing ground for up-scalable technologies" (Tjallingi and Schilperood 2001). In response to this suggested initiative, Princess Cruise Lines installed ballast water treatment (BWT) systems aboard several of their ships beginning in 2002. The BWT system installed aboard the *Coral Princess* in 2003 consisted of a disk filter manufactured by Arkal Inc., Tel Aviv, Israel (nominal cutoff: 55 μm) mounted upstream from a secondary treatment system consisting of a medium pressure Aquionics UV system nominally rated at 200 mW s^{-1} cm^{-2}. This was very similar to the system undergoing testing at the Baltimore dockside site. While the primary goal of the shipboard trial was to determine the efficacy of the installed BWT system, the tests performed were designed to inform both efficacy and compliance testing, both in terms of the sampling strategy and the feasibility of the biological end points employed.

The trial took place during the repositioning cruise from British Columbia, through the Panama Canal to Fort Lauderdale, Florida; a distance of approximately 3,500 miles (Fig. 3). The passage through the Panama Canal provided an opportunity to determine the efficacy of ballast water exchange (BWE) by exchanging freshwater from the Panamanian Lakes system with saline western Caribbean water. A comparison of the salinity shift with changes in biota determined the efficacy of BWE and provided a direct comparison with BWT. While detailed accounts of results from this shipboard trial have been published elsewhere (Wright 2007, Wright *et al.* 2005b, 2007a,b), major findings and conclusions are summarized below:

- Ballast water exchange as determined by measurement of salinity tends to overestimate the exchange of planktonic organisms.
- Once-through filtration + UV as a ballast water treatment would not meet IMO standards, as currently drafted for organisms in the >50 μm range but would probably meet or exceed the efficacy of ballast water exchange (~ 90%).
- Once-through filtration + UV as a ballast water treatment controlled coliform bacteria but gave mixed results with culturable, heterotrophic bacteria.
- Filtration + UV as a ballast water treatment would not meet IMO standards as currently drafted for organisms in the 10–50 μm range. For example, the two dominant genera *Guinardia* and *Dytilum*, both within the 10–50 μm size range, would leave mean live residuals of $508 \times 10^6\ m^{-3}$ and $289 \times 10^6\ mL^{-3}$, thereby exceeding IMO draft standards by factors of 51 and 29, respectively.
- Detailed taxonomic identification of phytoplankton is too time-consuming to be used as a universal biological end point for efficacy or compliance testing and microscopic examination of nonmotile organisms will not provide sufficient evidence of viability status (e.g. chloroplast integrity is unreliable).
- Growth potential (through grow-out) of dominant phytoplankton taxa will probably provide sufficient evidence of viability for this group.
- Several phytoplankton groups overlap in size characteristics into the >50 μm size range, leaving post-treatment residual densities in this size range as high as $>10^8$ cells m^{-3}. This creates significant problems for a standard of 10 viable organisms per ton.
- Preliminary data suggest that a treatment regime comprising filtration + UV irradiation at ballasting and a repeated UV treatment at de-ballasting will achieve current IMO standards for zooplankton, although some clarification is required. The interval between ballasting and de-ballasting treatment in this trial was >96 h longer than the definitive STEP test (24 h).

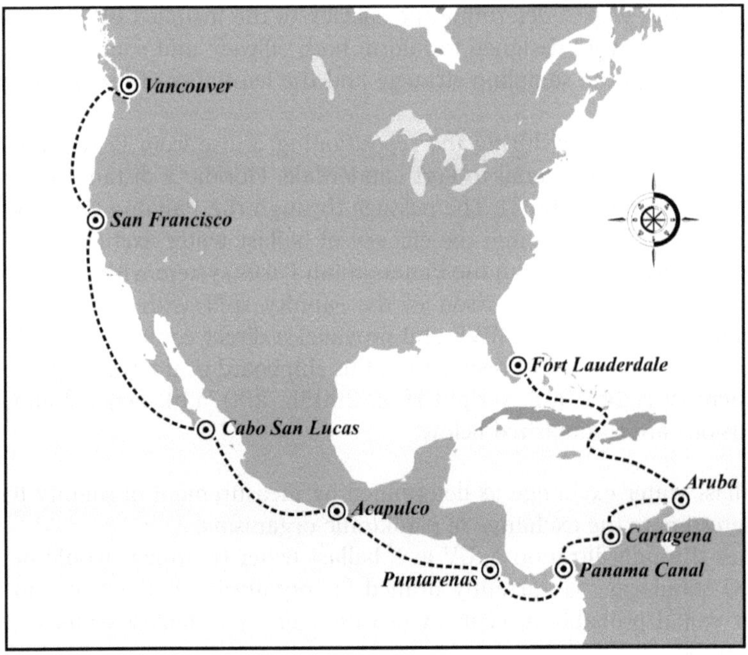

Fig. 3 Test cruise for STEP program aboard MV Coral Princess, September–October 2004.

CONCLUSIONS

Definitive STEP trials are anticipated for the *Coral Princess* in winter 2007–08 (Caribbean) and summer 2008 (NE Pacific) as part of a 5-year program culminating in a single shipboard trial in 2012. It is anticipated that further refinement will be made to biological end points throughout the course of these shipboard trials. These will include better live/dead assessment for nonmotile organisms, largely phytoplankton, and this in turn may lead to suggested changes in IMO standards, which are currently based on densities of specified size classes. Even with adjustments to standards that could, for example, account for large protists, such as dinoflagellates in the >50 μm range, the draft IMO standard of 10 live organisms, >50μm (zooplankton) per ton, remains difficult to meet for nonbiocide treatments. It is suggested that more stringent standards should be avoided pending more effective treatments and because they remain virtually unenforceable. A preliminary draft of the Ballast Water Management Bill (S 363) passed by the US Congress in August 2005 contained a post-treatment standard of 0.1 viable organisms per m^3 which would have required the filtration of a minimum of 100 tons of water aboard the vessel in order to accomplish such a measurement: not a feasible proposition aboard a working ship. Even with a minimum volume of 1 m^3 comprising each replicate sample, filtration, concentration, and examination of such a sample can be a challenge, particularly where high densities of organisms are concerned. Discrimination between live and dead organisms remains a high priority for nonmotile forms such as dormant stages and provides the impetus for investigations of vital stains such as Sytox Green and Cell Tracker Green that differentially stain cells with compromised cell membranes (Veldhuis *et al.* 2006, Wright *et al.* 2006, unpublished data).

The indication that ballast water treatment may be required during both ballasting and de-ballasting has crucial implications for land-based testing facilities, the most important being the necessity for large tanks to be available under such circumstances. Such tanks would be needed to store water to simulate ballast tank residence time (usually > 24 h) in sufficient quantity to avoid running the pump dry when the water is returned to the harbor during a simulated de-ballasting operation. Using a pump rated at 1500 gpm ($\sim 6 \times m^3\ min^{-1}$) it would require a minimum tank size of 100 m^3 to comfortably accommodate the 15 min period required for sample collection and filtration. Large tanks would also be needed in cases where significant storage time might be required prior to discharge to allow for chemical degradation in tests of biocide treatment.

It may also be argued that large tanks would be needed in cases where manipulations of water conditions (i.e. changes in water quality, additions of cultured organisms) would be required in order to conform to conditions of water quality and organism abundance stipulated by the IMO. Readers are directed to MEPC\53\24\Add.1 ANNEX 3 for a detailed account of these guidelines. However, there are inherent drawbacks to such an approach. Apart from the often prohibitively large effort involved with culturing the numbers of organisms required to satisfy density requirements for volumes of water likely to exceed several hundred m^3, there is a risk of creating an inherently unstable assemblage of organisms or set of water quality conditions (or both) that might be viewed as unrepresentative of the natural environment. For land-based test systems, a range of water quality and biological conditions is best achieved through the judicious selection of a variety of sites that cover a spectrum of environmental parameters. A compromise solution offering a greater degree of flexibility within the coastal environment is a barge-based test system combining a readily available sampling system with the ability to accommodate a variety of environmental conditions. Drawbacks to such a system, however, may include constraints regarding tank size. With respect of biogeochemical conditions, less flexibility is available to investigators involved with full-scale shipboard trials, although an improvement of test conditions can be achieved through the selection of more productive coastal water as the basis for tests, as opposed to poorly productive offshore water.

REFERENCES

Carlton, J. T. and J. B. Geller. 1993. Ecological roulette: the global transport of non-indigenous marine organisms. Science **261**, 78–82.

Carlton, J. T., D. M. Reid, and H. van Leeuwen. 1995. Shipping study. The role of shipping in the introduction of non-indigenous aquatic organisms to the coastal waters of the United States (other than their Great Lakes) and an analysis of control options. The National Sea Grant College Program/Connecticut Sea Grant Project R/ES-6. Department of Transportation, United States Coast Guard, Washington DC and Groton. Connecticut Report No. CG-D-11-95. Government Accession No. AD-A294809.

Tjallingi, F. and H. Schilperood. 2001. Global market analysis of ballast water treatment technology. October, 2001. Northeast Midwest Institute, Washington, DC.

Veldhuis, M. J. W., F. Fuhr, J. P. Boon, and C. C. ten Hallers-Tjabbes. 2006. Treatment of ballast water. How to test a system with a modular concept? Environmental Technology **27**, 909–921.

Waite, T. D., J. Kazumi, P. V. Z. Lane, L. L. Farmer, S. G. Smith, S. L. Smith, G. Hitchcock, and T. L. Kapo. 2003. Removal of large populations of marine plankton by a large-scale ballast water treatment system. Marine Ecology Progress Series **258**, 51–63.

Wright, D. A. 2006. Practical aspects of ballast water treatment, efficacy and compliance testing. Proceedings of the 2nd World Maritime Technology Conference. Maritime innovation - delivering global solutions. March 6–11, 2006, Institute of Marine Engineering, Science and Technology, London, UK.

Wright, D. A. 2007. Practical aspects of compliance testing for ballast water treatment. Proceedings of the Institute of Marine Engineering, Science and Technology, Journal of Marine Science and Environment, Journal of Marine Design and Operations **(B)**, in press.

Wright, D. A., R. Dawson, S. M. Moesel, and C. E. F. Orano-Dawson. 2005b. Shipboard trials of ballast water management options: Part 2. Efficacy of ballast water exchange. Proceedings of the Institute of Marine Engineering, Science and Technology, Journal of Marine Science and Environment **(C3)**, 15–20.

Wright, D. A., R. Dawson, and C. E. F. Orano-Dawson. 2005a. Shipboard trials of ballast water management options: Part 1. Efficacy of ultraviolet irradiation. Proceedings of the Institute of Marine Engineering, Science and Technology, Journal of Marine Science and Environment **(C3)**, 1–11.

Wright, D. A., R. Dawson, and C. E. F. Orano-Dawson. 2007a. Shipboard trials of menadione as a ballast water treatment. Marine Technology **44**, 68–76.

Wright, D. A., R. Dawson, C. E. F. Orano-Dawson, and S. M. Moesel. 2007b. A test of the efficacy of a ballast water treatment system aboard the Vessel Coral Princess. Marine Technology **44**, 57–67.

Wright, D. A., R. Dawson, C. E. F. Orano-Dawson, G. R. Morgan, and J. Coogan. 2006. The development of ultraviolet irradiation as a method for the treatment of Ballast water in ships. Proceedings of the Institute of Marine Engineering, Science and Technology, Journal of Marine Science and Environment **(C4)**, 3–12.

Chapter thirty-seven

Virtues and shortcomings of EU legal provisions for managing NIS: **Rana catesbeiana** *and* **Trachemys scripta elegans** *as case studies*

Riccardo Scalera

INTRODUCTION

Every year since 1928 a competition takes place in Angels Camp, California: the Jumping Frog Jubilee, a popular event inspired by Mark Twain's famous short story *The celebrated jumping frog of Calaveras County*. At the time – the story was first published in 1865 – the now endangered California red-legged frog (*Rana aurora draytonii* Baird and Girard) was very common in that area: thus this was probably the species used for the competition and to which the story refers. However, at the end of the 19th century a non-indigenous species (NIS) entered the scene: the American bullfrog (*Rana catesbeiana* Shaw). Soon after being introduced into the area, this species replaced the indigenous red-legged frog within the Calaveras competition, and now, thanks to "Rosie the Ribiter" and its jump over 6.5 m, the American bullfrog has held the world record since

Francesca Gherardi, Biological invaders in inland waters: Profiles, distribution, and threats, 669–678.

1986. Nevertheless the greatest ultimate jump that the species managed to perform, with substantial help from human agency, has been through the Atlantic Ocean, from the New to the Old World.

Today, *R. catesbeiana*, together with the red-eared slider *Trachemys scripta elegans* (Wied), is among the approximately 270 non-indigenous amphibians and reptiles known to be naturalized in the world (Lever 2003). Both taxa, native to North America, have been introduced into several countries throughout the world (Chapter 7).

Within the European Union (EU) their presence is documented in several member States (for a review see Adrados and Briggs 2002). *Rana catesbeiana* is known to occur and reproduce in eight countries (Belgium, France, Germany, Greece, Italy, the Netherlands, Spain, and the UK) but it seems to be established only in Italy (Sindaco *et al.* 2006), France (Pascal *et al.* 2006), Belgium (Jooris 2005), and Greece (Crete) (Adrados and Briggs 2002).

The situation is similar for *T. s. elegans*, whose presence is documented in at least 12 countries (Austria, Belgium, Denmark, France, Germany, Greece, Italy, the Netherlands, Portugal, Spain, Sweden, and the UK). However, breeding populations are known only in Spain (Pleguezuelos *et al.* 2002), Italy (Sindaco *et al.* 2006), France (Pascal *et al.* 2006, Cadi *et al.* 2004), and possibly the Balkan peninsula (Džukić and Kalezić 2004).

A STORY OF PETS TURNING INTO PESTS

Today a major cause of intentional and unintentional introductions is related to the growing trade in wildlife and the overall liberalization of international business practices. The main problem related to the international trade in wildlife is that certain NIS marketed over a period of time and frequently released in the wild have a greater chance to establish self-sustaining populations (Perrings *et al.* 2002). In the case of *R. catesbeiana* and *T. s. elegans*, the high number of specimens exported worldwide for the pet and food trade during the last century has increased the chance of the species becoming naturalized following intentional or unintentional releases. In practice, *R. catesbeiana* and *T. s. elegans* get established in the wild as a consequence of human mismanagement, in general following abandonment by owners who no longer could or wanted to take care of them or through escapes from farms, garden ponds, and other breeding facilities.

This is particularly true for *T. s. elegans*. According to Hoover (1998), up to eight million individuals were exported annually from the USA to the European and Asian markets (from up to 150 commercial farms; Moll 1995). In total, about 52 million specimens of *T. s. elegans* were exported from the USA between 1989 and 1997 (Telecky 2001). In countries like Italy (see Scalera 2001), Spain (Pleguezuelos *et al.* 2002), France (Pascal *et al.* 2006), and Poland (Najbar 2001), trade in *T. s. elegans* involved hundreds of thousands of individuals per year.

As reviewed by Adrados and Briggs (2002), both taxa are considered a serious ecological threat for indigenous species. Indeed they are voracious opportunistic predators, eating a wide range of prey, from insects and other invertebrates to several vertebrates, including amphibians and reptiles, small mammals, and birds. Competition dynamics with indigenous species are also known to occur. For instance, *R. catesbeiana* may compete for food with indigenous amphibians, at either adult or larval stages (Kupferberg 1997). Likewise, *T. s. elegans* may compete with other indigenous pond turtles (i.e. *Emys orbicularis* and *Mauremys* spp.) for food, basking, and nesting sites (Gasperetti *et al.* 1993, Miele 2001, Cadi and Joly 2003, 2004). Such species may also bring the inherent risk of the spread of diseases and parasites. For instance, *R. catesbeiana* seems to be involved in the spread of harmful pathogens, like Chytridiomycosis (Mazzoni *et al.* 2003), a fatal disease caused by the fungus *Batrachochytrium dendrobatidis* Longcore, Pessier and Nichols (see also Chapter 7). *Trachemys s. elegans*, as many other species, is considered a potential vector for *Salmonella* Lignieres (Moll 1995; see also Chapter 7), a gastrointestinal infection similar to typhoid, dangerous also to humans (that explains the US ban concerning the sale of hatchling turtles within this country since 1975).

Although extensive studies on the economic impact of these species are not available, figures related to local situations show that their management may be very expensive. In the UK, for instance, early efforts to eradicate the first breeding bullfrog population cost some €32,000 (Adrados and Briggs 2002). In Germany, according to Reinhardt *et al.* (2003), the annual cost for measures to control a few bullfrog populations (in only five ponds) is €270,000. The same authors assessed that this figure would rise to €4.4 billion if control measures were needed throughout the whole country.

FIRST STEPS WITHIN THE EU TO MANAGE NIS

In general, the identification of the intervention to be envisaged depends on several biological, social, and economic factors. Prevention is always the most environmentally desirable and cost-effective strategy. When prevention has failed, the best option is the eradication of the undesired species before it becomes invasive, provided that its presence is detected in time. When eradication is not feasible – i.e. for either technical or ethical constraints (Genovesi 1998, Kraus and Campbell 2002, Chapter 34) – the only valid alternative is to reduce the species' population density below an acceptable threshold, or to prevent its spread beyond a given geographical boundary (Wittenberg and Cock 2001). The support of GIS methodologies may contribute to the development of theoretical distribution models to predict the potential expansion range of NIS (Sutherst *et al.* 1996) and of their impact, i.e. eventual pathogens associated to their presence (Ron 2005). When control or eradication strategies are not feasible, the only option is to learn to "live with" the undesired NIS and to mitigate its impact on indigenous species and on the invaded ecosystems (Wittenberg and Cock 2001, Chapter 38) by, e.g., implementing habitat restoration plans or restocking and reintroduction programmes for the endangered indigenous species.

At the EU level, notwithstanding the formal adoption of the pan-European strategy developed under the Council of Europe (Genovesi and Shine 2004, Chapter 34), an *ad hoc* strategy has not yet been developed. As a consequence, a unified approach to address the problems caused by NIS is lacking, and most management initiatives undertaken so far were planned on a merely local basis. For instance, although both *T. s. elegans* and *R. catesbeiana* are recognized as undesired NIS within the EU, eradication programmes have been carried out only for *R. catesbeiana* in Germany, the Netherlands, and the U.K. (Adrados and Briggs 2002), while in France a control programme for this species is just being implemented (Détaint and Coïc 2006). Other relevant actions, concerning *T. s. elegans*, include the disposal of live specimen abandoned by amateur pet owners at rescue centres and zoological gardens, as reported for Italy, Spain, and France (Adrados and Briggs 2002, Pascal *et al.* 2006).

However, a more harmonized approach using the available tools is certainly possible at a regional level. Despite the lack of a strategy, some legal measures to manage NIS are envisaged within the EU by a few directives and regulations and their relative amendments. The main ones are:

1. Council Regulation (EC) No 338/97 of 9 December 1996 on the protection of species of wild fauna and flora by regulating trade therein – also known as the Wildlife Trade regulations.
2. Council Directive 92/43/EEC of 21 May 1992 on the conservation of natural habitats and of wild fauna and flora – also known as the Habitats directive.
3. Council Regulation (EEC) No. 1973/92 of 21 May 1992 establishing a Financial Instrument for the Environment (LIFE) – also known as the LIFE programme.

Directives and regulations are enforced by all EU member States. In particular, a regulation immediately takes on the character of national law within member States, and is directly implemented, while a directive is implemented by means of national legislations.

The Wildlife Trade regulations

The Wildlife Trade regulations were adopted by the EU so as to comply with the provisions of the Convention on International Trade in Endangered Species of Wild Fauna and Flora. This treaty, also known as CITES, was signed in Washington in 1973 and is currently implemented in almost 170 countries. It represents one of the most effective, though complex, international environmental agreements, whose strength is due to the continuous amendments and updates (under the form of resolutions) related to its interpretation, definitions, application, and changes in species to be protected. The EU, although not yet a party to the CITES, has implemented the convention through specific regulations, which are based on Council Regulation (EC) No. 338/97, as amended (see also Magel 2002). These regulations incorporate all CITES provisions as well as other stricter measures, so as to be consistent with the adoption of the single market and with the EU nature conservation policy (i.e. the Habitats directive, which also includes non-CITES listed species.)

According to the Wildlife Trade regulations, the European Commission may establish restrictions on the import of "live specimens of species for which it has been established that their introduction into the natural environment of the Community presents an ecological threat to wild species of fauna and flora indigenous to the Community".

Rana catesbeiana and *T. s. elegans* are very special taxa, since they have been for almost 10 years the only two species whose import has been suspended following this provision. However, the suspension of imports was not extended to subspecies other than *T. s. elegans*. As a consequence of this major inconsistency, attempts to disguise the head diagnostic colours of specimen belonging to *T. s. elegans* have been made in order to smuggle them (Fiori and Avanzo 2002). Moreover, trade in other subspecies, such as *Trachemys scripta scripta* (Schoepff), *Trachemys scripta ornata* (Gray), has been increased as well as trade in similar species such as *Trachemys decussata* Gray, *Chrysemys picta* Schneider, and *Pseudemys concinna floridana* (Le Conte). As a result, some of these subspecies and species have already been recorded in the wild from time to time – i.e. in Italy (see Bologna *et al.* 2000), Germany (Winkel *et al.* 2000), and Spain (Pleguezuelos *et al.* 2002).

It is clear that the Wildlife Trade regulations can play a pivotal role in preventing further introductions, reducing the movement of NIS which might become invasive, as well as those already introduced and considered for control or eradication programmes. Such regulations are elastic enough to be adapted to the contingencies typical of NIS introductions. However, following a specific study funded by the European Commission on the operation of the import suspensions and on the effectiveness of these measures (Adrados and Briggs 2002), problems with the use of the Wildlife Trade regulations as a mechanism for dealing with NIS were identified, and the need for a new instrument was suggested. As a consequence, the Commission has become rather reluctant to make additions to the list of species whose import is suspended on these grounds. This is a major shortcoming, because, although *R. catesbeiana* and *T. s. elegans* are some of the most harmful NIS within the EU – being also among the "least wanted" taxa at the global level, they are only a small sample of the taxa whose trade should be suspended to prevent further harmful introductions.

However, in their report to the Commission, Adrados and Briggs (2002) introduced a couple of interesting concepts. In particular, they emphazised the need (a) to focus attention towards the "replacement species" and (b) to foresee adequate measures to control their trade once they are found to occur on the market above a recommended "threshold". Indeed, in 2005 two more species were subjected to import suspension, including the North American painted turtle *C. picta*, which was identified as a potential replacement species for *T. s. elegans* [see Commission Regulation (EC) No. 252/2005]. In fact, the export quota of *C. picta* toward the EU increased sharply soon after enforcement of the import suspension of *T. s. elegans*, with specimens sold at the same trading point, similar price, and similar quantities. Moreover, field observations showed that successful reproduction in the wild may occur, at least in Germany (Fritz and Lehmann 2002).

The Habitats directive and the LIFE programme

The Habitats directive is one of the main legal instruments for the conservation of wild threatened species and habitats in the EU, the main objective being the creation of a regional network of protected areas. This network, called Natura 2000, is aimed at ensuring the long-term conservation of all fauna, flora, and habitats of EU concerns (including species listed in the Birds directive 79/409/EEC). This directive formally recognizes NIS as a conservation threat, and has been supported by a financial instrument called LIFE [from the French acronym "L'Instrument Financier pour l'Environement", see Council Regulation (EEC) No. 1973/92], which has been aimed at providing adequate resources for the implementation and development of the EU environmental policy and legislation.

The LIFE financial tool, adopted between 1992 and 2006, has been the main source of funding specifically aimed at the enforcement of legal provisions for nature conservation. It has been managed directly by the European Commission and consisted of three branches, one of which, called LIFE-Nature, was specifically aimed at the implementation of the Habitats directive and at the creation of the Natura 2000 network. For the period 2007–2013 the LIFE programme is likely to be replaced by LIFE+, a new financial programme which is being developed by the European Commission.

Although actions to manage NIS were not explicitly considered among the objectives of LIFE, between 1992 and 2002 the European Commission spent about €30 millions on more than 100 projects including control or eradication actions against NIS (Scalera and Zaghi 2004). While only a few projects targeted NIS as a primary objective, a higher number included various measures for monitoring and controlling NIS as a generic component of site management programmes, some of which also targeted *T. s. elegans* and *R. catesbeiana*. As emphasized by Scalera and Zaghi (2004), LIFE projects targeting NIS were selected for funding without any clear guiding strategy. However, the actual contribution of the LIFE programme and the Habitats directive, despite some inherent limits, is quite concrete compared to other existing legal tools dealing with NIS, which generally lack specificity and enforceability due to their vagueness and/or the lack of financial resources for implementation (Scalera 2004).

Of course, in the future more attention should be paid to the potentialities of the Habitats directives and LIFE+. If adequately planned, this programme could offer important opportunities. Although the rules do not foresee actions specifically directed at managing NIS, it is possible to address them whenever they represent a threat for the species and habitats of EU importance. Experience with the former LIFE programme emphasized the need for a national framework of programmes or multinational initiatives (Scalera and Zaghi 2004). Therefore, notwithstanding the lack of a comprehensive strategy or a specific legislation developed at EU level – only a pan-European strategy is available so far (Genovesi and Shine 2004) – member States should encourage actions to deal with NIS in the projects to be financed by LIFE+.

CONCLUDING REMARKS: THE WAY AHEAD

Within the EU, the lack of a comprehensive strategy to deal with harmful NIS and the absence of specific funds for the implementation of effective measures (EU funds are only partially available to deal with NIS), may represent a major limit for the sound management of the undesired populations of successfully established NIS.

The management of *R. catesbeiana* and *T. s. elegans*, in accordance with the Wildlife Trade regulations, has shown that such a tool does not fully provide the powers needed to deal effectively with NIS and has confirmed the need for a specific EU legislation (Adrados and Briggs 2002). However, the time needed to develop a new legal framework would be very long. In order to be effective, this framework should guarantee a sound harmonization of all existing legislations relevant to all sectors involved in this cross-cutting issue, for instance nature conservation, trade, agriculture, fisheries, health, and research. Enforcing a new legislation would be also very expensive. Regulating the trade in a number of NIS needs stringent border control, which in turn requires a framework of rules, training and capacity-building programmes, increased cooperation between the relevant authorities, technical protocols for actions to be undertaken, and reference lists of species and identification guides.

Although the mechanisms for implementing such regulations are neither adequate nor intended to control the movement of an unlimited number of species – the number of NIS likely to become invasive is potentially unlimited – they have provided some concrete contributions. For instance, the Wildlife Trade regulations have allowed a number of seizures that prevented further possible releases of NIS as a consequence of the typically questionable mismanagement which affects the species commonly kept as pets. For instance, following implementation of such regulations, in Italy between 1999 and 2000 the national authorities seized about 23,000 specimens of *T. s. elegans*, with an economic value of €296,000 (Fiori and Avanzo 2002). Such seizures can have an important effect in facing the spread of NIS. In fact, in Europe most populations of *R. catesbeiana* and *T. s. elegans* are likely to be present especially as a consequence of continuous releases, rather than to the existence of viable populations (see Luiselli *et al.* 1997, Bruekers and Van der Keijlen 1999, Adrados and Briggs 2002, Cadi and Joly 2003, Cadi *et al.* 2004, Hill *et al.* 2005).

In conclusion, before an *ad hoc* legislation is available, focusing on the potential role of the existing Wildlife Trade regulations and of the Habitats directive (European Commission 2003, Scalera and Zaghi 2004) could be key to preventing further introductions of NIS within the EU Member States and to enacting positive synergies with other international treaties and agreements. Such regulations and directives can contribute either to prevent novel introductions or to respond to the spread of species already established (Adrados and Briggs 2002). These contributions comply with the provisions which should be included in a sound legal framework to face NIS (Shine *et al.* 2000). The measures tested on *T. s. elegans* and *R. catesbeiana* may support the implementation of Recommendation No. 99 (2003) of the Standing Committee of the Council of Europe, on the *European Strategy on Invasive Alien Species* (Genovesi and Shine 2004), based on the Convention on Biological Diversity guiding principles (Decision VI/23).

REFERENCES

Adrados, L. C. and L. Briggs, editors. 2002. Study of application of EU Wildlife Trade regulations in relation to species which form an ecological threat to the EU fauna and flora, with case studies of American bullfrog (*Rana catesbeiana*) and red-eared slider (*Trachemys scripta elegans*). Study report to the European Commission. Amphi Consult, Denmark.

Bologna, M. A., M. Capula, and G. M. Carpaneto, editors. 2000. Anfibi e Rettili del Lazio. Palombi editore, Rome, Italy, 159 pp.

Bruekers, J. and H. Van der Keijlen. 1999. *Trachemys scripta elegans* in The Netherlands: a threat for Dutch flora and fauna? Report in Web NSV (Nederlandse Schildpadden Vereniging): http://www.trionyx.nl/english/1.htm

Cadi, A., V. Delmas, A.-C. Prévot-Julliard, P. Joly, C. Pieau, and M. Girondot. 2004. Successful reproduction of the introduced slider turtle (*Trachemys scripta elegans*) in the South of France. Aquatic Conservation: Marine and Freshwater Ecosystems **14**, 237–246.

Cadi, A. and P. Joly. 2003. Competition for basking places between the endangered European ond turtle (*Emys orbicularis galloitalica*) and the introduced red-eared slider (*Trachemys scripta elegans*). Canadian Journal of Zoology **81**, 1392–1398.

Cadi, A. and P. Joly. 2004. Impact of the introduction of the red-eared slider (*Trachemys scripta elegans*) on survival rates of the European pond turtle (*Emys orbicularis*). Biodiversity and Conservation **13**, 2511–2518.

Détaint, M. and C. Coïc. 2006. La Grenouille taureau *Rana catesbeiana* dans le sud-ouest de la France. Premiers résultats du programme de lutte. Bulletin de la Société Herpétologique de France **117**, 41–56.

Džukić, G. and M. L. Kalezić. 2004. The biodiversity of amphibians and reptiles in the Balkan Peninsula. Pages 167–192 in H. I. Griffiths, B. Kryštufek and J. M. Reed, editors. Balkan biodiversity: pattern and process in the european hotspot. Kluwer Academic Publishers, Dordrecht, The Netherlands.

European Commission. 2003. Thematic report on alien invasive species. Second Report of the European Community to the Conference of the Parties of the Convention on Biological Diversity.

Fiori, M. and C. Avanzo. 2002. Enforcement in Italy. Pages 69–74 in M. Anton, N. Dragffy, S. Pendry and T. R. Young, editors. Proceedings of the International Expert Workshop on the Enforcement of Wildlife Trade Controls in the EU. 5–6 November 2001, Frankfurt/Germany. TRAFFIC Europe, Brussels, Belgium and IUCN – The World Conservation Union, Gland, Switzerland and Cambridge, UK.

Fritz, M. and H. D. Lehmann. 2002. Fund von Schlüpflingen der nordamerikanischen Zierschildkröte, *Chrysemys picta belli*, an einem Gewässer in Baden-Württemberg. Elaphe **10**, 45–48.

Gasperetti, J., A. F. Stimson, J. D. Miller, J. P. Ross and P. R. Gasperetti. 1993. Turtles of Arabia. Fauna of Saudi Arabia **13**, 170–367.

Genovesi, P. 1998. Italy's grey squirrels: action vs opposition. World Conservation **1**, 25–26.

Genovesi, P. and C. Shine. 2004. European strategy on invasive alien species. Nature and Environment 137. Council of Europe, Strasbourg, France, 67 pp.

Hill, M., R. Baker, G. Broad, P. J. Chandler, G. H. Copp, J. Ellis, D. Jones, C. Hoyland, I. Laing, M. Longshaw, N. Moore, D. Parrott, D. Pearman, C. Preston, R. M. Smith and R. Waters. 2005. Audit of non-native species in England. English Nature Research Reports n 662, U.K., 81 pp.

Hoover, C. 1998. The US role in the international live reptile trade: Amazon tree boas to Zululand dwarf chamaleons. TRAFFIC North America. Washington, DC.

Jooris, R. 2005. De Stierkikker in Vlaanderen. Nieuwe inzichten in verspreiding, foerageergedrag en ontwikkeling. Natuur.focus **4**, 121–127.

Kupferberg, S. J. 1997. Bullfrog (*Rana catesbeiana*) invasion of a California river: the role of larval competition. Ecology **78**, 1736–1751.

Kraus, F. and E. W. Campbell. 2002. Human-mediated escalation of a formerly eradicable problem: the invasion of Caribbean frogs in the Hawaiian Islands. Biological Invasions **4**, 327–332.

Lever, C. 2003. Naturalized reptiles and amphibians of the world. Oxford University Press, Oxford, UK, xx + 318 pp.

Luiselli, L., M. Capula, D. Capizzi, E. Filippi, M. Trujillo Jesus and C. Anibaldi. 1997. Problems for conservation of Pond Turtles (*Emys orbicularis*) in Central Italy: is the introduced Red-eared Turtle (*Trachemys scripta*) a serious threat? Chelonian Conservation and Biology **2**, 417–419.

Magel, N. 2002. More than CITES: A EU wildlife trade regulation. In M. Anton, N. Dragffy, S. Pendry and T. R. Young, editors. Proceedings of the International Expert Workshop on the Enforcement of Wildlife Trade Controls in the EU, 5–6 November 2001, Frankfurt/Germany. TRAFFIC Europe, Brussels, Belgium and IUCN – The World Conservation Union, Gland, Switzerland, and Cambridge, UK.

Mazzoni, R., A. A. Cunningham, P. Daszak, A. Apolo, E. Perdomo and G. Speranza. 2003. Emerging pathogen of wild amphibians in frog (*Rana catesbeiana*) farmed for international trade. Emerging Infectious Diseases **9**, 995–998.

Miele, M. 2001. Ecologia di una specie introdotta di testuggine acquatica (*Trachemys scripta*) e possibili effetti di competizione con una specie autoctona minacciata (*Emys orbicularis*). Degree thesis. Università degli studi "Roma Tre", Rome, Italy, 107 pp.

Moll, E. O. 1995. The turtle *T. s. elegans* and the pet trade. Aliens **2**, 3.

Najbar, B. 2001. Żółw czerwonolicy *Trachemys scripta elegans* (Wied, 1839) w województwie lubuskim (zachodnia Polska) [The red-eared terrapin *Trachemys scripta elegans* (Wied, 1839) in the Lubuskie Province (western Poland)]. Przeglad Zoologiczny **45**, 103–109.

Pascal, M., O. Lorvelec, J.-D. Vigne. 2006. Invasions Biologiques et Extinctions: 11000 ans d'histoire des Vertébrés en France. Coédition Belin – Quæ, Paris, France, 350 pp.

Perrings, C., M. Williamson, E. B. Barbier, D. Delfino, S. Dalmazzone, J. Shogren, P. Simmons, and A. Watkinson. 2002. Biological invasion risks and the public good: an economic perspective. Conservation Ecology **6**, 1. [online] URL: http://www.consecol.org/vol6/iss1/art1

Pleguezuelos, J. M., R. Marquez, and M. Liziana, editors. 2002. Atlas y libro rojo de los anfibios y reptiles de España. Direction General de Conservacion de la Naturaleza-Asociation Herpetologica Española, Madrid, Spain, 584 pp.

Reinhardt, F., M. Herle, F. Bastiansen, and B. Streit. 2003. Economic Impact of the Spread of Alien Species in Germany. Federal Environmental Agency (Umweltbundesamt), Berlin, Germany, 229 pp.

Ron, S. R. 2005. Predicting the distribution of the amphibian pathogen *Batrachochytrium dendrobatidis* in the New World. Biotropica **37**, 209–221.

Scalera, R. 2001. Invasioni biologiche. Le introduzioni di vertebrati in Italia: un problema tra conservazione e globalizzazione. Collana Verde, 103. Corpo Forestale dello Stato. Ministero delle Politiche Agricole e Forestali, Rome, Italy, 368 pp.

Scalera, R. 2004. The legal framework for the protection of amphibians and reptiles in Italy: an annotated overview of the provisions at the international, European community, national and regional level. Pages 21–32 in M. A. Bologna and S. La Posta, editors. The conservation status of threatened amphibian and reptile species of the Italian fauna. Italian Journal of Zoology **71** (Supplement 1).

Scalera, R. and D. Zaghi. 2004. Alien species and nature conservation in the EU. The role of the LIFE program. LIFE Focus. European Commission, Bruxelles, Belgium, 60 pp.

Shine, C., N. Williams, and L. Gündling. 2000. A Guide to Designing Legal and Institutional Frameworks on Alien Invasive Species - IUCN, Gland, Switzerland, xvi + 138 pp.

Sindaco, R., G. Doria, E. Mazzetti and F. Bernini, editors. 2006. Atlante degli Anfibi e dei Rettili d'Italia. Societas Herpetologica Italica, Edizioni Polistampa, Firenze, Italy, 791 pp.

Sutherst, R. W., R. B. Floyd, and G. F. Maywald. 1996. The potential geographical distribution of the Cane Toad, *Bufo marinus* L. in Australia. Conservation Biology **10**, 294–299.

Telecky, T. M. 2001. United States Import and Export of Live Turtles and Tortoises. Turtle and Tortoise Newsletter **4**, 8–13.

Winkel, S., M. Kuprian, E. Flößer, M. Homeier, O. Homeier, and G. Möckel. 2000. "Volkszählung": Eingeschleppte Schildkröten erobern Hessens Gewässer – Erste Ergebnisse aus der Wetterau. Jahrbuch Naturschutz in Hessen **5**, 68–75.

Wittenberg, R. and M. J. W. Cock, editors. 2001. Invasive alien species: a toolkit for best prevention and management practices. CAB International, Wallingford, Oxon, UK, xvii + 228 pp.

Chapter thirty-eight

Problems and opportunities managing invasive Bullfrogs: is there any hope?

Michael J. Adams and Christopher A. Pearl

INTRODUCTION

The American Bullfrog (*Rana catesbeiana* Shaw) is a widely introduced and invasive anuran that is frequently blamed for population declines of indigenous species (Bury and Whelan 1984). Once established, Bullfrog populations are often either difficult or impossible to eradicate depending on habitat and landscape features (Schwalbe and Rosen 1988, Doubledee *et al.* 2003, Govindarajulu *et al.* 2005). Bullfrogs are representative of a large but neglected suite of non-indigenous species (NIS) that are characterized by: (1) a broad invasion that is well established in some areas; (2) a lack of obvious economic impacts compared to some other invasive species; and (3) a lack of reasonably feasible control methods. Despite demonstrated conservation concerns, invasive species like the Bullfrog do not tend to attract the resources necessary to attempt large scale management because of their lack of economic impact and the difficulty of control methods. This leaves biologists responsible for managing habitats invaded by such species with little hope and few options for promoting the persistence of sensitive indigenous species. With these issues in mind, we consider the case of the Bullfrog, review management options, and suggest directions for future research with this and similar species.

Bullfrogs are among the most successful vertebrate invaders and are considered by the IUCN Invasive Species Specialist Group to be among the 100 worst invaders in the world (http://www.issg.org/database/welcome). The

Francesca Gherardi, Biological invaders in inland waters: Profiles, distribution, and threats, 679–693.

native range of the Bullfrog covers much of eastern North America, roughly from the Mississippi River and Great Lakes east to the Atlantic Ocean and from the State of Florida north into southern Canada (Bury and Whelan 1984). Few anurans exhibit such a large native range. This broad native distribution is indicative of the adaptability and success of Bullfrogs and is dwarfed by their present range. They now occupy much of the western USA (Casper and Hendricks 2005), and parts of western Canada (Green and Campbell 1984), Mexico (Casas-Andreu *et al.* 2002), Brazil (Borges-Martins *et al.* 2002), Ecuador (Cisneros-Heredia 2004), Venezuela (Hanselmann *et al.* 2004), Cuba (Sampedro *et al.* 1985), Dominican Republic (Kairo *et al.* 2003), Jamaica (Mahon and Aiken 1977), Puerto Rico (Lopez-Flores *et al.* 2003), Hawaii (Viernes 1995), Japan (Hirai 2004), China (Wu *et al.* 2004), Korea (Kim and Ko 1998), Italy (Lanza 1962), France (Neveu 1997), the Netherlands (Stumpel 1992), and the UK (Banks *et al.* 2000), among other locations (Lever 2003).

Original introduction of Bullfrogs to many of these locations occurred more than 50 years ago for culturing as a food source, sometimes after the overharvest of indigenous anurans (Jennings and Hayes 1985, Negroni 1997, Mazzoni 1999). Escapees and intentional releases established naturalised populations that are often difficult to eradicate. Post-metamorphic stages are capable of dispersing long distances and are adept at colonizing new sites (>1200 m; Willis *et al.* 1956). A single female Bullfrog can produce 1,000–25,000 eggs with the largest females sometimes producing more than 40,000 eggs (Bury and Whelan 1984). Breeding sites can achieve notably high densities (>780 adults ha^{-1}; Schwalbe and Rosen 1988).

The conspicuousness (e.g. large size, high densities, and loud vocalizations) and natural history (e.g. high fecundity and broad diet) of Bullfrogs make their introduction an obvious hypothesis to explain declines in indigenous species. Early reports suggested displacement of indigenous amphibians (Moyle 1973, Hammerson 1982), but separating the influence of Bullfrogs from correlated factors has proven difficult (Hayes and Jennings 1986, Adams 1999). Some studies suggest that other factors associated with Bullfrog presence, like introduced fish or habitat alterations, may be more detrimental to indigenous species than the Bullfrogs themselves (Kiesecker and Blaustein 1998, Adams 1999, 2000). However, an increasing number of studies shows direct and indirect negative effects of Bullfrogs on indigenous anurans via competition, predation, and habitat displacement (Boone *et al.* 2004, Pearl *et al.* 2004, others reviewed in Kiesecker 2003). Bullfrog invasions may also affect other taxa such as aquatic snakes and waterfowl (Viernes 1995, Rosen and Schwalbe 2002, Lopez-Flores and Vilella 2003, Wylie *et al.* 2003). Recent work raises the possibility that Bullfrogs may serve as a reservoir of a chytrid fungus, *Batrachochytrium dendrobatidis* (Longcore *et al.* 1999), pathogenic to some amphibians (Hanselmann *et al.* 2004, Pearl and Green 2005, Garner *et al.* 2006). Despite some conflicting reports and regional differences in effects, Bullfrogs are clearly a conservation concern.

MANAGEMENT OPTIONS

Ongoing expansion of Bullfrogs to a wide variety of regions and habitats underscores the need for a suite of management approaches. No discussion of management options for invasive species is complete without mentioning prevention. The best way to control invasive species is to prevent their introduction or establishment in new regions. This is because it is often difficult to detect new invasions early and eradication is much more realistic for species with a limited distribution (Simberloff *et al.* 2005). A review of the papers cited in our introduction on the geographic extent of the Bullfrog invasion suggests that most Bullfrog introductions have been associated with aquaculture. Escape from such operations appears impossible to stop completely. Other vectors include the use of Bullfrog tadpoles as bait for recreational fishing and the availability of live Bullfrogs for pets, landscape ponds, research, and teaching. Efforts to reduce or eliminate these vectors are warranted to slow or prevent the spread of Bullfrogs, but we will not attempt a detailed analysis of such prevention options here. Instead, we focus on methods applicable to established populations.

Direct removal

We define direct removal of Bullfrogs as actions that have a proximate result of death or removal of Bullfrog individuals from the wild. This is in contrast to other options described below that seek to reduce the survival or effects of Bullfrogs by manipulating aspects of their environment. There are a few anecdotal accounts of efforts to directly control Bullfrogs but we are not aware of any publications that fully detail such efforts. Banks *et al.* (2000) installed fencing around the main ponds to limit dispersal and used lamps to collect adult frogs at dusk. They then drained the ponds and excavated the sediment to remove remaining frogs and larvae. This effort apparently did not result in complete eradication: limited breeding was detected the following summer, and post metamorphic bullfrogs were found in the vicinity two years after management (B. Banks 2006, personal communication). Another direct removal effort that has been partially documented in the literature is in ponds that are relatively isolated in a desert landscape in Arizona, USA (Schwalbe and Rosen 1988, Rosen and Schwalbe 1995). They used funnel traps, gigs, guns, and hand capture to remove Bullfrogs annually. Reductions in Bullfrog densities were said to be small and short-lived.

Direct removal techniques are hampered by strong density dependence. Bullfrog populations may exhibit density dependence in both the larval and post-metamorphic segments of their life history (Doubledee *et al.* 2003, Govindarajulu *et al.* 2005). Demographic perturbation analyses of data from invaded ponds in British Columbia suggest that removal efforts should target juveniles and tadpoles that transform after one instead of two winters (Govindarajulu *et al.* 2005). This is consistent with elasticity analyses for other pond-breeding anurans that predict that reducing survival of juveniles rather than other life stages should have the greatest effect on population growth rate (Biek *et al.* 2002, Vonesh and de la Cruz 2002). For Bullfrogs and some other temperate ranids, incomplete removal of eggs or larvae can boost growth and survival of remaining individuals via strong density dependence (Altwegg 2002, Govindarajulu 2004). Likewise, a reduction in the density of adult Bullfrogs can increase the survival of juveniles that would otherwise be prey for adults (Werner *et al.* 1995, Doubledee *et al.* 2003, Govindarajulu 2004). Doubledee *et al.* (2003) used population models to evaluate the potential effectiveness of shooting adult Bullfrogs. Their results suggest that efforts sufficient to increase adult mortality by 65% or greater every 2 years would be necessary to reduce Bullfrog densities enough to benefit California Red-Legged Frogs (*Rana draytonii* Baird and Girard). However, they also suggest that this level of mortality would be difficult to achieve and that the resulting fluctuations in the Bullfrog population might lead to dangerous instability in the California Red-Legged Frog population. Bullfrog life history and demography vary among sites and regions (e.g. Viparina and Just 1975, Cecil and Just 1979), and this variation will need to be accounted for in control prescriptions (Govindarajulu *et al.* 2005).

The high fecundity, density dependence, and evasiveness of Bullfrogs, along with the complexity of invaded wetlands, often make direct removal difficult. Even in small and relatively simple ponds, direct manual removal may need to be coupled with other activities to eradicate or control a population (Banks *et al.* 2000, Doubledee *et al.* 2003). Still, such actions are warranted in situations where Bullfrogs are threatening an endangered species and in the early stages of invasion. Direct removal will be more effective for small, isolated ponds where removal can be complete and reinvasion by overland dispersal is less likely.

Habitat manipulation

Given that direct removal is usually difficult, finding methods to indirectly control Bullfrogs or their effects is appealing. Opportunities to manage habitats present themselves in the course of other management activities. For example, wetland creation, restoration, and enhancement projects offer the chance to manipulate wetland characteristics in ways that promote indigenous versus invasive species. Preventing or controlling invasive plants is often a goal of wetland restoration or enhancement (e.g. Kentula *et al.* 1992). The role that wetland characteristics can play in managing invasive animals like the Bullfrog is less clear.

Some authors have suggested habitat and landscape characteristics that might be managed to limit the dispersal of Bullfrogs. For example, connections to permanent ponds in the form of streams, ditches, or flooding might increase the chance that Bullfrogs will invade a site (Pearl *et al.* 2005). Among other things, this suggests that mitigation of isolated wetlands that are lost to development should emphasize isolation as a desirable characteristic for new sites created. This has intuitive appeal, but the factors that influence Bullfrog movement through a landscape require more study. Moreover, the effects of connectivity and landscape patterns on the dynamics of indigenous species must also be considered.

Habitat management can be viewed as a technique to indirectly reduce or eliminate Bullfrogs. An obvious example is the alteration of hydroperiod. Bullfrogs overwinter as larvae in many regions and they generally depend on permanent waters for larval growth. Maret *et al.* (2006) found that drying could be used to eliminate Bullfrogs in some livestock watering ponds. Pond drying was also effective for local elimination of non-indigenous fish (Maret *et al.* 2006), which can interact with Bullfrogs in ways detrimental to indigenous anurans (Kiesecker and Blaustein 1998, Adams *et al.* 2003). Model comparisons of pond drying and adult removal caused Doubledee *et al.* (2003) to conclude that draining ponds every two years might reduce Bullfrog densities enough to allow the persistence of California Red-Legged Frogs. Their models suggest that a combination of adult removal and periodic pond draining can be an effective strategy to allow coexistence of California Red-Legged Frogs with Bullfrogs.

The use of pond drying to limit Bullfrogs and benefit natives requires additional research and will be region-specific. How to use drying rotations to reduce Bullfrogs without harming natives in groups of wetlands is poorly known (e.g. Maret *et al.* 2006). Drying effects on indigenous species must be considered fully prior to implementing such management plans. A case in point is the conservation of the threatened California Red-Legged Frog, which was recently confirmed to overwinter as larvae in some sites (Fellers *et al.* 2001). In warmer portions of their range, Bullfrogs are capable of reaching transformation in their first summer (Cohen and Howard 1958, Bury and Whelan 1984). Care must be taken to time draining such that there will not be selection for rapid development of larval Bullfrogs. This means draining the pond fast enough and early enough to prevent any rapidly developing portion of the population from reaching metamorphosis.

Whether there are habitat features other than hydroperiod that can be manipulated to control Bullfrog density is an open question. There is some evidence that Bullfrogs are less abundant in ponds in the Pacific Northwest with shallow sloping banks and extensive emergent vegetation (Adams *et al.* 2003). This may have less to do with Bullfrog habitat requirements than with associated patterns in the community (see Community Characteristics below), but suggests the possibility that pond characteristics other than hydroperiod can be manipulated to limit Bullfrogs.

Habitat characteristics can also mediate the interactions between two species. For example, it has long been thought that habitat complexity can facilitate prey survival (Huffaker 1958, Crowder and Cooper 1982, Sredl and Collins 1992). Habitat diversity can decrease encounter rates by increasing habitat segregation of predator and prey (Smith 1972). Habitat segregation could serve to reduce both predation and competition (Smith 1972). Structure such as vegetation can also reduce the effectiveness of some predators by reducing encounter rates within microhabitats (Savino and Stein 1982, Babbitt and Jordan 1996). It has been argued that short term measures that favor the indigenous species might allow natives to adapt in a way that allows their long-term persistence without further intervention (Schlaepfer *et al.* 2005). For example, when exposed to chemical cues from Bullfrogs, Red-legged frogs from populations that are syntopic with Bullfrogs can exhibit behavioral defenses that are enhanced relative to allotopic populations (Kiesecker and Blaustein 1997).

The demography of the typical pond breeding anuran is such that predation by Bullfrogs on recently transformed juveniles might be particularly detrimental to indigenous populations (Biek *et al.* 2002, Vonesh and de la Cruz 2002). This suggests a hypothesis that providing some form of cover in the portion of a pond where juvenile indigenous frogs emerge could promote survival of natives by reducing Bullfrog predation on natives. It has also been suggested that providing riparian cover and feeding areas around ponds and suitable streams can encourage indigenous species to leave the pond habitats where they are more likely to encounter Bullfrogs (Govindarajulu 2004). These hypotheses illustrate that the potential use of habitat to mediate Bullfrog interactions with indigenous species warrants further study.

Despite extensive theoretical evidence that habitat characteristics could influence the probability that indigenous species can coexist with Bullfrogs, there is little information upon which to base habitat guidelines. Indeed, there is currently a need for observations that identify habitat characteristics to test. Studies that quantify associations between Bullfrogs and various indigenous species (e.g. Adams 1999, Kiesecker *et al.* 2001) provide a ready source of data to further define habitat attributes that could increase persistence of natives if Bullfrogs are not eradicated. For example, a study conducted by Pearl *et al.* (2005) in the Willamette Valley of western Oregon takes the traditional approach of evaluating the potential for Bullfrogs to exclude indigenous amphibians but could, instead, investigate predictors of coexistence between indigenous species and Bullfrogs. We revisited these data to determine whether wetlands with coexistence had habitat characteristics that differed from wetlands where only Bullfrogs or only the indigenous species (Northern Red-Legged Frog, *Rana aurora* Baird and Girard) are found. Using Principal Components Analysis, we explored the explanatory value of variables related to wetland size, vegetation, depth, height of riparian vegetation, substrate slope, and road length within 200 m. We targeted these variables because large wetlands and shallow wetlands with extensive emergent vegetation might allow greater microhabitat segregation of Bullfrogs and indigenous species. Likewise, suitable riparian characteristics might reduce the amount of time that indigenous species spend in contact with the more aquatic Bullfrogs. Our analysis indicated that wetlands with greater portions of surface area with emergent vegetation might be more likely to support coexistence or to support Northern Red-Legged Frogs alone

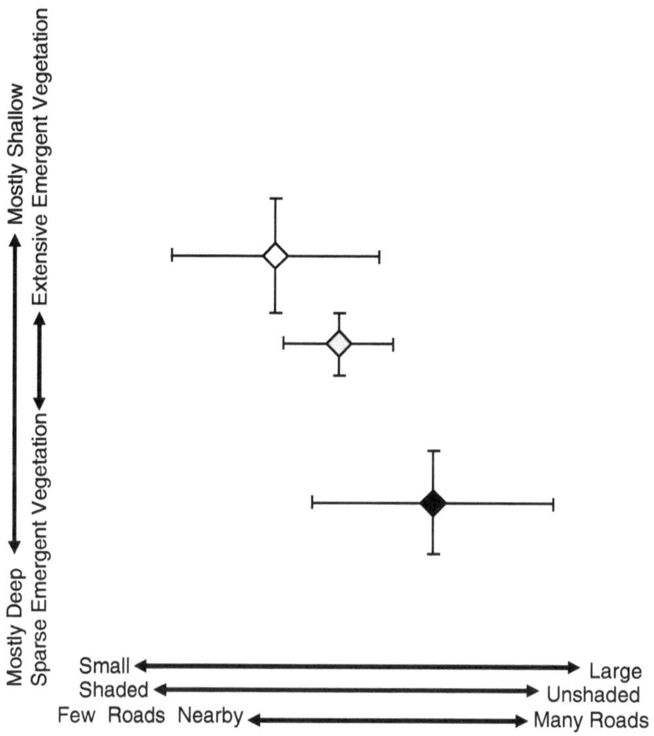

Fig. 1 Characteristics of wetlands with Northern Red-legged Frogs only (open diamond), Northern Red-legged Frogs and Bullfrogs (gray diamond), and Bullfrogs only (black diamond). The axes are principal components based on habitat variables measured during surveys for amphibians at 85 wetlands in the Willamette Valley, Oregon (Pearl *et al.* 2005).

(Fig. 1). As this was just an exploratory analysis, this pattern should be considered preliminary rather than conclusive. However, this approach can contribute to understanding coexistence by supplementing the more common approach of describing the occupancy and abundance of indigenous species relative to Bullfrog presence, Bullfrog abundance, and habitat characteristics. The pattern of coexistence in the Willamette Valley shows that further research seeking options to promote coexistence of indigenous amphibians with Bullfrogs via habitat management is warranted.

Community characteristics

Larval Bullfrogs differ from most other temperate ranids in several ecologically important ways. In their native range, Bullfrogs share permanent waters with a variety of warm water fish (Werner and McPeek 1994). In particular, Bullfrogs often co-occur with sunfish (family Centrarchidae), which include pumpkinseed, bluegill, crappie, and bass. Many fish avoid feeding on Bullfrog tadpoles which are somewhat unpalatable (Kruse and Francis 1977, Kats *et al.* 1988). These fish have an effect on pond communities that Bullfrogs exploit: they reduce the size and abundance of macroinvertebrates that can be major predators of Bullfrog larvae (Werner and McPeek 1994, Skelly 1996).

Research in the State of Oregon, USA, has shown that invasive bluegill (*Lepomis macrochirus* Rafinesque) increase the survival of Bullfrog tadpoles by reducing the abundance of indigenous aeshnid dragonfly larvae (Adams *et al.* 2003). Survival of Bullfrog tadpoles was 0% in experimental enclosures that lacked bluegill but had aeshnids; compared to 20% survival in enclosures with both bluegill and aeshnids. This suggests that reducing or eliminating bluegill and perhaps other similar centrarchids could be a way to reduce or eliminate Bullfrog populations. Moreover, limiting the spread or intentional introduction of such "facilitator" species may help limit the spread or abundance of Bullfrogs. This hypothesis is supported by field surveys in Oregon, showing that Bullfrogs are less likely to occur and appear to be less abundant at sites lacking introduced centrarchids compared to sites with centrarchids present (Adams *et al.* 2003).

This research also suggests a hypothesis that indigenous macroinvertebrates can resist Bullfrog invasion or help restrict Bullfrog populations to low enough densities that indigenous species can persist. Research is needed to understand the factors that regulate the abundance of predaceous macroinvertebrates and their effectiveness as Bullfrog predators. Such research might indicate features of wetlands that could be manipulated to manage the Bullfrog problem.

LIVING WITH INVASIVE SPECIES

Invasive species research and management have typically centered on prediction, prevention, and eradication (Mack *et al.* 2000, Simberloff *et al.* 2005). In the early stages of invasion, aggressive actions to eradicate the invader are warranted and, in some cases, intensive efforts to directly control invaders may be preferred even after broad establishment. However, many aggressive invaders like the Bullfrog, once established, are difficult or impossible to directly control or eradicate (e.g. Mack *et al.* 2000). Even if viable approaches for eradication exist, substantial resources are seldom available for any but the most economically damaging species. Moreover, intensive efforts to eradicate can sometimes have negative side effects (Zavaleta *et al.* 2001, Maret *et al.* 2006). These factors can leave few options for managing the problem.

The difficulty in eradicating Bullfrogs, particularly over large areas, is a common situation in invasive species management and may lead to a sense of futility. However, many of the non-eradication options discussed above have not been adequately explored and have the potential to promote the persistence of some indigenous species despite the invasion of Bullfrogs. In particular, while

Bullfrogs are a problem for a variety of indigenous species and there is also evidence that some otherwise vulnerable natives can sometimes coexist with Bullfrogs (Kiesecker and Blaustein 1997, Adams 1999, Govindarajulu 2004, Pearl *et al.* 2004). Coexistence suggests that other factors such as habitat conditions can mitigate the negative effects of Bullfrogs. Management options that focus on indigenous species persistence have potentially broad application but have received little research relative to more direct eradication and control measures.

We suggest that indigenous species persistence might be a primary goal of managers that have broadly established Bullfrog populations with little hope of eradication. However, managers must also consider the characteristics of the indigenous species when setting goals. It seems likely that some indigenous species simply may not be able to coexist with Bullfrogs while others, though vulnerable to negative effects, might benefit from efforts to promote their persistence despite the presence of Bullfrogs. Indigenous species are more likely to coexist with invaders such as Bullfrogs if natural history and microhabitat preferences of the former limit spatial and temporal overlap with the invader. This is likely to be particularly important during life history stages that are demographically influential.

Consider two pond-breeding ranid frogs that historically co-occurred in lowlands of north-western North America: Northern Red-Legged Frogs and Oregon Spotted Frogs (*Rana pretiosa* Baird and Girard). There is some evidence of decline for both species but Oregon Spotted Frogs have experienced the greatest losses. Bullfrogs have been implicated in population losses of both species, and both use the same general habitats as Bullfrogs in the region. Several factors indicate that Northern Red-Legged Frogs might be more likely than Oregon Spotted Frogs to coexist with Bullfrogs (Pearl *et al.* 2004). First, experimental trials in mesocosms showed that, given a choice between land or water, Oregon Spotted Frogs and Bullfrogs both chose water more often than Northern Red-Legged Frogs. This difference in microhabitat use could reduce contact between Bullfrogs and Northern Red-Legged Frogs. Second, this same study found that, when all three species are placed together in a mesocosm with both land and water available, juvenile Oregon Spotted Frogs do not survive as well as juvenile Northern Red-Legged Frogs. This supports the notion that differences in microhabitat use help protect Northern Red-Legged Frogs from Bullfrog predation. Third, larval Northern Red-Legged Frogs are able to modify risky behaviors in the presence of cues of invasive predators including Bullfrogs (Kiesecker and Blaustein 1997, Pearl *et al.* 2003). Fourth, juveniles of the Northern Red-Legged Frog tend not to linger around breeding ponds (Nussbaum *et al.* 1983, C. Pearl 2000, personal observation) and the adults spend much of their lives away from their breeding sites (Licht 1969, 1986). In contrast, postmetamorphic Oregon Spotted Frogs, like Bullfrogs, remain closely tied to aquatic habitats throughout their lives (Licht 1969, 1986, Pearl *et al.* 2004). Finally, field surveys at sites where both natives were known to occur found Northern

Red-Legged Frogs persisting after Bullfrog invasion more frequently than Oregon Spotted Frogs (Pearl *et al.* 2004).

Another example exists within the native range of the Bullfrog in Ontario, Canada, where Green Frogs (*Rana clamitans* Latreille), which share aquatic habitat with Bullfrogs, responded positively to local Bullfrog extinction (Hecnar and M'Closkey 1997). The relative abundance of Northern Leopard Frogs (*Rana pipiens* Schreber), which use aquatic habitats favored by Bullfrogs less than Green Frogs, decreased after the same Bullfrog extinction (Hecnar and M'Closkey 1997). In general, highly aquatic species such as the Oregon Spotted Frog may not be good candidates for coexistence with Bullfrogs whereas a species like the Northern Red-Legged Frog, despite having breeding sites in common with Bullfrogs, has greater potential to coexist with Bullfrogs. A detailed understanding of behavioral and life history characteristics of indigenous species may provide clues to their vulnerability to invasive species. Such insights can inform management prescriptions, including when to consider aggressive control efforts and when to focus on the persistence of natives.

CONCLUSIONS

There are no easy solutions to the Bullfrog problem but there is hope for progress. Given sufficient resources, eradication is an option for small isolated ponds and, if possible, should be emphasized when endangered indigenous species with vulnerable natural history characteristics are involved. When this fails, there is evidence that some indigenous amphibians can coexist with Bullfrogs in some habitats despite negative effects of Bullfrogs. This gives us reason to believe that managing habitat and community characteristics has potential to promote coexistence. Our primary thesis is that further research could lead to management options that promote the persistence of indigenous species despite the ongoing presence of Bullfrogs and that such options have been neglected. This conclusion is somewhat dependent on the assumption that observed coexistence between some indigenous species and Bullfrogs is not simply a transitory condition but rather is a function of species and site characteristics.

Several lines of inquiry need increased attention to develop management options for Bullfrogs in regions where they are well established or cannot otherwise be directly eradicated. These include:

1. Temporal dimensions of coexistence: Are cases of observed coexistence transitory or related, at least in part, to local conditions? This is a critical question whenever comparative studies based on patterns of coexistence are being used to suggest that certain habitat features might promote coexistence.
2. Habitat mediation: How do physical attributes of water bodies and upland habitats affect Bullfrog abundance and interactions with indigenous species?

Are there attributes that promote coexistence with some species? If so, are there ways to manage these attributes to benefit indigenous species?

3. Invasion resistance: What features of indigenous communities can help resist Bullfrogs? Options that prevent invasion are obviously desirable but there may also be options that help constrain Bullfrogs to low numbers. Despite evidence that some odonates are voracious predators of Bullfrog tadpoles, there is little information addressing interspecific differences in predation rates or habitat features that promote various odonate species. How can beneficial species be encouraged?
4. Mutualism: Are some NIS facilitating further invasion? Can knowledge of positive interactions among NIS be used to manage invasives? This is a new topic in invasive species research but a link has already been identified between non-indigenous fish and Bullfrogs (Adams *et al.* 2003). There is potential for other mutualistic interactions to be important.

In conclusion, we emphasize that while the Bullfrog invasion problem may seem discouraging, there are management options both existing and in need of further research that give reason to hope for future progress.

ACKNOWLEDGEMENTS

We thank J. Bowerman, J. Erickson, S. Galvan, F. Gherardi, and E. Muths for assistance and helpful suggestions on this manuscript.

REFERENCES

Adams, M. J. 1999. Correlated factors in amphibian decline: exotic species and habitat change in western Washington. Journal of Wildlife Management **63**, 1162–1171.

Adams, M. J. 2000. Pond permanence and the effects of exotic vertebrates on anurans. Ecological Applications **10**, 559–568.

Adams, M. J., C. A. Pearl, and R. B. Bury. 2003. Indirect facilitation of an anuran invasion by non-native fishes. Ecology Letters **6**, 343–351.

Altwegg, R. 2002. Trait-mediated indirect effects and complex life cycles in two European frogs. Evolutionary Ecology Research **4**, 519–536

Babbitt, K. J. and F. Jordan. 1996. Predation on *Bufo terrestris* tadpoles: effects of cover and predator identity. Copeia **1996**, 485–488.

Banks, B., J. Foster, T. Langton, and K. Morgan. 2000. British Bullfrogs? British Wildlife **11**, 327–330.

Biek, R., W. C. Funk, B. A. Maxell, and L. S. Mills. 2002. What is missing in amphibian decline research: insights from ecological sensitivity analysis. Conservation Biology **16**, 728–734.

Boone, M. D., E. E. Little, and R. D. Semlitsch. 2004. Overwintered Bullfrog tadpoles negatively affect salamanders and anurans in native amphibian communities. Copeia **2004**, 683–690.

Borges-Martins, M. M., M. Di-Bernardo, G. Vinciprova, and J. Measey. 2002. Geographic distribution. *Rana catesbeiana*. Herpetological Review **33**, 319.

Bury, R. B. and J. A. Whelan. 1984. Ecology and management of the Bullfrog. U.S. Fish and Wildlife Service, Resource Publication 155, Washington, DC.

Casas-Andreu, G., R. Cruz-Aviña, and X. A. Miguel. 2002. Geographic Distribution. *Rana catesbeiana*. Herpetological Review **33**, 63.

Casper, G. S. and R. Hendricks. 2005. *Rana catesbeiana* Shaw, 1802; American Bullfrog. Pages 540–546 in M. Lannoo, editor. Amphibian declines: the conservation status of United States species. University of California Press, Los Angeles, CA.

Cecil, S. G. and J. J. Just. 1979. Survival rate, population density and development of a naturally occurring anuran larvae. Copeia **1979**, 447–453.

Cisneros-Heredia, D. F. 2004. Geographic distribution: *Rana catesbeiana*. Herpetological Review **35**, 406.

Cohen, N. W. and W. E. Howard. 1958. Bullfrog food and growth at the San Joaquin Experimental Range, California. Copeia **1958**, 223–225.

Crowder, L. B. and W. E. Cooper. 1982. Habitat structural complexity and the interaction between bluegills and their prey. Ecology **63**, 1802–1813.

Doubledee, R. A., E. B. Muller, and R. M. Nisbet. 2003. Bullfrogs, disturbance regimes, and the persistence of California red-legged frogs. Journal of Wildlife Management **67**, 424–438.

Fellers, G. M., A. E. Launer, G. Rathbun, S. Bobzien, J. Alvarez, D. Sterner, R. B. Seymour, and M. Westphal. 2001. Overwintering tadpoles in the California red-legged frog (*Rana aurora draytonii*). Herpetological Review **32**, 156–157.

Garner, T. W. J., M. W. Perkins, P. Govindarajulu, D. Seglie, S. Walker, A. A. Cuningham, and M. C. Fisher. 2006. The emerging amphibian pathogen *Batrachochytrium dendrobatidis* globally infects introduced populations of the North American bullfrog, *Rana catesbeiana*. Biology Letters **2**, 455–459.

Govindarajulu, P. 2004. Introduced Bullfrogs (*Rana catesbeiana*) in British Columbia: impacts on native Pacific treefrogs (*Hyla regilla*) and red-legged frogs (*Rana aurora*). Ph.D. thesis. University of Victoria, British Columbia, Canada.

Govindarajulu, P., R. Altwegg, and B. Anholt. 2005. Matrix model investigation of invasive species control: Bullfrogs on Vancouver Island. Ecological Applications **15**, 2161–2170.

Green, D. M. and R. W. Campbell. 1984. The Amphibians of British Columbia. British Columbia, Canada, Provincial Museum Handbook Series 45, 120 pp.

Hammerson, G. A. 1982. Bullfrog eliminating leopard frogs in Colorado? Herpetological Review **13**, 115–116.

Hanselmann, R., A. Rodriguez, M. Lampo, L. Fajardo-Ramos, A. A. Aguirre, A. M. Kilpatrick, J. P. Rodriguez, and P. Daszak. 2004. Presence of an emerging pathogen of amphibians in introduced Bullfrogs *Rana catesbeiana* in Venezuela. Biological Conservation **120**, 115–119.

Hayes, M. P. and M. R. Jennings. 1986. Decline of Ranid frog species in western North America: are Bullfrogs (*Rana catesbeiana*) responsible? Journal of Herpetology **20**, 490–509.

Hecnar, S. J. and R. T. M'Closkey. 1997. Changes in the composition of a ranid frog community following Bullfrog extinction. American Midland Naturalist **137**, 145–150.

Hirai, T. 2004. Diet composition of introduced bullfrog, *Rana catesbeiana*, in the Mizorogaike Pond of Kyoto, Japan. Ecological Research **19**, 375–380.

Huffaker, D. B. 1958. Experimental studies on predation: dispersion factors and predator-prey oscillations. Hilgardia **27**, 343–383.

Jennings, M. R. and M. P. Hayes. 1985. Pre-1900 overharvest of California red-legged frogs (*Rana aurora draytonii*): the inducement for Bullfrog (*Rana catesbeiana*) introduction. Herpetologica **41**, 94–102.

Kairo, M., B. Ali, O. Cheesman, K. Haysom, and S. Murphy. 2003. Invasive species threats in the Caribbean Region. Report to The Nature Conservancy. [accessible via http://tncweeds.ucdavis.edu/products/library/carrassessment.doc]

Kats, L. B., J. W. Petranka, and A. Sih. 1988. Antipredator defenses and the persistence of amphibian larvae with fishes. Ecology **69**, 1865–1870.

Kentula, M. E., J. C. Sifneos, J. W. Good, M. Rylko, and K. Kunz. 1992. Trends and patterns in section 404 permitting requiring compensatory mitigation in Oregon and Washington, USA. Environmental Management **16**, 109–119.

Kiesecker, J. M. 2003. Invasive species as a global problem: toward understanding the worldwide decline of amphibians. Pages 113–126 in R. D. Semlitsch, editor. Amphibian Conservation. Smithsonian Institution, Washington, DC.

Kiesecker, J. M. and A. R. Blaustein. 1997. Population differences in responses of red-legged frogs (*Rana aurora*) to introduced Bullfrogs. Ecology **78**, 1752–1760.

Kiesecker, J. M. and A. R. Blaustein. 1998. Effects of introduced Bullfrogs and small mouth bass on the microhabitat use, growth and survival of native red-legged frogs. Conservation Biology **12**, 776–787.

Kiesecker, J. M., A. R. Blaustein, and C. L. Miller. 2001. Potential mechanisms underlying the displacement of native red-legged frogs by introduced Bullfrogs. Ecology **82**, 1964–1970.

Kim, H. S. and S. K. Ko. 1998. Distribution, food habit and seasonal cycles of germ cell activity in the introduced bullfrog, *Rana catesbeiana* in Korea. Journal of Forest Science **57**, 165–177.

Kruse, K. C. and M. G. Francis. 1977. A predation deterrent in larvae of the bullfrog, *Rana catesbeiana*. Transactions of the American Fisheries Society **106**, 241–252.

Lanza, B. 1962. On the introduction of *Rana ridibunda* Palla and *Rana catesbeiana* Shaw in Italy. Copeia **3**, 642–643.

Lever, C. 2003. Naturalized reptiles and amphibians of the world. Oxford University Press, New York, NY, 318 pp.

Licht, L. E. 1969. Comparative breeding behavior of the red-legged frog (*Rana aurora aurora*) and the western spotted frog (*Rana pretiosa pretiosa*) in southwestern British Columbia. Canadian Journal of Zoology **47**, 1287–1299.

Licht, L. E. 1986. Comparative escape behavior of sympatric *Rana aurora* and *Rana pretiosa*. American Midland Naturalist **115**, 239–247.

Longcore, J. E., A. P. Pessier, D. K. Nichols. 1999. *Batrochochytrium dendrobatidis* gen. et sp. nov., a chytrid pathogenic to amphibians. Mycologia **91**, 219–227.

Lopez-Flores, M. and F. J. Vilella. 2003. Predation of a white-cheeked pintail (*Anas bahamensis*) duckling by a bullfrog (*Rana catesbeiana*). Caribbean Journal of Science **39**, 240–241.

Mack, R. N., D. Simberloff, W. M. Lonsdale, H. Evans, M. Clout, and F. A. Bazzaz. 2000. Biotic invasions: causes, epidemiology, global consequences, and control. Ecological Applications **10**, 689–710.

Mahon, R. and K. Aiken. 1977. The establishment of the North American Bullfrog, *Rana catesbeiana* (Amphibia, Anura, Ranidae) in Jamaica. Journal of Herpetology **11**, 197–199.

Maret, T. J., J. D. Snyder, and J. P. Collins. 2006. Altered drying regime controls distribution of endangered salamanders and introduced predators. Biological Conservation **127**, 129–138.

Mazzoni, R. 1999. Bullfrog farming in South America. Infopesca Internacional **1**, 27–31.

Moyle, P. B. 1973. Effects of introduced Bullfrogs, *Rana catesbeiana*, on the native frogs of the San Joaquin Valley, California. Copeia **1973**, 18–22.

Negroni, G. 1997. Frog culture. World Aquaculture **28**, 16–22.

Neveu, A. 1997. The introduction of allochthonous green frog species into France, two dissimilar cases: *R. catesbeiana* and the foreign taxa of *esculenta* complex. Bulletin Francais de la Peche et de la Pisciculture **1997**, 165–171.

Nussbaum, R. A., E. D. Brodie Jr. and R. M. Storm. 1983. Amphibians and Reptiles of the Pacific Northwest. University of Idaho Press, Moscow, ID.

Pearl, C. A. and D. E. Green. 2005. *Rana catesbeiana* (American Bullfrog). Chytridiomycosis. Herpetological Review **36**, 305–306.

Pearl, C. A., M. J. Adams, G. S. Schuytema, and A. V. Nebeker. 2003. Behavioral responses of anuran larvae to chemical cues of native and introduced predators in the Pacific Northwestern United States. Journal of Herpetology **37**, 572–576.

Pearl, C. A., M. J. Adams, R. B. Bury and B. McCreary. 2004. Asymmetrical effects of introduced Bullfrogs (*Rana catesbeiana*) on native ranid frogs in Oregon, U.S.A. Copeia **2004**, 11–20.

Pearl, C. A., M. J. Adams, N. Leuthold and R. B. Bury. 2005. Amphibian occurrence and aquatic invaders in a changing landscape: implications for wetland mitigation in the Willamette Valley, Oregon, U.S.A. Wetlands **25**, 76–88.

Rosen, P. C., and C. R. Schwalbe. 1995. Bullfrogs: introduced predators in southwestern wetlands. Pages 452–454 in E. T. LaRoe, G. S. Farris, C. E. Puckett, P. D. Doran, and M. J. Mac, editors. Our living resources: a report to the nation on the distribution, abundance, and health of US plants, animals, and ecosystems. US Department of the Interior, National Biological Service, Washington, DC.

Sampedro, M. A., H. L. Montanez, and B. O. Suarez. 1985. Feeding of *Rana catesbeiana* in two capture zones of Cuba. Ciencias Biologicas **13**, 59–66.

Savino, J. F. and R. A. Stein. 1982. Predator-prey interactions between largemouth bass and bluegills as influenced by simulated, submersed vegetation. Transactions of the American Fisheries Society **111**, 255–266.

Schlaepfer, M. A., P. W. Sherman, B. Blossey and M. C. Runge. 2005. Introduced species as evolutionary traps. Ecology Letters **8**, 241–246.

Schwalbe, C. R. and P. C. Rosen. 1988. Preliminary report on effects of Bullfrogs on wetland herpetofaunas in southeastern Arizona. Pages 166–173 in R. C. Szaro, K. E. Steverson and D. R. Patton, editors. Management of amphibians, reptiles, and small mammals in North America. US Department of Agriculture, Forest Service, GTR-RM-166, Flagstaff, AZ.

Simberloff, D., I. M. Parker and P. N. Windle. 2005. Introduced species policy, management, and future research needs. Frontiers in Ecology and the Environment **3**, 12–20.

Skelly, D. K. 1996. Pond drying, predators, and the distribution of *Pseudacris* tadpoles. Copeia **1996**, 599–605.

Smith, F. E. 1972. Spatial heterogeneity, stability, and diversity in ecosystems. Transactions of the Connecticut Academy of Arts and Sciences **44**, 309–335.

Sredl, M. J. and J. P. Collins. 1992. The interaction of predation, competition, and habitat complexity in structuring an amphibian community. Copeia **1992**, 607–614.

Stumpel, A. H. P. 1992. Successful reproduction of introduced Bullfrogs *Rana catesbeiana* in northwestern Europe: a potential threat to indigenous amphibians. Biological Conservation **60**, 61–62.

Viernes, K. J. F. 1995. Bullfrog predation on an endangered common moorhen chick at Hanalei National Wildlife Refuge, Kaua'i. Elepaio **55**, 37.

Viparina, S. and J. J. Just. 1975. The life period, growth and differentiation of *Rana catesbeiana* larvae occurring in nature. Copeia **1975**, 103–109.

Vonesh, J. R. and O. de la Cruz. 2002. Complex life cycles and density dependence: assessing the contribution of egg mortality to amphibian declines. Oecologia **133**, 325–333.

Werner, E. E. and M. A. McPeek. 1994. Direct and indirect effects of predators on two anuran species along an environmental gradient. Ecology **75**, 1368–1382.

Werner, E. E., G. A. Wellborn, and M. A. McPeek. 1995. Diet composition in postmetamorphic bullfrogs and green frogs: implications for interspecific predation and competition. Journal of Herpetology **29**, 600–607.

Willis, Y. L., D. L. Moyle, and T. S. Baskett. 1956. Emergence, breeding, hibernation, movements and transformation of the Bullfrog, *Rana catesbeiana*, in Missouri. Copeia **1956**, 30–41.

Wu, Z. J., Y. P. Wang, and Y. M. Li. 2004. Natural populations of Bullfrog (*Rana catesbeiana*) and their potential threat in the east of Zhejiang province. Biodiversity Science **12**, 441–446.

Wylie, G. D., M. L. Casazza, and M. Carpenter. 2003. Diet of Bullfrogs in relation to predation on giant garter snakes at Colusa National Wildlife Refuge. California Fish and Game **89**, 139–145.

Zavaleta, E. S., R. J. Hobbs, and H. A. Mooney. 2001. Viewing invasive species removal in a whole-ecosystem context. Trends in Ecology and Evolution **16**, 454–459.

Concluding remarks

The need for more research should not be casually invoked as an excuse for inaction.
Daniel Simberloff (2003)

Chapter thirty-nine

A role for scientists

Francesca Gherardi

Species introductions and their ecological consequences have long fascinated scientists. In trying to explain patterns of species distributions and abundances, Charles Darwin (1859) was faced with the problems of non-indigenous species (NIS) (Cadotte 2006). He used them as a device to illustrate his theory of natural selection and descent with modifications, but he was also the first to note marked effects that these species had on the recipient communities. Invaders "from different quarters of the globe", as he wrote in *The Origin of Species* (1859), have greatly reduced in number the endemic species in the southeast corner of Australia (p. 124). So, caution should be taken, Darwin recommended, "in transporting animals from one district to another" (p. 136). It was a century later that Charles Elton's *The Ecology of Invasions by Animals and Plants* (1958) inspired much of the interest and understanding of invasions in our lifetime. Indeed, investigating the "ecological explosions" (p. 15) that occur "when a foreign species successfully invades another country" and analyzing the "mingling of thousands of kinds of organisms from different parts of the world" and their induced "changes in the natural population balance" (p. 18) are the major tasks of the contemporary invasion biologists.

The world today is much more "explosive" than in the 1950s. The volume of flora and fauna that is shuttled from one geographic realm to another has greatly expanded. The ballast water of cargo vessels alone can transfer up to 3,000–4,000 species per day from one continent to the other. Extrapolating these numbers to all the kinds of vessels at sea at any given time, a total of 7,000–10,000 species are translocated per any 24 h period (Carlton 1999). The increased complexity of international trade has opened new pathways and facilitated the ease with which potentially invasive species can move along these pathways. A wide diversity of vectors can transport invaders, including aquaculture, aquaria, ballast water, and even sushi and live seafood (Carlton and Geller 1993, Chapman *et al.* 2004). Almost every aquatic or wetland plant

Francesca Gherardi, Biological invaders in inland waters: Profiles, distribution, and threats, 697–702.

designated as a noxious weed can be ordered through e-commerce from an aquatic plant nursery somewhere (Kay and Hoyle 2001). The deregulation of national and international markets has reduced the barriers to trade and their surveillance, facilitating the movement, for instance, of thousands of species into North America by the aquarium industry alone (Welcomme 1984). In addition, the impact of biological invaders has been made more intense – and elusive – by the delay with which human behavior, social norms, and cultural traditions respond to the new risks that they pose (Perrings *et al.* 2002).

Indeed, much progress in the knowledge of invasion biology has been made within the five decades following the publication of Elton's book (Williamson 1996, 1999, Lonsdale 1999) – and this volume is the expression of the proliferation of scientific interest in biological invasions also in freshwater systems. The research has emerged on several fronts, especially following the series of invasion volumes published between 1986 and 1989 as the proceedings of the SCOPE (Scientific Committee on Problems of the Environment) symposia and workshops (e.g. Mooney and Drake 1986, Drake *et al.* 1989, Davis 2006). Since then, the scientific literature has moved on from its effort to describe the characteristics that make a species a good invader, or a community invasible, to the focus on the interactions between the invader and the target ecosystem. Qualitative studies have been replaced by quantitative assessments of the attributes of the invaders and of the invaded communities. Paleobiological studies, microcosm/mesocosm experiments, and modeling exercises have contributed to raise our awareness that history, chance, and determinism interact to shape ever-changing communities.

The increased concern about the rising economic and ecological costs inflicted by invasive species has induced many constituencies in several countries to seek to reduce their occurrence and impact. Problems caused by introduced species were included as a priority item (article 8h) in the 1992 "Rio" Convention on Biological Diversity and international organizations (Global Invasive Species Programme, Convention on Biological Diversity) began to implement this article. In the last decade, many nations have recognized the impact of some introduced species as a problem and have attempted to implement and improve administrative and legal solutions (e.g. New Zealand, South Africa, the USA, and European Union). National and international initiatives have been taken to assess the risk of future introductions, the potential for establishment and expansion, and the subsequent impacts (e.g. IUCN 1987, FAO 1995, 1996, US ANS Task Force 1996, NZ MAF 2002, UK Defra 2003). Finally, the control of NIS is part of the EU's policy approach taken to attain the Göteborg's target of halting biodiversity loss by 2010 (Commission of the European Communities 2006).

Due to the inevitable interplay in this issue among science, environmental ethics, and public policy (Lodge and Shrader-Frechette 2003), new areas of conflict have been opened in the recent times (cf. Simberloff 2003). A number of authors from different cultural fields (e.g. Sagoff 1999, 2005, Theodoropoulos

2003) joined with a few ecologists (e.g. Slobodkin 2001) in a "rearguard action" (Simberloff 2006, p. 915) to convince scientists and laypeople that the threat posed by introduced species is overblown. Criticisms were mainly directed towards the apparent "lack of adherence to sound scientific practices" (Thedoropoulos 2003) and to the formulation of unscientific generalizations or "reifications" (Slobodkin 2001). Among the other objects of contention, invasion biologists have been accused of a biased interpretation of the effects that invasive species exert on global biodiversity (Rosenzweig 2001, Gurevitch and Padilla 2004), of their adoption of "stipulative definitions" (Sagoff 2005) – such as the concept of biodiversity opposed to xenodiversity (in the meaning of Leppäkoski and Olenin 2000), and of their use of military and pejorative metaphors in their writings (Larson 2005). Finally, biological invasion research captured the attention of the media, which often magnified existing misunderstandings among scholars (e.g. Devine 1999).

A consequence is that invasion biologists are working today in an emotionally charged atmosphere where they are subject to the inevitable tension between the strong appeal that biological invasions elicit as ideal objects of study and their personal concern about the threats they pose to biodiversity. Davis (2006) coined the terms "Asilomar" (a town in California where the first Symposium of the International Union of Biological Sciences was held in 1964) and "Eltonian" (from Charles Elton) as attributes to label the two apparently contradicting attitudes assumed by scientists when they face the problem of invasions. On the one hand, some of them share the exclusive interest in ecological theories shown by the conveners in Asilomar and view invasions as very rich sources of information for capturing the complexities of ecological systems. In the same way that the physiology of an organism may be better studied during illness, anomalies of ecological systems – i.e. species introductions – are thus fundamental in understanding their functioning. On the other hand, an increased sense of social responsibility among scientists and their willingness to participate to the political forum made several of them believe that their findings on biological invasions might also assist resource managers in restoring and rebuilding the "ill" ecological systems.

As Soulé (1986) puts it, "fiddling" with ideas until the world is in ashes – like the Emperor Nero who is alleged to have continued playing his lyre while Rome burned – may lead to severe penalties (Gherardi 2006). Research is often "an unaffordable luxury that provides information only for the eulogy" (Coblentz 1990); on the contrary, efforts are to be directed "to construct ecological research program that dictate possibilities for managers, that investigate alternatives where the options used for management have failed, and that evaluate the processes of management" (Underwood 1995, p. 232). Research should be elevated to "a primary component of environmental decision-making", rather than being relegated to "an increasingly peripheral procedure" (Underwood 1995, p. 232), and in their turn scientists should allocate more time to practical concerns – rather than simply producing publications (Caro 1998). Real progress cannot be made in affecting environmental policy and management until scientists become more effective in the "processes" (the steps used to reach decisions, such as law, credibility, etc.) and in the "relationships" (related to human behavior, such as networks in communities), and not only in the understanding of objective and factual data ("substance") (Meffe 2002). In sum, "scientists need to make concerted efforts to learn these other skills and to become more active players in ensuring that good science is a real part of policy and management" (Meffe 2002, p. 367).

Particularly in the field of invasion biology, any attempt to manage the invasive species problem will greatly benefit from the role played by scientists of advocate on the one hand, and of their constructive partnership with resource managers on the other (Byers *et al.* 2002). Researchers are expected to identify and control pathways of accidental introductions, to promote measures that may prevent unwanted introductions, and to produce protocols for pre-introduction environmental risk assessment. They should stimulate cooperative actions among States, recognizing the risk – particularly high in Europe – that activities within their jurisdiction or control may pose to other States as a potential source of invasive species. By quantifying how invasive species affect native biodiversity, scientists also have the capacity to individualize effective systems for the early warning of the newcomer species and safe methods for the control/eradication of already established invaders.

Research has the potential to determine metrics reflecting all the biological changes that accompany any intervention; it can evaluate the role of NIS after their integration in the systems and suggest strategies that are flexible and in line with biogeographic and evolutionary realities (Cox 2004). Finally, research helps prioritize ecosystems at risk through assessing their invasibility and the duration of lag phases between the establishment and the spread of specific invaders. Risk assessment, decision theory, and epidemiology all offer useful insights for the development of policies to control NIS through a process that should involve both scientists and policy makers (Byers *et al.* 2002).

Certainly, much useful knowledge of NIS already exists – it merely requires reanalysis or reassembly into a form that managers can use (Byers *et al.* 2002). Despite the progress, to date much has still to be done – and scientists have not yet missed the boat (Puth and Post 2005).

REFERENCES

Byers, J. E., S. Reichard, J. M. Randall, I. M. Parker, C. S. Smith, W. M. Lonsdale, I. A. E. Atkinson, T. R. Seastedt, M. Williamson, E. Chornesky, and D. Hayes. 2002. Directing research to reduce the impacts of nonindigenous species. Conservation Biology **16**, 630–640.

Cadotte, W. W. 2006. Darwin to Elton: early ecology and the problem of invasive species. Pages 15–33 in M. W. Cadotte, S. M. McMahon, and T. Fukami, editors. Conceptual ecology invasion biology: reciprocal approaches to nature. Springer, Dordrecht, The Netherlands.

Carlton, J. T. 1999. The scale and ecological consequences of biological invasions in the world's oceans. Pages 195–212 in O. T. Sandlund, P. J. Schei, and Å. Viken, editors. Invasive species and biodiversity management. Kluwer Academic Publishers, Dordrecht, The Netherlands.

Carlton, J. T. and J. B. Geller. 1993. The ecological roulette: the global transport of nonindigenous marine organisms. Science **261**, 78–82.

Caro, T., editor. 1998. Behavioral ecology and conservation biology. Oxford University Press, Oxford, UK.

Chapman, J. W., T. W. Miller, and E. W. Coan. 2004. Live seafood as recipes for invasion. Conservation Biology **17**, 1386–1395.

Coblentz, B. E. 1990. Exotic organisms: a dilemma for conservation biology. Conservation Biology **4**, 261–265.

Commission of the European Communities 2006. Halting the loss of biodiversity by 2010 – and beyond. Brussels, Belgium, 22 May 2006.

Cox, G. W. 2004. Alien species and evolution. Island Press, Washington, DC.

Darwin, C. 1859. The origin of species. JM Dent, London, UK (1971).

Davis, M. A. 2006. Invasion biology 1958–2004: the pursuit of science and conservation. Pages 35–64 in M. W. Cadotte, S. M. McMahon, and T. Fukami, editors. Conceptual ecology and invasion biology: reciprocal approaches to nature. Springer, Dordrecht, The Netherlands.

Devine, R. S. 1999. Alien invasion: America's battle with non-native animals and plants. National Geographic Society, Washington, DC.

Drake, J. A., H. A. Mooney, F. di Castri, R. H. Groves, F. J. Kruger, M. Rejmanek, and M. Williamson. 1989. Biological invasions: a global perspective. SCOPE 37, Wiley, Chichester, UK.

Elton, C. S. 1958. The ecology of invasions by animals and plants. University of Chicago Press, Chicago, IL, Edition 2000.

FAO. 1995. Precautionary approach to fisheries, Part 1: Guidelines on the precautionary approach to capture fisheries and species introductions. FAO Fisheries Technical Paper No. 350, Part 1. FAO, Rome, Italy.

FAO. 1996. Guidelines for pest risk analysis. International standards for phytosanitary measures. Publ. No. 2, Secretariat of the International Plant Protection Convention, FAO, Rome, Italy.

Gherardi, F. 2006. Bioinvasions in fresh waters and the Nero dilemma. Polish Journal of Ecology, SEFS4 Special Issue **4**, 549–561.

Gurevitch, J. and Padilla, D. K. 2004. Are invasive species a major cause of extinctions? Trends in Ecology and Evolution, **19**, 470–474.

IUCN. 1987. The IUCN position statement on translocation of living organisms: introductions, re-introductions and re-stocking. Species Survival Commission, IUCN Commission on Ecology, Gland, Switzerland.

Kay, S. H. and S. T. Hoyle. 2001. Mail order, the Internet, and invasive aquatic weeds. Journal of Aquatic Plant and Management **39**, 88–91.

Larson, B. M. H. 2005. The war of the roses: demilitarizing invasion biology. Frontiers in the Ecology and the Environment **3**, 495–500.

Leppäkoski, E. and S. Olenin. 2000. Xenodiversity of the European brackish water seas: the North American contribution. Pages 107–119 in J. Pederson, editor. Marine bioinvasions. Massachusetts Institute of Technology, Cambridge, MA.

Lodge, D. M. and K. Shrader-Frechette. 2003. Nonindigenous species: ecological explanation, environmental ethics, and public policy. Conservation Biology, **17**, 31–37.

Lonsdale, W. M. 1999. Global patterns of plant invasions and the concept of invasibility. Ecology **80**, 1522–1536.

Meffe, G. 2002. Connecting science to management and policy in freshwater fish conservation. Pages 363–372 in M. J. Collares-Pereira, M. M. Coelho, and I. G. Cowx, editors. Conservation of freshwater fishes: options for the future. Blackwell, Oxford, UK.

Mooney, H. A. and J. A. Drake, editors. 1986. Ecology of biological invasions of North America and Hawaii. Springer, New York, NY.

NZ MAF 2002. Import risk analysis – animals and animal products. New Zealand Ministry of Agriculture and Forestry, Wellington, New Zealand.

Perrings, C., M. Williamson, E. B. Barbier, D. Delfino, S. Dalmazzone, J. Shogren, P. Simmons, and A. Watkinson. 2002. Biological invasion risks and the public good: an economic perspective. Conservation Ecology **6**, 1–7.

Puth, L. M. and D. M. Post. 2005. Studying invasion: have we missed the boat? Ecology Letters **8**, 715–721.

Rosenzweig, M. 2001. The four questions: what does the introduction of exotic species do to diversity? Evolutionary Ecology Research **3**, 361–367.

Sagoff, M. 1999. What's wrong with exotic species? Report from the Institute of Philosophy & Public Policy **19**, 16–23.

Sagoff, M. 2005. Do non-native species threaten the natural environment? Journal of Agricultural and Environmental Ethics **18**, 215–236.

Simberloff, D. 2003. Confronting introduced species: a form of xenophobia? Biological Invasions **5**, 179–192.

Slobodkin, L. B. 2001. The good, the bad and the reified. Evolutionary Ecology Research **3**, 1–13.

Soulé, M. E., editor. 1986. Conservation biology. The science of scarcity and diversity. Sinauer Associates, Sunderland, MA.

Theodoropoulos, D. I. 2003. Invasion biology: critique of a pseudoscience. Avvar Books, Blythe, CA.

UK Defra 2003. Review of non-native species policy – report of the working group. Product code: PB 8072. Department for Environment, Food and Rural Affairs, London, UK.

US ANS Task Force 1996. Report to the aquatic nuisance species task force. Generic nonindigenous aquatic organisms risk analysis review process (For Estimating Risk Associated with the Introduction of Nonindigenous Aquatic Organisms and how to Manage for that Risk). US Fish & Wildlife Service and National Oceanic & Atmospheric Administration; http://www.anstaskforce.gov/gennasrev.htm.

Underwood, A. J. 1995. Ecological research and (and research into) environmental management. Ecological Applications **5**, 232–247.

Welcomme, R. I. 1984. International transfers of inland fish species. Pages 22–40 in W. R. Courtnay Jr. and J. R. Stautter Jr, editors. Distribution, biology, and management of exotic fishes. Johns Hopkins University Press, Baltimore, MD.

Williamson, M. 1996. Biological invasions. Chapman & Hall, London, UK.

Williamson, M. 1999. Invasions. Ecography **22**, 5–12.

General keywords

P

R

Families, orders

Geographical names

C

D

E

Taxa common names

O

P

Q

R

S

T

U

V

W

Y

Z

Species index

A

D

E

MIX
Papier aus verantwortungsvollen Quellen
Paper from responsible sources
FSC® C105338

If you have any concerns about our products,
you can contact us on
ProductSafety@springernature.com

In case Publisher is established outside the EU,
the EU authorized representative is:
Springer Nature Customer Service Center GmbH
Europaplatz 3, 69115 Heidelberg, Germany

Printed by Libri Plureos GmbH
in Hamburg, Germany